The Combination Products Handbook

Combination products are therapeutic and diagnostic products that combine drugs, devices, and/or biological products. According to the US Food and Drug Administration (FDA), "a combination product is one composed of any combination of a drug and a device; a biological product and a device; a drug and a biological product; or a drug, device and a biological product." Examples include prefilled syringes, pen injectors, autoinjectors, inhalers, transdermal patches, drug-eluting stents, and kits containing drug administration devices co-packaged with drugs and/or biological products. This handbook provides the most up-to-date information on the development of combination products, from the technology involved to successful delivery to market. The authors present important and up-to-the-minute pre- and post-market reviews of combination product regulations, guidance, considerations and best practices.

This handbook:

- Brings clarity of understanding for combination products guidance and regulations
- Reviews the current state-of-the-art considerations and best practices spanning the combination product lifecycle, pre-market through post-market
- Reviews medical product classification and assignment issues faced by global regulatory authorities and industry

The editor is a recognized international Combination Products and Medical Device expert with over 35 years of industry experience and has an outstanding team of contributors. Endorsed by AAMI – Association for the Advancement of Medical Instrumentation.

The Combination Products Handbook

A Practical Guide for Combination Products and Other Combined Use Systems

Edited By

Susan W. B. Neadle

CRC Press is an imprint of the
Taylor & Francis Group, an **informa** business

First edition published 2023
by CRC Press
6000 Broken Sound Parkway NW, Suite 300, Boca Raton, FL 33487-2742

and by CRC Press
4 Park Square, Milton Park, Abingdon, Oxon, OX14 4RN

CRC Press is an imprint of Taylor & Francis Group, LLC

ISBN: 978-1-032-29162-8 (hbk)
ISBN: 978-1-032-29163-5 (pbk)
ISBN: 978-1-003-30029-8 (ebk)

DOI: 10.1201/9781003300298

Typeset in Times
by Deanta Global Publishing Services, Chennai, India

Contents

Preface

Increasingly, medicinal products[1] depend on medical devices for their administration, and likewise, there is growing use of medicinal substances to enhance medical device performance. Some countries/regions have adopted the classification "combination product" as a regulatory construct for combined use medical products, i.e., medical products in which two or more differently regulated medical products are to be used together (or are being studied for use together) to achieve an intended use, indication, or effect.[2] Other jurisdictions apply different terminology and regulatory constructs (Chapter 2).

Combination products are designed to offer greater benefits than medicinal products or devices acting alone. While there are terminology and classification differences across jurisdictions globally, our ultimate focus is aligned: Ensuring safety, efficacy, and usability of these products for intended users, intended uses and intended use environments, safeguarding, and supporting the patients we serve. Combination product development, manufacture, and post-market lifecycle management require an understanding of the risks and associated control strategies for each of the combination product constituent parts and their interactions, impacting the safety and efficacy of the combination product as a whole.

The objective of this book is to cover the breadth of topics needed for the novice to the experienced combination product practitioner to successfully navigate the dynamic global combination products space. The regulations, guidance, standards, and pre-market and post-market expectations in this space are evolving. They cover an expansive range of risks and complexity of medical devices, medicinal products, and their combined use. The rapid pace of technology changes – from the dynamic biotechnology space to connected health applications and beyond – and a growing propensity for the convenience and potential cost-effectiveness of at-home care, is driving increased recognition that the higher technical complexity of these combined use systems requires strengthening of risk analysis and controls strategies, authorization, and certification requirements.

This book is a comprehensive overview of the current state-of-the-art considerations and best practices spanning the combination product lifecycle, pre-market through post-market. The book is divided into two parts: Part 1 – Foundational Aspects, and Part 2 – Special Topics. There is a helpful appendix that includes a comparative summary of the Global Combination Product Landscape and also a combination products glossary of terms.

Part 1 steps through differences and commonalities in classification and terminology region to region; pre-market through post-market regulatory strategies; cGMP best practices; integrated drug/device product development and risk management – including essential performance requirements and human factors; submissions strategies; post-market lifecycle management; and inspection readiness considerations. Included in these discussions are critical success factors such as developing multilingual language skills (fluency in drug-, device-, biologic-, and combination product-speak), overcoming multicultural and multidisciplinary organizational challenges, and essential elements of supplier collaboration.

Part 2 delves more deeply into combination product material considerations and purchasing controls. Given that many combination product applicants are pre-disposed to purchasing at least one of their combination product's constituent parts, the partnership and collaboration model between the application holder and the constituent part supplier(s), component suppliers and contract manufacturing organizations is central to product success. Part 2 then provides insights into combination product considerations and best practices for analytical testing, biological products, connected health, and a review of the evolving global regulatory landscape.

NOTES

1. Some jurisdictions distinguish between "drugs" and "biological products"; others refer to and regulate them together as "medicinal products." The term "medicinal products" is used interchangeably with drugs/biological products throughout the book. In some countries, where drugs and biological products are regulated separately, a drug–biologic combination may be considered a "combination product." See Chapter 2 of the book, "What Is a Combination Product?"
2. ASTM International Combination Products Terminology Standard.

Foreword

My medical device career started in industry in the 1980s, where the company I worked for manufactured many different co-packaged combination products or convenience kits. Combination products have been around since the early 1970s with such products like radio-biologics, certain types of in vitro diagnostics, as well as a variety of convenience kits that combined drugs and devices. From 1970 through 1990, US FDA handled combination products on an Ad Hoc basis through Inter-Agency or Cross-Center Internal Agreements. I joined US FDA's Center for Devices and Radiological Health (CDRH) just after the issuance of the 1990 Safe Medical Devices Act which provided FDA their first legislative authority to formally develop regulations for combination products. Shortly thereafter in November of 1991, 21 CFR Part 3 was issued describing product jurisdiction and how FDA would designate which Center would take responsibility for pre-market review and regulation of combination products. At that time, the lead center ran combination products through their traditional programs with limited cross-center consultation(s).

As the CDRH member on the Center for Drugs "GMPs for the 21st Century Initiative" Steering Committee and as a co-chair for one of the Working Groups on Quality Systems, I worked in the early 2000s helping to draft some of the first guidance documents on quality systems/GMPs for combination products. Later I became a member of the Combination Productions Good Manufacturing Practice (GMP) Working Group which wrote, 21 CFR Part 4 "Current Good Manufacturing Practice Requirements for Combination Products" and its preamble.

As someone who was so involved in the development of the current regulations on combination products, I can confidently say that this book does an excellent job delving into the US requirements contained in both Part 3 and Part 4 in extraordinary depth to assist in understanding not only the requirements but the intent of the requirements.

The United States is not the only country that has a complicated history with combination product regulations; many international medical devices, pharmaceuticals, biologics, and combination product organizations or consortiums such as the Global Harmonization Task Force (GHTF), International Medical Device Regulators Forum (IMDRF), International Council for Harmonization of Technical Requirements for Pharmaceuticals for Human Use (ICH), Asia-Pacific Economic Cooperation (APEC), Asian Harmonization Working Party (AHWP), Pan African Harmonization Working Party (PAHWP) have tried to tackle the regulation of the ever-evolving nature of combination products. The different global regulatory agencies have different legislative basis under which to try to effectively manage the regulation of these diverse medical products, making harmonization or convergence efforts extremely challenging. There is currently a newly developed Combination Product standards work group, under ASTM Committee E55 working on global convergence of definitions with a draft document entitled "Guide for Definition of Combination Products (Drug/Device/Biologic Combinations Standard)."

This book will address some of the prominent global regulatory schemes for combination products beyond the United States. All the global efforts to converge regulatory requirements and expectations for combination products are vital now more than ever. The increase of combination products within advanced therapies and personalized medicines, as well as the increased use of digital health, artificial intelligence, mobile APPs, etc. calls for both the industry and the global regulatory agencies to work together. Industry and regulators need to find effective regulatory pathways that will encourage innovation and break down some of the barriers between our medical product sectors.

I want to thank all the extremely dedicated and experienced authors of this book for volunteering their time and wisdom. A special thank you to Susan W. B. Neadle, who gave many hours drafting chapters, coordinating content, and project managing the development of this comprehensive book. It is selfless dedication like this to improve the public health with new innovative combination products that make all our jobs so rewarding.

I am confident this book will be an invaluable resource for many years in understanding and complying with regulatory expectations for combination products.

Kim Trautman

Editor Biography

Susan W. B. Neadle, MS, BS, FAAO, is a recognized international Combination Products and Medical Device expert with over 35 years of industry experience. Networked, published, highly active in numerous industry groups, and with links to a number of teaching institutions, Susan brings deep knowledge and genuine passion for sharing that knowledge with others. Susan's leadership, innovation, and best practices have been recognized with multiple awards, including the Johnson Medal, Johnson & Johnson's highest honor for excellence in Research & Development, and most recently, the 2022 ISPE Joseph X. Phillips Professional Achievement Award for significant contributions as an integrator of industry and regulators, and as a Finalist in TOPRA's 2021 Awards for Regulatory Affairs Excellence. She serves as Principal Consultant and President of Combination Products Consulting Services LLC, applying her extensive leadership, technical skills, and complex program management experience to provide international quality & compliance, design excellence, regulatory, and executive advisory consulting services to the pharmaceutical, biotech, and medical device industries. Among her significant industry affiliations and contributions, she serves as Chair of the ISPE Combination Products Community of Practice, lead author on the ASTM International Combination Products Standard Committee, lead author on the AAMI Combination Products Steering Committee updating TIR 48, active member in Combination Products Coalition, and serves as faculty at AAMI and at University of Maryland Baltimore County Campus College of Natural and Mathematical Sciences for Combination Products Curricula. She is also active in PQRI and on AFDO/RAPS (previously Xavier Health) Combination Product Summit steering committee and speaks frequently at public venues on a variety of combination products topics.

Susan retired from Johnson & Johnson, where her distinguished 26+ years career included integral leadership roles in R&D, Quality Engineering, Risk Management, Design-to-Value, and Quality Systems Management, spanning pharmaceuticals/biotech, medical devices, and consumer health sectors, as well as strategic leadership at J&J corporate level. She served as Chair of J&J's Design Council, advancing world-class practices in product and process design and development to drive robust, customer-centric healthcare solutions across J&J. She also led J&J's cross-sector Combination Products Community of Practice. Among several achievements, Susan led the team that defined and implemented the globally integrated business model to meet Combination Products health authority regulations for Janssen, J&J's Pharmaceutical sector.

Susan also served as Executive Director and Head of Combination Products, Medical Devices, Digital Health & IVD Regulatory Affairs at Amgen, providing strategic leadership in global combination products/device regulatory development from initial clinical investigation through registration and commercial lifecycle. She was as an advisor for internal regulatory policy priorities, health authority engagement, and submission approaches through strategic engagement and mentoring of colleagues for individual projects and portfolio. She served as a catalyst for external consortium deliverables and strategic direction.

Susan holds an M.S. in Polymer Science & Engineering (University of Massachusetts Amherst) and B.Sc. in Biology/Chemistry (State University of New York at Albany). She is also a Fellow of the American Academy of Optometry. Susan is certified as a Black Belt in Design and Process Excellence and is a Design Thinking subject matter expert.

List of Contributors

Khaudeja Bano is VP, Combination Product Quality at Amgen. She is a Physician with a Masters in Clinical Research, Pharmaceutical engineering Certification, a Database Administrator, and a certified Project Management Professional.

Khaudeja has more than 25 years' professional experience, including clinical practice. She has held several global medical positions at Guidant, Abbott, AbbVie Inc., and now Amgen. Her career includes global medical/clinical and safety leadership roles in devices, diagnostics, pharmaceuticals, and combination products. She currently serves as the chair for the Post Marketing Safety working group for the Combination Product Coalition (CPC). Email: Khaudeja@gmail.com.

Daniel Bantz has over 25 years' practical experience in medical and analytical instrumentation development and testing. He is a subject matter expert in fluid metering product development and testing, combination product testing, and device reliability testing. Additionally, he is skilled in FMEA best practices and has developed product risk mitigation strategies and risk management platforms across multiple organizations. Daniel implements combination product functional testing strategies meeting ever-changing regulations in conjunction with customer needs and the competitive landscape. He earned an MBA in Operations and Technology from Aurora University in Illinois. Email: dbantz04@gmail.com

Dhiraj Behl is currently working as a Principal Regulatory Specialist professional, at PPD India Pvt. Ltd., Gurugram, India, and, a PhD scholar at Amity Institute of Pharmacy, Amity University, Noida, Uttar Pradesh, India. In total, Dhiraj has diversified working experience in the regulatory affairs field for over 9+ years to strategize, prepare, review, and submitting regulatory submissions to global health authorities. He managed R&D Projects from the Concept phase to complete the transfer to production and thereafter Lifecycle management, as positions from Specialist to Project Manager for products assortment include Pharmaceuticals, Medical Devices, and Combinational products (Drug-Device). In his most recent role, he is leading the group as Project Manager for client's global projects with responsibility for defining regulatory strategy and planning for global products, i.e., development to lifecycle management.

Dhiraj is actively engaged in several industry working groups on a range of Combination Products topics, including TOPRA, ISPE Community of Practice, and Xavier Peer Pros Network. Dhiraj earned an MS in pharmaceutical sciences and a Bachelor of pharmacy, as well as a registered pharmacist in Haryana, India. Before PPD, Dhiraj held roles in regulatory affairs at Nectar Lifesciences Ltd. and Jakob & Partners. Email: dhiraj1987@gmail.com

Ed Bills

During his career in medical devices, **Ed Bills** has held a number of quality and regulatory affairs positions for major medical device companies, including a period as Corporate Director of Risk Management. He has over 39 years' experience in the field of quality and regulatory affairs, including time as Director of Quality and Regulatory concurrently for four US sites. Currently, he consults and provides training in the area of medical device quality, regulatory and risk management. With Stan Mastrangelo, he co-edited Lifecycle Risk Management for Healthcare Products: From Research Through Disposal published by PDA.

ASQ has awarded Mr Bills with Fellow status as well as Certified Quality Engineer, Certified Quality Auditor, Certified Manager of Quality and Organizational Excellence, and he is a Regulatory Affairs Certified by the Regulatory Affairs Professionals Society.

Additionally, Mr Bills serves in international standards work, assisted in completing the revision of the third edition of ISO 14971 risk management standard and the guidance ISO TR 24971:2020 as an international member of the technical committee. He also serves on the US national committee for the medical devices quality system standard, ISO 13485, as well as the medical bed standards, IEC 60601-2-52, IEC 60601-2-89, and the AAMI technical committee CP contributing to the combination products risk management guidance, AAMI TIR 105. Email: elb@edwinbillsconsultant.com

Stephanie Canfield is a Quality Manager and Senior Consultant at Agilis Consulting Group and is active on standards committees and industry groups for medical devices and combination products.

In her current role, Stephanie manages projects of varying complexity for both medical devices and combination products, preparing various documentations, and executing HF evaluations. Stephanie integrates risk management and human factors to support successful regulatory submissions and ensure regulatory compliance. Stephanie has a Master of Science in Biology and over 10 years of experience in academia and industry. Stephanie has practical experience in a range of areas, from product feasibility, research and development, process validation, quality engineering, and human factors engineering throughout the IVD, medical device, and combination product space. Email: scanfield@agilisconsulting.com

Fran DeGrazio has 35+ years of experience in the pharmaceutical packaging and delivery industry with extensive expertise in sterile drug product systems, including vial container closure systems and prefillable systems for combination products. Fran is currently President & Principal Consultant at Strategic Parenteral Solutions LLC. Prior to this, she held numerous technical roles at West Pharmaceutical Services, including R&D, Quality & Regulatory, Technical Customer Support, Analytical Laboratories, and Scientific Affairs. Her final role prior to retiring from West was as Chief Scientific Officer.

Fran has published numerous technical articles and book chapters. She is the recipient of the PDA Packaging Science Award for 2021, Philadelphia Business Journal 2018 Healthcare Innovators of the Greater Philadelphia Region Award and the Healthcare Businesswoman's Association Luminary Award for West in 2017. Email: frandegraz@icloud.com

Stephanie Goebel is senior regulatory consultant for medical devices and combination products at Beyond Conception GmbH, with 15+ years of experience on implementation of legislative acts or standards (e.g., MDR/ ISO 13485), QMS remediation, documentation review, and auditing activities at global pharmaceutical and medical device companies. Stephanie also worked as lead auditor, technical documentations reviewer for active medical devices (covering MDD and ISO 13485) as well as product specialist on Article 117 (MDR) for a notified body and was a member of the Team NB (European Association of Notified Bodies) working group on Article 117 (MDR). She has combined her educational background as Medical Engineer with extensive experience in the device development, regulatory and quality fields, and is also an active contributor to conferences and working groups, e.g., at PDA or ISPE. Email: stephanie.goebel @beyond-conception.com

Shannon Hoste is the President of Agilis Consulting Group, an associate professor in the Quality Science Education program at Pathway for Patient Health and is active on several standards and conference committees for medical devices and combination products.
Formerly, Shannon worked as Team Lead for Human Factors in FDA/CDRH and as a CDER human factors reviewer within DMEPA. At the FDA, she led reviews of Human Factors data for pre-market submissions through 510(k), PMA, and De Novo as well as NDA, BLA, ANDA pathways, and IND and IDE requests. Outside of Shannon's work for the FDA, she continues to build on her 25+ year career where she has worked in the medical device, IVD and combination product industry as a Systems Engineer, R&D Manager, and Quality Director. Shannon has a Master of Science in Cognitive Systems Engineering and a Master of Science in Management. Email: shoste @agilisconsulting.com

Bjorg Hunter holds a BSc in Design and Innovation Engineering from Technical University of Denmark and an MSc in Biomedical Engineering from Aarhus University, Denmark. Bjorg started her career at GSK where she has held leadership positions with increasing responsibility in R&D Device Development and Regulatory Affairs. In 2020, Bjorg moved to NovoNordisk where she is Director of the RA Digital Health area within RA CMC and Devices. Bjorg has been very active in external advocacy especially related to combination products, including driving industry positions on behalf of EFPIA. Email: BJHT@novonordisk.com

Manfred Mäder is VP Global Device & Packaging Development in TRD (*Technical Research and Development*) at Novartis.
Prior to this, he held the position Head of Global Compliance & Audit for Devices & Combination Products overseeing all Alcon, Pharma, and Sandoz sites producing these types of products and Global QA Head of Technical Research and Development at Novartis Pharma starting in February 2011.

Prior to this position he was Senior VP of Quality Management & Regulatory Affairs, at Ypsomed, a company producing Medical Devices and Combination Products starting in 2007. Previously, he was responsible for Quality Assurance Management at Sanofi-Aventis for the Frankfurt Injectables site. Before then, being based in Kansas City, US, he had a global responsibility for Quality and Regulatory for one of the Aventis Blockbuster products. Prior to that, he held several positions in QA and QC.

By training he is pharmacist and holds a doctorate in pharmaceutical analytics and statistics by the University of Wuerzburg, Germany. Email: manfred.maeder@novartis.com

Cherry Malonzo Marty completed her medical degree in the Philippines and continued on to a second Masters in Biomedical Engineering in Switzerland. After several years of combined experience in science education and academia in international settings, Dr Malonzo Marty has been leveraging her clinical and engineering background to support her passion for medical and regulatory writing, and science communications. While covering a wide range of scientific fields from spine research and regenerative medicine, bio- and medical technologies, and pharmaceuticals, she developed a keen interest in combination products. Her fascination for regulatory writing and combination products has contributed to the success of an array of medical device and pharmaceutical companies by assuring the adequacy of their technical documentation and their compliance with global regulatory requirements. Email: cherry.marty@confinis.com

Ryan McGowan is the Director of Digital Devices and Combination Products in Regulatory Affairs at AstraZeneca. He has responsibility for developing regulatory strategies for the approval of digital health products including software as a medical device. Prior to joining AstraZeneca, Ryan was a pre-market reviewer and combination products team leader at FDA's Center for Devices and Radiological Health where he evaluated and influenced regulatory policy for drug delivery constituent parts of combination products. Email: ryan.mcgowan@astrazeneca.com

Susan W. B. Neadle
Principal Consultant & President, Combination Products Consulting Services, LLC. Email: sneadle@combinationprod.com

Meera Raghuram is currently the Director, Regulatory and Sustainability strategy for the Life Science Division of Lubrizol Corporation. She has over 20 years of experience in complex technical and strategic regulatory issues related to pharmaceutical excipients, drugs/devices, personal care products, and human health/environmental risk assessments. She has successfully used a holistic approach in assessing suitability of ingredients including consideration of safety, quality, environmental, and sustainability attributes ensuring minimization/management of risk enabling safe, compliant, and sustainable products. She is a member of the IPEC-Americas Executive Board and serves as the Chair of IPEC – Americas Regulatory Affairs Committee. She takes an active role globally on advocacy issues, understanding regulatory trends and finding innovative solutions for regulatory compliance. She is involved in advocacy and leadership on current issues related to novel excipients, polymer biodegradation, the European Union microplastics regulations, dietary supplements, and natural products. In her previous position, she was responsible for management of environmental remediation strategy and has successfully worked with US (state and federal) and international regulatory agencies to bring resolution to issues. Meera has a BS degree in Pharmacy from the Indian Institute of Technology and an MS in Pharmaceutical Sciences from Purdue University, Indiana. Email: Meera.Raghuram@lubrizol.com

Jennifer L. Riter is Senior Director, Business and Technical Operations for the Services and Solutions organization at West Pharmaceutical Services, Inc., Exton, PA. She joined West in August 1996 as an Associate Chemist in the Quality Control group. Jennifer has held many roles within the West organization and has experience and expertise in several areas from Technical Customer Support, Business Development, Sales and Marketing, and Quality. Her previous role was Senior Director, Analytical Services in the Scientific Affairs and Technical Services organization where she was responsible for the Analytical Services organization. Currently, Jennifer is responsible for the Business and Technology Operations in the Services and Solutions Organization in West. She is responsible for the Business Operations organization as well as the Technology Managers which have four primary areas of focus and thought leadership in Extractables & Leachables, Container Closure Integrity, Packaging and Device/Combination Product Performance, and Particle Analysis.

Her experience blends knowledge of West's components, containment, and delivery systems with hands-on experience of providing technical support and analytical solutions for packaging, delivery systems, and combination products to West's multinational customers. Ms Riter has also spoken at several symposiums on analytical testing of parenteral packaging components, devices, and combination products as well as on extractables and leachables analysis.

Ms Riter is an executive member of the Board of Directors for the Lock Haven University Foundation, American Association of Pharmaceutical Scientists, Parenteral Drug Association, American Association of Pharmaceutical Scientists, and Healthcare Businesswoman Association and ISPE. She earned a Bachelor of Science degree in Biology/Chemistry from Lock Haven University, Lock Haven, PA, and an MBA in Pharmaceutical Business at the University of the Sciences in Philadelphia, Philadelphia, PA. Email: Jennifer.Riter@westpharma.com

Suzette Roan, JD, is currently Associate Vice President and Head of Device Global Regulatory Affairs at Sanofi Inc., where she is responsible for the combination product and diagnostic device aspects of multiple development and commercial projects. Roan has more than 25 years of experience in the pharmaceutical industry, primarily focused on delivery systems and combination products. She is an active member of the Combination Products Coalition, PDA, and ISO/TC 84 working groups. She can be contacted at suzette.roan@sanofi.com.

Theresa Scheuble is Head of Design & Innovation for Johnson & Johnson. She is an expert on combination product requirements and considerations supporting all segments of Johnson & Johnson.
Prior to her current role, Theresa was Head of the Combination Product Drug Delivery Systems Development for the East Coast team in Janssen, a Johnson & Johnson Company. While at Janssen, Theresa was responsible for Combination Product Device development and led cross-sector initiatives. Theresa has been with Johnson & Johnson for greater than 25 years, working in all Segments of Johnson & Johnson with 18 years specifically in the Medical Device Segment. She was the project lead and technical lead for greater than 20 portfolio enhancing products by Johnson & Johnson companies. She also has experience in the Automotive and Aerospace industries. Email: TScheubl@its.jnj.com

Kimberly A. Trautman is an experienced Medical Devices, In Vitro Diagnostics, and Combination Product Expert with over 30 years of experience. She worked at the US Food and Drug Administration (FDA) for 24 years and continues to work with Regulatory Agencies around the globe. With industry and regulatory agency experience, she has a demonstrated history of working collaboratively with industry, regulators, and patient groups for the betterment of public health. She executes several medical device regulatory services and developed a formal Education/Training business. She also established an Authorized Medical Device Single Audit Program (MDSAP) Auditing Organization and a new Notified Body for EU IVDR/MDR Designation.

As an expert in global medical device regulations, she wrote and harmonized the current US FDA Quality System Regulation and was on the international authoring group of ISO 13485 since inception. She conceived and developed the Medical Device Single Audit Program and its consortium of five Global Regulators. She is a 20-year veteran of the Global Harmonization Tasks Force (GHTF) and foundational member of the International Medical Device Regulators Forum (IMDRF). She is also a recognized International Medical Device expert with a master's degree in Biomedical Engineering.

Kim holds an MS of Biomedical & Medical Engineering (University of Virginia, Charlottesville, VA) and a BSc of Molecular Cell Biology and Engineering Sciences (Pennsylvania State University, State College, PA). Email: Ktrautman_mdis@msn.com

As a Senior Consultant and Vice President at **confinis**, **Mr. Viky Verna** currently assists medical device and pharmaceutical companies with regulatory affairs challenges. His qualifications are firstly supported by his education; specifically, a BS and an MS in Biomedical Engineering from the University of Miami, an MS in Pharmacy and a Drug Regulatory Affairs Certificate from the University of Florida, and a Global Regulatory Affairs Certification (RAC) from Regulatory Affairs Professional Society (RAPS). Mr. Verna's experience with Combination Products started at the US Food and Drug Administration (FDA) as an investigator. Later, at the Center for Devices and Radiological Health (CDRH) of the FDA, Mr. Verna held several positions including (Acting) Branch Chief of the Respiratory, ENT, General Hospital, and Ophthalmic (REGO) devices branch, which handles the compliance activities of combination products among others, including drug delivery systems. During his time at CDRH, he also served as:

- A Subject Matter Expert (SME) reviewer in the quality system working group of the Office of Compliance, where he generated and reviewed the regulatory case reports (establishment inspection report review memos) for regulatory decisions and legal compliance actions.
- A combination product branch lead of the REGO branch. In this position, his responsibilities included training and reviewing the work of the team, as well as developing reviewing processes and techniques to be used by the office.

After joining confinis, Mr Verna has helped several companies of all sizes successfully comprehend, navigate around, and comply with the US regulatory requirements for medical devices and combination products including those involving drug delivery systems. By being an expert member of the ISO technical committee (TC 84), Mr Verna has also been leveraging his expertise and experience to help develop international standards for injection and respiratory products, infusion pumps, needles, catheters, and the likes. Email: viky.verna@confinis.com

Mike Wallenstein holds the position as Head Novartis MDR Implementation since April 2019. In this role, Mike oversees all activities related to the EU MDR implementation for Medical Devices & Combination Products at Novartis globally. Prior to this position, Mike was functioning as Executive Director QA/Senior Compliance Officer (GCA) since 2015 to oversee all compliance activities related to Medical Devices & Combination Products at Novartis globally.

He is leading the Medical Device and Combination Product Expert Network at Novartis, and member of several US and EU Expert Committees and Interest Groups on Combination Products.

Mike joined Novartis in 2010 as Global Auditor in Group Compliance and Audit.

Mike has over 25 years of experience in QA, R&D, and Manufacturing within the Medical Devices & Pharmaceutical Industry.

Before joining Novartis, he was Head Global Audit Systems at Gambro Renal Care (today Baxter) and European Head Quality Systems & Audits at 3M.

He studied Chemical Engineering and Plastic Technologies, in Münster, Germany. Email: mike .wallenstein@novartis.com

Richard (Rick) Wedge has spent nearly 25 years developing medical devices, diagnostics, and combination products since graduating with a degree in applied biology and a PhD in electrochemical biosensors. This has involved the development of prototype products in the laboratory, design and qualification of production instrumentation, establishing quality management systems, and submission of regulatory documentation. Products developed include blood cholesterol monitoring kits, illuminated endoscopes, and drug delivery devices.

Since joining Pfizer's Devices Centre of Excellence (DCoE) in 2016, Rick has been responsible for managing a team of subject matter experts concerned with the development and implementation of design control and risk management processes for both simple and complex drug delivery devices. Rick's team works closely with designers, manufacturers, QA, and Regulatory (both internally and externally) to ensure that products reaching the market are safe and effective for their intended use. In addition to the above, Rick is a member of the working group responsible for authoring ISO 14971 and ISO/TR 24971 and is a wider contributor to the combination products community. Email: Richard.Wedge@pfizer.com

John Barlow-Weiner, Esq. is the Associate Director for Policy in FDA's Office of Combination Products, which is tasked with ensuring the efficient, effective, and consistent regulation of combination products and the classification of medical products and their assignment to FDA components for regulation.

In addition to overseeing the policy program for OCP, Mr Weiner's responsibilities include providing guidance and direction for FDA policy initiatives and precedential regulatory actions concerning combination products and other medical products intended for combined use. His work involves engagement with components throughout FDA, with foreign counterparts, and with standard-setting bodies, as well as outreach to stakeholders and trade associations.

Mr Weiner was previously an Associate Chief Counsel for FDA, focusing on drug regulation and related innovation and competition issues, and on cross-cutting topics such as use of nanotechnology. He has served as FDA's liaison to the United States Trade Representative and other components of the United States Government on issues relating to pharmaceutical trade and competition, participating in multiple rounds of bilateral and regional trade negotiations as a technical advisor.

Before coming to FDA, Mr Weiner was in private practice, focusing on food and drug, environmental, and related public international and trade law. He has published and lectured in each of these areas. With clients in sectors including aviation, biotechnology, electronics, industrial chemicals and pesticides, information technology, and real estate, in addition to the healthcare and medical products sectors, his practice included regulatory counsel, engagement with regulatory agencies and Congress, litigation strategy, and representing clients before regional and multilateral organizations and in the negotiations of associated agreements.

Mr Weiner received a Bachelor of Arts degree from Princeton University and his *Juris Doctor* degree with honors from the Columbia University School of Law. Email: John.Weiner@fda.hhs.gov

Acknowledgments

I would like to acknowledge the countless time, effort, and passion invested by each of the contributing authors to this book. These authors represent globally recognized experts across the combination product industry, and like me, they are passionate to share their knowledge and expertise to benefit the combination products industry, and ultimately make positive impacts on the patients that we all serve. I want to specially recognize Amy Davis, of blessed memory, from DHI Publishing, who early on shared my vision for what this book could be and encouraged me in its development. Thank you to Hilary Lafoe, Taylor & Francis Group, for your incredible support to continue this work after Amy passed away. I am especially grateful to John Barlow-Weiner for his support in peer-review throughout the authoring and editing process. Special thanks to friends and colleagues Janine Jamieson, Amit Khanolkar, Doug Mead, Alireza Jahangir, and Renee Palmer for their parts in brainstorming and reviewing material. Thank you also to Dr Antonio Moreira, who encouraged this work, aligned to the Combination Products curriculum offered in-person and virtually at University of Maryland Baltimore Campus (UMBC). Finally, I dedicate this book to my family, who supported me throughout the authoring and editing of this book, and to Industry and Health Authority colleagues, who have shared (and continue to share!) in shaping and living the exciting journey of growth in the combination products space.

Dear Readers,
I hope you find this book a useful and lasting practical resource to inform and support your efforts to advance the development and delivery of innovative technologies in the combination products space! Enjoy!

Susan W. B. Neadle
Principal Consultant and President, Combination Products Consulting Services, LLC
(Retired, Johnson & Johnson)

Part 1

Foundation

1 Introduction: Key Considerations Advanced by this Handbook

John Barlow Weiner

CONTENTS

Since you are reading this book, you likely appreciate that developing and getting a combination product (or, more broadly speaking, a combined-use product – more on this below) to market can be complicated. You likely also appreciate that understanding regulatory expectations and applying best practices is a smart way to ensure safe and effective products get to market and stay there, and to minimize surprises along the way. When Ms. Neadle asked if I would contribute to this book, I happily agreed, seeing this practical guide as a meaningful way to enhance understanding of the challenges sponsors and regulators may face and how to address them successfully.

This practical guide is well timed. Policies and best practices have now developed to a point that such a guide can provide meaningful, wide-ranging insights to readers as they consider marketing of products in jurisdictions across the globe. Further, while this book largely focuses on questions relating to drug-led combination products (e.g., a drug packaged in or with a delivery device), it provides helpful information for all regulated entities who contribute to bringing safe and effective combination products to market, including device-led and biologic-led ones. The discussion presented here of current policies across the globe and of policy gaps also may help encourage further advancement and convergence on significant issues.

This introduction addresses a few significant considerations, which are reflected in this handbook and that this book may help advance: (i) core regulatory concepts and the critical role that a shared vision has played in their development and translation into regulatory policy and best practices; (ii) the importance of collaboration and coordination to successful shepherding of combination products by sponsors and regulators; (iii) two important advances in best practices and regulatory thinking, (a) embracing comprehensive, integrated risk management and (b) focusing on the safety and effectiveness questions raised by combined uses rather than the semantics of "combination product" status, and (iv) how efforts to date can be built upon to facilitate broader adoption of best practices and further convergence and harmonization of regulatory approach across jurisdictions.

DOI: 10.1201/9781003300298-2

A SHARED VISION – THE US POLICY PROGRAM FOR COMBINATION PRODUCTS AND ITS DEVELOPMENT

A shared vision among the regulatory community, FDA, and Congress has guided combination product policy development from the beginning in the United States and has been critical to its success. This vision has helped identify and strengthen core concepts from which a combination product regulatory program and best practices have been built and continue to be refined. These concepts include the application of a risk-based approach and relying on coordination and collaboration, to drive sound, consistent and predictable policies and practices as discussed below. This vision's maintenance and further propagation among regulated entities and regulators can facilitate continued progress. Its benefits to date are reflected throughout this book.

The combined use and marketing of different types of medical products are not new. Hypodermic needle syringes, for example, have been in use since the mid-19th century; prefilled syringes have been marketed for several decades. That said, the range and complexity of products intended for combined use have dramatically increased, and this can be expected to continue to accelerate in the coming years. The expanding use of software and connected health, for example, offers new and potentially potent tools (see Chapter 13 on connected health). The increasing use of biologically sourced materials offers extraordinary potential and raises practical challenges for product study and design (see Chapter 12 on some considerations for combination products that include biological product constituent parts).

Sponsors and regulators have had to consider over the decades how to ensure different types of medical products are safe and effective when used together. The first statutory recognition in the United States of a distinct category of products that combine a drug, biological product, and/or device together (referred to as the "constituent parts" of the combination product in the US FDA program) came in amendments made to the Federal Food, Drug, and Cosmetic Act (FFDCA) in 1990. These amendments added section 503(g) (21 USC 353(g)), which has remained the principal statutory authority in the United States for combination products. This statutory provision and the regulatory program that has followed have helped guide legislative and regulatory activity regarding combination products outside the United States as well. Chapters 2 and 14 introduce the reader to the concept of "combination product" across the globe (see also the appendix of this book).

Section 503(g) reflects a recognition that the regulation of such products warrants particular consideration to avoid confusion and ensure sound, predictable, efficient, and consistent regulatory treatment. The section has been amended since 1990, most substantially in 2016 by section 3038 of the 21st Century Cures Act. Greater specificity has been added over time regarding processes and requirements applicable to sponsors and the US Food and Drug Administration (FDA) alike. Section 503(g) began by addressing basic issues such as that combination products are a distinct category of medical product and the need to clarify what component of the FDA would have primary jurisdiction, i.e., the lead, for a given combination product's regulation. A now widely applied concept of "primary mode of action" (PMOA) was established to determine these jurisdictional assignments (see Chapter 2). From the beginning, shared regulatory responsibility among the Agency components with expertise in each constituent part type has been a core concept, with one component having the lead. PMOA has been relied on to inform not only lead component assignment but the application of regulatory programs and procedures, including premarket pathway availability. The PMOA construct has since been adopted and applied in other jurisdictions in a similar manner.

To promote sound implementation of regulatory expectations for combination products, Congress amended section 503(g) to mandate the establishment of an office in FDA's Office of the Commissioner to oversee the regulation of combination products. The office was to be tasked with making PMOA determinations as needed and with ensuring timely, effective premarket review, and consistent, appropriate postmarket regulation of combination products. This became the Office of Combination Products (OCP). Regulated entities and other regulators have pointed to OCP as an important aid to managing combination product regulatory activity and policymaking. At the same time, they have noted that it can be difficult to emulate, for example, in jurisdictions where regulatory bodies for drugs and devices are not within a single agency like FDA. (If you have questions about the classification of your medical product for the US market (is it a drug, a device, a biological product, a combination product?), OCP is designated by FDA to address these questions too.)

In subsequent amendments to section 503(g), among other adjustments, Congress elaborated upon the duties of OCP; codified a risk-based approach to regulation; called for leveraging existing knowledge of the products being combined; addressed what premarket pathways are available for combination products; and codified means for sponsors to enable the smooth running of the regulatory process.

In sum, section 503(g) calls for efficient, effective, and consistent regulation of combination products, relying on a risk-based approach and coordination and collaboration among centers to realize these expectations. As noted, a shared vision of regulators, regulated entities, and Congress has produced this construct, through iterative refinement captured through amendment of the section. The remarkable strength of this shared vision is demonstrated by the fact that these Congressional amendments, enacted over the past 30 years, have served to confirm and codify existing FDA practices and expectations, which were developed and refined by the Agency in light of experience and stakeholder input. The Cures Act amendments (for which I played a leading negotiating role for FDA) dramatically illustrate this dynamic. Proposed by the regulated industry, endorsed by its major trade associations, then refined with input from FDA, these amendments largely codify and thereby ensure the ongoing application of practices and policies developed by the FDA over the prior 25 years.

Along the way, stakeholders and FDA have also engaged in the development of rules and guidance to flesh out this framework. As this book reflects, there are a variety of specialized considerations specific to combination products for which regulatory requirements and policies and best practices have been developed. Such considerations are related, for example, to particular types of combination products, specific uses of them, particular types of constituent parts, or the use of products in conjunction with combination products. The expectations and practices focus on the distinct questions that arise when different types of medical products are studied, manufactured, and used together. Regulated entities and regulators have needed to determine how to reconcile, integrate, and apply expectations associated with drugs, devices, and biological products. Engagement by industry organizations such as AdvaMed, BIO, the Combination Products Coalition, and PhRMA, as well as individual sponsors, has been significant to this process.

The result is an increasingly comprehensive and robust collection of regulatory policies with others in development. The objective is the implementation of a coherent combination products regulatory program designed to apply consistent, appropriate, risk-based expectations to enable safe and effective products to come to market as efficiently as possible. There are, of course, differences of opinion on the specifics, for example, relating to premarket data expectations or postmarket product controls, but the shared vision of a coordinated, collaborative, risk-based regulatory approach to produce sound, consistent, efficient regulation continues to enable mutual understanding and efficient, substantive engagement to resolve issues. More work remains to be done, and policymaking is necessarily not a rapid process, but the shared core vision has enabled this work to progress in a coherent, linear manner, providing greater clarity by building on what already has been established.

This book addresses core topics that have generated significant questions and for which regulatory expectations and best practices have come into better focus. It also notes and explains where questions remain that may warrant further policy attention, indicating where proactive case-by-case discussion of products with regulators may be important in the interim. See, for example: Chapter 5 regarding overall product development; Chapter 7 on human factors analysis to assess user interfaces; Chapter 11 on how to approach analytical testing issues due to, for example, the different materials used in, and testing needs for, device and drug constituent parts, and Chapter 4 on current good manufacturing practices (CGMPs) and Chapter 8 on lifecycle management, regarding postmarket product management and reconciliation and integration of controls associated with devices and drugs.

While the history of the shared vision presented here is US centric, the concepts embraced have informed thinking in other jurisdictions enabling ongoing efforts to promote convergence and harmonization as discussed below under "Leveraging Efforts to Date to Advance Convergence and Harmonization".

COORDINATION AND COLLABORATION – KEYS TO EFFICIENT, EFFECTIVE PRACTICES

Turning from the policy realm to product-specific activities, perhaps the fundamental prerequisite of success is a collaborative mindset and approach. As reflected throughout this book, cross-cutting collaboration, bringing in additional areas of expertise as compared to that needed for drugs or devices alone, is critical throughout the lifecycle of a combination product. It ensures timely consideration of relevant issues with the benefit of all appropriate experience and expertise. It is essential to the work of sponsors and regulators alike. If the reader is not already convinced of the importance of such collaboration, this book makes a compelling case.

All along, regulated entities have faced practical questions of how to organize themselves and approach their activities as they have assessed, developed, and marketed products for combined uses, to ensure the products' safety and effectiveness for these intended uses. Not surprisingly, in considering such issues, sponsors have reported looking to regulatory expectations to guide their analysis and inform their decisions. They also have reported often seeking to leverage the experience and expertise of partners and suppliers.

Over time, however, sponsors have also increasingly reported recognizing the need to rethink themselves, for example, to embrace their responsibility for the combination product as a whole, rather than seeking to rely on others to address issues raised by one or the other of the constituent parts. Sponsors have come to realize that, while they may be subject to regulation and may collaborate and partner with others, it is their product and ultimately their responsibility to identify what questions need to be answered and to determine what measures need to be put in place to ensure safe and effective combined uses. Increasingly, sponsors have reported a conscious decision to refer to themselves as "combination product" companies as an expression of this cultural, psychological shift.

In relation to this change of mindset, sponsors have come to realize that they need to have appropriate expertise available and engaging at each step of the process from product development to postmarket management, to ensure that all issues relating to each constituent part and how the constituent parts interact are considered in a timely, coordinated, and collaborative manner. Sponsors have reported it becoming apparent that they needed to review and reconsider their processes and procedures to ensure that cross-cutting coordination would occur in an appropriate and timely way. Sponsor reports have also indicated that the complexity of the activities often demands a well-structured, comprehensive approach. This has led to development of new structures, systems, and procedures; sometimes tearing down silos; sometimes creating robust connections between them.

In the premarket space, this effort to ensure appropriate collaboration has manifested, for example, in expansion of the product team to include all relevant expertise relating to each constituent part, to help ensure timely, cross-cutting engagement, and that the team understands how issues need to be sequenced to avoid backtracking and surprises. This need and how to address it are addressed most comprehensively in Chapter 5 on integrated development and Chapter 6 on risk management.

Similarly, in the postmarket context, after publication of FDA's regulation on CGMPs for combination products (21 CFR 4.A), for instance, many sponsors reported efforts to reconceive their approach to product design control and transfer and to manufacturing controls, to integrate what had been isolated business components and/or to bring more comprehensive expertise in house. Similarly, sponsors have reported recognizing the need to establish stronger, appropriately managed, engagement between complaint management and product safety operations after promulgation of FDA's rule on postmarketing safety reporting for combination products (21 CFR 4.B). Chapter 8 on lifecycle management and Chapter 9 on inspectional readiness, for example, speak to the importance of coordination in the postmarket setting and how to approach it.

This greater focus on collaboration and coordination has extended to management of vendors and service providers. This element can be particularly complex and important because vendors have their own business priorities and concerns that may not readily align with those of the product sponsor (e.g., medical product support may not be the vendors' primary business and close control of proprietary knowledge and information may be of great importance to the vendor). Chapters 8 and 10 on supplier quality controls provide insights and useful resources to facilitate these efforts.

To be clear, companies that market combination products vary in their approaches, some drug-led combination product sponsors have bought device companies; some sponsors rely heavily on partners to supply components and support product design. Various approaches are viable. Failure to embrace and operationalize ultimate responsibility for the product from cradle to grave, however, carries real risks of delayed market entry and of failures of manufacturing controls and other challenges postmarket. This book offers help to those entering the field to avoid these pitfalls. It may help convince current sponsors who have not yet fully embraced ultimate responsibility for their products to make the necessary change of mindset, allocation of resources, and adjustments to training, vendor relations, and internal operating procedures and systems.

It will be obvious to readers of the various chapters of this book that the authors, all of whom are from industry, embrace this concept of ultimate responsibility and the associated need to ensure timely, robust coordination and collaboration, as foundational. This is reflected as they explain best practices throughout the product lifecycle from earliest stages of product viability assessment, through product development, to supporting manufacturing controls, and managing postmarket product safety and change controls. The importance of timely coordination and collaboration is reflected in the discussion of the specific, substantive topics of virtually every chapter. For a more general framing of the importance and application of such coordination and collaboration, see, for example, Chapter 3 on regulatory strategy, Chapter 5 on integrated development, and Chapter 6 on risk management.

Like sponsors, FDA has continued to enhance and support its efforts to ensure timely internal coordination and collaboration. These efforts include: adopting procedures calling for intercenter consultation as the norm for product review and regulation (in the absence of agreements that it is not necessary due to lead Center expertise in the particular case for example) (see, e.g., Staff Manual Guide 4101 on intercenter consultation requests); training programs for all steps of the regulatory process from jurisdictional determinations to premarket review and postmarket oversight; IT enhancements to support efficient consultation, activity tracking and assessment; policies for enabling resolution of differences of view between components including in rare cases, formal intercenter disputes (see, e.g., SMG 9010.2 on cross-center dispute resolution at FDA), and engagement by OCP to provide direction, facilitate, and otherwise engage, as requested or needed, on issues raised by FDA components or regulated entities.

As indicated above, a core purpose of coordination and collaboration for regulators is to ensure appropriate consistency of their regulatory actions. For FDA, this means, for example, that constituent parts are regulated consistently with how they would be regulated if marketed independently; i.e., establishing expectations among Agency components to ensure that if additional information or controls are sought, they are needed to address new questions raised by the combined use.

Regulators in other jurisdictions do not always have the same ability to collaborate as robustly across regulatory programs, associated disciplines, and areas of expertise. Nonetheless, norms have continued to develop favoring such engagement, whether directly or through the sponsor, in pursuit of sound, appropriately consistent regulatory expectations. Chapter 14 on the evolving global regulatory landscape, for example, speaks to enabling such engagement among other important topics for sponsors looking to understand and market in multiple jurisdictions.

ADVANCES IN BEST PRACTICES AND REGULATORY THINKING – EMBRACING HOLISTIC RISK MANAGEMENT AND A "COMBINED USE" MINDSET

Two recent trends reflected and discussed in this book warrant highlighting as they both offer meaningful opportunities to facilitate, strengthen, and align the efforts of sponsors and regulators. These are increasing (a) application of robust, holistic risk management and (b) focus on the safety and effectiveness of the "combined use" to determine what issues sponsors and regulators need to address.

RISK MANAGEMENT – COORDINATING AND INTEGRATING EFFECTIVE PRODUCT DEVELOPMENT AND MANAGEMENT

Recognition of the need to apply a comprehensive, well-structured, collaborative approach to address issues throughout the product's lifecycle has raised the question of how to operationalize such a goal. Increasingly, sponsors report using risk management programs to do so. Risk management applied holistically can be used to structure, integrate, and coordinate the full range of iterative analytical and investigational work needed from early concept to postmarket management, to enable design, development, pre-clinical and clinical study, manufacture, and marketing of safe and effective products. In sum, risk management, robustly conceived, can be a truly comprehensive activity and tool for combination products.

A well-developed, holistic risk management program enables timely, fully informed consideration and resolution of issues. At each stage from development and investigation to scale-up and marketing, such a program enables identification and articulation of precise issues, development of well-considered analyses, and application of appropriate responses, informed by all relevant areas of expertise. It can ensure issues are addressed in a timely, integrated, and appropriately sequenced manner. Because such a program enables both the clear and precise framing of issues and development and application of sound solutions, it can facilitate engagement within sponsors, with partners and vendors, and with regulators.

Risk management expectations are reflected in international standards and guidance such as the International Organization for Standardization (ISO) standard *14971: 2019 Medical devices – Application of risk management to medical devices* and the International Conference on Harmonisation of Technical Requirements for Registration of Pharmaceuticals for Human Use (ICH) *Q9 Quality Risk Management* guidance. What is distinct for combination product and other combined uses (see section Embracing a Combined Use Construct - Breaking Free of "Combination Product" Semantics) is the need to integrate risk management needs for each constituent part to address the full range of issues for their combined use. The range of questions to address may vary based on such factors as the complexity of the product, the intended use, the use environment, and the user population. The goal is to create a comprehensive, coordinated risk management program to capture and manage the full and interactive sweep of issues relating to the constituent parts, their interactions, and their combined usage, in a timely manner.

This concept and sound approaches to stand up and apply such a program for combination products have been presented, for example, in the Association for the Advancement of Medical Instrumentation (AAMI) Technical Information Report (TIR) *TIR105:2020 Risk Management Guidance for Combination Products*. Chapter 6 delves further into these issues, presenting the concept of risk management in the context of combined use of different types of medical products, and how to establish and maintain a robust risk management program for your product.

EMBRACING A COMBINED USE CONSTRUCT – BREAKING FREE OF "COMBINATION PRODUCT" SEMANTICS

Increasing attention to risk and embracing a comprehensive view of risk management have facilitated another important shift. In framing product development and evaluation expectations, sponsors report shifting away from reliance on regulatory classifications (as drugs, devices, combination products), to a focus on what questions need to be considered to ensure safe and effective combined use of the different types of medical products at issue. This shift in mindset has begun to be reflected in the use of the term "combined-use products" to embrace all medical products intended to be used with other types of medical products, whether the configuration satisfies a jurisdiction's definition of combination product or not.

Many sponsors market a type of product intended for use with a class of another type of product, an obvious example being the marketing of a type of syringe that can deliver a class of drugs or of a drug that can be delivered by a category of syringe. Risk management for the device needs to take into account with which drugs the device can (and cannot) be safely and effectively intended for use and *vice versa* for such drugs. Such assessment is necessary to guide product development, identify data needs, and inform whether design, packaging, labeling, or other mitigation measures may be needed to ensure safe and effective combined uses.

The "combination product" concept has been helpful as an indicator of where combined-use questions may arise, but it is incomplete and imprecise and can, therefore, be unreliable and misleading. Not all jurisdictions expressly recognize combination products as a distinct category of medical product, and those that do may define the term differently. For example, some may include only physically/chemically combined products, others may include co-packaged ones too, and some include certain separately distributed "cross-labeled" products as well. In addition, jurisdictions use different terms ("combination product," "borderline product," "reference product"). Further, jurisdictions apply these concepts in varying ways (e.g., to define a legally distinct category of products or as a more informal aid to regulation).

The underlying scientific and technical issues, however, aren't driven by such semantics. Combination product status may be an indicator that additional questions need to be addressed beyond those raised by a drug or device individually, but not always, and safety and effectiveness questions raised by their combined use arise for products that may not be classified as combination products in any jurisdiction. This disconnect between terminology and substantive questions was highlighted in discussions of when products may qualify for "cross-labeled" combination products status in the United States, but it reaches beyond this special case. What issues may need to be addressed depends on which drugs and devices are being used together, for what purpose, by what users, in what settings, as well as whether they are being manufactured and distributed together.

Adoption of the "combined use" reframing to guide what issues need to be considered has also helped shape more precise assessments of issues for which drugs and devices need to be studied together and which can be considered drug or device "agnostic" and evaluated without regard to which specific drugs might be used with the device or *vice versa*. As a consequence, among other things, combined use analytical framing can facilitate successful developmental planning and engagement with regulators regarding the use of "platform" technologies, as discussed in various chapters of this book. See, for example, Chapter 5 on integrated development and Chapter 6 on risk management.

More broadly, this reframing has the potential to clarify communications with partners and regulators by fostering a focus on what scientific and technical issues need to be addressed rather than what regulatory classification may apply to the products at issue. Support for adoption of a "combined use" construct to enhance analytical clarity and drive best practices and sound regulatory expectations has, for example, been reflected in standard-setting body activities, e.g., at ASTM International, an encouraging indicator of acceptance and adoption of the construct.

LEVERAGING EFFORTS TO DATE TO ADVANCE CONVERGENCE AND HARMONIZATION

Much has already been done to promote best practices and clarify regulatory expectations concerning combination and other combined-use products. Regulators and regulated entities continue to work together in regional and international fora, including ICH, the International Medical Device Regulators Forum, ISO, ASTM, and AAMI, to develop guidance and standards to facilitate efficient interaction with regulators, application of consistent information and data expectations, and best practices by regulated entities and regulators. Still this work remains incomplete, and it can be challenging due to such factors as differences in regulatory programs for drugs and devices and from jurisdiction to jurisdiction, in associated expectations and cultural characteristics of regulated entities and regulators, and in terminology and its interpretation.

This book explains and advances a number of important concepts and best practices. It reflects where regulatory expectations are well developed and aligned, and where further work may be needed. Understanding of where we are and need to go is, of course, valuable to improve understanding and promote sound approaches. From a practical perspective, it is encouraging to hear from regulated entities that challenges currently faced often have more to do with inefficiencies among processes than substantive regulatory differences of opinion across jurisdictions, as continuing to address such inefficiencies should be more straightforward, but both types of challenges exist.

I have seen shared understanding developing over time among sponsors and regulators of what needs to be considered and shown to ensure safety and effectiveness. A conference several years ago offered a good example. A panel of industry experts was discussing current good manufacturing practices for combination products. While the expectations being addressed were captured in FDA regulations (21 CFR Part 4), the panel members were promoting the substantive importance and value of the cross-cutting substantive expectations, rather than focusing on regulatory expectations as the driver for adoption of the practices. In standards bodies, conferences, sponsor meetings, and informal conversations since, with sponsors and fellow regulators alike – not to mention in this book – I've seen similar evidence of such shared understanding and general expectations. This shared thinking is not universally embraced, and specifics of premarket data needs and postmarket controls are, of course, discussed and debated, but general expectations for what questions need to be addressed and how across the product lifecycle are increasingly shared and advanced by industry members to one another, and among regulators.

As discussed above in regard to the development of the US regulatory program for combination products, shared expectations offer a foundation for further clarification, refinement, and convergence of practices and policies, to the benefit of regulated entities, regulators, and the consumers and patients that combined-use products may benefit. Shared expectations can facilitate making paths to market more consistent, predictable, and efficient across the globe. Propagation of the combined-use mindset reflects such an increasingly shared perspective and can help regulated entities and regulators to understand one another, avoid getting bogged down in semantics, and focus on the scientific and technical issues raised by the combined use, whether in development of standards and shared guidance or in case-by-case regulation of products. Greater use of comprehensive, well-structured risk management programs, as recommended in this book, could meaningfully contribute as well. By enabling more clear, consistent, and timely framing of issues as noted above, use of such risk management programs can enable clearer, more timely, and consistent discussions of products, from product to product and regulator to regulator. This can reasonably be expected to lead to more clear, consistent, and appropriate risk-based expectations.

CONCLUDING REMARKS

We have come a long way from the initial Congressional efforts in the United States in 1990 to address safety and effectiveness questions for a growing, not clearly defined class of products. There is plenty more to do, but we have some key concepts well developed and increasingly embraced, including the importance of collaboration; the critical role of comprehensive, well-structured planning and analysis, and the need for a risk-based focus on substantive safety and efficacy issues raised by combined uses. These shared concepts have enabled the ongoing development of efficient, effective, and consistent regulatory programs and of best practices for regulated entities and regulators.

The extent of collaboration and planning needed can vary, and regulatory expectations and best practices can be refined in light of the risks to be managed. But commitment to these core concepts helps ensure that we bring the right expertise to bear and identify and mitigate the risks that need to be managed. Our success in maintaining and building on what has been accomplished in the United States and other jurisdictions can be expected to depend on the ongoing adoption and application of these concepts. This raises the question of how we can best assure continued success in these efforts.

As for any medical products, ultimately, the success of our efforts to manage the complexities of bringing products to market for combined use will depend on the strength of our commitment to ensuring the safety and effectiveness of these products to meet patient and consumer needs. Collaboration takes time, and best practices to manage risk demand resources. There will always be the temptation to save time and money by making exceptions, discounting risks, but so long as we retain firm contact with the touchstone of protecting and improving patient wellbeing, our procedural and substantive choices can be expected to keep us on a sound path.

FDA and combination product sponsors do not always agree, but we have engaged authentically, bringing reliable, robust information to the process to inform both product-specific actions and policy development. These core concepts have enabled such engagement to occur. In a time when misinformation and disinformation about science and fact pose greater threats to trust in medical products and their regulation, our efforts to engage openly and rigorously are significant. Not only are they important to our immediate goals and efforts, but as an indicator of the capacity and commitment to maintain and expand authentic engagement, critical to the successful promotion and protection of public health.

A shared vision and the core concepts it has engendered have enabled the development and refinement of appropriate, consistent practices and expectations for combination and other combined-use products. Their continued application can help advance sound, efficient industry and regulatory practices, and can further convergence of regulatory expectations. I am pleased to introduce you to this book as a significant manifestation of this vision and a noteworthy contribution to its advancement.

2 What Is a Combination Product?

Susan W. B. Neadle

CONTENTS

Globally, health authorities establish regulatory frameworks and quality management system expectations to help assure quality, safety, efficacy, and availability of drugs, biologics, cell tissue therapies, and devices. The increasing convergence and interrelationship of life sciences, pharmaceutical, engineering, digital, and medical technologies, examples illustrated in Figure 2.1, has blurred traditional lines of product classification and created a range of challenges for premarket development, identification, and application of regulatory pathways, post-market product lifecycle management, and safety reporting. In response to the technological evolution, health authorities and industry have increasingly come to embrace the concept of "combination product," "drug-device combinations," or "combined-use systems" to capture at least some products that raise such considerations.

Currently, there are a variety of terminologies, definitions, and interpretations applied across international jurisdictions with respect to the combined use of medical products and technologies (see the Appendix[1] of this book, "Comparative Overview of Global Combination Product Regulatory Landscape"). In each case, regardless of the differences, the evolving regulatory and quality system landscape intends to ensure that, above and beyond the safety and efficacy of the drugs, devices and/or biological products alone, their combined use likewise meets quality, safety, and efficacy expectations, ultimately to promote and protect public health. In this chapter, we will review some high-level definitions and classifications for combination products as well as examples of what is – and is not – considered a combination product from jurisdiction to jurisdiction.[2]

COMBINATION PRODUCT: DEFINITION

Globally, the definition and interpretation of "combination product" and associated terminology varies (examples in Table 2.1).[3] In the United States, a combination product is a legally distinct type

DOI: 10.1201/9781003300298-3

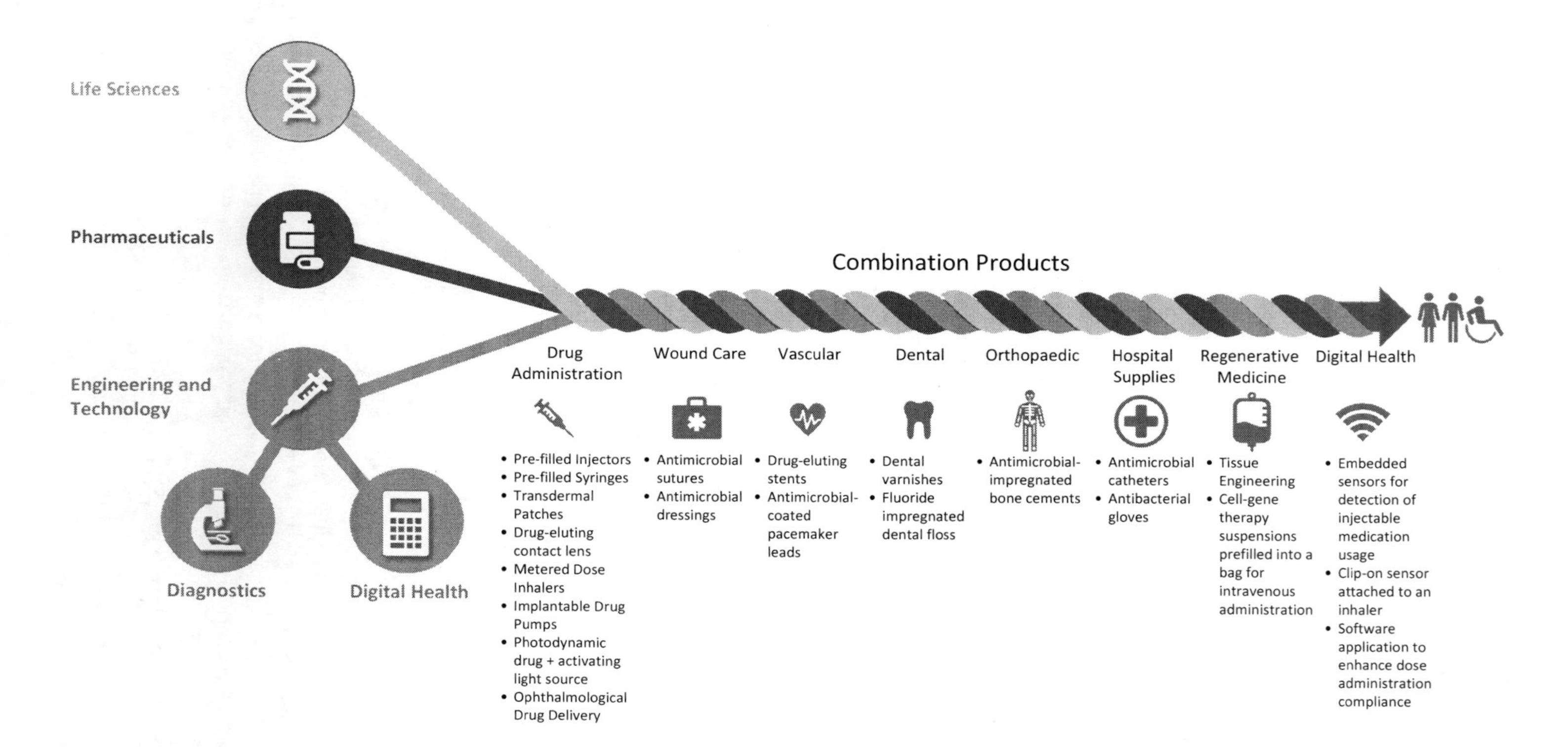

FIGURE 2.1 Combination products: The convergence of medical technologies.

TABLE 2.1

Examples of the Varied Jurisdictional Definitions for Combination Products. © 2023 Combination Products Consulting Services LLC.

Country	Definition of Combination Product
United States	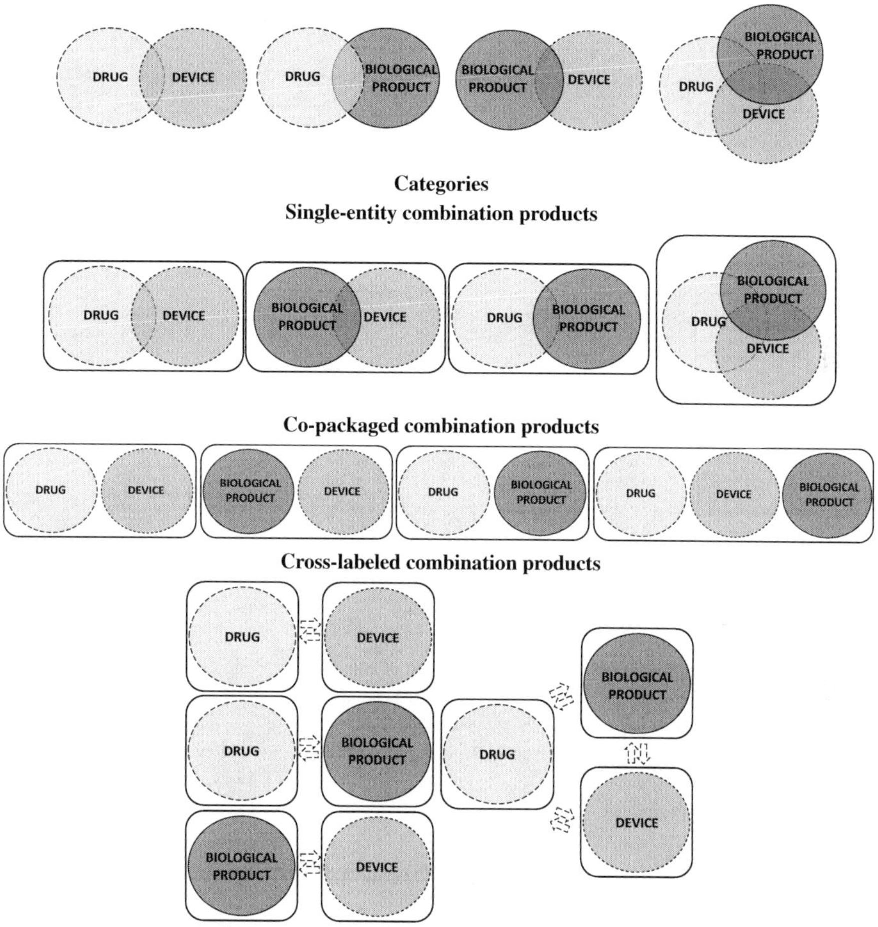 DRUG DEVICE; DRUG BIOLOGICAL PRODUCT; BIOLOGICAL PRODUCT DEVICE; DRUG BIOLOGICAL PRODUCT DEVICE **Categories** **Single-entity combination products** DRUG DEVICE; BIOLOGICAL PRODUCT DEVICE; DRUG BIOLOGICAL PRODUCT; DRUG BIOLOGICAL PRODUCT DEVICE **Co-packaged combination products** DRUG DEVICE; BIOLOGICAL PRODUCT DEVICE; DRUG BIOLOGICAL PRODUCT; DRUG DEVICE BIOLOGICAL PRODUCT **Cross-labeled combination products** DRUG DEVICE; DRUG BIOLOGICAL PRODUCT; BIOLOGICAL PRODUCT DEVICE; DRUG BIOLOGICAL PRODUCT DEVICE
European Union	Device administering drug: Medicinal Product MEDICINAL PRODUCT Drug and/or Biological Product DEVICE Single-Integral Non re-usable; Intended use only for given combination ***"Drug Device Combination Product"*** Device administering drug: Medicinal Product MEDICINAL PRODUCT Drug and/or Biological Product DEVICE Single Integral, Re-usable MEDICINAL PRODUCT Drug and/or Biological Product DEVICE **Co-packaged** (Or separately provided, but specifically indicated in the SmPC & package leaflet) MEDICINAL PRODUCT Drug and/or Biological Product DEVICE **Referenced product** Medical devices that are obtained separately by the user for use with medicinal products. Medical device with ancillary medicinal substance MEDICINAL PRODUCT Drug and/or Biological Product DEVICE

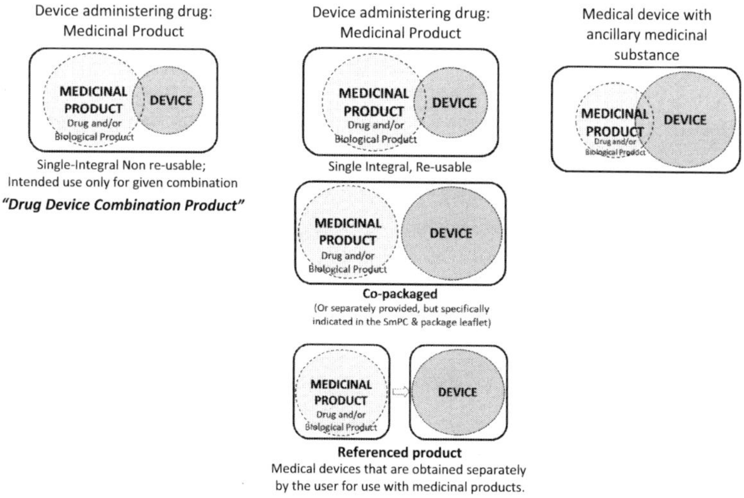

(*Continued*)

TABLE 2.1 (CONTINUED)
Examples of the Varied Jurisdictional Definitions for Combination Products. © 2023 Combination Products Consulting Services LLC.

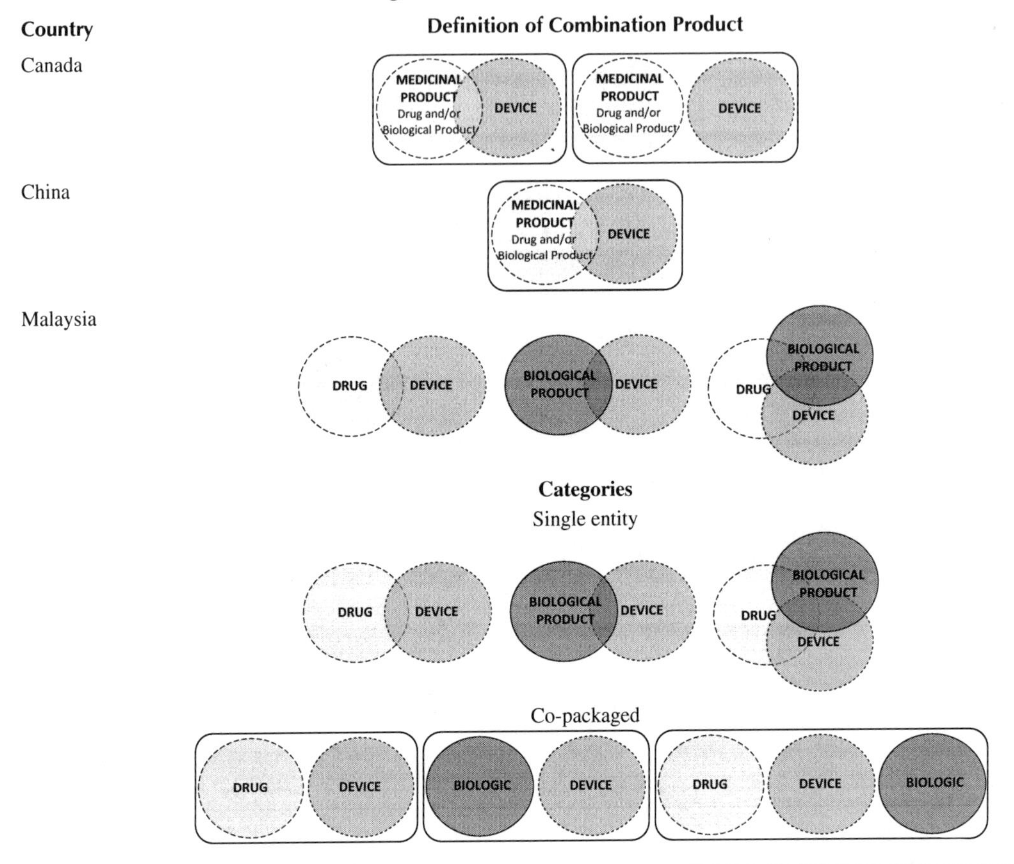

of product composed of two or more different types of medical products, i.e., any combination of a drug[4] and a device; a biological product and a device; a drug and a biological product; or a drug, device, and a biological product [21 CFR §3.2(e)]. When these different types of medical products are used in a combination product, they are referred to as constituent parts of the combination product. US regulations and policy statements further provide that the combination product classification includes medical products combined as a single entity, in a co-package, or cross labeled for use with one another. Of note, drug–drug and biologic–biologic combinations (otherwise known as "combination therapies" or "fixed dose combinations in the same category of product") (e.g., fibrin in a vial, co-packaged with thrombin in a vial, for combined use as a sealant), are considered in a separate category from "combination products" as defined under 21 CFR §3.2(e).

Contrast this US definition with Europe, where there is currently no formal legal definition given to the term "combination product." Under EU Medical Device Report (MDR) (2017/745) Article 117, combination products are generally referred to as medicinal products and medical devices that are placed on the market together. Under their draft "Guideline on the Quality Requirements for Drug-Device Combinations" (May 2019), EU's Competent Authority defines the term "drug-device combination product" (DDC) as a "medicinal product(s) with integral and/or non-integral medical device(s) necessary for administration, correct dosing or use of the drug product." Under EU MDR Article 117 regulatory framework, such products are ultimately classified either as medicinal products or as medical devices.

In Canada, a combination product is formally defined as a therapeutic product that combines a drug component and a device component (which by themselves would be classified as a drug or a device), such that the distinctive nature of the drug component and device component is integrated in a **singular product** (Health Canada, 2005; Kapoor et al., 2013).[5]

Similarly, in China, under the National Medical Products Administration (NMPA), Notification on Matters Concerning Registration of Drug-Device Combination Products (2021-10-28), a drug-device combination product is a product composed of any combination of a drug and a device, which is physically combined or mixed and produced as a single entity (NMPA, 2021).

The Ministry of Health Malaysia's (2020) definition of a combination product is more similar to the United States, where a combination product is comprised of (1) two or more regulated components, i.e., drug/device, biological/device, or drug/device/biological, that are physically, chemically, or otherwise combined or mixed and produced as a **single entity**, or (2) in which two or more separate products are **packaged together** in a single package or as a unit, and consist of drug and device products, or device and biological products. Convenience pack products (e.g., first aid kit consisting of medical devices and non-scheduled poison products), cross-labeled products, and health supplement products are excluded from this definition. Drug–biologic combinations do not fall under Malaysia's definition of the term combination product (Malaysia N., 2020).

The inconsistency and variety of interpretations for combination product terminology may lead to confusion and complexity for regulatory pathways, quality system expectations, and lifecycle requirements for these products.

Different interpretations mean that combination product regulatory, quality, and development professionals need to be multilingual: If they wish to work effectively in the combination products space, they must be able to speak the language of drugs, devices, and/or biological products included in the products with which they are charged and, they also need to become fluent in the language of the global combination products dictionary/thesaurus: different words that mean the same thing; the same terminology that means something different; and the expanse in between these two extremes.

In this chapter, for illustrative purposes, we'll compare terminology applied in the United States, with that in Europe, in Malaysia, and in China (*also see the Appendix and Chapter 14 in this book*). We also introduce "Primary Mode of Action," an important concept applicable to combination products across jurisdictions globally. Importantly, as we progress through the chapter, we'll observe that regardless of terminology differences globally, jurisdictions are striving to regulate the combined use of medical products in a way that is consistent with how they each approach their drugs, biologics, and devices, with varying levels of coordination within and/or across each jurisdiction's regulatory authorities responsible for the primary and secondary constituent parts that comprise a combination product, to assure the combination product is safe, effective, and usable for its intended use.

United States

As stated, in the United States, the term "combination product" is formally defined and includes any combination of a drug and a device; a biological product and a device; a drug and a biological product; or a drug, device, and a biological product [21 CFR §3.2(e)]. When these different types of medical products comprise a combination product, they are referred to as "constituent parts" of the combination product.

The US FDA goes on further to identify three categories of combination products in 21 CFR 3.2(e): (1) single entity, (2) co-packaged, and (3) cross labeled (Table 2.1 – high level, and Tables 2.2–2.5).

(1) **Single-entity combination product**: A product comprised of two or more regulated components that are physically, chemically, or otherwise combined or mixed and produced as a single entity. Examples of the myriad types of products considered single-entity combination products in the United States that have been, and continue to be, developed, approved,

TABLE 2.2
Combination Product Types Defined under 21 CFR 3.2(e). ©2023 Combination Products Consulting Services LLC.

Single-entity combination products

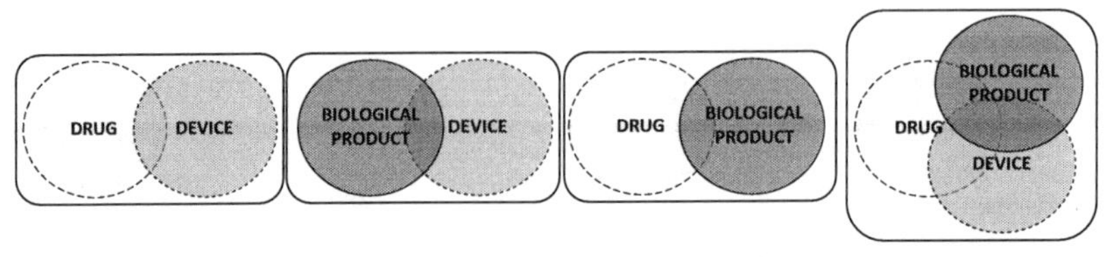

A product comprised of two or more regulated components that are physically, chemically, or otherwise combined or mixed and produced as a single entity.

Co-packaged combination products

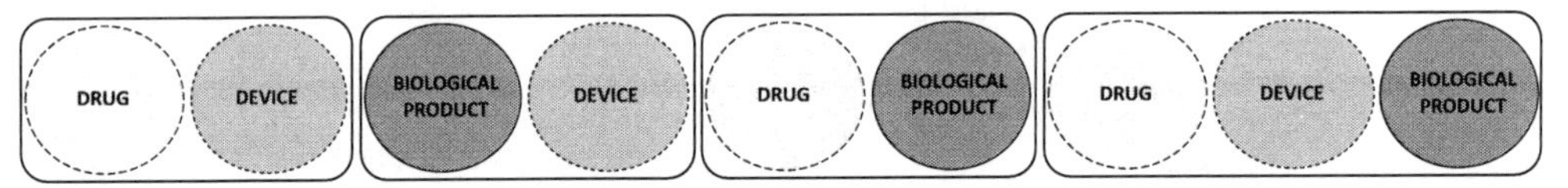

Two or more separate products packaged together in a single package or as a unit, and comprised of a drug and device; a device and biological product; a drug and biological product; or a drug, device and biological product.

Cross-labeled combination products

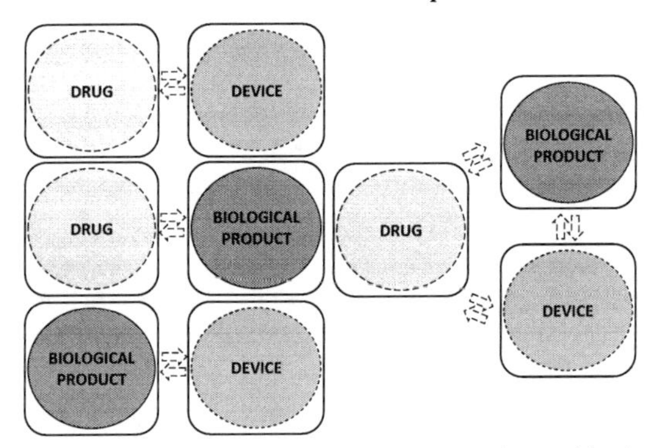

Constituent parts that are packaged separately, but specifically designated for combined use in order to achieve the intended use, indication, or therapeutic effect (see specific definition under 21 CFR 3.2(e)(3) and (4)).

and marketed are presented in Table 2.3. With the increasing ubiquity and vast array of single-entity combination products, few companies can afford to think of themselves any longer as simply manufacturers of drugs, devices, or biological products.

(2) **Co-packaged combination product**: Two or more separate products packaged together in a single package or as a unit, and comprised of a drug and device; a device and biological product; a drug and biological product; or a drug, device and biological product. Examples of products considered co-packaged combination products in the United States are presented in Table 2.4.

(3) **Cross-labeled combination product**: Constituent parts are packaged separately. This category of combination product embraces certain separately distributed medical products that are specifically designated for combined use in order to achieve the intended use, indication, or therapeutic effect. An example is listed in Table 2.5. Under 21 CFR 3.2(e)(3) and (4), a cross-label combination product includes:

1. A drug, device, or biological product packaged separately that according to its investigational plan or proposed labeling is intended for use only with an approved

TABLE 2.3
Examples of Single-Entity Combination Products

Example	Description
1) Prefilled drug-delivery device/systems	Prefilled drug syringes, disposable autoinjectors, transdermal patches with controlled release drug, nasal spray pump, metered-dose inhaler, prefilled iontophoresis system, drug-filled IV bag, multi-dose pen with a prefilled cartridge, on-body injector
2) Device coated/impregnated/ otherwise combined with drug	Drug-eluting stents, drug-eluting contact lenses; drug pills embedded with sensors, drug-eluting leads, condoms with spermicide, dental floss with fluoride, antimicrobial-coated sutures, bone cements impregnated with antibiotics, antimicrobial wound dressings, drug-releasing intrauterine device
3) Device coated or otherwise combined with biologic	Live cells seeded on or in a synthetic scaffold, extracorporeal column with column-bound protein
4) Prefilled biologic-delivery systems	Vaccine or other biological product in a prefilled syringe or autoinjector, prefilled nasal spray pump, transdermal systems, or microneedle patch preloaded with biological product, cell–gene therapy suspensions prefilled into a bag for intravenous administration
5) Drug/biologic combination	Progenitor cells combined with a drug to promote homing, antibody–drug conjugates
6) Drug/biologic/device combination	Syringe prefilled with drug–biologic conjugate

TABLE 2.4
Examples Co-packaged Combination Products

Example	Description
a. Drug or biological product vials packaged with device(s) or accessory kits	Empty syringes and/or transfer sets that are packaged together with a medicinal (drug or biologic) product
b. Medicinal product in a bottle, packaged with a delivery device	Oral dosing pipettes, medicinal cups, dosing spoons or piston syringe, packaged with a medicinal product
c. Convenience Kits[6]	First aid kit containing devices (e.g., bandages and gauze) and drugs (antibiotic ointments, pain relievers)
d. Surgical kit	Surgical tray with surgical instruments, drapes, and anesthetic or antimicrobial swabs
e. Kit containing two different single-entity combination product types	Drug-eluting device provided together with a biologic-prefilled syringe

TABLE 2.5
Example of Cross-Labeled Combination Products

Example	Description
Photodynamic therapy	Photosensitizing drug and activating laser/light source may comprise such a combination product

individually specified drug, device, or biological product where both are required to achieve the intended use, indication, or effect and where upon approval of the proposed product labeling of the approved product would need to be changed, e.g., to reflect a change in intended use, dosage form, strength, route of administration, or significant change in dose; *or*

2. Any investigational drug, device, or biological product packaged separately that according to its proposed labeling is for use only with another individually specified investigational drug, device, or biological product where both are required to achieve the intended use, indication, or effect.

Guidance on this category of combination products has been limited. It is a unique construct recognized by the US FDA. Other global jurisdictions have not recognized, nor adopted, "cross-label" under their definitions and classification of combination products at this time. Regardless of jurisdiction, however, on its face it may be helpful to recognize the spirit behind the concept of cross-labeling: This category speaks to questions of how best to ensure safety and efficacy of products intended for combined use (see discussion in ASTM International Combination Products Terminology Standard (2023)). This includes, for example, whether products need to be studied together, whether sponsors for the products at issue need to collaborate with one another, and how the products should be labeled to ensure safe and effective combined use. "Combined use," from a practical perspective, is a term for when to consider more than one product in your risk analysis and control strategy. The risk analysis and control strategy for combined-use products need to consider each medical product and their use together.

Importantly, FDA has stated that concomitant use of a drug, biologic, and/or device does not typically create a cross-labeled combination product. Just because a drug is labeled for combined use with devices, or vice versa, does not typically make the two cross labeled. Drug labeling, for example, might provide technical specification criteria for the devices to be used, without establishing a cross-labeled combination product. For example, devices such as syringes, labeled for general use with parenteral drugs, do not comprise cross-labeled combination products with these drugs.

One implication of whether one's combined-use products fall within the legal category of "cross-labeled combination product" is the possibility of having the products reviewed under a single application as opposed to having to submit separate ones for each product. Further, there is a requirement under 21 CFR §4B to share certain post-market adverse event information with the applicant for the other constituent part if you are not the sponsor for both. These considerations will be addressed further in Chapter 8 (section on PMSR).

Those with additional questions or product-specific questions are encouraged to contact the US FDA Office of Combination Products.

PRIMARY MODE OF ACTION (PMOA)

Generically, irrespective of global jurisdiction, a mode of action (MOA) is the means and/or methods by which a medical product achieves its intended effect(s) or action(s) (modes of action: drug, device, or biological product). Each combination product, therefore, has at least two modes of action, one for each of its constituent part types. Generally, all you need to know is what types of constituent parts a given combination product has, and you'll know what MOAs there are.

Understanding the mode of action is critical to determine the classification of a product, namely whether under medicines (drug and/or biologic) or medical device legislation. However, the definition of a medical device in relation to mode of action varies between regions and respective regulatory frameworks. For example, in the United States, the definition of a medical device states that it "does not achieve its primary intended purpose through chemical action within or on the body of man or other animals, and which is not dependent upon being metabolized for the achievement of it primary intended purposes." This contrasts with Europe under (EU)2017/745 on medical devices (MDR), as well as within the World Health Organization (WHO) Global Model Regulatory Framework for Medical Devices, as well as other regional guidance documents, where the wording used is "which does not achieve its principal intended action by pharmacological, immunological, or metabolic means, in or on the human body, but which may be assisted in its intended function by such means." While these definitions might be interpreted similarly, the use of different terms may increase the likelihood of similar products being classified differently.

The Primary Mode of Action (PMOA) is the single MOA that makes the greatest contribution to the combination product's overall intended use(s). Interpretation of which MOA is deemed as the primary one may differ across jurisdictions, but regardless, the PMOA is an important concept recognized across jurisdictions.[7] The PMOA designation is an important aspect of combination product status with implications across the product's lifecycle: PMOA designation can impact premarket pathway availability and other premarket regulatory processes, substantive requirements for marketing authorization and timing of review, cGMP expectations, and even post-marketing safety reporting obligations. Regardless of jurisdiction, the general intent is to apply a risk-based approach, using application types based on the Primary Mode of Action, while gaining endorsement and input from the regulatory body associated with the constituent part providing the secondary MOA of the combination product.

Following are the examples of how PMOA may be determined.

Consider a biologic product-prefilled syringe. The focus of this combination product is to provide the therapeutic effect of the biological product, while the syringe plays a support role to deliver the required dose of the biologic into the body. The PMOA in this case is therefore that of the biological product.

Note:
Just because a constituent part is not identified as providing the primary mode of action does not lessen the importance of taking that secondary constituent part seriously. In the example above, while the PMOA is that of the biological product, one should not discount the importance of the device design. If the device design, for example, results in painful dose administration, one possible unintended outcome might be non-compliance to the intended dose regimen; just because the device does not provide the PMOA, does not make it less important for user-needs focused design, functionality robustness, and reliability requirements.

Similarly, in the case of a device PMOA product with a drug or biologic secondary MOA, one would need to take the drug/biologic seriously. For example, the drug will rarely be formulated like a previously approved drug or biologic product that includes the same active (if there is such an approved product), and that approved drug or biologic often is not labeled for the same route or mode of administration, so one cannot simply rely on the fact that there is an approved NDA or BLA for a related use.

Consideration of both primary and secondary MOA critical performance requirements is discussed further in the subsequent chapters of this book.

Consider another example of an antibiotic-coated suture: Wound closure is the key impact needed, while the antibiotic coating is ancillary, preventing infection during and subsequent to wound closure. In this case, the medical device provides the PMOA.

Generally, PMOA for combination products is straightforward to determine. However, in some cases, it may not be possible to determine PMOA directly (e.g., due to lack of clear data), or the constituent parts may not be contributing to the same intended effect at all. Such products may be referred to as "Borderline Products," i.e., in the combined-use context, medical products that are covered by at least two legislations (e.g., both medical device and medicinal product[8]), whose lead legislation within a jurisdiction may be unclear. The term "Borderline" is used in a broader context in some jurisdictions. For example, MHRA Guidance (6 January 2021) "Borderline Products: How to Tell If Your Product Is a Medicine" indicates that "Some products are hard to distinguish from medicines, for example products that might be medical devices, cosmetics, food supplements or biocidal products." In Australia, Therapeutic Goods Administration (TGA) refers more broadly to medicinal product-device combined-use products as "device-medicine boundary products" (https://www.tga.gov.au/publication/device-medicine-boundary-products). In some countries, the designation as "borderline" is temporary. Identification of which products are considered "borderline," and the approach for determining a borderline product's lead legislation and applicable regulatory framework within a jurisdiction varies from country to country.

Consider the example of a drug-eluting contact lens. While the contact lens is intended to correct the vision of the wearer, the drug with which the lens is impregnated or coated may be intended to elute out in order to treat an eye health condition. In such cases, the designation of PMOA is not always straightforward. Issues are complex and product specific. Where there is ambiguity in PMOA designation, various regulatory strategies are applied across jurisdictions for risk-based classification or assignment to a lead entity for regulatory review and oversight. In this chapter, we will touch briefly on PMOA analysis in the United States, the European Union, China, and Malaysia just to illustrate the complexity.

United States

The US FDA has both informal and formal mechanisms to support requests for designation of the product [Pre-Request for Designation Guidance (2018), Request for Designation Guidance (2011), Classification of Products as Drugs and Devices and Additional Product Classification Issues (2017), and Draft Pathways Guidance (2019)] to support **both classification (as a combination product, drug, device, or biological product) and/or PMOA determination**.[9] When PMOA is in doubt, the US FDA applies a two-step assignment algorithm (21 CFR§3.4) to determine PMOA. As per this algorithm (illustrated in Figure 2.2), if there is a combination product that raises similar questions of safety and effectiveness that have been previously assigned, then US FDA will assign the product to that same Lead Center that regulates the similar combination product(s) ("Consistency"). If there is not such a previously assigned combination product, then FDA will assign the combination product to a Lead Center based on which constituent part raises the most significant questions of safety and effectiveness, so that the Center with the most expertise related to the determined safety and efficacy questions has oversight ("Safety & Effectiveness"). Reference the RFD Guidance (2011) for more information on issues FDA considers relevant to determining whether a combination product can be assigned on the first step or the second.

Table 2.6 lists examples of combination products based on their US FDA classification and PMOA.[10] [*Note*: There are exceptions. Not all drug PMOAs are assigned to the Center for Drug Evaluation and Research (CDER), nor all device PMOAs to the Center for Device and Radiological Health (CDRH), nor all biologic PMOAs to the Center for Biologics Evaluation and Research

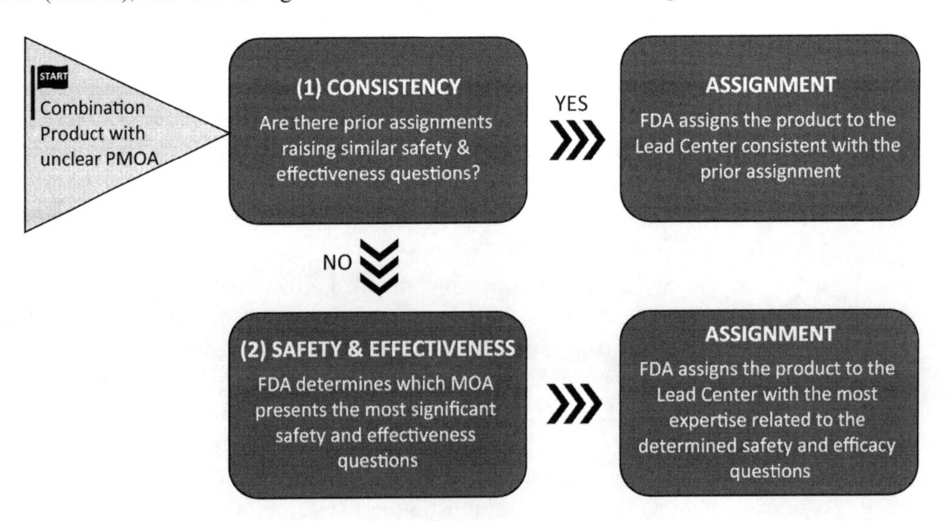

FIGURE 2.2 FDA assignment algorithm for combination products with non-obvious PMOA (adapted per 21 CFR §3.4).

TABLE 2.6
Combination Products Examples – US Classification and PMOA

Single entity: drug (or biologic) PMOA	• Cell-gene-tissue therapy prefilled IV bag • Drug-eluting beads • Drug-eluting contact lenses • Drug-eluting discs • Dry-powder inhalers • Iontophoretic delivery systems • Metered-dose inhalers • Prefilled dual-chamber syringe • Prefilled autoinjectors • Prefilled nasal spray • Prefilled on-body delivery system • Prefilled pen injectors • Prefilled syringes • Scaffolds for tissue engineering • Transdermal patches
Single entity: device PMOA	• Drug-coated catheters • Antibiotic-impregnated bone cements • Spermicidal condoms • Antibacterial-releasing dental restorative materials • Drug-coated pacemaker leads • Drug-coated stents • Antimicrobial surgical scrubs • Antibiotic-impregnated surgical mesh • Antimicrobial wound dressings
Co-packaged combination products (varied PMOAs based on intended uses)	• Liquid drug co-packaged with dosing dispenser (drug PMOA) • Syringe co-packaged with fibrin sealant (biologic PMOA)
Cross-labeled combination products (each constituent part is generally regulated based on constituent part type, however PMOA would matter for purposes of investigational submissions, and if a single marketing application is submitted for the combination product, and would be made clear)	• Photodynamic therapy + activating light source (drug PMOA) (PMOA would determine product assignment if both products are being studied or reviewed under a single application. If separate marketing applications are made for the drug and device, then those applications would be assigned to separate centers, though the centers would still coordinate on premarket review and regulation of the product.)

(CBER) (e.g., therapeutic proteins are assigned to CDER instead of CBER, so combination products made of a drug and such a protein product would always be assigned to CDER, regardless of PMOA; devices for use with blood and blood products are assigned to CBER, so a combination product comprised of such a device and biological product would always be assigned to CBER regardless of PMOA).] (*This is discussed in more detail in Chapters 3 and 14 of this book.*)

Examples follow of how other jurisdictions approach to PMOA designation, specifically Europe, Malaysia, and China.

Europe

In Europe, as stated earlier, products which combine a medicinal product or substance and a medical device are regulated either under EU MDR (2017/745) (Device PMOA) or under Medicinal Product Directive (MPD) 2001/83/EC of the European Parliament and of the Council (Drug PMOA) (Figure 2.3). There are very different approaches in regulatory approval systems for a medicinal product and a device. Similar to the "*Primary* Mode of Action" concept in the United States,

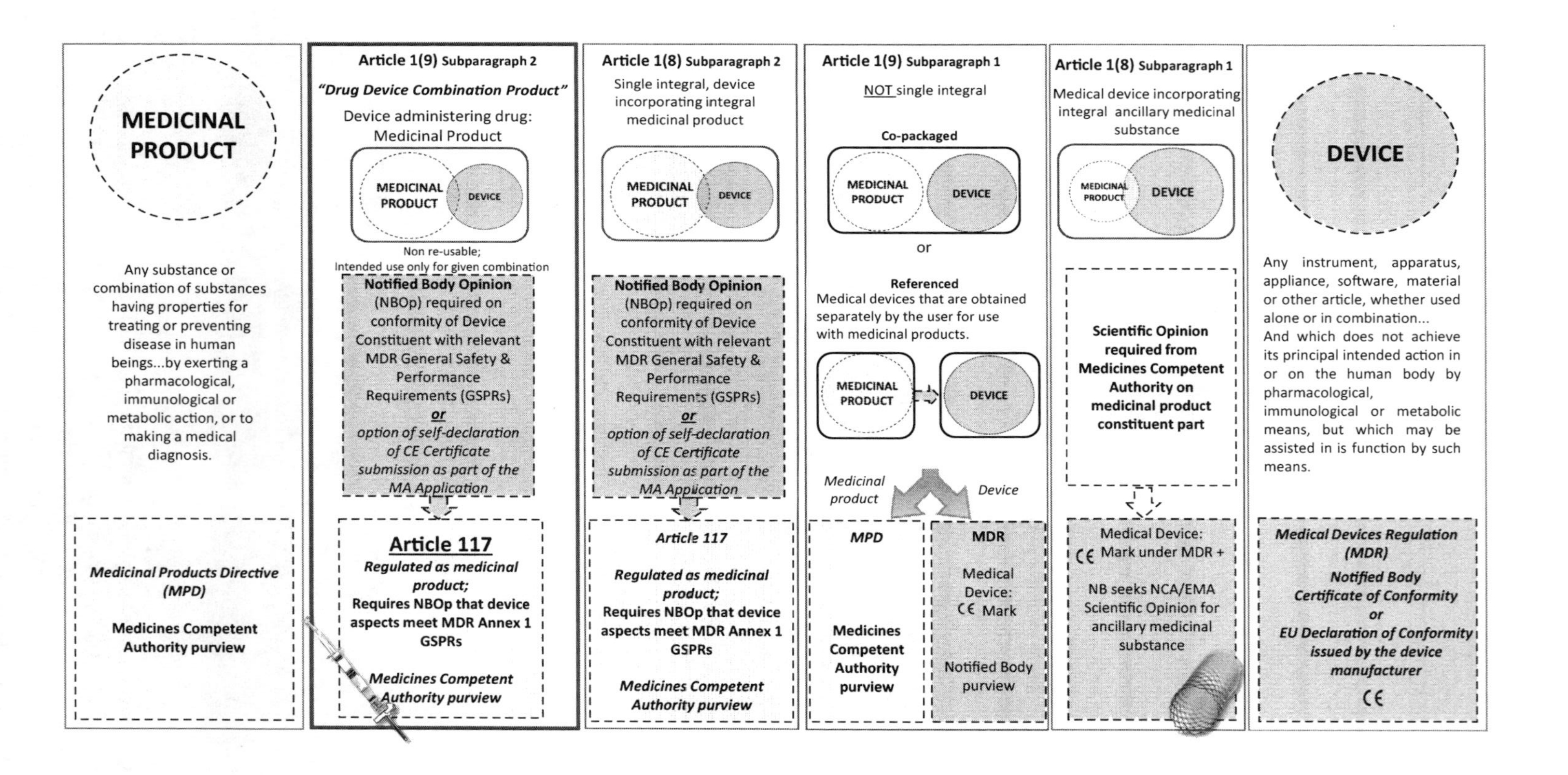

FIGURE 2.3 Combination products under EU MDR (2017/745) and PMOA designation (Leading Legislative Act). ©Combination Products Consulting Services, LLC. All Rights Reserved.

though, in Europe classification as either a medicinal product or as a medical device depends on the "*Principal* Mode of Action" and intended use of the product. Articles 1(8), 1(9), and Article 117 of the EU MDR articulate the regulatory expectations for drug and device combinations.

Under Article 117 of the EU MDR (2017/745), the legislation calls for coordination between Notified Body and Competent Authority to consider regulatory issues raised by each of the constituent parts of the combination product. The legislation calls for appropriate interaction in terms of consultations during premarket assessment, and of exchange of information in the context of vigilance activities involving such combination products. Under EU MDR Article 117, designation of PMOA as medicinal product or as device triggers the associated type and extent of coordination of combination product duties between Competent Authority and Notified Body.

Under Article 1(9): A drug PMOA drug-device combination product is

> a device intended to administer a medicinal product and the medicinal product are placed on the market in such a way that they **form a single-integral product** which is **intended exclusively for use in the given combination** and which is **not reusable**.

Also included under the designation of drug PMOA drug-device combination, under Article 1(8): A **device incorporating as an integral part** a substance considered to be a medicinal product, whose action is principal and not ancillary to that of the device.

So, under EU MDR Article 117, the analysis of PMOA, and trigger of associated DDC duties, centers on the following key question:

> Is the device intended to deliver a medicinal product?

If yes, to decide if it is a single-integral product subject to Article 117, ask the following questions:

1) Do the drug and device make a single unit that cannot be separated by the end user? (i.e., the drug and device are considered single integral)
2) Are the drug and device intended exclusively for use in the given combination?
3) Is the drug/device combination *not* reusable?

If the answer to all of these questions is "YES," the product is considered a single-integral "Drug-Device Combination" (DDC) product and will be treated as a medicinal product (**drug PMOA**), regulated under the medicinal product's framework 2001/83/EC, overseen by the Competent Authority (CA). In that case, DDC duties are triggered: The sponsor must demonstrate relevant General Safety and Performance Requirements (GSPRs) for the device constituent and submit a Notified Body Opinion for the device constituent part to the Competent Authority or European Medicines Agency (EMA), who approves the product as a medicinal substance. (GSPRs are discussed in more detail in Chapter 4 of this book.)

As illustrated in Figure 2.3, if the device part of a DDC is already a Conformité Européenne (CE)-marked medical device, and if this CE-marked medical device is used within its intended purpose in the DDC, evidence on conformity based on the Declaration of Conformity, and/or the EU certificates is sufficient. There is no need for a Notified Body Opinion. See also the official text in Article 117, EU MDR (2017/745):

> the conformity of the device part with the relevant general safety and performance requirements set out in Annex I to that Regulation contained in the manufacturer's EU declaration of conformity or the relevant certificate issued by a notified body allowing the manufacturer to affix a CE marking to the medical device.

Some examples of integral drug-device combination products (Medicinal Product PMOA) are single-dose prefilled syringes, single-dose dual-chamber syringes, prefilled pens, prefilled autoinjectors, and a preloaded on-body delivery system (OBDS). Other examples of integral devices that incorporate a medicinal substance as the PMOA include drug-releasing intrauterine devices and implants containing medicinal products whose primary purpose is to release the medicinal product.

If the answer to any of the above questions is "NO," the notified body provides oversight on product conformity through CE marking of the device, where a Notified Body (NB) would be required for CE marking of the medical device if it were a stand-alone product, i.e., a Class I device with measuring function (Im) or provided sterile (Is), Class IIa, Class IIb, and Class III devices (the new MDR Class Ir for reusable devices is unlikely to be relevant as this applies to surgical instruments and endoscopes).

Non-integral/Co-packaged Combination Products with Medicinal Product PMOA

Where a CE-marked device for the administration of a medicinal product is co-packaged or is referred to in the product information of a marketing authorization for a medicinal product, additional information on the device may need to be provided by the applicant with regard to any impact on the quality, safety, and/or efficacy of the medicinal product in the Market Authorization Application (MAA). The nuances of non-single-integral and reusable combination products are complex and guidance is currently limited. A draft guidance from the EMA Quality and Biologicals Working Parties "Guideline on the Quality Requirements for Drug-Device Combinations" was published in May 2019 and received over 800 comments. EMA held a workshop in November 2020 with Team NB and the industry to discuss questions and requests for clarification. Subsequently, EMA issued clarifications via a Questions & Answers document (23 June 2021 rev. 2) and a guideline on quality documentation for drug-device combination products in July 2021 that came into effect on January 1, 2022. Further updates are anticipated as this space evolves in Europe, e.g., European Commission is expected to update the current document Guidance on Classification (EU Commission Guidance MEDDEV 2.1.3 rev 3).

Examples of "non-integral combination products," regulated as devices, include injection needles and refillable pens and injectors that use cartridges (e.g., a drug cartridge co-packaged with a delivery device, where the user loads the drug cartridge into the device, even if the drug cartridge only works with the device component, and the device component is specifically designed for the drug cartridge; even though the two need to be together and the device serves merely to deliver the drug, it's still not considered a medicinal product combination product under Article 117) (Goebel (October 2019)).

However, before a Market Authorization (MA) can be granted, a CE certificate for the co-packaged device will be required by the medicines CA/EMA. Information on the device will be expected in the MAA according to the relevant sections of the EMA "Guideline on the Quality Requirements for Drug-Device Combinations" and further questions may be asked by the MA reviewer. For Class Is, Im, IIa, IIb, and III devices, a Notified Body will be involved in CE marking.

Given the broad range of non-integral devices, the information to be provided depends on the specifics of the device, its intended use, and associated risks to the quality, safety, and/or efficacy of the medicinal product. There are separate guideline sections for devices that are co-packaged and for those that are obtained separately and referred to in the product information.

Combination Products with Device PMOA and Companion Diagnostics

According to the October 2019 "Questions & Answers on Implementation of the Medical Devices and In Vitro Diagnostic Medical Devices Regulations ((EU) 2017/745 and (EU) 2017/746)" issued by the European Medicines Agency:

- For medical devices incorporating a medicinal substance (with action ancillary to the device) (Regulation 2017/745 Annex IX 5.2), the notified body shall seek a scientific opinion from either the National Competent Authorities (NCAs) or EMA.
- The notified body shall seek the opinion of EMA for medicinal products falling exclusively within the scope of centralized procedure (Annex I, Regulation (EC) No 726/2004), or that

incorporate human blood or plasma derivatives. [Regulation 726/2004 refers to products of biotechnology; new active substances for treatment of certain diseases including cancer, diabetes and viral diseases; and orphan medicinal products].

- For devices that are composed of substances, or of combinations of substances, that are systemically absorbed by the body in order to achieve their intended purpose, the notified body shall seek a scientific opinion from either the NCAs or the EMA (Regulation 2017/745 Annex IX 5.4).
- For companion diagnostics, the notified body shall seek a scientific opinion from either the NCAs or the EMA (Regulation 2017/746 Annex IX 5.2, Annex X 3(k)).

The Q&A goes on to state that the European Commission may consult EMA when deliberating on the regulatory status of products in borderline cases involving medicinal products (Regulation 2017/745 Article 4, Regulation 2017/746 Article 3).

To summarize, according to EMA's Q&A guidance (updated on October 2019), if the action of the medicinal substance is ancillary to the device, in other words, if it's a device PMOA, then the medical device must bear a CE mark and a scientific opinion must be obtained from the medicines authority for the medicinal substance before the notified body can issue the conformity certificate (see Figure 2.3).

Malaysia

In Malaysia, a **drug-medical device combination product** (DMDCP) (drug PMOA combination product) is one whose PMOA is driven by pharmacological, immunological, or metabolic action in/on the body where the National Pharmaceutical Regulatory Agency (NPRA) is the primary agency of the combination product. On the other hand, **a medical device-drug combination product** (MDDCP) (device PMOA combination product) is one whose PMOA in or on the human body is *not* driven by pharmacological, immunological, or metabolic means, but that may be assisted in its intended function by such means. MDDCPs are regulated under the Medical Device Authority (MDA) as the primary agency of the combination product. Malaysia does not include drug–biologic combinations under its definition of combination product.

Similar to the Notified Body strategy in the European Union, an Endorsement Letter, a document issued by the secondary agency after a satisfactory review of the ancillary component, is provided to the agency who has primary jurisdiction based on the combination product's PMOA. In the case of a DMDCP, MDA provides an endorsement letter to NPRA in the product review process. In the case of an MDDCP, the NPRA first reviews and endorses the ancillary drug constituent part, and sends an endorsement letter to the MDA for consideration in overall product approval (Ministry of Health Malaysia, Guideline for Registration of Drug-Medical Device and Medical-Device-Drug Combination Products, September 2020 "Guideline For Drug-Medical Device And Medical Device-Drug Combination Products, 5th Edition, January 2023"). This secondary agency "Endorsement Approach" is risk based with regard to device constituent parts. "Low-risk" ancillary medical device components (e.g., per Ministry of Health (MoH) Malaysia Guidance, asthma inhaler, syringe without a needle, measuring cup, measuring spoon, medicine dropper, dosing spoon, or pipette) are not required to apply an endorsement letter from MDA to NPRA prior to registration. Further, ancillary medical device components that have already obtained registration approval with MDA through a medical device registration application need only to present proof of medical device registration (certificate) when applying for DMDCP registration.

Of note in Malaysia, the applicant may submit an application for an endorsement letter and registration of their combination product concurrently to both the primary and the secondary agencies. However, the approval of the combination product registration by the primary agency is based both on fulfillment of all registration requirements, including receipt of the endorsement letter from the secondary agency. Significantly, Malaysia allows for an approach to leverage "platform" approaches (e.g., with respect to device-related considerations when a device technology is used for more than one application, process, or product).

For MDDCPs, a DMDCP endorsement letter can be used for different drug approval applications with respect to the same brand-named medical device. The applicant is required to list all the drug product applications to which the endorsement letter applies.

Of note, Malaysia distinguishes between a combination product which has already obtained regulatory clearance in reference countries (i.e., Therapeutic Goods Administration (Australia), Health Canada, EMA or other Competent Authorities from EU Member Countries, Pharmaceuticals and Medical Devices Agency (Japan), and US FDA) and those without approval from reference countries. A combination product that has already obtained reference country regulatory clearance has abridged documentation requirements compared to those without reference country approval.

China

In China, under the Chinese FDA, there are multiple institutes and divisions. Among these is the Center for Administration of Medical Device Standard (CAMDS). Consistent with other global jurisdictions that recognize combination products, the lead review center is determined by the product's PMOA. If a product PMOA is drug or biologic, the Center for Drug Evaluation (CDE) is the Lead Center. If the product PMOA is device ("medicated medical devices"), then the Center for Medical Device Evaluation (CMDE) serves as the Lead Center. If there is ambiguity on classification, a request for classification is submitted to CAMDS to determine classification as drug, device, or combination product.

As indicated earlier, in China, medicine and device combination products refer to products that are composed as a single entity. According to CMDE (April 2019), if the technology is relatively mature, and similar products are already on the market in China, with no major adverse reactions or adverse events, a single review Lead Center independently undertakes the technical review work. If a joint review is required after assessment, the lead unit can transfer the registration application materials to the cooperating unit for joint review; if necessary, the lead unit can organize the relevant persons in charge of the two review centers to convene work coordination meetings with major technical reviewers, or jointly convene expert consultation meetings. The joint review is based on "the rationality of the basis of the combination product," including "the meaning of the combination; identifying the new risks and risk-return ratios after the combination, and fully considering the interaction between the drug part and the device part," and the associated impact, including "the quality control measures and control indicators of the final product; the determination of the product technical requirements, application scope and contraindications of the final product." Joint review is therefore focused on evaluating new issues that arise from a combination of the constituent parts, e.g., drug administration, dosage, stability, toxicology, the impact of device materials on degradation or absorption of the drug, and risk-based overall evaluation of the combination product, assessing the significance for combining constituent parts, risk benefits, and quality control measures and indicators for the combination (NMPA, 2009–2019).

If a joint review is required, the two review centers will respectively issue supplementary information or review comments for the corresponding review part, which will be summarized by the lead unit, comprehensively the overall safety and effectiveness of the product, and issue a final review conclusion.

Products registered and managed as medical devices include drug-coated stents, catheters with antibacterial coatings, drug-containing condoms, and drug-eluting birth control rings. Bandages containing antibacterial and anti-inflammatory drugs are classified as drugs. Chinese medicines of external application products are also registered as drugs.

SUMMARY

Table 2.7 shows a comparison between PMOA assignment for the United States, the European Union, Malaysia, and China for a sampling of products. As stated earlier, PMOA designation is

TABLE 2.7
Classification of Product PMOA: United States vs European Union vs Malaysia vs China

Product	United States	European Union	Malaysia	China 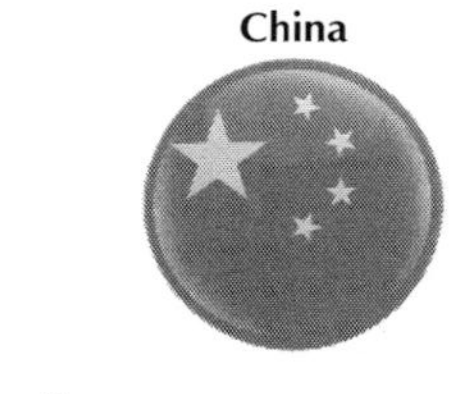
Syringe prefilled with drug product (single entity or single integral) 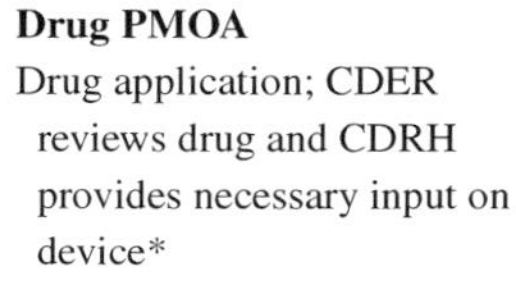	**Drug PMOA** Drug application; CDER reviews drug and CDRH provides necessary input on device*	**Medicinal Product (Drug PMOA)** Competent Authority (CA) or European Medicines Agency (EMA) oversight; NB opinion submitted to CA/EMA	**Drug PMOA** Drug-Device Combination Product Regulated as Drug (NPRA) NPRA oversight; Endorsement letter from MDA for device constituent submitted to NPRA	**Drug** Syringe components are filed as packaging components under Drug application; CDE reviews Drug and CMDE reviews device as needed; Joint review of combination product
Wearable sub-cutaneous injector with integrated drug-delivery system (single entity or single integral) 	**Drug PMOA** Drug application; CDER reviews drug and CDRH provides necessary input on device*	**Medicinal Product (Drug PMOA) (if non-reusable)** CA/EMA oversight NB opinion submitted to CA/EMA	**Drug PMOA** NPRA oversight; Endorsement Letter from MDA for device constituent submitted to NPRA	**Drug PMOA under Drug application** CDE reviews Drug and CMDE reviews device as needed; Joint review of combination product

Note: * indicates that in the absence of an agreement between Centers, consults are expected.

(Continued)

TABLE 2.7 (CONTINUED)
Classification of Product PMOA: United States vs European Union vs Malaysia vs China

Product	United States 	European Union 	Malaysia 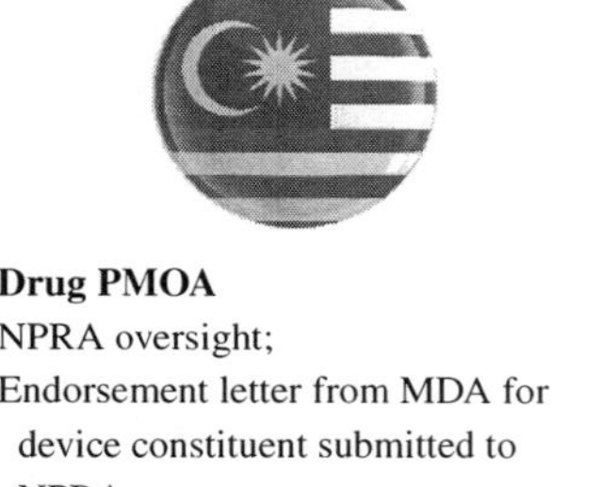	China
Non-integral Vial adapter co-packaged with drug product	**Drug PMOA** Drug application; CDER reviews drug and CDRH provides necessary input on device*	**Co-pack is not considered a combination product** CE mark for device through NB Drug approval through CA/EMA Currently unclear re: coordination between NB and CA /EMA – refer to EMA DDC quality guideline	**Drug PMOA** NPRA oversight; Endorsement letter from MDA for device constituent submitted to NPRA Abridged application if 510(k) or CE Mark exist	**Co-pack is not considered a combination product** Vial Adapter (Medical Device filed with CMDE) (licensed as a device) Drug product submitted under Drug Application
Bandages impregnated with antibacterial ointment	**Depending on the nature of claims, PMOA could shift** If Device PMOA; Device application; CDRH lead reviewer and CDER provides necessary input on drug* *Devices for use with blood and blood products are assigned to CBER, regardless of PMOA*	**Depending on the nature of claims, PMOA could shift** If Device PMOA with ancillary drug: Submission to NB for CE Mark and CA opinion for medicinal component (*Per Rule 14*) *However, if the claim is to treat infection, it could be classified as a Medicinal Product PMOA with bandage used to deliver the drug. The medical device component as a Class I would not require NB input*	**Depending on nature of claims, PMOA could shift** A medicated dressing is generally a device. For Device PMOA, MDA oversight; Endorsement Letter from NPRA to MDA for drug constituent submitted to MDA. If Drug PMOA, Endorsement Letter not required for low-risk ancillary medical device components prior to registration	Bandages containing antibacterial drugs are classified as **Drugs** CDE reviews. (Coordination, if necessary, with CMDE for any joint review, however, this is considered a drug.)

Note: * indicates that in the absence of an agreement between Centers, consults are expected.

product specific and can be complex, so these are just illustrative. This comparison highlights disparate interpretations of combination product definitions and PMOA designations country by country, which can lead to challenges. The differences in current regulations and regulatory body structures require that manufacturers be prepared to follow varying regulatory pathways and substantive product expectations in different regions.

CONCLUSION

"What is a combination product?" seems like a simple question, but it does not have a very simple answer. The term "combination product" is used widely, but its definition and interpretations vary across the globe. Increasingly there is an understanding that the term refers to a product composed of two or more types of medical products intended for combined use. Where there remains disparity in interpretation is in regard to whether the term captures only single-integral (single-entity) products or also applies to non-integral products co-packaged together, or to any products specifically labeled for combined use. The successful combination product sponsor must be aware of the gamut of interpretations in order to successfully navigate the regulatory expectations and processes among jurisdictions.

The Primary Mode of Action of a combination product is also a critical concept applied across jurisdictions. It often governs premarket review pathways, substantive product expectations, and other regulations throughout the combination product's lifecycle, including post-market modifications and safety reporting. While there are some differences across jurisdictions in defining PMOA, and even sometimes the conclusions of PMOA assignment, there are mechanisms being put in place to drive coordination between the regulatory bodies within jurisdictions, driving consideration of each constituent part and their potential interactions, ultimately with the intent to ensure safety and efficacy of these products. Health authorities are grappling with the best approach to ensure safe and effective combined use within their own regulatory frameworks.

Whether regulating these products as medical devices or as medicines (drugs, biologics, human cell tissue therapies) or as a separate legal category as in the United States, as technologies and therapies become more integrated to improve patient treatments and outcomes, the lines of traditional regulatory categorizations and approaches for medical products are increasingly becoming blurred. Ensuring product safety and efficacy necessitates enabling appropriate expertise in review, during development and throughout the product lifecycle. This accentuates the need for understanding the dictionary and thesaurus of combination products: Classification, PMOA, components, and constituent parts, as well as implications for regulatory pathways and other regulatory expectations across jurisdictions. Development professionals, regulatory teams, quality teams, supply chain personnel, reviewers, and investigators will all benefit from building an understanding of the vocabulary, and similarities and differences in interpretation (see Chapter 5 in this book for more on "Language opportunities"). Harmonizing the global combination product dictionary and improving the clarity of the thesaurus may help promote application of consistent risk-based approaches and get these products efficiently to the patients that need them (ASTM International Combination Product Terminology Standard (2023)).

NOTES

1. The content of the appendix of this book is intended to be a high-level overview of the complex global regulatory combination products' regulatory environment. The information gathered and reflected under this Appendix is a culmination of public domain information, regulations, directives, guidelines, and standards coupled with author experience for the regions across global jurisdictions as of the time of this writing (January 2023). Given the very dynamic evolution of the global environment with respect to combination products, the reader is strongly encouraged to refer to the quick references in the table for respective health authority websites for the most current updates.

2. In conjunction with review of this chapter and this book, readers should refer to relevant regulations, guidance and standards, and other relevant references cited in the bibliography for additional information. Regulatory requirements vary between jurisdictions, and may change. Those presented may differ from actual regulatory requirements imposed by Health Authorities for specific combination products and in different regions globally. Specific combination products and classes of combination products may raise distinct issues.
3. WHO "Global model regulatory framework for medical devices including in in vitro diagnostic medical devices (GMRF)"; WHO/BS/2022.2425 (July 2022).
4. Throughout this chapter, unless otherwise noted, the term "drug" is used generally to refer to both drugs and biological products; in some regions, the term medicinal products references both drugs and biological products.
5. In Canada, November 30, 2005 Health Canada Guidance "Drug/Medical Device Combination Products," accessible at https://www.canada.ca/en/health-canada/services/drugs-health-products/drug-products/applications-submissions/policies/drug-medical-device-combination-products.html, the phrase "integrated in a singular product" restricts the scope of applicable products to single entity and certain co-packaged combinations of drugs and devices. This is a much narrower scope of products than, for example, that considered by the US FDA, as Canada does not classify all co-packaged drugs and devices nor cross-labelled products as combination products. Additionally, Health Canada has a single set of regulations (Food & Drug Regulations) for drugs and biologics, so the regulation of drug–biologic product combinations can be accomplished from within one set of legislation.
6. "Convenience Kit" is a term that can have a specialized meaning for purposes of regulation in the United States. In the context of combination products, a kit includes two or more different types of medical products (e.g., a device and a drug), and is considered a co-packaged combination product under Part 4. If the kit includes only products that are (1) legally marketed independently and (2) packaged in the kit as they are when marketed independently, including labeling used for independent marketing, it is considered a "convenience kit" [21 CFR §4 Final Guidance, January 2017, and pre-amble to Part 4: 78 FR 4310].
7. The concept of PMOA is reviewed in more depth in Chapters 3 and 14 of this book. US FDA's December 2020 guidance "Requesting FDA Feedback on Combination Products," January 2022 guidance, "Principles of Premarket Pathways for Combination Products," and April 2011 guidance "How to Write a Request for Designation" (and other guidance per global health authorities cited in the references for this chapter) delve further into the concept of PMOA.
8. In some jurisdictions (e.g., EMA, MHRA), the term "medicinal product" generally refers to any drug and/or biologic product. Generally speaking, medicinal products achieve their primary intended purpose through chemical, pharmacological, immunological, or metabolic action, whereas medical devices do not. (https://www.ema.europa.eu/en/glossary/medicinal-product; EU MDR (2017/745) Article 2). Other jurisdictions (e.g., US FDA) treat drugs and biological products as separately regulated categories. Therefore, additional requirements may apply for drug/biological product combinations depending on the regional/country regulatory framework.
9. There is overlap in the definitions for medical device and drug under the Federal Food, Drug and Cosmetic Act [*US, § 201(g) of the FD&C Act (21 U.S.C. 321(g)] (drug definition) and §201(h) of the FD&C Act (21 U.S.C. 321(h))(device definition)*]. Genus Technologies vs. FDA (April 2021) Court Decision stemmed from this overlap. The take-away is that if a product meets both the device and drug definitions, it is regulated as a device. Subsequent to the court decision, legislation was passed indicating that the major categories of products implicated by the decision, i.e., radiopharmaceuticals, imaging agents for use with imaging equipment, and over-the-counter monograph drugs, are to be regulated as drugs. However, some products that have historically been regulated as drugs are now considered combination products, in particular ophthalmic dispensers packaged with drugs. Historically, under 21 CFR 200.50, dispensers packaged with ophthalmic drugs were regulated as drugs. FDA has declared 21 CFR 200.50 to be obsolete. The dispensers for ophthalmic preparations meet the "device" definition, and the ophthalmic preparations and dispensers are to be regulated as drug-led combination products. FDA has stated that it is evaluating its approach to post-market manufacturing requirements for low-risk device constituent parts, e.g., an eye dropper bottle. Sponsors are encouraged to reach out to FDA if they need clarification.
10. Sometimes people confuse PMOA analysis with drug vs device classification. These are separate issues. PMOA analysis is applied to determine Center assignment for combination products (e.g., in the United States and in China), while the other concerns whether a product can be classified as a device as opposed to a drug or biologic. For example, let's reflect on the antibiotic-coated suture. Our PMOA analysis

determined this product has a device PMOA, i.e., the suture providing wound closure is the PMOA. But what if the suture is not coated with a drug but is absorbable? In the United States, the language of the statutory definition for device states that a product cannot be a device if it achieves a "primary intended purpose" through chemical action. The primary intended purpose of an absorbable suture is still to rejoin tissue, and the suture is, therefore, a device even though it is designed to be resorbed by the body through a combination of chemical action and metabolic activity (Classification of Products as Drugs and Devices and Additional Product Classification Issues Guidance for Industry and FDA Staff, September 2017).

REFERENCES

ASTM International Standard Combination Product Terminology, 2023.

China Med Device, LLC, Key Points for Registering Drug-Device Combination Products in China, September 3, 2018, accessed at https://chinameddevice.com/key-points-for-registering-drug-device-combination-products-in-china/ on July 12, 2020.

EMA (Oct 2019) Questions & Answers on Implementation of the Medical Devices and In Vitro Diagnostic Medical Devices Regulations ((EU) 2017/745 and (EU) 2017/746), accessed at https://www.ema.europa.eu/en/documents/regulatory-procedural-guideline/questions-answers-implementation-medical-devices-vitro-diagnostic-medical-devices-regulations-eu/745-eu-2017/746_en.pdf on August 17, 2020.

EU MDR (2017/745) Article 117.

European Commission Guidance MED/DEV 2.1/3 rev 3 Borderline Products, Drug-Delivery Products and Medical Devices Incorporating, as an Integral Part, an Ancillary Medicinal Substance or an Ancillary Human Blood Derivative, accessed at http://www.meddev.info/_documents/2_1_3_rev_3-12_2009_en.pdf on December 24, 2020.

European Medicines Agency (EMA)/CHMP/QWP/BWP/259165/2019 (Draft, May 29, 2019) "Guideline on the Quality Requirements for Drug-Device Combinations."

European Medicines Agency (EMA)/CHMP/QWP/BWP/259165/2019 (Draft, April 8, 2020) "Guideline on the Quality Requirements for Drug-Device Combinations."

European Medicines Agency (EMA/37991/2019 Human Medicines Division (June 23, 2021, Rev. 2, Questions & Answers for Applicants, Marketing Authorization Holders of Medicinal Products and Notified Bodies with Respect to the Implementation of the Medical Devices and In Vitro Diagnostics Medical Devices Regulations ((EU)2017/745 and (EU)2017/746), accessed at https://www.ema.europa.eu/en/documents/regulatory-procedural-guideline/questions-answers-implementation-medical-devices-vitro-diagnostic-medical-devices-regulations-eu/745-eu-2017/746_en.pdf on July 9, 2022.

European Medicines Agency (EMA/CHMP/QWP/BWP/259165/2019) (22 July 2021) Guideline on Quality Documentation for Medicinal Products When Used with a Medical Device, accessed at https://www.ema.europa.eu/en/documents/scientific-guideline/guideline-quality-documentation-medicinal-products-when-used-medical-device-first-version_en.pdf on July 9, 2022.

Goebel, Stephanie, (2019, October) Notified Body Perspective on EU MDR and its Impact on Combination Products-TUV SUD presented at IQPC, Berlin, Germany.

Health Canada, (November 30, 2005) Drug/Medical Device Combination Products – Canada.ca.

Health Sciences Authority Therapeutic Products Guidance, December 2020; Guidance on Therapeutic Product Registration in Singapore (TPB-GN-005-006), accessed at https://www.hsa.gov.sg/docs/default-source/hprg/therapeutic-products/guidance-documents/guidance-on-therapeutic-product-registration-in-singapore_dec2020.pdf on June 2, 2020.

Kapoor, Vicky and Kaushik, Deepak, (2013) A Comparative Study of Regulatory Prospects for Drug-Device Combination Products in Major Pharmaceutical Jurisdictions. *Journal of Generic Medicines*, 10(2), 86–96.

MHRA Guidance, (2021, January 6) Borderline Products: How to Tell If Your Product Is a Medicine, accessible at https://www.gov.uk/guidance/borderline-products-how-to-tell-if-your-product-is-a-medicine.

Ministry of Health Malaysia, (2020, September 15) *Guideline for Registration of Drug-Medical Device and Medical Device-Drug Combination Products* (3rd Edition), https://www.mda.gov.my/documents/guideline-documents/1518-guideline-for-registration-of-drug-medical-device-and-medical-device-drug-combination-products-3rd-edition-15th-september-2020-003/file.html, accessed January 8, 2023.

Ministry of Health Malaysia, (2023, January 3) *Guideline For Drug-Medical Device And Medical Device-Drug Combination Products* (5th Edition), https://www.npra.gov.my/index.php/en/component/content/article/225-english/1527155-guideline-for-registration-of-drug-medical-device-and-medical-device-drug-combination-products-3.html?Itemid=1391, accessed January 8, 2023.

Ministry of Health Malaysia, Medical Device Authority, https://portal.mda.gov.my/industry/medical-device-registration/combination-product.html, accessed December 24, 2020.

NMPA, *Notice Regarding Drug-Device Combination Products Registration* (July 27, 2021) accessed at https://www.ccfdie.org/en/gzdt/webinfo/2021/10/1640588951989517.htm on August 7, 2022.

NMPA, *Announcement of Further Improving the Bundling Review, Approval and Supervision of Drug* (No.56 [2019]) http://www.nmpa.gov.cn/WS04/CL2138/339042.html.

NMPA, *Announcement on Adjusting the Definition of the Drug-device Combination Products* (No.28 [2019]) http://www.nmpa.gov.cn/WS04/CL2138/338096.html.

NMPA, *Announcement on Registration Items of Medical Device Master File (Draft)* https://www.cmde.org.cn/CL0101/18451.html.

NMPA, *Announcement on the Results of the Definition of Properties of Pharmaceutical Combination Products* (No.181 [2017]) http://www.nmpa.gov.cn/WS04/CL2439/353262.html.

NMPA, *Introduction to the Technical Review of Pharmaceutical Combination Products* (19 April 2019) https://www.cmde.org.cn/CL0033/19022.html.

NMPA, *Notice on Related Matters Concerning the Registration for Drug-Device Combination Products* (No.16 [2009]) http://www.nmpa.gov.cn/WS04/CL2138/299892.html.

Team NB, Position paper for the interpretation of device related changes in relation to a Notified Body Opinion as required under Article 117 of Medical Device Regulation (EU)2017/745, December 2020.

US FDA 21 CFR §3.2(e), §3.2(m).

US FDA 21 CFR §3.4.

US FDA 21 CFR §4A (2013).

US FDA 21 CFR §4B (2017).

US FDA, Classification of Products as Drugs and Devices and Additional Product Classification Issues Guidance for Industry and FDA Staff, September 2017.

US FDA, How to Prepare a Pre-Request for Designation (Pre-RFD) Guidance for Industry, February 2018.

US FDA, How to Write a Request for Designation (RFD) Guidance for Industry, April 2011.

US FDA, Principles of Premarket Pathways for Combination Products, January 2022 Final Guidance.

US FDA, Requesting FDA Feedback on Combination Products Guidance for Industry, December 2020.

US FDA Section 503(g).

WHO, Global Model Regulatory Framework for Medical Devices Including in Vitro Diagnostic Medical Devices (GMRF); WHO/BS/2022.2425 (July 2022).

3 Combination Products Regulatory Strategies

Suzette Roan

Contents

DOI: 10.1201/9781003300298-4

INTRODUCTION

Development and lifecycle management of combination products require integrated cross-functional teams, synergistically working together. The regulatory strategy for the combination product is critical for guiding these activities. Although it will be developed during the early stage of product development, the regulatory strategy for the product will evolve throughout development and during the commercial lifecycle. This chapter focuses primarily on the development and refinement of the regulatory strategy and submission planning and execution for drug Primary Mode of Action (PMOA) combination products which combine a drug or biologic with a medical device. Many of the concepts discussed can be applied for device PMOA combination products and for drug–biologic combination products, but they will need to be adapted accordingly for those product-specific considerations.

As described in Chapter 5, the development timelines for the combination product and the drug product are not always aligned. For some programs, the combination product is developed along with the drug product, but more often, the combination product development trails the drug product or the device constituent part development. For the purposes of this chapter, the development stages refer to the development of the combination product.

The global regulatory strategy typically includes an overview of the key regulatory interactions and submissions anticipated to support the target product profile (TPP) and quality target product profile (QTPP), considering the current regulatory expectations (FDA, 2007). Key inputs to the regulatory strategy are typically contained within the TPP, including, but not limited to route of administration, device needs, packaging configuration (e.g., prefilled system or separately provided), intended user and use environment, and intended commercial markets and countries to be included in the clinical studies. As the regulatory strategy is linked to the designation of the product in the target markets, documenting the strategy for designation of the product is an important first step, including whether product designation can be determined by the company based upon regulation and precedence or if there is a need to reach out to health authorities to determine or confirm the product designation and associated lead within the health authority which will be responsible for the review and approval of the product. The regulatory strategy also provides for a mechanism to document regulatory risks, requirements associated with the design, potential clinical study requirements, and associated timing, as well as identification of applicable regulatory requirements for the product. A simplified flow diagram of the inputs into the regulatory strategy and the steps during development are presented in Figure 3.1.

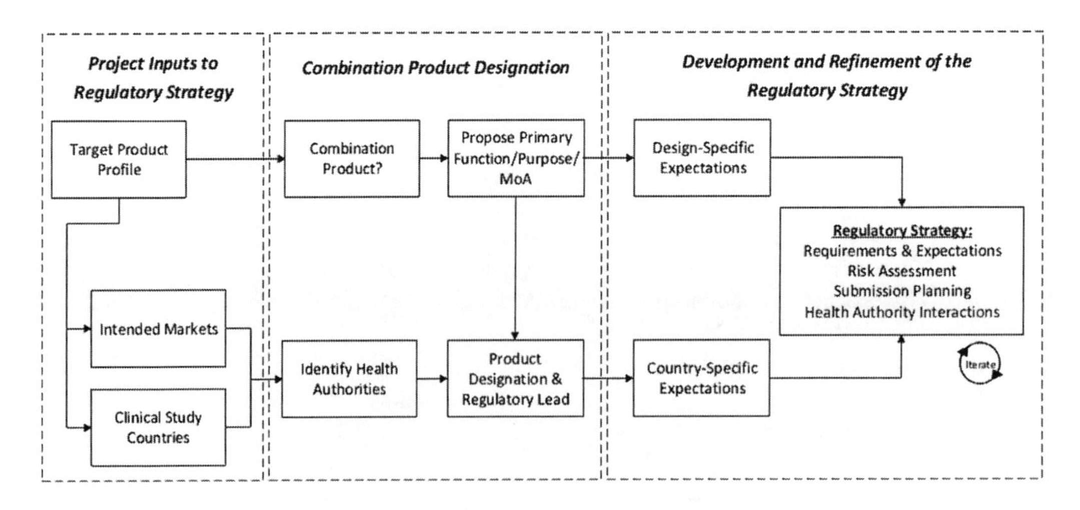

FIGURE 3.1 Regulatory strategy flow diagram.

This chapter is organized into three parts:

- **Part A: Combination Product Designation** provides an overview of the steps for defining the product as a combination product, identifying the health authorities and processes for determining product designation and the regulatory lead within the health authority.
- **Part B: Development and Refinement of the Regulatory Strategy** provide a question-based approach for the considerations related to the initial development of the regulatory strategy and management of the regulatory strategy throughout the product lifecycle.
- **Part C: Combination Product Regulatory Strategy Execution** provides practical recommendations related to the processes for identifying regulatory requirements, health authority engagements, submissions, and post-approval considerations.

PART A: COMBINATION PRODUCT DESIGNATION

Defining the Product as a Combination Product

When a product combines a drug or biologic constituent part with a medical device, a combination product may be formed, depending on the definition for combination products in the markets. It is important to understand what the intended commercial presentation(s) for the product will be. While it is not typically defined in early development, it is important for products which use both device and drugs or biologics to have the TPP include specifics regarding the intended commercial presentation. This becomes increasingly important when the constituent parts are not physically combined in an integral manner, but rather are combined by placing them in the same package (e.g., co-packaged), or are combined through the labeling for the products, as the commercial presentation can define whether the product is considered a combination product.

In some markets, when a medical device and a drug product are combined into the same package, a combination product is created (e.g., the United States, Malaysia, Japan, Saudi Arabia). This applies even when the constituents are individually packaged inside the combined package. In other markets, only products physically combined in an integral manner are considered combination products, and the packaging configuration does not define whether the product is considered a combination product (e.g., Brazil, Canada). In China, the combination products definition for integral products requires that the drug/biologic and the device be inseparable, which excludes prefilled drug delivery systems, as the drug/biologic is separated from the delivery device during dosing.

Even when the product is not officially considered a combination product in a given market, there are expectations to support the combined use of the medical device and the drug product. These expectations consist of demonstrating suitability for use together (ICH, 2002), including the relevant quality aspects of the device itself and its use with the particular medicinal product (EMA, 2021a). Additionally, requirements and guidance on labeling and instructions for use apply to these products (FDA, 2010; FDA, 2019h).

Identification of Intended Markets and Their Associated Health Authorities

Before the regulatory strategy can be developed, it is important to ensure that the intended (or likely, if not finalized) markets are identified, which is an output of the business strategy for the product. Knowledge of the intended markets will ensure that the global regulatory framework is considered when planning the regulatory strategy for the product. While combination products are not formally designated in all markets, many markets have developed regulation and guidance on how the combined product is regulated. Table 3.1 provides a non-exhaustive listing of representative health authorities with roles relevant for combination products and links to websites providing useful information. Notified bodies have been designated by the respective competent authorities to perform the conformity assessment for medical devices.

Refer to the chapter on Combination Products Evolving Global Regulatory Environment for additional details, including how countries without a formal "combination product" designation regulate

TABLE 3.1
Health Authorities with Combination Product Responsibilities

Market	Health Authority	Relevant Combination Product Website Links
Canada	Health Canada (HC)	Drug/Medical Device Combination Products – https://www.canada.ca/en/health-canada/services/drugs-health-products/drug-products/applications-submissions/policies/drug-medical-device-combination-products.html
China	National Medical Products Administration (NMPA)	Laws and Regulations – http://subsites.chinadaily.com.cn/nmpa/lawsandregulations.html
Europe	European Commission (EC)	Medical Device – https://ec.europa.eu/growth/sectors/medical-devices_en Notified Bodies Nando – https://ec.europa.eu/growth/tools-databases/nando/
Europe	European Medicines Agency (EMA)	Quality requirements for drug-device combinations – https://www.ema.europa.eu/en/quality-requirements-drug-device-combinations
Japan	Pharmaceutical and Medical Devices Agency (PMDA)	Regulatory Information – https://www.pmda.go.jp/english/safety/regulatory-info/0001.html
Japan	Ministry of Health, Labour and Welfare (MHLW)	Pharmaceuticals and Medical Devices – https://www.mhlw.go.jp/english/policy/health-medical/pharmaceuticals/index.html
Malaysia	National Pharmaceutical Regulatory Agency (NPRA)	Combination Products – https://www.npra.gov.my/index.php/en/combination-products-main-page.html
United States	Food and Drug Administration (FDA)	Combination Products – https://www.fda.gov/combination-products
Notified Bodies	The European Association for Medical devices of Notified Bodies (TEAM NB)	Coordination of Notified Bodies Medical Devices (NB-MED) documents – https://www.team-nb.org/nb-med-documents/

Source: Website links accessed on December 29, 2019.

these combined biologic or drug products with medical devices. (Note: Currently the United States and Malaysia also regulate drug/biologic combinations as combination products.)

Additionally, when the global aspects for development are considered early, it reduces the need to redesign or repeat testing (e.g., pack size, transport validation, clinical studies) to meet other market requirements, taking into account which markets are high priority for the company, which countries have more stringent requirements and/or long review times. Considering these global aspects in the early strategy can reduce the impacts later in development.

Product Designation and Health Authority Regulatory Lead Identification

The first purpose of the regulatory strategy is to define the designation of the product, e.g., whether the product is considered a drug, a device, a biologic, or a combination product. The designation is the determination of how the product will be regulated in the various markets, which drives the clinical and marketing submission pathway, reviewing agency(ies), and quality requirements. In general, most countries designate the primary function, purpose, and/or mechanism of action of the combination product to define the submission pathway and division or center in the health authority responsible for the review, with country-specific differences in terminology and/or approach (see Chapter 2).

In the United States, the Primary Mode of Action (PMOA) of the combination product is defined as the means by which a product achieves an intended therapeutic effect or action (FDA, 2005a). The PMOA is determined based on which constituent part of the combination product provides the greatest contribution to the product's intended effects. Where a combination product is a prefilled delivery system (e.g., prefilled syringe or autoinjector), the drug product contained within the delivery system provides the therapeutic effect and, thereby PMOA, with the device providing a supportive purpose to aid in administration. In such a case, the FDA center with jurisdiction of the review of the combination product will either be the Center for Drug Evaluation and Research (CDER) or the Center for Biologics Evaluation and Research (CBER), based on which Center is charged with review of such drugs or biologics (FDA, 2005b). Alternatively, where the combination product is a medical device which includes a drug product for an ancillary purpose, such as a drug-eluting stent, the PMOA is that of a medical device and the FDA center with jurisdiction of the review will generally be the Center for Device and Radiological Health (CDRH). With a drug-eluting stent, the physical effect of the stent propping open an artery is the primary role, with the drug providing a secondary role during healing. Generally, the submission pathway for the combination product is a suitable submission type for the constituent which has the PMOA: A drug-led combination product will be filed as a New Drug Application (NDA) (or an abbreviated new drug application if it is a generic of an already approved combination product); a biologic-led combination product will be submitted as a Biologics Licensing Application (BLA); and a device-led combination product will be submitted as a 510(k) premarket notification, a De Novo request or as a Premarket Application (PMA), depending on the review questions needing to be addressed. In the guidance on premarket pathways, the FDA has emphasized that the 510(k) pathway would not be successful for a device-led combination product if the proposed predicate device is not combined with the same drug or biologic constituent, since the addition of the drug or biologic constituent would likely result in a new intended use and/or constitute a different technological characteristic that raises different questions of safety and effectiveness as compared to the predicate. Instead, the FDA recommends that the applicant consider utilizing the De Novo process (FDA, 2022).

In 2017, the European Union published two device regulations introducing a new legal framework for medical devices, effective from May 20, 2021,[1] for devices, and May 20, 2022, for In Vitro Diagnostics. These regulations will be examined later in this chapter and in greater depth in Chapter 14 (Global Regulatory Environment). Let's first look briefly at the approach that the European Union has historically applied in determination of the submission pathway, outlined in the governing directives – Medicinal Product Directive (MPD) and in the device directives (Medical Device Directive (MDD), Active Implantable Medical Device Directive (AIMDD), and In-vitro Diagnostic Device Directive (IVDD)) and device regulations (Medical Device Regulation (MDR) and In-Vitro Diagnostic Device Regulation (IVDR)) issued by the EC (EC, 2001; EC, 1993; EC, 1990; EC, 1998). A "directive" is a legislative act that sets out a goal that all EU countries must achieve. However, it is up to the individual countries to devise their own laws on how to reach these goals (EU, 2019). These directives and their amendments provided the legal framework and established requirements for medicinal products and medical devices. Decision criteria were included in Article 1(5) of the MDD for determining whether a product falls under the MPD or the MDD, "particular account shall be taken of the principal mode of action of the product" (EC, 1993). This principal mode of action is deduced from the scientific data regarding the mechanism of action and the manufacturer's labeling and claims (EC, 2009).

The MDD further outlined in Article 1(3) where the product is a single integral product intended exclusively for use in the given combination and is not reusable, it is governed by the MPD, but the relevant essential requirements (ER) of Annex I to the MDD apply as far as safety and performance-related device features are concerned. The distinction of whether a product is a medical device or a medicinal product was further clarified through supplemental guidance to help understand how to classify these products, especially when the device is for administering a medicinal product (MHRA, 2011; EC, 2009).

Where a medical device incorporates an ancillary medicinal substance, the product is regulated as a medical device and is assessed by a notified body (NB) with consultation to a medicinal product competent authority. As part of the notified body assessment of the device-medicinal product combination, the scientific opinion is obtained from a competent authority on the quality and safety of the medicinal substance, including the clinical benefit/risk profile of the incorporation of the medicinal substance into the device (EC, 2009; MHRA, 2017a).

As indicated previously, the two EU device regulations published in 2017 introduced a new legal framework for medical devices, the Medical Device Regulation (MDR) and the In-Vitro Diagnostic Device Regulation (IVDR), with implementation periods of 5 and 7 years, respectively. A "regulation" is a binding legislative act. It must be applied in its entirety across the European Union (EU, 2019). The MDR supersedes the MDD and the AIMD (EU, 2017a). The IVDR came into effect on May 20, 2022, and superseded the IVDD (EU, 2017b). The MDR includes the same decision criteria in Article 1(6) regarding the principal mode of action of the product and designates the MPD as the governing directive in Article 1(9) where the product is placed on the market in such a way that they form a single integral product intended for use in the given combination and not reusable, including the need to apply the relevant general safety and performance requirements (GSPR) of Annex 1 to the MDR. In addition, Article 117 of the MDR included an amendment to the MPD requiring where a device part is not CE marked, a dossier for a marketing authorization application (MAA) needs to include results of a conformity assessment of the device part with the relevant GSPR issued by a notified body (EU, 2017a).

Some countries have mechanisms whereby the company can seek a determination of the designation of the product. In the United States, the pre-request for designation (pre-RFD) and request for designation (RFD) processes are avenues to seek a classification decision from the FDA. The RFD process is a formal mechanism to obtain a binding FDA determination from the Office of Combination Products (OCP) regarding product classification and/or which Center will have primary jurisdiction for the premarket review of their products (FDA, 2005c; FDA, 2011). In addition, the pre-RFD process was formalized in 2016 as a means for sponsors to obtain an informal designation for their products (Nguyen, T. and Sherman, R., 2016), and guidance was issued in 2018 providing more details on the pre-RFD process (FDA, 2018b). A summary of the similarities and differences between the RFD and pre-RFD processes is presented in Table 3.2. While both mechanisms are available to sponsors, the PDUFA VI Assessment reports that the pre-RFD process is used more often (Eastern Research Group, Inc., 2020). As per this report, sponsors have preferred obtaining combination product assessments by means of pre-RFDs much more than RFDs, as the pre-RFD provides more opportunities for interaction with FDA and the submission requirements are more flexible.

In 2018, there were 8 RFD decisions and 82 pre-RFD assessments (Nguyen, T. and Weiner, J., 2019). Whether a sponsor elects to utilize the RFD or pre-RFD process, it is important that the sponsor must ensure that the request is as clear and concise as possible to allow the FDA to assess how the product works for its intended use and to give the most informed view of classification/PMOA. The reader is highly encouraged to read the actual guidance documents and recognize that guidance may evolve over time.

As another example of a country designation process, in Malaysia, products which are not clearly defined as medical device or drug/cosmetic may apply for a classification decision to the National Pharmaceutical Regulatory Agency (NPRA) by submission of a product classification application (Ministry of Health Malaysia, 2017; NPRA, 2017). In China, a jurisdictional designation needs to be made to the Administrative Processing Service Center of the State Food and Drug Administration (SFDA) prior to the submission of a marketing application in China. This application takes 20 working days to process and requires information related to the composition, intended use, mode of action, usage and duration, summary of development, manufacturing process, and description of related products (SFDA, 2009).

TABLE 3.2
US RFD and Pre-RFD Process Overview

RFD Process*	Pre-RFD Process*
Formal	Informal, more flexible
Written request submitted to OCP including the following required information: • Description of the product • Proposed use or indication for use • Description of the manufacturing process • Supportive data/studies • Description of how the product achieves its intended therapeutic/diagnostic effect • Analysis of Classification, Primary Mode of Action (PMOA), and Jurisdictional assignment • Description of related products • Sponsor recommendations	Written request submitted to OCP including the following recommended information: • Description of product • Proposed use or indication for use • Description of how the product achieves its intended therapeutic/diagnostic effect Additionally, the following optional information may be provided in the pre-RFD, if available, as relevant: • Description of the manufacturing process • Supportive data/studies • Analysis of Classification, Primary Mode of Action (PMOA), and Jurisdictional assignment • Description of related products • Sponsor recommendations
15-page limit	No page limit, succinct summary encouraged
Letter of designation issued within 60 days	60-day assessment target
Binding	Preliminary, not binding
Request for reconsideration of decision may be made by Sponsor. Additionally, Sponsor may request review of the Agency decision.	No appeal process, as the decision is not binding. Sponsor may request a meeting to discuss, provide additional information, and/or submit a new pre-RFD or an RFD

Source: FDA (2005c), FDA (1991), FDA (1998), FDA (2011), and FDA (2018b).
*The reader is highly encouraged to read the actual guidance documents as they may continue to evolve.

PART B: DEVELOPMENT AND REFINEMENT OF THE REGULATORY STRATEGY

Once clarity on the product designation is established, the regulatory strategy needs to be defined. Development of the regulatory strategy for a combination product involves addressing several questions, which include:

- Regulatory Requirements: *What are the regulatory requirements for clinical use and marketing approval?*
- "Soft" Intelligence: *Are there any other expectations (e.g., "soft" regulatory intelligence) to consider?*
- Regulatory Risk Assessment: *What are the regulatory risks and what mitigations can be implemented to address?*
- Submission Approaches: *What submissions are needed to support clinical use, marketing, and lifecycle management?*
- Health Authority Interactions: *What are the plans for obtaining health authority feedback and what are the outcomes?*

The approach presented in Figure 3.2 is used to establish the initial regulatory strategy and iterates throughout the product lifecycle to refine the regulatory strategy.

In an ideal scenario, the regulatory strategy is initiated commensurate with the start of combination product development. It is important to develop the regulatory strategy early in the development process, as it has implications from premarket product development and regulatory pathways all

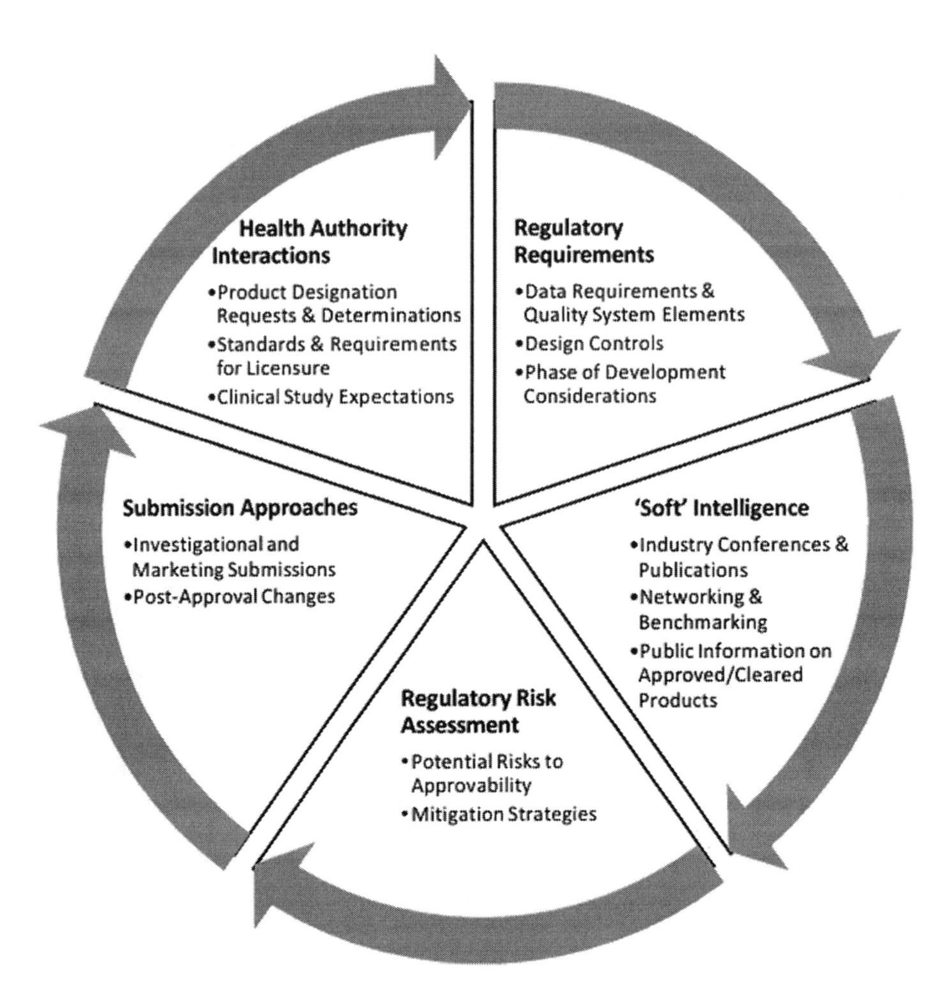

FIGURE 3.2 Iterative approach to developing and refining the regulatory strategy.

the way through postmarket lifecycle management. Involvement of quality and regulatory affairs colleagues with expertise in combination products on the team when that project team is considering combining a drug or biologic with a medical device is a best practice to ensure that the quality and regulatory aspects are considered. As the product development plans for the combination product are mapped out, the TPP can be a tool to identify and prioritize the intended claims for the combination product (e.g., sharps protection, in-use conditions), as the regulatory strategy will need to consider the requirements necessary to support the intended product claims.

As recommended by guidance, the manufacturer should consider the scientific and technical issues raised by a combination product and its constituents and develop a strategy accordingly, with the aim of addressing issues while minimizing duplication of effort. Even when a previously approved or cleared device is used with an approved drug product, additional considerations may arise when used in combination (FDA, 2006; FDA, 2022).

What Are the Regulatory Requirements for Clinical Use and Marketing Approval?

A critical step in developing the regulatory strategy is to identify applicable regulations, essential principles, guidance, and consensus standards that apply to the constituent parts and to the overall combination product. The regulatory strategy should consider the expectations for the individual constituents of the combination product (as if they were to be marketed independently), the expectations for those individual constituents of the combination product when marketed as a combination product and the expectations for the combination product, as additional regulatory questions may need to be addressed relating to the interaction between the device and the other constituent part(s). To emphasize these different expectations, consider a co-packaged combination product consisting of a drug product in a vial packaged with two separate medical devices. A syringe and a hypodermic needle:

- Constituent expectations (standalone): Drug product quality, safety and efficacy; syringe and hypodermic needle legally marketed.
- Constituent expectations in the combination product: Individual constituent design fit and function considerations, such as whether the connectors between the syringe and the hypodermic needle conform to the technical standards for compatibility. These fit and function requirements may not be required to legally market the individual medical devices but would come into play if the manufacturer intended to include the medical devices together in the kit.
- Combination product expectations: Drug product and medical device compatibility, as well as safe and effective use of the drug product with the syringe and hypodermic needle, demonstrated through verification and validation of the combination product for its intended use.

Figure 3.3 illustrates various sources of regulatory requirements, which are discussed in the following sections, with examples highlighted. With all of these input sources, there is the possibility that requirements may conflict from the various sources and the Sponsor will have to determine which requirements to apply considering the recommendations from the health authorities.

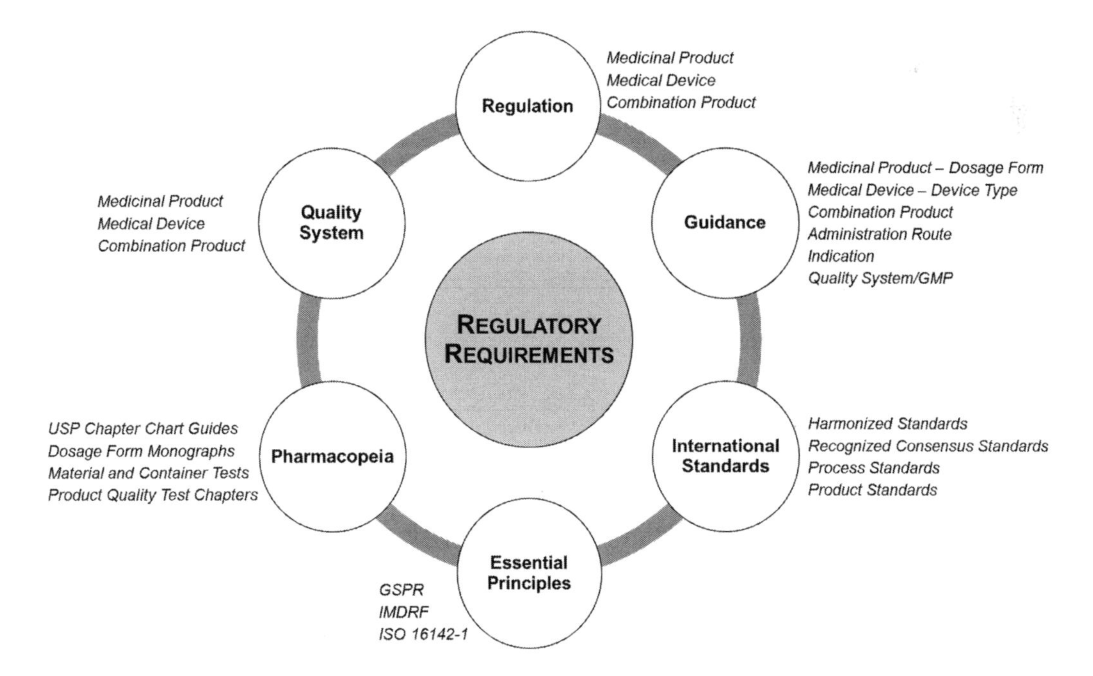

FIGURE 3.3 Sources of regulatory requirements.

Regulations

The regulations applicable to the combination product need to be identified holistically, considering the medicinal product and medical device regulations associated with the individual constituent parts, as well as any additional regulations applicable specifically to the combination product. The applicable regulatory approach for the constituent parts, if they were to be marketed independently for the same or a related intended use, offers a guide to the regulation of that medical device or medicinal product as a constituent part of the combination product, recognizing that **additional regulatory questions may apply relating to the interaction between the constituent part(s) and any differences in intended use as compared to the individual constituent parts if they were marketed independently**.

The regulations applicable to combination products are summarized in the chapter "Combination Products Evolving Global Regulatory Environment." For the purposes of developing the regulatory strategy, a few key regulations are highlighted. In the United States, the combination products regulations are promulgated in 21 CFR 4, subparts A and B, regarding how to apply drug, device, and biological product current good manufacturing (subpart A) and postmarketing safety reporting (subpart B) requirements (FDA, 2013a; FDA, 2016a), while other regulations not specific to combination products (i.e., those for constituent parts) may still apply. In the EU, where the single integral product is intended for use in the given combination and not reusable, conformity assessment of the device part with the relevant GSPR is needed for products marketed under the MDR.

In order to apply the regulations for combination products, it is important to understand the medicinal product and medical device regulatory frameworks. The regulations for drugs, biologics, and medicinal products require that the manufacturer demonstrate that the product satisfies three key criteria, namely quality, safety, and efficacy, to ensure that products administered to patients are of suitable quality and provide a positive benefit-risk balance (EC, 2020). Products are reviewed to determine whether it is safe and effective in their proposed use(s), whether the benefits outweigh the risks, whether the content of the labeling is appropriate, and whether the methods used in manufacturing and the controls used to maintain quality are adequate (FDA, 2020d).

The regulation of medical devices globally typically uses a risk-based paradigm under which devices are classified considering the risk presented by the medical device into three or more categories. The regulatory controls, requirements for marketing, and manufacturer obligations applicable to the device are based on these classification categories. In the United States, the device classification system is based on the level of control necessary to assure device safety and effectiveness based on intended use and indications for use, with Class I including the lowest-risk devices and Class III including those medical devices with the greatest risk (FDA, 2020c). Similarly, in the European Union, classification is achieved through application of a series of rules outlined in Annex VIII of the MDR, which considers the intended purpose of the device and its inherent risks, with four levels of classification (EU, 2017a). The classification drives the marketing requirements (e.g., whether a submission is needed and content to be included), as well as ongoing manufacturer obligations. In the United States, these medical device manufacturer obligations include regulatory controls, which are one of the sources to consider (e.g., unique device identification (UDI), labeling, good manufacturing practices (GMPs)) when defining the applicable regulatory requirements for the combination product. It is important to note that FDA does not simply apply device regulatory controls and mechanisms for device constituent parts of drug/biologic PMOA combination products, and issues addressed via medical device regulatory controls alone are not all that is relevant for combination products. (Note – see Part C for additional details on US medical device regulatory controls.)

Guidance

Research applicable guidance documents published by the relevant health authorities for the dosage form, administration route, indication, and/or device part, which may include references to standards and regulations that may also be relevant. Additionally, guidance documents on expectations for combination products may be available in some markets, including guidance on quality system expectations for the products (FDA, 2017b).

The guidance documents included in Table 3.3 represent a sampling of currently available documents commonly used during combination product development. As guidance documents are revised and finalized, and new guidances are issued on an ongoing basis, refer to the health authority websites (see Table 3.1) for contemporary guidances.

Other global consensus documents should be considered when developing the regulatory requirements for a given combination product and its constituent parts, including ICH guidelines (ICH, 2019a), International Medical Device Regulators Forum (IMDRF) documents (IMDRF, 2019a), and Parenteral Drug Association (PDA) technical reports (PDA, 2019). ICH guidelines on specifications, stability, pharmaceutical development, quality risk management, and pharmaceutical quality system should be reviewed for the elements which are applicable to the combination product. PDA technical report (TR)73 on prefilled syringes is a valuable resource that consolidates best practices and learnings from industry and regulators related to the use of glass prefilled syringes for biotechnology applications. TR73 includes practical and regulatory considerations for material selection and evaluation, syringe preparation and handling, human factors, drug product compatibility with syringe materials, and mode of delivery (PDA, 2015).

TABLE 3.3
Combination Product Guidance Documents

Issued by	Topic (Citation)	Summary
FDA	Principles of Premarket Pathways (FDA, 2022)	Guidance on the pathways for premarket submission and interactions with FDA for combination products.
FDA	Combination Product Feedback (FDA, 2020b)	Guidance on the types of meetings and interaction opportunities available throughout the lifecycle of combination products.
FDA	Technical Considerations for Demonstrating Reliability of Emergency-Use Injectors (FDA, 2020a)	Draft guidance for emergency-use injectors submitted under a BLA, NDA, or ANDA, explaining how combination product developers can demonstrate that their emergency-use injectors will reliably deliver drugs as intended in a life-threatening emergency.
FDA	Bridging (FDA, 2019j)	Framework for assessing changes to the combination product from that used in pivotal clinical trials to the product intended for marketing.
FDA	Transdermal and topical delivery systems (FDA, 2019m)	Expectations regarding transdermal and topical delivery system combination product design and pharmaceutical development, manufacturing process and control, and finished product control. The guidance also addresses special considerations for areas where quality is closely tied to product performance and potential safety issues, such as adhesion failure and the impact of applied heat on drug delivery.
EMA	Medicinal products which use a device (EMA, 2021a)	Expectations for integral and non-integral drug-device combinations for marketing application and variations. Expectations to demonstrate the suitability of a device for its intended purpose taking into account both the relevant quality aspects of the device itself and its use with the particular medicinal product.
EMA	Quality of Medicines – Part 1 (EMA, 2019c)	Harmonized position on issues, including uniformity of dosage units, variations.
EMA	Quality of Medicines – Part 2 (EMA, 2019d)	Harmonized position on issues, including administration via enteral feeding tube, packaging material sterilization, eye drops, DPI, graduation of measuring devices, needle safety systems, orally inhaled products, endotoxin, sterility, stability.
FDA	Metered Dose Inhaler (MDI) and Dry Powder Inhaler (DPI), draft (FDA, 2018g)	Revised draft reflecting current standards and requirements, with a detailed section on combination products, including clarifying that design controls apply to these products. Provides commentary to assist pharmaceutical manufacturers familiar with the Quality by Design (QbD) approach leverage those practices to satisfy the 21 CFR 4 obligations.

(Continued)

TABLE 3.3 (CONTINUED)
Combination Product Guidance Documents

Issued by	Topic (Citation)	Summary
MHRA	Human Factors and Usability Engineering (MHRA, 2017b)	Advisory guidance on application of human factors and usability engineering throughout the product lifecycle. Section on drug delivery devices and drug-device combination products considering the critical characteristics and nature of the medicinal products to be delivered. Considerations relating to bridging of differences between the device used in the pivotal trial and that proposed for marketing. Approaches to human factors for well-established platform drug delivery devices.
FDA	Combination Product Human Factors (FDA, 2016b)	Considerations on whether to submit combination products human factors study data. Considerations for design changes after human factors validation. Expectations for human factors information to submit in a combination product investigational application. Expectations for marketing application review of human factors studies and certain labeling.
EMA	Transdermal patch (EMA, 2014)	General requirements concerning the development and quality of transdermal patches for systemic delivery for marketing applications and variations.
FDA	Pen, Jet, and Related Injectors (FDA, 2013b)	Described content to be included in premarket submissions and outlined functional and performance testing expectations.
FDA	Glass syringes for delivery of drug and biological products, draft (FDA, 2013c)	Requirements relating to interconnectivity of devices and prefilled glass syringes, as well as performance and submission expectations.
FDA	Rheumatoid Arthritis, draft (FDA, 2013d)	Expectations for development of products for rheumatoid arthritis, including the expectation that the to-be-marketed drug-device combination product will be used in the pivotal studies supporting the efficacy and safety of the combination product for marketing approval. Expectations for ongoing evaluation of device performance incorporated into the pivotal studies for the combination product and included an example framework for bridging when transitioning from a prefilled syringe to an autoinjector delivery system.
EMA	Orally Inhaled Products Clinical Development (EMA, 2009)	Data requirements that are often dependent on the performance of the device. Relationship of the knowledge of *in-vitro* performance and the clinical development program. Considerations for spacers and nebulizers, including how and when these need to be listed in the labeling.
EMA and HC	Inhalation and Nasal Drug Products (EMA, 2006; HC, 2006)	Quality expectations for marketing and clinical trial use, including expectation for extensive characterization of batches used in pivotal clinical trials.
EMA	Plastic Immediate Packaging Materials (EMA, 2005a)	Specific requirements for plastic immediate packaging materials for active substances and medicinal products for marketing applications.
EMA	Graduation of devices for liquids (EMA, 2005b)	Recommendations regarding the graduation of delivery devices for liquid dosage forms. Expectations for the dosing accuracy and precision as well as the suitability of the measuring device.
FDA	Nasal Spray and Inhalation Solution (FDA, 2002)	Quality expectations for the combination product, including aerosol characteristics, device characteristics, material expectations, characterization, and product robustness studies.
FDA	Container Closure Systems (FDA, 1999)	CMC (chemistry, manufacturing, and controls) documentation expectations related to the container closure system (CCS) for investigational and marketing applications. CCS suitability expectations, which are relevant to drug delivery systems, e.g., protect the dosage form, be compatible with the dosage form, be comprised of materials safe for use, and perform as intended.

International Standards

International standards are recognized by some of the markets as appropriate for the products to be marketed in that market. The FDA publishes the status of the recognition in the CDRH recognized standards database. The database lists the version of the standard recognized and outlines the scope of the recognition (FDA, 2018e). In the European Union, a listing of harmonized standards is published and available online. A standard which is harmonized can be used to demonstrate that the product complies with the relevant EU legislation (EC, 2019a; EC, 2019b).

It is also important to recognize that the international standards are typically considered voluntary, even if recognized or harmonized. This means that the manufacturer is free to choose an alternate approach to demonstrate that the product is safe and effective, provided that the approach is supported by sound scientific and regulatory justification. Where a manufacturer elects to define an alternate approach, it is recommended to review these approaches with the relevant health authorities prior to the marketing submission. (Note: Importantly, while sometimes the case, just because a standard is recognized by one center (e.g., CDRH for devices), that does not always mean that a sponsor can rely on that standard for a product in another center (e.g., CDER for a drug/biologic-led combination product). Careful risk-based assessment of the specific combination product and its constituent parts must be applied based on the particular intended uses, intended users, and use-environments for a given product. The manufacturer is encouraged to engage with applicable health authorities to ensure alignment.)

As the listing of standards is developed, document whether the standard is recognized or harmonized for the intended markets, including considering the scope of such recognition or harmonization. (Note, see Part C for a recommended approach to identify potentially relevant international standards.)

Essential Principles

The GSPR for the device part should be considered during development to ensure that the development activities are appropriately scoped to meet the GSPR. Conformance to the GSPR necessitates a completely implemented design or product realization process, as requirements define the critical medical device assessments, such as risk, design, materials/construction, sterility, biocompatibility, clinical data, safety, labeling, and instructions for use. Internationally, the IMDRF published the Essential Principles of Safety and Performance of Medical Devices and IVD Medical Devices in 2018 to provide a set of globally harmonized essential principles which should be fulfilled in the design and manufacturing of devices to ensure that they are safe and perform as intended (IMDRF, 2018). The EU GSPR share many similarities with the IMDRF essential principles but do have some differences which should be reviewed when defining the relevant essential principles. Determine the essential principles applicable to the product, considering the published regulations of the various markets.

Utilize the recommendations included in ISO 16142-1 to guide the identification of standards appropriate to fulfill the essential principles identified as applicable to the medical device in the product (ISO, 2016b). Assess the applicability of the essential principles for the device to be developed and consider the standards tabulated in ISO 16142-1 for the essential principle. Additionally, consider the product-specific standards enumerated in ISO 16142-1 for applicability to the device to be developed.

Pharmacopeia

The United States Pharmacopeia (USP) includes chapters that relate to the container, device constituent, and combination product aspects of the product. The applicable chapters can be determined by referencing the USP chapter chart guides, which provide guidance to help determine which chapters should be used for specific drug products (USP, 2019a). Select the chart guides that are applicable to the product being developed, for example, use Chart 5 for a monoclonal antibody drug product to identify the applicable chapters to address topics for identification, characterization, equipment, miscellaneous tests, description, safety, assay, physiochemical, and impurities. Additionally, review the generally applicable chapters and charts included in the chart guide. An example helpful chapter is USP <1151> Pharmaceutical Dosage Forms, as it provides general descriptions and definitions for the drug products and dosage forms organized by route of administration and explains the general product quality tests (USP, 2019c). Also consider Chart 9 to ensure that the chapters relevant to drug product distribution are included. The chapter charts include references to the product quality test chapters for the drug product types in chapters through <5> and <771>, which provide the compendial expectations comprising the universal tests applicable to all products of the given type and specific tests to consider in addition to the universal tests, as well as the product quality tests for specific dosage forms (USP, 2019d; USP, 2019e; USP, 2019f; USP, 2019g; USP, 2019h; USP, 2019i; USP, 2019j). USP chapter numbers above <1000> are considered guidance that provides additional context and recommendations which facilitate implementing the applicable General Chapters (USP, 2019b).

European Pharmacopeia (Ph. Eur.) requirements can be identified for the combination product by reviewing the dosage form specific monographs, such as the monograph on parenteral preparations (Ph. Eur, 2019b). Included within these monographs is the definition of the dosage form, and requirements for materials, containers, production, and testing of the dosage form, and may also include storage and labeling requirements. The "Materials used in the manufacture of containers" and "Containers" chapters/subchapters of the Ph. Eur. provide the expectations for the material and containers described in the chapters, but the Ph. Eur. does not require that only those materials and containers may be used and alternates may be used subject to competent authority approval (Ph. Eur., 2019c; Ph. Eur., 2019f). Use of a material not described in the Ph. Eur. is an example of where the various regulatory design input sources should be used collectively, as the EMA's *Guideline on Plastic Immediate Packaging Materials* provides that the specifications for plastic primary packaging materials should refer to the Ph. Eur. monograph, but if the material is not described in the Ph. Eur., the manufacturer should develop an in-house specification and perform extraction studies (EMA, 2005a). All materials described or not described in the Ph. Eur. used in containers need to have a risk assessment for transmissible spongiform encephalopathies (TSE) carried out, with suitable measures taken to minimize risk (Ph. Eur., 2019c; Ph. Eur., 2019g; EMA, 2011).

The Ph. Eur. also contains general monographs, which cover classes of products, outlining requirements applicable to the products in the given class. Several of the general monographs cover the active ingredient or drug substance, such as *Substances for pharmaceutical use* and *Monoclonal antibodies for human use* (Ph. Eur., 2019d; Ph. Eur, 2019e). The manufacturer should consider whether any of the general monographs apply in addition to the dosage form monograph.

For pharmacopeial requirements, alternative methods of analysis may be used for control purposes, provided that the methods used enable an unequivocal decision to be made as to whether compliance with the standards of the monographs would be achieved if the official methods were used (Ph. Eur., 2019a). Where alternative methods of analysis are planned, consultation with the regulatory team member is needed.

Quality System

Typically, general quality system elements do not need to be specifically described and included in the regulatory strategy for a product, provided that the applicable quality system elements are incorporated into the quality policy for the organization. There are instances, however, where there are additional quality system requirements for the combination product, or an assessment of specific GMP provisions to define the extent of applicability of the requirement to the product is needed.

In these cases, it is useful to describe these regulatory requirements as part of the given product's regulatory strategy.

> **21 CFR 4:** The initial publication of 21 CFR 4 only included Subpart A – *Current Good Manufacturing Practice Requirements for Combination Products*, with Subpart B marked as [Reserved] and it became common for FDA and industry to refer to Subpart A as Part 4. Subpart B (§4.100-4.105), *Postmarketing Safety Reporting for Combination Products* was finalized in December 2016 and commonly referred to as PMSR (FDA, 2016a). Therefore, references in this chapter to 21 CFR 4 without the subpart designation refer to 21 CFR 4(A), §4.1-4.1, inclusive.

For example, in the United States, the combination product needs to be developed and manufactured in accordance with 21 CFR 4(A) (FDA, 2013a). The requirements under 21 CFR 820.30 (FDA, 1996b) and 21 CFR 820.50 (FDA, 1996c) are quite impactful for the combination product. Design controls are an interrelated set of practices and procedures applied to control design activities for making systematic assessments of the design as an integral and interactive part of development (FDA, 1997). Refer to Chapter 4, Combination Product cGMPs, and Chapter 5, Integrated Development, for additional details on best practices to implementing the 21 CFR 4 GMP requirements for combination products and for additional considerations relating to integrating contemporary requirements into the quality system for the organization.

GMPs for medical devices are expectations not just for device-led combination products, but now also for drug-/biologic-led combination products. Conformance to ISO 13485 and ISO 14971 can be used as evidence of a quality management system and a risk management system

Relationship to Design Controls

The identified applicable regulatory requirements, guidance, essential principles, and consensus standards, together with critical quality requirements and essential tasks associated with the drug, biologic, and/or device constituent parts and combination product as a whole, contribute to a combination product's Design Inputs under Design Controls (21 CFR 820.30 or ISO 13485) (see Chapter 5) (ISO, 2016a, 21 CFR 820.30 and 21 CFR Part 4A)). Therefore, another key element of the regulatory strategy is to outline which elements of the regulatory strategy should be treated as design inputs as part of the Design Controls development efforts.

The regulatory affairs colleague utilizes experience and expertise gained from previous projects in the company's portfolio to identify these regulatory design inputs, supplements with additional sources identified through research and identify any conflicts among them. Databases and resources published by the health authorities are particularly useful and the regulatory colleague should become experienced in how to navigate these sources (see Table 3.1). The applicable regulatory requirements should consider the regulatory requirements for the drug product if it were to be marketed not in combination with a device, the regulatory requirements for the device if it were to be marketed standalone, as well as the regulatory requirements which apply to the combination of the drug product and the device when used together. It is also recommended to include regulatory requirements if the product will be used as part of another product, for example, a prefilled syringe which will be also used in an autoinjector presentation.

Are There Any Other Expectations (e.g., "Soft" Regulatory Intelligence) to Consider?

Company-owned and publicly available contemporary information can be extremely useful when mapping out the regulatory expectations for a given combination product. It is imperative for the regulatory strategy to stay current with the evolving expectations. While surprises can never be completely prevented, incorporating approaches to maintain a contemporary view of the global regulatory landscape and anticipated upcoming new or changing regulatory requirements should help ensure changes in expectations are incorporated into the development and lifecycle of products. Some of the best practices to keep current include participating, researching, and networking, as outlined in Table 3.4. Parsing through the voluminous information that is available to separate the true regulator expectations and to identify trends and directionality of the expectations is best performed by a skilled combination products regulatory affairs colleague who understands the nuances and subtleties of the expectations and can distill those expectations into a concrete regulatory strategy. In addition to the published regulations and guidance, the regulatory strategy should consider the "soft intelligence" which includes contemporary regulator expectations – oftentimes long before these expectations are officially published.

TABLE 3.4
Strategies for Staying Current with the Regulatory Expectations

Participate	Research	Network
• Join standards development committees, such as American Association for Medical Instrumentation (AAMI) and International Standards Organization (ISO) • Get involved in industry groups, such as the Combination Products Coalition, Parenteral Drug Association (PDA), AdvaMed, PhRMA, and BIO Interest Groups • Attend and/or present at combination products conferences • Sit on an organizing committee for a combination products conference • Provide comments and feedback to draft guidance, rules, and standards	• Reading and understanding research presented in articles, posters, and publications • Subscribe to e-mail lists which include highlights of new and changing information • Become familiar with the wealth of information available on websites from health authorities and industry groups • Understand published regulatory guidance, standards, and regulations/rules, including reading the published preambles and responses to comments • Summaries of reviews (see Table 3.5)	• Build relationships with peers internal and external to your organization • Participate in virtual groups, such as LinkedIn groups and PDA interest group discussion boards • Engage in informal discussions at conferences • Contact colleagues at other companies to gain perspective and insights • Share perspectives and insights when colleagues from other companies reach out

One key source of public information is the published review documents supporting approval and/or for other similar products. Refer to Table 3.5 for an overview of several of the sources of public information on approved or cleared products, including where useful review summaries for combination products can be located.

These summaries from the health authority review of similar products provide useful information to inform the regulatory strategy for combination products. The following learnings may be available in the summaries described in Table 3.5:

- 510(k) summary, PMA Summary of Safety and Effectiveness Data (SSED), and PMA summary review memos usually identify the performance testing which was performed on the device, including references of the international standards and regulatory guidance used and can be useful to identify the applicable regulations and standards.
- CDER and CBER discipline review documentation for combination products usually includes the review memos from the CDRH and CDER Division of Medication Error Prevention and Analysis (DMEPA) consultative reviews in the "Other Reviews" section. The CDRH reviews provide expectations regarding device performance, human factors (HF) biocompatibility, and quality system expectations and may include references to international standards and regulatory guidance. The DMEPA reviews provide insights into the

TABLE 3.5
Public Information on Approved/Cleared Products

Source	Location	Content
CDER Summary Review and Discipline Review Memos for drug products	Drugs@FDA	Basis of FDA's decision to approve an application. It is a comprehensive analysis of clinical trial data and other information prepared by FDA drug application reviewers. A review is divided into sections on medical analysis, chemistry, clinical pharmacology, biopharmaceutics, pharmacology, statistics, and microbiology. Includes original approval and key supplements.
CBER Approval History, Letters, Reviews, and Related Documents	Licensed Biological Products with Supporting Documents Database	Summary basis for regulatory action, individual correspondence files, requests for information, meeting minutes, and discipline review memos. Includes original approval and key supplements.
510(k) Summary	510(k) Premarket Notification Database	510(k) applicant document that summarizes the device description, intended use, technological characteristics, and performance data (including references to standards which the device complies with).
PMA Summary of Safety and Effectiveness Data (SSED)	PMA Database	FDA document that summarizes the key content of the PMA, such as the Device Description, Preclinical Evidence, and Clinical Evidence, as well as FDA's analysis of the scientific evidence that served as the basis for FDA's decision regarding the reasonable assurance of the safety and effectiveness of the device.
PMA Summary Review Memo	PMA Summary Review Memos for 180-Day Design Changes Database	Review memo summarizing the change to the device, the data available to support the change, and the FDA reviewer assessment of the change.
European public assessment report (EPAR)	EMA Medicines Database	A set of documents describing the evaluation of medicine authorized via the centralized procedure and including the product information, and assessment history. Initial MAA and variations for major changes.
Australian Public Assessment Reports for prescription medicines (AusPARs)	Therapeutics Goods Administration (TGA) AusPAR Search Database	An AusPAR provides information about the evaluation of a prescription medicine and the considerations that led the TGA to approve or not approve an application.

Source: FDA (1994), FDA (2019a), FDA (2014c), FDA (1996a), FDA (2019b), FDA (2019c), (FDA) (2019d), FDA (2019e), FDA (2019f), EMA (2019a), EMA (2019b), Australian Government (2019a), and Australian Government (2019b).

expectations for HF and instructions for use (IFU). The "Administrative/Correspondence" section of the review may contain content related to information requests and feedback from meetings which can be useful to understand the contemporary expectations for product development, commercial manufacturing, and controls, as well as insights into strategies to bridge from one delivery system to another.

- EPAR and AusPAR documents include summaries of the clinical studies performed for the combination product, which can provide insights into the clinical dataset needed to bridge from one delivery system to another. These documents also refer to aspects of the review which were the subject of questions and interactions between the applicant and the health authority, and those summaries can provide insight into development data and/or manufacturing and control expectations for the combination.

What Are the Regulatory Risks and What Mitigations Can Be Implemented to Address?

One element of the regulatory strategy is to identify if there are any risks to clinical use and/or approvability. The impact of these risks can be a delay in health authority authorization of a clinical study approval, marketing approval, or aspects which may prevent the product from being approved. These regulatory risks should be tracked throughout development. As development progresses, many regulatory risks will be reduced because of the increased experience and confidence gained in the quality and performance of the product or through implementation of other mitigations. Mitigation strategies to address regulatory risks should be developed with the cross-functional team and the endorsed strategies included in the product development plan. Mitigations for regulatory risks may include data generation as well as reviewing the approach with health authorities. As the regulatory strategy is iterated throughout the product lifecycle, a review of the regulatory risks identified is recommended to ensure that the regulatory strategy reflects changes that have occurred in the risk profile.

What Submissions Are Needed to Support Clinical Use, Marketing, and Lifecycle Management?

The pathways for regulatory submissions for combination products vary across the global markets, so it is important to organize the documentation in a flexible manner to avoid significant rework for subsequent submissions to additional countries. It is also important to recognize that the submission pathways for investigational submissions may differ from those used for commercial combination products. (Note – see Part C for discussions related to the different submission types during the lifecycle of the combination product.)

Where content needs to align with medical device submissions, such as when a technical file or a device application is needed, the device documentation approaches should be considered. Medical device companies have been harmonizing the structure and format of the submissions for device licensing. The Global Harmonization Task Force (GHTF) developed the STED (Summary Technical Documentation) structure in 2008 to encourage and support the global convergence of regulatory systems, including harmonization of the documentation of evidence of conformity to the essential principles (GHTF, 2008). The IMDRF succeeded the GHTF and has developed the Regulated Product Submission (RPS), an internationally harmonized, modular, format for use when filing medical device submissions to regulatory authorities for market authorization (IMDRF, 2019b–d). Expectations from notified bodies, such as the British Standards Institute (BSI) and NB-MED, include recommendations to use the STED/RPS structure for document submissions (BSI, 2014; NB-MED, 2000).

Similarly, for drug or biologic applications, the use of the common technical document (CTD) as outlined in ICH M4 is expected for submissions. Additionally, many biopharmaceutical companies have developed approaches to utilize the granularity of the CTD to repurpose the documents for additional regions, e.g., to support the production of the CTD for the Association of Southeast Asian Nations (ASEAN), referred to as the ACTD (ASEAN, 2016).

Table 3.6 presents an overview of the marketing submission pathways for drug-led combination products across several key countries. As evidenced from the table, the same or similar information is needed for the countries, but the structure of the documentation may vary.

The FDA issued guidance on premarket pathways to help sponsors determine which type of premarket submission(s) are needed as part of efforts to implement the combination products provisions of the Cures Act (FDA, 2022). Regarding submission pathways, the Cures Act provided that FDA is to perform premarket review of any combination product under a single marketing application whenever appropriate per 21 United States Code (USC) §353(g)(1)(B), and that a Sponsor is not prohibited from submitting separate applications for the constituent parts of a combination product,

TABLE 3.6
Marketing Submissions for Drug-led Combination Products

Country	Integral Products	Packaged Together	Mutual Labeling
United States	BLA/NDA for combination product (note: device details often provided in a master file)	BLA/NDA for combination product; or BLA/NDA for drug product plus 510(k)/De Novo/ PMA for device	BLA/NDA for combination product[a] or BLA/NDA for drug product plus 510(k)/ De Novo/PMA for device
Europe	MAA for drug-device combination product plus Notified Body Art. 117 Assessment	MAA for medicinal product plus CE mark for device[b]	MAA for medicinal product plus CE mark for device[b]
Canada	New drug submission (NDS) for combination product	NDS for drug plus medical device license (MDL) for device[b]	NDS for drug plus MDL for device[b]
Japan	Japanese new drug application (JNDA) for combination product	Japanese new drug application (JNDA) for combination product	Japanese new drug application (JNDA) for combination product[a]
China	NDA for combination product	NDA for drug, plus device recorded or registered based on its classification[b]	NDA for drug, plus device recorded or registered based on its classification[b]
Malaysia	ACTD for combination product	ACTD for combination product	ACTD for drug, plus medical device registration

Source: SFDA (2009), FDA (2022), EU (2017a), EC (1993), EMA (2021a), Health Canada (2005), MHLW (2014), and NPRA (2019).

[a] When designated as a combination product in the country (e.g., in Japan, products which the constituents cannot be marketed individually are considered combination products).

[b] Not classified as a combination product in this country.

unless the FDA deems a single application necessary per 21 USC 353(g)(6). FDA clarified that their current thinking is that a single application would generally be appropriate for a combination product, but acknowledged that separate applications would generally be permissible for the constituent parts of cross-labeled combination products (FDA, 2022).

It is recommended to review the proposed submission approach with FDA during relevant agency engagements, such as end-of-phase or pre-submission meetings. Table 3.7 presents considerations for when sponsors are developing their proposed approach for the number of applications for marketing associated with their co-packaged and cross-labeled combination products in the United States. Even with multiple submissions, the PMOA submission needs to include reference to the other licenses (with right of reference provided in a letter of authorization), as well as evidence of suitability of the constituents (dosing device and drug product) for the combined intended use (ICH, 2009).

What Are the Plans for Obtaining Health Authority Feedback and What Are the Outcomes?

Once the initial regulatory strategy is developed (preferably early in the design process), if possible, it is highly advisable to engage in discussions on the plans for more complex, less well-understood combination products. Multiple pre-submission request avenues may be available for seeking advice from health authorities with feedback provided through face-to-face meetings and teleconferences and/or in written form.

TABLE 3.7
Considerations on Number of Applications for US Combination Products

Approach	Pro	Con
Single Submission (e.g., BLA)	• Including all content to be reviewed in a single submission is easier for the reviewer • Considered to be more streamlined for a single combination product • Enables simplified use of ICH Q12 for device content	• eCTD structure for device content can be difficult, especially with non-single-entity combination products • Often requires an MAF for device partner data • Post-approval change category selection can be challenging • Not feasible for off-the-shelf (OTS) constituents that have separate license • If the same device constituent is used in multiple combination products, the device-specific content is re-reviewed with each submission • Doesn't allow for a US clearance or approval designation for the device constituent filed under a single submission
Multiple Submissions (e.g., BLA + 510(k))	• Allows each manufacturer in a partnership to "own" filing for their product • Lifecycle management for each license type defined in regulation and guidance for the license type • Enables "platforms" where the constituent license supports multiple combination products • Aligns with other regions (see Table 3.6) • This is commonly the pathway for cross-labeled combination products • Similar approach used for companion diagnostics products, providing precedence for combination products	• Requires reviewer to assess multiple files to make determination • Sequencing of approvals can create logistical challenges for FDA • Post-approval changes may result in updates to multiple submissions • Separate device filing as a constituent part of a cross-labeled combination product would create "Device Constituent Applicant" responsibility under 21 CFR Part 4B, Combination Products PMSR if the sponsor is different than for the drug/biologic constituent part's application.

Source: FDA (2005a), FDA (2016c), FDA (2016a), FDA (2017c), FDA (2019l), and FDA (2022).

The process of interacting with health authorities for any product intended for marketing should be iterative throughout development, where the sponsor can obtain input from the regulators and incorporate that feedback into the approach and for combination products, a similar process is followed. There are a few elements specific to combination products to consider:

- Defining the participants from the health authorities who need to be involved in the interaction, to ensure appropriate expertise relating to each constituent part, their interactions and interrelationships are brought to bear
- Ensuring that the internal project team plans for health authority interactions include those able to address not only the pharmaceutical or biologic, but also the device/secondary constituent part and cross-cutting issues for the combination products
- Ensuring that the requests for interaction appropriately identify the expertise and/or location within the organization of such participants needing to be involved
- Whether the device and/or combination product-specific topics should be discussed in a cross-discipline interaction or in a separate interaction

There are multiple opportunities to obtain health authority feedback on specific topics during development of a combination product and the regulatory strategy should map out which topics will be covered during which interactions. The regulatory strategy should consider topics where feedback will be needed and outline the feedback avenues and sequencing. As more information becomes available during development, subsequent discussions with regulators may be warranted, including revisiting a given topic, so it is important for the regulatory affairs colleague to allow flexibility in negotiations and keep channels of communication open. The regulatory strategy typically includes a historical inventory of the interactions to date, including a summary of any commitments that have been made by the company, and planning for future feedback.

Common topics for feedback for combination products include, but are not limited to, the following:

- Human factors protocols/validation strategy
- Bridging strategy when "to-be-marketed" combination product is not used in pivotal trials
- Submission content and dossier structure
- Data package to support expiry, shipping
- Design verification test plan
- Drug/device compatibility studies
- Container closure integrity/sterility
- Performance/functionality requirements
- Control strategy

As an example, when there is a need to gain input on the control strategy for the combination product, it is recommended to take advantage of all opportunities to present the proposed development approach and control strategy and list the sequencing of the interactions planned along with the expected datasets which will be available to support the discussion in the regulatory strategy. Similar to meetings for a standalone medicinal product where feedback on the drug substance and drug product specifications and stability approaches are typically obtained multiple times during development, the combination product control strategy can be assessed during meetings throughout product development. Discussing the control strategy during multiple interactions will allow the company to consider the feedback early and provide opportunities for a dialogue where there are areas of disagreement. Control strategies can generally be discussed as part of pre-submission meetings with most regulators, as well as during end-of-phase (EOP) and Type C meetings with the FDA and scientific advice in the EU. EMA encourages applicants to seek advice within the EU Competent Authority (medicines) network early in development, particularly for new and/or emerging technologies (EMA, 2021a).

Considerations for Refining of the Regulatory Strategy during the Product Lifecycle

Documentation Considerations

The overall regulatory strategy should be vetted with the appropriate internal stakeholders and documented in a manner which can be updated simply, as the regulatory strategy is revised periodically throughout the product lifecycle.

Biopharmaceutical and device organizations typically have business processes for documenting the regulatory strategy to support product development. In addition to the product development aspects, elements of the regulatory strategy for a combination product feed into design controls (21 CFR 820.30 and ISO 13485:2016, Section 7, Product Realization). Each organization needs to define the appropriate structure for its regulatory strategy documentation, considering whether to combine the product development (business process) aspects into the same documentation with the GMP design control documentation or to maintain these aspects separately. Where the regulatory strategy is structured as a single document, the business-related product development aspects would then also be included in the design history file (DHF, also referred to as a design and development file under ISO 13485), even though these are not required elements of design controls. A single document is easier to maintain than having separate documents, often stored in separate locations (e.g., DHF for the design control aspects and project files for the business-related product development aspects). Some organizations choose to maintain separate documentation for the regulatory strategy, with only design control aspects included in the DHF, and the business-related product development aspects in a separate document. This approach allows for clear separation of the GMP content from the business-related content but does take additional effort needed to maintain separate documents. See Table 3.8 for a breakdown of the product development (business process) and design control aspects typically included as part of combination product regulatory strategies. The design control aspects presented in Table 3.8 relate to the expectation that the design inputs for product development include applicable regulatory requirements and standards, as specified in Section 7.2.1 (Determination of requirements related to product), Section 7.2.2 (Review of requirements related to product), and Section 7.3.3 (Design and development inputs) of ISO 13485:2016. The product-specific quality system requirements and applicable standards, regulations, and essential requirements (ER)/general safety and performance requirements (GSPR) are typically linked to the designation and classification of the product.

TABLE 3.8
Product Development and Design Control Aspects Included in Combination Product Regulatory Strategy

Product Development (Business Process) Aspects	Common Aspects	Design Control Aspects
Submission pathway(s) Submission timing and sequencing Planned health authority interactions and summary of prior interactions Commitments from prior reviews Submission strategies to support intended product claims Input from key opinion leaders Inspection planning	Intended markets Intended product claims (e.g., TPP/ Intended Use) Product designation and classification Analysis of new/revised regulatory expectations Regulatory risks and mitigation strategies Plans for significant changes to the product, including bridging approaches	Quality system requirements (product-specific) Identification of applicable standards and regulations Identification of applicable ER/ GSPR

Development Phase Considerations and Applying Learnings during Development

As development progresses, the data and information available on the product will increase. There will be product learnings which are aligned with the anticipated profile of the product that do not necessitate adjustments to the strategy. However, there may be product learnings that differ from the expectations, resulting in the need to modify or refine the regulatory strategy.

The regulatory strategy is an evolving one. Broad considerations for each stage of development – whether focused on clinical investigations, analytical testing, marketing requests, human factors studies, validation efforts, or even multi-generational planning for postmarket changes and bridging studies – are all in scope when it comes to developing the regulatory strategy. The regulatory professional is responsible for ensuring that the appropriate regulations, directives, guidance, and unwritten expectations/recommendations (which may be learned during public presentations) are incorporated into the regulatory strategy, inclusive of the elements described in Table 3.8. The regulatory strategy should consider the phase of development and would typically include more comprehensive content for the submission pathway(s), timing, and sequencing for investigational applications, as well as plans for health authority interactions during clinical trials. It would be preliminary for many of the elements relating to the marketing submissions. Early strategies should document the identified regulatory risks and include strategies to mitigate the risks during development. The regulatory strategy should comprehensively assess the intended product claims and map out the data requirements necessary to support those claims. It is recommended to ensure that the regulatory risks are linked to the overall product development plan, so that the activities to mitigate are included in the planned activities at appropriate stage(s) of development.

The regulatory strategy should be refined as the product moves into the late stage of development, by incorporating learnings during development. It is recommended to document the strategy commitments made or agreements reached during previous interactions and investigational submission reviews, to ensure that they are addressed in the marketing application. Building upon the initial regulatory strategy, comprehensively assess the intended product claims and design the marketing submission strategies that outline how the data generated during development will be presented in the submissions to support those claims. Consider obtaining input from external experts and key opinion leaders.

Continued monitoring of external and internal intelligence is important to ensure that as the regulatory strategy is refined during development, it remains consistent with contemporary expectations. Consider learnings from similar products under development in the company, as well as publicly available information (see Table 3.5). Additionally, where new or revised regulation or guidance is released, the regulatory strategy and product development plans should be assessed for impact and revised accordingly.

Planning Early for Post-approval Changes

Lifecycle management is a key consideration for the regulatory strategy. Continuous improvement plans should be considered and mapped out as early as possible in the regulatory strategy. This is especially important when there is a need to bridge from one delivery system to another. The expectation from regulators is for sponsors to use the "to-be-marketed" version of the product in pivotal clinical trials (FDA, 2013d; FDA, 2016b; EMA, 2021a; MHRA, 2017b). When there are plans to commercialize one delivery system, but alternate means of delivering the drug product are used during the Phase 3 or pivotal clinical trials, bridging is needed to secure approval of the intended delivery system either as part of the initial approval of the product or as a supplemental change.

Health authority interactions to discuss strategies for the data package requirements to support changes should be planned when there is sufficient information available to provide the health authorities with the necessary background to provide actionable feedback. The FDA has published two draft guidances that provide recommendations for bridging (FDA, 2013d; FDA, 2019j). In accordance with these guidances, the regulatory strategy can be built considering the potential bridging scenarios and mapping out the preliminary leveraging approach, which would be the subject of health authority interactions. When mapping out the scenarios and proposed leveraging approaches, an impact analysis is highly recommended to understand the differences between the product used in pivotal trials and "to-be-marketed" combination product through a side-by-side comparison followed by a risk assessment to understand the potential for impact on the existing dataset. The scenarios should consider the therapeutic window of molecule, patient tolerability effects, existing experience with delivery device and molecule. The proposed strategy to be discussed with health authorities should include an explanation for how the existing data can be leveraged and supplemented with studies on the new product filling the gaps in the dataset. These topics are commonly addressed during FDA EOP or Type C meetings and EU scientific advice.

PART C: COMBINATION PRODUCT REGULATORY STRATEGY EXECUTION

This part of the chapter provides practical information to support the execution of the regulatory strategy and includes:

- Strategies for identifying potentially applicable regulatory requirements: International standards and regulatory controls;

- Health authority engagement processes and interaction mechanisms; and
- Submission planning and execution: marketing submission structure and content, referencing other submissions, investigational submissions, applying ICH Q12, human factors submissions, dossier review, and post-approval considerations.

Strategies for Identifying Potentially Applicable Regulatory Requirements

Using FDA CDRH Databases to Identify International Standards

CDRH databases can be useful to identify international standards that may be relevant. A suggested approach is included in Figure 3.4.

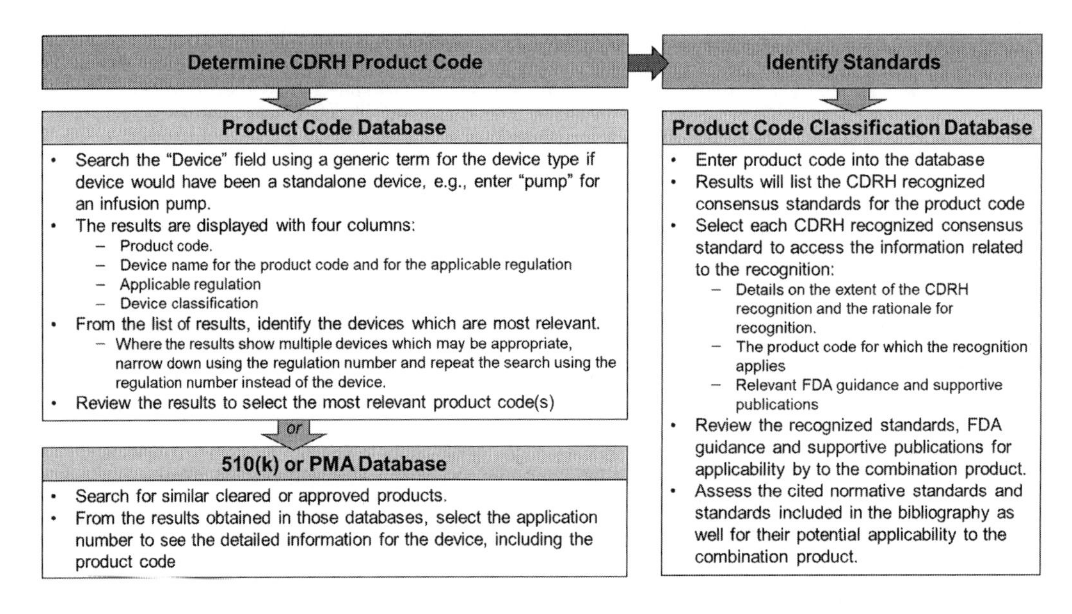

FIGURE 3.4 Suggested approach to identify standards which may be relevant using CDRH databases (FDA, 2018c).

US Medical Device Regulatory Controls

Recognizing that many biopharmaceutical colleagues are not as familiar with medical device regulations, the following summary of regulatory controls for medical devices is provided as additional background. As described earlier, device regulation in the United States is grounded in a risk-based paradigm under which devices are classified (Class I through III) in light of their risk and regulatory requirements that differ based on this classification. General controls apply to all device classifications. Devices that fall into Class II may also have special controls defined. Special controls are device specific and may include performance standards, postmarket surveillance, patient registries, special labeling requirements, premarket data requirements, and guidelines (FDA, 2018f; FDA, 2019i). General Controls are defined in the Food, Drug and Cosmetic (FD&C) Act and codified in Title 21 of the United States Code (USC) and apply to all classified devices, unless exempted by regulation. The general controls for US medical devices are listed in Table 3.9 with the FD&C Act section and 21 USC section references, as well as related 21 CFR section references.

Health Authority Engagement Processes and Interaction Mechanisms

While the process may be well understood by some readers of this chapter, the following is provided as an overview of the engagement processes. At a high level, health authorities have defined

TABLE 3.9
General Controls for Medical Devices (US)

FD&C Act Reference	Title	21 USC Reference	Related 21 CFR References
Section 501	Adulterated devices	351	800: General 801: Labeling 820: Quality system regulation
Section 502	Misbranded devices	352	800: General 801: Labeling
Section 510	Registration of producers of devices: Establishment registration and device listing; premarket notification (510k); reprocessed single-use device	360	807: Establishment registration and device listing for manufacturers and initial importers of devices
Section 516	Banned devices	360f	895: Banned devices
Section 518	Notification and other remedies: Notification; repair, replacement, or refund; reimbursement; mandatory recall	360h	810: Medical device recall authority
Section 519	Records and reports on devices: adverse event report; device tracking; unique device identification (UDI) system; reports of removals and corrections	360i	803: Medical device reporting 806: Medical devices; reports of corrections and removals 821: Medical device tracking requirements 830: Unique device identification
Section 520	General provisions respecting control of devices intended for human use: Custom device; restricted device; Good manufacturing practice (GMP) requirements; exemptions for devices for investigational use; transitional provisions for devices considered as new drugs; humanitarian device exemption	360j	812: Investigational device exemptions 820: Quality system regulation 814: Subpart H: Humanitarian use devices

Source: FDA (2018f).

mechanisms by which sponsors can request feedback during the product lifecycle, with the specifics of the process defined for each health authority. Generally, the processes have four main steps:

1. Request made by sponsor
2. Briefing materials submitted to health authority (if not required to be submitted at Step 1)
3. Feedback received (written, face-to-face, teleconference)
4. Minutes published (for interactions which are face-to-face or teleconference)

Typically, there are statutory timeframes for the steps in the processes, depending on the type of interaction and stage of development. The regulatory affairs colleague maps out the planned interactions and ensures that the product development team members participating in the interaction are informed of the timelines. When interactions are planned for multiple health authorities for the same or similar topics, typically, the request and briefing materials are prepared and reviewed in a merged/core format at the company, consolidating the requirements for the health authorities that the team will be interacting with. Once the content is finalized, the regulatory affairs team will produce the country-specific documents based on the content from the core document. This approach streamlines the preparation and review activities by minimizing the duplication of effort, as the team can work on core documents, instead of multiple regional versions in parallel.

The request for the interaction made by the sponsor should indicate when specific expertise from the health authority such as device reviewers or staff from OCP is desired (FDA, 2019k; FDA, 2020b). Making these requests for specific attendees ensures that the sponsor will have the opportunity to obtain the feedback that they request.

Many of the opportunities for health authority interactions for combination products are similar to those available for a standalone medical device or drug product. The combination product-specific interactions include product designation requests and determinations, as previously described in Part A of this chapter. Additionally, the US 21st Century Cures Act ("Cures Act") introduced a new meeting type specific to combination products, the Combination Product Agreement Meeting (CPAM) per 21 USC 353(g)(2), available to sponsors to:

- Address the standards and requirements for market approval or clearance of the combination product;
- Address other issues relevant to such combination product, such as requirements related to postmarket modification of such combination product and good manufacturing practices applicable to such combination product; and
- Identify elements to be discussed at a later date given that scientific or other information is not available, or agreement is otherwise not feasible regarding such elements, at the time a request for such meeting is made.

As described in the guidance on feedback for combination products, product development and lifecycle interactions should go through the lead center responsible for the review of the combination product and the lead center will coordinate with the appropriate consult review centers. The FDA recommends that interactions for a combination product use the application-based pathways available to all products as the preferred approach and to utilize these avenues prior to requesting a CPAM (FDA, 2020b).

The Cures Act included a few requirements relating to interactions with the FDA, as codified in 21 USC 353(g)(8). Sponsors may request in writing the participation of representatives of OCP in meetings regarding their products, or to have OCP otherwise engage in regulatory matters concerning the product. Additionally, Agency will ensure that meetings between the FDA and sponsors are attended by review staff from each center as appropriate in light of the topics and purpose of the meeting (Regulation of Combination Products, 2016).

The application-based mechanisms provide sponsors with the opportunity to obtain feedback on the questions posed by sponsors, but typically no agreements are made. A CPAM is an additional means for sponsors to obtain clarity and certainty, with intent of reaching agreement as the outcome. As agreements are difficult to be reached when there is uncertainty or limited data, FDA

> encourages sponsors to consider CPAMs only when they believe they have identified the indication for use and design of the combination product they will pursue and sufficient information can be provided to ensure an effective review by all relevant disciplines and centers. Accordingly, it may be helpful to interact through application-based mechanisms to provide FDA an opportunity to evaluate technical data or engage in scientific discussion before considering a CPAM.
>
> **(FDA, 2020b)**

In limited circumstances, it may be appropriate to engage with CDRH on the device constituent development engineering and technical requirements, such as when the sponsor is developing a device constituent as a "platform" intended to be used in one or more future drug or biologic PMOA combination product development programs or for feedback on device-only studies that are agnostic of the drug product, e.g., clinical study assessing volume/flow rate tolerability using a placebo.

While these CDRH device-specific interactions are informative and the feedback would be applicable to future combination products, as the interactions do not include the future to-be-assigned lead center, the feedback would not be binding on future combination products, as new considerations may arise when a sponsor seeks to use a platform technology with a particular drug or biologic product and/or for a use that may differ from that for which the platform may have been approved/cleared. These interactions would usually be through the Q-submission program, as informational meetings or pre-submission meetings (FDA, 2019g).

Additionally, sponsors may request meetings with OCP throughout development. If intended to support product designation, FDA recommends that the meetings be requested ahead of pre-RFD or RFD submissions (FDA, 2018b; FDA, 2011). The meeting requests should be submitted to OCP, defining purpose, issues to be addressed, FDA participants, and includes supportive information.

An overview of the available FDA engagement opportunities during the lifecycle of a combination product is illustrated in Figure 3.5.

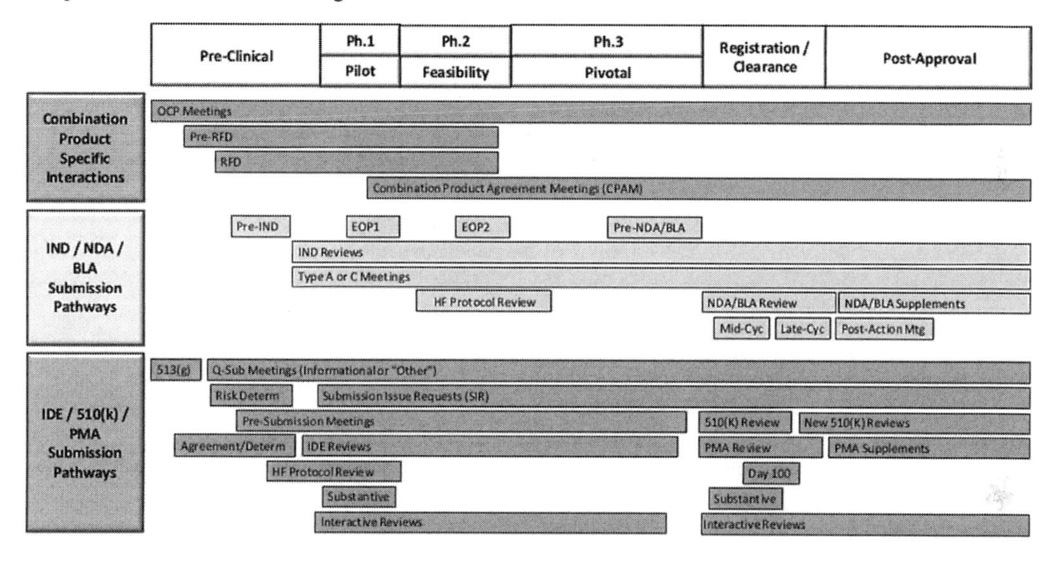

FIGURE 3.5 Opportunities for interaction with FDA during combination product lifecycle (FDA, 2001; FDA, 2019g; FDA, 2017a; Regulation of Combination Products; 2016, FDA, 2011; FDA, 2018b; FDA, 2014a; FDA, 2014b; FDA, 2018d).

Similarly, scientific advice may be sought on scientific and regulatory aspects of product development through national or centralized processes in the EU. In the early phases of development, the advice is typically sought direct from one or more national competent authorities to obtain feedback relating to program development, clinical study, CMC, and/or other aspects related to the development of a medicinal product. As part of the national scientific advice processes, the applicant is able to engage in dialogue with the national competent authority. These national scientific advice meetings may be product-specific or may be for broader scope topics, such as meetings to discuss "complex issues of drug/device combination products" (MHRA, 2018) and portfolio meetings with the German Federal Institute for Drugs and Medical Devices (BfArM), providing applicants with the opportunity to present an overview of different development projects (BfArM, 2019).

The centralized mechanism for scientific advice through the Scientific Advice Working Party (SAWP), which has the remit of providing scientific advice and protocol assistance to applicants, based on the documentation provided by the applicant considering the current scientific knowledge. Although the scientific advice feedback is not legally binding, the advice is taken into consideration during the MAA review, with any deviations requiring justification. In addition to providing scientific advice on medicinal products, the procedure may be used to obtain advice on the ancillary medicinal product components of medical devices, where the advice is focusing on the development needed to enable the Notified Body to submit the application needed for the EMA to provide its scientific opinion (EMA, 2017; EMA, 2012).

The process for obtaining centralized scientific advice includes the submission of a letter of intent and draft package for feedback, followed by the final package for validation which has been updated following EMA feedback. There is an opportunity to request a pre-submission meeting for scientific advice to gain input on the content and scope of the questions to be posed in the scientific advice and the supportive briefing materials to facilitate obtaining satisfactory answers. Where a pre-submission meeting for the scientific advice is requested, the letter of intent and draft package are submitted approximately 7 weeks ahead of the start of the scientific advice procedure. When a pre-submission meeting is not requested, the letter of intent and draft package are submitted approximately 3 weeks ahead of the start of the scientific advice procedure. The scientific advice procedure follows a timetable for evaluation, which enables the scientific advice coordinator to circulate the materials to experts on the working parties and the Committee for Medicinal Products for Human Use (CHMP) and consolidate the feedback received. The evaluation phase may be a 40-day procedure or a 70-day procedure, depending on whether the SAWP determines that a discussion meeting with the applicant is needed. Although discussion meetings are possible through the centralized scientific advice procedure, typically the scientific advice follows the 40-day procedure with only written feedback provided. In either scenario, the applicant may request follow-up scientific advice, which follows the same procedure as the initial scientific advice (EMA, 2017).

In addition to the scientific advice procedures with EU competent authorities and EMA, for integral drug-device combination products that require a notified body opinion per Article 117 of the MDR, it is recommended to request meetings with the selected notified body, applying the recommendations of that notified body.

Submission Planning and Execution

Marketing Submission Structure and Content

Once the pathway is determined, defining the approach to the content to be included in the filing(s) and the structure of the information is the next step. The regulatory strategy should map the structure and content of the submissions for each program. As each combination product is unique, the specifics relating to the content and structure for the regulatory filings can take a variety of approaches.

In addition to the content expectations for the safety and effectiveness of each type of constituent part, submissions need to include content related to the suitability of each of the constituents in the combination product for the combined intended use, including relating to their interrelationship and interactions. It is important to note that irrespective of the approach to the submission, the data/information needed to address safety and effectiveness questions related to the non-lead constituent part of a combination product may differ from the data/information needed to obtain approval/clearance for that article as a standalone product. For example, when the device constituent part has previously been 510(k) cleared, additional data/information may need to be provided to support safety and efficacy for the specific intended use with the drug product, as the data required under the 510(k) pathway may not be sufficient to address the safety and effectiveness questions for that combined use (FDA, 2022; Dorgan, 2018). These considerations are explained in the premarket pathways guidance, highlighting whether there are new questions of safety and effectiveness for the combination product (FDA, 2022). In these cases, the necessary additional data and information can be provided in the submission documents in accordance with existing guidance. For example, ICH M4Q(R1) includes expectations related to the suitability for use together, to answer the specific safety and effectiveness information on the dedicated combination product for the intended patient population (ICH, 2002). These expectations are aligned with the EMA expectations for both integral and non-integral drug-device combinations (DDC), wherein assessment of the suitability of a device for its intended purpose should take into account both the relevant quality aspects of the device itself and its use with the particular medicinal product. The complexity of the device, relevant patient characteristics, and the clinical setting in which the DDC is to be used are also important aspects of the review process. In the European Union, where a CE-marked device for the administration of the medicinal product is co-packaged or is referred to in the summary of product characteristics (SmPC), additional information may need to be provided by the applicant with regard to the device if the device may have an impact on the quality, safety, and/or efficacy of the medicinal product. The medicinal product dossier should include full evaluation of the impact of the device on the QTPP, CQA, and overall control strategy of the medicinal product (EMA, 2021a).

In the United States, a drug PMOA single-entity combination product may include device component information within the NDA/BLA or alternatively, the NDA/BLA may cross-reference a separate device clearance or approval (510(k), De Novo or PMA) or a master file (see next section on Referencing Other Submissions to Support Sponsor's Filing). A device PMOA single-entity combination product may include the drug information within the PMA/510(k)De Novo device filing, or it also may cross-reference an NDA/BLA registration dossier or master file. Even when cross-referencing other submissions, the combination product application needs to address the safety and efficacy of the combination product as a whole, and may require new data and information to support the approval. In the European Union, single-entity combination products governed by the MPD have historically included the device component information within the MAA to demonstrate conformity with the relevant essential requirements of Annex I to the MDD. This may be presented as a completed essential requirements checklist or a declaration of conformity. There is currently some uncertainty regarding the extent of device-specific information to be included in the MAA once the MDR is effective, as the device information will need to be in the documentation provided to the notified body for the Article 117 assessment of the evidence of the conformity of the device part with the relevant GSPRs.

For the countries which recognize co-packaging as a combination product, both device and drug information may be provided within one application or it can cross-reference other clearances or approvals or a master file. For example, a drug vial combined with a device (e.g., empty syringe) forms a co-packaged combination product. The NDA/BLA for the kit could have both the drug and device constituent information or it may cross-reference a 510(k) or master file for the syringe. In countries where co-packaging is not designated as a combination product, such as the European Union, the device is assessed through the CE marking process and appropriate references (e.g., CE mark certificate) are included in the medicinal product dossier to support the inclusion of the medical device in the packaging for the product.

For the cross-labeled combination product, the drug constituent and device constituent are packaged separately, with labeling specifically indicating use with the other constituent part. A single investigational application is appropriate, and either a single marketing application or separate ones for each constituent part is generally permissible. If separate marketing applications are used for each constituent part, what data and other information are included under which submission may vary, with cross-referencing to ensure access to such information as may be needed to support action on each submission (FDA, 2022).

Defining the information needed to be included and the most appropriate locations within the submission to place the device and combination product content to facilitate review, linkage to other sections and lifecycle maintenance continues to be an area where there is variability across industry and global regulators. This variability has made it challenging for reviewers to locate the content that they need to complete their reviews, resulting in information requests relating to providing the location of the specific data within the submission. Regulators have issued guidance intended to outline their expectations, particularly when the combination product content is included in a CTD structure (FDA, 2019l; EMA, 2021a). For example, EMA has clarified that Module 3, Section 3.2.P should contain information on the product-specific quality aspects related to the device relevant to the quality, safety, and efficacy of the medicinal product. In general, Module 3 of the MAA dossier should include appropriate information on the manufacture, control, and usability of the DDC as defined for the intended patient population (EMA, 2021a). Additionally, PDA TR73 provides recommendations on the types of content to be included in regulatory submissions, as well as tabulated descriptions of the elements related to prefilled syringes to consider for inclusion in Module 3 sections of the BLA or NDA (PDA, 2015).

Table 3.10 provides a summary of the currently available guidance on content and placement expectations for combination products, organized by the CTD hierarchy (ICH, 2016). Design and development information is included in the submission to provide support for the robustness of the design and rationale for the manufacturing and control approaches. It is recommended to include an overview of the combination product, including operating principles presented with illustrations explaining the mechanics of the device functions relevant for each key use step, provided to orient the reviewer to the design of the system, including the features and functions as a baseline to support the development approach and the proposed controls that are included in the submission. Another key aspect of the submission is to describe the approach used for determining the critical or key functions of the system and conveying how the controls applied across the supply chain provide assurance of the performance of the combination product (suppliers via purchasing controls, incoming inspection, in-progress testing, release testing, stability), including demonstrating why the controls are the right controls, why the acceptance criteria are appropriate, and why the control is placed at the right point.

TABLE 3.10
Summary of CTD Placement and Content Expectations for Device Information in Combination Product Submissions

Section	Content and Expectations from Guidance
1.1 [US]	**356h Form** Identify combination product manufacturing and testing facilities (FDA, 2019l) Include facilities involved in the design control activities in this list (FDA, 2019n)
1.2 [US]	**Reviewers Guide to Device Content** High-level overview of the submission's content listing the location in the eCTD (FDA, 2019l)
3.2.P.1	**Description of Combination Product** Information on each device, including description and function (EMA, 2021a) Description and intended use (FDA, 2013b)[a] Describe the device constituents/components (e.g., injector presentation or kit components) in a brief descriptive sentence and include them in a component table (PDA, 2015)
3.2.P.2	**Combination Product Development** Tabular summary of the changes in container closure components used throughout the development (FDA, 2018g) Overview of developmental process for the entire product including the device constituent part (FDA, 2018g) Information relevant to development of the device as integrated into the medicinal product summarized (EMA, 2021a) Rationale for the selection or design of the proposed CCS/device constituent (FDA, 2018g; EMA, 2021a) Include suitability of the device for its intended use, in the context of the device performing as intended (EMA, 2021a) A risk assessment summary for the DDC, aligned with suitable risk assessment principles in ICH Q9 (EMA, 2021a)
3.2.P.2.1	**Description** Design and safety features (FDA, 2013b)[a]
3.2.P.2.2	**Drug Product** The functional aspects of the medical device (part) should be qualified in line with its complexity and should include the rationale for the choice and optimization of the relevant aspects of design and performance (EMA, 2021a)

(Continued)

TABLE 3.10 (CONTINUED)
Summary of CTD Placement and Content Expectations for Device Information in Combination Product Submissions

Section	Content and Expectations from Guidance
	If there are changes in device design during development, include summary bridging data, with links to data in Module 4/5. Appropriate data should be provided to demonstrate and justify the equivalence of the overall performance of the DDC prototype(s) used during pivotal clinical development with the DDC intended for marketing (EMA, 2021a) Describe the device constituents/components briefly (PDA, 2015)
3.2.P.2.3	**Manufacturing Process Development** Summary of prior knowledge and risk assessments used to identify process parameters and material attributes that have the potential to impact product CQAs (FDA, 2018g)[b] Description of DDC manufacturing process development, including justification and suitability of sterilization processes of any devices or the DDC, where relevant. Include comparison of the manufacturing process of DDCs from pivotal or bridging clinical studies to the commercial DDC and describe development of the control strategy for the DDC manufacturing process (EMA, 2021a) Information related to the selection and optimization of the manufacturing process, as well as process elements that can influence the performance of the product (ICH, 2002) Describe how the device assembly process or manufacturing was developed and link to 3.2.P.3.3 (PDA, 2015)
3.2.P.2.4	**CCS Description** Brief description of the container closure system, including any non-integral medical devices needed for correct use of the DDC (EMA, 2021a)
	Brief details of critical functional components, description of controls and alarms and their instructions. Confirmatory signals for dose delivery, sharps injury prevention features, safety/lock-out features to prevent over-dosage, and safe disposal information. Brief description and rationale for any related technologies, e.g., software (EMA, 2021a) Materials of construction, such as chemical, grade, and brand name, and indicate which materials are in or will affect the fluid path (for injectors) (FDA, 2013b)[a] For implantable/transdermal devices, information on the matrix and reservoir, including mechanism of drug release (EMA, 2021a) **Suitability of CCS and Device** Suitability of materials (appropriateness and safety of materials) and compatibility with drug product (ICH, 2002) Evidence of suitability appropriate for the drug product and the actual use of the device and safety and functionality for the duration of direct contact (FDA, 2018g) Evidence to support lack of interaction between device materials, manufacture, and process residuals with drug product, extractables, and leachables (FDA, 2013b)[a] Provide "suitability information for the device constituent/components. For delivery devices, include a high-level overview of drug-specific data: e.g., accuracy, 510(k) history, summary of human factors studies of target population, shear studies, component aging, ISO 10993 data. Link to 3.2.R and cross-reference master files (PDA, 2015) **Performance** Dose accuracy and precision, mechanical functionality, and/or other functionalities directly related to the intended use of the device with the medicinal product and its impact on quality, safety, and/or efficacy (EMA, 2021a) Where the combination product is also the CCS, performance (such as reproducibility of the dose delivery from the device) (ICH, 2002) Results of performance testing, including stability, stress testing, and functional testing assessing mechanical specifications and safety features (FDA, 2013b)[a] For MDI/DPI, results of product characterization studies (FDA, 2018g)

(Continued)

TABLE 3.10 (CONTINUED)
Summary of CTD Placement and Content Expectations for Device Information in Combination Product Submissions

Section	Content and Expectations from Guidance
3.2.P.2.5	**Microbiological Information** For sterile products, the integrity of the container closure system to prevent microbial contamination (ICH, 2002) throughout the use and shelf life (EMA, 2021a) Demonstration of container closure integrity (FDA, 2013b)[a] Provide information regarding testing (e.g., development, stability, and shipping studies) to confirm that container closure integrity is maintained (PDA, 2015)
3.2.P.2.6	**Compatibility** Information on the compatibility of the drug product with dosage devices (e.g., sorption on injection vessels, stability) to support labeling (ICH, 2002) Evidence of physical and chemical compatibility of the drug product with the device(s), considering all materials in contact with the drug product results from interaction studies, including extractable and leachable studies (EMA, 2021a) Describe the compatibility of the device with the drug product if there is contact, and provide summaries of compatibility studies if conducted. Include human factors engineering summary report[c] with a cross-reference to the studies in 3.2.P.2 (PDA, 2015)
3.2.P.3	**Manufacturing** Information pertaining to manufacturing or assembly of the finished combination product as a whole (FDA, 2019l)
3.2.P.3.1	**Manufacturers** Names/addresses for DDC assembly, packaging, DDC sterilization, labeling, and quality control sites, as well as for the EU batch release site(s). Device part suppliers and sub-assembly manufacturers need not be stated (EMA, 2021a) Description of all manufacturing facilities; what activities occur and what constituents are at the site. For facilities subject to 21 CFR 4, identity quality system for the facility, including base regulation when following the streamlined approach (FDA, 2017b; FDA, 2019l) Provide names and addresses of the sites performing assembly, kitting, or packaging operations (PDA, 2015)
3.2.P.3.2	**Batch Formula** No device-related information (FDA, 2019l)
3.2.P.3.3	**Manufacturing Processes** Description of the manufacturing process of the DDC, including operations relating to the combination of device(s) and drug product. Critical processes, technologies, and/or packaging operations that directly affect product quality should be described in detail, together with critical process parameters, in-process controls, and acceptance criteria (for critical steps) (EMA, 2021a) Include a description of the final combination product assembly steps and testing, but not detailed component manufacturing (parts molding or machining). Include hold times, process flow diagrams, assembly or packaging equipment, etc. (PDA, 2015) **Quality System** Include general descriptions/summaries of quality documents demonstrating conformity with 21 CFR 4. Link to 3.2.R. to support manufacturing process (FDA, 2019l)
3.2.P.3.4	**Control of Critical Steps** Any critical steps should be justified, and any device-specific intermediates should be defined, along with relevant specifications, test methods, and their validation. Any holding times should be defined and justified (EMA, 2021a) For integral combination products, define intermediate checks, in-process controls, and critical process parameters of manufacturing or assembly process. Content is dependent on product complexity. This section is not applicable to kit components (PDA, 2015)

(*Continued*)

TABLE 3.10 (CONTINUED)
Summary of CTD Placement and Content Expectations for Device Information in Combination Product Submissions

Section	Content and Expectations from Guidance
3.2.P.3.5	**Process Validation** Process validation for the manufacture of the DDC, including the assembly and sterilization of the device(s) (if applicable) and any filling steps. Actual or simulated transportation studies (EMA, 2021a) Provide evidence that installation qualification, operational qualification, production qualification, and process validations were completed (e.g., device release test data on multiple representative clinical/stability/commercial lots) (PDA, 2015)
3.2.P.5	**Control of Combination Product** Release specifications for combination product and constituent parts; link to 3.2.R for design transfer information (FDA, 2019l) Description/appearance, performance/functional tests relevant to the intended use and other critical test parameters related to CQAs of the integral medicinal product (EMA, 2021a) Methods, justification of control strategy (FDA, 2018g)
3.2.P.5.1	**Specifications** Provide a table of release specifications for the combination product with limited, key device quality inspection or performance test specifications (ranges). Do not include device component design specifications (PDA, 2015)
3.2.P.5.2	**Analytical Procedures** Provide an adequate summary of the release test method/protocol for each reported device component specification[d] (PDA, 2015)
3.2.P.5.3	**Validation of Analytical Procedures** If noncompendial, provide summary validation information (e.g., gauge repeatability and reproducibility (GR&R) with data on measured parameters or cite standards. Descriptive device inspections may not require validation. (PDA, 2015)
3.2.P.5.4	**Batch Analysis** Provide release tests results for device characteristics for clinical and/or process validation lots[e] as part of overall combination product batch data (PDA, 2015)
3.2.P.5.6	**Justification of Specifications** Provide an overview of device design inputs that justify specifications (e.g., force to actuate). Supporting design verification data may be necessary (PDA, 2015)
3.2.P.7	**CCS/Device Component Information** If device is also the CCS, include content per CCS guidance (FDA, 1999), with links to 3.2.R. for device constituent testing (FDA, 2019l) Where device is also the primary container closure system, description of the CCS, quality control specifications, critical dimensions, test procedures, and materials of construction. For device parts not in direct contact with the drug product, information commensurate with its functionality is to be provided (EMA, 2021a) Provide a high-level description of device and device components. Link to 3.2.R. and cross-reference to master files (PDA, 2015)
3.2.P.8.1	**Stability Summary and Conclusions** Provide a summary of the device constituent aging test program and combination product functional stability test program, stability-indicating parameters, and proposed shelf life for the combination product. Present and discuss stability data (PDA, 2015) Stability studies for the integral medicinal product should include stability-indicating functionality and CQA tests, as well as in-use stability testing (EMA, 2021a)
3.2.P.8.2	**Post-Approval Stability Protocol and Commitment** Provide commitments and protocol summaries, if applicable (PDA, 2015)

(*Continued*)

TABLE 3.10 (CONTINUED)
Summary of CTD Placement and Content Expectations for Device Information in Combination Product Submissions

Section	Content and Expectations from Guidance
3.2.P.8.3	**Stability Data** Provide functional stability testing data (PDA, 2015)
3.2.A.1	**Facilities and Equipment** Describe the combination product (final assembly/kitting) facility, flow diagrams, principal equipment, other products, and cross-contamination procedures (PDA, 2015)
3.2.A.2	**Adventitious Agents Safety Evaluation** Identify all materials of human or animal origin, along with information assessing the risk with respect to potential contamination with adventitious agents should be provided (EMA, 2021a)
3.2.R [US]	**Device Constituent Information** Information pertaining to the manufacturing or assembly of device constituents (FDA, 2019l) Documentation, narrative explanations, plans, and reports, addressing design input requirements, design output specifications, design verification and validation, risk management, and traceability matrices (FDA, 2019l) 510(k)-like documentation for review by CDRH (PDA, 2015)
3.2.R [EU]	**Medical Device** Include relevant information related to the demonstration of compliance of the device(s) with MDR Annex 1 (the GSPRs) e.g. Notified Body (NB) Opinion, NB Certificate of Conformity, and/or device manufacturer's EU Declaration of Conformity (EMA, 2021a) Technical file-like document for review by notified body or competent authority (PDA, 2015)
5.3.5.4	**Human Factors**[e] HF validation protocols, reports, and other HF submissions, with links to Module 3; use HF-file tag (FDA, 2019l) Usability and human factor studies, with appropriate reference to Module 3 (EMA, 2021a)

[a] Guidance provides content expectations, without recommendation on CTD location.
[b] Guidance recommends Section 3.2.P.2, without specifying sub-section.
[c] Current recommendations from FDA and EMA are to include these reports in Section 5.3.5.4 (refer to the information for human factors in this table).
[d] Although this states "device component specification," it should have stated "device quality inspections/tests" as the section is intended to provide the analytical procedures for the tests included in Section 3.2.P.5.1.
[e] Additionally, it is recommended to include lots used for stability, design verification, and summative human factors, as appropriate to facilitate the review.
[e] As Module 5 is the Clinical module, some companies choose to instead place a document in Section 5.3.5.4. providing the links to the Module 3 location for the HF documents.

The following additional considerations apply to the US-specific information included in Table 3.10:

- For the reviewers guide to device content, it is recommended to structure this as a reviewer-friendly orientation to the content provided, with hyperlinks (but not presented as a table of contents).
- Regarding the 3.2.P.3.3 quality system information, as expectation is US-specific, some companies place this only in 3.2.R (CPC, 2018). The PMA quality system information guidance (FDA, 2003a) is referenced to guide the procedural information to be provided, but the Electronic Common Technical Document (eCTD) technical conformance guide (FDA, 2019l) allows for summaries for combination products. As quality system Standard

Operating Procedures (SOPs) are periodically reviewed and updated, manufacturers were concerned with the potential lifecycle management impact of including these procedures in the eCTD and whether the manufacturer would need to submit post-approval supplements if these procedures changed. Manufacturers sought to gain clarity that these procedural updates would not trigger the need for post-approval supplements, and in some cases, received feedback that changes to the procedures could require supplements (FDA, 2016d).
- Where there is not a logical location within 3.2.P for the device information, this information should be placed in 3.2.R (FDA, 2019l).
- When placing combination product files in 3.2.R for submissions to the United States, prefix the leaf title with "DEVICE" to facilitate differentiation between combination products and other categories of files (FDA, 2019l).
- Submission content and location expectations included in the 2018 Metered Dose Inhaler (MDI)/Dry Powder Inhaler (DPI) draft guidance were not aligned with the expectations from Industry, which are more aligned with the recommendations in the eCTD technical conformance guide that allows manufacturers the flexibility to include supportive information in the pharmaceutical development or regional sections.

Referencing Other Submissions to Support Sponsor's Filing

With partnerships or supplier relationships, the applicant may not have access to all necessary details to support registration and may need to rely on master files or other NDA/MAA/BLA/PMA/510(k) submissions. Where referenced submissions are required, it is important to ensure that the messaging included in the referenced submission and the combination product application is reviewed to ensure that there are no conflicts, as conflicts can cause confusion during the review and additional information requests.

Master files (MFs) are confidential filings to a health authority, which can be referenced by an applicant in support of their filing. A master file permits the master file holder (i) to incorporate the information by reference when submitting an application or an amendment or supplement to the application, or (ii) to authorize other individuals to rely on the information to support a submission to the FDA without having to disclose the information (Roan, 2009). The types of master files available vary by country (the European Union, the United States, Canada) and not all regions support the use of master files. In the United States, there are both drug master files (DMF) and master access files (MAF, or device master file). Additionally, CDER issued guidance in 2019 describing a new option for a Type V DMF that features a device constituent part with electronics and/or software that is planned to be used as a platform, for example, used in multiple CDER-led combination products (FDA, 2019o).

Master files are neither approved nor rejected by the FDA. The master file is only reviewed by the FDA when referenced by an applicant sponsor (typically a customer of the DMF holder) in a submission (investigational new drug (IND), NDA, etc.), through a letter of authorization to the master file. It is best practice to ensure that the letter of authorization for the master file includes the specific locations (section number, page number) for the referenced details to facilitate the review of the master file as it relates to the IND/NDA/BLA. This becomes increasingly important where the master file is used to describe multiple materials, as they can become very large filings.

The supply chain is often described using *n, n–x* nomenclature, where *n* is the biopharmaceutical company, *n–1* is the supplier that provides the packaging or device components and *n–2* are the suppliers of materials and sub-components procured by the *n–1* supplier, and so forth. As the sponsor of a submission referencing the *n–1* master file to support one or more elements in the submission, it is important to understand the hierarchy of *n–x* master files that may support the product. The review of the *n–1* master file may include a review of master files referenced by the *n–1* master file.

While master files are a useful regulatory tool, the preference from FDA reviewers is to ensure that the combination product submission is as complete as possible, supplemented by master file content. For example, review documentation for an initial BLA whereby FDA indicated that master files "should only be for confidential proprietary information that is not otherwise known to the BLA holder" (FDA, 2015). FDA has also expressed the importance of conveying the context of the information in the master file, as it relates to the intended use of the combination product (McMichael, 2017).

Investigational Submissions

During the clinical development phase, submissions to support the use of the drug or biologic PMOA combination products in clinical trials are typically filed as investigational new drug (IND) applications in the United States, Investigational medicinal product dossier (IMPD) or Clinical Trial Applications (CTA) in the European Union. While there are no general combination product guidances regarding the extent of content to be included for IND/IMPD submissions, most companies apply a phase-appropriate approach for submission content related to the device constituent parts of combination products and for separate cleared/approved delivery devices to be used to administer investigational medicinal products, aligned with the phase-appropriate expectations for the drug product. These phase-appropriate expectations involve emphasis in an initial Phase 1 CMC submission placed on providing information that will allow evaluation of the safety of subjects in the proposed study (FDA, 1995). For Phase 2 and 3 studies, the amount and depth of CMC information that would be submitted to the Agency depends, in large part, on the phase of the investigation, the type of testing proposed in humans, and whether the information is safety related (e.g., has the potential to affect the safe use of the drug). Corroborating information is described as the additional supportive information that is less likely to affect the safe use of the drug but should be submitted in annual reports to ensure the quality of the product. In general, the corroborating information should focus on summaries and analyses of data rather than extensive compilations of data (FDA, 2003b). In the author's experience, more extensive corroborating information is expected and/or requested during the consultative review for the device constituent in investigational products than for the drug or biologic information.

Applying ICH Q12

Lifecycle management tools, such as those outlined in ICH Q12 (ICH, 2019b), can be applied to drug-device combination products with a drug primary mode of action. From a practical standpoint, this would include combination products submitted as BLAs/NDAs/MAAs using the ICH CTD structure (ICH, 2002). The ICH Q12 guideline is intended to demonstrate how increased product and process knowledge can contribute to a more precise and accurate understanding of which post-approval changes require a regulatory submission as well as the definition of the level of reporting categories for such changes. One goal of ICH Q12 is to shift from "Tell then Do" to "Do and Tell" allowing for industry and regulators to focus resources on the critical changes and increase reliance on the quality systems for the minor changes. ICH Q12 provides a framework for lifecycle management comprising categorization of changes, established conditions (EC), post-approval change management protocols (PACMP), effective pharmaceutical quality system, and utilization of product lifecycle management strategy (PLMS). The regulatory strategy should address how to implement the ICH Q12 regulatory tools and enablers to support lifecycle management of the combination product. While the application of these tools drug-device combination products is not described in extensive detail in the guideline, the examples provided in the Q12 Annexes are informative (ICH, 2019c). Table 3.11 provides an overview of opportunities for the use of these tools and enablers for combination products.

At the time of publication of this chapter, experience is limited regarding the application of ICH Q12 to combination products. As a means to provide guidance to readers, the following is provided as a starting point to consider when applying ICH Q12 to future combination products. It is recommended to research contemporary recommendations and to utilize health authority engagement avenues to confirm the approach for a given product. FDA has included guidance related to

TABLE 3.11
Opportunities from Implementing ICH Q12 Tools and Enablers for Combination Products

Q12 Tools and Enablers	Description	Potential Applicability for Combination Products
Categorization of Changes	Risk-based categorization of changes for reporting to regulators	Risk-based categorization for combination product change reporting can focus reporting on those changes with potential for impact rather than minimal risk changes
Established Conditions (EC)	Legally binding elements for a product necessary to assure product quality. ECs require reporting if changed	Increased clarity regarding which elements are ECs for the device aspects would be beneficial for sponsors and regulators to enable the right level of change reporting
PACMP	Protocol submitted ahead of the change which defines studies and acceptance criteria, typically with lower reporting category	PACMP can be used for changes in materials/components, combination product assembly sites, and processes for device aspects
Pharmaceutical Quality System (PQS)	An effective PQS and increased confidence changes are supported by data obtained through application of patient-centric, risk-based principles	The effective PQS expectation from ICH Q12 for the device part can be demonstrated through certification to international standards. ISO 13485/21 CFR 820 is consistent with ICH Q10, and ISO 14971 is consistent with ICH Q9
Product Lifecycle Management Strategy (PLCM)	Strategy document which outlines the EC, reporting categories for EC changes, PCAMPs, post-approval commitments	The PLCM strategy is the repository for all of the Q12 aspects, so this tool would also be appropriate to address the above aspects which are applicable to combination products

identification of ECs for device constituent parts of combination products in the draft ICH guidance (FDA, 2021), which includes the following:

1. Elements unique to drug product-device combination products that may be ECs include the device identification, description, design features, and performance specifications. The regulatory strategy should map out a process for selection of ECs for device elements, including assessing the "characteristics of the product that are essential for its safe and proper use" (ISO, 2016a) and identifying the ECs associated with these characteristics (referred to as primary characteristics). Aspects of the device which may be considered as primary characteristics include:
    - Functions of the delivery device identified as essential for safe use utilizing risk management (e.g., ICH Q9 and/or ISO 14971).
    - Design features essential to achieve delivery of the dose, as they are considered essential for proper use.
    - Characteristics which impact and/or ensure that the drug product maintains its critical quality attributes (CQAs).
2. The following may be considered potential ECs:
    - Design features that are primary characteristics.
    - Manufacturing process elements for the device constituent part that need to be controlled to ensure a primary characteristic.
    - Other control strategy elements for the device constituent part that ensure a primary characteristic.
3. The extent of product and process understanding, knowledge from design and development to manage the risks and evidence to support designation, and justification of the primary characteristics of the device are expected to support the proposed ECs.

4. Appropriate justification should also be provided to support the proposed reporting categories for future changes to the EC. The reporting categories for potential future changes are based on potential for risk to safe and proper use, as determined using the risk management principles in ICH Q9 and/or ISO 14971 (FDA, 2021).

Human Factors-Specific Submissions

FDA has issued guidance with human factors submission expectations for IND products. The expectation is that the combination product's specific use-related risk analysis (URRA) is submitted to the IND. If the applicant determines from the URRA that a HF study is not needed, the justification for this conclusion should be included along with the URRA. If the URRA indicates that a HF study is necessary, FDA encourages applicants to submit before commencing the HF validation study: The HF protocol; a summary of HF formative study results and analysis; a summary of changes made to the product user interface after the HF formative studies, including how the results from the HF formative studies were used to update the user interface, and intend-to-market labels and labeling (including instructions for use if any are proposed) that will be tested in the HF validation study for feedback (FDA, 2016b).

The 2017 Food and Drug Administration Reauthorization Act (FDARA) incorporates by reference the sixth reauthorization of the Prescription Drug User Fee Act (PDUFA VI) performance goals and procedures for fiscal years 2018–2022 (FDA, 2017d). One of the performance goals relates to the process for human factors protocol submissions and the timelines for protocol review. In response, FDA issued guidance to define the submission expectations for the following HF use-related risk analysis (URRA), HF validation protocols, and reports, as well as threshold analyses (FDA, 2018d).

Dossier Review

The review of the combination product dossier by the health authority follows the same general approach as drug or biologic products, where the dossier is first assessed for its completeness, e.g., does the submission contain all of the elements required to consider it a "fileable" dossier. For an MAA in the European Union, this is assessed during the validation phase. In the United States, the first 74 days of the review for the NDA and BLA processes involve this assessment of the sufficiency of the filed package. Similarly, both PMA and 510(k) submission reviews include a period where the FDA is assessing whether the submission can be accepted for review.

Where the review includes a non-lead component of the health authority, typically there is a coordination process between the lead and non-lead reviewers. In the United States, the review by the non-lead centers are considered consult reviews (FDA, 2018a). In 2018, FDA issued Staff Manual Guide 4101 which details FDA's process for when and how to request, receive, process, and track the progress of Inter-Center Consult Requests (ICCRs) between FDA review centers (FDA, 2018h). As part of the consultation process, Centers are expected to coordinate as appropriate prior to issuance of communications regarding the review, as Agency communications from the lead center can be considered communications on behalf of all centers involved in the review (FDA, 2022).

For some markets, the non-lead review is conducted by separate bodies, such as notified bodies for EU for drug/biologic-led combination products and conformity assessment bodies in Malaysia. In these cases, the applicant is interacting with both the competent authority as well as these separate review bodies. Further discussion of the international processes is included in the chapter "Combination Products Evolving Global Regulatory Environment."

Post-approval Considerations

It is important to build the submission strategy with the final commercial product in mind, but also with future post-approval iteration considerations, as the approaches for describing the studies and the content included in the initial submission can pay dividends when incorporating post-approval changes years later. For example, the device portion of the combination product may experience more frequent updates (e.g., iterations on a device design) throughout the product lifecycle than the drug component. The lifecycle plan may warrant incorporation of study designs and comparability protocols included in the initial submission aimed at helping facilitate post-approval continuous improvement and innovation.

Once a Sponsor has received approval for their combination product, the challenge of maintaining the license during the lifecycle begins. Assessing changes for their impact and identifying the appropriate reporting category can be challenging for combination products, as the majority of the available guidance does not address combination products. In the United States, the FDA issued a draft guidance regarding submissions for post-approval modifications to combination products when 21 CFR 4 was published (FDA, 2013e). This guidance outlined FDA expectations for a manufacturer to holistically assess changes to combination products approved under a BLA, NDA, or PMA, including changes made to any of the individual constituents (i.e. drug/biologic constituent and device constituent), and provided manufacturers with a framework upon which to assess the appropriate reporting category for changes to combination products. This framework provides that the sponsor should assess the change to the constituent part based on the submission type that would have been required if the constituent were a standalone product and then, using the translation tables provided in the guidance, identify the corollary submission type based on the combination product original application type.

The guidance does not expressly address how to approach postmarket changes to devices that might be considered equivalent to 510(k) rather than PMA devices however. Yet, the majority of device constituents, which are part of drug delivery system combination products, would be classified as low or moderate risk if they were standalone devices, requiring 510(k) clearance and not a PMA. FDA has stated that it plans to issue revised guidance addressing these issues.

When changes to sterile parenteral products are considered, there are specific change types from the drug constituent guidances that apply. These include a prior approval supplement for any change that could impact drug product sterility assurance; a changes being effected (CBE) supplement for change to or in a container closure system that does not affect the quality of the drug product, and minor changes such as changes to glass supplier and changes to crimp caps being annually reportable (FDA, 2004). In the European Union, the following medicinal product post-authorization change categories might be utilized:

- B.II.e.1 Change in immediate packaging of the finished product – Type II variation for sterile products.
- B.II.e.4 Change in shape or dimensions of the container or closure (immediate packaging) – Type Ib variation for a sterile product and a Type II variation when the change concerns a fundamental part of the packaging material, which may have a significant impact on the delivery, use, safety, or stability of the finished product.
- B.IV.1 Change of a measuring or administration device – ranges from Type IA notification (Type IAN) to Type II for devices not integrated with the primary packaging, but is a Type II when the device is integrated with the primary packaging (EC, 2013).

CONCLUSION

As part of product development and lifecycle management of combination products, it is critical to have a solid regulatory strategy to guide the activities. As described in this chapter, the regulatory strategy is initially developed during the early stages of product development and is refined throughout the product lifecycle. This chapter has outlined the key considerations for developing the strategy, including product designation and jurisdiction, implications of the commercial presentation, and identification of the regulatory design inputs.

The global regulatory strategy typically includes an overview of the key regulatory interactions and submissions anticipated, considering the current regulatory expectations. The regulatory strategy also provides for a mechanism to document regulatory risks, requirements associated with the design, potential clinical study requirements, and associated timing, as well as identification of applicable regulatory requirements for the product.

Importantly, the regulatory strategy needs to evolve along with the development of the combination product to integrate contemporary external and internal intelligence, as well as feedback from clinical filings and interactions with health authorities. It is best practice for the identified regulatory risks to be managed along with the revisions to the strategy, with the aim of mitigating those risks prior to the marketing submission.

As combination products provide for unique application of the expectations for clinical and marketing submissions based on the varied nature of the designs, the regulatory strategy can serve to guide the submission pathways, content, and structure. Finally, the regulatory strategy should be used for prospective planning of post-approval changes to the combination product, including anticipated filing categories, submission expectations, and planned health authority interactions to review the changes.

NOTE

1. Implementation date delayed by one year (EU, 2020) due to COVID-19; original date of implementation was May 2020.

REFERENCES

ASEAN (2016) The ASEAN Secretariat. *ASEAN Common Technical Dossier (ACTD).* Accessed on January 15, 2020 at https://asean.org/storage/2017/03/68.-December-2016-ACTD.pdf.

Australian Government (2019a) TGA Australian Public Assessment Reports for Prescription Medicines (AusPARs). Accessed on November 7, 2019 at https://www.tga.gov.au/australian-public-assessment-reports-prescription-medicines-auspars.

Australian Government (2019b) TGA AusPAR Search Database. Accessed on November 7, 2019 at https://www.tga.gov.au/ws-auspar-index.

BfArM (2019) *Scientific and Regulatory Advice by the Federal Institute for Drugs and Medical Devices (BfArM) Guidance for Applicants (Revision 9).*

BSI (2014) Documentation Submissions – Best Practices Guidelines. Accessed on January 15, 2020 at https://www.bsigroup.com/globalassets/meddev/localfiles/ja-jp/documents/bsi_best_practice_guidelines.pdf.

CPC (2018) *CPC Comments to Docket No. FDA-2018-D-1098: Metered Dose Inhaler (MDI) and Dry Powder Inhaler (DPI) Drug Products—Quality Considerations.* Document ID: FDA-2018-D-1098-0019. Accessed on January 15, 2020 at https://www.regulations.gov/document?D=FDA-2018-D-1098-0019.

Dorgan, C (2018) *Regulatory Considerations for Complex Container Closure System*, presented at the 2018 PDA Container Closure Performance and Integrity Conference in Bethesda, MD.

Eastern Research Group, Inc. (2020) *Assessment of Combination Product Review Practices in PDUFA VI.*

EC (1990) Active Implantable Medical Device Directive (as Amended), 90/385/EEC. Will be Superseded by MDR on May 20, 2020 (EU, 2017a).

EC (1993) Medical Device Directive (as Amended), 93/42/EEC. Will be Superseded by MDR on May 20, 2020 (EU, 2017a).

EC (1998) In-vitro Diagnostic Medical Devices Directive (as Amended), 98/79/EC. Will be Superseded by IVDR on May 20, 2022 (EU, 2017b).

EC (2001) Medicinal Product Directive (as Amended), 2001/83/EC.

EC (2009) *Medical Devices: Guidance Document – Borderline Products, Drug-delivery Products and Medical Devices Incorporating, as an Integral Part, an Ancillary Medicinal Substance or an Ancillary Human Blood Derivative*, MEDDEV 2. 1/3 rev 3.

EC (2013) *Guidelines on the Details of the Various Categories of Variations, on the Operation of the Procedures Laid Down in Chapters II, IIa, III and IV of Commission Regulation (EC) No 1234/2008 of 24 November 2008 Concerning the Examination of Variations to the Terms of Marketing Authorisations for Medicinal Products for Human Use and Veterinary Medicinal Products and on the Documentation to be Submitted Pursuant to Those Procedures*, 2013/C 223/01.

EC (2019a) Harmonised Standards. Accessed on December 4, 2019 at https://ec.europa.eu/growth/single-market/european-standards/harmonised-standards_en.

EC (2019b) Medical Devices (Harmonised Standards). Accessed on December 4, 2019 at https://ec.europa.eu/growth/single-market/european-standards/harmonised-standards/medical-devices_en.

EC (2020) Legal Framework Governing Medicinal Products for Human Use in the EU. Accessed on December 20, 2020 at https://ec.europa.eu/health/human-use/legal-framework_en.

EMA (2005a) Committee for Medicinal Products for Human Use (CHMP), Committee for Medicinal Products for Veterinary Use (CVMP). *Guideline on Plastic Immediate Packaging Materials*. CPMP/QWP/4359/03 and EMEA, CVMP/205/04.

EMA (2005b) CHMP. *Guideline on the Suitability of the Graduation of Delivery Devices for Liquid Dosage Forms*. EMEA/CHMP/QWP/178621/2004.

EMA (2006) CHMP. *Guideline on the Pharmaceutical Quality of Inhalation and Nasal Products*. CHMP/QWP/49313/2005.

EMA (2009) CHMP. *Guideline on the Requirements for Clinical Documentation for Orally Inhaled Products (OIP) including the Requirements for Demonstration of Therapeutic Equivalence between Two Inhaled Products for use in the Treatment of Asthma and Chronic Obstructive Pulmonary Disease (COPD) in Adults and for the use in the Treatment of Asthma in Children and Adolescents*. CPMP/EWP/4151/00 Rev. 1.

EMA (2011) *Note for Guidance on Minimising the Risk of Transmitting Animal Spongiform Encephalopathy Agents via Human and Veterinary Medicinal Products – Revision 3*. EMA/410/01 rev. 3.

EMA (2012) *European Medicines Agency Recommendation on the Procedural Aspects and Dossier Requirements for the Consultation to the European Medicines Agency by a Notified Body on an Ancillary Medicinal Substance or an Ancillary Human Blood Derivative Incorporated in a Medical Device or Active Implantable Medical Device*. EMA/CHMP/578661/2010.

EMA (2014) *Guideline on Quality of Transdermal Patches*. EMA/CHMP/QWP/608924/2014.

EMA (2017) *European Medicines Agency Guidance for Applicants Seeking Scientific Advice and Protocol Assistance*. EMA/4260/2001 Rev. 9.

EMA (2019a) *European Public Assessment Reports: Background and Context*. Accessed on November 9, 2019 at https://www.ema.europa.eu/en/medicines/what-we-publish-when/european-public-assessment-reports-background-context.

EMA (2019b) *Medicines Database*. Accessed on November 9, 2019 at https://www.ema.europa.eu/en/medicines.

EMA (2019c) *Quality of Medicines Questions and Answers: Part 1*. Accessed on November 9, 2019 at https://www.ema.europa.eu/en/human-regulatory/research-development/scientific-guidelines/qa-quality/quality-medicines-questions-answers-part-1.

EMA (2019d) *Quality of Medicines Questions and Answers: Part 2*. Accessed on November 9, 2019 at https://www.ema.europa.eu/en/human-regulatory/research-development/scientific-guidelines/qa-quality/quality-medicines-questions-answers-part-2.

EMA (2021a) *Guideline on Quality Documentation for Medicinal Products When Used with a Medical Device*. EMA/CHMP/QWP/BWP/259165/2019.

EU (2017a) Regulation (EU) 2017/745 of the European Parliament and of the Council of 5 April 2017 on Medical Devices, Amending Directive 2001/83/EC, Regulation (EC) No 178/2002 and Regulation (EC) No 1223/2009 and Repealing Council Directives 90/385/EEC and 93/42/EEC. *Official Journal of the European Union*, L117 1-175.

EU (2017b) Regulation (EU) 2017/746 of the European Parliament and of the Council of 5 April 2017 on In Vitro Diagnostic Medical Devices and Repealing Directive 98/79/EC and Commission Decision 2010/227/EU. *Official Journal of the European Union*, L117 176-332.

EU (2019) *Regulations, Directives and Other Acts*. Accessed on January 15, 2020 at https://europa.eu/european-union/eu-law/legal-acts_en.

FDA (1991) *Letter of Designation*. 21 CFR 3.8.

FDA (1994) *Content and Format of a 510(k) Summary*. 21 CFR 807.92.

FDA (1995) FDA, CDER, CBER. *Content and Format of Investigational New Drug Applications (INDs) for Phase 1 Studies of Drugs, Including Well Characterized, Therapeutic, Biotechnology-Derived Products: Guidance for Industry*.

FDA (1996a) *Confidentiality of Data and Information in a Premarket Approval Application (PMA) File*. 21 CFR 814.9.

FDA (1996b) *Design Controls*. 21 CFR 820.30.

FDA (1996c) *Purchasing Controls*. 21 CFR 820.50.

FDA (1997) FDA, CDRH. *Design Control Guidance for Medical Device Manufacturers*.

FDA (1998) *Internal Agency Review of Decisions*. 21 CFR 10.75.

FDA (1999) U.S. Department of Health and Human Services (HHS), FDA, CDER, CBER. *Container Closure Systems for Packaging Human Drugs and Biologics, Chemistry, Manufacturing and Controls Documentation – Guidance for Industry.*

FDA (2001) FDA, CDRH, *Early Collaboration Meetings Under the FDA Modernization Act (FDAMA); Final Guidance for Industry and for CDRH Staff.*

FDA (2002) U.S. Department of HHS, FDA, CDER. *Nasal Spray and Inhalation Solution, Suspension, and Spray Drug Products guidance— Chemistry, Manufacturing, and Controls Documentation – Guidance for Industry.*

FDA (2003a) U.S. Department of HHS, FDA, CDRH. *Quality System Information for Certain Premarket Application Reviews.*

FDA (2003b) U.S. Department of HHS, FDA, CDER. *INDs for Phase 2 and Phase 3 Studies Chemistry, Manufacturing, and Controls Information: Guidance for Industry.*

FDA (2004) U.S. Department of HHS, FDA, CDER. *Changes to an Approved NDA or ANDA – Guidance for Industry.*

FDA (2005a) *Definitions.* 21 CFR 3.2.

FDA (2005b) *Designated Agency Component.* 21 CFR 3.4.

FDA (2005c) *Request for Designation.* 21 CFR 3.7.

FDA (2006) U.S. Department of HHS, Office of the Commissioner, OCP. *Guidance for Industry and FDA Staff: FDA Early Development Considerations for Innovative Combination Products.*

FDA (2007) U.S. Department of HHS, FDA, CDER. *Guidance for Industry and Review Staff Target Product Profile — A Strategic Development Process Tool (Draft Guidance).*

FDA (2010) U.S. Department of HHS, FDA, CDER, CBER. *Guidance for Industry: Dosage and Administration Section of Labeling for Human Prescription Drug and Biological Products — Content and Format.*

FDA (2011) U.S. Department of HHS, FDA, Office of the Commissioner, OCP. *Guidance for Industry: How to Write a Request for Designation (RFD).*

FDA (2013a) *Current Good Manufacturing Practice Requirements for Combination Products.* 21 CFR 4.1-4.4.

FDA (2013b) U.S. Department of HHS, FDA, CDRH, CDER, CBER, OCP in the Office of the Commissioner. *Technical Considerations for Pen, Jet, and Related Injectors Intended for Use with Drugs and Biological Products – Guidance for Industry and FDA Staff.*

FDA (2013c) U.S. Department of HHS, FDA, CDRH, CDER, CBER, OCP in the Office of the Commissioner. *Glass Syringes for Delivering Drug and Biological Products: Technical Information to Supplement International Organization for Standardization (ISO) Standard 11040-4 – Guidance for Industry and FDA Staff*, Draft.

FDA (2013d) U.S. Department of HHS, FDA, CDER, CBER, CDRH. *Rheumatoid Arthritis: Developing Drug Products for Treatment – Guidance for Industry*, Draft.

FDA (2013e) U.S. Department of HHS, FDA, OCP, Office of Special Medical Programs, Office of the Commissioner. *Guidance for Industry and FDA Staff: Submissions for Postapproval Modifications to a Combination Product Approved Under a BLA, NDA, or PMA – Guidance for Industry and FDA Staff*, Draft.

FDA (2014a) FDA, CDER. *21st Century Review Process Desk Reference Guide.*

FDA (2014b) U.S. Department of HHS, FDA, CDRH, CBER. *Types of Communication during the Review of Medical Device Submissions - Guidance for Industry and FDA Staff.*

FDA (2014c) *Procedures for Review of a PMA.* 21 CFR 814.44.

FDA (2014d) *PMA Supplements.* 21 CFR 814.39.

FDA (2015) Administrative Documents and Correspondence for BLA 125522 Initial Review, page 99. Accessed on January 15, 2020 at https://www.accessdata.fda.gov/drugsatfda_docs/nda/2015/125522Orig1s000AdminCorres.pdf.

FDA (2016a) *Postmarketing Safety Reporting for Combination Products.* 21 CFR 4.100-4.105.

FDA (2016b) U.S. Department of HHS, FDA. CDRH, CDER, CBER, OCP in the Office of the Commissioner. *Human Factors Studies and Related Clinical Study Considerations in Combination Product Design and Development – Guidance for Industry and FDA Staff*, Draft Guidance.

FDA (2016c) U.S. Department of HHS, FDA. CDRH, CDER, CBER. *Principles for Codevelopment of an In Vitro Companion Diagnostic Device with a Therapeutic Product – Guidance for Industry and FDA Staff*, Draft Guidance.

FDA (2016d) Administrative Documents and Correspondence for BLA 761029 Initial Review, page 202. Accessed on January 15, 2020 at https://www.accessdata.fda.gov/drugsatfda_docs/nda/2016/761029Orig1s000AdminCorres.pdf.

FDA (2017a) U.S. Department of HHS, FDA, CDER, CBER. *Formal Meetings Between the FDA and Sponsors or Applicants of PDUFA Products Guidance for Industry (Draft guidance).*

FDA (2017b) U.S. Department of HHS, FDA, OCP in the Office of the Commissioner, CDER, CBER, CDRH, Office of Regulatory Affairs (ORA). *Current Good Manufacturing Practice Requirements for Combination Products – Guidance for Industry and FDA Staff.*

FDA (2017c) Master Files. Accessed on January 20, 2020 at https://www.fda.gov/medical-devices/premarket-approval-pma/master-files.

FDA (2017d) FDA, PDUFA VI Commitment Letter. Accessed on January 20, 2020 at https://www.fda.gov/downloads/ForIndustry/UserFees/PrescriptionDrugUserFee/UCM511438.pdf.

FDA (2018a) *Intercenter Agreements.* Accessed on September 3, 2019 at https://www.fda.gov/combination-products/classification-and-jurisdictional-information/intercenter-agreements.

FDA (2018b) U.S. Department of HHS, FDA, OCP in the Office of the Commissioner. *How to Prepare a Pre-Request for Designation – Guidance for Industry (Pre-RFD).*

FDA (2018c) *Product Code Classification Database.* Accessed on August 18, 2019 at https://www.fda.gov/medical-devices/classify-your-medical-device/product-code-classification-database.

FDA (2018d) U.S. Department of HHS, FDA, CDER, CBER. *Contents of a Complete Submission for Threshold Analyses and Human Factors Submissions to Drug and Biologic Applications - Guidance for Industry and FDA Staff.*

FDA (2018e) *Recognized Consensus Standards Database.* Accessed on December 3, 2019 at https://www.accessdata.fda.gov/scripts/cdrh/cfdocs/cfStandards/search.cfm.

FDA (2018f) *Regulatory Controls.* Accessed on December 3, 2019 at https://www.fda.gov/medical-devices/overview-device-regulation/regulatory-controls.

FDA (2018g) U.S. Department of HHS, FDA, CDER. *Metered Dose Inhaler (MDI) and Dry Powder Inhaler (DPI) Products – Quality Considerations – Guidance for Industry.* Revision 1 (Draft).

FDA (2018h) FDA Staff Manual Guides, Volume IV – Agency Program Directives. *Combination Products Inter-center Consult Request Process.* SMG 4101. June 2018.

FDA (2019a) *510(k) Premarket Notification Database.* Accessed on August 30, 2019 at https://www.accessdata.fda.gov/scripts/cdrh/cfdocs/cfpmn/pmn.cfm.

FDA (2019b) *Premarket Approval (PMA) Database.* Accessed on August 30, 2019 at https://www.accessdata.fda.gov/scripts/cdrh/cfdocs/cfPMA/pma.cfm.

FDA (2019c) *PMA Summary Review Memos for 180-Day Design Changes Database.* Accessed on August 30, 2019 at https://www.accessdata.fda.gov/scripts/cdrh/cfdocs/cfpma/pmamemos.cfm.

FDA (2019d) Drugs@FDA Database. Accessed on September 6, 2019 at www.fda.gov/drugsatfda.

FDA (2019e) *Drugs@FDA Glossary.* Accessed on September 6, 2019 at https://www.accessdata.fda.gov/scripts/cder/daf/index.cfm?event=glossary.page.

FDA (2019f) *Licensed Biological Products with Supporting Documents Database.* Accessed on September 6, 2019 at https://www.fda.gov/vaccines-blood-biologics/licensed-biological-products-supporting-documents.

FDA (2019g) U.S. Department of HHS, FDA, CDRH, CBER. *Requests for Feedback and Meetings for Medical Device Submissions: The Q-Submission Program; Guidance for Industry and Food and Drug Administration Staff.*

FDA (2019h) U.S. Department of HHS, FDA, OCP, CDER, CBER. *Guidance for Industry: Instructions for Use – Patient Labeling for Human Prescription Drug and Biological Products and Drug-Device and Biologic-Device Combination Products – Content and Format*, Draft.

FDA (2019i) *Search for FDA Guidance Documents.* Accessed on December 22, 2019 at https://www.fda.gov/regulatory-information/search-fda-guidance-documents.

FDA (2019j) U.S. Department of HHS, FDA, CDER, CBER, CDRH. *Bridging for Drug-Device and Biologic-Device Combination Products Guidance for Industry*, Draft.

FDA (2019k) *Frequently Asked Questions About Combination Products.* Accessed on December 22, 2019 at https://www.fda.gov/combination-products/about-combination-products/frequently-asked-questions-about-combination-products#pre-IND.

FDA (2019l) U.S. Department of HHS, FDA, CDER, CBER. eCTD Technical Conformance Guide – Technical Specifications Document.

FDA (2019m) U.S. Department of HHS, FDA, CDER. *Transdermal and Topical Delivery Systems – Product Development and Quality Considerations – Guidance for Industry*, Draft.

FDA (2019n) U.S. Department of HHS, FDA, CDER, CBER. *Identification of Manufacturing Establishments in Applications Submitted to CBER and CDER Questions and Answers – Guidance for Industry.*

FDA (2019o) U.S. Department of HHS, FDA, CDER. *Type V DMFs for CDER-Led Combination Products Using Device Constituent Parts With Electronics or Software Guidance for Industry*, Draft.

FDA (2020a) U.S. Department of HHS, FDA CDRH, CDER, CBER, OCP. *Technical Considerations for Demonstrating Reliability of Emergency-Use Injectors Submitted under a BLA, NDA, or ANDA: Guidance for Industry and FDA Staff*, Draft.

FDA (2020b) U.S. Department of HHS, FDA, OCP, CBER, CDER, CDRH. *Requesting FDA Feedback on Combination Products – Guidance for Industry and FDA Staff.*

FDA (2020c) *Classify Your Medical Device.* Accessed on December 30, 2020 at https://www.fda.gov/medical-devices/overview-device-regulation/classify-your-medical-device.

FDA (2020d) New Drug Application. Accessed on December 30, 2020 at https://www.fda.gov/drugs/types-applications/new-drug-application-nda.

FDA (2021) U.S. Department of HHS, FDA, CBER, CDER, CDRH, OCP. *ICH Q12: Implementation Considerations for FDA-Regulated Products*, Draft.

FDA (2022) U.S. Department of HHS, FDA, OCP, CBER, CDER, CDRH. *Principles of Premarket Pathways for Combination Products – Guidance for Industry and FDA Staff.*

GHTF (2008) *Summary Technical Documentation for Demonstrating Conformity to the Essential Principles of Safety and Performance of Medical Devices.* GHTF/SG1/N011:2008.

Health Canada (2005) *Drug/Medical Device Combination Products Policy.*

Health Canada (2006) *Guidance for Industry – Pharmaceutical Quality of Inhalation and Nasal Products.* File Number 06-106624-547.

ICH (2002) International Conference on Harmonisation of Technical Requirements for Registration of Pharmaceuticals for Human Use (ICH). *ICH Harmonised Tripartite Guideline: The Common Technical Document for the Registration of Pharmaceuticals for Human Use: Quality – M4Q(R1) Quality Overall Summary of Module 2; Module 3: Quality.*

ICH (2009) *ICH Harmonised Tripartite Guideline: Pharmaceutical Development – Q8(R2).*

ICH (2016) *ICH Harmonised Tripartite Guideline: The Common Technical Document for the Registration of Pharmaceuticals for Human Use – M4(R4).*

ICH (2019a) *ICH Guidelines.* Accessed on December 2, 2019 at https://www.ich.org/page/ich-guidelines.

ICH (2019b) *ICH Q12 – Technical and Regulatory Considerations for Pharmaceutical Product Lifecycle Management.*

ICH (2019c) *ICH Q12 Annexes – Technical and Regulatory Considerations for Pharmaceutical Product Lifecycle Management.*

IMDRF (2018) *Essential Principles of Safety and Performance of Medical Devices and IVD Medical Devices.* IMDRF/GRRP WG/N47 FINAL: 2018.

IMDRF (2019a) *IMDRF Documents.* Accessed on December 2, 2019 at http://www.imdrf.org/documents/documents.asp#imdrf.

IMDRF (2019b) *Non-In Vitro Diagnostic Device Market Authorization Table of Contents (nIVD MA ToC).* IMDRF/RPS WG/N9 FINAL:2019 (Edition 3).

IMDRF (2019c) *Assembly and Technical Guide for IMDRF Table of Contents Submissions.* IMDRF/RPS WG/N27 FINAL:2019.

IMDRF (2019d) *IMDRF Communication Regarding Regulated Product Submission (RPS).* Accessed on January 15, 2020 at http://www.imdrf.org/docs/imdrf/final/procedural/imdrf-proc-190124-rps-communication.pdf.

ISO (2016a) ISO 13485:2016. *Medical Devices — Quality Management Systems — Requirements for Regulatory Purposes.*

ISO (2016b) ISO 16142-1:2016. *Medical Devices – Recognized Essential Principles of Safety and Performance of Medical Devices Part 1: General Essential Principles and Additional Specific Essential Principles for All Non-IVD Medical Devices and Guidance on the Selection of Standards.*

McMichael, J. (2017) *Regulatory Challenges for Combination Products*, presented at PDA Combination Product Interest Group Meeting, 11 May 2017.

MHLW (2014) *Handling of Marketing Application for Combination Products.* PSFB/ELD Notification No. 1024-2.

Ministry of Health Malaysia (2017) *Guideline for Registration of Drug-Medical Device and Medical Device-Drug Combination Products.* Accessed on August 7, 2019 at https://www.npra.gov.my/index.php/en/component/content/article/146-english/guidelines-combo-products/1723-guideline-for-registration-of-drug-medical-device-and-medical-device-drug-combination-products.html?highlight=WyJtZWRpY2FsIiwiZGV2aWNlIiwibWVkaWNhbCBkZXZpY2UiXQ==&Itemid=1391.

MHRA (2011) Medicines and Healthcare Products Regulatory Agency (MHRA). *Bulletin No. 17: Medical Devices and Medicinal Products.*

MHRA (2017a) *Guidance Note 31: Guidance for Notified Bodies – Devices Which Incorporate an Ancillary Medicinal Substance.* Accessed on December 6, 2019 at https://assets.publishing.service.gov.uk/government/uploads/system/uploads/attachment_data/file/760519/Guidance_Note__31_September_2017__3_.pdf.

MHRA (2017b) *Human Factors and Usability Engineering – Guidance for Medical Devices Including Drug-device Combination Products.* Version 1.0.

MHRA (2018) Guidance - Medicines: Get Scientific Advice from MHRA. Accessed on December 2, 2019 at https://www.gov.uk/guidance/medicines-get-scientific-advice-from-mhra.

NB-MED (2000) *Chapter 2.5.1. Conformity Assessment Procedures; General Rules. Recommendation: Technical Documentation.* NB-MED/2.5.1/Rec5.

Nguyen, T. and Sherman, R. (2016) Making Continuous Improvements in the Combination Products Program: The Pre-RFD Process. *FDA Voice.* Accessed on August 4, 2019 at http://wayback.archive-it.org/8521/20180926000006/https://blogs.fda.gov/fdavoice/index.php/2016/08/making-continuous-improvements-in-the-combination-products-program-the-pre-rfd-process/.

Nguyen, T. and Weiner, J. (2019) *Office of Combination Products Update*, presented at the Xavier Health Combination Products Summit 2019, Cincinnati, OH.

NPRA (2017) Product Classification Application NPRA 300.1. Accessed on August 7, 2019 at https://npra.gov.my/images/Application_Form/General_Forms/2017/BORANG_NPRA_3001_BORANG_PENGKELASAN_PRODUK.pdf.

NPRA (2019) Ministry of Health Malaysia. *Drug Registration Guidance Document.* Second Edition, September 2016, Revised January 2019.

PDA (2015) *Technical Report No. 73: Prefilled Syringe User Requirements for Biotechnology Applications.*

PDA (2019) *Technical Reports Portal.* Accessed on December 6, 2019 at https://www.pda.org/publications/pda-publications/pda-technical-reports.

Ph. Eur. (2019a) *1. General Notices.* Ph. Eur. 9.2 page 4275

Ph. Eur. (2019b) *Monograph on Parenteral Preparations.* Ph. Eur. 9.0 page 871.

Ph. Eur. (2019c) *3.1. Materials Used for the Manufacture of Containers.* Ph. Eur. 9.0 page 391.

Ph. Eur. (2019d) *Substances for Pharmaceutical Use.* Ph. Eur. 9.3 page 4777.

Ph. Eur. (2019e) *Monoclonal Antibodies for Human Use.* Ph. Eur. 9.0 page 826.

Ph. Eur. (2019f) *3.2. Containers.* Ph. Eur. 9.0 page 423.

Ph. Eur. (2019g) *5.2.8. Minimising the Risk of Transmitting Animal Spongiform Encephalopathy via Human and Veterinary Medicinal Products.* Ph. Eur.

Regulation of Combination Products (2016) 21 U.S.C. 353(g).

Roan, S (2009) *Use of Type III Drug Master Files in Product Registrations.* Regulatory Focus, 2009, 14(12), p. 40.

SFDA (2009) *Circular on Issues Relevant to Registration of Drug and Medical Device Combination Products* No. 16.

USP (2019a) *Chapter Charts.* USP42-NF37 2S page 9131.

USP (2019b) *General Notices and Requirements applying to Standards, Tests, Assays and Other Specifications of the United States Pharmacopeia.* USP42-NF37 2S page 9081.

USP (2019c) *<1151> Pharmaceutical Dosage Forms.* USP42-NF37 1S page 9037.

USP (2019d) *Injections and Implanted Drug Products (Parenterals) – Product Quality Tests.* USP42-NF37 page 6339.

USP (2019e) *Oral Drug Products – Product Quality Tests.* USP42-NF37 1S page 9007.

USP (2019f) *Topical and Transdermal Drug Products – Product Quality Tests.* USP42-NF37 2S page 9463.

USP (2019g) *Mucosal Drug Products – Product Quality Tests.* USP42-NF37 page 6356.

USP (2019h) *<5> Inhalation and Nasal Drug Products – Product Quality Tests.* USP42-NF37 page 6360.

USP (2019i) *<771> Ophthalmic Drug Products – Quality Tests.* USP42-NF37 page 6918.

4 Combination Products Current Good Manufacturing Practices (cGMPs)

Susan W. B. Neadle and Mike Wallenstein

CONTENTS

DOI: 10.1201/9781003300298-5

INTRODUCTION

Combination products and other combined use systems are comprised of two or more differently regulated medical products[1] (e.g., drug/device; biological product/device, drug/biological product, or drug/device/biological product). Their safety, efficacy, and usability inherently depend on considerations related to the design, monitoring, and control of each constituent part and their combined use. Increasingly health authorities, recognizing this, are imposing regulatory constructs and requirements for product development, risk management, and control strategies for the respective constituent parts, constituent part components, and the overall combined use product. The trend is that these regulatory constructs take into account the scientific and technological considerations raised by the intended combined use of medical products.

Good manufacturing practices[2] are regulations requiring manufacturers to take proactive steps to ensure that their products are consistently produced and controlled according to quality standards, with an aim to ensure product safety, efficacy, and usability. cGMPs, or Current Good Manufacturing Practices, are a framework for the *minimum* requirements to assure product quality. cGMPs have been established for medical products that also serve as constituent parts of combination products. These include drug cGMPs, biological product cGMPs, and medical device cGMPs. In many jurisdictions around the globe, drug and biological product cGMPs are the same, treated together as medicinal product cGMPs. Where biological product cGMPs are called out separately, they generally still apply the drug cGMPs, and supplement them with specific requirements distinctly applicable to the biological product.

With regard to drugs and biological products, adherence to cGMPs assures the identity, strength, quality and purity of drug and biological products by requiring that manufacturers of medications adequately control manufacturing operations such that the product meets quality standards. The cGMPs emphasize a strong quality management system, appropriate raw materials, robust operating procedures, detecting and investigating product quality deviations, and maintaining reliable testing laboratories. Adherence to the cGMPs should prevent instances of contamination, mix-ups, deviations, failures, and errors. For the United States, drug cGMPs are codified at 21 CFR 210/211 and additional requirements for biological products are codified at 21 CFR Parts 600-680.

The cGMPs for medical devices in the United States, under 21 CFR 820, have been referred to as the Quality System Regulations (QS Regs). ISO 13485:2016 serves as a similar framework of quality system expectations for medical devices (under the 2/22/2022 FDA Proposed Rule, the QS Regs are being harmonized to the extent possible with ISO 13485:2016, by incorporating ISO 13485:2016 clauses by reference in the US regulation (discussion later within this chapter)). There are a wide range of device types that fall along a continuum of risk – from low-risk tongue depressors to high-risk implantable medical devices, like pacemakers and artificial hearts. Given the variety of device types, the regulations for medical devices are not prescriptive on how a manufacturer must produce a device, but rather, the regulations provide a framework that all manufacturers must follow – by requiring that manufacturers establish and follow procedures, and fill in the details appropriate to a given device according to the current state-of-the-art manufacturing for that specific device. Adherence to the Quality System Regulations is intended to assure that any device – including an accessory to any device – is safe and effective for its intended use, by requiring that manufacturers of medical devices adequately control manufacturing operations such that the product meets quality standards.

Across the gamut of medical products, at the core, cGMPs are quality system requirements to assure proper design, monitoring and control of products, manufacturing processes, and facilities. When it comes to combination products, the cGMPs need to address the risk-based considerations of each constituent part, as well as those raised by their being combined.

This chapter delves into cGMP expectations for combination products, and more specifically, reviews the US FDA cGMP expectations for Combination Products under 21 CFR Part 4.[3,4,5] While this regulation is US-based, it is a risk-based approach that may be helpful conceptually to ensure sound manufacturing controls that can help inform cGMP best practices outside the United States, too. The analysis and approach would be essentially the same under ISO standards and International Conference on Harmonization (ICH) guidance. That said, while there are differing expectations among markets (see Chapters 2 and 14 in this book), the practices are robust, and it makes sense to apply an approach consistent with the best practices being presented in this chapter even if you aren't intending to enter the US market, or if you are planning to market in the United States as well as abroad.

The chapter also reviews the impactful expectations (and still evolving understanding) of combination products under EU MDR and touches on UK post-Brexit.

In summary, this chapter includes:

- Responsibilities for Combination Product cGMPs;
- The Streamlined Approach to demonstrating compliance, with case studies; [6]
- cGMPs for "combination products" under EU MDR (2017/745);[7] and
- cGMPs for "combination products" in the United Kingdom post-Brexit.

21 CFR PART 4

Who Is Responsible for Combination Product cGMPs?: The Roles of Sponsors, Applicants, Component Suppliers, and Contract Manufacturers

Chapter 2 of this book, "What Is a Combination Product?" generally describes a combination product as a product composed of two or more *differently* regulated types of medical products, e.g., a combination of a medicinal product and a medical device.[8] The drug, device (including Software as a Medical Device), or biological product (including HCT/Ps[9]) that is part of the combination product is referred to as a constituent part. From a scientific and technological perspective, it is logical that if one is combining two or more constituent parts for use together, risk-based controls required to ensure the safety and efficacy of each constituent part on its own would still be relevant if these constituent parts are intended for combined use. It also makes scientific and technological sense that, in addition to the risk-based considerations of each standalone constituent part, one would need to consider if additional controls might be needed to address any potential interactions of those constituent parts when used together, to ensure the safety and efficacy for the combined use application is not compromised. Fundamentally, this is the foundation for combination product cGMPs. The cGMPs that apply to each constituent part of a combination product also apply to the combination product, and necessitate an assessment of, and controls for, any potential interactions of the constituent parts (*see also Chapters 5 and 6, Integrated Product Development and Combination Product Risk Management*).

So, we've established that cGMPs for each constituent part apply to the combination product. Who is responsible for adhering to cGMPs for combination products, given they are comprised of more than one constituent part?

Importantly, the user and/or patient see the entire combination product system, not just the device or the drug or the biological product. So, the Sponsor/Applicant needs to take a system view of the combination product they are putting on the market. Ultimately, the Sponsor/Applicant is responsible for complying to all the applicable cGMPs for their specific combination product.

We need to unpack this a bit further, though: What is "Combination Product Manufacturing" and who is considered a "Combination Product Manufacturer"?

Combination Product Manufacturing and Combination Product Manufacturer

The US FDA interpretation of the term "Manufacturer" differs for medical devices, drugs, biological products, and combination products. For combination products, a combination product manufacturer is an entity that engages in any of the activities for a combination product within the scope of manufacture of any of the constituent parts of the combination product.

> **Medical Device Manufacturer:**
> Any person who **designs,** manufactures, fabricates, assembles, or processes a finished device. Manufacturer includes but is not limited to those who perform the functions of contract sterilization, installation, relabeling, remanufacturing, repacking, or **specification development,** and initial distributors of foreign entities performing these functions. (21 CFR 820.3(o))
>
> **Drug Manufacturer:**
> Manufacture, processing, packing, or holding of a drug product includes packaging and labeling operations, testing, and quality control of drug products. (21 CFR 210.3(12))
>
> **Biological Product Manufacturer:**
> Any legal person or entity engaged in the manufacture of a product subject to license under the act; "Manufacturer" also includes any legal person or entity who is an applicant for a license where the applicant assumes responsibility for compliance with the applicable product and establishment standards. (21 CFR 600.3(t))
>
> **Combination Product Manufacturer:**
> An entity (facility) engaged in activities for a combination product that are considered within the scope of manufacturing for drugs, devices, biological products, and HCT/Ps. Such manufacturing activities include, but are not limited to, specification development/designing, fabricating, assembling, filling, processing, sterilizing, testing, labeling, packaging, repackaging, holding, and storage*, including a contract manufacturing facility.* (21 CFR §4.2 and June 2020 FDA Combination Products Compliance Program)

A unique aspect of many combination products is that Sponsors often purchase the constituent part that is secondary to their organizations' core product type. For example, pharmaceutical and biotech companies often source the device constituent part used for delivery of the drug or biological product from a third party. These third parties have historically considered themselves to be mere "packaging component" or "container-closure" suppliers. Container-closures that are used to deliver a metered dose of a medicinal product, or that serve a transfer function for a medicinal product, may now ***also*** be considered ***medical devices***! This means that companies that previously considered themselves as solely component suppliers may now actually be device constituent part manufacturers. The distinction in terminology is important. The cGMP regulations do not apply to manufacturers of components or parts of finished devices,[10] but the cGMPs *do* apply for constituent parts of a combination product. Even if they are "only" supporting design or specification development for the device constituent intended for use in the combination product, the cGMPs for that device development activity (Design Controls) now apply. If their scope of activities includes assembly, filling, or processing the constituent parts together at one facility, they are **combination product manufacturers**, *even though they are not the sponsors for the combination product.*

This interpretation issue is a bit of a blind spot for many, and has confounded many in the pharmaceutical and biotech industries as they work to comply with 21 CFR Part 4 and are dependent upon container-closure/device constituent part suppliers for drug-agnostic Design Control and risk management supporting information.

Sponsors frequently employ Contract Manufacturing Organizations (CMOs) to support activities ranging from constituent part design through to manufacturing processes like filling, assembly, packaging, testing, labeling, and sterilization. ***These CMOs are considered combination product manufacturers. They are obligated to comply with combination product cGMPs aligned to the activities they perform at their facilities.***

In Preamble Comment 15 (78 FR 4313-4314) to 21 CFR Part 4, US FDA states:

> ***We reject the proposal that the CGMP requirements applicable to a constituent part come into effect only after that constituent part has been formed.*** Such an approach would be inconsistent with the application of the underlying CGMP regulations listed in § 4.3. The trigger is whether the facility is conducting manufacturing operations that would be subject to the underlying CGMP requirements.
>
> For example, if a facility is manufacturing only device components, it might not be subject to CGMP requirements under the QS regulation. However, ***a facility that is manufacturing a finished device from such components is subject to the QS regulation. Therefore, for example, if a facility is manufacturing a finished combination product, a prefilled syringe for instance, from device components and drug components, that facility is subject to both the QS regulation and drug CGMPs.***

FDA is clearly indicating that component suppliers that also serve as contract manufacturing facilities have an obligation to augment their own quality management systems as combination product manufacturers under 21 CFR Part 4.

Combination Product cGMPs: What Are They, and How to Demonstrate Compliance

Table 4.1 summarizes applicable cGMP regulations for products marketed in the United States, including:

- 21 CFR 210/211 defines the cGMP expectations for drugs. International Conference on Harmonization (ICH) has similarly established regulatory tools to guide pharmaceutical companies in development. These guidance documents include ICH Q10 (Pharmaceutical Quality System), which aligns well with 21 CFR 210/211.[11]
- Biological products and HCT/Ps are also required to comply with the cGMPs for drugs. Additionally, there are unique cGMP expectations applicable for biological products under 21 CFR 600-680.
- HCT/Ps are subject to 21 CFR 1271 and may also be subject to other cGMPs.
- For medical devices, 21 CFR 820 applies.

21 CFR Part 4 introduces the phrase "cGMP operating system" to mean the quality system established within an organization that is designed and implemented in order to address and to meet the cGMP requirements applicable to the manufacture of a combination product. In the United States, combination product cGMP requirements apply to all combination products.

- ***The constituent parts of combination products retain their regulatory status, even after they are combined.*** For combination products, as summarized in Table 4.1, the cGMPs applicable for each of the given combination product's constituent parts apply to the combination product.
- For cross-labeled[12] combination product constituent parts, if the constituent parts are entirely manufactured at separate facilities, the cGMPs that will apply are the same as those that would apply if the given constituent part was not part of a combination product.

TABLE 4.1
Applicable US cGMP Regulations for Drugs, Biological Products, Medical Devices, and Combination Products

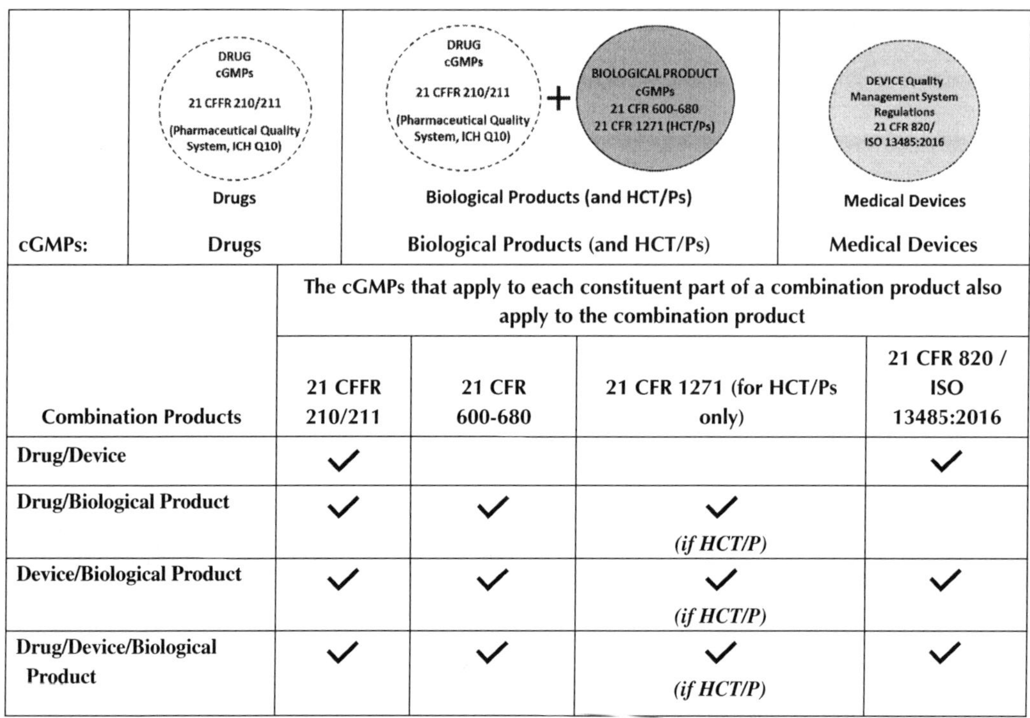

	DRUG cGMPs 21 CFFR 210/211 (Pharmaceutical Quality System, ICH Q10) Drugs	DRUG cGMPs 21 CFFR 210/211 (Pharmaceutical Quality System, ICH Q10) + BIOLOGICAL PRODUCT cGMPs 21 CFR 600-680 21 CFR 1271 (HCT/Ps) Biological Products (and HCT/Ps)	DEVICE Quality Management System Regulations 21 CFR 820/ ISO 13485:2016 Medical Devices
cGMPs:	Drugs	Biological Products (and HCT/Ps)	Medical Devices

	The cGMPs that apply to each constituent part of a combination product also apply to the combination product			
Combination Products	**21 CFFR 210/211**	**21 CFR 600-680**	**21 CFR 1271 (for HCT/Ps only)**	**21 CFR 820 / ISO 13485:2016**
Drug/Device	✓			✓
Drug/Biological Product	✓	✓	✓ *(if HCT/P)*	
Device/Biological Product	✓	✓	✓ *(if HCT/P)*	✓
Drug/Device/Biological Product	✓	✓	✓ *(if HCT/P)*	✓

Note: ©2022 Combination Products Consulting Services LLC.

Table 4.2 compares key elements of a pharmaceutical quality system cGMP requirements vs that of a medical device quality system.

Table 4.2 reveals several common elements between pharmaceutical and medical device cGMP quality system expectations. These include risk-based requirements for:

- Management responsibilities, including organization and personnel;
- Document Controls;
- Robust Facilities and Equipment Controls;
- Materials Controls;
- Production and Process Controls;
- Change Control;
- Ongoing Assessment of Systems; and
- Corrective and Preventive Action.

While there are many common elements, there are also some key differences that readily stand out between the quality system expectations (see items in ***bold italics*** in Table 4.2). One clear difference is that, in the United States, cGMP requirements for biological products are distinct compared to drug and device requirements.[13,14] Another key difference is that for medical devices, there are mandatory expectations to apply Design Controls (which also includes risk management under ISO 14971:2019 and Human Factors), whereas for drugs and biological products, Quality by Design (ICH Q8(R2)) and Quality Risk Management (ICH Q9) are considered best practices, not

TABLE 4.2
Comparison of Key Elements of Pharmaceutical vs. Medical Device Quality Systems

Quality System Overview	
Pharmaceutical Quality System (21 CFR 211/ICH Q10)	**Medical Device Quality System 21 CFR 820/ISO 13485:2016**
Assures identity, strength, quality, and purity of drug and biological products by requiring that manufacturers of medications adequately control manufacturing operations such that product meets quality standards. HCT/Ps are subject to 21 CFR 1271 and may also be subject to other CGMPs. Emphasizes a strong quality management system including:	Provides a framework of controls commensurate with risk according to the current state-of-the-art manufacturing for each specific device type. Adherence assures any device or accessory to any device is safe and effective for its intended users, intended use and intended use environment. Emphasizes the following key elements:
 	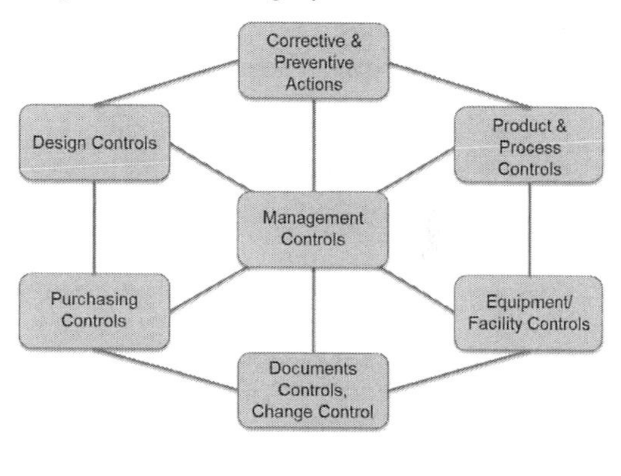
• Management responsibilities, including organization, personnel • Records and reports • Appropriate raw materials; appropriate labeling and packaging • Robust facilities, equipment, and operating procedures • Robust product and process monitoring and control • Ongoing assessment of systems • Detecting and investigating product quality deviations; corrective and preventive actions • Change management • Maintaining reliable testing laboratories • ***For biological products and HCT/Ps: Assures controls for the unique risk-based considerations for these product types***	• Management controls, including organization and personnel • Document controls • Purchasing Controls and appropriate raw materials • Robust facilities, equipment, and operating procedures • Robust product and process monitoring and controls • Ongoing assessment of systems • Detecting and investigating product quality deviations; corrective and preventive actions • Change management • ***Design controls***

cGMP requirements (we'll discuss this further later in the chapter). Importantly, there is a frequent mishap that presents itself here. Many pharma companies mistakenly assume that Design Controls are the only thing they need to pay attention to because the terminology applied between drug cGMPs and device cGMPs is so similar. This is a critical mis-step. As we will discuss in the coming sections, while the terminology applied is similar, some of the interpretations under device quality system expectations and drug cGMPs are distinct! We will address this further as we progress in the chapter.

Notably, US FDA has historically exempted some low-risk devices from all, or certain provisions of, the device cGMPs, including Design Controls. Exemption is not a consideration relevant for most types of devices that may be used as constituent parts of combination products. If, however, such an exemption *does* apply to a device type, it is also considered to apply to the device constituent part and the combination product of which they are a part, if the device constituent part *does not have a new intended use and does not otherwise raise different device performance-related safety and/or effectiveness questions.* The combination product manufacturer must have documentation to support the exempt status for the constituent part *as used in the combination product*. *If exemptions for a device constituent part of a drug-device combination product cover the device-called-out provisions (see section in this chapter on the "Streamlined Approach"), then FDA will consider the combination product manufacturer cGMP – compliant so long as the cGMP operating system is compliant with 21 CFR Part 211.*[15]

Under ISO 13485:2016, there are no such exemptions. Application of Design Controls and Device cGMPs is commensurate with the complexity and risk of each product. With FDA's Quality Management System Regulation proposed rule (2/2022) to include ISO 13485:2016 by reference, it is yet to be determined if any exemptions will still apply. With respect to combination products, FDA states that the proposed rule is not intended to change any of the cGMP expectations for combination products under 21 CFR Part 4.

Combination Product cGMPs under 21 CFR Part 4A: Streamlined Approach

Under 21 CFR Part 4, US FDA gives options on approaches to demonstrate combination products cGMP compliance for:

- Single-entity combination products;
- Co-packaged combination products; or
- Cross-labeled combination products whose constituent parts are manufactured at the same facility.

One option to demonstrate compliance with the cGMPs for a combination product is to demonstrate full compliance with each of the quality systems that apply to the constituent parts of the combination product. Given the breadth of common elements between the device and drug quality system cGMPs, the US FDA recognized an opportunity – and gives industry that opportunity – to apply a streamlined and risk-based approach to demonstrating cGMP compliance for combination products. The FDA vetted through the pharmaceutical, biological product, and medical device cGMP requirements to identify which provisions under each of these quality systems are uniquely interpreted versus which aspects are commonly interpreted. As mentioned earlier, the "devil is in the details!" Table 4.3 summarizes the distinctly interpreted provisions identified by the US FDA based on their assessment and comparison of medical device, pharmaceutical, and biological product cGMP operating systems, as well as the FDA streamlined approach for demonstrating compliance under 21 CFR Part 4.

On February 22, 2022, US FDA proposed a long-awaited rule to amend its Quality System Regulation (QSR) 21 CFR 820 for the cGMP standards for medical devices and Part 4 to align more closely with the international consensus standard for device quality management systems set by the International Organization for Standardization (ISO), ISO 13485:2016. Through this proposed rule, FDA also proposes conforming edits to 21 CFR Part 4 to clarify the device cGMP requirements for combination products. The proposed rule explicitly states that these edits do not impact the cGMP requirements for combination products. Figure 4.1 highlights the called-out provisions under the proposed rule, reflecting the applicable referenced clauses in ISO 13485:2016.

TABLE 4.3

Called-Out Provisions: cGMP Regulations that Are Distinctly Interpreted under Each of the cGMP Operating Systems and US FDA Streamlined Approach for Demonstrating Compliance to Combination Products cGMPs

Foundation for Streamlined Approach under 21 CFR Part 4: Many common cGMP elements; some elements are distinctly interpreted	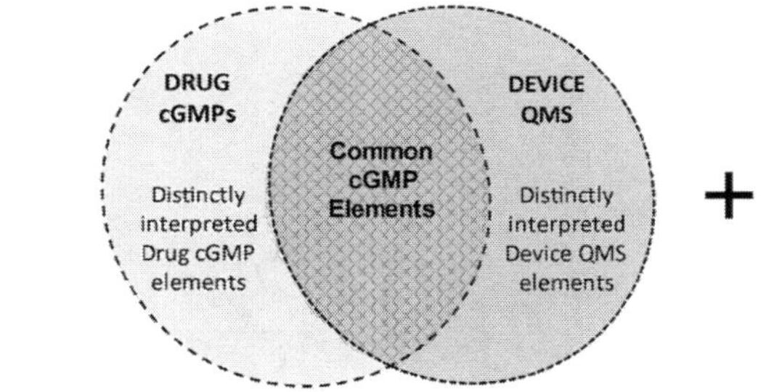 	+	BIOLOGICAL PRODUCT cGMPs 21 CFR 600-680 21 CFR 1271 (HCT/Ps)
Distinctly interpreted "Called-Out Provisions"	**Medical Devices (21 CFR 820)** 820.20 Management Controls 820.30 Design Controls 820.50 Purchasing Controls 820.100 CAPA 820.170 Installation 820.200 Servicing	**Drugs (21 CFR 210/211)** 211.84 Testing/approval/ rejection of components 211.103 Calculation of Yield 211.132 Tamper-Evident Packaging 211.137 Expiration Dating 211.165 Testing & Release for Distribution 211.166 Stability Testing 211.167 Special Testing Requirements 211.170 Reserve Samples	**Biological Products (21 CFR 600-680) (21 CFR 1271 for HCT/Ps)** All biological product cGMP requirements 21 CFR 600-680 + 21 CFR 1271 for HCT/Ps
Drug-Device Combination Product DRUG DEVICE	**<u>Drug cGMP-based Operating System:</u>** Demonstrate compliance to all Drug cGMPs + Distinctly interpreted Medical Device QMS Called-out Provisions		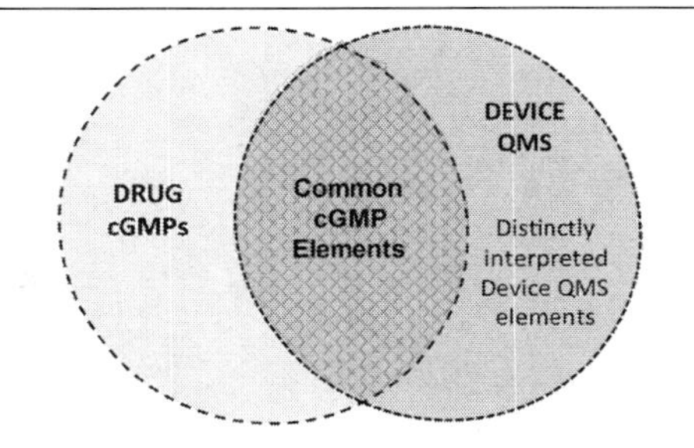

(Continued)

TABLE 4.3 (CONTINUED)

Called-Out Provisions: cGMP Regulations that Are Distinctly Interpreted under Each of the cGMP Operating Systems and US FDA Streamlined Approach for Demonstrating Compliance to Combination Products cGMPs

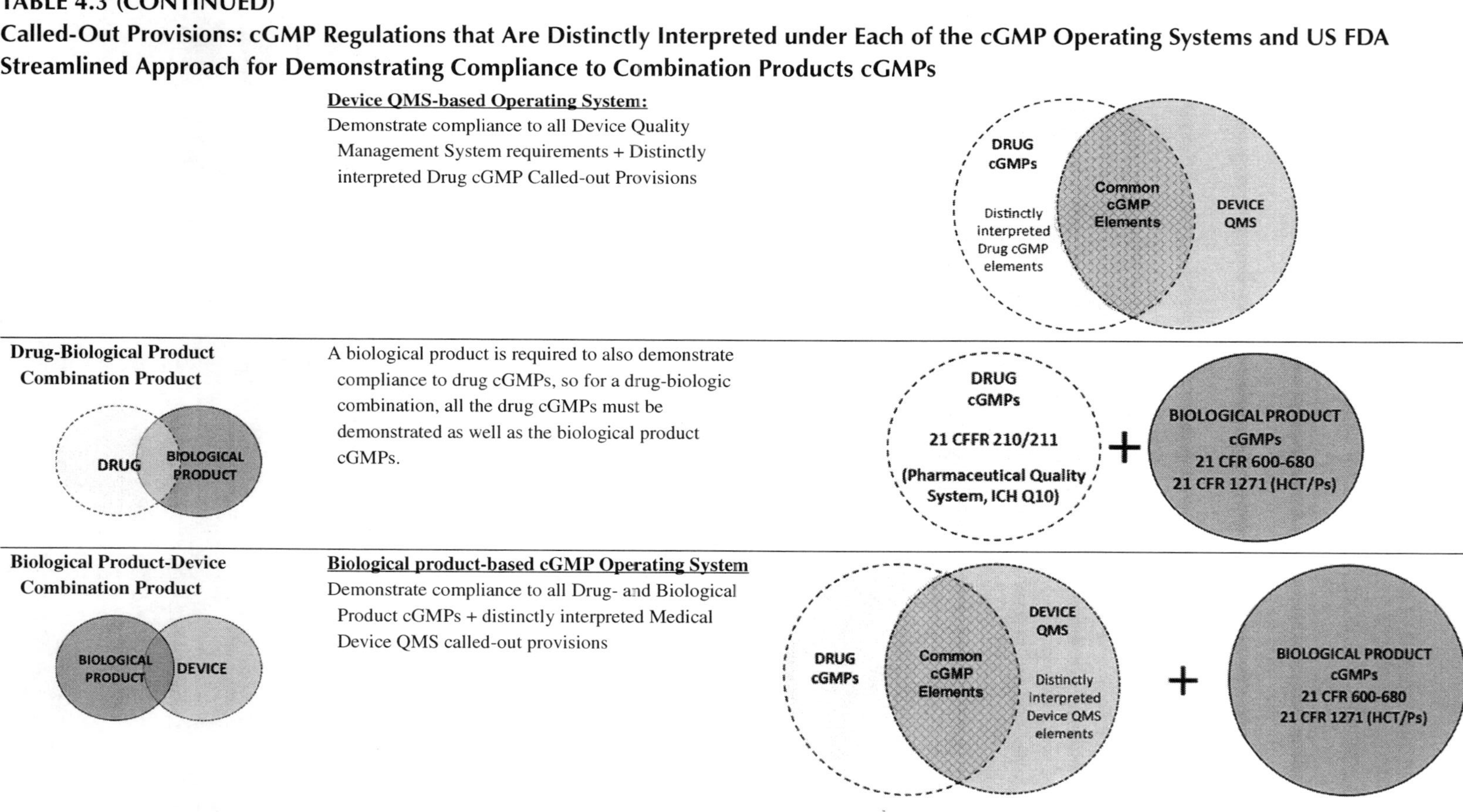

	Device QMS-based Operating System: Demonstrate compliance to all Device Quality Management System requirements + Distinctly interpreted Drug cGMP Called-out Provisions	
Drug-Biological Product Combination Product	A biological product is required to also demonstrate compliance to drug cGMPs, so for a drug-biologic combination, all the drug cGMPs must be demonstrated as well as the biological product cGMPs.	
Biological Product-Device Combination Product	**Biological product-based cGMP Operating System** Demonstrate compliance to all Drug- and Biological Product cGMPs + distinctly interpreted Medical Device QMS called-out provisions	

(Continued)

TABLE 4.3 (CONTINUED)

Called-Out Provisions: cGMP Regulations that Are Distinctly Interpreted under Each of the cGMP Operating Systems and US FDA Streamlined Approach for Demonstrating Compliance to Combination Products cGMPs

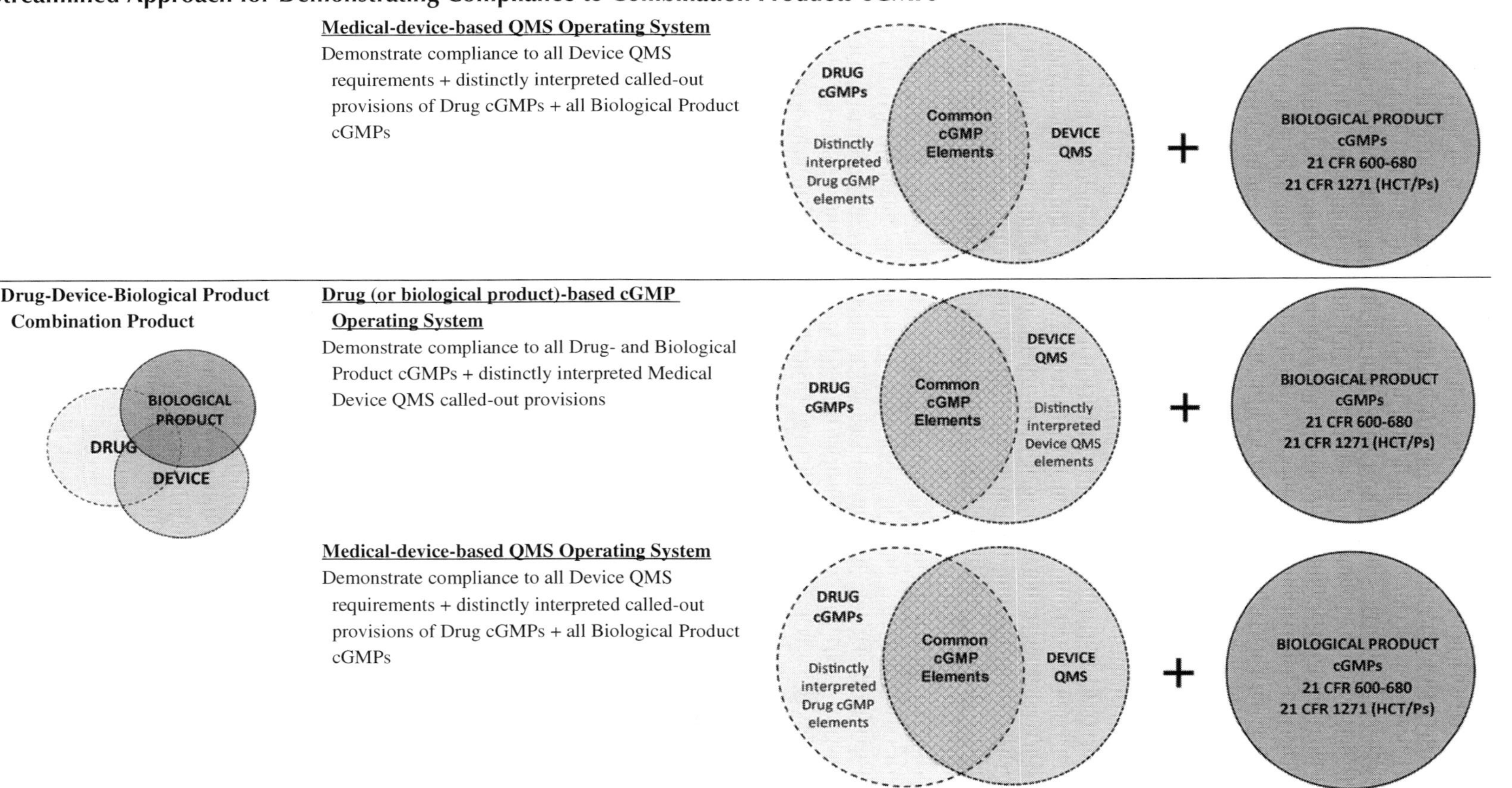

	Medical-device-based QMS Operating System Demonstrate compliance to all Device QMS requirements + distinctly interpreted called-out provisions of Drug cGMPs + all Biological Product cGMPs	
Drug-Device-Biological Product Combination Product	**Drug (or biological product)-based cGMP Operating System** Demonstrate compliance to all Drug- and Biological Product cGMPs + distinctly interpreted Medical Device QMS called-out provisions	
	Medical-device-based QMS Operating System Demonstrate compliance to all Device QMS requirements + distinctly interpreted called-out provisions of Drug cGMPs + all Biological Product cGMPs	

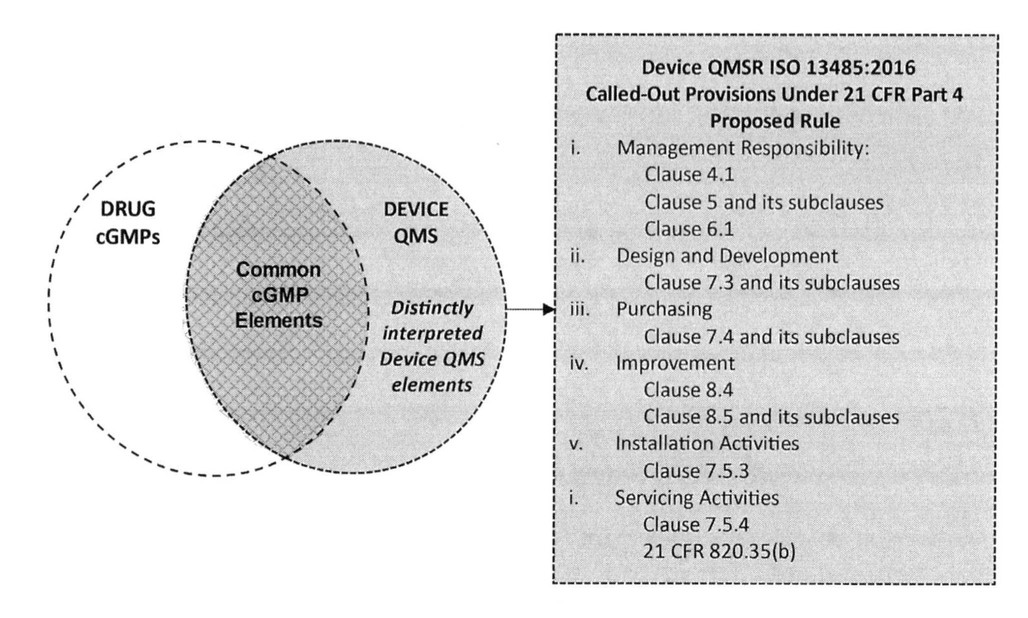

FIGURE 4.1 Called-out provisions under 2/22/22 FDA Quality System Regulation Proposed Amendments (ISO 13485: 2016 standard has been made available in the ANSI Incorporated by Reference (IBR) Portal at https://ibr.ansi.org/). ©2022 Combination Products Consulting Services, LLC. All Rights Reserved.

Under 21 CFR Part 4.4(c) and (d):

> During any period in which the manufacture of a constituent part to be included in a co-packaged or single-entity combination product occurs at a separate facility from the other constituent part(s) to be included in that single-entity or co-packaged combination product, the current good manufacturing practice operating system for that constituent part at that facility must be demonstrated to comply with all current good manufacturing practice requirements applicable to that type of constituent part.

When two or more types of constituent parts to be included in a single-entity or co-packaged combination products have arrived at the same facility, or the manufacture of these constituent parts is proceeding at the same facility, the combination product manufacturer needs to demonstrate compliance with the cGMPs of their base cGMP operating system plus the called-out provisions for the other constituent part. A combination product sponsor needs to demonstrate cGMP compliance for the combination product, as does a CMO, based on the activities the CMO performs.

Now let's take a closer look at the called-out provisions for device quality management system expectations and drug cGMPs. In the coming section, particularly for the device called-out provisions, we will include "what is different between the device and drug cGMP interpretation" given that often the same words are used, but the interpretations and expectations are distinct.

Device Called-Out Provisions

Let's start with the device called-out provisions. Under 21 CFR §4.4(b)(1), if the current good manufacturing practice operating system has been shown to comply with the drug cGMPs, and the combination product includes a device constituent part, in addition to the drug cGMPs, a combination product manufacturer needs to demonstrate compliance to the device called-out provisions summarized in Table 4.4.

Management Controls

Companies with a drug cGMP-based quality management system are subject to cGMP provisions related to Management Responsibility and quality management systems (Section 501(a)(2)(B) of the

TABLE 4.4
Device cGMP Called-Out Provisions

21 CFR Citation	Description	Comments
820.20	Management Controls	• Top-level corporate commitment to product quality • Documentation includes quality policy, management review procedures, quality plan • Management representative appointed
820.30	Design Controls	• Proactive consideration of product intended users, intended uses, and associated requirements: Design in Quality! • Documentation includes design control procedures, including risk analysis, and product design information, including Design History File (DHF)
820.50	Purchasing Controls	• Manage suppliers of products, components, and services • Define type and extent of control over suppliers (e.g., supplier quality agreements, notification of changes
820.100	Corrective & Preventive Action (CAPA)	• Evaluate sources of quality data and respond when needed to correct existing and/or prevent potential problems • Ensure corrective actions taken are effective and do not create a new problem
820.170 820.200	Installation Servicing	• Requirements that apply if the combination product requires installation and/or servicing (not applicable to most drug-led products) • *(Installation does NOT refer to implantation into a patient!)*

FD&C Act [21 USC 351(a)(2)(B)]), but there are distinct requirements under 21 CFR 820.20 that are not explicitly required under drug cGMPs. Combination Product Manufacturers and Sponsors of a combination product with a device constituent part have to satisfy *all* the elements of 21 CFR 820.20.

Under Management Controls, the expectation is that there is demonstrated top-level (Management with Executive Responsibility[16]/Top Management) corporate commitment to product quality. This includes (1) providing adequate resources for device/drug/combination product design, manufacturing quality assurance, distribution, installation, and servicing activities; and (2) assuring that the quality system is functioning properly through ongoing monitoring of the quality system and making necessary adjustments. Elements that are used to demonstrate compliance with Management Controls expectations are summarized in Table 4.5. At the bottom half of the table, there is also a comparison with the 21 CFR 211 interpretation of Management Responsibility, to help bring further clarity to what is distinct compared to the drug cGMPs.

CASE STUDY 1:

FDA is inspecting a manufacturer of a drug-eluting stent. The FDA investigator requests the management review procedures and an agenda for the most recent management review. The investigator notices that all agenda items are related to the device constituent part. Should the investigator be concerned?

Yes, there is a cause for concern. Management reviews should cover *all* the aspects of the quality system related to the combination product, *not* just the device constituent part.

Of note, detailed management review documents are typically not made available to the investigator, but they may review CAPAs and other quality records to ensure the company is considering the drug constituent part, the device constituent part, and the combination product as a whole, when addressing product problems. The adequacy of the procedure for conducting management reviews might also be reviewed.

TABLE 4.5
Management Controls 21 CFR 820.20

<table>
<tr><th colspan="2">Key Elements of Management Controls</th></tr>
<tr><td colspan="2">Key Question: Does Management with Executive Responsibility(820.3(f)) (Top Management) ensure that an adequate and effective quality system is established and maintained?</td></tr>
<tr><td>Documentation</td><td></td></tr>
<tr><td>• Quality Policy (820.3(u))</td><td>Overall intentions and directions of an organization with respect to quality</td></tr>
<tr><td>• Management Review & Quality Audit Procedures</td><td>At defined intervals, ensure the quality policy and associated objectives are met.</td></tr>
<tr><td>• Quality Plan (820.20(d))</td><td>Quality Manual, Device Master Record(s), Production Procedures;
Defines quality practices, resources, and activities relevant to the devices/ combination products being designed and manufactured at that facility.</td></tr>
<tr><td>• High-Level Quality System Procedures</td><td></td></tr>
<tr><td>• Quality System Procedures and Instructions</td><td></td></tr>
<tr><td>Management Representative (820.20(b)(3))</td><td>An individual is appointed (e.g., through org chart and job description) as responsible for ensuring requirements of the quality management system regulation are established effectively and maintained.</td></tr>
<tr><td>Management Reviews (QS Regulation Preamble, Comment 53)</td><td>Management reviews are conducted, including a review of suitability and effectiveness of the applicable quality system elements. For combination products, the Management Review needs to address the applicable quality system elements for each constituent part and for the combination product.
Failure to conduct management reviews is a source of significant 820.20 violations.</td></tr>
<tr><td>Quality Audits</td><td>Quality audits, and re-audits of deficiencies, are conducted.
NOTE: The agency's policy relative to the review of quality audit results is stated in CPG 7151.02 (CPG Manual subchapter 130.300). This policy prohibits FDA access to a firm's audit results. Under the Quality System Regulation, this prohibition extends to reviews of supplier audit reports and management reviews. However, the procedures and documents that show conformance with 21 CFR 820.50, Purchasing Controls, and 21 CFR 820.20(3)(c), Management Reviews, and 21 CFR 920.22 Quality Audit, are subject to FDA inspection.</td></tr>
<tr><td colspan="2">How this compares to Management Responsibility under 21 CFR 211</td></tr>
<tr><td colspan="2">21 CFR 211.22 Quality Control Unit: Does not specifically require all elements of 820.20, but does include some quality oversight responsibilities.
For example, under drug cGMPs, there is no explicit requirement for conducting management reviews to assess the suitability and effectiveness of the quality system at defined intervals, but this is a requirement under 21 CFR 820.20.
Section 501(a)(2)(B) of the FD&C Act requires oversight and controls to ensure product quality, including to manage risks and to establish the safety of raw materials and in-process materials, as well as the finished product:

a drug (including a drug contained in a medicated feed) shall be deemed to be adulterated if the methods used in, or the facilities or controls used for, its manufacture, processing, packing, or holding do not conform to or are not operated or administered in conformity with current good manufacturing practice to assure that such drug meets the requirement of the act as to safety and has the identity and strength, and meets the quality and purity characteristics, which it purports or is represented to possess.

ICH Q10 has elements of Management Responsibility.
• Quality Systems Approach to Pharmaceutical CGMP Regulations, Sept 2006 http://www.fda.gov/downloads/Drugs/.../Guidances/UCM070337.pdf
• Guidance for Industry Q10 Pharmaceutical Quality System, Apr 2009 http://www.fda.gov/downloads/Drugs/Guidances/ucm073517.pdf</td></tr>
</table>

Design Controls

Design Controls (21 CFR 820.30 and ISO 13485:2016, Clause 7.3) is a mandatory methodology to control the design process, to design in quality, and to incorporate relevant expertise throughout the development process. The scope of Design Controls includes both the product design and the process for manufacturing that product. The intent is to assure that the device constituent part and the combination product meet user needs, intended uses, and specified requirements.

Combination product design controls under 21 CFR 820.30 apply to the combination product, including design considerations for *each* constituent part specifically used in the combination product. Combination product manufacturers may leverage pre-existing design and development information for any constituent part as part of the overall Design Control process for the combination product. Design Controls for a combination product composed of a device constituent part to be used with an already developed drug product, for example, could leverage the drug properties (CQAs, CMAs, CPPs[17]) as inputs for Design Control activities focused on ensuring the device constituent delivers the drug per dosing requirements, and that the drug quality is not negatively impacted by its contact with the device constituent part.

Design Controls activities could take place at separate facilities from other manufacturing activities for the combination product. Design, including specification development, is considered in scope in the definition "combination product manufacturer." Design and/or specification development facilities may maintain a Design History File for the combination product. According to FDA Compliance Program 7356.000,

> The combination product DHF may include cross-references to relevant information rather than be a direct repository for all the information it needs to include. Regardless of where DHF information is maintained, the combination product manufacturer should be able to access necessary DHF information during an inspection to demonstrate compliance with design control requirements.[18]

Notably, if a specification developer, or some other site, is responsible for Design Controls for the combination product, the FDA may decide to inspect that site. This is an important point for combination product manufacturers who may, for example, be outsourcing design or purchasing devices that have been designed by a third party. Per FDA Compliance Program 7356.000:

> In cases where the combination product manufacturer is purchasing the device constituent part (or its components to be assembled) from another entity (e.g., buying syringe components to be combined and filled with the drug product), the combination product DHF may reference design information from the supplier to support design of the device constituent part. However, the DHF should demonstrate how the constituent part's specifications are appropriate for its use in the combination product, addressing any interaction of the constituent parts when combined. Similarly, if the combination product manufacturer designed the entire product, previously developed constituent part information can also be referenced in the DHF (e.g., using development information on a previously approved drug product that is now being developed in a pre-filled delivery device configuration).

For pharmaceutical and biotech companies entering into the combination products space, Design Controls requirements may seem daunting. Quality by Design (QbD) (ICH Q8(R2))[19] is a similar methodology used by pharma companies to control the design process for pharmaceutical products, to design in quality, and to incorporate relevant expertise throughout the development process (see Chapter 5). Some of the QbD information generated may be useful in supporting the Design Controls efforts for the combination product. As discussed in Chapter 5, your ability to effectively articulate/communicate to an investigator how your design practices align with Design Controls requirements is critical.

Key elements of Design Controls, and a high-level comparison to drug cGMPs, are listed in Table 4.6 and illustrated in Figure 4.2A. (Figure 4.2B illustrates the interplay of QbD and Design Controls together; *see also Chapter 5*.)

Given that Design Controls is a cGMP element that many pharma companies are less familiar with, and its critical role in combination products development, this chapter reviews combination product Design Controls expectation in some detail. (Risk management is an underpinning of the

TABLE 4.6
Design Controls (21 CFR 820.30 (also ISO 13485:2016, Clause 7.3)

Key Elements of Design Controls
• Identifying user needs • Design and development planning • Translating the user needs into technical design requirements, referred to as "design inputs" • Risk analysis and risk management (ISO 14971:2019)[a] • Generating "design outputs," i.e., the results of the design process • Clearly identifying which design outputs are essential to the proper functioning of the device constituent part and combination product (Essential Performance Requirements) • Verifying that the design outputs meet the design input requirements • Validating that the design meets the user needs acceptance criteria (including Human Factors[b] studies, where appropriate) • Controlling design changes, including verification or validation where appropriate • Reviewing design results, including support of an independent reviewer • Generating an end-to-end trace matrix (tracing user needs to design inputs to design outputs to verification and validation activities) • Transferring the design to production, ensuring the design is translated effectively into production specifications • Compiling a Design History File (DHF) • Appropriate Use of Statistical Techniques throughout (21 CFR 820.250) It is important to note that the scope of Design Controls includes *both* product *AND* process. Some people mistakenly believe it is only about the device constituent physical design and labeling. Design Controls includes Design Transfer (820.30(h), and under 820.30(h) is reference to Process Validation *(820.75 (a)–(c))* requirements, which precede or may be done concurrently (at risk) with Design Validation. Design Validation (21 CFR 820.30(g)) establishes that the product specifications meet the user needs and intended uses. "Where the results of a process cannot be fully verified by subsequent inspection and test, the process shall be validated." "When changes or process deviations occur, the manufacturer shall review and evaluate the process and perform revalidation where appropriate." *The device constituent part/combination product design and process validations are completed prior to distribution of any combination product.* (See additional discussion on combination product process validation in Chapter 8 of this book).
How this compares to Expectations under 21 CFR 211
21 CFR 211.10 Production & Process Controls includes provisions for controlling process design and developing appropriate specifications, but there are no overall design controls obligations. ICH Q8(R2) Quality by Design[c] and ICH Q9(R1)[d] Quality Risk Management have similar foundational concepts, but are not mandated, and are not the same. Similar to Human Factors expectations for the device constituent and combination product under Design Controls, CDER has issued Guidance on Safety Considerations for Product Design to Minimize Medication Errors[e] (April 2016). Process Validation is an enforceable requirement for finished drug products: • *Drug cGMP regulations are designed to assure that quality is built into the design and the manufacturing process at every step (i.e., facilities are in good condition, equipment is properly maintained and calibrated, employees are qualified and fully trained, and processes are reliable and reproducible linking CGMPs, quality safety, and efficacy).* • 21 CFR 211.100(a): "written procedures for production and process control designed to assure that the drug products have the identity, strength, quality, and purity they purport or are represented to possess." • 21 CFR 211.110(a): "procedures shall be established to monitor the output and validate" Under drug cGMPs, effective Process validation is essential to connect clinical significance to the quality of the drug product the patient will receive once the application is approved. Connection of the filed exhibit and clinical batch data to the conditions under which the drug product is manufactured during commercial production. **(Process validation for devices/combination products is generally done at scale before marketing authorization while it may occur after for drugs; process validation is needed prior to distribution in any case**).

[a] Also reference AAMI TIR 105:2020, "Combination Products Risk Management" and Chapter 6 of this book.
[b] US FDA Draft Guidance "Human Factors Studies and Related Clinical Study Considerations in Combination Product Design & Development" (February 2016), accessed August 27, 2022 at https://www.fda.gov/regulatory-information/search-fda-guidance-documents/human-factors-studies-and-related-clinical-study-considerations-combination-product-design-and.
[c] ICH Q8(R2) Pharmaceutical Development, November 2009 accessed August 27, 2022 at http://www.fda.gov/downloads/Drugs/Guidances/ucm073507.pdf.
[d] ICH Q9(R1) Quality Risk Management, June 2022, accessed August 26, 2022 at https://www.fda.gov/regulatory-information/search-fda-guidance-documents/q9r1-quality-risk-management.
[e] Safety Considerations for Product Design to Minimize Medication Errors Guidance for Industry (DMEPA), April 2016, accessed August 26, 2022 at https://www.fda.gov/regulatory-information/search-fda-guidance-documents/safety-considerations-product-design-minimize-medication-errors-guidance-industry.

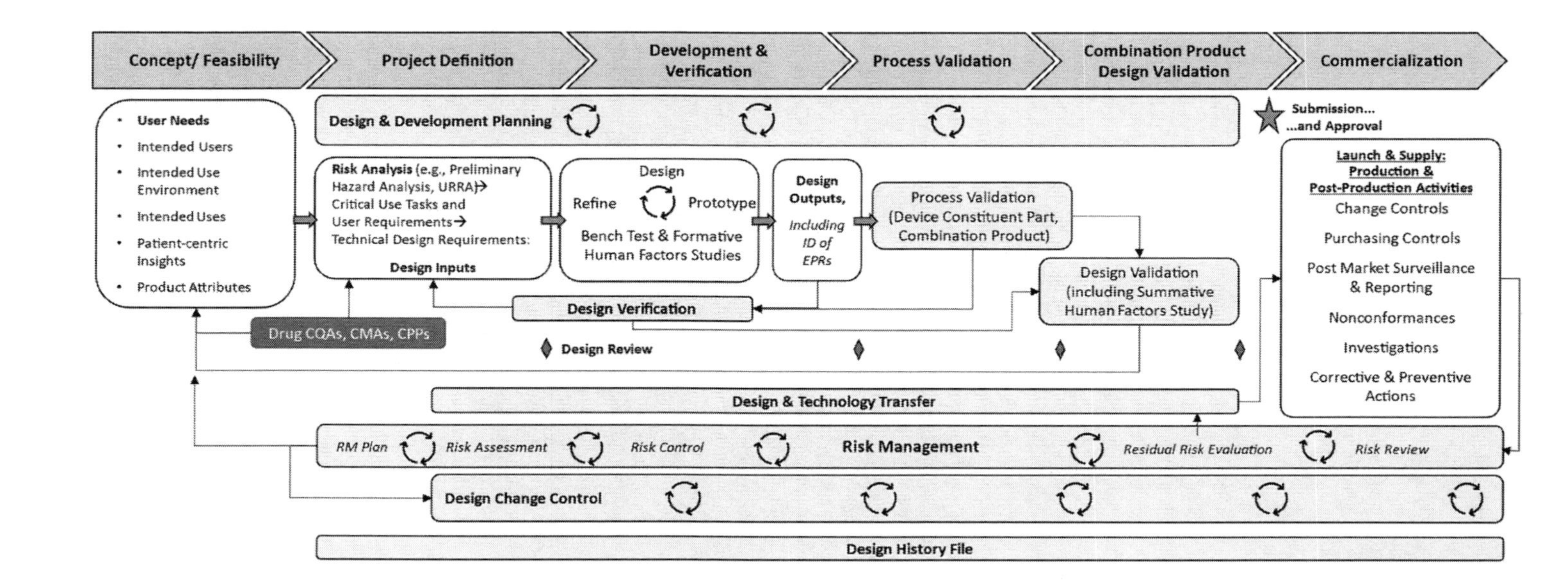

FIGURE 4.2A Design Control process.

"ID of EPRs": Identification of Essential Performance Requirements
CQA: Critical Quality Attribute
CMA: Critical Material Attribute
CPP: Critical Process Parameter
URRA: Use-Related Risk Analysis

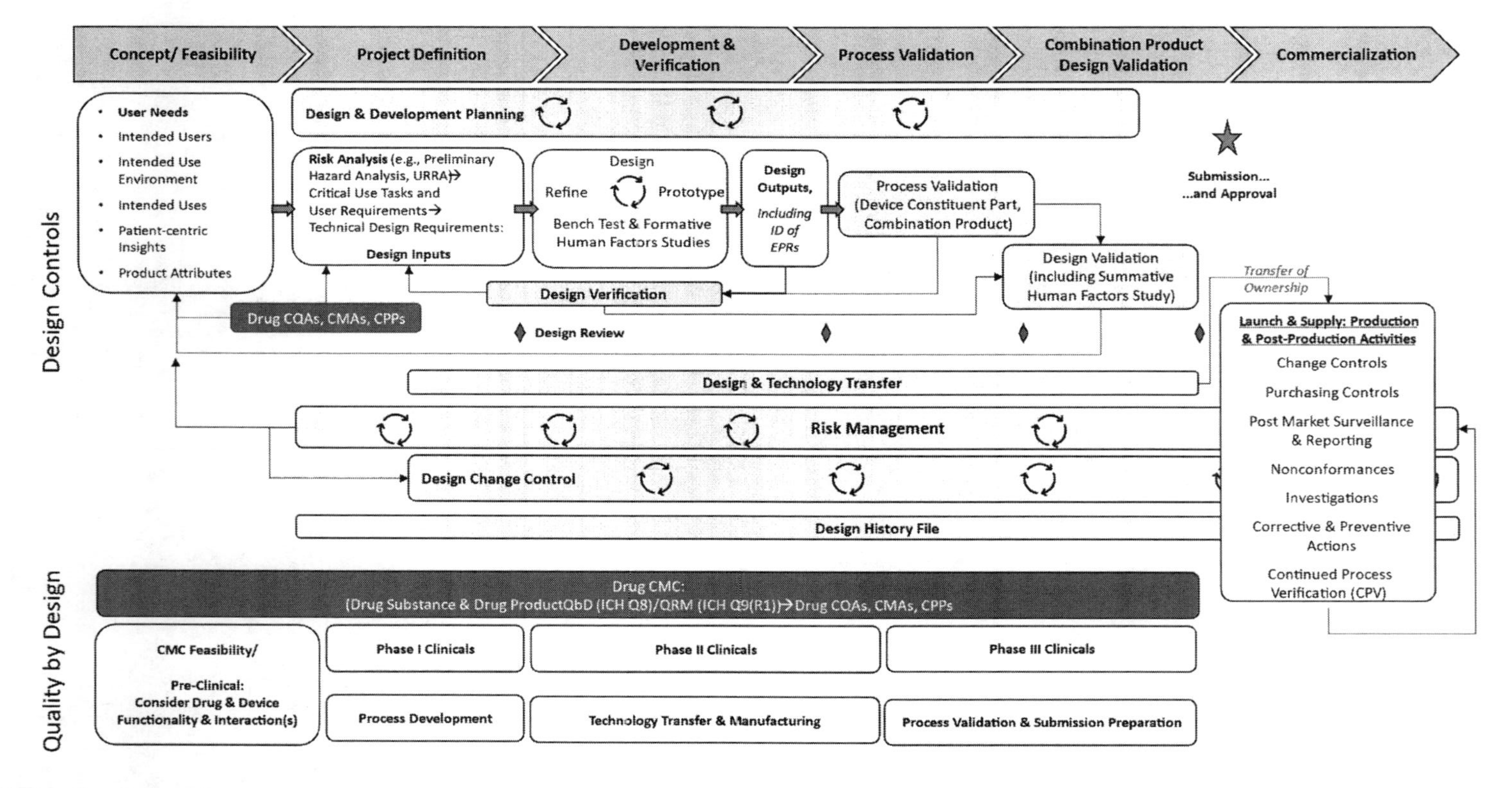

FIGURE 4.2B Design Control process with QbD. © 2022 Combination Products Consulting Services, LLC. All Rights Reserved.

"ID of EPRs": Identification of Essential Performance Requirements
CQA: Critical Quality Attribute
CMA: Critical Material Attribute
CPP: Critical Process Parameter
URRA: Use-Related Risk Analysis
CMC: Chemistry, Manufacturing & Controls

entire Design Controls process, and will also be discussed, but is more extensively reviewed in Chapter 6, "Risk Management," in this book.)

Examples of significant 820.30 violations – found in 483[20] and Warning Letter citations – include:

- Lack of Design Controls procedures;
- Inadequate design verification procedures and processes;
- Inadequate design validation, including lack of risk analysis and (where applicable) software validation; and
- Design changes implemented without validation/verification activities conducted prior to implementation.

As noted in Table 4.6, in pharmaceutical development, Quality by Design (QbD) and Quality Risk Management are applied. Quality by Design is a systematic approach to development that begins with predefined objectives and emphasizes product and process understanding and process controls, based on sound science and quality risk management (see Chapter 5 for discussion on integrated combination product development). QbD can be a parallel process to Design Controls, but the outputs of QbD and pharmaceutical Quality Risk Management (i.e., therapeutic needs, Critical Quality Attributes (CQAs), Critical Material Attributes (CMAs), and Critical Process Parameters (CPPs)) are generally inputs to user needs and Design Input requirements.

Design Controls can be thought of as project management best practices that have been mandated for assuring the development of safe, effective, and usable device constituents and combination products. Table 4.7 compares QbD for pharmaceutical development to Design Controls, aligning both to Project Management best practices. It is important to understand this alignment, particularly for companies who may have developed or purchased their device constituent part as a "container-closure," not recognizing that Design Controls obligations apply. A company with 21 CFR 211-based operating system likely has generated QbD documentation to support its development activities. Some of these documents may help support building the required medical device/combination product Design History File. The proverbial "devil is in the details," as while there is general alignment between the principles of Design Controls and Quality by Design, the specifics of the deliverables are distinctly interpreted.

Following is a closer look at each of the Design Control elements.

When Do Design Controls Apply?

Premarket, Design Controls apply after feasibility/"proof of concept" prototyping. Importantly, though, they should be applied before a human clinical investigation. A mechanism of change control should also be in place during any clinical investigation. Post-market, change control, and risk management may trigger the need to revisit the Design Controls process.

Design and Development Plan (D&DP) 21 CFR 820.30(b)

Design and Development Planning is the set of approved activities for the design and development team. The D&DP is used to determine the adequacy of the design requirements, and to ensure that the design will eventually be released to production meeting the approved requirements. The D&DP is a helpful communication tool between organizational groups/functions that are providing input into design and development, or are on the receiving end of the design outputs generated. Once it is decided that a design will be developed, a plan needs to be established (Preamble Comment #62) (e.g., applied to clinical trial activities for Phase 1 and 2 Human Trials). The D&DP describes the flow of the development process, describing which tasks should be completed first, and which tasks can be completed in parallel. Typically, the D&DP includes proposed quality practices; assessment methods; recordkeeping and documentation requirements; resources; describes or references activities; defines responsibilities for implementation; and identifies and describes interfaces with different groups or activities. The D&DP is under revision control, and gets reviewed, updated, and

TABLE 4.7
Comparison of Pharmaceutical Development under QbD to Device and Combination Product Development under Design Controls*

Inputs

Project Management	**Drug and/or Biological Product Constituent Part Quality by Design (ICH Q8(R2))/ Quality Risk Management (ICH Q9(R1))**	**Device and Combination Product Design Controls [21 CFR 820.30 (ISO 13485:2016, clause 7.3 and its subclauses] and Risk Management (ISO 14971:2019)**
Best Practices Project Management Planning; assigning roles and responsibilities	*Best Practices* Project Plan	*Mandatory* Design and Development Planning (820.30(b))
Who are the users and what are their needs?	User Needs and Intended Uses	User Needs and Intended Uses
Ensure safe and effective use; avoid predictable mis-use	Human Factors for Medication Error Reduction	Human Factors/Usability Engineering
Translate the user needs to design requirements	Target Product Profile (TPP) and QTPP	Design Inputs (820.30 (c))
Translate product requirements into specifications	Product Specifications	Design Outputs (820.30(d))
Reviews at development milestones, addressing problems	CMC Stage Gate Reviews	Design Review (820.30(e))
Ensure product meets design specifications	Product Characterization	Design Verification (820.30(f))
Ensure product meets user needs and intended uses	Clinical Studies	Design Validation (820.30(g))
Scale-up and commercialization of the product	Technical Transfer Studies	Design Transfer (820.30(h)
Manage change to maintain performance	Change Control	Design Change Control (820.30(i))
Document the development data and activities	Product Dossier and Development Reports	Design History File (820.30(j))
Recipe for repeatable and reproducible product	Master Batch Record (MBR)	Device Master Record (DMR) (820.181)
Ensure benefits of product use outweigh risks	Quality Risk Management (ICH Q9(R1))	Device & CP Risk Management (ISO 14971:2019/A11:2019)

*The correspondence between QbD and Design Control elements is not necessarily one-to-one, but gives a sense for how the steps of the two processes relate.

is generally approved as a project evolves by a cross-functional team during a Design Review. It also reflects who is on the Design Team and Interfaces, describing the expertise, skills, and quality system expertise needed.

Following is an example of what might be included in a typical D&DP "Table of Contents":

1. **Revision Date and Number**
2. **Purpose and Scope**
3. **Definitions/Abbreviations**
4. **Responsibilities**
5. **Team Roster**
6. **Organizational interfaces**
7. **Project Schedule/Milestones**
8. **Deliverables**
9. **Design Reviews:** *Development lifecycle – The D&DP specifies the minimum number of design reviews that will be held and who will participate*
10. **Manufacturing Strategy and Design Transfer Plan**: *Design transfer approach*
11. **Design Verification Plan**
12. **Design Validation Plan**
13. **Software Validation Plan**: *Software validation approach*
14. **Human Factor/Usability Engineering Plan**
15. **Clinical Strategy**
16. **Regulatory Strategy: Target markets and requirements for each market**
17. **Stability Plan**
18. **Risk Management Strategy:** *Risk Management approach and responsibilities*
19. **Relevant Standards and Guidance:** *Consensus standards, guidance, special controls, voluntary, or mandatory standards*
20. **Trainer Device (if applicable)**
21. **Other**
22. **References**
23. **Attachments**

User Needs and Intended Uses and Human Factors → Design Inputs

The combination products user interface is a critical driver for the safe and effective use by intended users for intended uses in intended use environments. Clearly defining the target patient population, intended users, uses, and use environments is essential for proactive risk management and effective combination product design. Human Factors[21] is a key component of proactive risk management (ISO 14971:2019, Annex C), and foundational to addressing potential use-related hazards for the combination product (*see Chapters 5–7 on Combination Products Integrated Development, Risk Management, and Human Factors*). Proactively defined user needs, intended uses, and human factors use-related risk analysis (URRA) are all sources of Design Inputs requirements. Table 4.8 is an example template for recording User Requirements Specifications.

Example considerations of users, user needs, and intended use environments include:

User

- Who is the user? A Health Care Provider (HCP)? A patient? A caregiver/ layperson? Their experience, training, and education can shape their ability to safely and effectively interpret and execute tasks with the user interface (e.g., instructions for use and other labeling, operation of the device constituent part, reconstitution steps).
- What is the age of the user? Is it a pediatric patient? An adolescent? A senior citizen? What unique risks arise due to the user's age? Cognitive skills, dexterity, physical abilities, ergonomic factors, sensory abilities (e.g., low vision) are all considerations.

- Mental or emotional condition can also impact safe and effective product use. For example, if someone is experiencing a life-threatening emergency, the stress of that situation may influence their actions. User interface design considerations need to help off-set new potential risks that arise from the stressful environment.

Intended Use

- Target condition/therapeutic effect: Is the condition one that requires acute care? Or is it for treatment of a chronic condition? Such considerations can impact the expectations for the reliability of functionality of the device constituent part as well as the drug efficacy.
- *How* the device constituent/Combination Product (CP) is used (per instructions for use (IFU)): The more complex the use, the more opportunities for use error. Even with few steps, if a user interface is confusing to interpret, results might not be safe and effective use.
- Predictable Mis-use: Are there uses of the device constituent part or combination product that are not aligned with the instructions for use, but that one might be able to readily predict? Design risk analysis and technical design requirements should take that into account to prevent unwanted and unsafe outcomes.

TABLE 4.8
Potential Template for Recording User Requirement Specifications

Product Name			
Review Date			
Project Team Leader		**URS version number**	
URS number	**URS**	**Source**	
U001			
U002			
U003			
U004			

Use Environment

- Self-administration/home-use: The COVID-19 pandemic shifted much of health care away from clinical settings into home-use environments. Where some therapies might have historically been offered in a clinical setting, based on emergency use and compromised immune systems, many turned to trying to receive home treatment. One needs to consider what might be unique in terms of design considerations suitable for a home-use setting based on the specific medicinal product/device combination. Self-administration raises the bar on human factors considerations given the range of users, user experience, and capabilities.
- Dose regimen/when the device/CP is used: How frequently is the patient to receive their dose? Will they remember how to do it each time? Is the user interface so complex or perhaps administration is cumbersome or painful, such that the patient won't want to comply with their dose regimen? The use-related risk assessment (see Chapter 7, Human Factors) is an important component of Design Controls.

- Where the device/CP is used – light, noise, distraction, motion/vibration, workload – these all figure into safe and effective use. A healthcare worker in a busy hospital jumping from patient to patient all day long – or treating someone in a medivac for emergency transport – or a child injured on a playground – the range of settings for care is diverse, and the combination product design needs to accommodate safe and effective use, based on intended use-settings.
- Other devices interfaced with: Is the device constituent/combination product designed to work with other devices? Will excipients of the drug formulation interact with any of the components in the fluid path?

Risk analysis, human factors, standards, regulations, and guidance – along with other quality data sources are valuable to inform establishing design inputs (see Chapters 5–7).

Design Inputs (21 CFR 820.30(c))

The user needs and intended uses are translated into technical design requirements, i.e., the physical and performance characteristics of the device constituent part and CP that are used as the basis for product design. These technical design requirements are known as Design Inputs. Design Inputs ensure the requirements are appropriate. They should be measurable (quantitative), testable (clear and complete), unambiguous, and traceable.

Design Inputs are informed by consensus standards, regulations, and guidance, and importantly, by risk management. QbD and Pharmaceutical Risk Management are used to identify drug CQAs, CMAs, and CPPs. (Additional CQAs, CMAs, and CPPs may be identified through the Clinical phases; see Chapter 5). These may inform user needs and design input requirements for a combination product. Further, User Requirements Risk Analysis (URRA) (see Chapter 7, Human Factors) and Hazard and Harm Analyses (ISO 14971:2019/ ISO 24971:2020/ AAMI TIR 105:2020) are used to proactively identify design input requirements necessary to ensure the combination product will be safe, effective, and usable. Preliminary Hazard Analysis (PHA) is also a useful tool to support the identification of safety requirements, which are Design Inputs. PHA can be used as early as Research Phase to identify possible hazards even before a design is solidified (see Chapter 6).

Design Inputs should address incomplete, ambiguous, or conflicting requirements. For example, if a patient has low dexterity, but they are to use an autoinjector for self-administration of a drug, the actuation force for the autoinjector would need to be sufficiently low to help the patient actuate their dose, but not so low that actuation occurs prematurely.

Design Inputs are under revision control over the course of development. They are reviewed, approved, and may be updated under design change control throughout the development effort.

<u>Some categories of combination product Design Inputs</u>

- Functions to be performed
- Features/options
- Performance
- Safety requirements
- Reliability
- Energy source(s)
- Physical characteristics
- Biocompatibility
- Manufacturability
- Installation, maintenance, and servicing

- Human factors/user interface
- Use environment
- Environmental limits
- Applicable standards (mandatory and optional)
- Regulatory requirements
- Compatibility with other devices
- Labeling and packaging

Drug-specific Design Input examples:

- Fill volume
- Rheological properties, e.g., viscosity
- Dosage strength
- Route of administration
- Sensitivities, e.g., to light or oxygen
- Known interactions, e.g., silicone, tungsten
- Dosage form
- Shelf life
- Storage conditions

Device/manufacturing-specific Design Input examples:

- Bulk vs prefillable syringe
- Incoming goods specifications and controls
- Internal and/or external manufacturing capabilities
- Sterile filtration
- Aseptic vs terminal sterilization
- Assembly process for safety system/delivery device
- Visual/automated inspection
- Labeling/packaging

Design Outputs (21 CFR 820.30(d))

Design outputs are the complete set of approved documentation that captures the product specifications. They are established at each design phase, and at the end of the total design effort. They are described in terms that enable design verification, i.e., objectively demonstrating that the results of the design effort meet the technical design requirements (Design Inputs). The total finished design outputs include packaging, labeling, and the Device Master Record (DMR) (21 CFR 820.181). The design outputs are documented, reviewed, and approved prior to release.

One of the requirements under Design Controls is that the combination product manufacturer identify which design outputs are considered essential for the safe and proper functioning of the device constituent part and the combination product. These are called **Essential Performance Requirements** (EPRs) (*see Chapter 6 for more discussion on EPRs*).

Notably, for combination products where the device is a constituent part in a drug-cGMP-based operating system, it is likely that the organization is using a Master Batch Record (MBR). As long as the required DMR elements are embedded within the MBR, there is no requirement to have a separate DMR.[22] The DMR is a compilation of records containing the procedures and specifications for the finished device constituent part/combination product. The records include all the instructions, drawings, and other records that must be used to produce a product, covering product design and production. If a manufacturer chose to "lift and shift" their operations to a new facility in another part of the world, the DMR includes all the information required for "copy exact" and reliably producing the same constituent parts/combination product.

The DMR includes:

- Device and sub-assembly specifications, drawings, raw material and component specifications, composition, formulation and data sheets, Bill of Materials (BOM), and Parts list;
- Device software specifications (source code, build instructions, libraries, executables);
- Production process specifications, including appropriate equipment specifications, production methods, production procedures, and production environment specifications and flowcharts;
- Quality Assurance (QA) procedures, specifications, sampling plans, and acceptance criteria, including quality assurance equipment to be used (with capable measurement systems and test methods);
- Packaging, labeling and user guides specifications, including methods and processes used; and
- Installation, maintenance, and servicing procedures and methods, where applicable.

Design Review (21 CFR 820.30(e))j

A Design Review is a formal, documented, comprehensive, and systematic examination at appropriate stages of development to evaluate the adequacy of design requirements; evaluate the capability of the design to meet those requirements; identify any problems over the course of development; and to confirm that risks have been mitigated where necessary such that benefits outweigh the risks, and the level of risk is appropriate.

A design review is *not* a business or project management review. (This is a distinction between a CMC Stage Gate Review and a Design Review; the CMC Stage Gate Review is often a mix of business review and project review. Design Reviews are focused on identifying and addressing product and process design problems.)

- Design reviews occur, at a minimum, at a frequency/at milestones as predefined in the Design & Development Plan.
- Design reviews are also held as significant changes occur in project scope, resources, or timelines, e.g., a product design change or update to the D&DP could trigger the need for a design review.
- Design reviews are conducted with two or more people, and include representatives from those functions concerned with the specific design change being reviewed, as well as any specialists or subject matter experts needed.
- At least one independent observer should participate in each design review. That individual should be designated in meeting minutes as a participant in the design review. The independent observer is a person who does not have direct responsibility for the design stage being reviewed, but has the know-how to constructively challenge/ask questions during the design review.
- The results of design reviews should be documented, along with approvals to proceed, in the Design History File. Design review documentation should include identification of the device constituent part/combination product design, the date of the design review, and a list of the individuals who perform the review, including specifically designating who was the independent observer.

Design Verification (21 CFR 820.30(f))

Recall that design inputs are measurable (quantitative), testable, unambiguous, and traceable. Design outputs are traced to the uniquely identified design inputs; design verification then confirms

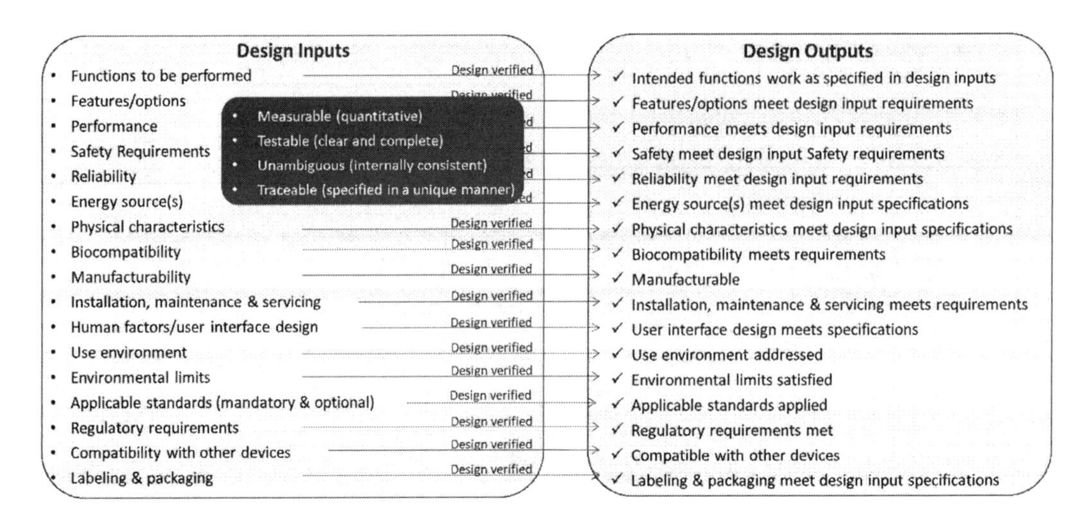

FIGURE 4.3 Design Controls verification. ©2022 Combination Products Consulting Services LLC. All Rights Reserved.

by objective evidence that the design outputs meet the design input requirements. *All design inputs and all Essential Performance Requirements require verification.* Many test reports associated with Design Verification are included in submissions.

Consider the example of a prefilled syringe. Pharmaceutical requirements that would likely require verification include drug/container interactions, extractables and leachables, degradation, absorption, or change of stability indicating parameters, the effect of shear forces on the drug product, biologic requirements, and delivered volume and particles. When considering functional requirements, design verification would likely include establishing performance data on break loose and extrusion forces, burst resistance, closure system forces and torques, dosing accuracy, residual volume, sharps injury protection requirements, liquid leakage beyond the plunger, and markings.

Design verification is documented in the Design History File. Figure 4.3 illustrates the traceability of design verification.

The following are warning letter excerpts with respect to Design Verification:

> Your firm claims conformance to a standard and you have identified requirements and sample sizes for the design verification testing based on Acceptable Quality Levels (AQLs) from this standard. While the sampling plan ensures that lots having a quality level equal to the AQL are consistently accepted, it does not ensure that lots accepted will consistently achieve this quality level. In other words, your selection of sampling plans based on your specified AQL means that you would accept the design if it had a defect level equal to the AQL. However, it would not necessarily ensure that the design would be rejected if it had a defect level exceeding the AQL, i.e., a worse defect level.

> **Failure to adequately establish and maintain procedures for verifying the device design, as required by 21 CFR 820.30(f). Design verification shall confirm that the design output**

meets the design input requirements. The results of the design verification, including identification of the design, method(s), the date, and the individual(s) performing the verification, shall be documented in the DHF.

Your firm did not have adequate analysis to show that design verification ensured that the outputs for the [redacted combination product] conform to the defined inputs for the products. The occurrence of multiple serious component and product failures for your [redacted combination product] indicates a need to review the adequacy of your outputs to ensure conformance with the defined inputs. For example, your firm claims conformance to the ANSI/ASQ Z1.4-2003 (R2013) standard in your design verification testing, and you have identified requirements and sample sizes for the design verification testing based on Acceptable Quality Levels (AQLs) from this standard. However, using the standard's AQLs for design verification does not confirm that design outputs meet design inputs.

The ANSI/ASQ Zl.4-2003 (R2013) standard states, in section 4.3, that "the AQL alone does not describe the protection to the consumer for individual lots or batches, but more directly relates to what is expected from a series of lots or batches provided the provisions of this standard are satisfied." While ANSI/ASQ Z1.4 sampling plan ensures that lots having a quality level equal to the AQL are consistently accepted, it does not ensure that lots accepted will consistently achieve this quality level. In other words, your selection of sampling plans based on your specified AQL means that you would accept the design if it had a defect level equal to the AQL. However, it would not necessarily ensure that the design would be rejected if it had a defect level exceeding the AQL, i.e. a worse defect level. Therefore, your firm's particular use of the ANSI/ASQ Z1.4 standard does not confirm that the design meets your particular quality requirements. Thus, you have not demonstrated that you adequately establish and maintain procedures for verifying your device design.

We reviewed your firm's response and conclude that it is not adequate. In your firm's response, you state that AQL is used to define sample sizes for testing, that no critical defects are allowed, and that you will **(b)(4)**. However, this does not address the use of the ANSI/ASQ Z1.4-2003 (R2013)'s AQLs for design verification, and as such, your design verification does not necessarily confirm that design outputs meet design inputs. Therefore, in response to this letter, provide information that demonstrates your sampling plans are written and based on valid statistical methods. In addition, please clarify how you determined the unacceptable quality level in design verification.

Design Validation (21 CFR 820.30(g))

Design Validation ensures through objective evidence that the finished device constituent part and combination product meet the user needs and intended uses. Said otherwise: Design validation proves that the product that has been created is the "right product" to satisfy the user needs and intended uses. Design Validation helps complete the risk analysis that began as part of the user needs/design input stage. It validates information for safety and claims for risk control, and supports the determination that the benefits of using the combination product outweigh the risks. Design Validation generally follows successful design verification.

Design Validation is conducted using final production – or production-equivalent – units (see Chapter 8 on Design Transfer). Preamble Comment 81 to the QS Regulation states that when equivalent devices are used in the final design validation, the manufacturer must document in detail how the device was manufactured, and how the manufacturing is similar to – and possibly different from – the initial production. Where there are differences, the manufacturer has to justify why design validation results are valid for production units, lots, or batches. A detailed justification should consider all aspects of a Device Master Record (Tech File).

Design outputs may need to be revised in light of design validation. These final outputs are what need to be brought over into manufacturing in design transfer. (The Design Control process is iterative, and this is an example of its iterative nature.)

Design validation can be accomplished through multiple means. These include:

- Clinical evaluation via analysis of relevant scientific literature;
- Comparison to marketed same or similar products; and
- Simulated or Actual Use Testing.

For simulated or actual use testing, Design Validation is conducted using initial production units, lots, or batches (or their equivalents)[23] under defined operating conditions:

- Simulated use of the device/combination product by users, but not on human subjects
- Usability testing by users, with a focus on the user interfaces, labels, and labeling
- Clinical trials, where there is actual use of the device constituent part/combination product by users, on human subjects (under IND[24] or IDE[25]) to ensure safety and efficacy for study subjects

Design Validation also includes software validation, where appropriate.

The Design Validation is reviewed, approved, and documented in the Design History File. Results of the Design Validation are typically included in premarket submissions.

Design Validation: Human Factors

Human factors studies often play a central role in Combination Products Design Validation, by demonstrating the safety and efficacy of the device constituent part and combination product. People sometimes mistakenly believe that the Human Factors Summative Study is the entire Design Validation. The scope of Design Validation is broader than Human Factors. Design Validation is based on *all* the user needs and intended uses. *Human Factors are focused on the suitability of the user interface, since performance can depend on attributes that are not part of, and are not managed through, the user interface.* Human Factors formative studies are done during development as the design of the user interface is refined and finalized to promote safe and effective use. A Human Factors Summative Study is used as part of Design Validation to:

- Demonstrate that users can operate the device/user interface successfully for the intended use(s);
- Provide evidence that users can operate the combination product/device constituent part without injury or negative clinical consequences to either the user or the patient; and
- Demonstrates that potential use errors and failures have been eliminated or limited to the extent possible through appropriate application of risk management and human factors.

Figure 4.4 illustrates a case study example that shows a range of Design Control activities, from user needs identification, translation of user needs into technical design requirements, design outputs and Essential Performance Requirement (EPR) identification, through design verification and a range of validation activities, including Human Factors studies and various others.

Table 4.9 is an example of a Combination Product Traceability Matrix Template to support required end-to-end traceability under Design Controls.

Product Name	"RADrug"		
Review Date	MM/DD/YYYY		
Project Team Leader	S. Neadle	URS version number	1
URS number	URS	Source	
U001 *(intended users)*	Target User Population 19-60+ yr. olds	Market Research Study RADrug	
U002 *(intended use environment)*	Home use/self-administration	Market Research Study RADrug	
U003 *(intended uses)*	Subcutaneous administration of 'RADrug' with viscosity 15 centipoise ±2 cp and volume 2 ml ±0.2 ml to relieve symptoms of RA	Target Product Profile; Clinical Research Team Report XXXX	
U004 *(patient centric insights → product attributes)*	Positive experience (low pain... 0 or 2 on pain scale) during dose administration.	Market Research Study RADrug Literature Search Summary Report XXX	

Product Name	"RADrug"		
Review Date	MM/DD/YYYY		
Project Team Leader	S. Neadle	TDR version number	1
URS number	URS	Technical Design Requirements	
U001	Target User Population 19-60+ yr. olds	Dimensions/ Ergonomic grip design	
U002	Home use/self-administration	Handling Steps Labeling requirements Label Font / Font size	
U003	Subcutaneous administration of 'RADrug' with viscosity 15 centipoise ±2 cp and volume 2 ml ±0.2 ml to relieve symptoms of RA	Injection depth Needle length Needle bore Barrel dimensions Dose volume Dose completion signals Dose Accuracy Shelf life/ Stability Extractables & Leachables Biocompatibility	
U004	Positive experience (low pain... 0 or 2 on pain scale) during dose administration.	Needle tip design/ bevel Needle bore Injection time Handling steps Reliability Cap Removal Force Grip Force Activation Force Spring Force	

FIGURE 4.4 Hypothetical design controls cascade for an autoinjector to treat rheumatoid arthritis (RA) (for illustration purposes only).

Design Input ID	Design Input *measurable, testable, unambiguous, traceable*	Design Output ID	Design Output *clinically relevant specifications*	Design Verification *Design outputs meet input requirements*
DI 1	Grip Dimensions	DO 1 DO 2	Drawings Specifications	✓ Grip Dimensions circumference x' mm± 0.05 mm
DI 2	Labeling requirements	DO 3	Labeling requirements per standards	✓ reading level 3rd grade
DI 3	Label Font	DO 4	Font per standards	✓ Label Font– 8 pitch font, high contrast
DI 4	Subcutaneous injection depth	DO 5	Angle and depth specifications	✓ Injection depth– z mm ± 0.1mm
DI 5^{EPR}	**Needle length**	DO 6 DO 7	Drawings Specifications	✓ Needle length– z' mm ± 0.1mm
DI 6	Needle bore	DO 8 DO 9	Drawings Specifications	✓ Needle bore- 1.27 mm ± 0.05 mm
DI 7	Barrel dimensions	DO 10 DO 11	Drawings Specifications	✓ Barrel diameter x mm± 0.1mm
DI 8 ^{EPR}	Dose volume/ Dose Accuracy	DO 12	Quality Target Product Profile (QTPP) specifications	✓ Dose volume- 2.0 ml ± 0.2 ml
DI 9^{EPR}	**Dose completion signal**	DO 13 DO 14	Drawings Specifications	✓ audible (decibels) and visual signal
DI 10	Shelf life/ Stability	DO 15	QTPP specifications	✓ Shelf life/ Stability x months for drug/device CP
DI 11	Extractables & Leachables	DO 16	E&L per standards	✓ E&L meets standards
DI 12	Biocompatibility	DO 17	Biocompatibility per standards	✓ Biocompatibility meets standards
DI 13	Needle tip design/ bevel	DO 18 DO 19	Drawings Specifications	✓ Needle tip / bevel dimensions
DI 14^{EPR}	**Injection Time target**	DO 20	QTPP specifications	✓ Injection Time (8 seconds± 1 second)
DI 15	Handling Steps per URRA	DO 21	Handling Steps (IFU)	✓ IFU (*design validation required*)
DI 16	Reliability criteria	DO 22	QTPP specifications	✓ Acceptable reliability for chronic use (95% CI)
DI 17^{EPR}	**Cap Removal Force target**	DO 23 DO 24	Drawings Specifications	✓ Cap Removal Force y Newtons (N)± 5N
DI 18	Grip Force target	DO 25 DO 26	Drawings Specifications	✓ Grip Force y' N±5N
DI 19^{EPR}	**Actuation Force target**	DO 27	Drawings Specifications	✓ Actuation Force 20N± 5N
DI 20^{EPR}	**Needle Safety Lockout Force**	**DO 28**	Specifications per standards	✓ Force meets standards
DI 21	Spring Force target	DO 29 DO 30	Drawings Specifications	✓ F_{spring}40N ±5N

User Needs	Design Validation Activities
Target User Population 19-60+ yr. olds home use/self-administration	"RADrug" HF Summative Study: HF Summative Study on representative users demonstrated through simulated use testing with initial production units (IFU + packaging + combination product) that patients are able to safely and effectively self-administer the dose of medication in a home-use environment, and that predictable mi-uses are mitigated.
Subcutaneous injection with 'RADrug' to relieve symptoms of RA	'RADrug' Phase III Clinical Trial using initial production units or equivalent
Positive experience (low pain... 0 or 2 on pain scale) during dose administration.	Design Validation clinical study (integrated question into Phase III Study), study participants rated administration of dose on scale and 95% rated pain at level "0" or "2", meeting pre-established acceptance criteria .

FIGURE 4.4 (Continued)

TABLE 4.9
Example Traceability Matrix Template

User Requirement (URS) Unique Identifiers	URS	Design Input Unique Identifiers	CtQ (Y or N)	Stability Indicating Attribute (Y or N)	Design Input Requirement	Design Output Specifications	Design Verification Report (Related to TDR)	Design Validation Report (Related to URS)	Comment	Meets Requirements (Y or N)
Design Inputs						Design Outputs	Design Verification	Design Validation		
U001	The device components shall be provided sterile	DI –001.1	Y	N	The vial meets standards for aseptic filling (e.g., USP 797 for Pharmaceutical Compounding, Sterile Preparations)	TR-XXX Document Title	TR-XXX Technical Rationale (Confirmation of Vendor Specifications)	TR-XXX Technical Rationale (Confirmation of Vendor Specifications)	N/A	Y
							TR-XXXX (Interim Design Verification Summary Report)	TR-XXXX Human Factors Summative Report		

The following are FDA Warning Letter excerpts with respect to Design Validation:

> Your firm has not completed any testing of the combination products in order to ensure that the products conform to the defined intended uses. Although you inform investigators that you have a risk assessment document, you state that you have not reviewed or updated the risk analysis.

> **Failure to adequately establish and maintain procedures for validating the device design. Design validation shall ensure that devices conform to defined user needs and intended uses. Design validation shall include risk analysis, where appropriate, as required by 21 CFR 820.30(g).**

USER NEEDS AND INTENDED USES

The occurrence of multiple serious components and product failures for your [redacted combination products] indicates issues with the ability of your product to conform to the defined user needs and intended uses. Your firm's standard operating procedure (SOP) for design verification and validation describes the process for execution of design validation. However, your firm has not completed any validation testing of the design of [redacted combination products] in order to ensure that the products conform to the defined intended uses. Your firm's representatives confirmed this, during the inspection, when they stated that no design validation testing has been conducted for the [redacted combination products].

We reviewed your firm's response and conclude that it is not adequate. Your firm has not demonstrated that you have performed design validation. Therefore, in response to this letter, provide design validation of the finished combination product that ensures that the products conform to defined user needs and intended uses, and include risk analysis, where appropriate. Your firm's analysis of design validation must address the finished combination product.

Design Transfer (21 CFR 820.30(h))

Design Transfer ensures that the combination product and device constituent part design is effectively translated into production specifications. During Design Transfer, processes are validated, and we prove the adequacy of the manufacturing process through simulated or actual use testing. The process specifications and procedures are reviewed and approved, and overall risk acceptability against pre-established acceptance criteria is evaluated (acceptability criteria are pre-established in the Risk Management Plan; see Chapter 6). ISO13485:2016 Clause 7.3.8 shares similar expectations for Design Transfer.

As illustrated in Figure 4.2A, Design Transfer is not a point in time, but rather, it spans across much of development – it is a transition period. Design Transfer occurs throughout the iterative design process. There is a final stage of development intended to ensure all design outputs are adequately transferred. This closes out the Design Transfer process. Process validation (21 CFR 820.75) is a deliverable under Design Transfer, and sometimes a blind spot for those with drug-led combination products. Process validation under 21 CFR 820.75 answers the question "Do we understand this process, so that we can reliably and reproducibly produce product that conforms to predetermined specifications?" Process validation for devices/combination products is generally done at scale before marketing authorization while it may occur after for drugs; process validation is needed prior to distribution in any case. We'll discuss Process Validation more extensively in Chapter 8, Lifecycle Management.

Key Design Transfer elements include, for example:

(1) Release of design outputs to production in the Device Master Record (or integrated into the Master Batch Record);
(2) Training and qualification of operators and other relevant production staff by the development team;
(3) Qualification of suppliers, and inclusion on the Approved Suppliers List (to be discussed further under Purchasing Controls);
(4) Qualification of new production equipment and process validation;
(5) Calibration and qualification of inspection and testing equipment;
(6) Test method validation;
(7) Qualification and/or process validation of new facilities, utilities, and environmental controls; and
(8) Design validation (21 CFR 820.30(g)), demonstrating that the device constituent part and combination product meet user needs and intended uses are typically done subsequent to or in parallel with process validation under Design Transfer.

Table 4.10 is an example of a Design Transfer Checklist for a drug-device combination product.

Design Change Control (21 CFR 820.30(i))

Design Change Control includes identification, documentation, validation, or where appropriate, verification, review, and approval of design changes ***before*** their implementation, to ensure that the changes are appropriate and that the device constituent part and combination product will continue to perform as intended. According to Preamble Comment #87, all design changes made after the design review that approves the design inputs for incorporation into the design, and those changes made to correct design deficiencies once the design has been released to production, must be documented. This documentation is valuable not just to satisfy a requirement, but to inform and to expedite later activities, by showing what has previously been considered and rejected or what has previously changed, and why.

Design change control starts at the very beginning of Design Controls. Once you have approved *any* of the design inputs, a mechanism to manage changes needs to be in place. At a minimum, Design Change Control requirements include: (1) establishing a design change control procedure; (2) determining, for each change, the risk presented by that potential change; (3) implementing Design Control requirements *commensurate with the risk* presented by the proposed change; and (4) conducting an impact analysis on design output activities.

Change control is tightly integrated with risk management (see Chapter 6), as there may be unintended consequences of a proposed change, and for combination products, the risk assessment and application of controls, and Design Controls, need to include *each* constituent part, their components, and the combination product, to protect against any potential negative interactions. (There is a case study at the end of Chapter 5 that speaks to just how critical this is! A seemingly innocuous change to a drug formulation can have dramatic consequences for device functionality or interactions of drug with the device. Likewise, "minor" changes to a device component or constituent part might have significant impacts on its functionality or reliability, which could interfere with the accurate and reliable delivery of a drug). The scope and impact of a proposed change should be assessed against the entirety of the Design Controls process. If the change impacts intended users, uses, or use environment, it might trigger the need for new design inputs, design outputs, verification, and validation. An existing device that is intended for a new user population, new indication for use, or a new use environment would be such a significant change. Depending on the scope and impact of a change, it may require a new premarket submission, supplement, or study. Changes must be

TABLE 4.10
Example of a Potential Design Transfer Checklist

Combination Product Design Transfer Checklist		
Category	**ITEM**	**Reference Document No.**
1	**Planning, Manufacturing Strategy, Site Assessment**	
1.1	Manufacturing Strategy and Design Transfer Plan	
1.2	Manufacturing Site Assessment	
1.3	Regulatory Plan	
1.4	Purchasing Controls	
1.5	Supplier Qualification (Approved Supplier List)	
2	**Design Output, Production Process**	
2.1	Device Component Specifications	
2.2	Labeling and Packaging Specifications	
2.3	Storage and Manufacturing Environmental Requirements	
2.4	Control Strategy/Control Plan	
2.5	Test Methods	
2.6	Receiving Inspection Procedures	
2.7	Finished Product Acceptance Criteria, Test, and Inspection Procedures Combination Product Release Testing	
2.8	Device Master Record (DMR) elements integrated into Master Batch Record (MBR)	
3	**Process Development & Qualification**	
3.1	Manufacturing Procedure including Process Flow Chart/Production Layout/Material Flow	
3.2	Product Risk Analysis, URRA, dFMEA, pFMEA, and Criticality Analysis If applicable, eFMEA for each equipment; aFMEA for Software, if applicable[a]	
3.3	Process Characterization studies	
3.4	Preventative Maintenance and calibration procedures for Manufacturing	
3.5	FATs, SATs, Equipment URS[b]	
3.6	Manufacturing Process Training	
3.7	Facility, Equipment, Utility Design and Qualification (IQ, OQ[c])	
3.8	Performance Qualification (PQ) Protocol & Report	
3.9	Final Process Validation (PPQ) Plan & Report	
3.10	Continued Process Verification (CPV)	
3.11	Process Development & Validation/Manufacturing Data Review	
4	**Process Validation**	
4.1	Validation Master Plan	
4.2	Cleaning Validation	
4.3	Sterilization Validation	
4.4	Packaging & Labeling Validation	
4.5	Stability Studies Report/Stability Monitoring Plan	
5	**Other**	
5.1	Business Systems Interface Confirmation & Review (SAP/LIMS[d] etc.)	
5.2	Update to Supply Chain information	
5.3	Complaint Handling Readiness	
5.4	Post-Market Safety Reporting (PMSR) Readiness	
5.5	Malfunction Assessment Reportability List Readiness	
5.6	Review of open Change Controls, Events, and CAPAs	

[a] URRA: Use-Related Risk Analysis; dFMEA: design Failure Modes & Effects Analysis; pFMEA: process FMEA; eFMEA: equipment FMEA; aFMEA: applications FMEA.

[b] FAT: Factory Acceptance Test; SAT: Site Acceptance Test; URS: User Requirements Specification (technical requirements in all aspects of any equipment that the organization decides to purchase for its productivity).

[c] IQ: Installation Qualification; OQ: Operational Qualification.

[d] SAP: System Applications and Products in Data Processing; LIMS: Laboratory Information Management System.

communicated to the FDA if a device constituent part and/or combination product are under review of investigational use (IND or IDE) or premarket review.

Design Control applies both before and after Design Transfer. Post-transfer, design changes for the device constituent part/combination product are covered under 21 CFR 820.30(i), Document Controls (820.40),[26] and Production and Process Controls (820.70) (similarly, subpart F under 21 CFR 211). (This is discussed further in Chapter 8, Lifecycle Management.)

The following is an excerpt from a warning letter with respect to Design Change Control:

> By switching from a [redacted] to a [redacted], the design of the combination product was changed, which, according to your data, resulted in a significant increase in product complaints. Between the release of the redesigned combination product in January, the complaint rate for the combination product increased from [redacted] complaints to [redacted] complaints.

Design History File (21 CFR 820.30(j))

The Design History File (DHF) is a compilation of all the records which describe the design history of the finished device constituent part and combination product. The DHF tracks the evolution of the device constituent part and combination product design, including all the versions of the Design & Development Plan. It is a summation of all design actions from initiation of Design Controls through Design Transfer, including design changes. The DHF is a history of design practices, and not a file of all Design documents. It is objective evidence that the Design & Development Plan has been executed as planned. It is a valuable source of device design information throughout the product lifecycle.

For a combination product, the DHF must address all design requirements resulting from the combination of the constituent parts, and that the drug or biological product is appropriate for use with the device constituent part, i.e., that there are no negative interactions noted with the drug, that the delivery system will deliver the drug properly, and that the container-closure integrity and shelf life can be maintained. (For a Conformité Européenne (CE)-marked device, according to EU MDR 92017/745), a Device Technical File would be generated.)

Following is a summary of an example Combination Product DHF Table of Contents:

1. **Purpose/Scope**
   a. *The Combination Product Name*
   b. *Constituent Parts*
2. **Definitions/Abbreviations**
3. **Responsibilities**
   a. Specific groups/functions involved in the plan with specific roles and responsibilities
4. **Design and Development Planning**
   a. Include all revisions of the D&DP
5. **Design Inputs**
   a. User Needs/Intended Uses/Intended Use Environment
   b. Design Input Requirements
6. **Design Outputs**
   a. Device Master Record Documents (or Master Batch Record including DMR elements)
   b. Identification of Essential Performance Requirements
   c. Traceability Matrix
7. **Risk Management**
8. **Design Reviews**

9. **Test Methods**
   a. Test Method Validation Documents with Statistical Rationale
   b. Test Method Equipment Documents
10. **Design Verification**
    a. Protocols
    b. Reports
    c. Investigations
    d. Drug/Device Interactions
11. **Design Validation**
    a. Human Factors Formative Study Protocols
    b. Human Factors Formative Study Reports
    c. Human Factors Summative Study Protocol
    d. Human Factors Summative Study Report
    e. Design Validation Protocol
    f. Design Validation Report
12. **Design Transfer**
    a. Manufacturing Equipment Specifications
    b. Process Validation Documents
    c. Protocols and Reports
13. **Design Changes**

A DHF is established for *each* device constituent part/combination product. While it may be possible to establish a "platform" DHF for a device constituent part that is used in multiple combination products, a DHF should be created for each combination product. Pre-existing data can be evaluated for applicability to the combination product DHF, but the point of the DHF is to address both the device constituent part *and* the specific drug/device considerations and risks for the specific combined use application. (The use of platform technology is discussed further in Chapter 5, Integrated Development and Chapter 6, Risk Management.)

The DHF should include or reference records and information necessary to demonstrate that the design was developed in accordance with the Design and Development Plan (D&DP) and the quality management system requirements. The DHF contains or references all the revisions of documents, new documents, and the Device Master Record from project initiation to product obsolescence. This last point is important. The DHF is not a "check-the-box" activity that is simply done once the product launches. Ownership for the DHF post-market should be designated in Design Controls procedures. In some organizations, Research & Development maintains ownership of the DHF for the product life; in others R&D transfers ownership to a technical operations function post-market. Generally, Quality provides oversight. There is a strong linkage between the DHF and post-market change control and risk management.

Risk Management [ICH Q9(R1) +ISO 14971:2019/ ISO 24971:2020/ AAMI TIR 105:2020]

Risk management is the foundation for Design Controls (actually, as will be discussed, it is the foundation for all the called-out provisions). Risk is a combination of the probability of occurrence of harm and the severity of that harm (ISO 14971 and ICHQ9). Safety is freedom from unacceptable risk (ISO 14971, ICH Q9). Risk management serves as a framework to structure, integrate, and coordinate the full range of iterative analytical and investigational work needed from early concept through post-market management, to enable design, development, pre-clinical and clinical study, manufacture, and marketing of a safe and effective product. What is distinct for combination products is the need to integrate risk management needs for drugs and devices to address the full range of issues for a combination product.

Specific to Design Controls, under 21 CFR 820.30(g) Design Validation, the FDA states, "Design validation shall include risk analysis, and where appropriate, software validation."[27] In 61 Federal Register at 52620, Comment 82, it states

> When conducting a risk analysis, manufacturers are expected to identify possible hazards associated with the design in both normal and fault conditions. The risks associated with the hazards, including those resulting from use error, should then be calculated in both normal and fault conditions. If any risk is judged unacceptable, it should be reduced to acceptable levels by appropriate means.

In Chapter 6, Table 6.11 illustrates the inter-relationship of Risk Management and Design Controls. The chapter delves further into combination products risk management, presenting the concept of risk management in the context of combined use of different types of medical products, and how to establish and maintain a robust risk management program for your product.

Notably, under the Proposed Rule Quality System Regulations Amendment (2/23/2022), where ISO 13485:2016 is incorporated by reference into the Quality Management System Regulation, FDA states:

> The substance of the ISO 13485 requirements and the activities and actions required for compliance are primarily the same as under the current part 820. ISO 13485 has a greater emphasis on risk management activities and risk-based decision making than the current part 820. ... the explicit integration of risk management throughout the clauses of ISO 13485 more explicitly establishes a requirement for risk management to occur throughout a QMS and should help industry develop more effective total product life-cycle risk management systems. Effective risk management systems provide the framework for sound decision making within a QMS and provide assurance that the devices will be safe and effective (see section 520(f) of the FD&C Act).

In summary, Design Controls and risk management are foundational to Combination Product Development of safe and effective products. Table 4.11 summarizes the Design Controls elements and expectations.

The following is an excerpt from a warning letter with respect to risk analysis:

> **Failure to adequately establish and maintain procedures for validating the device design. Design validation shall ensure that devices conform to defined user needs and intended uses. Design validation shall include risk analysis, where appropriate, as required by 21 CFR 820.30(g).**

RISK ANALYSIS

> Your firm did not include risk analysis related to the design validation, where appropriate. Your firm's SOP for design verification and validation for new products states that product risk assessment is an input that is required to start design validation. Although you informed investigators that you have a risk assessment document, you stated that you have not reviewed or updated the risk analysis since YYYY. Therefore, you have not provided adequate risk analysis in your design validation to ensure that the products conform to their defined intended uses.

The adequacy of your firm's response cannot be determined at this time. Although your firm's response has discussed an updated risk assessment, your firm has not provided this document or the procedures that guide routine review of the document. Without evidence that risk analysis has been adequately performed, we are unable to assess whether design validation has been properly completed.

TABLE 4.11
Summary of Design Controls Elements

Design Controls Element	Key Questions
Design & Development Planning	• Written plans for each design are established & maintained? • All design activities are described/referenced? • Are responsibilities defined? • Are internal and external interfaces identified & described? • Are design plans reviewed, updated & approved? • Do the design plans encompass the design controls project?
Design Input	• What is the need for this device/ combination product? • Who will use this device/combination product? • Where will this device/combination product be used? • How will this device/combination product be used? • What will it be used with? • How long will it be used? • Other questions related to the specific device/combination product being considered? **Typical considerations related to Design Inputs:** – Risk Analysis – Human Factors – Intended Use/Applications for Use – User Interfaces – Regulatory Requirements/Standards – Performance Characteristics – Compatibility with accessories & auxiliary devices.
Design Output	• Do Design Output procedures define and document design outputs in terms that allow an adequate evaluation of conformance to design input requirements? • Do Design Output procedures contain or make reference to acceptance criteria, and ensure that those design outputs that are essential for the proper functioning of the device/ combination product are identified? • Are Design Output(s) documented, reviewed, and approved before release? **Design Outputs are used to:** – Document the Design – Verify & Validate Performance – Manufacture & Test the Product – Install & Service the Product – Obtain Market Clearance – Satisfy Regulatory Requirements – Inform Users how to Use the Device
Design Reviews	• Are Design Reviews Documented – following an established Review Procedure? • Are formal reviews of design results planned and conducted per development stage – as a systematic exam? • Are the review participants – concerned with stage of design, but with no direct responsibility for the design stage being reviewed? • Are the review results recorded in the DHF? • *Team meetings do not constitute a Design Review.*
Design Verification	*Evaluates conformance by examination and provision of objective evidence that specified requirements have been fulfilled.* *Verification and validation of a process-dependent component may only be completed on a finished device/combination product.* • Are design verification procedures established & maintained? • Is the device/combination product design verified?

(*Continued*)

TABLE 4.11 (CONTINUED)
Summary of Design Controls Elements

Design Controls Element	Key Questions
	• Is it confirmed that design outputs meet design input requirements? • Are design verification results documented in the Design History File (e.g., identification, date, methods, individual(s)? **Typical Design Verification Activities**: – Bench testing – *In vitro* Measurements – Tolerance stack-up analysis – Trace Matrix
Design Validation	*Measures the appropriateness of the finished device/combination product for its intended use, including the interactions between the verified subsystems.* *Verification and validation of a process-dependent component may only be completed on a finished device/combination product.* • Are design validation procedures established & maintained? • Is the device/combination product design validated? • Is design validation performed under defined operating conditions? • Is conformance to user needs ensured? • Are production units (or equivalent) tested under actual or simulated use conditions? • Was software validation and software-related risk analysis included, where appropriate? • Are design validation results documented in the Design History File (e.g., identification, date, methods, and individuals)? **Typical Design Validation Activities:** – Clinical Trials – Usability Studies – "Black box" testing – Scientific Literature Research – Benchmarking – Emissions/Susceptibility testing – Sterilization Validation – Trace Matrix, Test ID vs Functional Requirements
Design Transfer	• What will be transferred? • When will the transfer take place? • How will the transfer take place? • What are the Regulatory Requirements? **Typical considerations related to Design Transfer:** – The product – The size of the company – The Development Process – The Manufacturing Process – The Skill of Production Personnel – Consider Process Differences – Engineering Process – Manufacturing Process

Design Changes	Design Change Control Covers:
	• The Design
	• The Device
	• All Subsystems
	  – Accessories
	  – Packaging and Labeling
	  – Processes

Purchasing Controls (21 CFR 820.50)

Purchasing Controls are intended to control products and services procured from suppliers for use in a combination product. Many combination product Sponsors leverage third-party suppliers for design and/or manufacture of at least one of their constituent parts, making Purchasing Controls a critical part of assuring combination product safety and efficacy (Table 4.12). Purchasing Controls are important for acceptance activities. They are required for products received at the facility for use in combination product manufacturing. They are also required for *all* the suppliers of these products and for the suppliers of services (e.g., a service provider for terminal sterilization who is different from the combination product manufacturer).

TABLE 4.12
Purchasing Controls Considerations under 21 CFR 820.50

Key Purchasing Controls considerations for components, constituent parts, and combination product

- Are Subcontractors evaluated and qualified?
- Are Quality Agreements in place?
- Are Quality Parameters specified in orders?
- Is pertinent information included in orders?
- Are Consultants covered?
- Is (are) "Sister" Facility(ies) covered? (Sister facilities within a company, that are each considered to be separate legal entities, also need to have intra-company quality agreements and controls.)
- Is the performance of subcontractors reviewed?

The type and extent of controls needed, and accordingly management of suppliers of products, components, and services should be based on the criticality of and risk associated with the use of the specific product or service purchased. Specific to combination products, Purchasing Controls should include suppliers of constituent parts and/or components of the combination product. Under 21 CFR 820.50, components are not just those of the device constituent part. They *also* include drug components, as well as containers and closures subject to 21 CFR 211.84.

Under Preamble Comment 115 of 61 Federal Register at 52626, FDA states

> **the need for specifications should be based on the criticality of and risk associated with the use of the specific manufacturing material**.... The extent of the specification detail necessary to ensure that the product or service purchased meets requirements will be related to the nature of the product or service purchased, taking into account the effect the product or service may have on the **safety or effectiveness of the finished device**.

Key expectations for Sponsors and Combination Product Manufacturers under purchasing controls include the following:

- Establishing purchasing controls procedures and related incoming acceptance procedures.
- Documenting purchase records, including the approach to accepting suppliers and communicating specifications for purchased products and services (e.g., **Approved Supplier List, Supplier Qualifications, and Supplier Quality Agreements**, to manage products and components supplied, as well as services provided by contractors).
- **Supplier qualifications and controls, commensurate with the significance and risk of the supplied product or service.**[28]
- Establishing and objectively, consistently applying performance triggers and actions in response to findings from supplier assessments for increasing supplier controls and/or acceptance criteria.
- Establishing relationship(s) and agreements (e.g., Supplier Quality Agreements)[29] between the Sponsor, combination product manufacturer (if not the same as the Sponsor), constituent part manufacturers and component suppliers (here, these are all together referred to as

"entities" involved in combination product manufacture). A critical part of these agreements is change management. Changes to a raw material, component, constituent part, or processing might seem innocuous, but could have inadvertent significant impacts on the joint use of the medical products in the combination product, so there needs to be a documented mechanism that reflects how changes made by any of the entities are communicated to the combination product manufacturer, and, if applicable, to the other entities.
- Testing and controls of received products to ensure specifications are met. Under the drug cGMP streamlined approach, related acceptance activities are included in 21 CFR 211.82, "Receipt and storage of untested components, drug product containers, and closures" and 21 CFR 211.84, "Testing and approval or rejection of components, drug product containers and closures" and 21 CFR 211.34, "Qualification of consultants." Purchasing controls under 21 CFR 820.50 do not apply to the API (active pharmaceutical ingredient) and drug components at the same facility, if they are not used to manufacture a combination product.[30]
- Qualifying suppliers, regardless of third-party certifications.
- Be aware of language and interpretation differences (see Chapter 5), and some suppliers who may perceive themselves as out of scope of Part 4 (this can be a big undertaking, requiring education and alignment on responsibilities).

GHTF/SG3/N17:2008 – Quality Management System – Medical Devices – Guidance on the Control of Products and Services Obtained from Suppliers (specifically called out in the FDA combination product compliance program) http://www.imdrf.org/docs/ghtf/final/sg3/technical-docs/ghtf-sg3-n17-guidance-on-quality-management-system-081211.pdf is a helpful guidance in effectively applying Purchasing Controls for medical devices. The principles of this guidance, 21 CFR 820.50 and ISO 13485 Clause 7.4 and its subclauses, are well-aligned and serve as a sound foundation for sponsor coordination and collaboration to help ensure the safe and effective combined use of the medical products that comprise the combination product. Collaboration and coordination considerations may include specification development, compatibility, design verification and validation, risk management, change control, product supply, complaint handling, and potentially, post-marketing safety reporting for each of the combination product constituent parts. It is critical to ensure that the medicinal product and device function together for the intended use(s), user(s), and use environment(s) safely and efficaciously, as well as sustaining market supply.

Examples of significant violations against 21 CFR 820.50 include:

- Deficiencies in both purchasing controls and acceptance activities;
- Inadequate or lack of supplier agreements (particularly for supplied products/services that introduce higher risks to the product if not of sufficient quality);
- No supplier qualifications or qualified suppliers list; and
- Deficiencies in documentation.

The following are excerpts from Warning Letters with respect to Purchasing Controls:

> With regards to the [redacted] located in your firm's facility, upon request, your firm could not certify that the contractor that serviced the equipment had been evaluated.

> You and your customer, [redacted], have a quality agreement regarding the manufacture of [combination] products. You are responsible for the quality of combination products you produce as a contract facility, regardless of agreements in place with [redacted] or with any of your suppliers. You are required to ensure that your combination products are compliant with the CGMP requirements applicable to each manufacturing process that occurs at your facility.

See Chapter 10 on Supplier Quality Considerations for more in-depth discussion of Purchasing Controls considerations.

Corrective and Preventive Action (CAPA) (21 CFR 820.100)

CAPA is considered one of the most important quality system elements. CAPA helps ensure that identified and potential product and quality problems are systematically identified and investigated. The CAPA process ensures a comprehensive review of activities is undertaken to determine the cause of existing or potential problems. These could include manufacturing problems, deviations (including issues with product yield), or nonconformities for a constituent part or the combination product. CAPA is considered essential for dealing effectively with product and quality problems, preventing their recurrence, and preventing or minimizing device constituent part and combination product failures. A combination product manufacturer must ensure that applicable 21 CFR 820.100 requirements are met for their facility.

In 61 FR 52633-52634, Comment 159, US FDA states "the degree of CAPA taken to eliminate or minimize actual or potential nonconformities must be appropriate to the magnitude of the problem and commensurate with the risks encountered." FDA expects combination product manufacturers to develop procedures for assessing the risk of each constituent part and their combined use, the actions that need to be taken for different levels of risk, and how to correct or prevent the problem from recurring, based on the risk assessment.

Combination Product Manufacturers CAPA[31] includes the following:

- Considering implications of CAPA to ***all constituent parts and the combination product***.
- Collecting and analyzing information, identifying and investigating product and quality problems, and taking effective corrective and/or preventive action to prevent their recurrence.[32] A combination product manufacturer should compare the results of analyses from different quality data sources to identify and develop the extent of product and quality problems.
- Taking appropriate measures, which may include CAPAs, with regard to all relevant manufacturing activities, including coordinating with other manufacturers, as necessary, to correct problems with the combination product and to prevent or mitigate them going forward.
- Verifying that there is control to prevent the distribution of nonconforming product (21 CFR 820.90[33]).
- Ensuring information regarding quality problems and CAPA has been properly disseminated, including dissemination for management review.

The CAPA process for combination products should consider implications of corrective and preventive actions to **all constituent parts and the combination product**. *Effectiveness checks might need to consider the combination product even if the corrective action is only done to a single constituent part.* CAPA documentation, for example, needs to reflect how changes may impact constituent parts *and* the product as a whole, including adequate testing or evaluation of those changes. (Notably, under the June 2020 Combination Products Compliance Program 7356.000, compliance with Medical Device Reporting (MDR), Corrections and Removals, and Medical Device Tracking is typically not evaluated in an inspection for a Center for Drug Evaluation and Research (CDER)-led combination product). Take care, though. While MDR is part of QSR inspection for devices, and not cGMP inspections for drugs, this does not mean that FDA won't inspect for Post-Market Safety Reporting (PMSR)/MDR compliance. It just is not part of a CDER cGMP inspection.

The device quality system interpretation of CAPA is a frequent blind spot for pharmaceutical companies. While pharma companies aligned with 21 CFR 211 and ICH Q10 implement "CAPA," 21 CFR 211 does not expressly require all elements in 21 CFR 820.100. Drug CGMPs that are relevant, though, include 21 CFR 211.192 and 21 CFR 211.180(e). Quality data from out-of-specification investigations and any related corrective actions should be addressed through the combination

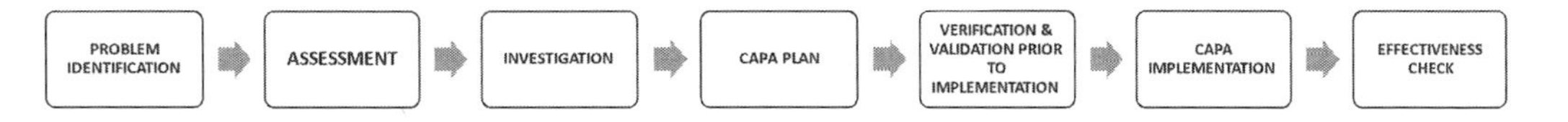

FIGURE 4.5 Typical CAPA process under 21 CFR 820.100.

TABLE 4.13
Example Quality Data Sources[a] That Can Be Used to Inform CAPA

- Regulations, Standards, and Guidance
- Clinical Trials
- Predicate devices
- Same-similar products/combination products
- Human Factors Evaluations
- Reliability analysis
- Risk Management Tools: Harm/Hazard Analyses (e.g., URRA, user-, design-, process- Failure Mode and Effects Analyses (FMEA), Preliminary Hazard Analysis (PHA), Fault Tree Analysis (FTA), Ishikawa Diagrams, Hazard Operability Analysis (HAZOP), Risk Ranking, and Filtering, etc.) and supporting statistical tools
- Management Review
- Supplier (performance/controls)
- Complaint Handling Adverse Event Reporting
- Process Controls
- Finished Product Quality
- Audits (internal/external)
- Product Recall
- Spare Parts Usage
- Returned Product
- Market/Customer Surveys/Competitive Intelligence
- Scientific Literature
- Media Sources
- Product Realization (design, purchasing, production and service and customer information)

[a] The data sources are likewise used to inform risk management. See Chapter 6.

product manufacturer's quality system. Many pharma companies implement CAPA as a repository of all their nonconformances and deviations. In contrast, under the device interpretation, CAPAs are generally opened in response to significant and/or systemic issues. Figure 4.5 illustrates the typical CAPA process under 21 CFR 820.100. Table 4.13 lists some common data sources that can be used to inform the CAPA process.

GHTF/SG3/N18:2010 – Quality management system – Medical Devices – Guidance on corrective action and preventive action and related QMS processes http://www.imdrf.org/docs/ghtf/final/sg3/technical-docs/ghtf-sg3-n18-2010-qms-guidance-on-corrective-preventative-action-101104.pdf is a helpful guidance in effectively applying Corrective and Preventive Action. In the combination product context, one should consider application of this guidance, 21 CFR 820.100 and ISO 13485 Clauses 8.4, 8.5, and its subclauses for each constituent part and the combination product.

Examples of 820.100 violations include failure to implement CAPA procedures. This includes having inadequate corrective and preventive actions; not evaluating all quality data sources; not verifying or validating CAPA; and/or having inadequate documentation, e.g., of verification activities. If procedures do not address all the required elements of CAPA under 21 CFR 820.100, that is likewise an example of a violation, and would be considered a failure to establish a CAPA system.

CASE STUDY 2:

During a review of a drug-eluting stent manufacturing operations, an FDA investigator observes what appears to be a reduced yield for the drug constituent part at one of the manufacturing steps, as compared to the theoretical yield. Should these data be considered as part of a CAPA?

Yes*. Data from the manufacturing process are a source of quality data that should be analyzed. Even if the process is within acceptable limits, that data gives insights on existing* *or potential nonconformities. This does not mean the reduced yield results in the opening of a CAPA. The determination of whether a CAPA is needed should be considered in light of the company's CAPA procedures "action limits" and the implications of the reduced yield as it relates to the overall manufacturing process.*

The following is an excerpt from a warning letter with respect to CAPA:

> Your firm does not distinguish between the different failure modes of rejected components/ units that are collected in reject bins on the combination product manufacturing assembly line. For example, the *Packaging and Inspection Master Specification* instructs that [redacted] which leads to commingling of different types of rejected components. Your firm does not assess the types or causes of rejects, and instead only records the total number of rejects.
>
> Your firm does not use appropriate statistical methodology for process capability in order to analyze the quality of production machinery output at critical process steps and to detect recurring quality problems. Your firm's Process Capability Report for the combination products states that you performed capability analysis on [redacted] test results to determine process capability of the manufacturing operations involved in production. However, various specifications were only analyzed at the finished product attribute-level. Since capability is not determined at the [redacted], and since the capability calculations were performed using final batch data collected after some defective units were removed, this analysis does not adequately demonstrate the ability to detect recurring quality problems.

The following is an excerpt from a warning letter with respect to CAPA:

STATISTICAL METHODOLOGY AND CAPA

Your firm does not employ appropriate statistical methodology for analyzing complaint trends to identify recurring quality problems and/or existing and potential causes of nonconforming product. *Your SOP* specifies how many complaints constitute a trend, and it requires trends to be investigated to identify the need for a corrective and preventive action (CAPA) plan. However, you have not used statistical analysis to justify your definition of a trend. Consequently, we note that CAPAs were not adequately implemented to address several recurring issues seen in complaints and quality reports.

We reviewed your firm's response and conclude that it is not adequate. Your response describes statistically based alert limits for similar complaints within the same lot; however, it does not discuss alert limits for recurring quality problems that are not associated with a specific lot. Additionally, you do not discuss how complaint trends will be addressed by your firm's CAPA system and trigger the requirements for implementing corrective and preventive actions.

The following is an excerpt from a warning letter, including a drug provision related to CAPA.

> **Your firm failed to thoroughly investigate any unexplained discrepancy or failure of a batch or any of its components to meet any of its specifications, whether or not the batch has already been distributed (21CFR211.192)**
>
> Among other things, you manufacture two [redacted] auto-injectors at your facility ... These products are intended to deliver a lifesaving drug ... during emergency treatment of serious allergic reactions .... If your auto-injectors do not operate as expected and deliver the intended

amount of … drug when deployed in emergencies, patients can die or suffer serious illness. You failed to thoroughly investigate multiple serious component and product failures for your [combination] products, including failures associated with patient deaths and severe illness. You also failed to expand the scope of your investigations into these serious and life-threatening failures or take appropriate corrective actions, until FDA's inspection.

In response to this letter, provide:

- A comprehensive review of all your manufacturing investigations, including an evaluation of any other failures or discrepancies of a batch or any of its components that could potentially affect other products, whether or not they have been distributed or recalled; and
- Your plans for addressing the patient safety and product quality risks for product still in distribution.

The following is an excerpt from a warning letter, including a drug provision related to CAPA.

Your firm failed to establish and follow adequate written procedures describing the handling of all written and oral complaints regarding a drug product (21 CFR 211.198(a))

COMPLAINT CLASSIFICATIONS

Your procedures for handling complaints are inadequate. Your complaint classification scheme, listed in your standard operating procedure describes three classifications – expedite, high, and normal – for customer complaints. This complaint scheme is deficient because it does not prioritize complaints based on risk to patients.

For example, you classify complaints for products that fail to activate when the patient has followed the proper sequence as "expedite." However, you classify complaints for products that dispense the drug spontaneously prior to patient use as "normal," your lowest priority classification. Both problems result in the patient not receiving the needed drug in a life-threatening situation.

In response to this letter, provide your revised complaint classification scheme that prioritizes complaints commensurate with potential harm to patients, and your updated standard operating procedure for complaint classification and handling. Also provide your interim plan for addressing complaints you received before implementing your revised classification scheme and procedure to ensure that you have reviewed and handled complaints commensurate with the potential risks to patients.

The following is an excerpt from a warning letter, including a drug provision related to CAPA.

Your firm failed to establish and follow adequate written procedures describing the handling of all written and oral complaints regarding a drug product (21 CFR 211.198(a)).

TREND ANALYSIS

As part of your complaint handling procedure, you define a trend as "complaints of a similar nature on the same lot." You have no scientific or statistical basis for defining a trend as similar complaints, and you stated to our investigators that you had no rationale for using this value.

In response to this letter, provide your procedure that includes a statistical trend analysis with both intra- and inter-batch bases for complaints received. Also provide a detailed analysis of complaint trends across all lots distributed within the last two years.

Installation and Servicing (21 CFR 820.170 and 21 CFR 820.200)

Installation[34] and Servicing provisions ensure that unique challenges associated with these activities are adequately controlled. Generally, these provisions are only applicable to durable medical devices. Typically, they are reviewed only when there are CAPA indicators of a problem with installation/servicing, or as a specific item in a directed inspection. A combination product manufacturer who makes such device constituent parts is expected to have procedures governing installation, inspection, and testing, if applicable.[35] Typically these provisions are significant only when implicated as part of field failures and/or recalls, or in combination with other significant CAPA issues.

FDA references ISO 9001:1994 "Quality Systems- Model for Assurance in Design, Development, Production, Installation and Servicing" in QS Preamble comments 178–179 and 199–202.

US FDA's February 2022 Proposed Rule for Quality Management System Regulations (QMSR) to Harmonize 21 CFR 820 and ISO 13485:2016

February 2022, FDA published a proposed update to the long-standing Quality System Regulation, intended to support harmonizing US-specific regulatory standards for facilities and controls for medical device manufacturing, packaging, storage, and installation with ISO 13485:2016. ISO 13485:2016 standard has been made available in the ANSI Incorporated by Reference (IBR) Portal at https://ibr.ansi.org/. The FDA indicates that

> globally harmonizing the regulation of devices will help provide consistent, safe, and effective devices, contributing to public health through timelier access to patients.... This proposal does not fundamentally alter the requirements for a QS that exists in the current part 820.

The two systems are perceived to already be closely aligned. The proposed rule incorporates linkages to ISO 13485:2016 by reference in the US FDA regulatory framework, while keeping minimum requirements necessary to remain in alignment with the Food Drug & Cosmetics (FD&C) Act. Under the proposed rule, the updated version of 21 CFR 820 will be referred to as the Quality Management System Regulation (QMSR).

Under the proposed rule, the cGMP requirements for combination products are not intended to change.[36] Figure 4.1 summarizes the Combination Product cGMP called-out provisions under the proposed rule. Given that the cGMPs for combination products, i.e., the device called-out provisions, are not fundamentally changing, this chapter does not delve deeply into this topic. Table 4.14 highlights combination products call-outs based on shifts to ISO 13485:2016 referenced clauses.

TABLE 4.14
Highlights of Device-Called-Out Provisions under the Proposed Quality Management System Regulation Amendment under 21 CFR Part 4

If the combination product includes a device constituent part and a drug constituent part, and the current good manufacturing practice operating system has been shown to comply with the drug CGMPs, the following clauses of ISO 13485 within the QMSR requirements for devices must also be shown to have been satisfied; upon demonstration that these requirements have been satisfied, no additional showing of compliance with respect to the QMSR requirements for devices need be made:	
Management Responsibility (21 CFR 820.20)	(i) Management responsibility. Clause 4.1, Clause 5, and its subclauses and Clause 6.1 of ISO 13485
Design Controls (21 CFR 820.30)	(ii) Design and development. Clause 7.3 and its subclauses of ISO 13485
Purchasing Controls (21 CFR 820.50)	(iii) Purchasing. Clause 7.4 and its subclauses of ISO 13485
CAPA (21 CFR 820.100)	(iv) Improvement. Clause 8.4, Clause 8.5 and its subclauses of ISO 13485 (subclauses referenced include analysis of data, statistical techniques, and CAPA)
Installation (21 CFR 820.170)	(v) Installation activities. Clause 7.5.3 of ISO 13485
Servicing (21 CFR 820.200)	(vi) Servicing activities. Clause 7.5.4 of ISO 13485 and § 820.35(b)

Of note, the FDA states that they will retain inspectional authority and will have to update and/or replace the QSIT (Quality Systems Inspection Technique). FDA inspections will not result in the issuance of certificates of conformity to ISO 13485:2016. Further, manufacturers with a certificate of conformance to ISO 13485:2016 are not exempt from FDA inspections.

Drug Called-Out Provisions

When a combination product manufacturer has a 21 CFR 820-based quality management system, their quality system needs to also address specific called-out drug cGMP provisions (see Table 4.3). In contrast to the 820 call-outs discussion, typically fewer issues have been raised to the attention of the authors on application of the 211 call-outs, as well as the biologic and HCT/P sections, as generally, these called-out provisions are prescriptive. For completeness, this section includes excerpts from the FDA Combination Products cGMP Guidance, and largely tracks with the guidance, focusing extensively on the drug constituent part.

If a third party is manufacturing the drug for the combination product, that third-party manufacturer is responsible for complying with 21 CFR 211 requirements for the drug manufacturing process it performs. The combination product manufacturer, however, is responsible for ensuring that the supplier satisfies these requirements, as a part of the combination product manufacturer's purchasing controls for its combination product under 21 CFR 820.50 Purchasing Controls.

The following summarizes the drug cGMP called-out provisions.

Testing/Approval/Rejection of Components (21 CFR 211.84)

Under 211.84, samples of each shipment of each lot are examined, specifically identified and tested to written specifications, and are then released by the quality control unit with all relevant information recorded/documented. *If the specifications are not met, the component should be rejected, and all relevant testing information included in documentation of the rejection of the component.*

Sample collection for testing or examination should be based on statistical principles. *Samples shall be collected in such a manner as to prevent introduction of contaminants into the component and consideration should be given to sampling different portions of the lot (i.e., beginning, middle, and end) and individual testing of these subdivisions of the container.*

- *Facilities operating under a device QS-based Streamlined Approach can augment 21 CFR 820.80* (Receiving, in-process, and finished device acceptance) *to incorporate 211.84 compliant measures.*
- *For any API and/or drug product intended for further processing into the drug constituent part that is supplied to the CP manufacturer, 211.84 should include confirming that at least one test is done to verify the identity of incoming material.*
- Combination product manufacturers do not need to comply with 211.84 for device constituent parts or materials used in the manufacture of a device constituent part *unless the constituent part is also the drug container-closure or a part thereof.* Syringe components prefilled with the drug constituent part, for example, would be in-scope, because they are container-closures. But materials used solely for manufacture of the device constituent part that is not part of the drug container-closure system, e.g., a co-packaged syringe that is not prefilled with the drug, would be out of scope of 211.84, and rather would be controlled under 21 CFR 820.50.
- If the facility relies on the supplier's Certificate of Analysis (C of A), the facility should have established the reliability of the supplier's analysis through appropriate validation of the supplier's test results at appropriate intervals, and conduct additional testing.
- 21 CFR 211.84 may be related to other purchasing controls activities (e.g., 21 CFR 820.50).

Calculation of Yield (21 CFR 211.103)

Calculation of yield ***for the drug constituent part of a combination product*** must be performed during manufacture of the combination product to ensure process control:

- Yield calculation at the conclusion of each appropriate phase of manufacturing, processing, packaging, and holding for the drug constituent part(s) and for the combination product.
- Determined at each phase at which drug component, in-process material, or product loss may occur, including during formulation of the drug, during incorporation of the drug into the combination product (e.g., filling or coating), and, where applicable, during the packaging process. For single-entity combination products, the calculation of yield should be done on every batch of the combination product. For co-packaged combination products, the calculation of yield is on every batch of the drug constituent part(s).[37]
- The final calculation of yield is done on the finished combination product lots.
- Although data on the number of device constituent parts and components used and lost during combination product manufacturing may be needed for manufacturing process controls (21 CFR 82070), yield calculation under 21 CFR 211.103 is *not* required for device constituent parts. (See note 37 for important exception.)

If a third party is manufacturing the drug for the combination product, that third-party manufacturer is responsible for complying with the calculation of yield requirement at the appropriate phases of the drug manufacturing process it performs; and the combination product manufacturer, is responsible for ensuring that the supplier satisfies these requirements under 21 CFR 820.50 Purchasing Controls.

CASE STUDY 3:

What if there are problems with the device constituent part? Could those affect drug yield?

Yes*. Consider the example of prefilled syringes rejected because of a nonconformity of the syringe needle. This nonconformity could lead to a loss of the corresponding drug. Any loss would be captured as part of the yield calculations for the drug constituent part. An investigation into the cause of that loss should identify the manufacturing problem that led to the device nonconformances.*

Other regulations to consider relevant to 21 CFR 211.103 calculation of yield include:

*211.68**	*Equipment**
*211.115**	*Reprocessing**
*211.111**	*Time Limitations on Production**
*211.125**	*Labeling Issuance **
820.100	CAPA
*820.70**	*Production & Process Control**
*820.75**	*Validation**

*Non-called-out provision.

Non-called-out provisions are italicized in the above list. As mentioned earlier in this chapter, non-called-out provisions are helpful to consider, but FDA does not enforce compliance with them. They may be helpful to review to inform understanding for best practices, and as an aid to help in understanding how FDA may interpret the corresponding base set of provisions/requirements for the combination product. The reader should take seriously what FDA can require under the base set of cGMPs, given the breadth of expectations, even if specifics of a non-called-out provision aren't replicated.[38] Needless to say, you need to be able to demonstrate that your approach is sound. (Throughout this section, non-called-out provisions are italicized for clarity.)

Tamper-Evident Packaging for Over-the-Counter (OTC) Human Drug Products (21 CFR 211.132)

If a combination product is accessible as a non-prescription drug to the public, it should have tamper-evident packaging. The tamper-evident feature needs to remain intact when the product is handled in a reasonable manner during manufacturing, distribution, and display on a retail shelf. Manufacturers marketing an OTC combination product need to notify regulators of changes to tamper-evident packaging and labeling. For single-entity combination products, tamper-evident packaging applies to the combination product. For co-packaged combination products, tamper-evident packaging applies to the drug constituent part(s). This requirement can be met if the entire combination product, including the drug constituent part, has tamper-evident packaging. Some exemptions apply.[39]

Expiration Dating (21 CFR 211.137)

Expiration dating ensures that the label for the combination product reflects the date through which the product has been shown to meet applicable standards of identity, strength, quality, and purity (drug CQAs). Stability testing is used to determine expiration dating for combination products (refer to 21 CFR 211.166, and also see Chapter 11, Analytical Testing Considerations). One must ensure that batches placed on stability for purposes of establishing expiration date are representative of the proposed marketed product (21 CFR 211.166).

Consideration should be given to expiration dating implications from *each constituent part*, including environmental factors, e.g., temperature and humidity, that may impact the constituent part. For the medical device constituent part, shelf-life testing may be done under accelerated and real-time conditions at a specific temperature or multiple temperatures. This is usually based on Design Controls and risk analysis considerations (21 CFR 820.30). Stability studies for drug constituent parts generally take into account additional relevant environmental factors, such as humidity, light exposure, and regional temperature zones.[40]

If the product is to be reconstituted at the time of dispensing, its labeling must identify expiration information for both the reconstituted and un-reconstituted forms. Shipping and handling conditions and parameters (during manufacturing and post-release, to include storage) need to be validated through appropriate testing. *When constituent parts of a co-packaged combination product can be used independently, expiration dating, when required, should be listed separately for each constituent part. If a single expiration date is listed for a co-packaged combination product, the date should be the earliest expiration date/shortest shelf life for any constituent part.*

The expiration date of a combination product may be shorter than the expiration date or shelf life of its constituent part)s, if marketed independently, due to (1) interactions between constituent parts when combined, (2) effects of additional manufacturing steps (e.g., sterilization), or (3) one constituent part having a shorter expiration date than the other.

Stability Testing (21 CFR 211.166)

The United States Pharmacopoeia (USP) defines stability as "the extent to which a product retains, within specified limits, and throughout its period of storage and use, i.e., its shelf life, the same properties and characteristics that it possessed at the time of manufacture." Testing to support the stated expiration date and storage conditions for the combination product is stability testing. For combination products, one needs to consider the stability of the drug constituent part, the stability (typically referred to as "shelf life") of the device constituent part,[41] and the stability of the combination product (e.g., interactions between the constituent parts and/or components for the combined use configuration).

Stability criteria generally focus on chemical performance (e.g., degradation, interaction, device packaging and interaction, decay, manufacturing), physical considerations (e.g., physical and/or functional performance characteristics, manufacturing process, storage conditions), microbiological characteristics (e.g., sterility, environmental control, antimicrobial effectiveness), therapeutic effect, toxicological performance, and biocompatibility testing. Combination product stability considerations should also include component interactions or degradation which can cause the device operating characteristics or the drug performance to fall outside of the prescribed tolerances. Stability should also address external environment impacts, e.g., the shipping or storage conditions can cause a breakage in the device, a failure of the barrier properties of a sterile package or degeneration of the device itself. Preconditioning of constituent parts and/or components may be necessary, e.g., for functional stability testing, representing the overall lifetime of the product (including the time that elapses from point of manufacture of the final finished product, its release into supply chain, through the point of first use and actual use of the product in the hands of users, to the point of last operational use and disposal). [42]

The materials, components, and packaging that are to be used in the manufacture of the drug, device, and combination product need to be examined for their individual stability/shelf life characteristics in addition to their effect on the stability of the finished device/drug/combination product. Some materials and components may need special handling to maintain their characteristics within the desired specifications. Also, the intended use of the combination product is an important consideration because this will greatly influence the tolerance level of stability or other failures of the drug/device/combination product.

For both single-entity and co-packaged combination products, stability testing is performed on the drug constituent part as incorporated into the finished combination product. For co-packaged combination products, if the drug product is purchased from another manufacturer for inclusion in the combination product, the combination product manufacturer is still responsible for ensuring the stability of the drug constituent part as marketed in the co-packaged combination product through, e.g., Purchasing Controls (820.50), including documentation of oversight in the cGMP records. The combination product manufacturer has to ensure the adequacy of the drug product manufacturer's stability testing or conduct additional stability testing.

Combination product manufacturers can leverage bracketing and matrixing approaches, or stability data from a previously marketed product, if a new combination product is a modification of one that is already marketed, and the change does not impact the drug constituent part stability (21 CFR Part 4, §VI.B.6).[43,44] If a combination product manufacturer *does* leverage previously existing data, they need to document objective evidence and the rationale to support the approach. FDA provides additional clarification on stability testing for combination products under FR 87 FR 56066, "Alternative or Streamlined Mechanisms for Complying with the Current Good Manufacturing Practice Requirements for Combination Products; List Under the 21st Century Cures Act" (September 13, 2022).[45]

The batches placed on stability to establish expiration date need to be representative of the marketed product. Given that, process validation requirements should be considered prior to producing units for formal stability testing. Further, process and design changes made after stability units have been produced and may result in the need to repeat stability testing or to conduct supplemental testing. When significant changes are made to the product or manufacturing process, it is important for manufacturers to consider whether new or supplemental stability testing is needed.

Other considerations include addressing out-of-specification (OOS) results during stability testing. Quality data from investigations of OOS results and related corrective actions should be addressed through the combination product manufacturers' CAPA system under 21 CFR 820.100. Sampling, testing, and release should include the application of appropriate statistical techniques (21 CFR 820.250), and Labe Controls should be applied to ensure analytical results obtained are accurate.

As mentioned under the overview of 211.137, stability testing is used to determine expiration dating for combination product.

Generally, medical devices are conditioned at specific temperatures for accelerated aging or recommended storage temperature for real-time aging to establish expiration dates. Typically, accelerated aging data is attained first to establish expiration dating of medical devices with real-time data attained later to confirm this dating/shelf life. Manufacturers should consider using the same production lots for both accelerated and real-time aging testing. Discussing and aligning upon stability plans (e.g., lots, times, conditions, proposed functional tests) with the regulator early in development is recommended.

FDA Guidance on Stability Testing of Drug Substances and Drug Products refers to the ICH Q1A which involves exposure of product to multiple temperature and humidity combinations. In some cases, accelerated aging data can be used together with available real-time aging data to extrapolate the combination product expiry date. This should be confirmed with real-time data on the finished combination product.

Manufacturers are responsible for establishing and managing their stability program. If a combination product manufacturer purchases a drug product from another manufacturer for inclusion in its co-packaged combination product, the ***combination product manufacturer*** is responsible for ensuring the stability of the drug product ***as marketed in the co-packaged product*** through appropriate mechanisms. This could include, for example, applying purchasing controls (21 CFR 820.50) to ensure the adequacy of the drug product manufacturer's stability testing or by conducting additional stability testing. These documents should be included in cGMP records. (*See also Chapter 11, Combination Product Analytical Considerations, for additional discussion on stability testing.*)

Testing and Release for Distribution (21 CFR 211.165)

Testing to support the stated expiration date and storage conditions for the combination product is done under 21 CFR 211.165. The following considerations apply to single-entity combination products or to the drug constituent part of any co-packaged or cross-labeled combination product. The combination product manufacturer must ensure the following:

- Each batch meets specifications including the identity and strength of each active ingredient prior to release.
- Each batch is laboratory tested as necessary to be free of microorganisms.
- Written procedures exist for sampling and testing.
- Acceptance Criteria used by Quality Control Unit (or equivalent) are adequate for approval and release.
- Accuracy, sensitivity, specificity, and reproducibility of the test method being used are validated.
- Reprocessed rejected material meets standards, specifications, and other relevant criteria.

Functionality of the combination product is considered for assessment as part of release.

Other regulations to consider relevant to 21 CFR 211.165 Testing for Release and Distribution include:

211.167(a):	Sterile or Pyrogen-free
211.167 (c):	Controlled Release
211.137:	Calculation of Yield
211.192:	*Production Record Review**
211.173:	*Laboratory Animals**
211.176:	*Penicillin Containment**
820.72:	*Inspection, measuring, and test equipment**
820.80:	*Receiving, in-process, and finished device acceptance**
820.90:	*Nonconforming product**
820.160:	*Distribution**

*Non-called-out provision (see Note 32 of this chapter).

Special Testing Requirements (21 CFR 211.167)

Special testing requirements apply only if a combination product or the drug constituent part is supposed to be sterile and/or pyrogen-free, e.g., for sterility, pyrogenicity, ophthalmic ointments, and controlled release dosage forms. After identifying product requirements, some special testing for batches may be needed to confirm that the combination product requirements have been met, e.g., sterility or pyrogenicity, or controlled release rate of the drug). The term "batch" can be defined based on the drug constituent part rather than the finished combination product for purposes of special testing requirements for pyrogens and endotoxins (evidence and rationale need to be documented). Special testing may be required for the primary packaging and/or other parts of the combination product.

Parametric release may be acceptable for some terminally sterilized combination products (this is more common for CDRH-led combination products). Use of parametric release requires approval from FDA for combination products approved under an NDA, BLA, ANDA, or PMA. For 510(k) cleared products, adopting parametric release may be allowed for some well-established sterilization methods like ethylene oxide (EtO) or gamma irradiation. Per Compliance Program 7356.000, such changes could be accomplished under change controls under a combination product manufacturer's QMS without FDA review. For terminally sterilized CDRH-led combination products that are labeled sterile, generally Sterility Assurance Level of (SAL) 10^{-6} is considered appropriate.[46]

Other relevant provisions that should be considered in relation to 21 CFR 211.167 include:

211.160(a), (b)	*General Requirements**
211.165(e)	Testing for Release and Distribution
211.137	Calculation of Yield
211.192	*Production Record Review**
211.194	*Laboratory Records**
610.12	Sterility testing of Biological Products
820.72	*Inspection, measuring and test equipment**
820.80	*Receiving, in-process, and finished device acceptance**
820.90	*Nonconforming product**
820.160	*Distribution**

*Non-called-out provision (see Note 32 of this chapter).

Reserve Samples (21 CFR 211.170)

21 CFR 211.170 requires retain sample to support any potential post-distribution drug investigations. Single-entity and co-packaged combination product manufacturers are expected to keep reserve samples of each lot of the active ingredient, if any, that they receive, in whatever form it arrives at their facility. For example, if the active is received as bulk API, the combination product manufacturer should retain representative samples of the bulk API. If, on the other hand, the active is received already incorporated into an in-process material, the representative retain sample would be of that in-process material. Active ingredient samples are typically kept for 1 year after the expiration date for the last lot of the combination product containing the active ingredient. Likewise, samples from each lot of bulk drug substance would be retained for 1 year after the expiration date of the last lot of the combination product that uses that lot of the active ingredient.

Given that retain samples are for the purpose of any potential post-distribution ***drug*** investigations, for co-packaged combination products, the combination product manufacturer is expected to maintain reserve samples of the drug constituent part in its immediate container-closure system, *without* retaining samples of the device constituent part from the same package. Reserve samples are typically maintained for 1 year after the expiration date for the drug product (radioactive and OTC products may have different requirements under 211.170). A sample of the device constituent(s) may also need to be kept if, for example, the device constituent part is needed to perform any of required tests. For example, a combination product consisting of an injector system (device constituent part) into which the user inserts a prefilled cartridge containing the drug, reserve samples of the prefilled cartridge alone would generally suffice to comply with the drug product sample retention requirements. An injector may need to be available, though, to enable testing of reserve samples.

For single-entity combination products, generally the finished combination product, ***including*** the device constituent part (and/or components of which the device constituent part is comprised) that come into contact with the drug constituent part as packaged for distribution ***would*** be retained. For example, in the case of a prefilled syringe, you'd need to retain the entire combination product, or a separable portion like a cartridge used in an injector system, or the complete packaged combination product, e.g., drug-eluting contact lenses packaged in blisters and their carton.

Reserve samples may present challenges, particularly for device manufacturers less familiar with retain sample requirements under Laboratory Controls (211.170). For a drug-coated implant, e.g., a drug-coated hip, retaining samples of entire hips could become an unwieldy expectation. US FDA's Combination Product cGMPs and their Compliance Program 7356.000 make allowances for that, giving options such as the possibility for combination product manufacturers to retain reserve samples representative of, but not identical to, a finished drug constituent part or combination product. Alternatively, a combination product manufacturer could maintain validated surrogates for some of the necessary testing, while also retaining complete samples of the combination product for other testing. The FDA goes on to say that retention of samples representative of lots of a larger batch might also be considered acceptable. Consider, for example, a range of sizes of drug-eluting stents, drug-coated catheters, or drug-coated screws. Representative samples of each size from within a broadly defined batch that includes multiple sizes of the same family (a "device design space") of such coated combination products could be acceptable. Importantly, appropriate evidence and an explanation of the rationale to support the reserve sample approach should be accessible at the manufacturing facility for review during an inspection. Adequate justification and data to support this approach requires documenting any differences in the manufacturing process for the reserve sample and the finished combination product, ensuring they do not affect the drug constituent part. Further, the manufacturer needs to assure the immediate container/closure has essentially the same characteristics as the immediate container/closure for the drug as packaged in the combination product for distribution (in the case where the actual immediate container-closure is not being used). Rationale must also show that the proposed representative samples are suitable for all required testing of the drug constituent part for which those reserve samples are being kept.

The reader is strongly encouraged to read FDA's Guidance, Current Good Manufacturing Practice Requirements for Combination Products,[47] as it goes into much more specific detail on this called-out provision. Some additional key points are summarized as follows:

- The combination product manufacturer must maintain twice the quantity of active ingredient and combination product reserve samples necessary to perform required tests, except for sterility and pyrogen testing (see 21 CFR 211.167 Special Testing Requirements).
- The quantity of product kept as reserve samples should be aligned with the batch definitions in any related premarket submission for the combination product.
- Reserve samples for the API and finished drug product should be retained under labeled storage conditions for a specified time, to enable conduct of any necessary tests.
- Manufacturers need to examine, investigate, and maintain associated records of any drug product sample deterioration.
- In the event that reserve samples of API, drug/biological product constituent part or finished combination product are stored at a facility other than that of the combination product manufacturer, the manufacturer has to ensure appropriate controls like storage conditions, sample access, ability to analyze samples, and ability to maintain adequate samples for each lot.

Biological Products Considerations

The CGMP requirements for biological products in 21 CFR parts 600 through 680 address the unique challenges biological products pose. A biological product regulated under §351 of the Public Health Service (PHS) Act is also, by definition, a drug or a device, so in addition to the requirements in parts 600 through 680, a biological product is always either subject to the drug CGMPs or subject to the device QS regulation, regardless of whether the biological product is a constituent part of a combination product. For biological products, consistency of manufacturing procedures can be a principal way of ensuring product safety, purity, and potency. The CGMP requirements for biological products applicable to a given product can vary based on the specific considerations applicable to a particular biological product. The specific requirements in parts 600 through 680 that must be met to comply with 21 CFR Part 4 for a combination product that includes a biological product constituent part therefore depend upon the specific type of biological product it includes. Combination product manufacturers whose product contains a biological product constituent part are encouraged to reach out to the product's lead Center or Office of Combination Products with any questions. Specific combination products policy for inspection of CBER-led products has yet to publish.

Human Cellular and Cellular-Tissue-Based Products (HCT/Ps)

21 CFR Part 1271 distinguishes between HCT/Ps regulated solely under section 361 of the PHS Act (42 U.S.C. 264) and 21 CFR Part 1271 vs those that are ***also*** regulated as drugs, devices, and/or biological products. The reader should refer to 21 CFR 3.2(e), 1271.10, 1271.15, and 1271.20 to determine whether an HCT/P is regulated as a drug, device, or biological product constituent part of a combination product. An HCT/P that is *not* regulated solely under section 361 of the PHS Act and Part 1271 is *also* regulated as a drug, device, or biological product. The drug CGMPs, device QS regulation, and the requirements in parts 600 through 680 may apply to an HCT/P depending on whether the product is regulated as a drug, device, or biological product. Current Good Tissue Practices (cGTPs)[48] apply to combination products that include an HCT/P.

An HCT/P is regulated solely under §361 of the PHS Act if:

(1) It is minimally manipulated;
(2) It is in intended for homologous use[49] only, as reflected by the labeling, advertising, or other indications of the manufacturer's objective intent;
(3) The manufacturer of the HCT/P does not involve the combination of the cells or tissues with another article (other than water, crystalloids, or a sterilizing, preserving, or storage agent) provided that the addition of water, crystalloids, or the sterilizing, preserving, or storage agent does not raise new clinical safety concerns with respect to the HCT/P; and
(4) Either:
   - The HCT/P *does not* have a systemic effect *and is not dependent* upon the metabolic activity of living cells for its primary function;

   or
   - The HCT/P *has* a systemic effect or *is dependent upon* the metabolic activity of living cells for its primary function, *and*
     - It is obtained from the same individual as it is intended for use (i.e., for autologous use);
     - It is for allogeneic use (involving tissues or cells that are genetically dissimilar and hence immunologically incompatible, although from individuals of the same species) in a first-degree or second-degree blood relative; or
     - It is for reproductive use.

At the time of this writing, there is limited additional guidance on cell-gene therapies in combination products.

COMBINATION PRODUCTS' CGMPS UNDER EU MDR

To this point, our analysis and approach on cGMP best practices have been through the lens of 21 CFR Part 4. The best practices are essentially the same under ISO standards and ICH guidance. The practices are risk-based and sound, and it makes sense to apply these best practices for combination products even if you aren't intending to enter the US market. With the understanding of combination product cGMPs and best practices in the United States, let's now turn our focus to the European Union, where impactful regulations have recently gone into effect and are still evolving. We necessarily will first review background on how combination products are interpreted under the EU MDR (2017/745), and will then turn to Annex 1 for specific considerations for single integral drug-device combinations.

BACKGROUND

Before we can effectively delve into combination product cGMP expectations in the European Union, it is important to understand the Medical Devices Regulation (2017/745/EU)(MDR), adopted May 2017. As such, we will first review the EU MDR that serves as context for combination product expectations in EU.

The EU MDR regulation replaced the Medical Devices Directive (93/42/EEC)(MDD) and the Active Implantable Medical Device Directive (90/385/EEC)(AIMDD).[50] The new MDR is intended to create a robust, transparent, and sustainable regulatory framework, which improves clinical safety and creates fair market access conditions for manufacturers. In contrast to directives, the regulations are directly applicable. They do not need to be transposed into national law. The MDR therefore reduces the risk of discrepancies in interpretation across the EU.

The publication of the MDR in May 2017 marked the beginning of a four-year period of transition from the MDD and the AIMDD to the EU MDR. The publication of the In-vitro Diagnostic Device Regulation (IVDR) in May 2017 marked the beginning of a five-year period of transition from the In-vitro Diagnostic Device Directive (IVDD). This transition was aimed to allow manufacturers and other economic operators to prepare for the implementation of the regulation, while also giving healthcare professionals and health institutions time to learn what will be required from them, notably in terms of the traceability of devices. To avoid market disruption and allow a smooth transition from the MDD to the MDR, several transitional provisions are also in place. Some devices with certificates issued under the MDD may continue to be placed on the market until 27 May 2024, and made available or put into service until 27 May 2025 (see Figure 4.6).

In general, no requirements from the MDD have been removed; the new regulations emphasize a lifecycle approach to safety, backed up by clinical data. Figure 4.7 illustrates the Annexes of the new EU MDR.

Risk Classification of Devices and Scope of the Regulations

Under the EU MDR medical devices, as previously under the MDD, are categorized into four classes (Classes I, IIa, IIb, and III), but the MDR reclassifies certain devices and has a wider scope. These classes are summarized in Table 4.15 and Figure 4.8.

Clinical Investigations (MDR Articles 62 to 82)

The rules on clinical investigations for medical devices and performance have been reinforced in the EU MDR. The new rules describe clearly how these investigations shall be designed, notified and/or authorized, conducted, recorded, and reported. If you are a Sponsor or take part in clinical investigations or performance studies, please read the relevant articles carefully so that you are informed of all the new obligations.

- Devices with **valid MDD certificates** can continue to be placed on the market during the transition period only **if there is no significant change to the device and PMS, vigilance and registration according to MDR is followed**

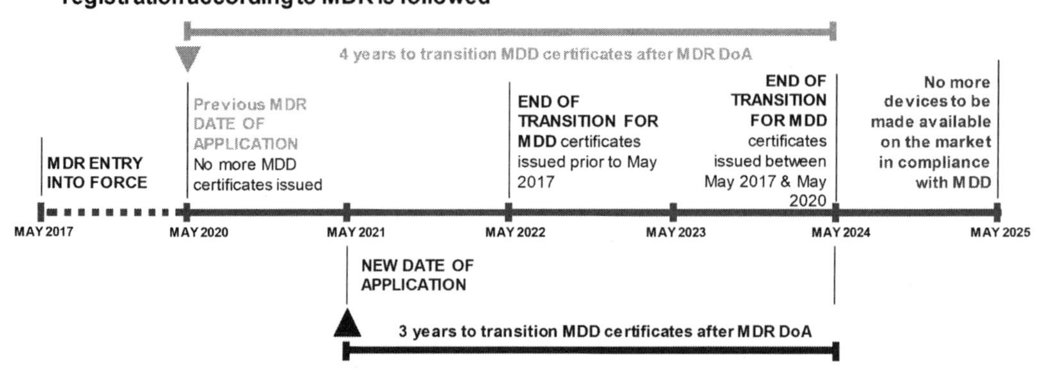

FIGURE 4.6 Transition from MDD to MDR.

Note: In February 2023 the EU Parliament and Council approved a delay in the MDR transition. (They also removed a "sell-off date" that would have prevented the sale of products already on the market, but still in the supply chain, but not yet received by users before the transition period ends.)

The new transition dates are: May 2026 for Class III implantable, custom-made devices; 31 December 2027 for Class IIb implantable and Class III devices; and 31 December 2028 for Class IIb (all other), Class IIa, Class I sterile/measuring and Class I self-certified devices

This delay doesn't modify the MDR's current safety and performance requirements, but is intended to give manufacturers additional time to move from the old rules to the new requirements.

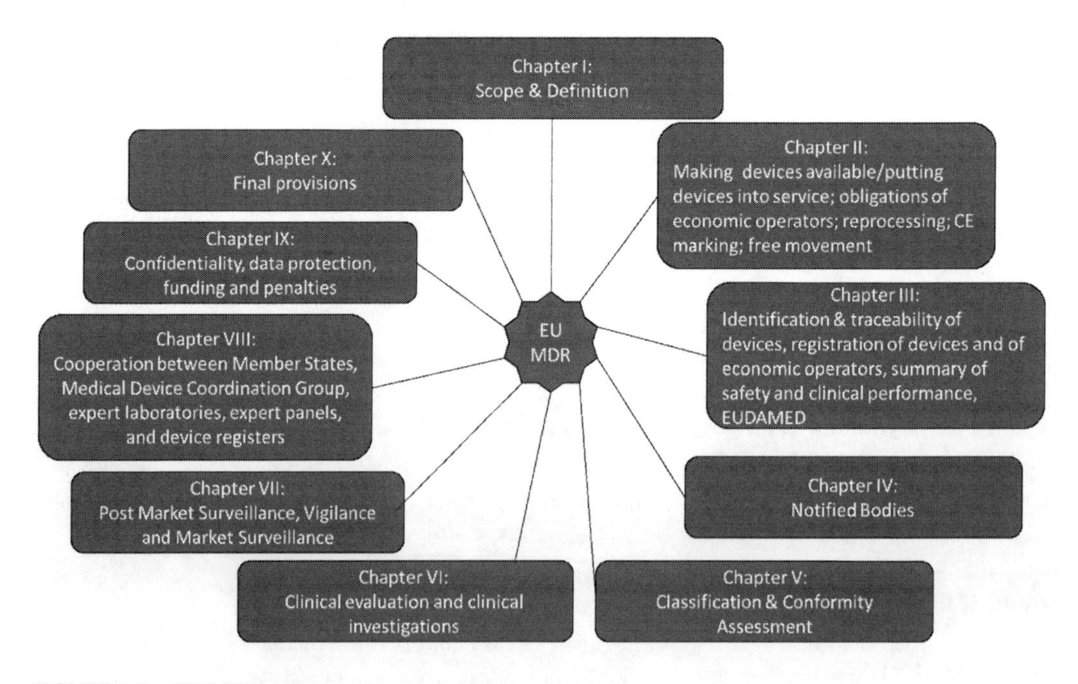

FIGURE 4.7 EU MDR annexes.

TABLE 4.15
Classification and Conformity Assessment Route

LOWEST RISK	**Class I**	Self-declaration for CE marking Notified Body involved in case of sterile or measuring devices and for reusable surgical instruments
↓		
↓	**Class IIb and IIa**	Full quality assurance system, self-declaration for CE marking
↓	**Class IIb implants**	Notified Body assessment of clinical evaluation before CE marking
↓	**Class III**	Full quality assurance system and Notified Body assessment of dossier and issue EC product certificate before CE marking. Scrutiny procedure for Class III implantable devices
↓		
↓		
HIGHEST RISK		

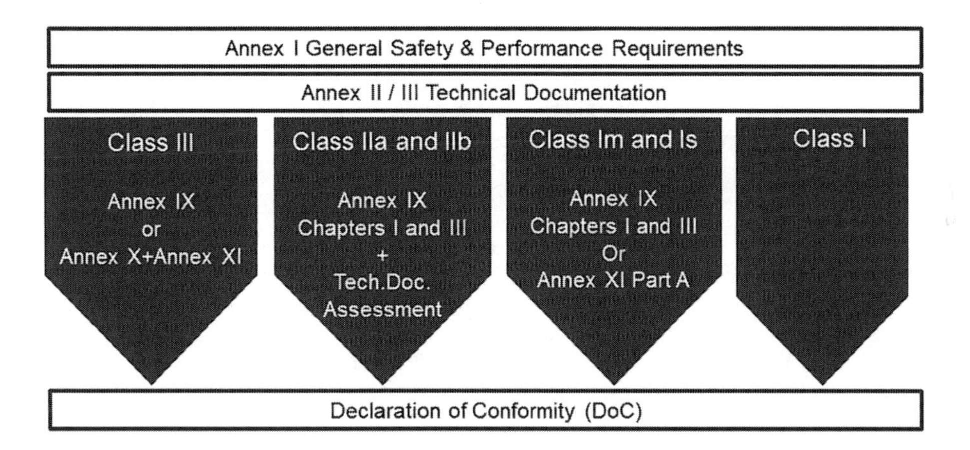

FIGURE 4.8 Expectations for risk-based medical device classifications under EU MDR.

Obligations and Regulatory Requirements of Economic Operators

MDR Article 2(35) defines the term "Economic Operator." An Economic Operator is a manufacturer, a distributor, an authorized representative, an importer, or a system/procedure pack producer. The MDR clarifies the respective obligations of manufacturers, distributors, authorized representatives, importers, and system/procedure pack producers (MDR Articles 10 to 16). See Table 4.16 for a summary.

TABLE 4.16
Summary of Economic Operators and Their Responsibilities under EU MDR

Economic Operator	Obligations
Manufacturer	Manufactures a device or has a device designed/manufactured, and markets that device under its name or trademark
Distributor	Makes a device available on the market
Authorized Representative	Mandated to act on behalf of extra-EU manufacturer for specific tasks of the Regulation
Importer	Places a device from a third country on the EU market
System/Procedure Pack Producer	Person combining devices bearing a CE-marking with other devices or products, in order to place them on the market as a system or procedure pack

For **all economic operators**, the regulation adds new requirements and reinforces existing requirements:

Manufacturers

Manufacturers have to put systems in place for risk and quality management, conduct clinical or performance evaluations, draw up technical documentation, and keep all of this up to date. Manufacturers are also required to apply conformity assessment procedures in order to place their devices on the market. The level of clinical evidence needed to demonstrate the conformity of a device depends on its risk class. Once they have completed their obligations, manufacturers should draw up a declaration of conformity and apply the CE mark to their devices.

The Regulations require manufacturers to implement post-market surveillance follow-up plans. This includes compiling safety reports and updating the performance and clinical evaluation throughout the lifecycle of a device. Manufacturers outside the EU market should have a contract with an authorized representative inside the EU.

Distributors

- Verify CE marking and Declaration of Conformity
- Provide labeling and instructions for use in the local language
- Provide name and address of importer (where applicable)
- Ensure the presence of unique device identifier (UDI)
- May perform checks on a sampling basis
- Shall not place defective devices on the market
- Notify the local authorities in case of serious risk
- Keep a register of complaints, nonconforming products, product recalls, and withdrawals
- Forward complaints to manufacturer, Authorized Representative (AR), and/or importer
- Provide samples of devices to authorities
- Comply with the manufacturer's storage and handling instructions

Importers

- Verify CE marking and Declaration of Conformity
- Check AR on the label and in EUDAMED
- Ensure appropriate labeling and instructions for use
- Ensure presence of UDI
- Register devices in EUDAMED, including UDI (within one week after placing on the market)
- Comply with the manufacturer's storage and handling instructions
- Put the name and address of the importer on the label or accompanying documents
- Shall not place defective devices on the market
- Notify the local authorities in case of serious risk
- Keep a register of complaints, nonconforming products, product recalls, and withdrawals
- Forward information to the manufacturer, AR, distributor
- Provide samples of devices to authorities
- Ensure storage and handling conditions preserve device safety and performance requirements
- Comply with the manufacturer's storage and handling instructions
- Register as importer in EUDAMED

System or Procedure Pack Producer (SPPP)

A System or Procedure Pack Producer:

- Draws up a statement declaring that
  - It has verified the mutual compatibility of the devices and if applicable other products;
  - It packaged the system or procedure pack and supplied relevant information to users incorporating the information to be supplied by the manufacturers of the devices or other products which have been put together;

  - The activity of combining devices and, if applicable, other products as a system or procedure pack was subject to appropriate methods of internal monitoring, verification, and validation.
- Assigns a Basic UDI-DI (unique device identifier) and UDI-PI (production identifier).[51]
- Registers as SPPP and obtains a Single Registration Number (SRN) in EUDAMED (when available).
- Registers the device on EUDAMED and submits all information of Annex VI A&B.

Vigilance and Post-market Surveillance

The MDR also clarify the distinction between vigilance and post-market surveillance. The former includes identifying and reporting serious incidents and conducting safety-related corrective actions. It requires direct and efficient cooperation between healthcare professionals, health institutions, manufacturers, and national competent authorities for medical devices. Post-market surveillance involves monitoring the available information to periodically reconfirm that the benefits of the device continue to outweigh its risks.

CE Marking of Conformity (MDR Article 20)

Devices, other than custom-made or investigational devices, that are considered to be in conformity with the requirements of the MDR shall bear a CE mark (European Certificate of Conformity). Medical devices in Class I, which are the less risky devices, generally do not require the involvement of a Notified Body (NB) for their placement on the market. All other devices need a certificate issued by a Notified Body. In these cases, the CE mark is followed by the number of the Notified Body.

Notified Bodies

Notifies Bodies' tasks include:

- Assessing the manufacturer's quality management system;
- Evaluating the technical documentation – sometimes together with product sample verification; issuing CE marking certificates;
- Announced annual surveillance audits;
- Unannounced audits at least every 5 years, with sample testing; and
- Post-market surveillance review.

The list of designated Notified Bodies can be found on the NANDO database.[52] In addition to the evaluation made by the Notified Bodies, certain high-risk devices are subject to additional scrutiny of their clinical files by an independent expert panel with clinical, scientific and technical expertise (MDR Article 54). The new Regulations reinforce the responsibilities of national competent authorities and the Commission in terms of controlling and monitoring devices on the market.

Traceability

A completely new feature of the MDR is the unique device identification (UDI) system (MDR Article 27), which applies to all devices placed on the EU market. The UDI is a barcode, a QR code, or any other machine-readable code. The intent of the UDI is to enhance the identification and traceability of devices and the effectiveness of post-market safety-related activities through targeted field safety corrective actions and better monitoring by competent authorities.

Economic operators shall be able to identify any health institution or healthcare professional to which they have directly supplied a device (MDR Article 25). The UDI should also help to reduce medical errors and fight against falsified devices. Usage of the UDI system should also improve purchasing, waste disposal, and stock management by health institutions and other economic operators;

where possible, UDI should be compatible with other authentication systems already in place in those settings (MDR recital 41).

Identification

Unique device identifiers (UDIs) are used to uniquely and unambiguously identify devices, both individually and when packaged, or in the case of reusable devices by direct marking of the device itself. Each medical device and, when applicable, each level of their packaging will have a UDI that will be indicated on the labels. UDIs are to be added to labels in stages, but will be complete by 2027, depending on the risk class of the device. For Class III implantable devices, health institutions are expected to store and keep – preferably by electronic means – the UDIs of the devices they have supplied, or with which they have been supplied (MDR Article 27(9)). With each implantable device the manufacturer will have to deliver an implant card-carrying-appropriate information. This card, including the patient's identity, shall be supplied to each patient fitted with an implant. Health institutions are expected to allow rapid access to the information contained on the implant card to any patient fitted with a device, unless the type of implant is exempt from this obligation (currently this includes, for example, staples and dental hardware) (MDR Article 18).

EUDAMED Database

The MDR is intended to increase transparency by making the UDI the key to publicly available information on devices and in studies. EUDAMED, the new European database for medical devices and in vitro diagnostic medical devices, will play a central role in making data available and increasing both the quantity and quality of data (MDR Article 33). The EUDAMED database is to be central to EU MDR application. Figure 4.9 illustrates the role of EUDAMED.

The central European database will allow all stakeholders to access basic information on MDs and IVDs, such as the identity of the device, its certificate, the manufacturer, the Authorized Representative, and the importer. The EUDAMED database (MDR Article 92) will adequately inform the public, including healthcare professionals, about:

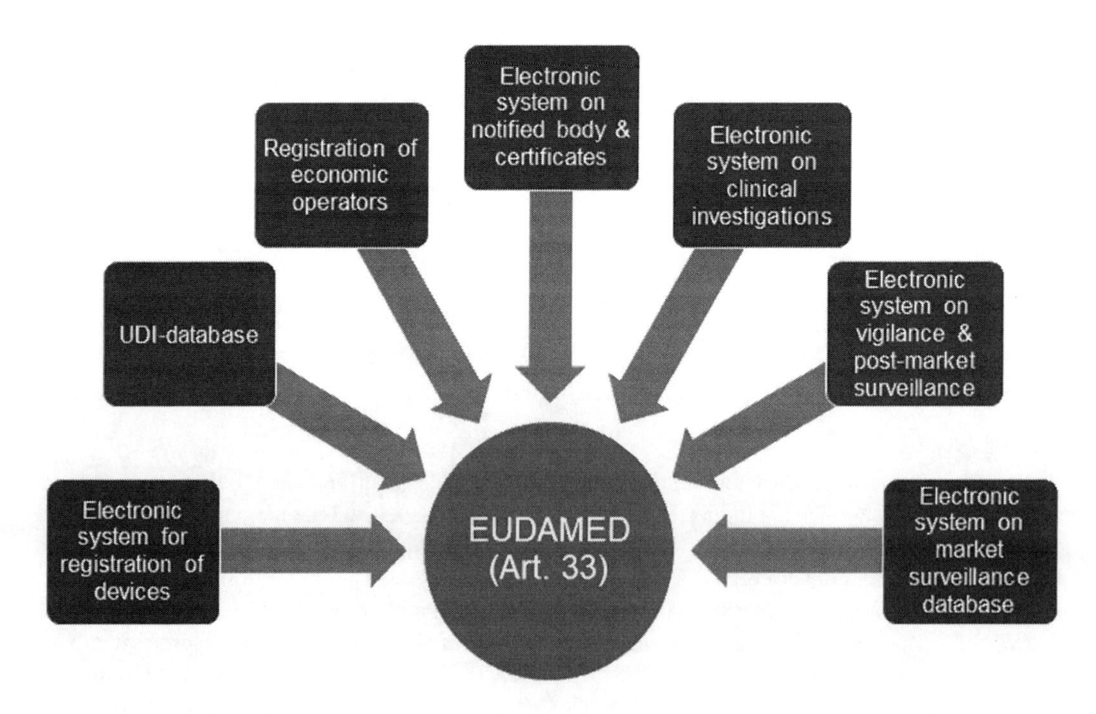

FIGURE 4.9 EUDAMED database.

- Clinical investigation reports on medical devices and performance: study reports on in vitro medical devices; the summaries of the main safety and performance aspects of the device and the outcome of the clinical/performance evaluation; and
- Field safety notices by manufacturers and certain aspects of serious incident reports.

Healthcare professionals can use this information and may receive questions from patients about what they have read in EUDAMED. In addition, member states are expected to take appropriate measures, like organizing targeted information campaigns, to encourage and enable healthcare professionals, users, and patients to report to the competent authorities on suspected serious incidents occurring with devices (MDR Article 87(10)).

Labeling and Instructions for Use

In addition, the MDR improves labeling. New requirements intend to simplify identification of products, finding instructions for use, and getting information about the safety and performance of devices. For example, labels will contain new information, along with symbols showing the presence of hazardous or medicinal substances (MDR Annex I Chapter III(23)). In general, each device is to be accompanied by the information needed to identify it and its manufacturer, and by any safety and performance information relevant to the user, or any other person, as appropriate. The information may appear on the device itself, on the packaging, or in the instructions for use, and, if the manufacturer has a website, is to be made available and kept up to date on the website.

Drug-Device Combination Products under the EU MDR

Under EU MDR Articles 1(8) and (9), there are two types of "combination products" considered under the MDR: Integral and co-packaged. The regulatory requirements for the medical device differ depending on whether or not it is integral.

- Integral: The medicinal product and device form a single integrated product, e.g., prefilled syringes and pens, patches for transdermal drug delivery, and prefilled inhalers.
- Co-packaged: The medicinal product and the device are separate items contained in the same pack e.g., reusable pen for insulin cartridges, tablet delivery system with controller for pain management.

> Article 1(8):
> "Any device which, when placed on the market or put into service, incorporates, as an integral part, a substance which, if used separately, would be considered to be a medicinal product as defined in point 2 of Article 1 of Directive 2001/83/EC, including a medicinal product derived from human blood or human plasma as defined in point 10 of Article 1 of that Directive, and that has an action ancillary to that of the device, shall be assessed and authorized in accordance with this Regulation.
>
> - However, if the action of that substance is principal and not ancillary to that of the device, the integral product shall be governed by Directive 2001/83/EC or Regulation (EC) No 726/2004 of the European Parliament and of the Council, as applicable. In that case, the relevant general safety and performance requirements set out in Annex I to this Regulation shall apply as far as the safety and performance of the device part are concerned."
>
> Article 1(9):
> "Any device which is intended to administer a medicinal product as defined in point 2 of Article 1 of Directive 2001/83/EC shall be governed by this Regulation, without prejudice to the provisions of that Directive and of Regulation (EC) No 726/2004 with regard to the medicinal product.
>
> - However, if the device intended to administer a medicinal product and the medicinal product are placed on the market in such a way that they form a single integral product which is intended exclusively for use in the given combination and which is not reusable, that single integral product shall be governed by Directive 2001/83/EC or Regulation (EC) No 726/2004,

> as applicable. In that case, the relevant general safety and performance requirements set out in Annex I to this Regulation shall apply as far as the safety and performance of the device part of the single integral product are concerned."

As with combination products in the United States, in the European Union, the Primary (Principal) Mode of Action (PMOA) is important in the regulatory framework and cGMP expectations for combination products. As discussed in Chapter 2, though, under EU MDR, drugs and biological products are considered to be medicinal products, and not governed under separate regulations. Medicinal products (or "medicines") are those whose primary mode of action is pharmacological, immunological, or metabolic. Medical devices are those that perform a physical function.

The level of integration between the drug and device constituent parts, together with their PMOA, significantly influences the cGMP framework for "combination products" under EU MDR (See Figure 4.10).

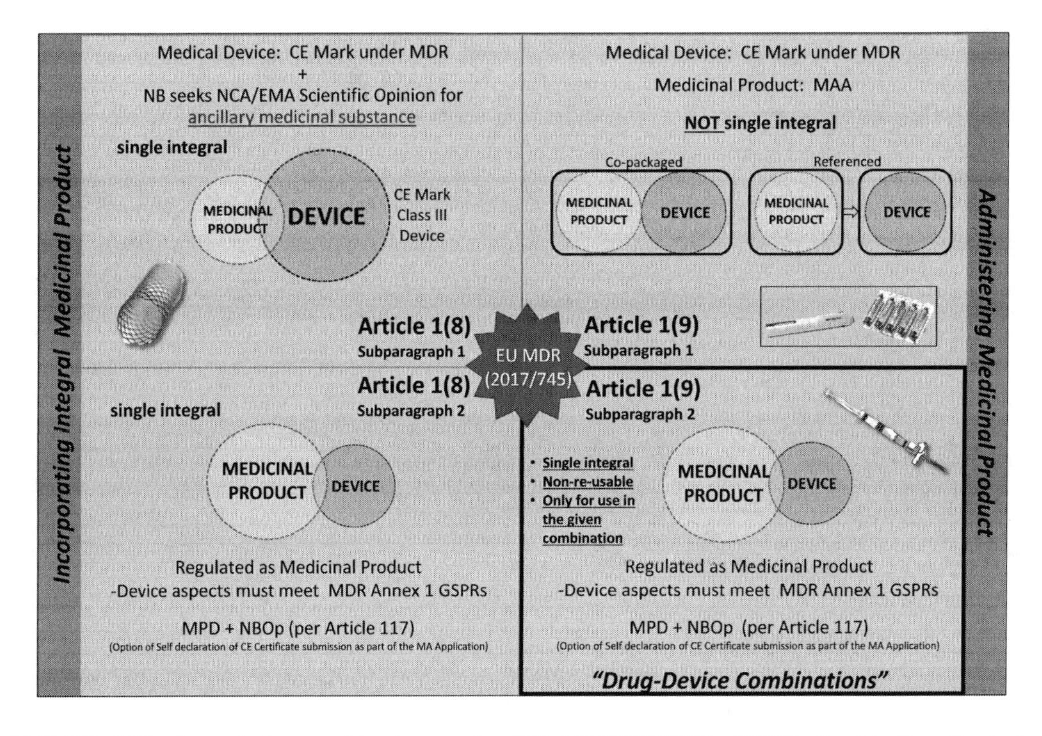

FIGURE 4.10 Drug-device combinations under EU MDR(2017/745).

Single Integral Products MDR Article 117

Drug-Device Combinations (MDR (Art.1(8),(9)) are either categorized/registered as medicinal product ("MP," in Figure 4.10) or as medical device and thus regulated either by the Directive on Medicinal Products (Directive 2001/83/EC)("MPD" in Figure 4.10) or the EU MDR.

However, the medical device part of such a product must at least fulfill the relevant General Safety and Performance Requirements (GSPRs (Annex I of the MDR)) despite the combined product being considered a medicinal product. This is an impactful aspect of EU MDR. ***Compliance to GSPRs now needs to be assessed by a Notified Body for Class Is, Im, IIa, IIb, or III device parts of single integral products.***

- These are **medicinal products** with a medical device part.
- Conformity must be shown either by CE marking of the whole medical device part or by a notified body opinion (NBOp) – as per Article 117.

- NBOp is needed for all new marketing authorization application and submissions for substantial changes (component replacements, intended use extension).
- No UDI or EUDAMED registration/submission is required for these products.

For all single integral products incorporating a device part of Class Is, Im, IIa, IIb, or III, Notified Body involvement is required to confirm compliance of the device part of single-use integral products with relevant General Safety and Performance Requirements (GSPRs):

1. **CE mark certificate and/or Declaration of Conformity (DoC)**

Or, if not available, a

2. **Notified Body Opinion (NBOp)**

Non-integral Drug-Device Combinations (Co-packed)

The term "Non-Integral Drug-Device Combinations" was defined in EMA's Draft Guidance on Quality Requirements for Drug-Device Combinations (DDCs) and adopted in EMA's Q&A on Article 117.[53] A typical example of a non-integral DDC includes a vial kit, composed of a vial with medicinal product, filter/injection needle, vial adaptor, empty syringe, syringe with diluent, alcohol swabs, etc.

These products are not per se subject to Notified Body assessment, they are considered medicinal products subject to Directive 2001/83/EC. This means that the labeling of the DDC itself is not in the scope of the MDR. However, **the devices within the medicinal product packaging are expected to comply *in full* with the MDR.** In addition, the relevant output from the risk management, including the relevant safety information included by the manufacturer of the devices in the IFU, shall be incorporated in the medicinal product labeling, unless it is in conflict with Directive 2001/83/EC.

Tables 4.17–4.20 are intended be a helpful resource for the reader in support of cGMP compliance (and inspection readiness) efforts.

TABLE 4.17
Product Registration and Device Declaration of Conformity Assessment

Format, Content	Format and content specified in Annex IV: • Manufacturer identification including registration number • Issued under sole responsibility of the manufacturer • Device UDI, name, trade name, product code • Device intended use • Risk classification • Conformity statement to EU MDR, but also to other EU laws that require a DoC • Reference to relevant harmonized standards and CS used • Notified Body identification • Conformity assessment procedure • Certificate numbers • Name/place/date/signature
Updates	• DoC must be continuously updated • DoC must be available in local language

TABLE 4.18
Traceability and UDI Requirements

<table>
<tr><td>Trace</td><td colspan="6">• Distributors and importers must cooperate with the manufacturer or AR to achieve the appropriate level of traceability
• Must be able to identify the previous and next economic operator or HCI/HCP to whom they have supplied a device</td></tr>
<tr><td>UDI</td><td colspan="6">• UDI must be available on the label and all higher levels of packaging (excl. shipping containers)
• UDI must be used when reporting AE and FSCA, and provided to the patient for implanted devices
• Basic UDI must be on the DoC
• UDI must be part of the Technical Documentation
• UDI must be part of the traceability data kept by the economic operators</td></tr>
<tr><td rowspan="2">UDI Dates</td><td>MDR Date of Application</td><td>UDI CLASS III & IMPLANTABLE</td><td>EUDAMED REGISTRATION</td><td>UDI CLASS IIA & CLASS IIB</td><td></td><td>UDI CLASS I</td></tr>
<tr><td colspan="2">May 2021</td><td>May 2022</td><td>May 2023</td><td>May 2024</td><td>May 2025</td></tr>
<tr><td>UDI Database</td><td colspan="6">• Will not contain the Production Information part of the UDI or commercially confidential information
• Core data of UDI will be publicly available and can be downloaded from EUDAMED
• Device UDI has to be entered in the UDI Database before the device is placed in the market
• Member states may implement national databases on distributors and importers</td></tr>
</table>

TABLE 4.19
Registration of Devices and Economic Operators

Devices	• Class lll: UDI needs to be assigned before submitting the dossier for approval to the Notified Body • Devices UDI and related information need to be uploaded in EUDAMED before placing on the market
Manufacturer/ Authorized Representative	• Register themselves and devices in EUDAMED before placing devices on the market • The CA will issue a registration number (SRN) to the manufacturer or AR (fee!!) • Changes to registration data have to be updated within 1 week, data accuracy check after 1 year and then every second year • Put the registration number on the Declaration of Conformity • Upload the summary of safety and clinical performance (SSCP) in EUDAMED for implants and Class lll
SSCP (available to the public)	• Device and manufacturer information, UDI • Intended purpose, (contra-)indications, target population • Device description, including parents and accessories • Possible diagnostic and therapeutic alternatives • Reference to harmonized standards and CS • Summary of the clinical evaluation and post-market clinical follow-up • Suggested profile and training for users • Information on residual risks, undesirable side-effects, warnings and precautions

The EU MDR includes expectations around clinical evaluations and clinical investigations. Clinical evaluation is a systematic and planned process to continuously generate, collect, analyze, and assess the clinical data pertaining to a device in order to verify the safety and performance, including clinical benefits, of the device when used as intended by the manufacturer; clinical evaluation is required for all devices. Clinical investigation is systematic investigation involving one or more human subjects, undertaken to assess the safety or performance of a device. A clinical investigation is not always required (depending on the device class and the available safety and performance data). Requirements are summarized in Figure 4.11 and Table 4.21. Clinical data gathered must be scientifically valid, reliable, and robust. There is a strong focus on protection of (vulnerable) subjects, informed consent, and risk minimization. The sponsor and the investigator need to have a

TABLE 4.20
Conformity Assessment Procedure

Full QMS	• Class III: Full quality assurance system and Notified Body assessment of dossier and issue EC product certificate before CE marking • Scrutiny procedure for class III implantable devices
Production QMS	• Class IIb and IIa: Full quality assurance system, self-declaration for CE marking • Class IIb implants: **Notified Body assessment of clinical evaluation before CE marking**
Class I	• Class I: Self-declaration for CE marking Notified Body involved in case of sterile or measuring devices and **for reusable surgical instruments**
NB Power	• Notified Bodies can withdraw the certificate • Notified Bodies may **impose restrictions** to the intended use or **require** manufacturers to initiate **PMCF studies**

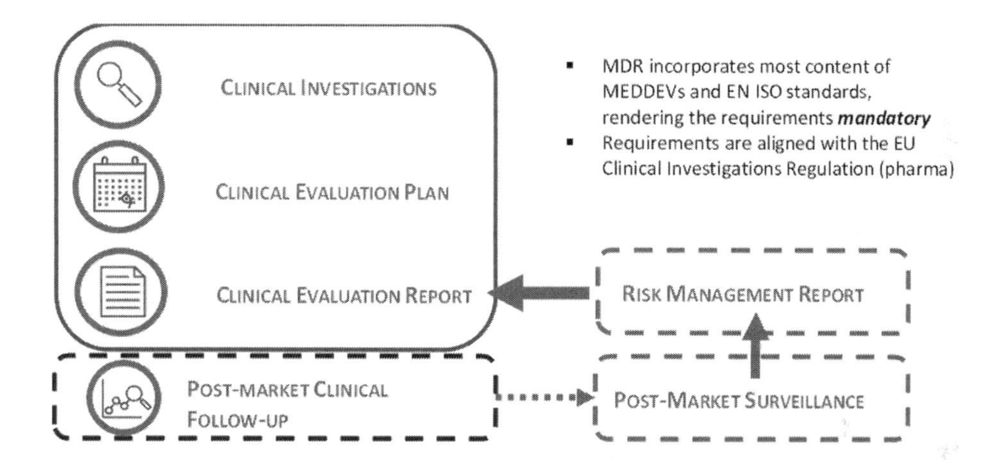

FIGURE 4.11 Clinical evaluation and clinical investigation.

TABLE 4.21
Clinical Evaluation and Investigation

Class	Clinical Evaluation	Clinical Investigation	Expert Panel Involvement	PMCF Plan	PMCF Report
Class I	X	(X)		X**	X**
Class Is/Im	X	(X)		X**	X**
Class IIa	X	(X)		X**	X**
Class IIb	X	(X)		X**	X**
Class IIb active	X	(X)	X*	X**	X**
Class III	X	X¥		X**	X**
Class III implant	X	X¥	X	X**	X**
Implant	X	X¥		X**	X**

() provided that the Clinical Evaluation Report (CER) determines that there is no need for additional clinical data and is duly justified.

¥ Summary of safety and clinical performance (SSCP) also required.

*With exceptions.

**Unless duly justified in technical documentation.

system in place for damage insurance in line with local law. Electronic application in EUDAMED (once it is operational) will deliver a unique number. Changes to the application or to the clinical investigation need to be updated within 1 week.

There are specific notification requirements for clinical termination. Early termination requires notification to the member state(s) and must be justified within 15 days. If there is a safety reason, notification to the member state(s) is required within 24 hours. Normal termination notification is 15 days. The clinical investigation report has to be submitted within 1 year, or within 3 months in case of early termination or halt.

Clinical investigations for "other" reasons (e.g., user preference studies) on CE-marked devices also require ethics committee approval and protection of subjects.

Post-marketing Surveillance (PMS) is also a requirement under EU MDR.[54] PMS requirements are summarized in Figure 4.12 and Table 4.22. A planned post-market surveillance system (PMSS) is part of the Quality Management System to gather data on the quality, safety, and performance of the device during the entire lifecycle. Post-market surveillance plan per device (group) is part of the Technical File.

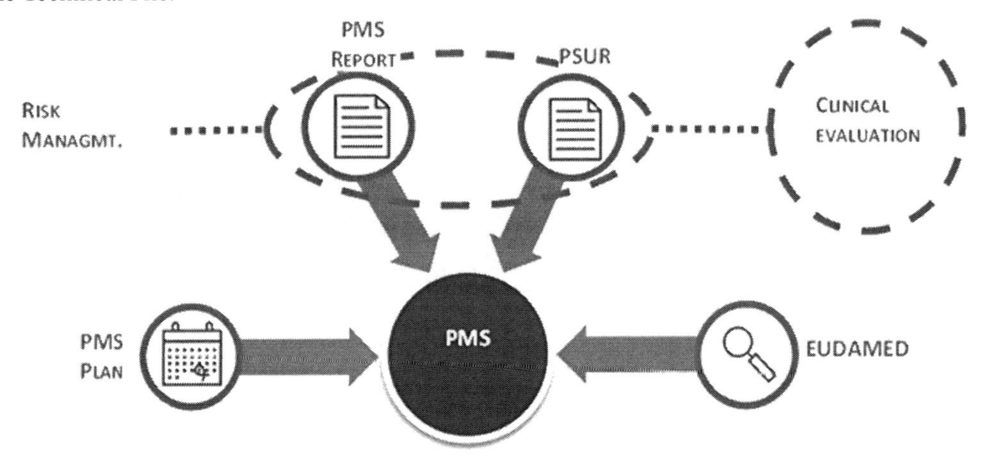

FIGURE 4.12 Post-marketing surveillance (PMS).

TABLE 4.22
Post-market Surveillance Requirements under EU MDR

Device Class	PMS Plan	PMS Report	PSUR	EUDAMED PMS Module
Class I	X	X		
Class Is/Im	X	X		
Class IIa	X		X (minimum every 2 years)	
Class IIb	X		X (minimum annually)	
Class IIb active	X		X (minimum annually)	
Class III	X		X (minimum annually)	X
Class III implant	X		X (minimum annually)	X
Implant	X		X (minimum annually)	X

Data gathered must feedback into:

- The Risk Management Report, Benefit/Risk Assessment, Design, Manufacturing, Instructions for Use (IFU), and labeling;
- Clinical evaluation;

- Summary of safety and clinical performance; and
- Trend reporting.

A Post-Market Surveillance Update Report (PSUR) is expected per device/device group with risk-based update frequency. For Class I devices, the PMS report is updated when necessary and according to manufacturer's plan. The periodic safety update report (PSUR) for Class IIa devices is to be updated when necessary. For Class IIb and Class III devices, the periodic safety update report (PSUR) is to be updated annually and sent to the Notified Body for evaluation. For Class III and implantable devices, the PSUR has to be sent to the Notified Body for assessment. The PSUR includes analysis of post-market surveillance data, with description of preventive and corrective actions, conclusion of the benefit/risk evaluation, main findings of post-market clinical follow-up (PMCF) report, as well as sales volumes, estimate of the population using the device, and usage frequency of the device.

Table 4.23 summarizes Vigilance reporting requirements.

TABLE 4.23
Vigilance Reporting

Report	• Serious incidents (except known side-effects) • Field Safety Corrective Action (FSCA)
Timing	• Serious incident: 15 day reporting • Serious public threat: 2 day reporting • Death or unanticipated serious deterioration in state of health: 10 day reporting
Summary Reports	When agreed with the coordinating competent authority: • For similar incidents with known root cause or FSCA implemented • For common, well-documented incidents
Report To	• Centralized electronic reporting in EUDAMED

With respect to Trend Reporting and Field Safety Corrective Actions (FSCA), there is mandatory reporting of statistically significant increase in frequency or severity of non-serious incidents or expected side-effect(s) that could impact risk–benefit ratio. The Sponsor needs to define "statistically significant increase" proactively in the Technical File as part of the Post-Market Surveillance Plan for the device. Competent Authorities may ask manufacturers for corrective actions, and will inform the Notified Body, other member states, and the EU Commission.

In case of FSCA, the Competent Authority (CA) may perform their own risk assessment, and the manufacturer has to provide the supporting documentation. The CA may intervene in the manufacturer's investigation. The Field Safety Notice needs to contain the UDI and the manufacturer's SRN. The information needs to be uploaded in EUDAMED. Of note, the EU Commission will perform trending and signal detection based on the data in EUDAMED.

Uploads in EUDAMED need to happen by the end of the transition period, or, if EUDAMED is not finalized by then, within six months after EUDAMED is released:

- Economic operator information should be checked after 1 year and then every second year;
- Summary of Safety and Clinical Performance (class III and implants) should be uploaded annually; and
- Application for clinical investigation, clinical investigation report and summary, and post-market clinical studies should all also be uploaded.

UDI and device registration in EUDAMED need to happen within 18 months after the end of the transition period, or, if EUDAMED is not finalized by then, within 18 months after EUDAMED is released.

The EU MDR has multiple Annexes. These are summarized in Table 4.24.

TABLE 4.24
EU MDR Annexes

Annex	Content	New	Significant Change Compared to MDD
I	General Safety and Performance Requirements (GSPRs)		X
II	Technical Documentation (content)	X	
III	Technical Documentation on Post Market Surveillance	X	
IV	EC Declaration of Conformity (content)	X	
V	CE Mark of Conformity		
VI	Registration of Economic Operators, Devices, and UDI	X	
VII	Requirements to be Met by Notified Bodies		X
VIII	Classification Criteria		
IX	Conformity Assessment QMS and Review of Technical Documentation		X
X	Conformity Assessment Based on Type Examination		
XI	Conformity Assessment Based on Product Conformity Verification		
XII	Procedure for Custom-Made Devices		
XIII	Certificates Issued by a Notified Body		
XIV	Clinical Evaluation and Post-Market Clinical Follow-up (PMCF)	X	
XV	Clinical Investigations		X
XVI	List of Products (devices without medical purpose)	X	

Demonstrating conformance to Annex 1, GSPRs, is a major focus of drug-device combination product manufacturers. GSPRs are summarized in Table 4.25. (The reader should observe that, like the device-called-out provisions under US Combination Product cGMPs under Part 4, there is a strong emphasis on risk management in the GSPRs.)

If your organization is preparing for a cGMP inspection under EU MDR, the following suggestions may be helpful in your inspection preparation. Prior to starting, you should put a plan in place. The steps below will guide you through the main topics:

1. General Safety and Performance Checklist:
   Prepare a checklist for the General Safety and Performance Requirements (GSPRLs).
2. Technical File:
   Technical file update according to the MDR requirements.
3. Risk management according to ISO 14971:
   The risk management activities (plan, analysis, and report) should be aligned with your PMS and PMCF activities.
4. UDI System:
   Each medical device needs a UDI-DI (Unique Device Identification – Device Identifier) and UDI-PI (Unique Device Identification – Production Identifier) and must be submitted and transferred to the UDI database (see document from the EU Commission).
5. Post-Market Surveillance (PMS):
   Post-market surveillance is defined in chapter VII of the MDR.
6. Post-Market Clinical Follow-Up (PMCF) & 5.7 Clinical Evaluation:
   Device Manufacturers shall conduct a clinical evaluation in accordance with the requirements set out in Article 61 and Annex XIV, including a PMCF as described in MEDDEV 2.7.1 Rev.4.

7. Labeling:

   Device manufacturers shall ensure that the device is accompanied by the information set out in Section 23 of Annex I in an official Union language(s) determined by the Member State in which the device is made available to the user or patient. The particulars on the label shall be indelible, easily legible, clearly comprehensible to the intended user or patient.

8. EUDAMED registration:

   Currently, the EUDAMED is not online but a first FUNCTIONAL SPECIFICATION version is available. The initial release was March 2020, but the European Commission has decided to change the initial release up to May 2022 (see decision letter). Any changes to the release date will be communicated with our Regulatory Intelligence Paper.

9. Common Specifications:

   Take care about upcoming specification updates. We keep you informed with our Regulatory Intelligence Paper. Furthermore, this paper helps to address regulatory activities according to ISO 13485:2016, Clause 5.6 Management Review.

TABLE 4.25
General Safety and Performance Requirements

GENERAL REQUIREMENTS
• Devices must be safe and effective • Devices must be "state of the art" • Risks must be reduced as far as possible, without adversely affecting the risk/benefit ratio
Risk management system is a continuous process throughout the lifecycle of the device, requiring regular systematic update • Risk management plan per device • Identify hazards, associated risks taking into account production and post-production phases • Implement actions to reduce risk where necessary
Risk control measures priority: • Design or manufacturing changes to eliminate or reduce risk • Build in protection measures for risks that cannot be eliminated • Provide information for safety and training Information on residual risk is not a risk control measure
Use error risk reduction has to take both the type of user and the use environment into account
HAZARDOUS SUBSTANCES
Devices containing hazardous substances in a concentration over 0.1% need justification • Carcinogenic, mutagenic or toxic to reproduction (CMR) • Endocrine disruptors
Justification needs to consider: • Patient exposure • Evaluation of possible alternatives • Need for the particular substance for performance of the device • Discussion of target patients • Scientific Committee guidelines
• EU Commission has to issue guidelines on the use of phthalates by end of transition period, and update every 5 years • Scientific Committee has to issue guidelines on the use of other CMR
Labeling • Identification of hazardous substances on the label • Residual risks and appropriate measures for vulnerable patients in Instructions for Use
DEVICE LABELING
When the manufacturer has a website, the relevant safety and performance information from the labeling has to be posted on the website, in addition to the labels provided with the device

(*Continued*)

TABLE 4.25 (CONTINUED)
General Safety and Performance Requirements

New (additional) labeling requirements: • Hazardous substances • UDI Indication that the device is a medical device
Qualitative composition and quantitative information for devices dispersed/absorbed in the body
Information on the sterile packaging: • Indication that the package is sterile • Declaration that the device is sterile • Method of sterilization • Name and address of manufacturer • Device description • Month and year of manufacture • Time limit for using the device safely • Instruction to check IFU if package is damaged
DEVICE INSTRUCTIONS FOR USE (IFU)
• All elements of 23.4 need to be addressed in the IFU, unless the requirement is not applicable to the device • IFU needs to bear date of issue and identifier of revision
New (additional) requirements for IFU: • Specification of target groups • Indications and contra-indications • Expected clinical benefit • Link to the summary of safety and clinical performance (SSCP) • Performance characteristics • Levels of disinfection required and all available methods for achieving those levels of disinfection • Requirements for facilities, user training, or qualifications • Reusable devices: – Method of resterilization "suitable to the Member State" – Information to identify when the device should no longer be reused
New (additional) requirements for IFU: • Warnings and contra-indications: – Malfunction or changes in performance – Exposure to external influences or environmental conditions – Interference with other devices – Precautions regarding CMR and endocrine disruptors – Interaction with other devices, drugs or other substances for dispersed/absorbed devices, and undesirable side-effects and risks related to overdose – Overall qualitative and quantitative information on materials and substances to which the patient is exposed in case of implantable devices • Safe disposal of the device – Microbial hazards – Physical hazards (sharps) • Circumstances when a lay user needs to consult an HCP • Notice that any serious incident needs to be reported to the manufacturer and the competent authority • Patient information leaflet

COMBINATION PRODUCTS' CGMPS POST-BREXIT: UNITED KINGDOM

Classification of Drug-Device Combinations into integral and non-integral "combination products" is similar in the UK post-Brexit. Integral DDC products are regulated as medicines under the Human Medicines Regulations 2012 (as amended). In addition to this, under regulation 8 of the UK Medical Device Regulations 2002 (UK MDR 2002), integral DDCs must meet the relevant

Essential Requirements (ERs) set out in Annex I of EU Directive 93/42/EEC (as modified by Part 2 of Schedule 2A to the UK MDR 2002) with respect to the safety and performance-related features of the device component. Evidence of compliance of the device component of the integral DDC with relevant essential requirements (ERs) should be provided in a Sponsor's Market Authorization Application (MAA). It is up to the Applicant to decide how these ERs will be demonstrated to be met. A Notified Body opinion from an EU Notified Body can be accepted in the GB MAA to demonstrate evidence of compliance with the essential requirements of the UK MDR.

SUMMARY

Good manufacturing practice regulations center on proactive, risk-based approaches to ensure that manufacturers' products are consistently produced and controlled according to quality standards, with an aim to ensuring product safety, efficacy, and usability. Adherence to cGMPs emphasize a strong quality management system, appropriate raw materials, robust operating procedures, detecting and investigating product quality deviations, and maintaining reliable testing laboratories. Specific to combination products, key factors to success include (1) collaboration with suppliers; (2) ensuring engagement of all relevant expertise throughout the product lifecycle; and (3) compliance to obligations for cGMPs under Part 4 for all facilities where the whole combination product/multiple constituent parts are brought together to create the combination product. When it comes to combination products, best practices warrant that the cGMPs address the risk-based considerations of each constituent part, as well as those raised by the intended combined use of each of the constituent parts.

In this chapter, we reviewed cGMP expectations and best practices for combination products, highlighting those in the United States and the European Union. In both cases, cGMPs are risk-based. There are opportunities to harmonize cGMP expectations across regions. Fundamentally, as regulators and industry strive to ensure safe and effective products, the scientific and technological considerations that underlie the controls required should not differ substantially region to region and country to country for the very same combined use application. US FDA is making a step toward harmonizing the device cGMPs by referring to ISO 13485:2016 by reference in its 2022 Quality Management System Requirements Proposed Rule. International recognition of ISO 14971:2019 as a harmonized standard for Risk Management is another positive step forward. Perhaps, building on that, there will be harmonized recognition of AAMI TIR 105:2020, Combination Products Risk Management, and ideally, not far behind, harmonization of broader Combination Product cGMPs.

NOTES

1. Medical product is a generic term for any product used to diagnose, cure, mitigate, treat, prevent, or manage a disease, including any medical device, drug, or biological product (or medicinal product).
2. World Health Organization (Nov. 2015) https://www.who.int/medicines/areas/quality_safety/quality_assurance/gmp/en/).
3. 21 CFR Part 4, at https://www.ecfr.gov/current/title-21/chapter-I/subchapter-A/part-4?toc=1.
4. FDA Final Guidance "Current Good Manufacturing Practice Requirements for Combination Products"(January 2017) at https://www.fda.gov/regulatory-information/search-fda-guidance-documents/current-good-manufacturing-practice-requirements-combination-products.
5. The FDA issued a elaboration and clarification of some flexibilities also addressed in the final guidance, as well as how/when to engage the US FDA in an update to the Federal Register: "Alternative or Streamlined Mechanisms for Complying with the Current Good Manufacturing Practice Requirements for Combination Products: List under the 21st Century Cures Act" (September 13, 2022) https://www.federalregister.gov/public-inspection/2022-19713/alternative-or-streamlined-mechanisms-for-complying-with-the-current-good-manufacturing-practice.
6. Based on 21 CFR Part 4, subsequent US FDA Guidance, "Current Good Manufacturing Practice Requirements for Combination Products" (January 2017) and Technical Report AAMI TIR 48, "Quality Management System (QMS) Recommendations on Application of the US FDA's cGMP Final Rule on Combination Products."

7. *See also Chapter 14, "the Global Evolving Regulatory Landscape" for additional helpful discussion on this topic.*
8. Several jurisdictions globally regulate drugs and biological products together, under the umbrella of 'medicinal products.' In these jurisdictions, e.g., EU, drug/biological products are not considered combination products, because they are not regulated differently. Recall a combination product includes two or more *differently* regulated constituent parts. (*See also Chapter 2, "What's a Combination Product," Chapter 14 on "Global Evolving Regulatory Landscape", and the Appendix of this book).*
9. Human cellular, tissue, and cellular tissue-based products.
10. Component manufacturers are encouraged to use appropriate provisions of the cGMP regulations as guidance.
11. ICH Q8, Pharmaceutical Development, and ICH Q9, Quality Risk Management, together with ICH Q10, Pharmaceutical Quality System, form a robust foundation for pharmaceutical lifecycle management.
12. A cross-labeled combination product is defined under 21 CFR 3.2(e)(3) and (4) as: (3) A drug, device, or biological product packaged separately that according to its investigational plan or proposed labeling is intended for use only with an approved individually specified drug, device, or biological product where both are required to achieve the intended use, indication, or effect and where upon approval of the proposed product the labeling of the approved product would need to be changed, e.g., to reflect a change in intended use, dosage form, strength, route of administration, or significant change in dose; or (4) Any investigational drug, device, or biological product packaged separately that according to its proposed labeling is for use only with another individually specified investigational drug, device, or biological product where both are required to achieve the intended use, indication, or effect.
13. Under 21 CFR 210, it states, "If a person engages in only some operations subject to the regulations in this part, in parts 211, 225, and 226 of this chapter, in parts 600 through 680 of this chapter, and in part 1271 of this chapter, and not in others, that person need only comply with those regulations applicable to the operations in which he or she is engaged."
14. Under 21 CFR 210, it states that the biological product cGMPs are intended to supplement the drug cGMPs, unless the regulations explicitly indicate otherwise. In the event there is a conflict between applicable regulations, the regulation specifically applicable to the drug product in question supersedes the more general expectations.
15. "Alternative or Streamlined Mechanisms for Complying with the Current Good Manufacturing Practice Requirements for Combination Products: List under the 21st Century Cures Act" (September 13, 2022): https://www.federalregister.gov/public-inspection/2022-19713/alternative-or-streamlined-mechanisms-for-complying-with-the-current-good-manufacturing-practice.
16. Management with Executive Responsibility are the senior employees of a manufacturer who have the authority to establish or make changes to the manufacturer's quality policy and quality system.
17. Critical Quality Attributes, Critical Material Attributes, Critical Process Parameters. See Chapters 5 and 6 on integrated development and risk management for more on this topic).
18. "Special Instructions Concerning Design Controls" in Compliance Program 7383.001 (https://www.fda.gov/media/82616/download) and Compliance Program 7382.845 (https://www.fda.gov/media/80195/download) accessed September 3, 2022.
19. "Guidance for Industry on Q8(R2) Pharmaceutical Development" (November 2009), accessed September 3, 2022 at https://www.fda.gov/regulatory-information/search-fda-guidance-documents/q8r2-pharmaceutical-development
20. An FDA Form 483 is issued to an organization's management at the conclusion of an inspection when an investigator(s) has observed any conditions that in their judgment may constitute violations of the Food Drug and Cosmetic (FD&C) Act and related Acts. Each observation noted on the FDA Form 483 is made when in the investigator's judgment, conditions or practices observed would indicate that any food, drug, device, or cosmetic has been adulterated or is being prepared, packed, or held under conditions whereby it may become adulterated or rendered injurious to health. Lack of procedures for required cGMP activities is considered as such a condition whereby a combination product may become adulterated.
21. ANSI/AAMI 75:2009, ISO IEC 62366: 2015, and FDA Guidance on Combination Products Human Factors.
22. The same holds true for an organization who may apply a device cGMP-based QMS operating system. The combination product manufacturer does not have to have both a DMR and MBR. The MBR information can be embedded within the DMR.
23. Product from Product/Process Qualification stage is sometimes used "at risk"; if the PQ fails, the design validation activity likely requires re-manufacture and a re-do of the design validation effort.
24. Investigational New Drug Application.

25. Investigational Device Exemption; IDE (21 CFR §812) must comply with Design Controls (21 CFR 820.30), but are exempt from other QS Regulations; ISO 14969:2004 briefly discusses Clinical Evaluation (section 7.3.6). It refers to additional guidance in ISO 14155-1, *Clinical investigation of medical devices for human subjects – Part 1: General requirements:* "Testing by users" could be performed by actual users, or by people with knowledge /abilities /training that is representative of users (i.e., not typically by developers).
26. Under the 2/22/2022 proposed amendments to 21 CFR 820, ISO 13485:2016 Clauses 4.2.4 and 4.2.5 similarly address control of documents.
27. The phrasing in the regulation is a bit confusing; this is what it means.
28. During a combination products inspection (*see chapter on Inspection Readiness*), performance or quality issues identified based on customer complaints or CAPAs that are supplier related may be pointers to potential gaps in Purchasing Controls. The authors has observed FDA investigators audit not just the direct supplier of a Sponsor, but even the Sponsor's supplier's supplier being subject to inspection, where the criticality and risk of a given supplier's product or services warrant the extra scrutiny. Qualifying a supplier as having a quality system that is suitable and capable includes the supplier having appropriate controls and vetting of their own suppliers.
29. FDA Guidance "Contract Manufacturing Arrangements for Drugs: Quality Agreements" (November 2016) accessed September 22, 2022, at https://www.fda.gov/media/86193/download.
30. Per Compliance Program 7256.000, syringe components that are prefilled with the drug constituent part would be subject to 21 CFR 211.84 *and* to 21 CFR 820.50. Materials used solely for manufacture of a device constituent part *that is not part of the drug container or closure* (e.g., a co-packaged syringe with a drug vial) would not be subject to 21 CFR 211.84. These materials that are used *solely* for manufacture of the device constituent part would be subject to 21 CFR 820.50.
31. Guide to Inspections of Quality Systems, Quality System Inspection Technique (QSIT). Refer to the QSIT section on CAPA; Compliance Program 7382.845 – Inspection of Medical Device Manufacturers; Compliance Program 7383.001 – Medical Device PMA Preapproval and PMA Post-market Inspections; Section IV.A.4, FDA Guidance for Industry and FDA Staff – Current Good Manufacturing Practice Requirements for Combination Products.
32. Non-called-out provisions are helpful to consider, but FDA doesn't enforce compliance with them. They may be helpful to review to inform understanding of how FDA may interpret corresponding base set requirements for the combination product, and the reader should take seriously what FDA can require under the base set of cGMPs, given the breadth of expectations, even if specifics of a non-called-out provision aren't replicated.Needless to say, you need to be able to demonstrate that your approach is sound.

    Under CAPA, three such helpful non-called-out provision include 21 CFR 820.250 Statistical Techniques, 21 CFR 820.70, Production & Process Controls, and 21 CFR 820.90, Nonconforming Product. Application of appropriate statistical methods is used to detect recurring quality problems, as needed. Production and Process Controls are useful to support, among other things, the device constituent part and the combination product process validation activities. The Nonconforming Product provision is one that combination product manufacturers with drug cGMP-based Quality Systems should consider within their quality system to support effectively implementing CAPA for their combination products. 21 CFR 820.90 states that each manufacturer must establish and maintain procedures to control product that does not conform to specified requirements.
33. https://www.ecfr.gov/current/title-21/chapter-I/subchapter-H/part-820/subpart-I/section-820.90.
34. An important clarification: "Installation" is not "implantation." Installation refers to the process for making hardware and/or software ready for use. Implantation is the act of inserting or fixing (tissue or an artificial object) in a person's body, e.g., by surgery.
35. Preamble to the Quality System Regulation, Comments 178–179 (61 FR 52636 – 52637), Comments 199–202 (61 FR 52640) accessed September 3, 2022 at https://www.fda.gov/medical-devices/quality-system-qs-regulationmedical-device-good-manufacturing-practices/medical-devices-current-good-manufacturing-practice-cgmp-final-rule-quality-system-regulation.
36. 87 FR 10119: 10119-10134 (16 pages) at https://www.federalregister.gov/documents/2022/02/23/2022-03227/medical-devices-quality-system-regulation-amendments.
37. Some medical devices are produced using continuous manufacturing processes. If the drug constituent part is introduced to the medical device in the midst of the continuous manufacturing process, calculation of yield may apply to the entire continuous manufacturing process, not just the step where the drug is introduced. A related control strategy should therefore take into account the entire process. If a Sponsor has specific questions on how "calculation of yield" applies to their specific product and production process, they are encouraged to seek clarification from FDA proactively.

38. See example warning letter citations throughout this chapter.
39. [211.132(b)(1)], e.g., toothpaste co-packaged with a toothbrush, dermatological drug prefilled into a delivery device; and [211.132(c)(1)], e.g., propellant-based aerosols and saline nasal sprays are exempt from tamper-evident features on the package.
40. Exemptions apply: [21 CFR 211.137 (e)–(h)], e.g., homeopathic drug products and investigational use products.
41. FDA Guidance "Shelf Life of Medical Devices" (April 1991), accessed September 13, 2022 at https://www.fda.gov/media/72487/download.
42. For example, ISO 11608-1:2022 Section 10.3.8 and Annex D make reference to preconditioning for functional stability testing.
43. "Guidance for Industry Q1D Bracketing and Matrixing Designs for Stability Testing of New Drug Substances and Products," January 2003. https://www.fda.gov/media/71720/download.
44. "Guidance for Industry Quality of Biotechnological Products: Stability Testing of Biotechnological/Biological Products" Q5C, July 1996. https://www.fda.gov/media/71441/download.
45. https://www.federalregister.gov/documents/2022/09/13/2022-19713/alternative-or-streamlined-mechanisms-for-complying-with-the-current-good-manufacturing-practice.
46. Quality Systems Inspection Technique: Product & Process Controls Subsystem and Sterilization Process Controls, accessed September 8, 2022 at https://www.fda.gov/media/76038/download.
47. FDA Final Guidance, "Current Good Manufacturing Practice Requirements for Combination Products" (January 2017), accessed September 8, 2022 at www.fda.gov/media/90425/download.
48. Current Good Tissue Practice (CGTP) and Additional Requirements for Manufacturers of Human Cells, Tissues, and Cellular and Tissue-Based Products (HCT/Ps) (CGTP Guidance) (December 2011), accessed September 8, 2022 at https://www.fda.gov/media/82724/download.
49. Repair, reconstruction, replacement, or supplementation of a recipient's cells or tissues with an HCT/P that performs the same basic function or functions in the recipient as in the donor (21 CFR 1271.3(c)).
50. The IVDR (2017/746) was also adopted in May 2017, replacing the In Vitro Diagnostic Medical Devices Directive (98/79/EC)(IVDD). The IVDs are not the focus of this book.
51. https://health.ec.europa.eu/medical-devices-topics-interest/unique-device-identifier-udi_en.
52. https://economie.fgov.be/en/themes/commercial-policy/technical-barriers/database-notified-bodies-nando.
53. "Guideline on quality documentation for medicinal products when used with a medical device" (effective 01/2022) accessed September 8, 2022 at https://www.ema.europa.eu/en/quality-documentation-medicinal-products-when-used-medical-device.
54. Chapter 8 includes discussion of 21 CFR Part 4B, Post-Marketing Safety Reporting for Combination Products in the United States.

5 Combination Products Integrated Development

Susan W. B. Neadle

CONTENTS

INTRODUCTION

Medicinal product[1] prefilled syringes, nasal sprays, metered-dose inhalers, antimicrobial wound dressings, scaffolds for tissue engineering, antibody–drug conjugates, transdermal and iontophoretic drug delivery systems, antibacterial-releasing dental restorative materials, medicinal product-coated stents, medicinal-eluting contact lenses, antibiotic-loaded bone cements, liquid medication co-packaged with dose dispensers, photodynamic therapy (e.g., a light-emitting device and a drug activated by that light) – these are just a few examples of the vast array of combination products that have been, and continue to be, developed, approved, and marketed to benefit the patients that we serve, given the increasing convergence of life sciences, pharmaceutical sciences, engineering, and technology (Figure 5.1). Couple these with connected health, the promise of improved healthcare delivery through digital technologies and the possibilities become mind-boggling. Combination products are designed to offer greater benefits than drugs or devices acting alone. Their development and manufacture require an understanding of each of the combination product constituent parts and their interactions, impacting the safety and efficacy of the combination product as a whole. Companies engaged in this space can no longer afford to think of themselves simply as manufacturers of a singular type of medical product. This mindset shift applies across the lifecycle of combination products, from user needs identification through product development and

DOI: 10.1201/9781003300298-6

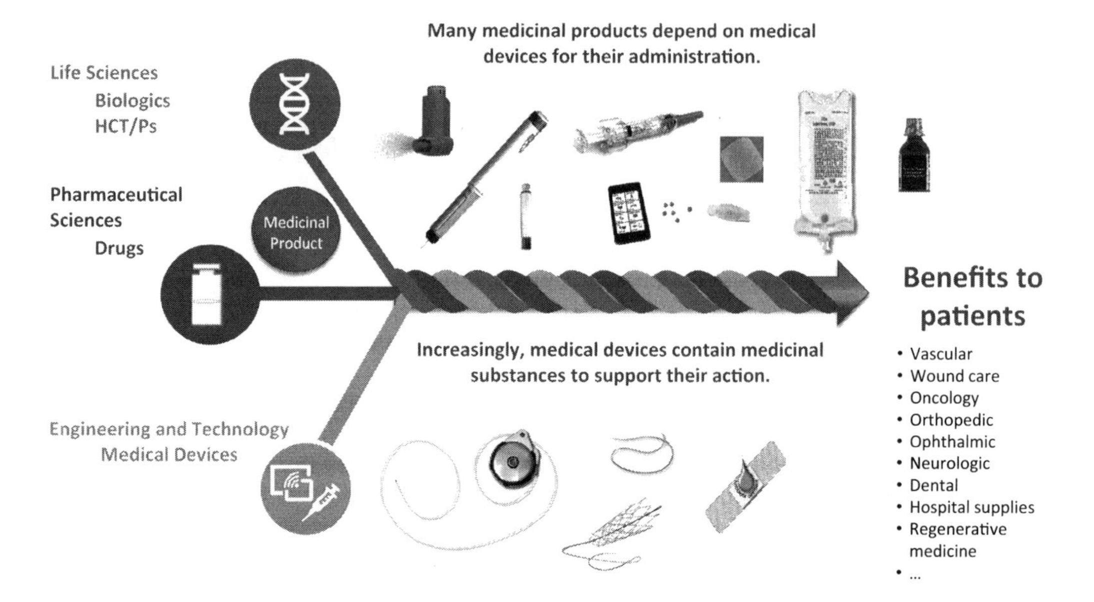

FIGURE 5.1 Combination Products: Convergence of Life Sciences, Pharmaceutical Sciences, Engineering and Technology to benefit the patients that we serve (©2022 Combination Products Consulting Services, LLC. All Rights Reserved).

post-marketing considerations. In this chapter, we focus on practices for successful combination product development.

KEYS TO COMBINATION PRODUCT DEVELOPMENT SUCCESS

Above and beyond the foundational need for technology that works, entering into the development of combination products requires awareness of a range of hurdles. These hurdles might simply be categorized as **language**, **process**, **culture**, **and supplier collaboration**. The simplicity of these category labels can be deceiving, however. The proverbial saying that the "devil is in the details" applies. Let's explore these hurdles, or more positively framed, these "keys to success" for combination product development.

LANGUAGE

The concept of "Forming, Storming, Norming and Performing" (FSNP) describes the typical stages of team development as they work on a project.[2] Anyone who has experienced FSNP as teams move through each stage to learn to work together, overcome challenges, and eventually focus on accomplishing a shared goal, is likely well-versed in the opportunities project teams face with respect to effective communication. These challenges are magnified in organizations accustomed to developing only one type of medical product, now engaged in combination product development, because there is a factor at play that many do not initially recognize: Vocabulary, i.e., the different dictionary definitions for device and medicinal product professionals. Cross-functional Teams need to become keenly aware that there is inconsistent interpretation of terminology by those viewing development through the lens of "pharma" and those viewing development through the lens of "device." We'll refer to these lenses here as the "Pharma World" and the "Device World." The inconsistency in interpretations extends beyond the internal stakeholders to external stakeholders, like suppliers, and even health authorities, who have different interpretations of the same vocabulary words between "device" and "drug" within regulatory bodies and across jurisdictions. The differences in interpretation can lead to breakdowns throughout product development and even post-market lifecycle management. **Communication with a focus on ensuring consistency of and alignment on interpretation is key to establishing a high-performing team**. Table 5.1 includes some important examples of the "Combination Products Dictionary."

In summary, the people engaged in combination product development – and post-market lifecycle management – need to become "multilingual." They need to develop fluency in the "languages" of drug, biologic, and medical devices. The varying interpretations of vocabulary can lead to confusion, poor planning, missed deliverables, delay in programs, and even duplication of effort.

Consider the example where one is defining procedures around packaging and supplier qualification, only to realize that the device development project sub-team interprets those procedures completely differently than does the pharmaceutical project sub-team, because they each believe the procedure applies to a different part of the product!

Suppliers likewise may struggle with vocabulary differences. Does the device constituent part supplier view themselves as "only" a component supplier, when in fact, that supplier is now considered the manufacturer of your device constituent part? Such a situation requires supplier education, collaboration, and likely updates to Supplier Quality Agreements. Is the combination product market authorization holder using a supplier to do design specification development – which means that supplier meets the definition of "combination product manufacturer"? That has implications for the quality management system expectations for that supplier, and in turn, the supplier qualification.

Recognition of the differences in the interpretation of "Product Development Process" and even just the term "product" will help avoid confusion. Often, project sub-teams working on one constituent part of the combination product tend to underestimate the scope of work required to develop the other constituent part. Updates to project timelines to address and integrate the different interpretations of "Product Development Process" deliverables for the drug constituent part, the device constituent part, and the combined use system are essential (this will be addressed further within the body of this chapter).

The confusion in language leads to shifts in establishment registration,[3] to differences in substantive product expectations, to varying premarket review pathways, and to lack of harmonized submissions formats and expectations. It can even lead to confusion during audits and health authority inspections. Being aware of, and attuned to, the language differences is the first step in the path to successful combination product development.

Integrated Process

Combination products are comprised of two or more differently regulated constituent parts for combined use (e.g., a medicinal product and a medical device).[4] In a pharma-centric organization, the development scientists and engineers eagerly focus on the technical challenges of developing the drug within their area of expertise, while less focused on the device, or perhaps relying on a third party for potential purchase of the device constituent part. In a device-centric organization, the engineers jump in to develop a device design and may very well purchase the drug constituent part. Some organizations manufacture and market both constituent parts. Each organization is faced with the challenge of bringing the constituent parts together for safe and effective combined use.

A successful combination product organization will benefit from establishing a development model where innovation for each constituent part and their combined use deliverables are aligned throughout an integrated development process, for timely realization of the overall intended therapeutic effect/intended use of the combination product.

Take the case of a biotech- or pharma-centric organization: It will benefit from having a dedicated device team, an accompanying device development process model, and device quality system. This device infrastructure should not be siloed from the rest of the organization, nor treated as an afterthought. Proactive integration of the device development deliverables and open device/pharma

TABLE 5.1
A Sampling of the Combination Products Dictionary

Terminology	"Device World" Interpretation	"Pharma World" Interpretation	"Combination Product" Interpretation
What we are looking at:	ON BODY INJECTOR		
Packaging:	The configuration of materials designed to: • avoid physical damage, biological contamination, and any other external disturbance to the medical device or assembly components, from time of assembly to point of use; • support the proper identification and use of the device by means of labeling, such that at any time point before the end of shelf life, the device can be used safely and efficaciously.	The combination of components necessary to contain, preserve, protect and deliver a safe, efficacious drug product, such that at any time point before expiration date of the drug product, a safe and efficacious dosage form is available[a].	In the pharma world, the "packaging" (container-closure) is considered part of the dosage form; In the combination products/device world, if the medicinal product container-closure both holds and protects the medicinal product ***and*** delivers a metered dose of (or serves a transfer function for) that medicinal product, it may ***also*** be considered a device. Be careful of the implications of confusion on "packaging" terminology; some may not perceive the criticality of a "package" to be as significant as a medical device, impacting, e.g., supplier audit frequency. Others may not realize their "packaging component" is actually a medical device constituent part!
Container-Closure	If it delivers a metered dose of the medicinal product or serves as a transfer function for the medicinal product, it's a **medical device**.	Holds and protects the medicinal product; also known as "**primary package.**"	

(Continued)

TABLE 5.1 (CONTINUED)
A Sampling of the Combination Products Dictionary

Terminology	"Device World" Interpretation	"Pharma World" Interpretation	"Combination Product" Interpretation
Blister	**Primary Package** for the medical device	**Secondary Package (or tertiary)** for the medicinal product	The blister serves a primary package for the combination product.
Component	"Any raw material, substance, piece, part, software, firmware, labeling or assembly which is **intended to be included as part of the finished, packaged and labeled device**." (21 CFR §820.3(c))	"Any ingredient intended for use in manufacture of a drug product, **including those that may not appear in such drug product." (21 CFR §210.3)**	Each definition of component still applies to the constituent parts of a combination product. Just be aware there are differences for drug versus device interpretation. Container-closure manufacturers may have historically considered themselves as packaging "component suppliers." But if that packaging "component" is used to deliver a metered dose of the drug, the component is now considered a *device* constituent part, and that supplier is a device supplier.
Product Development Process	Design Controls (21 CFR 820.30 or ISO 13485 Clause 7.3 and its subclauses) and Risk Management (ISO 14971:2019) are **mandatory**.	ICH Q8(R2) Quality by Design and ICH Q9 Quality Risk Management are **best practices**.	For combination products, design controls and combination product risk management are mandatory. The critical quality attributes, critical material attributes, and critical process parameters for the drug constituent part are design inputs for the combination product design controls and risk management processes.
Product	**Finished device** means any device or accessory to any device that is suitable for use or capable of functioning, whether or not it is packaged, labeled, or sterilized.[b,c]	A finished dosage form, generally, but not necessarily, in association with inactive ingredients.[d]	The combination product includes: Each constituent part (device, medicinal product) + the combination product as a whole.

(Continued)

TABLE 5.1 (CONTINUED)
A Sampling of the Combination Products Dictionary

Terminology	"Device World" Interpretation	"Pharma World" Interpretation	"Combination Product" Interpretation
Manufacturer	Any person who **designs,** manufactures, fabricates, assembles, or processes a finished device. Manufacturer includes but is not limited to those who perform the functions of contract sterilization, installation, relabeling, remanufacturing, repacking, or **specification development**, and initial distributors of foreign entities performing these functions. (21 CFR 820.3(o)).	**Drug (21 CFR 210.3(12))** Manufacture, processing, packing, or holding of a drug product includes packaging and labeling operations, testing, and quality control of drug products. **Biologic (21 CFR 600.3(t))** Any legal person or entity engaged in the manufacture of a product subject to license under the act; "Manufacturer" also includes any legal person or entity who is an applicant for a license where the applicant assumes responsibility for compliance with the applicable product and establishment standards.	**Combination Product Manufacturer**[e]: An entity (facility) engaged in activities for a combination product that are considered within the scope of manufacturing for drugs, devices, biological products, and HCT/Ps. Such manufacturing activities include, but are not limited to, designing, fabricating, assembling, filling, processing, sterilizing, testing, labeling, packaging, repackaging, holding, and storage, ***including a contract manufacturing facility*** (21 CFR §4.2).

[a] H. Bulchandani, "Pharmaceutical Packaging, Components and Evaluation" accessed July 2, 2022 at http://pharmaquest.weebly.com/uploads/9/9/4/2/9942916/hitesh-pharmaceutical_packaging_component_and_evaluation.pdf#:~:text=Pharmaceutical%20Packaging%3A%20Pharmaceutical%20packaging%20means%20the%20combination%20of,product%2C%20a%20safe%20%26%20efficacious%20dosage%20form%20is.

[b] 21 CFR 820.3(l).

[c] Design controls cGMP requirements apply to the full scope of "product," including the "finished device," the production process, the final medical device (packaged, labeled, sterilized) and its labeling.

[d] Manufacturing, Processing, or Holding Active Pharmaceutical Ingredients FDA Guidance.

[e] *October 2019 FDA guidance Identification of Manufacturing Establishments in Applications to CBER and CDER Q&A.*

subject matter expert communication throughout the Chemistry Manufacturing & Controls (CMC) process will help avoid late development or post-market breakdowns.

We can extend this example further by examining how the device infrastructure is resourced. The device development team in a biotech/pharma organization typically is developing/engineering device constituents[5] that potentially can serve as a basis for administering more than one therapeutic application. For traditional biotech/pharma companies, resources get assigned to individual molecules or therapeutic business units. In such a model, resourcing of the device development may be sub-optimal, as the pharmaceutical and biotechnology organization's primary focus and expertise are centered on drug development. Given the potential breadth of applications of a given device constituent part to multiple therapeutic applications, a more optimal approach is to resource the complex device engineering development and supporting infrastructure such that funding and resources are independent of a given therapeutic area, i.e., across therapeutic areas, with associated visibility across therapeutic areas, and with integration into multiple applicable drug development strategies.

Figure 5.2 illustrates the concept of integrated device and drug development teams, and considerations of drug and device functionality and interaction(s) *together* throughout product development. (QbD and Design Controls and their integration are also addressed in Chapter 4 (Combination Products cGMP) of this book.)

Culture and Organization

One of the bigger hurdles, at least for organizations entering into – or recognizing that they are already operating in – the combination product space, is the cultural element foundational to integrated product development. Like the need to consider the unique characteristics and performance requirements of each constituent part, and their interactions as they are brought together for combined use, when it comes to combination products development, there needs to be awareness of the unique perspectives of the device- and pharma-"sides" of the organization, both internally and externally. Effective collaboration, bringing together the diverse skillsets, knowledge, and perspectives of the multidisciplinary stakeholders is crucial for successful combination product development (Figure 5.3).

Medical devices, by definition, generally are designed to address physical phenomena, while drugs are designed to address chemical and/or biological phenomena. A device developer tends to have a strong focus on technology and engineering competencies, while a drug developer often has an increased focus on the discovery and research of new molecular entities or chemical syntheses and biological processing. Drug developers often do not have strong insights into or appreciation for what it takes to effectively develop a medical device, and likewise, device developers may view the development of drugs as outside their bailiwick and scope of understanding. These disconnects can lead to difficulties beyond the technical development hurdles. If not addressed proactively, through "multilingual-competent" communication and education, a combination product development team is likely to encounter breakdowns that have little to do with their technical skills.

One needs to realize there may be blind spots. It may not even occur to one group (e.g., medicinal product developers) that there are questions they need to be asking of, or consulting with, the developers of the other constituent part (e.g., device developers), and vice versa. In a pharma organization, for example, where historically the therapeutic effect of the drug has been, and continues to be, of utmost importance, incorporating device development, engineering, quality, and regulatory considerations may be perceived as an additional burden on their already challenging product development timelines. Their Chemistry Manufacturing & Controls (CMC) stage-gate business process might not reflect much, if any, of the considerations and deliverables relative to the device constituent part and combination product as a whole. It might not even occur to medicinal product developers that there is a nuance associated with the medical device or its components that need to be taken into account as part of their development activities. They may be unaware of differences in competencies and deliverables required. With their focus on the therapeutic effect of the drug, the

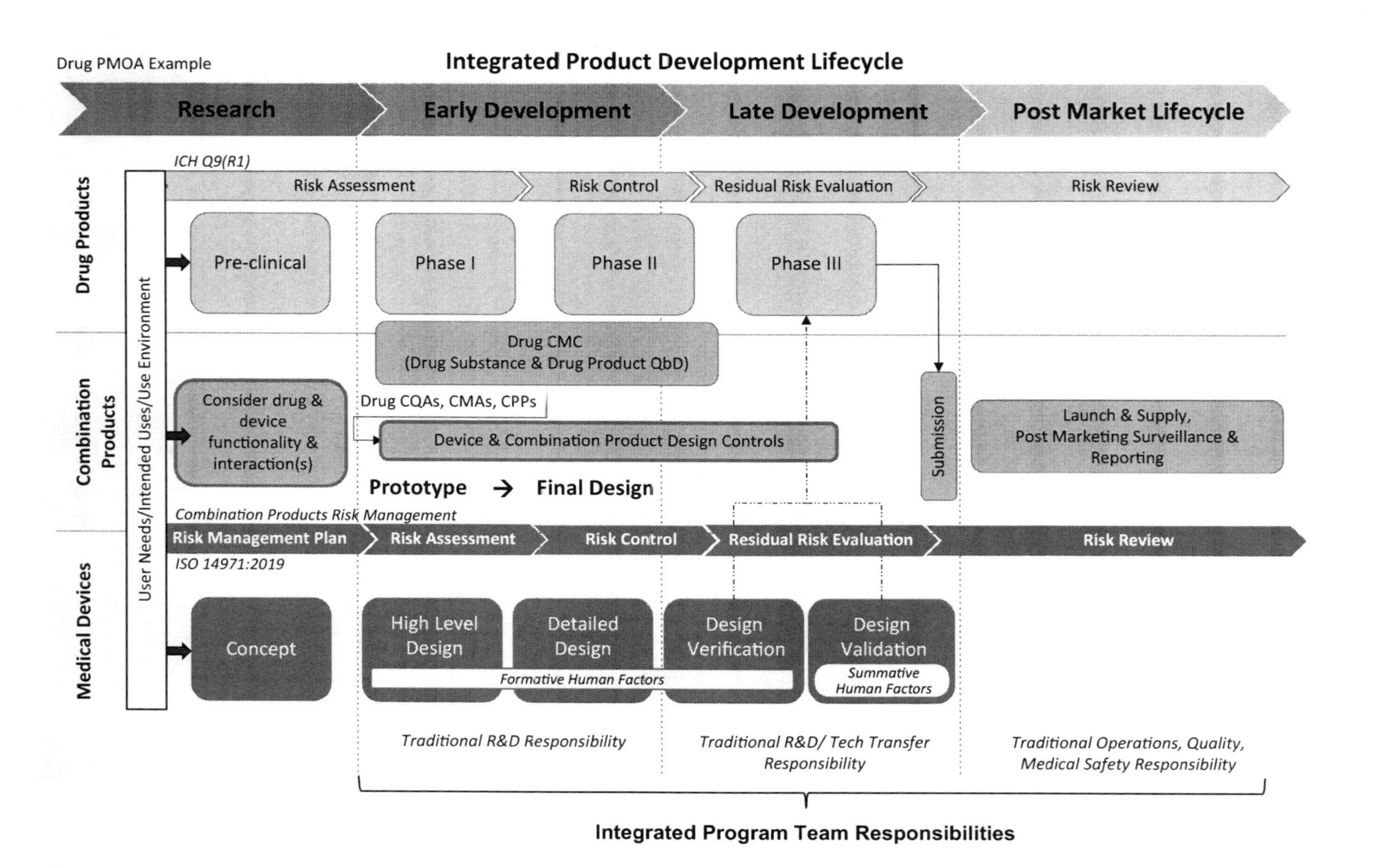

FIGURE 5.2 Example integrated drug/device development process (©2022 Combination Products Consulting Services, LLC. All Rights Reserved).

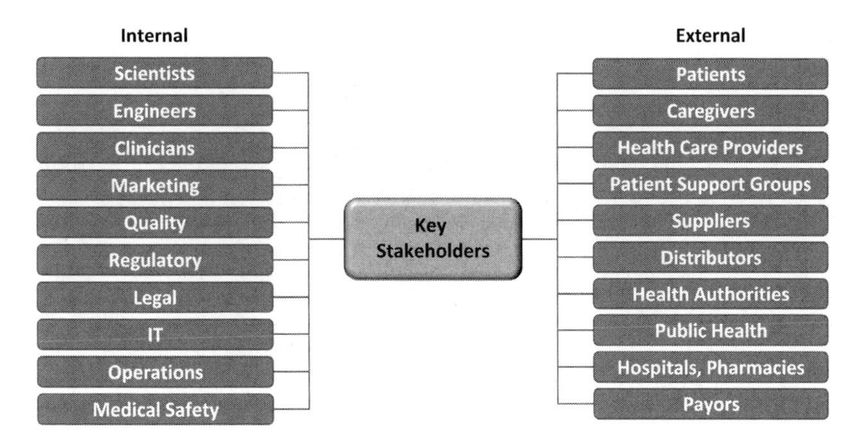

FIGURE 5.3 Collaboration among a diverse set of stakeholders is crucial for successful combination product development.

device might be perceived as a "necessary evil" that "those device people just need to deal with." This is a theme frequently heard across a variety of industry venues.

With no malintent whatsoever, medicinal product developers and device developers just don't know what they don't know with respect to the constituent part that is out of their comfort zone – they do not even know a question needs to be asked!

On the other hand, when some organizations discover that they have a product that is considered a combination product, they may go through a phase of denial: "No, my transdermal drug delivery patch is NOT a combination product! It has always been successfully filed as a drug only!" or "My IV bag is not a device constituent part of a combination product! It is a simple container closure!" "My prefilled syringe is just considered a drug in China … what do you mean it is a combination product?!" In at least some jurisdictions, e.g., in the United States under 21 CFR 3.2(e), these products clearly ARE considered combination products. Change management includes educating these colleagues and helping them move along the change curve from shock and denial to acceptance and action, to incorporate the practices and expectations of each constituent part and the combination product as a whole, from development throughout the entire product lifecycle.

A successful combination product developer needs to understand that each constituent part development sub-team probably knows some things that the other does not, and then ensure that they hold hands the whole way through the development journey to get the best outcome.

Supplier Collaboration

Suppliers play a significant role in most combination product organizations. Most purchase at least one constituent part (the one that is outside the "comfort zone" of the given combination product applicant) from a third-party supplier. Further, given a lack of device expertise, some pharma organizations may choose not only to purchase the device constituent part, but they may also outsource device development, relying heavily on medical device design partners (suppliers). Likewise similar challenges may exist in a medical device organization with respect to the supply of the medicinal product constituent part they purchase for combined use with their medical devices. This makes Purchasing Controls and effective supplier collaboration an essential success factor for combination product developers.

The relationship with suppliers can be complex.[6] When some suppliers learn that their component is now considered a constituent part of a combination product, they may go through their own phase of denial. Supplier education, partnering, and update of expectations in Supplier Quality

Agreements become important aspects of the Combination Product Market Authorization Holder's change management efforts.

Using the example of a medicinal-led combination product Market Authorization Holder (MAH) who is purchasing the device constituent part from a third party, Table 5.2 summarizes a few of the deliverables expected of the device constituent part supplier compared to the MAH with respect to the overall combination product.

Ultimately the Combination Product Market Authorization Holder (MAH) (drug-led) or Legal Manufacturer (device-led) is accountable to make sure each constituent part and their combined use supports the safe and effective performance of their combination product. Challenging situations arise when suppliers/third parties refuse to share their development information, e.g., design controls or risk management files. This presents development, submissions, and potentially post-market lifecycle management challenges. A Sponsor[7] may want to provide all the necessary information in one marketing application. Premarket, investigational, or marketing applications often contain trade-secret or confidential information. In some cases, where a supplier/third-party manufacturer does not want to share proprietary information directly with the MAH or Legal Manufacturer, some jurisdictions have mechanisms to support preservation of each manufacturer's proprietary information. US FDA, for example, will allow a MAH to submit a letter authorizing cross-reference from the owner of the referenced material (e.g., from an existing application or from a Master File)[8] for their current market application. The letter grants FDA permission to consider the referenced

TABLE 5.2
Device Supplier vs Combination Product Market Authorization Holder Comparison of Typical Development Considerations and Deliverables

Device Constituent Part Supplier	Combination Product Market Authorization Holder
Identifies and evaluates failures associated with the device components and constituent part design, manufacturing process and use, including foreseeable misuse, of the device.	Assesses and addresses unique interactions of the component or constituent part with the finished product for **each specific medicinal product, intended use, intended users, and use environment**.
Evaluates potential harms associated with the medical device, **agnostic of the medicinal product**.	Links (potential) failures to hazardous situations and harms for the constituent parts and their combined use.
Identifies, implements, and verifies risk controls **independent of a specific medicinal therapy**.	• Identifies, implements, and conducts final verification of risk controls, assessing effectiveness for each **constituent part and their combined use**. (Leveraging of some supplier data, with appropriate rationale/justification, may be possible.) • Residual risk assessment, including benefit/risk analysis for the constituent parts and the combination product system.
Lifecycle management (e.g., change control, distribution supply availability, investigations, post-marketing safety reporting) for their device. Information sharing with their customers (the MAH).	Lifecycle Management (e.g., change control, distribution supply availability, investigations, post-marketing safety reporting) for the combination product and its constituent parts.[a]
Maintain a risk file for the device.	Maintain a risk file for a combination product and its constituent parts.
Maintain Quality System that ensures suitability of their products for intended applications.	Qualify supplier, and provide supplier quality oversight (e.g., audits), under a Supplier Quality Agreement.

[a] In the event of a cross-labeled combination product, where the Legal Manufacturer for the device constituent part is different from the MAH of the drug constituent part, information sharing requirements exist in some jurisdictions, e.g., in the United States under 21 CFR Part 4B.

material in its review of the current application. Not every supplier is willing to provide such a letter of authorization, and some jurisdictions do not have such a mechanism to preserve proprietary information. The long and short of it: Close collaborative relationships with supplier(s) are an essential element of combination product success (Combination Product Purchasing Controls are discussed more extensively in Chapter 4, cGMPs, and Chapter 10, Supplier Quality Considerations).

THE DEVELOPMENT PROCESS

The potential therapeutic advantages of combined use of medicinal products (including drugs and biologics)[9] and medical devices are evident (Figure 5.1). Realizing those benefits comes with increased complexity, though, as combination products raise a variety of development, regulatory, and review challenges compared to individually categorized products (drug-only, biological product-only, or medical device-only). The controls under the main categories of drugs, devices, and biologics often share commonalities, but they also have distinct differences. When it comes to combination products, the separate "Pharma World" and "Device World," which differ in fundamental principles, "language," approaches, and scientific bases, need to be considered and reconciled. Regulatory and cGMP expectations for each combination product, its constituent parts, and any issues that may be raised through their combined use, need to be vetted, and addressed. As mentioned, a key to combination product development success is establishing a core infrastructure – e.g., in a pharma organization, a dedicated device team, an accompanying device development process model, and device quality system – integrated into the overall therapeutic/intended use product's development process. In this section of the chapter, we will discuss elements of a successful integrated combination product development process further. Elements of successful development include:

- Regulations, Guidance, and Standards; and
- Aligned Development Models – Drug, Device, and their Integration.

Regulations, Guidance, and Standards

Several health authorities have recognized that the higher technical complexity associated with bringing constituent parts together for combined use requires strengthening of authorization and certification requirements. Regulators are establishing regulatory frameworks and coordination mechanisms between medicinal/drug/biologic authorities and medical device authorities within jurisdictions. Their focus is on driving successful practices and control strategies throughout combination product development and post-market to assure public health, i.e., ensuring safety,[10] efficacy, and usability of the constituent parts and their combined use by intended users, for intended uses, and in intended use environments. Currently, regulations for combination products (albeit not all regions use the nomenclature "combination products") are enforced, for example, in the United States,[11,12] Europe,[13] Malaysia,[14] Canada,[15] Switzerland,[16] and China,[17] with regulations evolving elsewhere, e.g., in Australia[18] and Saudi Arabia.[19]

Specific to combination product development, (See Chapter 14 and Appendix) impactful cross-cutting regulations, standards, and guidance have been issued to promote risk-based regulatory frameworks for the development of innovative products that combine devices, drugs, and/or biological products. These aim to provide requirements, specifications, guidelines, and/or characteristics to be used to ensure fit-for-purpose, safe and effective materials, products, processes, and services.

Among the impactful regulations and guidance are the US FDA final rule 21 CFR Part 4 (January 22, 2013), and final Guidance (January 2017), that define Combination Product cGMP expectations, founded on principles of risk management, as well as impactful EU.[20] Impactful EU regulation EU MDR (2017/745) Article 117, EU (2017/746)[21] and associated Q&A Guidance.[22] US FDA also issued helpful draft guidance in September 2006 on "Early Development Considerations for Innovative Combination Products"[23] and continues to issue insightful product-specific draft and final guidance documents (Table 5.3) intended to provide context for development and current state-of-the-art thinking for a range of combination products and practices supporting successful combination product development. A complete list of the most recent version of US FDA guidance documents can be found at https://www.fda.gov/RegulatoryInformation/Guidances/default.htm.

TABLE 5.3
Cross-cutting Regulations and Guidance, and Product-Specific Guidance Documents Issued by US FDA to Support Combination Product Development (Partial List)

cGMPs	
Date (YYYY.MM)	**Regulation or Guidance**
2022.09	Alternative or Streamlined Mechanisms for Complying with the Current Good Manufacturing Practice Requirements for Combination Products: List under the 21st Century Cures Act
2022.03	21 CFR Part 600. Biological Products: General.
2022.02 (Draft Rule)	21 CFR Parts 4 and 820 Medical Devices; Quality System Regulation Amendments
2017.01 (Final)	Current Good Manufacturing Practice Requirements for Combination Products (Part 4A)
2003.02	Quality System Information for Certain Premarket Application Reviews - Guidance for Industry and FDA Staff)
1997.03 (Final)	Design Control Guidance for Medical Device Manufacturers
1996.10 (under revision)	Medical Device Quality System Regulation and Preamble 21 CFR 820
1985.02*	21 CFR Part 314 For FDA approval to market a new drug.
1978.09*	21 CFR Part 210. Current Good Manufacturing Practice in Manufacturing Processing, packing, or Holding of Drugs.
1978.09*	21 CFR Part 211. Current Good Manufacturing Practice for Finished Pharmaceuticals.
**Updates noted within the regulation*	
Jurisdictional	
Date (YYYY.MM)	**Regulation or Guidance**
2018.02 (Final)	How to Prepare a Pre-Request for Designation
2017.09 (Final)	Classification of Products as Drugs and Devices and Additional Product Classification Issues
2011.04 (Final)	How to Write a Request for Designation (RFD)
2007.07 (Final)	Devices Used to Process Human Cells, Tissues, and Cellular and Tissue-Based Products (HCT/Ps)
**Updates noted within the regulation*	
Product Development (Premarket)	
Date (YYYY.MM)	**Regulation or Guidance**
2022.07 (Draft)	Evaluation of Therapeutic Equivalence
2022.01 (Final)	Principles of Premarket Pathways for Combination Products
2020.12 (Final)	Requesting Food and Drug Administration Feedback on Combination Products
2021.09 (Draft)	Investigator Responsibilities – Safety Reporting for Investigational Drugs and Devices
2019.12 (Draft)	Bridging for Drug-Device and Biologic-Device Combination Products
2006.09 (Final)	Early Development Considerations for Innovative Combination Products
2008.02 (Final)	Guidance: Container and Closure System Integrity Testing in Lieu of Sterility Testing as a Component of the Stability Protocol for Sterile Products
1997.03 (Final)	Design Control Guidance for Medical Device Manufacturers
Human Factors	
Date (YYYY.MM)	**Guidance Document**
2022.12 (Draft)	Content of Human Factors Information in Medical Device Marketing Submissions
2022.07 (Final)	Instructions for Use – Patient Labeling for Human Prescription Drug and Biological Products – Content and Format Guidance for Industry

(*Continued*)

TABLE 5.3 (CONTINUED)
Cross-cutting Regulations and Guidance, and Product-Specific Guidance Documents Issued by US FDA to Support Combination Product Development (Partial List)

cGMPs	
Date (YYYY.MM)	**Regulation or Guidance**
2022.05 (Final)	Safety Considerations for Container Labels and Carton Labeling Design to Minimize Medication Errors
2018.09 (Draft)	Contents of a Complete Submission for Threshold Analyses and Human Factors Submissions to Drug and Biologic Applications
2016.04 (Final)	Applying Human Factors and Usability Engineering to Medical Devices
2016.04 (Final)	Safety Considerations for Product Design to Minimize Medication Errors
2016.02 (Draft)	Human Factors Studies and Related Clinical Study Considerations in Combination Product Design and Development
Product Specific	
Date (YYYY.MM)	**Guidance Document**
2022.03 (Final)	Certain Ophthalmic Products: Policy Regarding Compliance With 21 CFR Part 4 Guidance for Industry
2021.12 (Draft)	Technical Considerations for Medical Devices with Physiologic Closed-Loop Control Technology
2021.11 (Draft)	Content of Premarket Submissions for Device Software Functions Draft Guidance for Industry and FDA Staff
2020.04 (Draft)	Technical Considerations for Demonstrating Reliability of Emergency-Use Injectors Submitted under a BLA, NDA, or ANDA
2019.11 (Draft)	Transdermal and Topical Delivery Systems – Product Development and Quality Considerations
2018.04 (Draft)	Metered-Dose Inhaler (MDI) and Dry Powder Inhaler (DPI) Drug Products – Quality Considerations (Related article (May 2022) https://www.fda.gov/drugs/regulatory-science-action/effects-realistic-in-vitro-test-factors-aerosol-properties-metered-dose-inhalers)
2015.06 (Draft)	Draft Guidance on Budesonide; Formoterol fumarate dihydrate (metered dose inhaler)
2014.12 (Final)	Infusion Pumps Total Product Life Cycle Guidance for Industry and FDA Staff
2013.06 (Final)	Technical Considerations for Pen, Jet, and Related Injectors Intended for Use with Drugs and Biological Products
2013.04 (Draft)	Glass Syringes for Delivering Drug and Biological Products: Technical Information to Supplement International Organization for Standardization (ISO) Standard 11040-4
2011.05 (Final)	Dosage Delivery Devices for Orally Ingested OTC Liquid Products
2009.12 (Draft)	New Contrast Imaging Indication Considerations for Devices and Approved Drug and Biological Products
2008.03 (Draft)	Coronary Drug-Eluting Stents – Nonclinical and Clinical Studies -Companion Document

Product-specific guidance and standards within a jurisdiction may also be helpful across jurisdictions globally, as many of these standards and guidance documents focus on the technical considerations that should be addressed to ensure the safety and efficacy of these combination products – a common goal for all combination products developers.

There are also several impactful international standards and guidance that help institutionalize the critical aspects of the constituent parts and their combined use. These standards and guidance are established through committees of industry and regulatory experts. Table 5.4 summarizes some of these important documents that developers may (also) need to reference in support of their design and development activities.

TABLE 5.4
Impactful International Guidance and Standards (partial list) in the Combination Product Space[a]

cGMPs	
ISO 13485:2016	Medical Devices – Quality Management Systems – Requirements for Regulatory Purposes.
ICH Q10	Pharmaceutical Quality System
AAMI TIR 48:2015 (under revision)	Quality Management System (QMS) Recommendations on The Application of The U.S. FDA's CGMP Final Rule on Combination Products.
Standards on Product Development & Risk Management	
EN ISO 14971	Medical devices – Application of risk management to medical devices.
EN ISO 14971:2019+A11:2021 (December 2021)	The amendment replaces the European Foreword and adds two new Annex Zs (ZA and ZB) to show the relationship between the clauses of the standard and the requirements of the EU MDR (2017/745) and EU IVDR (2017/746).[b]
AAMI TIR 105:2020	Combination Products Risk Management
ICH Q9 [ICH Q9(R1)]	Quality Risk Management
ICH Q8(R2)	Quality by Design
ISO 10993 series	Biological evaluation of medical devices
ISO 20069	Guidance for assessment and evaluation of changes to drug delivery systems.
Sterilization:	
ISO 11135	Sterilization Of Health Care Products – Ethylene Oxide – Requirements for Development, Validation and Routine Control of a Sterilization Process for Medical Devices.
Sustainability:	
AAMI TIR 65:2015	Sustainability of Medical Devices – Elements of a Responsible Product Lifecycle
Clinical Investigation:	
ISO 14155:2020	Clinical Investigation of medical devices for human subjects – Good clinical practice
Standards on Human Factors/Usability Engineering	
AAMI/IEC 62366-1:2015(R2021)	Medical devices – Part 1: Application of usability engineering to medical devices, including Amendment 1
AAMI HE 75:2009/(R)2018	Human factors engineering – Design of medical device
AAMI TIR 59: 2017	Integrating Human Factors into Design Controls
Standards on Symbols	
ISO 15223-1:2021	Medical Devices – Symbols to Be Used with Information to Be Supplied by The Manufacturer – Part 1: General Requirements
Standards for Needle-based Injection Systems	
ISO 11608 SERIES (Needle-based injection systems)	
11608-1 (Parent document)	Key requirements for a Needle-based injection system, including: General requirements Risk management Free-fall testing Dose accuracy Determining and testing essential functions Visual inspection, Markings & Labeling
11608-2	Requirements and Test Methods for Needles
11608-3	Requirements and Test Methods for Finished Containers (including fluid path)
11608-4	Pen injectors; electro-mechanical components, electronics, firmware, software, and/or batteries

(*Continued*)

TABLE 5.4 (CONTINUED)
Impactful International Guidance and Standards (partial list) in the Combination Product Space[a]

11608-5	Needle-based injection system with an automated function (e.g., needle insertion) or the medical device limits access to a function
11608-6	Requirements and Test Methods for On-Body Delivery System (OBDS)
11608-7	Requirements for NIS claimed to be appropriate for the visually impaired
ISO 23908	Sharps injury protection for single-use hypodermic needles, catheters, and needles for blood sampling; Needle shielding
ISO 8537	Sterile single-use syringes, with or without needle, for insulin
Standards re: Syringes	
ISO 9626	Stainless Steel Tubing
ISO 7886-1	Single-Use Syringes
ISO 7886-2	Syringes for Syringe Pumps
ISO 11040 -5, -6, -8	Prefilled Syringes
ISO 11607-1:2019	Packaging for terminally sterilized medical devices – Part 1: Requirements for materials, sterile barrier systems, and packaging systems
ISO 11607-2:2019	Packaging for terminally sterilized medical devices – Part 2: Validation requirements for forming, sealing, and assembly processes
ISO/TS 16775:2021	Packaging for terminally sterilized medical devices – Guidance on the application of ISO 11607-1 and ISO 11607-2
ISO 21881:2019	Sterile packaged ready-for-filling glass cartridges
AAMI TIR 101:2021	Fluid delivery performance testing for infusion pumps
Additional Potentially Applicable Standards:	
IEC 60601-1	Medical Electrical Equipment
IEC 60601-1-8	Alarms
IEC 60601-1-11	Home Healthcare
EN 62304 and IEC 800002-1	Software
IEC 15026-1, -2, and -4	Systems and Software Engineering
IEC gogo1-2, IEC 61000-4 and CISPR 11-14-1	Electro-mechanical compatibility (EMC)
IEC 60812	Dependability
EN 61078:2016	Reliability Block Diagrams
Delivery requiring a connector or catheter:	
EN 1615 and 1618	Catheters, Tubing
ISO 594-2 and ISO 80369-1	Connectors
ISO 11070	IV introducers + guide wires
Pump-Specific Standards:	
IEC 60601-2-24	For pump-based, needle-based injection systems
Medicinal Administration without a Needle:	
ISO 20072	Aerosol Drug Delivery Devices
ISO 21649	Needle-free injection systems

[a] ISO = International Organization for Standardization; IEC = International Electrotechnical Commission; EN = European Norm; ICH = International Council on Harmonization.

[b] Medical Devices Regulation ((EU) 2017/745) (MDR) and In vitro Diagnostic Medical Devices Regulation ((EU) 2017/746) (IVDR).

In addition to the plethora of international standards available, there are also a number of country-specific technical standards. These can present industry challenges, as the regulations are dynamic and at times not communicated widely. Awareness of those standards and timely access to them during product and process development are among the hurdles companies/product development teams face. One such problematic standard has been the Korean needle standard, where companies have developed products without awareness of such standard, only to be faced with breakdowns during the submission process. China's regulatory construct, regulations, and standards have likewise presented challenges for developers, given a combination of document access issues and interpretation challenges. In an attempt to help those who seek to develop products for the China market, Table 5.5 illustrates some of the consensus standards required at the time of this writing for product development and submissions in China.

TABLE 5.5
Comparison of select US FDA and China Recognized Standards Applicable to an Autoinjector

	FDA Recognized Consensus Standards	China Standards
Packaging	ISO 11607-1: 2019	GB/T 19633.1-2015 (ISO11607-1:2006, IDT)
Biocompatibility	ISO 10993-1: 2018	GB/T 16886-1:2011 (ISO10993-1:2009, IDT)
	ISO 10993-4: 2017	GB/T 16886-4:2003 (ISO10993-4:2002, IDT)
	ISO 10993-5: 2009	GB/T 16886-5:2017 (ISO10993-5:2009, IDT)
	ISO 10993-10: 2010	GB/T 16886-10:2017 (ISO10993-10:2010, IDT)
	ISO 10993-11: 2017	GB/T 16886-11:2011 (ISO10993-11:2006, IDT)
	ISO 10993-17: 2002	GB/T 16886-17:2005 (ISO10993-17:2002, IDT)
	ISO 10993-18: 2020	GB/T 16886-18-2011 (ISO10993-18:2005, IDT)
Sterility	ANSI/AAMI ST67: 2011	ChP 2020 sterility
	ANSI/AAMI/ISO 10993-7:2008/(R)2012	GB/T 16886-7-2015 (ISO10993-7:2008, IDT)
Needle-Based Injection System	ISO 11608-1: 2014	The YY/T standard referring to ISO 11608-1 is being drafted YY/T 1768.1-2021
	ISO 11608-4: 2006	
	ISO 11608-5: 2012	
Sharp Injury Protection	ISO 23908: 2011	GB/T standard is currently being drafted. It is IDT ISO 23908: 2011
Sterile Hypodermic Needle	ISO 7864: 2016	GB 15811-2016 (refers to ISO 7864:1993, modified)
Battery	IEC 60086-1: 2015	GB/T 8897.1-2013 (refers to IEC 60086-1:2011, modified)
EMC	IEC 60601-1: 2012 ANSI/AAMI ES60601-1:2005/(R) 2012 + A1:2012 + C1:2009/(R)2012 + A2:2010/(R)2012, ED 3.1	GB 9706.1-2007 Medical electrical equipment – Part 1: General requirements for safety (IEC 60601-1:**1988**, IDT) Note: GB 9706.1-2020 will be effective 1 May 2023)
	IEC 60601-1-2: 2014 AAMI/ANSI/IEC 60601-1-2:2014	YY 0505-2012 Medical electrical equipment – Part 1-2: General requirements for safety – collateral standard: electromagnetic compatibility – requirements and tests (IEC 60601-1-2: **2004**, IDT)
	IEC 60601-1-8: 2012 (May update to 2020) AAMI/ANSI/IEC: 60601-1-8:2006 and A1:2012	YY 0709-2009 (IEC 60601-1-8:2003, IDT) (YY 9706.108-2021 will be effective 1 May 2023)
	IEC 60601-1-11: 2015 AAMI/ANSI HA60601-1-11:2015	YY 9706.111-2021 (effective 1 May 2023)

Combination Product Development Strategy

Introduction

Given the wide gamut of types and complexities of combination products, there is not a one-size-fits-all approach to combination product development. The backdrop of regulations, guidance, and standards helps to inform the process, but needs to be scrutinized and adapted for each combined use application based on the unique questions raised and risks associated with, each of the constituent parts intended functionality and their potential interaction(s) when used together. The concept of Quality by Design (ICH Q8(R2)) is a best practice applied for medicinal product development. Design Controls (21 CFR 820.30 or ISO 13485 §7.3) is the current state-of-the-art expectation for development of medical devices. Risk management (ICH Q9(R1) and ISO 14971:2019) underpins both drug and device development. These approaches for developing each of the constituent parts in isolation, however, may not address the scientific and technical issues of their combined use. As illustrated in Figure 5.2, based on insights into user needs, intended uses, and intended use environment for a given combination product, innovation starts with consideration of the functionality and potential interactions between the constituent parts for the intended combined use application. These interactions may lead to a range of risk management and regulatory approaches based on where the combination product use sits along a continuum of risk. Even if a combination product might be comprised of an already-approved drug and an already-approved device, new scientific and technical issues may arise when these constituents are combined or used together (see examples, Table 5.6). In some cases, the constituents may actually have synergistic effects that need evaluation. Consider, for example, technology that enables previously inaccessible parts of the body to now receive localized/targeted delivery. This provides an opportunity to reduce the drug dose used, but also has the potential to lead to higher exposure to that target than when the drug has been administered systemically.

TABLE 5.6
Examples of Constituent Part Interactions in Drug/Device Combination Products

Combination Product	Examples of Potential Interactions
Drug-eluting stent	The mechanical attributes of the polymer coating system contain the drug substance; they are important for stent deployment, drug release, biocompatibility, and stability.
Drugs/biologics in a prefilled syringe	Metals and manufacturing materials that may be components of a syringe support mechanical performance of the syringe (e.g., siliconized barrel for enhanced break-loose/glide force) may react with certain drugs/biologics (e.g., creating protein aggregates). Biosurfactants, which may have antimicrobial and antiviral activity in a biological product formulation, might interact with the syringe component materials, leading to product immunogenicity.[a]

[a] Locatelli F, Aljama P, Barany P, et al. Erythropoiesis-stimulating agents and antibody-mediated pure red-cell aplasia: where are we now and where do we go from here? *Nephrol Dial Transplant* 2004;19: 288–293.

In short, the scientific and technological considerations raised by the intended combined use of the medical products need to be addressed in development, risk analysis, and control strategy, for each constituent part and for the combined use system as a whole. Founded on a framework of combination products risk management, integration of drug/biologic Quality by Design (QbD) with device Design Controls underpins successful combination product development.

In the following, we will review combination product development considerations, focused on the process to establish safe and effective combination products. Let's begin with those combined use applications that leverage one or more constituent parts that are already marketed.

Development Strategy: Using Already-Marketed Constituent Parts for Development of New Combination Products

Developers may choose to leverage a constituent part in their combination product that has previously received marketing authorization approval. For example, a pharma company may choose to use an already 510(k) approved syringe for administration of their drug. In order to achieve that prior approval, safety and efficacy data for the constituent part had to have been vetted for the syringe. Importantly, though, it was vetted agnostic to the pharma company's specific combined use application, i.e., drug-agnostic. The *combination product* that is being developed can possibly leverage the pre-existing drug-agnostic data for the syringe, but additional information will likely be needed to address scientific and technical considerations for the new use of that constituent part for the pharma company's specific drug. The considerations include not just the combined use, functionality, and interactions; they also include considerations of intended users, intended uses, and intended use environments. Is the treatment emergency use or chronic care? Is this a new route

TABLE 5.7
Considerations for Constituent Parts of a Combination Product

For the proposed indication:	
	Medical Device-led or Medicinal-Product-led Combination Product
1	Are either of the constituent parts already approved for an indication?
2	For the constituent part with an already-approved indication, is it similar to that proposed for the combination product? What are the differences, if any?
3	Does the combination product expand the indication or intended target population beyond that previously approved for the already-approved constituent part? Is the intended use environment altered?
4	Is the combination product presenting a new route of administration for the patient?
5	Is there a new local or systemic exposure profile for an existing indication?
6	Is a regenerative medicine and/or HCT/Ps being used in the combination product? What new questions arise for the delivery system intended for use with such biologically sourced materials?
7	If the drug is already approved, is the drug formulation in the combination product different from that used in the already approved drug?
8	Is there a need to update the device constituent design to accommodate the new use?
9	Is the device constituent being used in an area of the body that differs from that included in its existing approval?
10	Are the constituent parts chemically, physically, or otherwise combined into a single entity?
11	How do the drug and device constituent parts interact? (i.e., How does the device impact the drug properties? How does the drug impact the device functionality?)
12	Does the device constituent part function as a delivery system?
13	Is the device constituent part used to prepare a final dosage form?
14	Does the device constituent part provide active therapeutic benefit?
15	Is there any other change in design or formulation that could impact the safety and/or efficacy of any existing constituent part or the combination product as a whole?
16	Is the **drug** constituent part a new molecular entity (regardless of whether the device is already marketed)?
17	Is the **device** constituent part a complex, new device (regardless of whether the drug is already marketed)?

of administration for the drug? Is there a shift from systemic exposure to local exposure of the drug? Is the target population changing? Adults? Adolescents? Pediatric patients? Is the drug being administered by a trained professional or is it to be administered by lay-persons? Is drug administration in a clinician's office, a hospital, in a helicopter, or at home? Table 5.7, drawn from US FDA Guidance,[24] lists some of the many questions for consideration. These same questions apply to whether a combination product is device-led or drug-led. Further, these questions are applicable regardless of jurisdiction, as they are focused on the scientific and technological considerations of combined use. The range of questions may vary substantially based on the type and complexity of product and the context of intended use. The reader is encouraged to review "FDA Guidance for Industry and Staff: Early Development Considerations for Innovative Combination Products" (September 2006).

Generally speaking, product development is founded on risk management. As discussed in Chapter 6 (Combination Products Risk Management) of this book, combined use considerations such as those listed in Table 5.7 may be indicative of new or different hazards,[25] hazardous situations,[26] harms,[27] or harm severities that emerge from a particular combined use application.[28] Combination product development and risk management should address the risk-based questions applicable to each unique combined use context (intended uses, intended users, and intended use environment) for each constituent part and for the combination product as a whole.

Development Strategy: Medicinal Product (Drug and Biological Product) Constituent Part Considerations

Development of the drug constituent part of a combination product is generally accomplished using Quality by Design (ICH Q8(R2)) in conjunction with Pharmaceutical Risk Management (ICH Q9) best practices (see Chapter 6 on Combination Products Risk Management), in conjunction with a series of Clinical Phases[29,30] (summarized in Figures 5.2, 5.4, and 5.5, and Table 5.8).

Yu et al. (2014)[31] summarized the goals of Quality by Design:

1) Achieve clinically relevant product quality specifications
2) Enhance product and process knowledge, design, understanding and control, driving robust process capability, reduced product variation, and reduced defects
3) Increase efficiencies throughout product development and manufacturing
4) Improved root cause analysis and post-approval change control

> Under QbD, these goals can often be achieved by linking product quality to the desired clinical performance, and then designing a robust formulation and manufacturing process to consistently deliver the desired product quality.
>
> **Yu et al. (2014)**

Prior to testing a new molecular entity (NME) on people [First in Human (FIH)], the NME's safety and effectiveness, i.e., the NME's potential to cause serious harm (toxicity), is assessed through preclinical *in vitro* and *in vivo* studies (Table 5.8). For a combined use application, information is needed not just for the NME; considerations of the interactions between NME and device constituent part must also be assessed and addressed. Risk analysis can be used to identify applicable characterization information required, e.g., certain preclinical pharmacology and toxicology studies (including, for example, genotoxicity, mutagenicity, immunotoxicity, and local tolerance) for each constituent part and their combined use (Figures 5.2 and 5.6) to establish the safety profile of the NME with the particular device constituent part.

TABLE 5.8
Drug Development Clinical Phases[a]

In This Table, Those Clinical Phase Elements Specific to Combination Products Are italicized within Each Phase			
Clinical Phase	**Primary Goal**	**Clinical Subjects**	**Comments**
0	Understanding pharmacodynamics and pharmacokinetics of drug	Preclinical[b]	Prior to testing a drug on people, assess the potential for a drug to cause serious harm (toxicity) through *in vitro* and *in vivo* studies.
		For a combination product with a medicinal product constituent part that is already approved, the preclinical efforts might be tailored to address the safety considerations raised by the new context of use. *In vivo pharmacokinetic (pK) studies may be needed to assess changes in formulation, strength, route of administration, dosing, population, or other factors that might change the extent or duration of systemic exposure. These studies may be used to evaluate drug release kinetics (e.g., release rate, local peak concentrations of the drug, drug distribution, and systemic bioavailability).* *Acute and repeat dose toxicity studies using the new route of administration or method of delivery may be appropriate to determine the NOAEL (no observed adverse effect level) and toxicity profile of the combination product. These studies typically evaluate the intended clinical formulation and dosing regimen that will closely approximate its use in clinical settings.*	
I	Dose-ranging study (sub-therapeutic and ascending doses)	20–80 healthy subjects[c]	Gather information about how a drug interacts with the human body: • How much of a drug the body tolerates and have acute side effects? • Side effects associated with increased dosage • How it works in the body • Early information about how effective the drug is • Determine how best to administer the drug in order to limit risks and maximize potential benefits
		For combination products, dose-ranging or dose-finding studies in humans may be appropriate to determine dose adjustments for safety/effectiveness when a therapy is targeted to a local site.	
II	Assess the efficacy and side effects of the drug at therapeutic dosing levels	Up to several hundred people with the disease/ condition	Provide researchers with additional safety data to: • Refine research questions • Develop research methods • Design new Phase III research protocols
		For combination products, special safety studies may be appropriate for certain patient populations or specific risks, e.g., hepatoxicity, QT prolongation *Specific safety monitoring to obtain data on the novel aspects presented by the combination product, e.g., local toxicity for a new route of administration, may be appropriate.[d]*	
III (Pivotal)	Testing on target population to assess efficacy and adverse reactions	300-3000 volunteers who have the disease/ condition	• Demonstrate whether or not a product offers a treatment benefit to a specific population • Provide most of the safety data (including long-term or rare side effects)
		For combination products, specific safety monitoring to obtain data on the novel aspects presented by the combination product, e.g., local toxicity for a new route of administration, may be appropriate.[e]	

(*Continued*)

TABLE 5.8 (CONTINUED)
Drug Development Clinical Phases[a]

In This Table, Those Clinical Phase Elements Specific to Combination Products Are italicized within Each Phase			
Clinical Phase	**Primary Goal**	**Clinical Subjects**	**Comments**
IV	Post-market; long-term safety *Not always a requirement. May be a post-market commitment*	Several thousand users of the approved/marketed product who have the disease/ condition	Carried out once the drug or device has been approved by FDA during the Post-Market Safety Monitoring

[a] Adapted from The Drug Development Process, http://www.fda.gov/ForPatients/Approvals/Drugs/defualt.htm.

[b] 21 CFR 58 "Good Laboratory Practice for Nonclinical Laboratory Studies" at https://www.accessdata.fda.gov/scripts/cdrh/cfdocs/cfcfr/CFRSearch.cfm?CFRPart=58.

[c] If a new drug is intended for use in cancer patients, researchers conduct Phase I studies in patients with that type of cancer.

[d] FDA Guidance "E1A The Extent of Population Exposure to Assess Clinical Safety: For Drugs Intended for Long-Term Treatment of Non-Life-Threatening Conditions" (1995), accessed July 2, 2022 at https://www.fda.gov/regulatory-information/search-fda-guidance-documents/e1a-extent-population-exposure-assess-clinical-safety-drugs-intended-long-term-treatment-non-life.

[e] FDA Guidance "E1A The Extent of Population Exposure to Assess Clinical Safety: For Drugs Intended for Long-Term Treatment of Non-Life-Threatening Conditions" (1995), accessed July 2, 2022 at https://www.fda.gov/regulatory-information/search-fda-guidance-documents/e1a-extent-population-exposure-assess-clinical-safety-drugs-intended-long-term-treatment-non-life.

If a combination product's drug or biological product constituent part is already approved for another indication, as reflected in the section "Development Strategy: Using Already-Marketed Constituent Parts for Development of New Combination Products", consideration needs to be given to the potential for change to the previously established safety, effectiveness, and/or dosing requirements relative to the new combination product configuration, intended use, users, and/or use environment. Depending on the specific context of use (intended use, intended users, intended use environment), additional preclinical or clinical safety information or even clinical studies may be needed for the drug/biological product constituent part and/or for the combined use product. Some *context of use* examples that could trigger the need for additional data/studies are as follows:

Change Type	**Examples**	**Considerations: Why Care?** ***Changes to intended uses, intended users, and intended use environment raise questions as to the safety and efficacy of the combination product: Are there new hazard (potential sources of harm)s, hazardous situations (exposure to harm), or harms (injury or damage to people, property or the environment) possible as a result of the proposed change? (See Chapter 6 on Risk Management in this book.)***
Change in intended use from the approved indication	Treatment of a different disease	A few examples for consideration: • Are the performance requirements for the administration device consistent for the treatment of the new disease? Is the administration of the drug now intended for a different part of the body? (The drug/device combination might be safe and effective for administration, e.g., for the epidural space, but not for use in the more constrained intrathecal space.) Is the required dose volume the same? Regimen change? Dose accuracy requirements? • Are the reliability expectations of the combined use system aligned to the new intended use? (e.g., acute life-saving care versus chronic care)

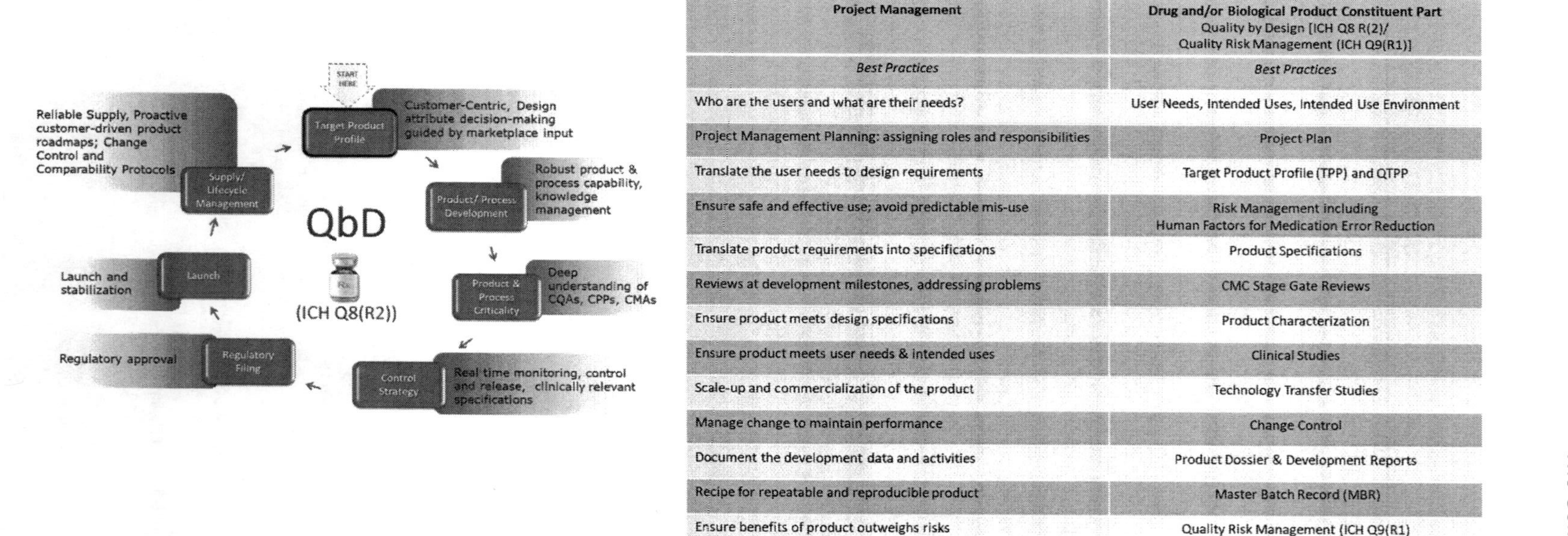

Project Management	Drug and/or Biological Product Constituent Part Quality by Design [ICH Q8 R(2)/ Quality Risk Management (ICH Q9(R1)]
Best Practices	*Best Practices*
Who are the users and what are their needs?	User Needs, Intended Uses, Intended Use Environment
Project Management Planning: assigning roles and responsibilities	Project Plan
Translate the user needs to design requirements	Target Product Profile (TPP) and QTPP
Ensure safe and effective use; avoid predictable mis-use	Risk Management including Human Factors for Medication Error Reduction
Translate product requirements into specifications	Product Specifications
Reviews at development milestones, addressing problems	CMC Stage Gate Reviews
Ensure product meets design specifications	Product Characterization
Ensure product meets user needs & intended uses	Clinical Studies
Scale-up and commercialization of the product	Technology Transfer Studies
Manage change to maintain performance	Change Control
Document the development data and activities	Product Dossier & Development Reports
Recipe for repeatable and reproducible product	Master Batch Record (MBR)
Ensure benefits of product outweighs risks	Quality Risk Management (ICH Q9(R1)

FIGURE 5.4 Quality by design in drug development process (© 2022 Combination Products Consulting Services, LLC. All Rights Reserved) CQA = Critical Quality Attribute; CPP = Critical Process Parameter; CMA = Critical Material Attribute.

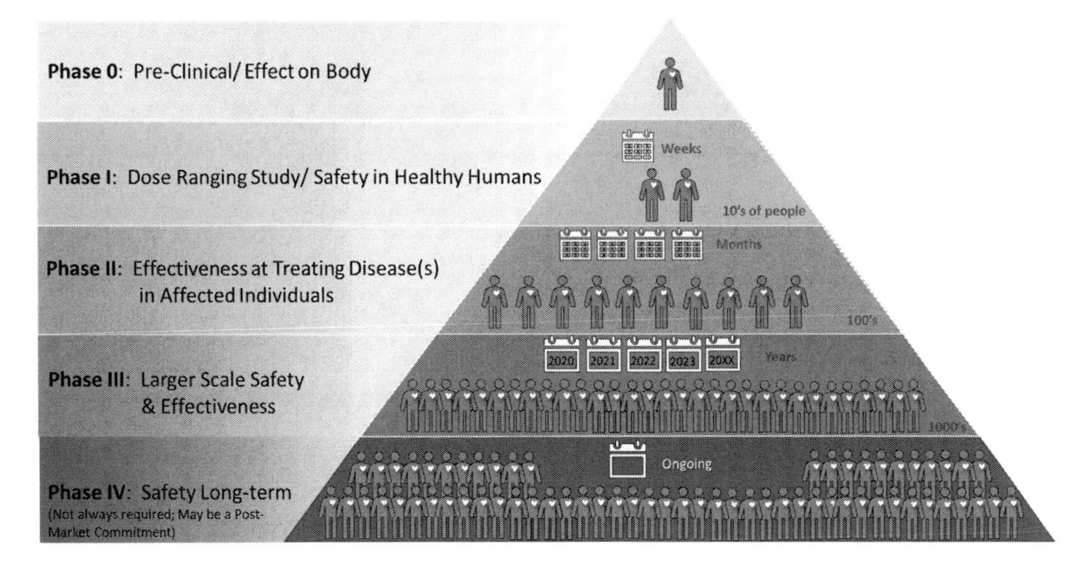

FIGURE 5.5 Drug development clinical phases.

Change in intended users/ target patient population	Neonates, pediatric patients, geriatric patients, nursing mothers, home care vs healthcare professionals	Changing an intended user population raises questions on the new user's ability to safely and effectively interact with the user interface. Device design changes may also be needed, e.g., to address differences in tissue depth, metabolism, and physical dimensions of the intended target population. Some examples: • Cognitive skills: Ability to understand and apply instructions for use; e.g., patient ability to safely navigate the user interface, of a patient-controlled analgesia system, where failure to perform the tasks correctly could result in missed dose, overdose, or underdose. • Physical skills; differences of intended users to execute critical tasks due to dexterity or ergonomic constraints (e.g., hand grip size and strength, low vision). • Experience and education impacting the ability to execute complex tasks, e.g., preparation and administration of a reconstituted drug under sterile conditions.
Change in intended use environment/ dosage	Dose volume, dose duration, dosing regimen, or local vs systemic total exposure	• Changes in the local adverse reaction profile could be a consideration, e.g., including those related to a change in drug viscosity, formulation, or change in injection rate. • Is the same device that was used for chronic care now to be used for emergency use? This triggers the need to have greater confidence in functional reliability, and may also require updates to the user interface to ensure safe and effective use under emergency/high stress conditions. • If administering a dose in a clinical environment vs home-use vs on a playground or in an emergency evacuation helicopter, there will likely be differences in user experience and emotional state that need to be addressed.
Change in the approved drug or biological product formulation, strength, route of administration or delivery method	Prefilled syringe to autoinjector	• Changes in the drug formulation might lead to shifts in bioavailability of the drug and/or its metabolic profile due to changes in the device, formulation or route of administration, e.g., changes in needle depth, tissue plane, or rate of infusion. • Changes in the drug formulation may lead to shifts in drug-device interactions, e.g., impacting the leachable and extractable profiles and drug stability of the combination product. Consider, for example, interactions of excipients of the drug formulation with drug-contacting surfaces of the fluid path. • Changes in dose accuracy of the same device constituent part when the drug formulation is changed, e.g., with change in drug viscosity.

Data needs to be generated to substantiate the overall safety and effectiveness of each constituent part and their specific combined use. See combination product highlights in Table 5.8 with respect to other possible considerations when creating a development plan for a combination product that incorporates a drug/biologic constituent part. Developers may be able to provide relevant information from the literature or may rely on prior agency findings to address issues raised by combined

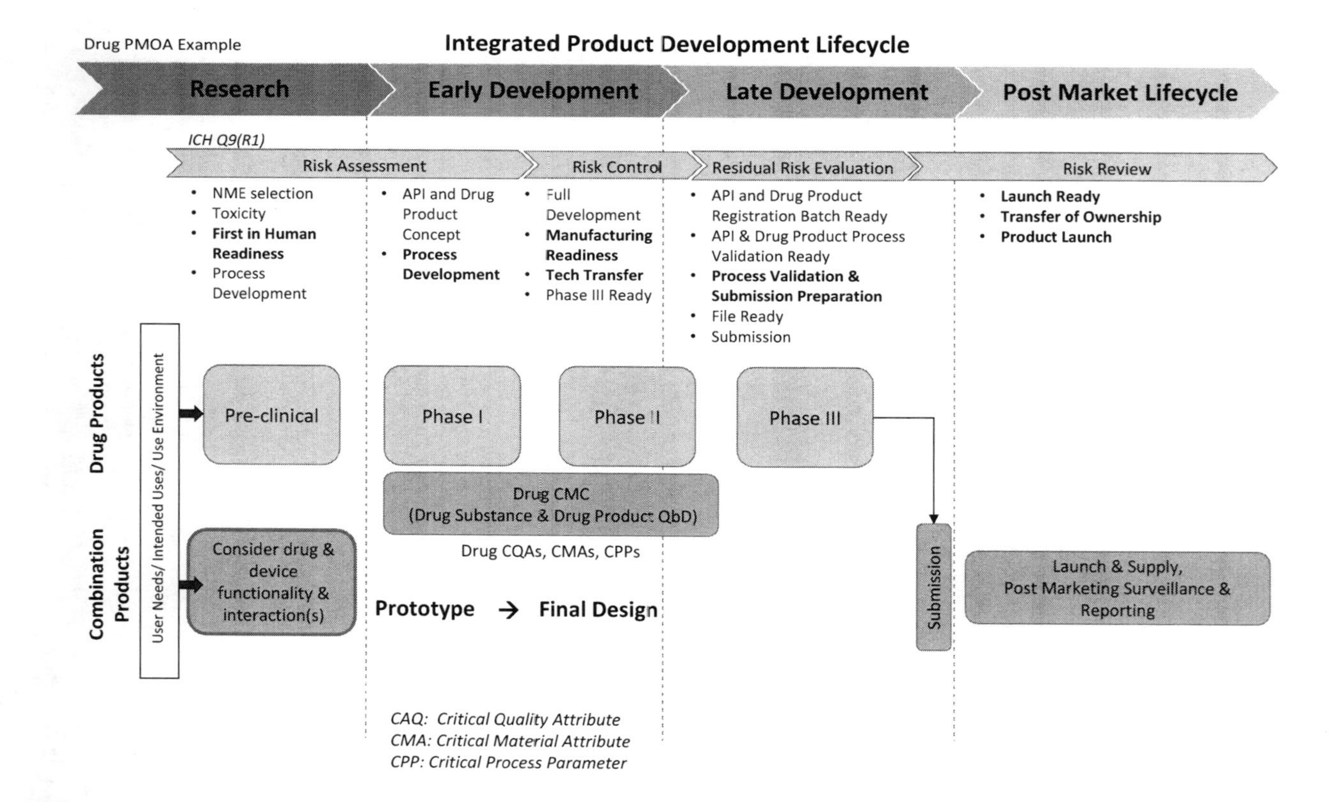

FIGURE 5.6 Drug product development pathway process (©2022 Combination Products Consulting Services, LLC. All Rights Reserved).

use, or changes to drug context of use. When this is not possible, then additional studies may be necessary.

FDA Draft Guidance "Bridging for Drug-Device and Biologic-Device Combination Products" (December 2019)[32] addresses how sponsors can bridge data from earlier stages of development, or from other development programs, to support an application. Bridging is founded on conducting an effective gap assessment, understanding scientific and technical similarities and differences between products, and doing risk analysis to determine what pre-existing data can be leveraged. The amount of information that can be leveraged depends on the specific combination product, and in some cases bridging may not be possible. A change in the route of administration for a complex biological product may raise additional safety and/or efficacy considerations, which preclude the ability to bridge to a proposed combination product. A change in drug or biological product formulation could likewise raise safety or efficacy considerations through undesirable interactions with the device constituent part. Any potential leveraging of pre-existing data necessitates robust vetting of similarities, differences, and risks.

Development Strategy: Device Constituent Part Considerations

Development of the device constituent part and the combination product as a whole is accomplished using Design Controls and founded upon risk management (Figures 5.2 and 5.7). Design Controls serve as a framework to control the design process, made up of quality practices and procedures incorporated into the design and development to ultimately assure the device (and combination product) meet user needs and intended use(s) and are safe, effective, and usable. Design Controls generally follows project management best practices. In Design Controls:

(1) Insights on user needs, intended uses, and use environment;
(2) Requirements of the medicinal product indication, critical quality attributes, critical material attributes, and critical process parameters; and
(3) Insights from analysis of the risks of joint combined use of each of the constituent parts in a combination product are together translated into technical design requirements (design inputs) for the device constituent and combination product as a whole. Through a mandated,[33] risk-based, methodical, planned, documented, and iterative design-build-test process that may include clinical and human factors studies, the device constituent part and combination product design outputs are developed, reviewed, verified, validated, and transferred for commercialization (Figure 5.7 and Table 5.9) (see Chapter 4, Combination Product cGMPs, in this book for more discussion of the Combination Products Design Controls process).

Compare and contrast Tables 5.8 and 5.9. The preclinical evaluation of drugs/biologics[34] is different from preclinical or nonclinical studies for devices (also see Table 5.12 for a further comparison). Both drug and device development processes are focused on building a deep understanding of critical attributes, critical material attributes, and critical process parameters of the finished product. Drug development is focused on reduction of the potential for medication errors; device development brings focus to the user interface of the product, identifying applicable critical use tasks for safe and effective product use. When developing a combination product, it is likely that neither isolated approach would fully address the relevant preclinical development questions for both constituents and those for the combination product as a whole. Consideration of the scientific and technical questions raised by the joint/combined use of the constituent parts should inform the approach an organization takes to integrate the product development activities for each of the constituent parts and for the combination product as a whole.

Development of a new device constituent part for use in a combination product may require some drug-agnostic safety and/or effectiveness testing of the device (and, if applicable, its components)

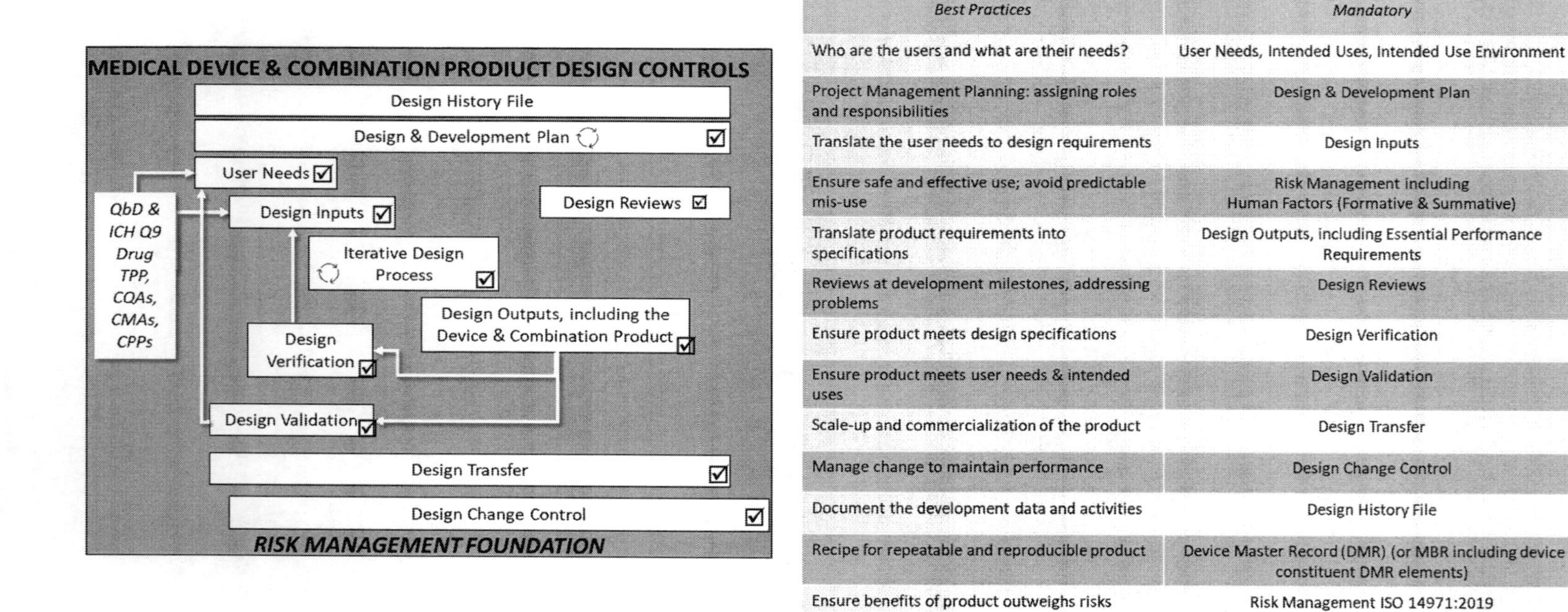

Project Management	Design Controls (21 CFR 820.30 or ISO 13485:2016, Clause 7.3 and its subclauses)
Best Practices	*Mandatory*
Who are the users and what are their needs?	User Needs, Intended Uses, Intended Use Environment
Project Management Planning: assigning roles and responsibilities	Design & Development Plan
Translate the user needs to design requirements	Design Inputs
Ensure safe and effective use; avoid predictable mis-use	Risk Management including Human Factors (Formative & Summative)
Translate product requirements into specifications	Design Outputs, including Essential Performance Requirements
Reviews at development milestones, addressing problems	Design Reviews
Ensure product meets design specifications	Design Verification
Ensure product meets user needs & intended uses	Design Validation
Scale-up and commercialization of the product	Design Transfer
Manage change to maintain performance	Design Change Control
Document the development data and activities	Design History File
Recipe for repeatable and reproducible product	Device Master Record (DMR) (or MBR including device constituent DMR elements)
Ensure benefits of product outweighs risks	Risk Management ISO 14971:2019 (for combination product: ISO 14971:2019 + ICH Q9(R1); AAMI TIR 105: 2020)

FIGURE 5.7 Design controls. (c)2023 Combination Products Consulting Services, LLC.

TABLE 5.9
Device Constituent Part Development Clinical Studies

In This Table, Those Clinical Phase Elements Specific to Combination Products Are italicized within Applicable Phases

Study Type	Primary Goal	Clinical Subjects	Comments
Device Discovery & Concept[a]	Understand device risk-based classification given impact on development process	Not applicable	Identification of unmet medical need; creation of product concept for a new device
Preclinical Research	Prototyping and testing in controlled laboratory settings	Not for human use	Prototype process attempts to reduce risk of harm in people
Early feasibility	Define and refine device constituent part technical design requirements	Small number of subjects; approval may be based on less nonclinical data than would be needed to support initiation of a larger clinical study on a more final medical device design	Device constituent part may be early in development, before final device/combination product design[b]
Feasibility and early development studies	Preliminary safety and effectiveness	Small number of subjects	Typically informs pivotal study *Risk analysis for specific combined use application may give insights into additional technical design requirements.*
Formative Human Factors studies	User interface assessments: Critical task identification and assessments. Identify the user interface's strengths and weaknesses, potential use errors that may result in harm to the patient/intended user	Varying degrees of formality and sample sizes depending on (1) how much information is needed to inform device design; (2) complexity of the device and its use; (3) variability of the user population; and/or (4) specific conditions of use	Typically done iteratively over the course of design development, with a focus on the user interface: Medical device design, systems, tasks, user documentation, and user training to enhance and demonstrate safe and effective use
	For combination products that include a device constituent part, evaluation of human factors of the device constituent use on the safety and effectiveness of the combination product may be needed. Based on user requirements/risk assessment of critical tasks, where needed, studies assess how users operate the combined use system in realistic conditions. Human Factors studies include an evaluation of all components and accessories needed to operate and properly maintain the device constituent part, e.g., controls, displays, software, logic of tasks, labels, instructions for use, use error hazard, and risk analysis. Human factors evaluations are ideally done throughout the iterative design-build-test process to identify design features that may need adjustment prior to conducting pivotal studies to establish the safety and effectiveness of the combination product (see Chapter 7 on Combination Products Human Factors in this book).		
Pivotal	Collect definitive evidence on safety and effectiveness for specific intended use	Statistically justified number of subjects	
Design Validation	*Demonstrate the medical device constituent part and combination product conforms to defined user needs and intended uses*	Statistically justified number of subjects	*Includes testing of production units under actual- or simulated use conditions; also includes risk analysis and software validation if applicable.*

(*Continued*)

TABLE 5.9 (CONTINUED)
Device Constituent Part Development Clinical Studies

In This Table, Those Clinical Phase Elements Specific to Combination Products Are italicized within Applicable Phases			
Study Type	**Primary Goal**	**Clinical Subjects**	**Comments**
Summative Human Factors Study	*Demonstrate the final device constituent and combination product user interface ensures safe and effective use.*	Minimum of 15 users per distinct user population. Tested by representative users under realistic conditions	*Part of Design Validation*

[a] US FDA (January 2018) "The Device Development Process" at https://www.fda.gov/patients/learn-about-drug-and-device-approvals/device-development-process.

[b] US FDA Guidance (October 2013) "Investigational Device Exemptions (IDEs) for Early Feasibility Medical Device Clinical Studies, Including Certain First in Human (FIH) Studies" at https://www.fda.gov/regulatory-information/search-fda-guidance-documents/investigational-device-exemptions-ides-early-feasibility-medical-device-clinical-studies-including.

alone prior to studies to establish the safety, efficacy, and usability profile of the combination product as a whole.

If a device constituent part has already been approved or cleared for a different purpose, again applying the principles of US FDA's Draft Bridging Guidance,[35] the extent of preclinical testing will focus on the new use/context of use of the device constituent part as part of the combination product. A gap analysis of the differences between the device constituent's approved indication and the new use, coupled with risk analysis on any new or different hazards, hazardous situations, and harms that could arise from the new combined use is necessary. For example, if an autoinjector was previously approved for use with a non-urgent care drug, and is now to be re-purposed for use in an emergency-use indication, human factors and design reliability engineering become critical design aspects. Risk analysis and consideration of a range of potential medicinal product/device interactions, whether desirable or undesirable, is an important aspect of effective combination product development. (See Chapter 7, Combination Products Human Factors, in this book for more detailed discussion on the important topic of human factors.)

Each step of the way, combination product development requires consideration of each constituent part and how its combined use with one or more other constituent parts may impact the safety, effectiveness, and usability of the constituent part and combination product. Consideration should be given to how the device might adversely (or even desirably) impact drug performance, as well as how the drug might impact the device constituent part. For example, excipients in a biological product formulation could interact with the material properties of a delivery catheter, creating byproducts that interact with an infusion pump mechanism, causing the device to stop performing. The interactions need to be considered for each constituent part and for the combination product as a whole.

Table 5.10 summarizes some example performance considerations for each constituent part and potential interactions associated with the combination product as a whole for a self-injection system employing a prefilled syringe.[36] Studies may be needed to evaluate leachables and extractables of the device components/constituent materials with the drug/biologic substance or final combination product. Changes in stability or functionality of the drug constituent part should be assessed when delivered by the device, or when used as a coating on a device (e.g., the drug-eluting stent example in Table 5.6) or, vice versa, changes in the stability or activity of a drug constituent when used together with an energy-emitting device. The stability of a combination product

TABLE 5.10
Example for a Self-Injection System Employing a Syringe

Drug/Biologic	Device (Protection)	Device (Compatibility)	Device (Performance)
• Purity • Stability • Dose Volume • Viscosity • Structure • Elution Profile • Particle Size • Concentration • Acute/Chronic Exposure • Dose-Ranging/Finding • Chemistry, Manufacturing & Controls (CMC) • Pharmacokinetics (Pk) • Local/Regional/ Systemic Toxicities	• Container-Closure Integrity • Permeability • Microbial Ingress • Dimensional Tolerances • Component inter-connectivity	• Biocompatibility • Particles • Extractables/Leachables • Loss of API via Absorption or Adsorption • Contaminants (Tungsten, Silicone Oil) • Surface Interactions, Adverse Chemical Reactions, Particle Formation	• Drug viscosity • Injection Time • Permeability • Siliconization • Shelf life • Coating Integrity • Clarity of Units of Measure • Break-loose/Extrusion/ Activation Force • Back-pressure • Needle Gauge/ Extended Length/ Insertion Depth • Clarity of Dose Completion

as a whole may also differ from that of each of its individual constituent parts. Drug adhesion/ or absorption to device materials, or even the device packaging (e.g., a drug-eluting contact lens that is in direct contact with its package) could alter the delivered dose. Manufacturing process or storage conditions should be studied as they could affect the safety and stability of the drug, or device functionality. For example, a biological product is temperature sensitive and requires storage at refrigerated (2–8°C) conditions. The device constituent part in which the biological product is prefilled may become brittle under cold storage conditions, and potentially break/malfunction at the time of use.

Based on insights drawn from user needs, intended uses/therapeutic effect/use environment, coupled with risk management and development activities, the device and combination product developer should identify those critical requirements essential to the safe and effective operation/ functioning and use of the device constituent and combination product as a whole, e.g., those performance characteristics needed for safe and effective drug delivery. As discussed in Chapter 6, Combination Products Risk Management, "*safety*" is defined as "freedom from unacceptable risk" (ISO 14971:2019, ICH Q9). These *Essential Performance Requirements* (EPRs) need to perform at an acceptable level of risk. In considering which characteristics of the combined use product are considered as EPRs, the following risk-based questions may prove helpful:[37]

- Is the characteristic essential for safe use of the constituent part or combination product based on risk management?
- Is the characteristic essential to achieve delivery of the labeled dose?
- Does the device characteristic impact the drug product's CQAs?
- Does the drug characteristic impact the device's functionality?

Specific guidance on EPRs has been issued for medical electrical equipment[38] and, at the time of this writing, is expected from US FDA with respect to drug/biological product-device combination products. The safety and performance risks that emerge from the combined use of the constituent parts, including clinical and technology risks and complexity of the device constituent, influence which characteristics are considered essential, and their associated reliability, manufacturing control strategy, verification, and validation expectations.

Table 5.11 summarizes some common EPRs by product type.[39] (Also see Chapter 6, Risk Management, in this book.) Other device-specific requirements may include:

- Biocompatibility, ensuring components are biocompatible, commensurate with the level and duration of patient contact. See ISO 10993-1;
- Sterilization and Packaging – demonstrating minimum SAL (10-6) of the fluid path & providing packaging testing of primary sterile barrier after shipping and aging;
- Bacterial endotoxin testing (e.g., USP <85> or ANSI/ AAMI ST 72:2019);
- Software/cybersecurity;[40]
- Electromagnetic compatibility (IEC 60601-1-11:2015);
- Reliability (level determined from risk, i.e., emergency use requires higher reliability); and
- Particulate testing per USP <788> on fluid-contacting portions.

Development Strategy for Generics

An NDA is generally the appropriate pathway for drug-led combination products in the United States, other than generic versions of already-approved drug-led combination products. An Abbreviated New Drug Application (ANDA) is generally the pathway for a drug-led combination product that has the same active ingredient(s), dosage form, strength, route of administration, conditions of use, and labeling as a product (with some differences permitted with respect to Reference Listed Drugs).[41] If the device constituent varies much from the reference device, one can expect approval challenges, so it is helpful to keep the device as similar as possible to the reference to avoid challenges. There can also be issues around combined use of separately distributed products, e.g., where a company wants to bring their own device to market; when can a reference device (or drug) be used; or one vs two application scenarios. If these are scenarios the reader is contending with, the authors encourage your engagement with regulators to avoid surprises.

The proposed generic combination product and its Reference Listed Drug (RLD) do not need to be identical in all respects, though.[42] What is important is ensuring adequate analysis and scientific justification that the user interface of the generic combination product to confirm that any difference in device constituent part and labeling for the proposed generic combination product can be substituted for – yet still produce the same clinical effect and safety profile as – the RLD under the conditions specified in the labeling.

Consistent with the discussion throughout the chapter, the analysis and data/information needed include risk-based assessment of each constituent part and the combination product as a whole. Table 5.12 lists an example of the risk-based information/analysis likely required for various dosage forms using a prefilled syringe (note: this information is not prescriptive).

USP<1> and other relevant USP chapters provide common guidelines for packaging systems or interactions, irrespective of whether or not the product is a combination product. Extractables and leachables information is important, particularly for fluid-path contacting materials (see Chapter 11, Analytical Considerations, in this book). Supporting toxicity studies may be required.

For the device constituent part, functional attribute and performance testing (e.g., for Essential Performance Requirements) comparable to the RLD are important. In our example of a drug-prefilled syringe, that would include assessing dose accuracy and ensuring glide force and break-loose force are comparable to the RLD. FDA 2013 Guidance "Technical Considerations for Pen, Jet and Related Injectors Intended for Use with Drugs and Biological Products" is a helpful resource in identifying performance considerations.

In addition to the technical aspects of the constituent parts and the combination product as a whole, there is a need to ensure the quality systems for applicable facilities aligned to 21 CFR Part 4 requirements (see Chapter 4, cGMPs, in this book).

TABLE 5.11
Common Essential Performance Requirements by Product Type[a]

Essential Performance Requirement (EPR)	Infusion Pump	On-body Injector (OBI)	Prefilled Syringe (PFS)	Auto-injector (AI)	Pen Injector (PI)	Piston Syringe	Metered-Dose Inhaler[b] (MDI)	Nasal Spray[c]
Total Volume Delivered		✓						
Delivery Rate	✓	*						
Time/Phase Dependent	✓	*						
Dose Accuracy or Dose Efficiency (user-filled)	✓	✓	✓	✓	✓	✓	✓	✓
Delivered Dose	✓	✓	✓	✓	✓	✓	✓	✓
Occlusion Detection	✓	✓						
Injection Time		✓		✓				
Actuation Force		✓		✓				
Breakloose/Glide Force			✓		✓	✓		
Extended Needle/Canula Length	✓	✓		✓	✓			
Cap Removal Force (if applicable)		✓	✓	✓	✓			
Needle Safety Lockout Force (if applicable)			✓	✓	✓			
Adhesion/Adhesion Peel Force per Area		✓						
Audible/Visual Feedback (if applicable)	✓	✓	✓	✓	✓			
Aerodynamic Particle Size Distribution (APSD)							✓	
Spray Pattern							✓	✓
Specific Resistance to Air Flow							✓	✓
Plume Geometry							✓	✓
Spray Velocity							✓	✓

*For some more recent technology evolutions of OBI's, they are able to deliver high volume doses, and as such, based on intended use, delivery rate and/or time/phase dependence may be applicable.

[a] Alan Stevens, CDRH, Xavier Combination Products Summit, 2019, 2020, 2021.

[b] FDA Draft Guidance (2018) "Metered Dose Inhaler (MDI) and Dry Powder Inhaler (DPI) Drug Products- Quality Considerations" accessed August 3, 2022 at https://www.fda.gov/regulatory-information/search-fda-guidance-documents/metered-dose-inhaler-mdi-and-dry-powder-inhaler-dpi-drug-products-quality-considerations.

[c] FDA Draft Guidance (2002) "Nasal Spray and Inhalation Solution, Suspension, and Spray Drug Products – CMC Documentation," accessed August 3, 2022 at https://www.fda.gov/regulatory-information/search-fda-guidance-documents/nasal-spray-and-inhalation-solution-suspension-and-spray-drug-products-chemistry-manufacturing-and.

TABLE 5.12
Example Analyses Required for a Drug-Led Combination Product ANDA relative to RLD[a]

Device	Drug	Combination Product (Injectable Drug Example)

Quality
Labeling
Biocompatibility
Comparative Analysis
Functional attributes/ performance testing:
• ***Dose Accuracy for drug design space (expelled volume – USP <697>)***
• ***Glide force comparable to RLD***
• ***Break-loose force comparable to RLD***
	Drug Substance:
	• Physicochemical properties: Description, pH, solubility, chirality, hygroscopicity, etc.
	• Impurities: Organic, inorganic, residual solvents
	• Specifications (Type II DMF or USP, as applicable)
	Drug product:
	• Microbiology
	• Manufacturing process
	• Facilities
	• Critical Quality Attributes, e.g., under USP
	  • Identification
	  • Assay
	  • Impurities
	  • Container content
	  • Container-closure integrity
	  • Particulate matter in injection
	  • Visible particulates in injections
	  • Sterility
	  • Bacterial endotoxin tests or pyrogen tests
	  • Residual solvents
	  • pH (if applicable)
	  • Osmolarity (if applicable)
	  • Viscosity (if applicable)
	  • Conformance to USP, ISP <232>/ ICH Q3D
	  • Packaging- USP <381>, USP <659>, <660>, <661>, <661.1>, <661.2>, <1663>, <1664>
		Quality
		Labeling
		Bioequivalence
		Comparative Analyses[b]
		Functional attributes/performance testing
		• Injection Solutions
		  For prefilled syringes:[c]
		• ***Dose Accuracy (expelled volume – USP <697>)***
		• ***Glide force comparable to RLD***
		• ***Break-loose force comparable to RLD***
		  For multiple-dose containers
		• Antimicrobial preservatives
		Sterile Powders (Lyo or Powder Fill) for Injections
		• Uniformity of dosage units, USP<905>
		• Completeness and Clarity of Solution
		• Water or Solvent Content
		• Reconstitution Time
		Injection Suspensions:
		• Uniformity of Dosage Units <905>
		• Particle Size Distribution
		• Dissolution
		• Re-suspendability
		• Sedimentation Rate and Volume
		• **Syringeability**[d]
		Injection Emulsions
		• Globule Size Distribution in Lipid Injectable Emulsions, USP<729>
		Packaging System:
		• Sterility/Container-Closure Integrity Testing (CCIT) of Primary CCS
		• Qualification of CCS: USP <381>, USP<660>, USP <661>
		• Glass delamination
		• Extractables and Leachables

[a] Adapted from presentation by Ashish Rastogi, PhD, FDA, "Generic Combination Products: Assessments and Regulatory Update" at Generic Drugs Forum 2020 (April 16, 2020), accessed August 9, 2022.

[b] See draft guidance for industry Comparative Analyses and Related Comparative Use Human Factors Studies for a Drug-Device Combination Product Submitted in an ANDA (January 2017). When final, this guidance will represent the FDA's current thinking on this topic.

[c] FDA Guidance "Technical Considerations for Pen, Jet, and Related Injectors Intended for Use with Drugs and Biological Products" (June 2013).

[d] Syringeability is the ease of withdrawing of a solution from vial to syringe, whereas injectability is "a term related to the ease of parenteral administration of a dosing solution, and includes dose preparation, dose administration, ergonomics related to these procedures, pain of injection, and other adverse events at the injection site" (E. Reid (2021), "Syringeability vs. Injectability" at https://blog.rheosense.com/syringeability-vs-injectability#:~:text=Syringeability%20is%20the%20ease%20of%20withdraw%20of%20a,other%20adverse%20events%20at%20the%20injection%20site%20%22.

Development Strategy: Manufacturing Considerations[43]

The scientific and technical aspects of manufacturing each constituent part of a combination product, and then bringing them together as a combination product, are likewise important to consider as part of the combination product development strategy. Table 5.13 summarizes the high-level differences between manufacture of biological product, drug and device constituent part, and ultimately combination product. Manufacturing Practices[44] are important for premarket development and post-market lifecycle management of combination products. For example, exposing a temperature-sensitive drug-prefilled syringe to the heat of a blister-sealing process in manufacturing can lead to drug degradation and shortened stability for the drug contained in that syringe. Post-assembly processing, such as application of certain terminal sterilization techniques, can alter the critical quality attributes of a drug or biological product. Manufacturing methods and controls need to address such drug/device interactions.

TABLE 5.13
Critical Process Considerations for Constituent Parts and Combination Product

Biological Product Manufacturing Process		Drug Manufacturing Process		Device Manufacture & Combination Product Assembly Process
Starting Materials	• Master cell bank • Media • Disposables	**Starting Materials**	• Precursors • Solvents • Reactors	• Resin Materials (or Glass or Metals) • Plasticizers, Coatings • Mold Tooling • Injection Molding or Machining
Drug Substance	• Preculture → expansion → harvest/capture → viral inactivation → chromatography → concentration → storage	**Drug Substance/ API**	• Chemical Synthesis storage	⇩ • Components • Subassemblies • Fill/Finish Assemblies integral (or not) with the drug and/or biological product
Drug Product	• Thawing → pooling/ mixing → filtration → aseptic filling → storage	**Drug Product**	• Blending → granulating → drying → excipients/fillers → tableting or vials → sealing	⇩ • Finished Combination Product • Potential Post-processing steps (e.g., sterilization, packaging)
Combination Products Manufacturing Process Development & Validation ⟷				

Once preclinical and clinical studies begin, potential changes to the design, formulation, or manufacturing process for the drug, biological product or device constituents, and/or their integration or post-processing, may affect the safety or effectiveness of the combination product as a whole. For example, changes in concentration, inactive ingredients, software, or in methods to combine two constituent parts, could affect the performance characteristics of the combination product. Inconsistencies in manufacturing processes can likewise impact combination product safety, efficacy, and usability. A few examples: (1) increased variation in component parts from injection molding with increased mold/tooling cavitation can lead to sub-assembly mismatches impacting seal integrity; (2) label material changes require consideration of label adhesives, which might migrate and interact with a drug constituent part, impacting drug stability and performance (see Chapter 11, Analytical Chapter in this book); (3) inconsistent cellular incubation time and methods before application of cellular constituent parts to a device can cause variation in performance characteristics.

A case in point highlighting the criticality of manufacturing process and design change control, pre-dating combination product-specific cGMP regulations, is that of epoetin-associated pure red-cell aplasia.[45] According to Bennet et al., "a confluence of factors related to the production, handling, and route of administration of epoetin may account for ... increased incidence of epoetin-associated pure red-cell aplasia." Multiple changes occurred – to the manufacturing process and to the constituent parts associated with epoetin production and administration – which, together, had deadly consequences:

1. Manufacturing processes used, like freeze drying, facilitated oxidation or aggregation of protein, enhancing immunogenicity.
2. A shift from intravenous administration to subcutaneous administration, with problems in storage and handling, also had the potential to induce antibody formation.
3. The biological product formulation was changed to incorporate a different biosurfactant, inadvertently interacting with/solubilizing organic leachables from rubber plungers.
4. Silicone oil was used as a lubricant in prefilled syringes, another potential source of increased immunogenicity.

Design controls for device constituents and the combination product as a whole could anticipate changes to manufacturing as part of a multi-generational plan during investigational development. Development of new manufacturing methods, in-process controls/characterization testing, test methods and specifications, and other approaches to evaluate changes in the constituent parts and for the combination product as a whole are risk-based underpinning to good manufacturing practices. For certain developmental changes, as mentioned in the section "Development Strategy: Medicinal Product (drug and biological product) Constituent Part Considerations", *in vitro-*, preclinical, and clinical bridging studies[46] may be appropriate.

Changes associated with purchased constituent parts are a critical aspect of manufacturing controls. Arrangements (documented) with the manufacturers of constituent parts are important to ensure sufficient awareness of manufacturing changes to constituent parts that may occur – pre-market or post-market – to ensure safety and effectiveness of the combination product. The combination product manufacturer needs enough time to assess the impact(s) of a proposed change in a way commensurate with the risk and complexity of change, and given the particular stage of combination product development. Post-market manufacturing changes require careful review prior to implementation. Dependent on the scope and nature of a change, design verification and design validation may need to be repeated, and prior approval may be needed before marketing.[47] *Note that regulatory filing requirements associated with such changes are different and evolving across jurisdictions globally, so the reader is encouraged to familiarize themselves with the applicable country and regional requirements (see chapters on Combination Products Regulatory Strategies (3), Combination Products Global Regulatory Frameworks (14), and Combination Products Lifecycle Management (8) in this book).*

CONCLUSIONS

Innovative drug, biological product, and device combined use applications hold great promise for advancing patient care, making treatments safer, more effective, or more convenient and acceptable to patients. Achieving these potential benefits is not without hurdles:

(1) Lack of consistent interpretation of key terminology between pharmaceutical, biotech, and medical device development practitioners – coupled with similar "language" barriers with suppliers and health authorities within and across jurisdictions – makes this an important factor to overcome.
(2) Integration of drug/biological product Quality by Design and device Design Controls product and process development approaches, founded on risk management, is critical to address distinct/additional questions relative to the joint compatibility, safety, efficacy, and usability of each constituent part and their combined use. If medical products are intended to be used together, their combined use needs to be vetted through risk analysis, and needed scientific and technological controls defined to ensure their joint compatibility, safety, efficacy, and usability.
(3) Pharmaceutical, biotech, and medical device practitioners/developers each bring unique perspectives, skillsets, and know-how to the development efforts; they are each accustomed to operating under different regulatory paradigms, and certain critical developmental issues, like the interaction(s) of the drug/biologic and device constituents, might not be readily apparent to individual practitioners. Effective collaboration, bringing together the diverse skillsets, knowledge, and perspectives of the multidisciplinary stakeholders, is crucial for successful combination product development.
(4) Often constituent parts, components, or services are outsourced for combination products. Outsourcing the design, or using an existing platform, changes the "design" process, necessitating robust collaboration with suppliers to assure rigor during technology evaluation, selection, development, transfer, and lifecycle management.

There are a plethora of combination product regulations, standards, and guidance available across global jurisdictions to inform and address the added complexity, and continuum of risks, presented by the combined use of drug/biological product and devices. A systems approach that consistently vets the potential interactions of these drug/device and biological product constituent parts for their specific combined use application will help ensure the safety, effectiveness, and quality of the combination product.

NOTES

1. Medicinal products are drugs and/or biologics; definitions for medicinal products and devices vary from region to region. Generally medicinal products achieve their primary intended purpose through chemical, pharmacological, immunological, or metabolic action, while medical devices do not. [US FDA §201(g); FDCA §201(h); PHSA §351; https://www.ema.europa.eu/en/glossary/medicinal-product; EU MDR (2017/745) Article 2).
2. Six Sigma Daily (August 17, 2020) "What is Forming, Storming, Norming and Performing?" at www.sixsigmadaily.com/what-is-forming-storming-norming-performing/.
3. FDA Guidance: Identification of Manufacturing Establishments in Applications Submitted to CBER and CDER Questions & Answers (October 2019), accessed July 2, 2022 at https://www.fda.gov/regulatory-information/search-fda-guidance-documents/identification-manufacturing-establishments-applications-submitted-cber-and-cder-questions-and.
4. Definitions vary globally. See ASTM International Combination Product Terminology Standard.
5. Or partnering with a third party to ensure a purchased device constituent part can serve such purpose.
6. Importantly, in the United States, FDA has stated that Contract Manufacturing Facilities fall in scope of the term "Combination Product Manufacturer" (see Table 5.1 and US FDA Compliance Program 7356.000 Inspections of CDER-led or CDRH-led Combination Products). In 78 FR 4313-4314 Comment 15, FDA states that cGMP requirements applicable to a constituent part come into effect even from the components that form the constituent part. The trigger is whether the facility is conducting manufacturing operations that would be subject to the underlying cGMP requirements.
7. The term "Sponsor" is used interchangeably with "Market Authorization Holder (MAH)," "Market Authorization Applicant," and/or "Legal Manufacturer."

8. US FDA Guidance "Principles of Premarket Pathways for Combination Products" (January 2022), accessed July 2, 2022 at https://www.fda.gov/regulatory-information/search-fda-guidance-documents/principles-premarket-pathways-combination-products.
9. In the United States, drug–biological product combinations also meet the definition of "combination product" under 21 CFR 3.2(e). While drug-biologics are an important category of products, in most jurisdictions, drugs and biological products are regulated under the one regulatory framework, i.e., medicinal products. In this chapter, we will focus on medicinal product (including both drug and biologics) – medical device combinations. The term drug and medicinal product are used interchangeably for simplicity.
10. Safety is defined as "freedom from unacceptable risk." [ISO 14971:2019 (3.26)]; Risk is defined as "the combination of the probability of the occurrence of harm and the severity of that harm."
11. US FDA Current Good Manufacturing Practice Requirements for Combination Products Guidance (January 2017), at https://www.fda.gov/regulatory-information/search-fda-guidance-documents/current-good-manufacturing-practice-requirements-combination-products#:~:text=This%20guidance%20describes%20and%20explains%20the%20final%20rule,4%29%20that%20FDA%20issued%20on%20January%2022%2C%202013.
12. AAMI TIR 48:2015 Quality Management System (QMS) Recommendations on the Application of the U.S. FDA's CGMP Final Rule on Combination Products.
13. EU MDR(2017/745 Article 117), at https://eur-lex.europa.eu/eli/reg/2017/745/2017-05-05.
14. Malaysia Medical Device Authority "Guideline for Registration of Drug-Medical Device and Medical Device-Drug Combination Products" at https://portal.mda.gov.my/industry/medical-device-registration/combination-product.html; National Pharmaceutical Regulatory Agency MoHM "Medical Device-Drug-Cosmetic Interphase Products" at https://www.npra.gov.my/index.php/my/classification-guideline/product-classification-guideline-medical-device-drug-cosmetic-interphase-products.html.
15. Health Canada "Drug/Medical Device Combination Products" (2006) at https://www.canada.ca/en/health-canada/services/drugs-health-products/drug-products/applications-submissions/policies/drug-medical-device-combination-products.html.
16. In its publications from 2017 (SMJ 06/2017), 2019 (SMJ 10/2019), and 2020 (SMJ 05/2020), Swissmedic provided information on integral and non-integral combination products. For the practical implementation of Art. 117 MDR, Swissmedic aligns itself to the EU's specification documents, in particular: (1) Guideline on the quality requirements for drug-device combinations (EMA/CHMP/QWP/BWP/259165/2019) (currently in draft form) and (2) Questions & Answers on Implementation of the Medical Devices and In Vitro Diagnostic Medical Devices Regulations ((EU) 2017/745 and (EU) 2017/746) (EMA/37991/2019).
17. NMPA Announcement No. 3 of 2022 at https://www.nmpa.gov.cn/xxgk/ggtg/qtggtg/20220117145645132.html; NMPA Notice on Matters Concerning the Registration of Drug-device Combination Products (July 23, 2021), at http://english.nmpa.gov.cn/2021-07/27/c_661024.htm.
18. Australian TGA "Biologicals packaged or combined with another therapeutic good" (2018) at https://www.tga.gov.au/biologicals-packaged-or-combined-another-therapeutic-good.
19. Saudi Arabia FDA Draft Guidance for Combination Products (February 2020) at https://old.sfda.gov.sa/en/oper/Documents/GuidanceCombinationProductsClassification.pdf.
20. See recognized consensus standards, ISO 13485:2016 Clause 7.3 and its subclauses, AAMI/ANSI/ISO 14971, ISO 24971, IEC TR80002, ICH Q9 (and ICH Q9(R1)), and AAMI TIR 105. In the United States, also see 21 CFR 820 (Device Quality Management System Regulations), 21 CFR 210 & 211 (Drug cGMPs); and 21 CFR 600-680 (Biologic cGMPs).
21. Medical Devices Regulation ((EU) 2017/745) (MDR) and In vitro Diagnostic Medical Devices Regulation ((EU) 2017/746) (IVDR).
22. Questions & Answers for applicants, marketing authorization holders of medicinal products and notified bodies with respect to the implementation of the Medical Devices and In Vitro Diagnostic Medical Devices Regulations ((EU) 2017/745 and (EU) 2017/746) (June 23, 2021) at https://www.ema.europa.eu/en/documents/regulatory-procedural-guideline/questions-answers-implementation-medical-devices-vitro-diagnostic-medical-devices-regulations-eu/745-eu-2017/746_en.pdf.
23. Early Development Considerations for Innovative Combination Products *Guidance for Industry and FDA Staff (September 2006)* at https://www.fda.gov/regulatory-information/search-fda-guidance-documents/early-development-considerations-innovative-combination-products.
24. "FDA Guidance for Industry and Staff: Early Development Considerations for Innovative Combination Products" at https://www.fda.gov/regulatory-information/search-fda-guidance-documents/early-development-considerations-innovative-combination-products.
25. Hazard: Potential source of harm (ISO 14971:2019; ICH Q9(R1).

26. Hazardous situations: Circumstances in which people, property or the environment are exposed to one (or more) hazard(s) (ISO 14971:2019).
27. Harm: Injury or damage to the health of people, or damage to property or environment (ISO 14971:2019; ICH Q9(R1)); Injury includes physical or mental injury that can occur from loss of product quality or availability (ISO/IEC Guide 6.3:2019, Clause 3.1 and ICH Q9(R1)).
28. AAMI TIR 105:2020, "Combination Products Risk Management," is a helpful resource.
29. Relative to the clinical development phases, clinical investigations for combination products are typically done under one investigational application for the combination product as a whole. In the United States, an Investigational New Drug (IND) application or an Investigational Device Exemption (IDE) application is submitted.

    [*Note: Clinical Guidance documents that may be helpful for combination products developers include FDA "Guidance to Industry, Investigators & Reviewers: Exploratory IND Studies" (January 2006)(https://www.fda.gov/regulatory-information/search-fda-guidance-documents/exploratory-ind-studies) and "Changes or Modifications During the Conduct of a Clinical Investigation; Final Guidance for Industry and CDRH Staff" (May 2001) (https://www.fda.gov/media/72429/download). The IND Guidance provides approaches for evaluating candidate products during research and development before selecting the composition for further development, while the "Changes" Guidance clarifies how to manage changes that may occur during investigational development of a device constituent part.*]

30. One of the challenges underlying combination product development is determining the clinical development strategy. Common questions surround trial design, sample size, statistical techniques, clinical endpoints, number of clinical studies, and appropriate indications for use or claims. Decisions should generally be commensurate with the risk and complexity of the science and technology for each of the constituent parts and the combination product as a whole. For example, approaches to evaluate drug-device interactions may raise questions or be of be of concern. The reader is encouraged to reach out to FDA or other pertinent health authority to address questions or concerns associated with combined use clinical trials.
31. Yu, L.X. et al. (2014) Understanding Pharmaceutical Quality by Design. *The AAPS Journal* 16 (4):771-783. Epub 2014 May 23.
32. FDA Draft Guidance "Bridging for Drug-Device and Biologic-Device Combination Products" (December 2019, accessed on July 2, 2022 at https://www.fda.gov/regulatory-information/search-fda-guidance-documents/bridging-drug-device-and-biologic-device-combination-products.
33. US FDA 21 CFR 820.30 or ISO 13485:2016 §7.3 and its subclauses, in conjunction with ISO 14971:2019/A11:2021 and ISO/IEC 62366-1:2015.
34. FDA "Preclinical Research," accessed on July 2, 2022 at https://www.fda.gov/patients/drug-development-process/step-2-preclinical-research and 21 CFR Part 58 "Good Laboratory Practice for Nonclinical Laboratory Studies" at https://www.accessdata.fda.gov/scripts/cdrh/cfdocs/cfcfr/CFRSearch.cfm?CFRPart=58.
35. FDA Draft Guidance "Bridging for Drug-Device and Biologic-Device Combination Products" (December 2019, accessed on July 2, 2022 at https://www.fda.gov/regulatory-information/search-fda-guidance-documents/bridging-drug-device-and-biologic-device-combination-products.
36. Neadle, S.; Riter, J.; McAndrew T.P. (Pharmaceutical Online, Nov. 30, 2020) "*Considerations in Combination Product Risk Management*"; https://www.pharmaceuticalonline.com/doc/considerations-in-combination-product-risk-management-0001.
37. Questions derived from ICH Q12 "Draft Guidance – Technical and Regulatory Implementation Considerations for FDA-Regulated Products" (May 2021).
38. IEC 60601-1-11:2015: Medical electrical equipment – Part 1–11: General requirements for basic safe and essential performance – Collateral standard: Requirements for medical electrical equipment and medical electrical systems used in the home healthcare environment.
39. See FDA Guidance titled "*Guidance for Industry and FDA Staff: Technical Considerations for Pen, Jet, and Related Injectors Intended for Use with Drugs and Biological Products*" – for list of additional potential EPRs and relevant standards; Also reference Alan Stevens, CDRH, Xavier Combination Products Summit, 2019, 2020, 2021.
40. FDA "Draft *Guidance for the Content of Premarket Submissions for Software Contained in Medical Devices" (November 2021)* and "*Content of Premarket Submissions for Management of Cybersecurity in Medical Devices" (October 2014).*
41. FDA Draft Guidance "Evaluation of Therapeutic Equivalence" (July 2022) https://www.fda.gov/media/160054/download states that "the evaluation of whether drug products have the same clinical effect and

safety profile is product-specific. For example, whether a proposed generic drug-device combination product with a user interface that contains differences from that for the Reference Listed Drug (RLD) can be substituted with the full expectation that the generic combination product will produce the same clinical effect and safety profile as the RLD under the conditions specified in the labeling is a product-specific determination, and additional information and/or data relating to the user interface may be appropriate to support approval and to perform this evaluation." Therapeutic equivalence evaluations are made between an ANDA and its RLD at time of approval, including for ANDAs for drug/device combination products. A generic combination product classified as therapeutically equivalent to the RLD can be expected to produce the same clinical effect and safety profile as the RLD under the conditions specified in the labeling.

42. See the Office of Combination Products guidance for industry and FDA staff Principles of Premarket Pathways for Combination Products (January 2022).
43. See also Chapter 8, Lifecycle Management.
44. FDA Guidance "Current Good Manufacturing Practice Requirements for Combination Products" (January 2017), accessed July 2, 2022 at https://www.fda.gov/regulatory-information/search-fda-guidance-documents/current-good-manufacturing-practice-requirements-combination-products.
45. Charles Bennet, MD, PhD, et al., "Pure Red-Cell Aplasia and Epoetin Therapy" New England Journal of Medicine 351 (September 30, 2004).
46. FDA Draft Guidance "Bridging for Drug-Device and Biologic-Device Combination Products" (December 2019, accessed on July 2, 2022 at https://www.fda.gov/regulatory-information/search-fda-guidance-documents/bridging-drug-device-and-biologic-device-combination-products.
47. FDA Draft Guidance "Submissions for Postapproval Modifications to a Combination Product Approved Under a BLA, NDA, or PMA" (January 2013), accessed July 2, 2022 at https://www.fda.gov/media/85267/download.

6 Combination Products Risk Management

Susan W. B. Neadle, Richard Wedge, and Ed Bills

CONTENTS

DOI: 10.1201/9781003300298-7

INTRODUCTION

What Is "Risk"?

What is "risk?" As defined in both ISO 14971:2019 "Medical devices – Application of risk management to medical devices" and ICH Q9 drug "Quality Risk Management," "**risk**" is "a combination of the probability of occurrence of harm and the severity of that harm."[1,2] The term "**harm**" is interpreted as (physical, emotional, or mental) injury or damage to the health of people or damage to property or environment. Under ISO/IEC Guide 63:2019, Clause 3.1 and ICH Q9, *damage to health* includes damage that can occur from loss of product quality or availability. Safety is defined as "freedom from unacceptable risk."

In these definitions, we are talking about more than just the patient – the definitions apply to all people that may come in contact with the medical product or its use – patients, caregivers, healthcare providers, etc. Manufacture and use of medical devices, medicinal products (drug products and/or biological products), and combination products – including their components – necessarily entail some degree of risk. Appropriate risk-based decision-making across the product lifecycle and actions commensurate with the risk and complexity of a given product help ensure the **safety** and effectiveness of these medical products for intended uses, by intended users, and in intended use environments.

What Is Risk Management?

"Risk management" is the systematic application of management policies, procedures, practices, insight/judgment and experience to the identification, analysis/evaluation, monitoring, and subsequent control/mitigation of risk. It serves as a framework for the identification and control of critical requirements essential to the safe and effective functioning and use of medical products. Risk management underpins the entire product lifecycle (Figure 6.1).

The application of risk management during premarket phases helps serves as a framework for product developers. It guides thinking, encourages questions, and brings discipline to help in the elimination or reduction of risks, proactively ensuring product safety, efficacy, and usability. Post-market, active surveillance and monitoring enable reactions commensurate with risk – e.g., through Change Control and/or Corrective and Preventive Actions, Supplier/Purchasing Controls, and Post-Market vigilance and reporting – ensuring that risks continue to be acceptable.

Risk management drives focus and standardization of health and safety considerations from development through post-market to final disposal of the device. It is also the foundation for communication among key stakeholders with respect to that product. Risk management acts as an interface between individuals, departments, manufacturing sites, or operating units within an organization, manufacturers and vendors and with regulators, users, and the public. Done effectively, risk management helps sponsors demonstrate their understanding of the science and technology and can serve to help give regulators confidence and likewise be mutually grounded in the science and technology of what you are out to achieve.

Chapter Scope

The focus of this chapter is specifically on risk management for combination products and other combined-use systems. The chapter begins with a high-level overview of combination products risk management considerations, and then delves more deeply into the topic, including:

- Key terminology as a basis for understanding;
- Considerations for a successful cross-functional risk management team;
- High-level process;
- Combination product-specific unique risk elements, including human factors and essential performance requirements (*see also Chapter 5 (Product Development) and Chapter 7 (Human Factors) in this book*);

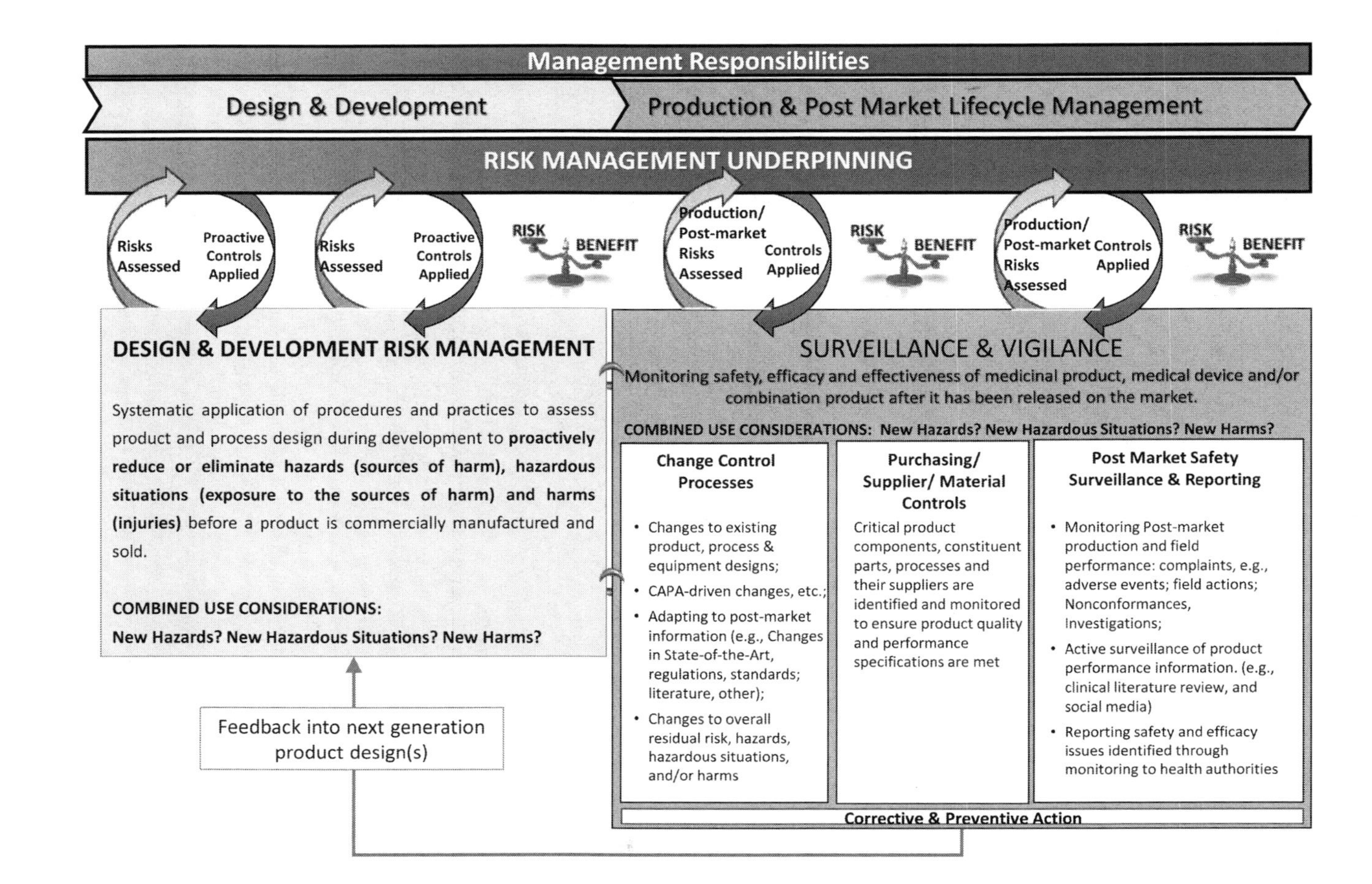

FIGURE 6.1 Combination products risk management: An underpinning of the combination products lifecycle (©2022 Combination Products Consulting Services, LLC. All Rights Reserved).

- A deeper dive into a modular stepwise process for combination product risk management, including key expectations and best practices;[3]
- Considerations for use of third-party suppliers;
- Risk management for components/constituent parts not developed under ISO 14971; and
- Finally, we will review and discuss the possibilities for a drug-agnostic risk assessment approach to help drive efficiencies in the development of medicinal-product-led combination products without compromising the critical proactive consideration of joint use of the differently regulated constituent parts.

Throughout the chapter, the authors underscore the critical connections between combination product development and post-market risk management activities to ensure safe and effective use of the combination product by intended users, for intended uses, in intended use environments. The authors hope that this chapter brings lasting value to those engaged in the combination products – and more specifically, combination products risk management space! Enjoy!

RISK MANAGEMENT FOR COMBINATION PRODUCTS AND OTHER COMBINED-USE SYSTEMS

High-Level Overview

Systematic application of procedures and practices to assess the product and process design and risks of joint use of the constituent parts during Design & Development enables proactive elimination or mitigation of (1) potential sources of harm; (2) circumstances where people, property or the environment might be exposed to the sources of harm; and (3) harms – before a product is commercially manufactured and sold. After launching the product, the manufacturer monitors the results over the life of the product to ensure the risks continue to be acceptable. Systematic risk management practices continue through post-market lifecycle management monitoring and control of the safety and effectiveness of each constituent part and the combination product as a whole. The development and manufacturing activities are inextricably linked to post-marketing safety reporting, where safety and efficacy issues identified through monitoring are reported upon to health authorities (see Chapter 8, Post-Market Lifecycle Management), and may trigger post-market design or process changes, preventive and/or corrective actions, or even new supplier controls. Risk management and weighing benefits versus risks is central to the combination products we develop and market to make a positive impact on peoples' lives.

Importantly, risk management isn't just about the ***safety*** *of products. It also includes not missing out on their* ***effectiveness****. When done properly, risk management is a truly comprehensive activity in relation to medical product development, study, manufacture, and marketing. Harm includes not just experiencing a negative effect of a product but also not receiving the intended benefit of that product.*

Globally, regulatory frameworks intend to drive successful practices and control strategies throughout the medical product lifecycle to assure public health, including safety, efficacy, and usability for intended uses, by intended users, in intended use environments. Risk Management relies on the use of available knowledge and science to identify harms and hazards and to evaluate risks. Where, after considering product benefits, the risk is determined to be unacceptable, manufacturers must use methods to reduce that risk, i.e., the manufacturer should eliminate harms if possible, and then look to mitigations.

> *Importantly, risk management isn't just about the* ***safety*** *of products. It also includes not missing out on their* ***effectiveness****. When done properly, risk management is a truly comprehensive activity in relation to medical product development, study, manufacture, and marketing. Harm includes not just experiencing a negative effect of a product, but also not receiving the intended benefit of that product.*

In the combination products space, the combined use of two or more differently regulated constituent parts with one another inherently raises questions about potential interactions of constituent parts with one another and may lead to a range of new and significant risks and risk management considerations based on where a product, or its use, sits along a continuum of risk. There is a need to consider more than one medical product in risk analysis and control strategy.

Importantly, one should restrain themselves from the temptation to treat drug risk management as a complete separate track of activities from device risk management, converging them only "after-the-fact." This creates a reactive situation. Proactive thinking about joint use risks is warranted, asking the right questions early enough in the development process (see also Chapter 5, Integrated Development). A unique aspect of many combination products is that Sponsors often purchase the constituent part that is secondary to their organizations' core competencies. In the Pharma/Biotech world, for example, pharmaceutical or biotech manufacturers often purchase the device constituent part of their combination product from a third party. The device constituent already exists. Caution should be taken: Even the selection of a constituent part for purchase should trigger integrated thinking about joint-use risks and approaches. While a Sponsor can leverage the risk management data/information from the third-party supplier, the Sponsor should not over-rely on that data, which was very likely established independent of any drug-device interaction considerations.

Rather, a Sponsor *must* proactively evaluate the unique elements and considerations for the specific combination product application. And if you, as Sponsor, *are* developing your own device constituent part for use in a combination product, you really *want* an integrated approach of development and risk management of the device with the drug. The questions raised by the joint use of drug and device should be proactively addressed throughout the development process.(Even for the section of this chapter, "Drug-agnostic Combination Products Risk Management," one needs to ensure proactive consideration of risks for combined use early on to avoid unwelcome surprises late in development – or worse, post-market.)

The Association for Advancement of Medical Instrumentation (AAMI) issued a helpful standard, AAMI TIR 105:2020, "Combination Products Risk Management" that highlights best practices for combined-use applications. As discussed in this Technical Information Report (TIR), requirements imposed on each constituent part and the overall combined-use product should be based on the scientific and technological considerations raised by the intended combined use of the medical products. Risk management, therefore, serves as the underpinning throughout combination product development and post-market lifecycle management. AAMI TIR:105:2020, in conjunction with ISO 14971:2019, ISO TR 24971:2020 "Medical Devices-Guidance on the Application of ISO 14971," and ICH Q9 (and ICH Q9(R1)) are foundational standards for this chapter.

ISO 31000

Notably, some organizations confuse broader business risk management with that focused specifically on product risk management. ISO 31000:2018 "Risk Management – Guidelines" is an international standard that provides principles, a framework, and a process for managing uncertainties on business objectives, i.e., business risks. ISO 31000 is applicable and adaptable to a wide gamut of industries and businesses – for "any public enterprise, private enterprise, association, group or individual." Its scope is broad, including, but not limited to: compliance with legal and regulatory requirements and international standards; financial management; business management; business continuity; stakeholder confidence; robust decision-making and planning; improving controls; resource allocation; operational effectiveness and efficiency; improving prevention and incident management; minimizing losses; improving organizational learning and resilience.

The umbrella of broader business risks does include aspects of product risk management. For example, business continuity is a key consideration for a business – whether in the face of severe weather conditions or ability to continue operations in the face of raw material supply shortages.

Mitigations a business might implement include having secondary source suppliers, secondary storage facilities, back-up generators, "cushioned" raw material inventory levels – all aimed at ensuring sustained business operations and product supply in the event of extenuating circumstances. Under ISO 14971 and ICH Q9, harm includes injury or damage to health that could occur from lack of product quality or availability, so the ISO 31000 risk mitigations a business takes to ensure business continuity may very well overlap with risk controls needed for product risk management. Product risk management activities such as qualification of alternate suppliers and materials, the associated product development, bridging, post-market change management and regulatory activities to support such supply continuity demonstrate some of the overlap between ISO 31000 and the specific focus of product risk management under ISO 14971 and ICH Q9.

A significant difference, though, is that ISO 31000 does not define "harm" and rather focuses on cost-effective risk management decision-making. In the product risk management-realm of ICH Q9 and ISO 14971, cost is not the consideration. In medical device/medicinal product/combination product risk management, broadly speaking, judgments are made as to whether the expected benefits of a medical product outweigh the potential risks associated with the product's intended use. This chapter is not intended to delve into the broader umbrella of ISO 31000 business risk management. Rather, this chapter is focused specifically on combination product risk management, a product safety risk activity.

KEY TERMINOLOGY

Table 6.1 summarizes key combination products risk management terminology that are used throughout this chapter.

TABLE 6.1
Key Terms and Definitions[a]

Term	Definition
Benefit	Positive impact or desirable outcome of the use of a medical product on the health of an individual, e.g., positive impacts to clinical outcome, quality of life, or diagnosis or positive impact on patient management or public health.
Control Strategy	A planned set of controls based on product and process understanding that ensure a process performs as it should, to ensure product quality is maintained.
Critical Material Attribute (CMA)	A physical, chemical, biological, or microbiological property or characteristic of an input material that should be within an appropriate limit, range, or distribution to ensure the desired quality of output material.
Critical Process Parameter (CPP)	A process parameter whose variability impacts a CQA or CtQ and therefore should be monitored or controlled to ensure the process produces the desired quality. (ICH Q8(R2))
Critical Quality Attribute (CQA)	A physical, chemical, biological, or microbiological property or characteristic that should be within an appropriate limit, range, or distribution to ensure the desired product quality (e.g., identity, strength/potency, purity, and safety). (ICH Q8(R2))
Critical to Quality (CtQ)	The quality of a product or service in the eyes of the voice of the customer; typically used to refer to the critical functionality/features of a medical device.
Critical Use Tasks	CDRH: "A user task which, if performed incorrectly or not performed at all, would or could cause ***serious*** harm to the patient or user, where harm is defined to include compromised medical care."[b]
	CDER: "User tasks that, if performed incorrectly or not performed at all, would or could cause harm to the patient or user, where harm is defined to include compromised medical care."[c] 1. Consider harm to both the patient or user, such as a caregiver or healthcare professional. 2. Dosing, such as dialing a proper dose using a pen injector. 3. Administration, e.g., the step of selecting the correct injection site. 4. Urgency/time. In some cases, e.g., emergency use-administration of epinephrine to treat a severe allergic reaction, drugs must be administered as quickly as possible to be effective.

(*Continued*)

TABLE 6.1 (CONTINUED)

Key Terms and Definitions[a]

Detectability	The ability to discover or determine the existence, presence or fact of a hazard.
Essential Performance Requirements	Critical requirements, achieved at an acceptable level of risk, that are essential to the safe and effective operation/functioning and use of a medical product, i.e., those performance characteristics needed for safe and effective dose delivery (based on IEC 60601-1-11:2015).
Harm	Injury or damage to the health of people, or damage to property or environment; injury includes physical or mental injury, including the damage that can occur from loss of product quality or availability. (ISO/IEC 63:2019)
Hazard	Potential source of harm. (ISO/IEC Guide 51)
Hazard Identification	The systematic use of information to identify potential sources of harm (hazards).
Hazardous Situation	Circumstances in which people, property or the environment are exposed to one (or more) hazard(s). (ISO/IEC Guide 63:2019).
Intended Use	Use for which a product, process or service is intended according to its specifications, instructions and information provided by the manufacturer. (ISO/IEC Guide 63:2019). Included in intended use includes intended users, target patient population, intended use environment, and operating principles of a given product.
Predictable Misuse (also called "Reasonably Foreseeable Misuse")	Use of a product or system in a way not intended by the manufacturer, but which can result from readily predicted human behavior. (ISO/IEC Guide 63:2019)
Product Lifecycle	All phases in the life of the product from the initial development through marketing until product discontinuation and disposal. (ISO/IEC Guide 63:2019)
Quality	The degree to which a set of inherent properties of a product, system, or process fulfills requirements.
Residual Risk	Risk remaining after risk control measures have been implemented. (ISO/IEC Guide 63:2019)
Risk	Probability of the occurrence of harm and the severity of that harm. (ISO/IEC Guide 63:2019)
Risk Acceptance	The decision to accept risk. (ISO Guide 73)
Risk Analysis	Systematic use of available information to identify hazards and to estimate the risk. (ISO/IEC Guide 63:2019)
Risk Assessment	Overall process including risk analysis and risk evaluation. (ISO/IEC Guide 51:2014)
Risk Communication	Sharing information about risk and risk management between stakeholders and decision maker.
Risk Control	Process in which decisions are made and measures implemented to reduce or maintain risks within specified levels; actions implementing risk management decisions. (ISO/IEC Guide 63:2019 and ISO Guide 73)
Risk Estimation	Process used to assign values to the probability of occurrence of harm and the severity of that harm. (ISO/IEC Guide 63:2019)
Risk Evaluation	Comparison of estimated risk to given risk criteria to determine the acceptability of the risk (ISO/IEC Guide 63:2019); process of comparing the estimated risk against a given risk criteria to determine the significance of a risk, using a qualitative or quantitative scale (ICH Q9).
Risk Management	Systematic application of management policies, procedures and practices to the tasks of analyzing, evaluating, controlling and monitoring risk across the product lifecycle. (ISO/IEC Guide 63:2019). It is a framework that supports identification and control of critical requirements essential to the safe and effective functioning and use of medical products.
Risk Management File	Set of records and other documents that are produced via risk management. (ISO 14971:2019 3.25)
Risk Management Plan	The iterative guide to risk management activities throughout the entire product lifecycle of a given product.
Safety	Freedom from unacceptable risk. (ISO/IEC Guide 63:2019)

(*Continued*)

TABLE 6.1 (CONTINUED)
Key Terms and Definitions[a]

Severity[d]	Measure of the possible consequences of a hazard (ISO/IEC Guide 63:2019); the intensity of a specific event (e.g., mild, moderate, or severe)
Use Error	User action or lack of user action while using the medical product that leads to a different result than that intended by the manufacturer or that expected by the user.[e] (ISO/IEC 62366-1:2015)
Verification	Sometimes also referred to as "qualification"; Confirmation, with objective evidence, that specified requirements have been fulfilled.

[a] Definitions, unless otherwise specifically noted, are based on ICH Q9(R1), ISO 14971:2019, ISO TR 24971:2020, and AAMI TIR 105:2020.

[b] US FDA CDRH Draft Human Factors Engineering Guidance, "Applying Human Factors and Usability Engineering to Medical Devices" (February 2016), accessed July 31, 2022 at https://www.fda.gov/media/80481/download

[c] Note that CDER does not include the word "serious" in their definition of critical use task, so the scope of tasks that are considered critical is broader than in the definition from CDRH. US FDA CDER Draft Human Factors Engineering Guidance, "Human Factors Studies and Related Clinical Study Considerations in Combination Product Design and Development"(February 2016, accessed July 31, 2022 at https://www.fda.gov/regulatory-information/search-fda-guidance-documents/human-factors-studies-and-related-clinical-study-considerations-combination-product-design-and

[d] *This is not the same as* seriousness, *which is based on patient/event outcome or action criteria usually associated with events that pose a threat to a patient's life or functioning. ("US FDA Guideline for Industry: Clinical Safety Data Management- Definitions and Standards for Expedited Reporting" ICH-E2A (March 1995), accessed July 31, 2022 at* https://www.fda.gov/media/71188/download). *The terminology "severity" versus "seriousness" sometimes raises confusion. Recall that harm is injury or damage to people, property or the environment. When conducting risk analysis, the ratings for severity of harm to people are, in fact, more aligned to pharmacovigilance interpretations (seriousness); or the use of severity terms used in vigilance/adverse event reporting; when applied to property or environment, harm severity, e.g., cybersecurity breaches, may better align to the traditional definition of severity (measure of possible consequences of a hazard).*

[e] Use error includes the inability of the user to complete a task and can result from a mismatch between the characteristics of the user, user interface, task, or use environment. Users may or may not be aware that there has been a use error.

COMBINATION PRODUCTS RISK MANAGEMENT: A CONSOLIDATED PROCESS

Interdisciplinary Risk Management Team

ISO 14971:2019 Clause 4.3 requires the manufacturer documents the competence of the persons performing the risk management activities. The standard also states in A.2.4.3 that "*this usually requires several representatives from various functions or disciplines, each contributing their specialist knowledge. The balance and relation between those representatives should be considered.*" The statement points out that we need a broad spectrum of experts. This is especially true in a combination product with its various constituents, components, and subsystems parts of the combination product system.

It is recommended that manufacturers establish a cross-functional team early in combination product development to holistically consider and evaluate risks across the combination product lifecycle. Individuals comprising a successful interdisciplinary risk management team represent a broad range of subject matter expertise, including, for example, quality, business development, designers, formulators, engineering, regulatory affairs, production operations, sales and marketing, supply chain, legal, statistics, medical professional/clinical and risk management. The broad core competencies of this interdisciplinary team support the necessary comprehensive consideration of risks for each constituent part and the combined-use system as a whole. Constituent part and combined-use risk considerations should include those based on the intended use(s) of the product, target user populations, and intended use environment (including product disposal), as well as those risks associated with raw materials, suppliers, manufacturing, and distribution (see Chapters 8 (Lifecycle Management), 10 (Supplier Quality), and 11 (Analytical Considerations) in this book). The risk management team may engage early in the development process with regulatory authorities to consider potential new or significant risks of the combination product, or perhaps even situations where risks from one constituent part are potentially outweighed by the benefits from another.

Process Overview

The degree of rigor and formality of product risk management should reflect available knowledge and be commensurate with the complexity and/or criticality of the issue to be addressed.[4] Table 6.2 provides a comparison of the risk management process and associated terms as described in ISO 14971:2019 and ICH Q9(R1). While the terminology may differ slightly, the basic concepts and activities, i.e., Risk Identification and Assessment, Control, Residual Risk Evaluation, and Risk Review, are essentially the same.

A key aspect of each of these product risk management processes is the identification of potential ***sources of harm***, i.e., ***hazards***. A hazard is a hazard whether an individual is exposed to it or not. A circumstance, i.e., a ***hazardous situation***, may occur such that ***people, property or the environment are exposed to one or more hazards***. Such exposure to the hazard may result in ***harm***, with a range of seriousness from negligible impact/injury to the health (physical or mental) of people, or damage to property or environment, to catastrophic results including death. Recall from earlier in the chapter that "risk" is the probability that a harm will occur, and the severity (seriousness) of that harm. Figure 6.2 illustrates this concept.

TABLE 6.2
Comparison of Medical Device and Pharmaceutical Product Safety Risk Management Processes and Associated Terms

	ICH Q9(R1) (Medicinal Product Quality Risk Management)		Basic Concepts		ISO 14971:2019 (Medical Device Risk Management)
Risk Communication	Risk Assessment • Risk Identification • Risk Analysis • Risk Evaluation	Risk Management Tools	Risk Identification & Assessment	Risk Management Plan	Risk Assessment • Risk Analysis ○ Intended Use ○ Identify Hazards ○ Estimate Risk • Evaluate Risk
	Risk Control • Risk Reduction • Risk Acceptance		Risk Control		Risk Control • Option Analysis • Risk Control Implementation • Residual Risk Evaluation • Benefit-Risk Analysis • Manage Risks Arising from Risk Controls • Assure Completeness of Risk Control
	Process Output/ Result of the Quality Risk Management Process		Residual Risk Evaluation		Evaluation of Overall Residual Risk
					Risk Management Review
	Risk Review • Review Events		Risk Review		Production and Post-Production Activities • General • Information Collection • Information Review • Actions

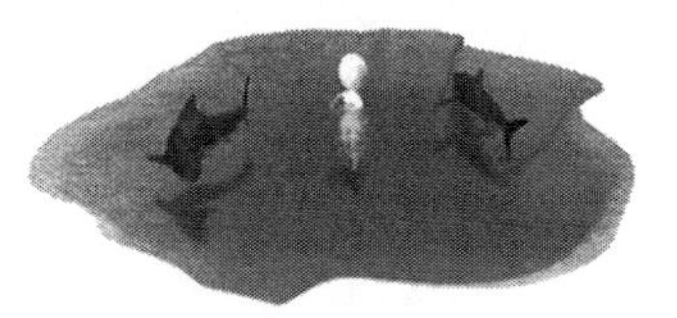

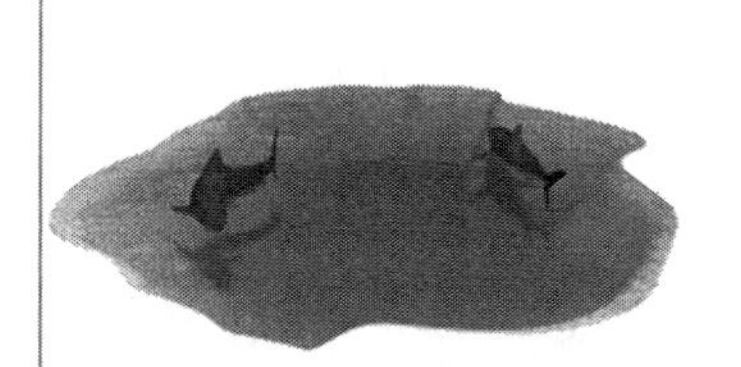

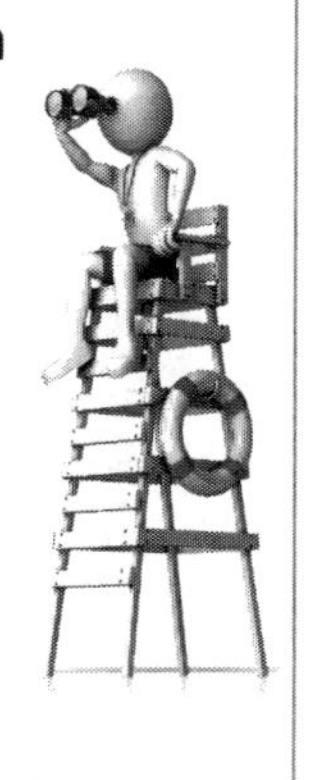

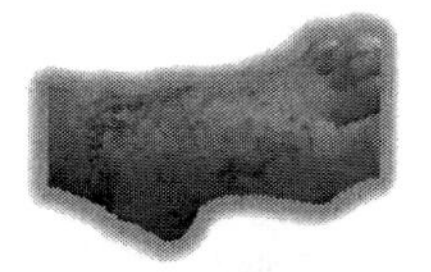

FIGURE 6.2 Hazard → hazardous situation → harm (©2022 Combination Products Consulting Services, LLC. All Rights Reserved).

Consider the example of a hungry shark swimming just offshore at a crowded beach. The shark swimming in the water represents a potential source of harm (HAZARD). A person who goes swimming in the "shark-infested" water is *exposed to that potential source of harm* (HAZARDOUS SITUATION) and could very well experience harm, i.e., get injured. The likelihood of that happening is probably higher if the shark is hungry, or the person has stubbed their toe on some coral and is bleeding, attracting the shark. If the injury results (shark bite or worse!), HARM has occurred. On the other hand, if that person, instead of going swimming, decides to stay on the beach sand, they are not exposed to the potential source of harm. The hazard (shark) is still there, but there is no exposure to that hazard, so no hazardous situation, and consequently no likelihood of harm. In risk management, we work to proactively identify and eliminate, where possible, potential hazards. Where we cannot eliminate them, we work to reduce the likelihood of exposure to those hazards, and ultimately, to eliminate or reduce the likelihood and severity of harm. In this example, what might you do to eliminate the hazard? What would you do to limit exposure to the hazard? What could you do, if you were exposed to the hazard, to minimize the likelihood and severity of harm?[5] More to come on that as we progress in this chapter.

When it comes to combination products, the manufacturer has to ensure each constituent part and the combined-use system as a whole are safe, effective, and usable for the product's intended use, by intended users, in the intended use environment. Applying combination products risk management therefore entails identifying and managing the risks of each constituent part, and then also paying attention to the joint risks of combined use of those constituent parts. ICH Q9 (ICH Q9 (R1)) is applied to identify and control the risks of the drug constituent part alone, identifying critical quality attributes, critical material attributes, and critical process parameters associated with the drug constituent part.[6] ISO 14971:2019 is applied to identify and control the risks of the device constituent part alone, determine what the critical to quality performance requirements, critical material attributes, and critical process parameters associated with the device constituent part and its components. Generally, the drug Critical Quality Attributes (CQAs), Critical Material Attributes (CMAs), and Critical Process Parameters (CPPs) can be considered **design inputs** (see Chapter 4 (cGMPs) and Chapter 5 (Integrated Product Development) in this book) for combination products design controls and risk management. For the device constituent part, risk control measures and their parameters are considered to be *essential design outputs* and should be documented in design outputs as such for communication of the criticality to downstream processes.[7]

The combination products risk management process centers on the risks of the drug and device combined-use system, including any risk change based on the interaction effects of one constituent part with the other(s). In short, combination products risk management asks the questions:

1) Are there new hazards arising from the combined use of the constituent parts?
2) Are there any new hazardous situations from combined use of the constituent parts?
3) Is there any increase or change in the severity of harm due to the combined use of the constituent parts?

Figure 6.3 illustrates the combination products risk management concept. Consider a potent drug that has historically been dispensed in a topical ointment formulation:

(1) The drug alone is a potential source of harm, e.g., formation of tiny pink bumps and acne, coupled with a major treatment downside of likely poor adherence to the dose regimen (frequency and dosage applied) may be inconsistent, leading to the harm of patient drug resistance (among other possible potential hazardous situations).

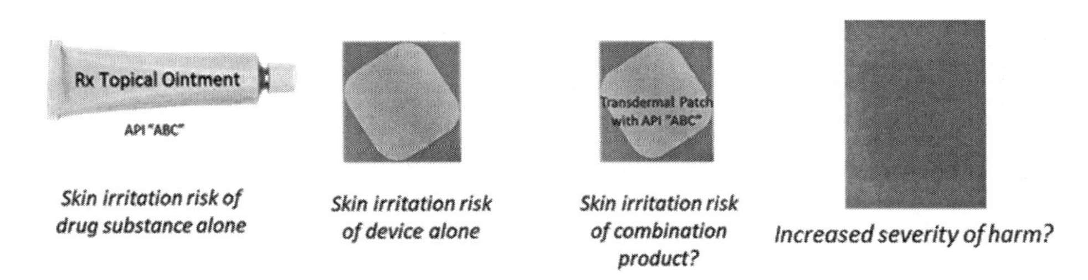

FIGURE 6.3 Combination product risk management example (©2022 Combination Products Consulting Services, LLC. All Rights Reserved).

The manufacturer decides to deliver the drug through a transdermal patch to overcome the dose regimen adherence hurdle. This enables a specific dosage of medication to be delivered through the skin and directly into the bloodstream of a patient. The transdermal patch manufacturer makes the patch using a "medical grade" adhesive (see Chapter 10, Supplier Quality Considerations in this book).

(2) The manufacturer does a risk assessment of the transdermal patch device alone to identify any potential harms from device use itself. The transdermal patch materials of construction/adhesive are a potential source of harm (i.e., a potential hazard), as they can cause skin irritation and sensitization.
(3) The manufacturer then assesses whether there are new hazards/hazardous situations or increased severity of harm associated with the combination of patch with the specific drug. Are there interaction effects between the drug and the device delivery system? There are specific drugs that cannot be delivered using the transdermal delivery method because interactions of drug and device might exacerbate skin irritation and sensitization from the drug by itself or device by itself. There can also be issues caused by environmental impacts (e.g., high humidity) on the adhesion of the patch due to different types of skin. Such adhesion lapses (hazards) might interfere with the delivery of the appropriate dose of the drug (hazardous situation), resulting in harms due to the underdose.

The manufacturer would apply risk controls to address identified hazards of each constituent part and the combination product as a whole (e.g., selection of adhesive materials for the transdermal patch design optimize adhesion without promoting skin sensitization; or drug formulation changes to minimize risks of skin irritation; or instructions for use (IFU) that specify skin preparation procedures to maximize skin adherence and/or an IFU contraindicating use of the transdermal patch under certain environmental conditions (e.g., showering)).[8,9]

Some regulatory schemes[10] and ISO 14971 Clause 7.1 prescribe a fixed hierarchy of risk controls that should be examined in the following order, where labeling is seen as least effective:

(1) Inherent safety by design
(2) Protective measures in the device constituent/combination product or its manufacture
(3) Information for safety, such as warnings, maintenance schedules, etc., and in some cases, training

Subsequent to risk mitigations, the combination product manufacturer would assess whether any potential new hazards/hazardous situations or harms have been introduced. A benefit-risk analysis of the combination product residual risks[11] would also be performed. Improved patient compliance to the intended dose regimen might outweigh the hazardous situations of under- or overdosing with the topical treatment alone.[12]

Combined-Use Hazards, Hazardous Situations, and Harms

Table 6.3 summarizes a few examples of potential new hazards and hazardous situations that could result from combined use of drug and device constituent parts. We will discuss these in more depth later in the chapter. The severity of the harm resulting from a given combination product hazard and/or hazardous situation is often drug- and dose-error dependent. Accordingly, risk tolerance often differs based on specific drugs, intended uses, intended users, and intended use environment. For an injectable drug or biological product intended to treat emergent, life-threatening conditions, failure of a device, preventing adequate delivery of a life-saving drug to a patient, can have fatal consequences, whereas the severity of harm might not be as alarming for a missed dose of a non-life-sustaining drug.[13]

TABLE 6.3
Examples of Potential New Hazards and Hazardous Situations from Combined Use

New Hazards	New Hazardous Situations
• Drug formulation/device material interactions • Device (container/closure) integrity • Device functionality and reliability • Ease of use • Production hazards • Cybersecurity risk to connected health element (AAMI TIR 57)	• Impact(s) to dose safety, accuracy, and/or device performance ◦ Drug stability profile impacted by device material interaction ◦ Device shelf life and reliability impacted by combination product production hazards ◦ Toxic impurities in the drug product due to drug-device interaction, extractables & leachables from drug contact materials in the device ◦ Physical deterioration and failure • Use error (see Chapter 7 on Human Factors in this book) • Dose regimen error due to software malfunction

Table 6.4 summarizes the overall risk management standards, related key standards, process outputs, and key considerations. The risk management activities for drug constituent part are design inputs to the device and combination product design controls and risk management process.[14] We will discuss this in more depth.

Human Factors Considerations

A critical aspect of combination products risk management is the consideration of human factors, the scientific discipline concerned with the understanding of interactions among humans and other elements of a system, and applying theory, principles, data and methods to design in order to optimize human well-being and overall system performance[15] (see Chapter 7 on Human Factors in this book). The user interface of a combination product refers to all components of a product with which a user interacts, e.g., labels, packaging, the device delivery constituent part, and any associated controls and displays.[16] The user interface supports their safe and effective use of that product – by intended users for intended uses and in intended use environments. A use error is defined as a

> user action, or lack of user action, while using the medical device (or combination product) that leads to a different result than that intended by the manufacturer or expected by the user ... and was not caused solely by device failure and did, or could, result in harm.[17]

A *use error* is a hazardous situation that may lead to exposure to hazards and should be considered and controlled. Any preventable event that may cause or lead to inappropriate medication use (use error) or patient harm while the medication is in the control of the health care professional, patient, or consumer is considered a *medication error*. A goal of combination product human factors is to develop and demonstrate that the product user interface supports safe and effective use of the combination product. Use Related Risk Analysis (URRA) or user Failure Modes and Effects Analysis are common tools to assess use tasks and to identify which of those tasks is considered critical. Interestingly, Center for Devices & Radiological Health (CDRH) and Center for Drug Evaluation & Research (CDER) apply different interpretations to the term "Critical Use Task".

CDRH draft guidance defines a critical use task as "a user task which, if performed incorrectly or not performed at all, would or could cause ***serious*** harm to the patient or user, where harm is defined to include *compromised medical care (i.e., the product cannot be used as intended).*"[18]

CDER does not include the word "**serious**" in their definition. In FDA Draft Guidance "Human Factors Studies and Related Clinical Study Considerations in Combination Product Design and Development," a critical use task is defined as a "user task that, if performed incorrectly or not

TABLE 6.4
Combination Products Risk Management Process Overview

ICH Q9(R1) (Medicinal Product Quality Risk Management)			AAMI TIR 105: 2020 Combined-Use Risk Management	ISO 14971:2019/ISO TR 24971:2020 (Medical Device Risk Management)	
Risk Communication	Risk Assessment • Risk Identification • Risk Analysis • Risk Evaluation	Risk Management Tools	Risk Identification & Assessment	Risk Management Plan	Risk Assessment • Risk Analysis ◦ Intended Use ◦ Identify Hazards ◦ Estimate Risk
					• Evaluate Risk
	Risk Control • Risk Reduction • Risk Acceptance		Risk Control		Risk Control • Option Analysis • Risk Control Implementation • Residual Risk Evaluation • Benefit-Risk Analysis • Manage Risks Arising from Risk Controls • Assure Completeness of Risk Control
	Process Output/Result of the Quality Risk Management Process		Residual Risk Evaluation		Evaluation of Overall Residual Risk
					Risk Management Review[a]
	Risk Review • Review Events		Risk Review		Production and Post-Production Activities • General • Information Collection • Information Review[b] • Actions

(Continued)

TABLE 6.4 (CONTINUED)
Combination Products Risk Management Process Overview

	ICH Q9(R1) (Medicinal Product Quality Risk Management)	AAMI TIR 105: 2020 Combined-Use Risk Management	ISO 14971:2019/ISO TR 24971:2020 (Medical Device Risk Management)
Additional Key Standards	FDA Guidance "Safety Considerations for Medication Error Reduction" ICH Q8(R2) Quality by Design	FDA Guidance "Human Factors Studies and Related Clinical Study Considerations in Combination Product Design and Development"	IEC 62366-1/2015 (Usability Engineering) FDA Guidance "Applying Human Factors and Usability Engineering to Medical Devices" 21 CFR 820.30 Design Controls ISO 13485:2016 Clause 7.3 Medical device design controls
Outputs	• Drug CQAs • Drug CMAs • Drug CPPs • Safety considerations to minimize medication errors for medicinal constituent part	**Joint Risks of Combined Use**: • New hazards? • New hazardous situations? • Increased severity of harm? Critical Use Tasks for the Combination Product	• Device CtQs • Device CMAs • Device CPPs • Critical Use Tasks for the device constituent part
Key Considerations	• User Needs & Intended Uses • Therapeutic Effect • Safety Considerations to Minimize Medication Errors	**Joint Risks of Combined Use**: • Essential Performance Requirements (see §3.3 of this chapter), e.g., - Ability to delivery labeled dose - Impact of device on drug CQAs - Impact of drug on device functionality - Usability/critical tasks accomplished without injury or issue/safety considerations to minimize medication errors	• User Needs & Intended Uses • Intended Use Environment • Functionality • Critical Use Tasks for the device constituent part

[a] This Risk Management Review assesses the correct execution of the Risk Management Plan and assures the overall residual risk (including Benefit Risk Assessment) is acceptable prior to the launch of the product (or the change being implemented). This Risk Management Review occurs prior to product release (pre-launch).

[b] This includes review for events occurring post-launch and is comparable to "Review Events" in ICH Q9.

performed at all, would or could cause harm to the patient or user, where harm is defined to include *compromised medical care*."[19] This is a broader interpretation than that for device alone. When it comes to combination products, more tasks will be classified as "critical" than they would be for a device alone.

Consider the example of removing a needle safety cap from a non-acute care drug-prefilled syringe (PFS). What is the impact if the patient trying to self-administer their dose is unable to remove the needle safety cap? Is it simply missing a dose or delaying therapy of a non-acute care drug? Is removal of that safety cap considered to be a "critical task"? The answer is yes: needle safety cap removal is considered a critical task. While the drug in the PFS might not be emergency-use, the intended use of that combination product is to enable patients to successfully self-administer their medication. So, if the patient is unable to remove the needle cap, they cannot use the product as intended, so they would experience compromised medical care. Even if that event happens only one time, or even if that event does not lead to patient injury or need for medical treatment, the inability to deliver their dose as expected could still be of concern. Removal of the needle safety cap is a critical use task.

For combination products, CDER recommends the manufacturer consider:

1. Harm to both the patient or user, such as a caregiver or healthcare professional.
2. Dosing, such as dialing a proper dose using a pen injector.
3. Administration, e.g., the step of selecting the correct injection site.
4. Urgency or time. In some cases, drugs must be administered as quickly as possible to be effective, e.g., emergency use-administration of epinephrine to treat a severe allergic reaction.

Essential Performance Requirements

Proactive consideration of the risks of combined use, applying risk management as a framework to guide development, helps to ensure safety and efficacy of combined-use products. The joint risks of combined use are pointers to Essential Performance Requirements (EPRs), i.e., critical requirements essential to the safe and effective operation/functioning and use of a medical product.[20] "ICH Q12 Implementation Considerations for FDA-Regulated Products Draft Guidance for Industry" (May 2021)[21] Appendix B describes key questions to ask in identifying established conditions for the device constituent part of a combination product. These questions put a focus on performance characteristics needed for safe and effective drug dose delivery, i.e., **essential performance requirements**:[22]

- Is the device performance characteristic essential for the safe use based on risk management?
- Is the device performance characteristic essential to achieve delivery of the labeled dose?
- Does the device performance characteristic impact the drug product's CQAs?

A fourth question, while absent from the Draft Guidance, should also be added:

- Does the drug characteristic impact the device's functionality/Critical to Quality (CtQs)?

Consider the case where excipients in a drug formulation interact with the medical device, causing the device to malfunction. For example, an infusion device was designed for use with a particular drug formulation in mind. A manufacturer decides at a later point to use that same device platform for another drug formulation. Excipients (e.g., surfactant) from the new drug formulation might interact with the tubing of the device, releasing oxidants that cause the motor to malfunction, ultimately leading to missed doses, and patient harm.

Device and drug characteristics, manufacturing process elements that need to be controlled to ensure the product design and functionality meets acceptance criteria, and other control strategy elements that ensure the required safe and effective, proper use of the combination product are Essential Performance Requirements. These may be characteristics of one or more constituent parts, or of the combination product as a whole. Considerations include, for example, the principles of device functionality (e.g., mechanical and/or electrical), mechanism of operation for delivery of the drug product (e.g., spray, mixing), materials of construction that are in direct contact with the drug fluid path or the patient. FDA EPR Guidance includes specific recommendations regarding preconditioning and stability testing, reliability expectations, and emphasis on selecting the correct specifications particular to a given combination product. Risks include clinical risks from the product and technology risks based on the complexity of the device. EPR control strategy is commensurate with the risks presented. All EPRs should be verified and/or validated (see also Chapter 8, Process Controls).

Potential EPRs for a self-injection system employing a syringe are listed in Table 6.5.[23] Table 6.6 includes potential EPRs for a range of commonly used device types, including infusion pumps, on-body injectors, prefilled syringes, autoinjectors, pen injectors, piston syringes metered dose inhalers, and nasal sprays.

Other device-specific requirements may include (but might not be limited to):

- Biocompatibility – ensure components are biocompatible and commensurate with the level and duration of patient contact. See ISO 10993-1;
- Sterilization and Packaging – demonstrate minimum SAL (10-6) of the fluid path & provide packaging testing of the primary sterile barrier after shipping and aging;
- Bacterial endotoxin testing (e.g., USP <85> or ANSI/AAMI ST 72:2019);
- Software/cybersecurity;[24]
- Electromagnetic compatibility (IEC 60601-1-2);
- Reliability (level determined from risk, i.e., emergency use requires higher reliability); and
- Particulate testing per USP <788> on fluid-contacting portions.

Combination Products Risk Management Process: A Deeper Look

For each specific combined-use system, the recommendation is that manufacturers/applicants use prior knowledge coupled with risk assessment tools such as those listed in ICH Q9, ISO 14971/ISO TR 24971 and AAMI TIR 105 (e.g., Failure Modes Effects Analysis, Criticality Analysis, Fault Tree Analysis, Ishikawa Diagram, Preliminary Hazard Analysis[25]) starting from early product development specific to their formulation, manufacturing process, and device constituent part, to identify CQAs, CMAs, and CPPs of the drug constituent part; CtQs, CMAs, and CPPs for the device constituent part, and then joint risks of combined use, all of which have the potential to impact product quality. The identified factors can be further studied (e.g., through experiments and/or modeling) to define an appropriate control strategy that assures the design and manufacturing process consistently produces a product of the desired quality, functionality, and usability.

Tactically, executing this joint-use assessment can be a bit overwhelming and confusing. Such confusion may lead some to fall into a mindset of treating the risk management activities as a paperwork exercise to slog through in order to support mandatory regulatory documentation. In reality, even if the risk management activities were not mandated from a regulatory perspective, from a science and technological perspective, scientists, engineers, and management would still want to incorporate risk management into the development and post-market lifecycle management of safe, effective, and usable products. This section aims to bring understanding and clarity to the execution of combination product risk management expectations and execution across the combination products lifecycle.

TABLE 6.5

Potential EPRs for a Self-Injection System Employing a Syringe[a]

Drug/Biologic (CQAs)	Device (Protection)	Device (Compatibility)	Device (Performance)
• Purity • Stability • Dose Volume • Viscosity • Structure • Elution Profile • Particle Size • Concentration • Acute/Chronic Exposure • Pharmacokinetics (Pk) • Local/Regional/Systemic Toxicities	• Container-Closure Integrity • Permeability • Microbial Ingress • Dimensional Tolerances • Component Inter-Connectivity	• Biocompatibility • Particles • Extractables & Leachables • Loss of API via Absorption of Adsorption • Contaminants (e.g., Tungsten, Silicone Oil) • Surface Interactions; Adverse Chemical Reactions; Particle Formation	• Drug Viscosity • Injection Time • Permeability • Siliconization • Shelf Life • Coating Integrity • Clarity of Units of Measure • Break-loose/Extrusion/Activation Force • Back-Pressure • Needle Gauge/Extended Length/Insertion Depth • Clarity of Dose Completion

[a] Neadle, S.W.; Riter, J.; McAndrew T.P.; (Pharmaceutical Online, Nov. 30, 2020) "*Considerations in Combination Product Risk Management,*" 23; https://www.pharmaceuticalonline.com/doc/considerations-in-combination-product-risk-management-0001

TABLE 6.6
EPRs for Common Drug Delivery Devices[a]

Essential Performance Requirement (EPR)	Infusion Pump	On-body Injector (OBI)	Prefilled Syringe (PFS)	Autoinjector (AI)	Pen Injector (PI)	Piston Syringe	Metered Dose Inhaler[b] (MDI)	Nasal Spray[c]
Total Volume Delivered		✓						
Delivery Rate	✓	*✓						
Time/Phase Dependent	✓	*✓						
Dose Accuracy or Dose Efficiency (User-Filled)	✓	✓	✓	✓	✓	✓	✓	✓
Delivered Dose	✓	✓	✓	✓	✓	✓	✓	✓
Occlusion Detection	✓	✓						
Injection Time		✓		✓				
Actuation Force		✓		✓				
Break-loose/Glide Force			✓		✓	✓		
Extended Needle/Canula Length	✓	✓		✓	✓			
Cap Removal Force (If Applicable)		✓	✓	✓	✓			
Needle Safety Lockout Force (If Applicable)			✓	✓	✓			
Adhesion/Adhesion Peel Force per Area		✓						
Audible/Visual Feedback (If Applicable)	✓	✓	✓	✓	✓			
Aerodynamic Particle Size Distribution (APSD)							✓	
Spray Pattern							✓	✓
Specific Resistance to Air Flow							✓	✓
Plume Geometry							✓	✓
Spray Velocity							✓	✓

*Recent technology advancements in on-body injectors enable delivery of high dose volumes; as such, for these combination products, some of the EPRs are similar to those of infusion pumps. Infusion pumps, in and of themselves, are generally stand-alone devices, typically marketed independently of a specific drug, whereas most OBIs are marketed for a specific drug-device combination product intended use.

[a] Alan Stevens, CDRH, Xavier Combination Products Summit, 2019, 2020, 2021.

[b] FDA Draft Guidance (2018) "Metered Dose Inhaler (MDI) and Dry Powder Inhaler (DPI) Drug Products- Quality Considerations" accessed August 3, 2022 at https://www.fda.gov/regulatory-information/search-fda-guidance-documents/metered-dose-inhaler-mdi-and-dry-powder-inhaler-dpi-drug-products-quality-considerations

[c] FDA Draft Guidance (2002) "Nasal Spray and Inhalation Solution, Suspension, and Spray Drug Products – CMC Documentation," accessed August 3, 2022 at https://www.fda.gov/regulatory-information/search-fda-guidance-documents/nasal-spray-and-inhalation-solution-suspension-and-spray-drug-products-chemistry-manufacturing-and

In Tables 6.2 and 6.4, we reviewed the general risk management framework for medicinal products [ICH Q9 (ICH Q9(R1)], medical devices (ISO 14971:2019/ISO TR 24971:2020), and combination products (AAMI TIR 105:2020). This framework is summarized in Figure 6.4. We apply outputs of Quality by Design (QbD) (see Chapter 5 on Integrated Product Development in this book) and criticality analysis (ICH Q9) for the drug constituent part (CQAs, CMAs, CPPs) as inputs into the device and combination products risk management process (ISO 14971/ISO TR 24971 and AAMI TIR 105), working our way through risk management planning, through risk assessment, risk control, residual risk evaluation, and risk review. Let's walk through it.

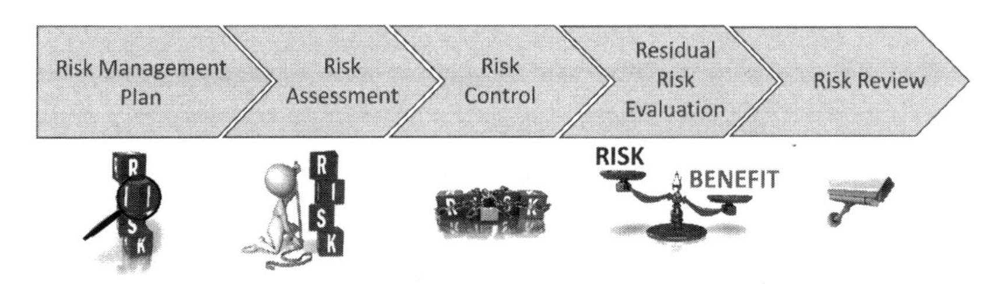

FIGURE 6.4 Combination product safety risk management framework (©2022 Combination Products Consulting Services, LLC. All Rights Reserved).

Risk Management Plan

As with any process, a documented plan, in this case typically called the **Risk Management Plan** (RMP), is required that defines the responsibilities, activities to be conducted, and timing of these activities. The RMP is an iterative guide to risk management activities throughout the product lifecycle. The plan also proactively defines the risk acceptability criteria (safety criteria "guardrails") used in all risk acceptability decisions throughout the process. "The objective of risk management is rarely to eliminate all risk, but rather to reduce risk to an acceptable level while maintaining feasibility and functionality."[26]

The plan is the first of many risk management documents that will be created as the risk management process continues. The RMP enables risk management decisions relative to reasonably foreseeable harms and benefits of various actions. The Risk Management Plan:

- Is a forward-looking document; it identifies activities, responsibilities, and authorities before activity occurs;
- Establishes relationships between hazards, foreseeable sequences of events, hazardous situations, and harms that occur;
- Includes risks associated with product design and manufacturing process, use, and disposal;
- Supports assessment of the acceptability of identified risks and specific control measures;
- Incorporates strategy for leveraging third-party vendor risk assessments for components or constituent parts;
- Incorporates the strategy for leveraging pre-existing information or data, e.g., platform technology or a vendor's drug-agnostic information for a device constituent part, ensuring gaps are identified, assessed, and addressed relative to the combination product manufacturer/applicant's specific intended uses, users, and use environment for the particular combined-use application;
- Identifies appropriate production and post-production information to capture;
- Identifies feedback into risk management activities throughout the product lifecycle as a revisable document; and
- Is under revision control, and included in a Risk Management File (RMF) for the combination product.

Risk Management File

The risk management process helps to tighten up one's thinking; it can help a regulator understand *your* thinking. Conducting and documenting effective risk management can help you – and a regulator – understand something that one might otherwise not have. A file called a **Risk Management File** (RMF) is created to manage all of the risk management documents during the entire lifecycle of the product, including the Risk Management Plan (and all its versions). The more robust the Risk Management File, the more confidence there will be in what you are doing. The RMF helps both you and the regulator be mutually grounded in the science and technology of what you are out to achieve.

New information may be received at any point in the life of the product that requires the updating of documents in the file or the creation of new documents. The RMF may exist in a variety of formats, e.g., with an index referring to the location of documents in an organization. Traceability is a required element of the Risk Management File under ISO 14971 Clause 4.5. (For a discussion of Traceability, see AAMI TIR 105 Annex B and later discussions in this chapter §3.4.3.6.)

There are many internal users of the RMF to consider, including the design and development team, manufacturing personnel, post-market surveillance personnel, and product investigational groups. Given that team members may be situated in a number of different locations, the responsibility and authority for managing the RMF needs to be identified in the Risk Management Plan at the beginning of the product, and again, when changes are made in the responsibility. Due to the critical nature of the RMF, these documents should undergo review and approval by a cross-functional team of experts, including those with current medical expertise, to assure that all identified hazards, harms, and severities are accurate. The cross-functional team of experts should also include those with current expertise in combination products as well as those who specialize in the drugs or devices that are component/constituent parts of the combination product. The review of risk management documents should also determine that the RMF is documented in such a manner useful to all of the interested parties. Importantly, the RMF can be subpoenaed in product liability cases. Consideration of this fact should be carried into the creation of documents that meet regulatory and legal requirements in the RMF.[27]

Risk Assessment

Risk Assessment is the systematic use of available information to identify hazards, i.e., potential sources of harm, in light of intended uses, intended users, and intended use environment, and to estimate risk. In the risk assessment stage, the manufacturer mines risk data, continually assessing quality data sources to inform risk management assumptions and living risk files. Examples of data sources can be, but are not limited to:

- Regulations, Standards, and Guidance;[28]
- Clinical Trials;
- Predicate devices;
- Same-similar products/combination products;
- Human Factors Evaluations;
- Reliability analysis;[29]
- Risk Management Tools:[30] Harm/Hazard Analyses (e.g., URRA, user-, design-, process- Failure Mode and Effects Analyses (FMEA), Preliminary Hazard Analysis (PHA), Fault Tree Analysis (FTA), Ishikawa Diagrams, Hazard Operability Analysis (HAZOP), Risk Ranking and Filtering, etc.), and supporting statistical tools;
- Management Review;
- Supplier (performance/controls);
- Complaint Handling;
- Adverse Event Reporting;
- Process Controls;
- Finished Product Quality;

- Audits (internal/external);
- Product Recall;
- Spare Parts Usage;
- Returned Product;
- Market/Customer Surveys/Competitive Intelligence;
- Scientific Literature;
- Media Sources; and
- Product Realization (design, purchasing, production and service and customer information).

The number of data sources and amount of data to inform decisions can be overwhelming for established companies or downright intimidating for those with limited experience with their innovative combination product technology. There are several risk management tools that one can draw from to support risk analysis. A very common risk assessment tool used by medical device and combination product manufacturers is FMEA. The FMEA format provides a structured approach for assessing risks and verifying that the controls implemented are effective. However, FMEAs may also prove challenging to execute. It is not the intent of this chapter to walk through every nuance of FMEAs. There are several resources that do just that (e.g., Carl Carlson's book "Effective FMEAs.")[31]

One source of frustration for teams is confusion between failure analysis and risk analysis (**Figure 6.5**).[32] FMEAs are a means of vetting through *failures*, an inability to perform a function under normal or single-fault conditions. Where those failures are also a potential source of harm, they are *also* considered hazards. A safety-related effect of a failure is transferred from the FMEA to the Risk Analysis as a Hazard. (No other information from the FMEA is transferred from the FMEA to the Risk Analysis, to avoid unnecessary confusion of what is a safety risk.)

Failure analysis focuses on identifying risk controls that will prevent the failures/hazards from being an issue at all. Detectability is ***not*** part of the definition in ISO 14971 or ICH Q9 as a separate component of risk. Detectability is useful, however, in determining where in production processes inspection and test points should be established. An inspection or test point should be able to detect the failure cause that leads to a hazard. So detectability may be considered in a pFMEA.

Where there is a potential source of harm (i.e., a hazard), the next opportunity is to prevent a sequence of events from occurring that would expose a person to that potential source of harm (i.e., prevent the hazardous situation), therefore reducing the likelihood of the impact of that exposure to the hazard actually resulting in a harm (physical or mental injury).

The harm may occur at varying levels of severity based on the level of an individual's exposure to the source of harm. We refer to the hazard-hazardous situation-harm analysis as the "**Risk Analysis.**"[33]

Note: *Detectability should* ***not*** *be used to allow hazardous situations to be passed to the user with the hope of detectability of the hazard prior to use of the product. While detectability is useful in process FMEAs to detect failure causes and to eliminate or reduce their impact, detectability is not a component of ISO 14971 or ICH Q9 product risk analysis.*

We'll discuss this in more detail, but first let's consider a case study:

Consider a patient who is self-administering a dose of insulin to treat their diabetes using a pen injector. The injector has a visual indicator feedback mechanism to signal dose completion. If the visual indicator malfunctions such that the patient receives premature feedback of dose completion, not only might the patient have an underdose of the drug, the sharp needle might still be protruding when the device is lifted from the skin, and a needle stick injury could occur.

Let's unpack this example a bit further. The visual indicator feedback malfunction, i.e., premature signaling of dose completion, is a design-related failure that is a potential source of harm – said differently, the visual indicator malfunction is a hazard. The exposed sharp needle is another potential source of harm, i.e., a second hazard.

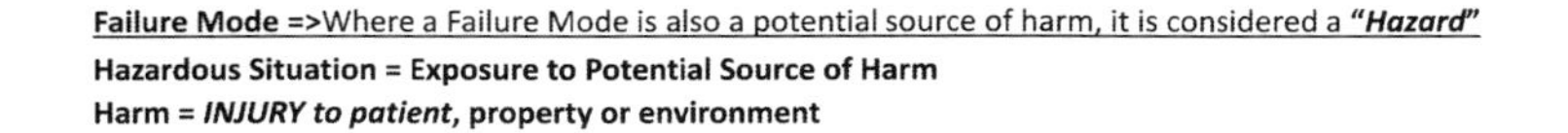

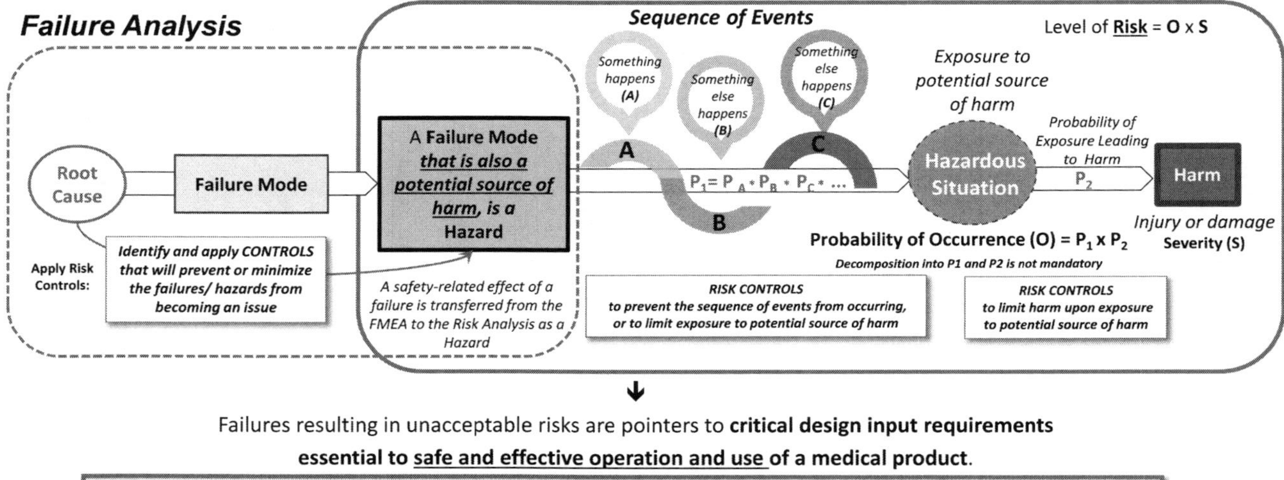

Cascade through:

Intended uses	Use environments (incl. disposal)	Suppliers
User populations	Raw materials	Manufacturing & Distribution

FIGURE 6.5 Failure analysis versus risk analysis (©2022 Combination Products Consulting Services, LLC. All Rights Reserved).

A sequence of events then occurred for this patient. The hazards (visual indicator and sharp needle) exist. The patient was exposed to the visual indicator hazard and thought the dose was complete, so they lifted the autoinjector from their skin prematurely. That led to immediate exposure to one potential source of harm: underdose. The visual indicator malfunction was also part of a sequence of events that led to exposure to the second potential source of harm – the sharp needle.

Circumstances where an individual is exposed to a potential source of harm are hazardous situations. So, the underdose and exposure to the sharp needle are each separate hazardous situations that derive from the one potential source of harm – the visual indicator hazard.

P1 and P2

The harm that results from exposure to the hazards, i.e., the injury to the person administering their medication, will have a given severity based on the given drug/intended function of the combination product. In the earlier case, the drug is intended to treat diabetes, so underdose would likely lead to some level of hyperglycemia[34]. The needle stick injury might likewise lead to a range of harm severities. These could include minor percutaneous skin injury to local or even systemic infection. Each of those injuries described has a different level of harm severity. (This example and others are illustrated in Table 6.9.) Each of these injuries needs to be considered in the risk analysis, not just the one with the greatest severity of harm. Why? Let's deconstruct this a bit:

- Risk is the probability of occurrence of harm and the severity of that harm.
- The risk management team would assess the likelihood of the potential source of harm (i.e., the hazard) undergoing a sequence of events that exposes the patient to that harm (hazardous situation) ("P1").
- The team also assesses the likelihood of harm (of a given severity) if a patient is exposed to the source of harm ("P2").

Teams sometimes make the mistake of only focusing on the harms with the greatest severity. That is inappropriate, as the probability of occurrence of the harm (P1 × P2), even with a harm of less severity, could lead to an Unacceptable Risk (see Figure 6.7). Additionally, the probability of each harm may be different; there is not one P2 for all harms from a single hazardous situation.

- As illustrated in Figure 6.5, the team would apply risk controls to minimize the likelihood of the sequence of events that exposes the patient to the harm (P1), and applies further controls if needed, to reduce the likelihood of harm if indeed an individual is exposed to the source of harm (P2).

Recall the earlier example of hazard/hazardous situation and harm, based on the sharks swimming just off-shore at a busy beach? Let's revisit that example here to elucidate P1 and P2, Probability of Occurrence of Harm and Severity of Harm. Figures 6.6A, 6.6B, and 6.6C are helpful for reference.

The shark swimming in the water represents a potential source of harm (HAZARD). A person who goes swimming in the "shark-infested" water is *exposed to that potential source of harm* (HAZARDOUS SITUATION) and could very well experience harm, i.e., get injured. If injury results (shark bite or worse!), HARM has occurred. Each harm outcome from a shark bite is usually a different probability, e.g., loss of a limb, severe lacerations, or minor abrasion.

We closed the earlier example with a few questions. The first was: **What might you do to eliminate the hazard?** One might get creative to eliminate the potential source of harm. Elimination of the hazard might take some effort (think "*JAWS*"). We could do a failure analysis to identify potential approaches. Let's assume, for the sake of this illustration, we do not have the option to kill the sharks, relocate the sharks, or even build a barrier against the sharks. Perhaps in real life those might be options, among others. So, for our illustration, the shark hazard exists.

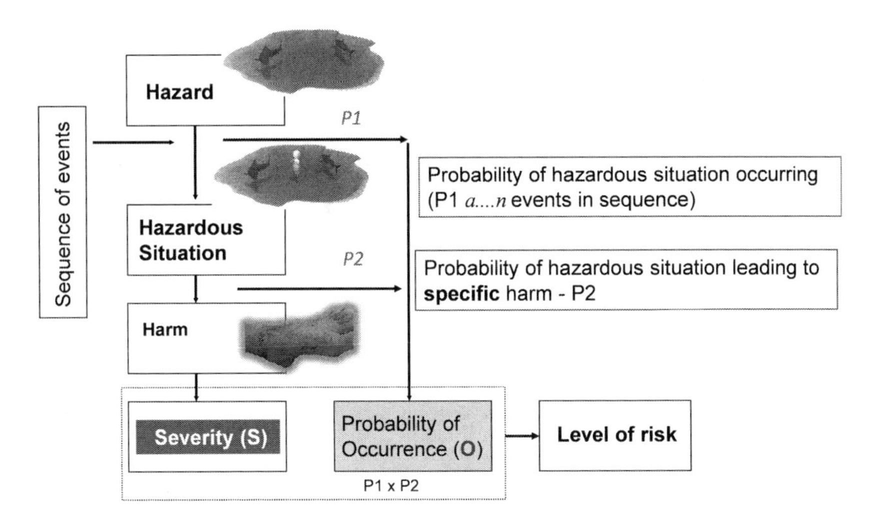

FIGURE 6.6A Risk analysis: Based on ISO 14971:2019 annex C.1 (©2022 Combination Products Consulting Services, LLC. All Rights Reserved).

So now we have to evaluate the probability that the shark hazard is going to lead to a circumstance where someone is exposed to the hazard. If it is a super-hot day out, and the person is off from work and eager to go to the beach, the chain of events is set in motion. The person gets to the beach, and there is no lifeguard on duty. There are no signs warning of the shark situation. The person decides to go snorkeling in the water. At this point, we have exposure to the potential source of harm (P1).

Oops! The person forgot their swim flippers. As they enter the water, they cut their foot on some coral, and their foot starts bleeding. Uh oh. The sharks are hungry and sense the blood. This is increasing the likelihood of the hazardous situation leading to harm (P2).

The sharks come by, and "chomp!" Off comes the foot. Harm has happened! The severity is critical.

Risk = (Probability of a Hazard Leading to a Hazardous Situation) (Probability of the Hazardous Situation Leading to Harm) (Severity of that Harm) = P1 × P2 × S

P1 was high, P2 was high, and the severity of harm was critical. Based on this sequence of events, we've concluded that the risk in this situation was very high/unacceptable.

Proactive risk analysis on this potential sequence of events could lead us to a different outcome.

- What would you do to limit exposure to the hazard? (Lower P1)
- What could you do, if you were exposed to the hazard, to minimize the likelihood and severity of harm? (Lower P2)
- What might you do to limit the severity of harm? (S)

Assuming the shark hazard exists, we could methodically review the sequence of events to limit exposure to the hazard. No swimming signs, a shark alert system, lifeguards on duty to ensure beachgoers stay on the beach and not in the water – these are all layers of defense (i.e., *risk controls*) that one could eliminate or reduce the probability of exposure to the source of harm (P1).

Reducing P2 is a bit more challenging. If the person has decided to go swimming despite all the layers of defense trying to mitigate that, then the person is in the hazardous situation. Ensuring the person's feet are protected (maybe wearing a spare pair of flippers, or some other foot covering, would avoid the foot injury that led to the cut and bleeding), and perhaps wearing some sort of ultrasonic shark repellant might keep the sharks at bay, even though the person is exposed to them. And even if the sharks did approach, they might be more timid in the face of the shark repellant

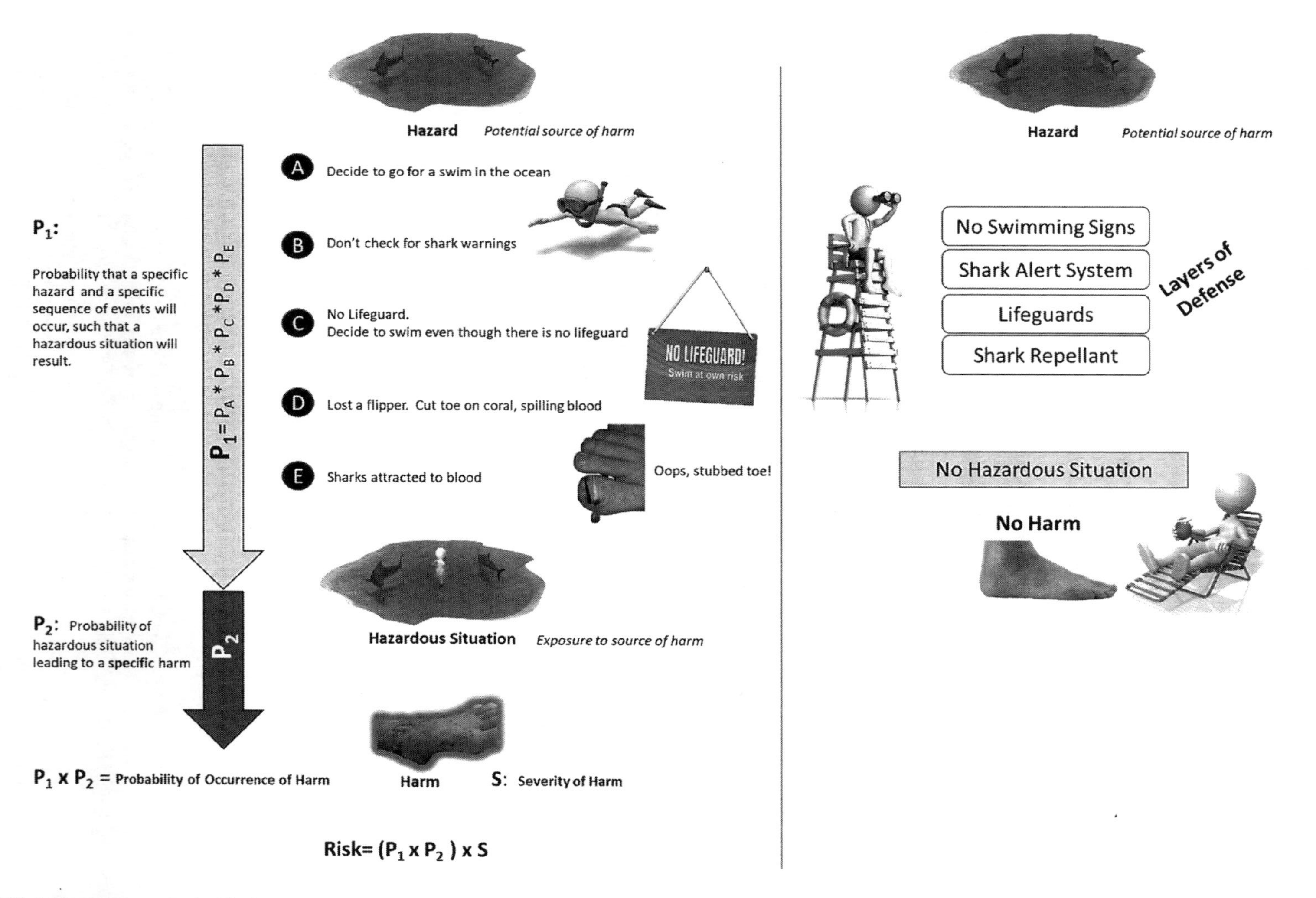

FIGURE 6.6B Risk analysis: Understanding sequence of events (©2022 Combination Products Consulting Services, LLC. All Rights Reserved). *(Also see ISO TR 24971:2020, Figure 1 in clause 5.4.7).*

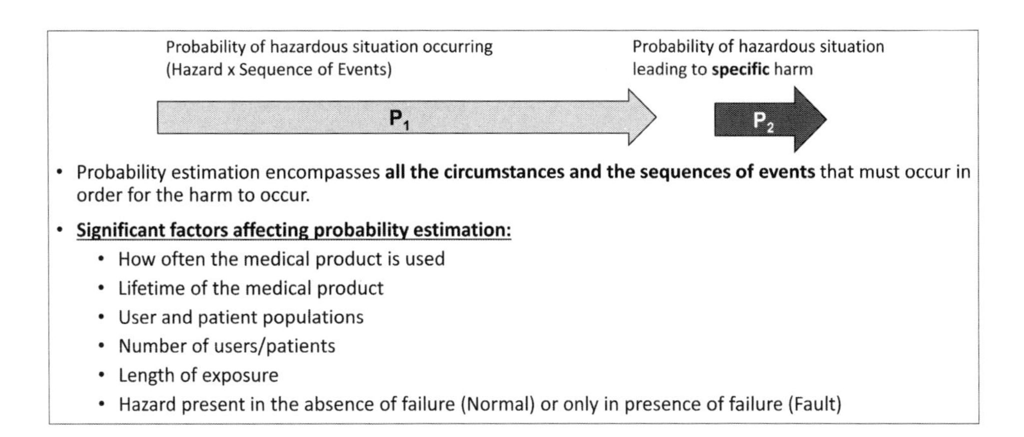

FIGURE 6.6C Risk analysis: Probability considerations.

than they might otherwise be. Another last-ditch opportunity could be that there is a boat out in the water that feeds the sharks, keeping them well-sated, so they are more attracted to the easily obtained shark food than to the person snorkeling. These are measures (i.e., more *risk controls*) that *might* lower the probability of P2.

It is hard to limit the severity of harm once a person is exposed to it. We'll discuss this further later in the chapter (Risk Controls section). For our shark scenario, maybe a horn that startles the shark upon exposure might scare it away, limiting the severity of damage done. The ideal scenario is that the person who wanted to swim heeds the signs and warnings, and stays on the beach, enjoying a day in the sun without exposure to the hazard in the first place.

Going beyond our shark example, consider how one might approach estimating probability of P1 and P2 for combination products. Figure 6.6C summarizes some factors affecting probability estimation. These factors likewise hold true for stand-alone medical devices and medicinal products.

Decomposition into P1 and P2 can be useful to estimate the probability of occurrence of harm (O), but it is not mandatory. According to ISO TR 24971:2020,

> When the probability of occurrence of *harm is decomposed into P1 and P2, it could be the case that one of them can be estimated and the other not. In such cases, a conservative approach can be used by setting the unknown probability equal to 1. Such approach can be useful when the estimated probability is either so low that the resulting risk becomes clearly insignificant or negligible, or so high that it is clear the resulting risk should be reduced.*

When there is enough data available to estimate the probability of occurrence of harm with adequate confidence, one should use a quantitative approach. If probability of occurrence of harm cannot be estimated, a qualitative approach, based on expert judgment, should be applied and may necessitate evaluating the risk on the basis of the severity of harm alone. This might be the case, for example, for a new combination product technology, where there is not yet suitable quantitative data available until later in design development/design validation, or even post-market. (One should have statistical confidence in data to use quantitative data; until then, the use of qualitative data is preferred.)

The manufacturer's risk control measures should then focus on reducing the probability of the hazardous situation entirely, or on reducing the probability that the hazardous situation (exposure to harm) leads to harm. If this is not possible, the risk control measures should focus on reducing the severity of the harm. Examples of such controls might include alarms on a device constituent to alert the user that there is an issue, so they can try to quickly address it. Other controls might include the use of a trainer device prior to dose administration, coupled with labeling (recognizing labeling is often least effective).

As previously indicated in the chapter, risk controls to be examined are preferably (1) inherent safety by design (ensuring the visual indicator feedback functions reliably); (2) protective measures in the device constituent/combination product or its manufacture (e.g., a failsafe that causes the needle to automatically retract on premature lift, and/or perhaps creating a back-up failsafe alarm signal); and lastly (and generally least effective) (3) information for safety, such as warnings, in the IFU to guide the patient.

Hazard Identification

The initial process step in risk management is to "*identify hazards.*" This requires the use of information such as knowledge of the composition of the product and the characteristics of the product to identify possible hazards. Hazard identification also applies to manufacturing processes, where hazards may be inserted into the product during manufacturing, such as by improper settings or not following process instructions properly. All of the discussions that follow apply to hazards that may be associated with the design and manufacture of the combination product either in its subsystem components/constituents or the combination (combined use) product. Table 6.7 summarizes some common hazard categories.

TABLE 6.7
Examples of Hazard Categories for Consideration

Device Hazard	Drug/Biologic Hazards	Production Hazards
Electrical	Purity	Incapable processes
Mechanical	Excipients	Inaccurate procedure
Thermal	Content Uniformity	Operator non-compliance
Biocompatibility	Sterilization	Cleaning
Usability	Stability	Labeling
Electromechanical		Test Methods
Software		
Interactive Hazards Between the Device and Drug/Biologic		

Hazardous Situations

Following the hazard identification step using the tools described in the previous section (also, see Chapter 7 on Human Factors), we need to determine how each hazard may result in harm. Just because a hazard is present does not mean harm will automatically occur. There must be exposure to the hazard for harm to occur. It is also possible that several different severities of a single type of harm may occur, each at a different rate of occurrence, this creating a distribution of harms. Hazardous situation identification assures that we understand situations and contributing factors that may occur which expose the hazard and may lead to different harms. This knowledge is important because once the mechanism leading from hazard to harm is fully understood, and given the goal of eliminating harms, we can first identify solutions to redesign the product to eliminate the hazard, or if that is not possible, take steps to reduce the probability of occurrence of harm. If these actions are taken early in product and process development, then risk can be reduced at a lower cost than later in the development process when a hazard may have been identified during final validation or late-phase clinical trials. It is even more expensive to identify hazards after product release when product actions such as recalls may be required.

Qualitative versus Quantitative Risk Management

While we have all utilized both qualitative and quantitative analyses at some point, due to our scientific and engineering training, most of us prefer quantitative data analyses versus qualitative. It is not uncommon during modern education that statistics professors will teach students how to convert qualitative data to quantitative so quantitative methods may be applied. Qualitative data may include intangibles such as colors, preferences, or comfortable feel. Small sets of numeric data may actually be considered to be semi-quantitative or semi-qualitative.

Qualitative and semi-qualitative risk management is commonly performed. There is no stigma in using qualitative analyses *per se,* even though ideally, we could have better decision-making with quantitative *sets of high confidence data.* Most often early in the design/development processes the available high confidence data sets are qualitative and quantitative data is lower confidence and often does not support good decision-making. Much can be learned by performing simple Preference Studies and Human Factors Tests. A well-designed, scientifically valid qualitative analysis is better than a numerically complex, possibly flawed, quantitative analysis. The use of quantitative data that has a low level of confidence should be avoided for data of higher confidence levels which may mean the use of qualitative analysis to be performed until a point when quantitative data of high confidence becomes available. Usually, this occurs later in the development cycle and perhaps even after the combination product is released for use. Even after release if *higher confidence* quantitative data become available, post-production quantitative data should be fed back into the risk management system during the post-market surveillance process.

It has been observed that complex schemes have been established by some manufacturers to perform Pareto analyses for various product failure modes. These complex schemes often mask the ability to make informed decisions because it requires mathematical calculations to determine acceptability rather than using common sense heuristics. The goal of Risk Management is not a scientifically sophisticated report, the goal is a safe and effective combination product.

Risk Estimation

After the hazardous situation(s) is identified, then the risk can be estimated. As previously discussed, **risk is defined as the *combination of the probability of occurrence of harm and the severity of that harm.*** The term, *harm* is then defined as *injury or damage to the health of people or damage to property or the environment.* We must continue to consider that we are trying to improve the health of people with our combination product and therefore we should not be damaging the health of people. We should be reducing risks. We can reduce risks by reducing the probability of occurrence of the hazardous situation, or we can reduce the severity of the harm associated with the hazardous situation. Most often the approach is to reduce the probability of occurrence, as when the event occurs it is difficult to reduce the severity of the harm.

Severity of Harm First, the severity of harm should be established for each hazardous situation, and it is possible that different harms could occur for each hazardous situation, each with its own severity. Each of these severities is listed and explored in the risk management process. Table 6.8A includes examples of Harm Severity classifications for the assessment of risks that result in harm to an individual.

TABLE 6.8A
Example Harm Severity Classifications for Assessment of Risks That Result in Harm to an Individual

Harm Classification		Description of the Harm to an Individual
Critical	**S-4**	**Life-threatening injury/illness (death has or could occur); Threat to life that requires immediate professional medical intervention.**
Serious	**S-3**	**Injury/illness that necessitates significant medical or surgical intervention to preclude permanent injury.**
Minor	**S-2**	**Injury/illness that is transient and that *does not* require medical intervention.**
Negligible	**S-1**	**No adverse health consequences, illness, or injury.**

Severity is usually a qualitative term, especially when referring to harm to people. While the nuclear and chemical industries have adopted quantitative values equating human life in financial terms, the medical products industries do not define severity in this manner. This may largely be due to the fact that the other industries may cause a large number of casualties in a single incident, where the health products usually only result in a harm to a single patient or person per product use. These other industries also use the term "catastrophic" to define the highest level due to the large number of casualties, such as the crash of a single aircraft resulting in a large number of casualties.

Each manufacturer defines those terms. It is advisable to apply the language used within applicable regulations for your products, such as 21 CFR 803, Medical Device Reporting, the FDA Adverse Event Reporting System (FAERS/VAERS) system for drugs, vaccines, and biologics or the European Union (EU) Vigilance terms. Using those terms provides alignment in the product development process and post-market events. This helps in communication across the company and with regulatory agencies. In that case, there might be only four levels. This may provide sufficient information to investigate events and determine reporting requirements. Note that even though numbers are used to identify levels, it is still a qualitative system. The numbers are not useful in "calculations" as they have no quantitative meaning. An example of such a table is presented in Table 6.8A.

It is important in this discussion to note that the FDA has adverse event reporting expectations for combination products (21 CFR Part 4), requiring Individual Case Safety Reports, and while they do not have severity levels identified, the manufacturer must comply with all appropriate event reporting requirements.

Probability of Occurrence of Harm The probability of occurrence of harm is then estimated for each hazardous situation. Probability of occurrence is not a purely quantitative term, although the term "probability" suggests quantitative use. The intention of the authors of the ISO risk management standard and the ICH guidance is to allow either qualitative or quantitative values. If the data available are of lower confidence, then qualitative should be used, until a higher confidence data set becomes available.

For disposable, or single-use products, the qualitative or quantitative levels should be defined based on the expected Occurrence of Harm per number of units distributed. Table 6.8B illustrates one possible example. An example of a semi-quantitative probability scale for a reusable product is shown in Table 6.8C based on the estimated product distribution and its use. There is no "one-size-fits-all" approach; the manufacturer is supposed to choose the proper terminology and frequencies that make sense for their specific combination product application. There is similarly no one "correct" number of levels to apply.

Both severity and probability are estimates until data with high confidence is available to confirm the estimates. Often the confirmation does not occur until the product has sufficient field experience due to limitations of small data sets with limited patient/user exposure to the product.

TABLE 6.8B
Example Probability of Occurrence of Harm Levels (e.g., For a Single-Use, Disposable Combination Product)

Level	Common Term	Qualitative Description (e.g., Time-Based[a])	Examples of Probability Range
O-4	Frequent	<Monthly	$\geq 10^{-3}$
O-3	Probable	>Monthly but ≤ Yearly	$\geq 10^{-5}$ and $<10^{-3}$
O-2	Occasional	>Yearly but ≤ 5 years	$\geq 10^{-6}$ and $<10^{-5}$
O-1	Remote	>5 years	$<10^{-6}$

[a] Typically occurs, or occurs on average within the time frames displayed.

TABLE 6.8C
Example Probability of Occurrence of Harm Levels (e.g., For a Reusable Combination Product)

Level	Common Term	Qualitative Description	Examples of Probability Range
O-4	Frequent	Occurs multiple times in life of a single unit	$\geq 10^{-3}$ of total installed base
O-3	Probable	Occurs once in life of single unit	$\geq 10^{-5}$ and $< 10^{-3}$
O-2	Occasional	Possible to occur in life of product model	$\geq 10^{-6}$ and $< 10^{-5}$
O-1	Remote	Not expected to occur in product model	$< 10^{-6}$

TABLE 6.8D
RISK: Probability of Occurrence of Harm × Severity of that Harm
4 × 4 Risk Chart Before Risk Levels Are Overlaid

Probability of Occurrence of Harm						
	Frequent	**O-4**				
	Probable	**O-3**				
	Occasional	**O-2**				
	Remote	**O-1**				
			S-1	**S-2**	**S-3**	**S-4**
			Negligible	Minor	Serious	Critical
			Severity of Harm			

Risk Chart With the establishment of the probability of occurrence of harm and the severity of the harm, we have defined the two essential elements of risk: (1) probability of the occurrence of harm and (2) severity of that harm. We have provided the basis for *risk estimation.*

A Risk Chart may be used to define risk acceptability using the levels of severity of harm and probability of occurrence of harm from the previous tables. The example Risk Chart shown in Table 6.8D depicts a system with four levels of probability of occurrence of harm and four levels of severity of harm. The levels chosen by the manufacturer typically vary between four and five, though three levels may be used for qualitative analysis (such as early in product development). That is up to an organization's discretion. Exceeding five levels usually leads to lengthy discussions regarding which level of severity or probability should be assigned to a hazardous situation, and typically does not result in any useful benefit. The more complex the analysis, the more documentation is necessary to justify how the data were parsed, thus requiring more time to be spent on the rationale and less time on the actual decision-making.

Risk Acceptability Risk acceptability criteria are required to be established as part of the Risk Management System and documented in the Risk Management Plan at the start of the product development and risk management processes. The criteria for risk acceptability are based on the specific intended uses, intended users, and intended use environment of a given combination product. ISO TR 24971:2020 Annex C.2 has an extensive discussion of risk acceptability. This establishes the acceptable risk by balancing the severity of harm with the probability that harm would occur. These criteria most often require the input of medical and clinical experts that are competent in the use of the combination product in its various uses, as well as the health professionals and patient population expected to use the combination product. The document recommends that manufacturers use the following to establish Acceptable Risk:

- Applicable regulatory requirements in markets served
- Relevant international standards (ISO and IEC) and guidance (ICH) for example
- Generally acknowledged *state of the art*[35]
- Validated concerns from stakeholders such users, clinicians, patients, or regulatory bodies (it is important to consider perception and understanding of risk acceptability can vary between different stakeholders)

A Risk Chart may be used to define risk acceptability, according to the company's Risk Management Policy, using the levels of severity of harm and probability of occurrence of harm from the previous tables (examples in Tables 6.8A through C). In evaluating risk acceptability, one looks at the intersection of "probability of occurrence of harm" with "severity of harm." The intersection of the two defines the level of risk. Figure 6.7 shows an example Risk Chart with three levels of risk:

Broadly Acceptable Risk
Moderate Risk
and Unacceptable Risk

In this example, "Unacceptable" risks are considered intolerable; the "Moderate" risks require investigation of further risk reduction; and the third level "Broadly Acceptable" is considered to be acceptable risks. The purpose of the "Moderate" (investigate) region for many is to identify those risks that may result in good return for establishing additional control measures. The particular manufacturer might require the product team to take action depending on the level of risk resulting from the severity and probability of harm. At the end of the risk management process, there are only two levels of risk: "Acceptable and Unacceptable."

In Figure 6.7, the intersection of an "Occasional" (O-2) probability of occurrence of harm, where the harm has a Serious severity (S3), results in a "Moderate" risk level. The manufacturer would be expected to investigate and work to reduce the probability of harm as low as possible. In the case where there are sufficient resources, a team may work to reduce all risks.

Probability of Occurrence of Harm						
	Frequent	O-4	Moderate Risk	Unacceptable Risk	Unacceptable Risk	Unacceptable Risk
	Probable	O-3	Broadly Acceptable Risk	Moderate Risk	Unacceptable Risk	Unacceptable Risk
	Occasional	O-2	Broadly Acceptable Risk	Broadly Acceptable Risk	Moderate Risk	Unacceptable Risk
	Remote	O-1	Broadly Acceptable Risk	Broadly Acceptable Risk	Broadly Acceptable Risk	Moderate Risk
			S-1	S-2	S-3	S-4
			Negligible	Minor	Serious	Critical
			Severity of Harm			

FIGURE 6.7 Example risk chart with risk levels. *Not for use without valid rationale.*

Risk and EU MDR (2017/745)

Notably, under Article 10.2 of the Medical Device Regulations EU MDR (2017/745), it is actually the expectation that *all* risks need to be reduced "as far as possible." Manufacturers are required to establish, document, implement, and maintain a product safety risk management system. The detailed requirements of the system are listed in Annex I Chapter I, points 2–9. Risk decisions need to remain focused on safety risk benefits. When a risk is identified, the product developer/manufacturer has to be able to demonstrate that all practical risk control measures, which reflect the "current state of the art" have been implemented, and any further risk controls would not further reduce risks without adversely affecting the benefit-risk ratio.[36] "State of the art" refers to "generally accepted state of the art," which relates to current best practices in the industry. It does not mean the most advanced or latest technology, just what is widely seen as current best practices. When making risk reduction decisions, and identifying and implementing risk controls, cost is not part of the decision-making process.[37] The benefits to the intended users/society should outweigh the residual risks associated with the device constituent part. Where residual risks remain, addition risk controls would not result in lower risk levels.

Risk Chart Example

Table 6.9 illustrates a more comprehensive risk chart example. In this example, the Risk Chart used depicts a system with six levels of probability of occurrence of harm and five levels of severity of harm. As indicated earlier, the number of levels of "Probability of Occurrence of Harm" and "Severity of Harm" are defined at the discretion of a manufacturer.

Often a risk can result in more than one severity of harm. A simple example, depicted in Table 6.9, rows 1.3.2 through 1.3.7, would be the range of harms derived from the hazard "sharp needle without a needle guard." Depending on the specific sequence of events, the critical task of handling a prefilled syringe, for example, with its sharp needle without a needle guard, could lead to harms of negligible severity all the way to critical severity.

The potential harms associated with this hazardous situation vary in their severity and can include "critical" injury (S-4) such as transmission of blood-borne pathogens or a systemic infection, to "negligible" (S-1) with minor percutaneous injury. Each of the severities potentially has a different probability of occurrence of harm. It is possible that the same event results in varying harm severities. For example, the "Sharp Needle with no needle guard" hazard/"Inadvertent Needle Stick with Unsterile Needle after Being Used by Same Patient" hazardous situation could lead to either a harm of "minor" S2 severity (minor percutaneous injury/wound) (Line 1.3.2) or it could lead to a "serious" S3 harm (local infection).

It is possible that a less serious event may have an unacceptable risk due to a high rate of occurrence of harm (i.e., probability of occurrence of harm). A more severe event may have such a low rate of occurrence it may not rate as an unacceptable risk, depending on the risk acceptability criteria established by the manufacturer (see ISO TR 24971:2020 for how to establish risk acceptability criteria).

In our "Sharp Needle with no needle guard" example, Hazard/Hazardous Situation→Harm in line 1.3.4 has a lower severity of harm than does the Hazard/Hazardous Situation→Harm in line 1.3.5, but the probability of occurrence for Hazard/Hazardous Situation→Harm in line 1.3.4 is more frequent ("Occasional O-3"), making them both "Moderate Risk," requiring investigation and controls to reduce the risk as low as possible.

It is important that each of the possible severities be assessed to determine the acceptability of the risk in the associated hazardous situation, not just the most severe ones. Risk, and acceptability of risk, is based on probability (rate of occurrence) of harm and the severity of that harm.

Traceability

ISO 14971 requires the use of **traceability** as part of the RMF. In defining traceability requirements, the combination product manufacturer should consider all those constituent parts, components, materials, and conditions which might cause the combination product or its constituent parts to not meet its specified requirements, including its safety requirements. Among considerations for traceability are: intended use (e.g., emergency use, life sustaining, or implantable), intended users and use environment for the combination product; origin of components and materials; processing history; probability of failure; and impacts of failure on patients, users, or other persons.

TABLE 6.9
Example Risk Traceability Summary (All Ratings Are "Mock" for Illustrative Purposes Only)

A: Harm and Probability of Occurrence of Harm Ratings Applied[a]						
Severity Level	*S Term*	*Description*[b]	*Probability of Occurrence of Harm*	*O Term*	*Probability Range*	*Qualitative Description (e.g., Time-Based)*
S-5	*Catastrophic/Fatal*	*Results in death*	*O-6*	*Always*	$\geq 10^{-2}$ and ≤ 1	Expected to occur
S-4	*Critical*	*Results in permanent impairment or irreversible injury*	*O-5*	*Frequent*	$\geq 10^{-3}$ and $< 10^{-2}$	3 days
S-3	*Serious/Major*	*Results in injury or impairment requiring medical or surgical intervention*	*O-4*	*Probable*	$\geq 10^{-4}$ and $< 10^{-3}$	>3 days but Monthly
S-2	*Minor*	*Results in transient injury or impairment, not needing medical or surgical intervention*	*O-3*	*Occasional*	$\geq 10^{-5}$ and $< 10^{-4}$	>Monthly but Yearly
S-1	*Negligible*	*Results in inconvenience or temporary discomfort*	*O-2*	*Remote*	$\geq 10^{-6}$ and $< 10^{-5}$	>Yearly but 5 years
			O-1	*Improbable*	$< 10^{-6}$	>5 years

[a] GHTF.SG3.N15-R8: Implementation of Risk Management Principles and Activities Within a Quality Management System (2005) accessed August 3, 2022 at https://www.imdrf.org/sites/default/files/docs/ghtf/final/sg3/technical-docs/ghtf-sg3-n15r8-risk-management-principles-qms-050520.pdf

[b] See ISO TR 24971:2020, Table 4: Example of Five Qualitative Severity Levels, and Table 5: Example of Five Semi-Quantitative Probability Levels.

B: Risk Chart (6 × 5)							
Probability of Occurrence of Harm	Always	**O-6**	Moderate Risk	Unacceptable Risk	Unacceptable Risk	Unacceptable Risk	Unacceptable Risk
	Frequent	**O-5**	Moderate Risk	Moderate Risk	Unacceptable Risk	Unacceptable Risk	Unacceptable Risk
	Probable	**O-4**	Broadly Acceptable Risk	Moderate Risk	Moderate Risk	Unacceptable Risk	Unacceptable Risk
	Occasional	**O-3**	Broadly Acceptable Risk	Broadly Acceptable Risk	Moderate Risk	Moderate Risk	Unacceptable Risk
	Remote	**O-2**	Broadly Acceptable Risk	Broadly Acceptable Risk	Broadly Acceptable Risk	Moderate Risk	Moderate Risk
	Improbable	**O-1**	Broadly Acceptable Risk	Broadly Acceptable Risk	Broadly Acceptable Risk	Broadly Acceptable Risk	Moderate Risk
	None observed	**O-0**	Broadly Acceptable Risk	Broadly Acceptable Risk	Broadly Acceptable Risk	Broadly Acceptable Risk	Broadly Acceptable Risk
			S-1	**S-2**	**S-3**	**S-4**	**S-5**
			Negligible	Minor	Serious/Major	Critical	Catastrophic
			Severity of Harm				

(Continued)

TABLE 6.9 (CONTINUED)
Example Risk Traceability Summary (All Ratings Are "Mock" for Illustrative Purposes Only)

	C: Risk Management Traceability Table Illustrative Example (©2022 Combination Products Consulting Services, LLC)						
	Risk Assessment						
	Risk Analysis						Risk Evaluation
	Hazard List	Reference/Source Document	Hazardous Situation	Harm	Risk Estimation		Risk Acceptability
Unique Identifier	**Hazard** *(Potential Source of Harm)*	*Identify Document Number and Line Item of Reference*	*(Potential Exposure of People, Property or Environment to the Source of Harm)*	*[Injury (Mental or Physical) to People, Property or the Environment]*	**Probability of the Occurrence of Harm**	**Severity of Harm**	**Risk Level**
1.1	**Functional Hazards**						
1.1.1	Premature signal of dose completion	dFMEA 243, line 22	Visual indicator feedback malfunction →Incomplete dose/underdose (e.g., insulin delivery)	Hyperglycemia	Probable O-4	Serious S-3	Moderate
1.2	**Mechanical Hazards**						
1.2.1	Motor failure	pFMEA 45, line 850	Motor malfunction →Under dose/missed dose (e.g., of insulin)	Lack of drug effect (drug-dependent harm) (e.g., diabetic shock)	Remote O-2	Critical S-4	Moderate
1.3	**Use Hazards**						
1.3.1	Complexity of Use/ Use Error	URRA HHA line 373	Incorrect orientation of emergency use injector (e.g., to treat severe allergic reaction) → **missed dose**	Anaphylactic shock	Remote O-2	Catastrophic S-5	Moderate
1.3.2	Sharp Needle (no needle guard)	URRA HHA Line 37	Inadvertent Needle Stick with Sterile Needle	Minor percutaneous injury/ wound	Probable O-4	Negligible S-1	Broadly Acceptable
1.3.3	Sharp Needle (no needle guard)	URRA HHA Line 41	Inadvertent Needle Stick with Unsterile Needle after Being Used by Same Patient	Minor percutaneous injury/ wound	Probable O-4	Minor S-2	Moderate
1.3.4	Sharp Needle (no needle guard)	URRA HHA Line 42	Inadvertent Needle Stick with Unsterile Needle after Being Used by Same Patient	Local infection	Probable O-4	Serious S-3	Moderate

(Continued)

TABLE 6.9 (CONTINUED)
Example Risk Traceability Summary (All Ratings Are "Mock" for Illustrative Purposes Only)

	C: Risk Management Traceability Table Illustrative Example (©2022 Combination Products Consulting Services, LLC)						
	Risk Assessment						
	Risk Analysis						Risk Evaluation
	Hazard List	Reference/Source Document	Hazardous Situation	Harm	Risk Estimation		Risk Acceptability
1.3.5	Sharp Needle (no needle guard)	URRA HHA Line 43	Inadvertent Needle Stick with Unsterile Needle after Being Used in Another Patient	Transmission of blood-borne pathogens	Remote O-2	Critical S-4	Moderate
1.3.6	Sharp Needle (no needle guard)	URRA HHA Line 44	Inadvertent Needle Stick with Unsterile Needle after Being Used in Another Patient	Systemic infection	Remote O-2	Critical S-4	Moderate
1.3.7	Sharp Needle (no needle guard)	URRA HHA Line 45	Inadvertent Needle Stick with Unsterile Needle after Being Used in Another Patient	Local infection	Remote O-2	Serious S-3	Broadly Acceptable
2	**Biological Hazards**						
2.1.1	Non-biocompatible materials	dFMEA 16 Line 567	Skin contact with non-biocompatible materials	Minor Allergic Reaction	Improbable O-1	Minor S-2	Broadly Acceptable
3	**Chemical Hazards**						
3.1.1	Excipients of drug formulation	dFMEA 2435 Line 152	Interaction of excipients from drug formulation with drug-contacting fluid path of device → extractables & leachables in drug administered to patient	Anaphylactic shock/death	Remote O-2	Catastrophic S-5	Moderate
4	**Manufacturing/Production Hazards**						
4.1.1	Heat seal process	pFMEA 123 Line 300	Heat-sensitive drug (e.g., insulin) exposed to excessive heat duration due to blister seal process downtime→ drug degradation→ lack of drug effect	Hyperglycemia	O-4	S-3	Moderate
5	**Software Hazards**						
5.1.1	Cybersecurity compromise	sFMEA SSS Line 261	Delayed cardiac treatment		O-2	S-5	Moderate

While *GHTF SG3/N15R8 Implementation of risk management principles and activities within a Quality Management System* provides an example of such a document, this example could be expanded to include columns referring to source documents as shown in the attached traceability summary seen in Table 6.9. The traceability summary is especially useful for products that have risk management information collected from a variety of sources, locations, and sub-processes, and especially for Combination Products which as a system consist of several major components or subsystems. Identified severity and occurrence categories are defined and justified. The traceability summary becomes a system index to all documents created by the various elements of the combination product system. The Traceability Summary allows for a complete view of the Combination Product as a system and includes the documentation of risks of all components, constituent parts, subsystems, and the complete Combination Product System.

A traceability summary may be created for each constituent part of the combination product by the manufacturer of that component and/or constituent part, and then combined into a System Risk Traceability Summary for the entire combination product system. Then, it could be used during the Overall Residual Risk Analysis for the entire combination product and, potentially, as a part of the product submission. The Risk Traceability Summary provides a single location to document all the hazards and harms associated with the entire combination product, as well as a location to document the results of the risk analyses and subsequent risk control activities. The Risk Traceability Summary is considered part of the Risk Management File and serves as an index to the documents listed. In the "Risk Assessment" of the Summary shown in Table 6.9, the "Risk Evaluation" column identifies the acceptability of each risk.

The Risk Traceability Summary is built from hazard categories (see Table 6.7, for example, categories to consider.

Risk Control

All risks that are rated as "Unacceptable or Intolerable" must result in risk control actions. Recall the shark example, when we discussed "layers of defense." These are basically "risk controls." In ISO 14971, a technique called "Option Analysis"[38] is used to choose a method of risk control *in the following order of preference* (see Figure 6.8):

(a) *Inherently safe design and manufacture*; Modify the product or process design (including the manufacturing process design) to eliminate the risk or reduce it to an acceptable level.
(b) *Protective measures in the medical device itself or in the manufacturing process;* Provide a method or safeguard to reduce the risk. For example, in a drug-delivery syringe, a needle guard may be incorporated to reduce the risk of a needle stick.
(c) *Information for safety, and where appropriate, training to users;* Provide information for safety (including training materials) so decisions on appropriate use of the combination product can be made by user and patient.

Note: Italics in (a). (b). and (c), above are quoted from ISO 14971:2019 7.1. Normal font text is explanation for combination products.

The third (and weakest) method of risk reduction is to provide information regarding the risk to the user so that action may be taken to reduce the probability of harm. This information gives the healthcare professional (and the patient) information to make informed decisions about either accepting or avoiding the risk. One method is to provide Warnings and Cautions at both the beginning part of instruction manuals and later in the text where an operational step is explained. The main problem with instructions is their availability at the time of use of the product and the ability of the instructions to convey the safety message regarding the condition and the steps to avoid the risk.

A common practice in both the pharmaceutical and medical device industries is to address issues through labeling. *Information for safety* is important and is a requirement of ISO 14971:2019 Clause

8 Evaluation of overall residual risk to disclose significant[39] residual or remaining risk after risk controls are implemented and to provide this necessary information for informed decisions on the appropriate use of the product. Warnings and Cautions may also appear on screens on electronically controlled devices to inform the user of a hazard in addition to those shown in labeling and documents accompanying the product. Identification of contraindications for the product is also important to assist with decisions on the appropriate use of the combination product.

This method of risk reduction is the least effective of the three options but is usually necessary in addition to other options. Information for safety is also a requirement for reducing product liability by giving warnings, cautions, and contraindications for use in decisions on appropriate product use. That said, addressing potential harms or issues in product labeling may be appropriate, but ***only*** if those issues cannot be otherwise eliminated. What that means may be different for drugs, devices, or combination products.

Other relevant control measures include change control processes, purchasing controls, CAPA systems, and management responsibilities (see Figure 6.1 in this chapter, and also Chapter 4 on Combination Products cGMPs in this book). These are tightly related to post-market safety surveillance and reporting. It is important to realize that risk reduction typically only reduces the probability of harm occurring and not the severity of the harm. Only in cases where the hazard is removed by a design solution is severity of harm possibly modified.

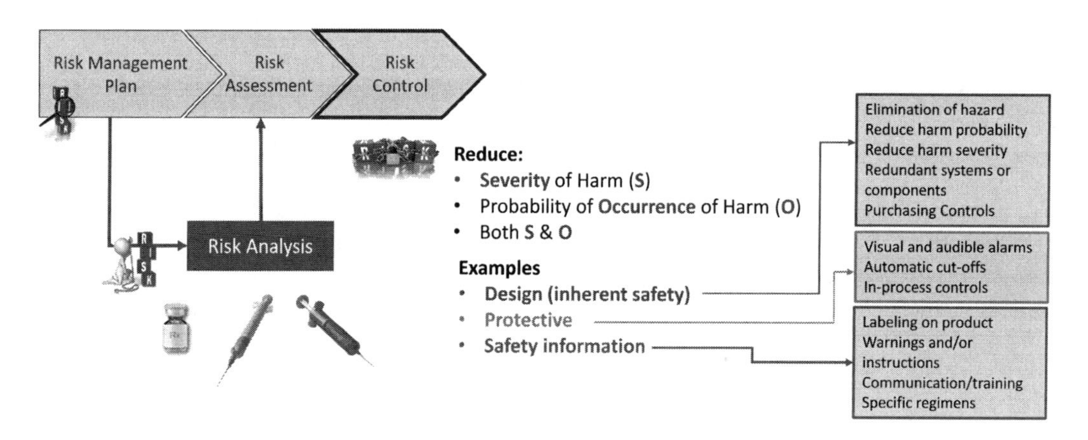

FIGURE 6.8 Risk controls: "Option analysis" (©2022 Combination Products Consulting Services, LLC. All Rights Reserved).

Risk Management Plan → Risk Assessment → Risk Control

Risk Analysis

Examples

Hazard	Control Measures	Verification of Effectiveness
Loss of sterility	Container closure design	Integrity testing; Packaging validation
Drug in PFS exposed to excessive heat at blister seal	In-process controls	Process Validation/ Testing
Drug contamination from materials of syringe construction (e.g., silicone oil)	Purchasing controls; In-process testing; Finished product testing	Verification of purchasing controls; Design controls; Test method validation
Syringe breakage during use	Syringe design and material specifications; Storage specifications; Reliability requirements	Reliability testing 1st article inspection Design verification Design validation

FIGURE 6.9 Examples of hazards and potential control measures (©2022 Combination Products Consulting Services, LLC. All Rights Reserved).

Eliminating a potential issue for a drug could include changes to their formulation and dosing, etc.; for devices, it might include design changes; for combination products – addressing a potential issue might be handled altogether differently, as combination products combine types of medical products and, thereby, offer not just another layer of complexity, but also potential synergies to manage harms that the "one medical product-type" alone would present, or not be able to eliminate or successfully mitigate alone. Consider, for example, a device-led combination product, where the drug supports the function of the device.[40] A drug may be coated on a device to mitigate undesired local physiological responses associated with the implantation procedures or the use of the product (e.g., an anti-inflammatory drug on a cardiac lead to reduce inflammation at the implantation site, or an anti-coagulant bound to the inner lumen of a catheter to prevent clot formation within the catheter, thereby maintaining catheter patency). The drug coatings often involve a lower dose and/or primarily local, rather than systemic, exposure to a drug as compared to what might otherwise be experienced for a stand-alone drug product. As such, the combination product may successfully eliminate certain issues that the stand-alone type drug or device could not. That said, though, the combination product configuration may raise different safety and effectiveness concerns, e.g., due to interactions of the drug with the device, that need to be addressed.

Figure 6.10 illustrates applicable sections of the Risk Traceability Summary document that would be used to document Risk Controls, the associated documents and actions. Each of the hazards identified in the illustration are ones whose risk was "Moderate." Per the Manufacturer's Risk Acceptability Criteria, risk rates at that level require risk control measures to be applied to reduce the risk. The following column is an identification of the Product Requirements documentation and line item which informs the design team of the product or the process requirement that must be fulfilled to reduce the risk.

In row 1.1.1 for the "premature signal of dose completion" hazard, by placing risk control measures in the product Specification, labeling, and syringe specification document, we also assure that the Design Verification process confirms that the risk controls are actually implemented. Risk controls that include information for safety as the method of risk reduction must also appear in the Product Requirements documents and be verified as appearing in the product documentation. Usability Validation is required to cover content and effectiveness of Labeling as a Risk Control. Any information for safety implemented through software would appear in the software requirements and should be verified as implemented during software verification.

Risk control measures could include manufacturing controls to reduce the probability of occurrence of causes of hazardous situations. These controls are documented in manufacturing process documents such as process instructions and verified as implemented and effective during process validation activities. Purchasing Controls might also be applied, where a constituent part or component is purchased from a third party, to ensure quality processes are in place to ensure parts meet acceptance specifications.

The Risk Traceability Summary shows columns for "Verification of Implementation ID" and "Verification of Implementation Results." In the first of these, we should identify both the Verification Protocol document and the section of the document containing the method and acceptability criteria used to verify the implementation of the Risk Control Measure. The second item identifies the Verification Report section and results that demonstrate the verification of the risk control. This verification is to assure that the risk control concepts identified in the Risk Control portion of the Risk Traceability Summary actually appear in the design of the product, including the manufacturing process, as appropriate.

The last section of the Risk Traceability Summary, shown in Figure 6.11, documents the verification of the effectiveness of the risk controls. The column "Verification of Risk Controls" refers to the objective evidence document(s) that contain the results of the verification of effectiveness protocol identified in the last column of Figure 6.10. If the Estimate is initially documented as Quantitative values, then the Verification of Effectiveness results must be documented as Quantitative values.

When verifying the effectiveness of the risk control, we need to reassess the residual risk that remains after the control is applied. The product or the process should achieve the risk estimated during the consideration of the risk control measure. The remaining or residual risk is documented

	Risk Management Traceability Table											
	Risk Assessment											
	Risk Analysis						Risk Evaluation					
	Hazard List	Reference/ Source Document	Hazardous Situation	Harm	Risk Estimation		Risk Acceptability	RISK CONTROLS				
Unique identifier	Hazard *(Potential Source of Harm]*	Identify document *number and line item of reference*	*(Potential exposure of people, property or environment to the source of harm)*	*[Injury(metal of physical) to people, property of the environment]*	**Probability of the Occurrencet of Harm**	**Severity of Harm**	**Risk Level**	**Risk Control Measures**	**Product Requirement or Characteristic ID**	**Verification of Implementation ID Verification of**	**Verification of Implementation Results**	**Verification of Effectiveness (or Validation) ID**
1.1	**Functional Hazards**											
1.1.1	Premature signal of dose completion	dFMEA 243, line 22	Visual indicator feedback malfunction→ Incomplete dose/ underdose (e.g., jnsulin delivery)	hyperglycemia	Probable 0-4	Serious S-3	Moderate	-Re-design visual indicator -add fail-safe backup alarm -IFU	Syringe spec 12.12 Labeling 12.12 Product **Specification**	**Verification** Protocol 12.12	Test Pass Protocol 12.12 Line 351	**Verification** Protocol 12.12 Human Factors Summative Study 12.12
1.2	**Mechanical Hazards**											
1.2.1	Motor failure	dFMEA 45, line 850	Motor malfunction→Under dose/Missed dose (e.g., of insulin)	Lack of drug effect (drug-dependent harm) (e.g., diabetic shock)	Remote 0-2	Critical S-4	Moderate	Incoming inspection; Win-process check Purchasing Controls	W123&Assy Equipment 22	**Verification** Protocol 22.1	Test Pass Protocol 22.1 Line 762	**Verification** Protocol 22.1
1.3	**Use Hazards**											
1.3.1	Complexity of Use/Use Error	URRA HHA line 373	Incorrect orientation of emergency use injector (e.g.,to treat severe allergic reaction)→ missed dose	Anaphylactic Shock	Remote 0-2	Catastrophic S-5	Moderate	IFU Training Model	Labeling3M Product **Specification**	**Verification** Protocol 345	Test Pass Protocol 345 Line 47	**Verification** Protocol 345 Human Factors Summative Study 345

FIGURE 6.10 Examples of risk controls in risk traceability summary (©2022 Combination Products Consulting Services, LLC).

	Risk Management Traceability Table											
	Risk Assessment											
	Risk Analysis						Risk Evaluation					
	Hazard List	Reference/ Source Document	Hazardous Situation	Harm	Risk Estimation		Risk Acceptability					
Unique identifier	**Hazard**	*Identify document number and line item of reference*	*(Potential exposure of people, property or environment to the source of harm)*	*[Injury (mental or physical) to people, property or the environment]*	**Probability of the Occurrence of Harm**	**Severity of Harm**	**Risk Level**		**POST CONTROL RISK**			**Comments**
	(Potential Source of Harm)							**Verification of Effectiveness Results**	**Probability of Occurrence of Harm**	**Severity of Harm**	**Residual Risk**	**Comments and Identify, and Affected Item**
1.1	**Functional Hazards**											
1.1.1	Premature signal of dose completion	dFMEA 243, line 22	Visual indicator feedback malfunction Incomplete dose/ underdose (e.g., insulin delivery)	hyperglycemia	Probable 0-4	Serious S-3	Moderate	Protocol 12.12 Pass HF Study 12.12 Pass	Remote 0-2	Serious S-3	Broadly Acceptable	**Pass, see final** report 12.12 Section 4
1.2	**Mechanical Hazards**											
1.2.1	Motor failure	pFMEA 45, line 850	Motor malfunction → Under dose/ Missed dose (e.g., of insulin)	Lack of drug effect (drug-dependent harm) (e.g., diabetic shock)	Remote 0-2	Critical S-4	Moderate	Protocol 22.1 Pass	Improbable 0-1	Critical S-4	Broadly Acceptable	**Pass, see final** report 22.1 Section 3
1.3	**Use Hazards**											
1.3.1	Complexity of Use/Use Error	URRA HHA line 373	Incorrect orientation of emergency use injector (e.g., to treat severe allergic reaction) → missed dose	Anaphylactic Shock	Remote 0-2	Catastrophic S-5	Moderate	Protocol 345 Pass HF Study 345 Pass	Improbable 0-1	Catastrophic S-5	Moderate	Passed testing, see **final report 345,** Lines 15–18; and **Benefit-Risk** Analysis 345 Section 7, Lines 759–800

FIGURE 6.11 Example risk traceability summary – Risk controls (©2022 Combination Products Consulting Services, LLC).

as Post-Control Severity of Harm and Probability of Occurrence of Harm. Recall from earlier in the chapter, that the risk acceptability of each constituent part alone may not be the same as when they are used together in the combined-use context. For example, an emergency-use drug for use with a given device may have a more stringent level of acceptance criteria than a non-emergency-use drug intended for use with that same device constituent part. Likewise, a combination product intended for use by pediatric patients might have more restrictive criteria than that same type of drug-device combination for an adult user population based on differences in cognitive and dexterity skills. The intended uses, intended users and intended use environment, or potential constituent part interactions, all are key factors influencing the level of controls, and acceptability of those controls applied.

Following the implementation of control measures, the manufacturer should reassess the combination product and its constituent parts for any new potential hazards/hazardous situations that may have been introduced to the combined-use system and then update risk estimates. Verification of effectiveness relative to acceptability criteria in the Risk Management Plan would then be repeated. A Benefit-Risk analysis of any residual risks is then needed to ensure benefits of the combination product outweigh its risks. One then needs to link risks and controls for the combination product to verification and validation documents.

Residual Risk Evaluation

Residual risk is the risk that remains after all risk control measures have been implemented. The residual risks may relate to the possible occurrence of side-effects or after-effects related to the use of the combination product or its constituent parts (Figure 6.12). The Benefit-Risk Analysis (BRA)[41,42] evaluates residual risks, if any, against pre-established acceptance criteria, answering the question: "Does the medical benefit outweigh the residual risk(s)?" In this step of the process, an independent review, using competent medical experts in the use of the product (independent of the product development team), is performed to evaluate the total risk associated with the combination product to determine if it is safe to place on the market. Comparisons are based on medical data, literature reviews, etc., based on the specific intended purpose and use of the combination product. The BRA applies to each individual constituent part and the combination product overall risks.

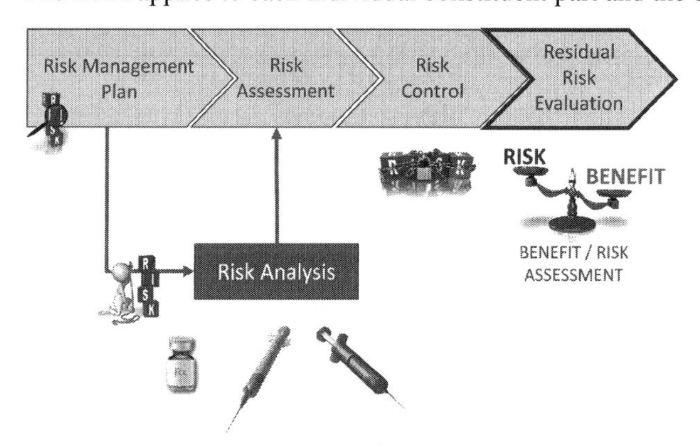

FIGURE 6.12 Residual risks (©2022 Combination Products Consulting Services, LLC. All Rights Reserved).

It is necessary for the manufacturer responsible for the combination product to look at all risks for the product (device and medicinal product) to look at all risks of the individual constituents and the risks of the entire product, such as from interactions between the various elements of the combination product. This is where the systems approach considers each element and the total system all together to evaluate the total overall safety performance. The evaluation process should also include a detailed review of the accompanying information (such as instructions for use, warnings, cautions, and contraindications) to determine if appropriate information on risks has been included.

It is possible that the risk is acceptable for each individual risk when assessed one at a time. But when all risks are evaluated together as a total, the overall risks package may be unacceptable.

For instance, each individual risk of one serious injury per 100,000 may be acceptable, but when 15 individual risks rolled up each have a potential of one serious injury, taken together 15 serious injuries per 100,000 may no longer be acceptable. It is also possible that a risk may be acceptable for one constituent when used by itself, but when used as part of a combination product, it may no longer be acceptable. The use of the Risk Traceability Summary in Overall Residual Risk Evaluation provides a vehicle to consider all of the risks for each constituent part of the combination product when assembled as a whole product. During the Overall Residual Risk Evaluation, a Benefit-Risk Evaluation is performed to assure that the benefit to the user and patient for the entire combination product outweighs the risk of use of the product.

This is to assure that the user (professional and non-professional) can make informed decisions on the appropriate use of the product and how to avoid unsafe conditions. For a combination product, this step includes all of the risks arising from the device and the drug/biologic. The final product manufacturer must consider how the drug/biologic impacts the device and vice versa. In other words, all risks are associated with the product individually and in combination. These include the risks of the supplied components of the product and the manufacturing process risks. The goal is to understand the risk profile of the product as would be experienced in the intended use and all stages of the product lifecycle (patient, health care professional, and other users).

If the results are acceptable, a manufacturer can document the results and proceed. If the residual risks are deemed unacceptable, and further controls impossible (see example in row 1.3.1, Figure 6.11), the manufacturer will need to conduct further BRA or revisit the design cycle. If the results continue to be unacceptable, there may not be an acceptable product for the market. The results of the Benefit-Risk evaluation should be included in the Risk Management File.

Risk Review

The final activity prior to the release of the combination product to market is to review the results and prepare a report. The report should document that all risk management activities in the Risk Management Plan have been successfully completed. Additionally, this report should confirm that the Overall Residual Risk Evaluation indicates the product meets the manufacturer's Risk Acceptability Criteria. Finally, this report should document that processes are established and implemented to assure identification and action on issues associated with the production and post-production experience with the product. This process is the monitoring and feedback activities of post-market surveillance. The risk management report is not a lengthy document but is intended to be a report to executive management, and usually contains a short executive summary with references to the Risk Traceability Summary and other important documents such as the Overall Residual Risk Evaluation Report and any Benefit-Risk Analysis. The report may refer to procedures and activities of production quality monitoring and procedures for post-market surveillance. For combination products (and devices), the Report documents the output of the Review activities as defined in the Risk Management Plan. The Report is part of the Risk Management File (Figure 6.13).

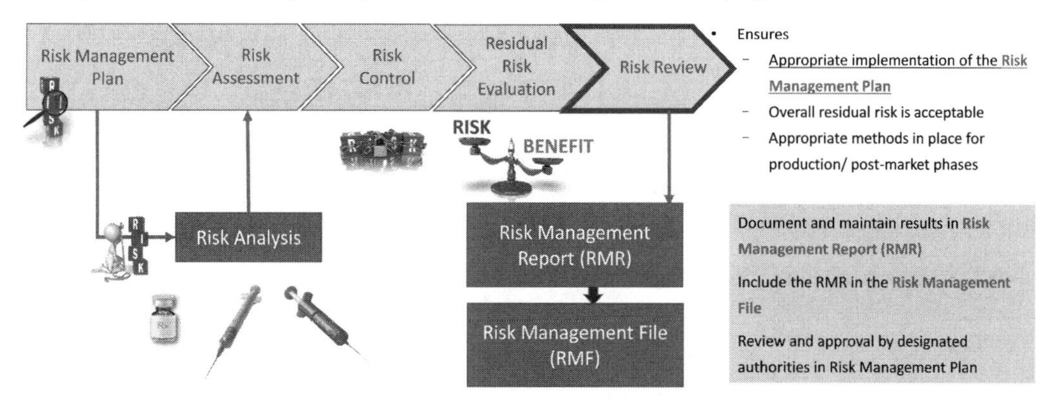

FIGURE 6.13 Risk review (© 2022 Combination Products Consulting Services, LLC. All Rights Reserved).

At this point in the preparation for initial product release, the RMF is ready for the regulatory submission process. Some regulatory bodies might request portions of the information contained in the Risk Traceability Summary, the entire RMF or just the Risk Management Report. In the case where a product is being submitted to a Certifying Body (CB) for a safety mark, the CB may also request all or a portion of the RMF.

However, the only part of the product lifecycle complete at this point is the development portion. The RMF must continue to be reviewed and updated throughout the product lifecycle. ISO 14971:2019 requires in Clause 10 *Production and post-production activities* that the manufacturer continues to monitor product performance throughout the product lifecycle and evaluate any information regarding product safety performance for possible actions that may be necessary by the manufacturer to address any product safety issues. The risk management file is a part of the post-market review process to assure not only is the devices continuing to be safe to use but that it continues to meet any post-market regulatory requirements, including any reporting requirements.

Figure 6.14, while based on device regulation 21 CFR 820, and ISO 13485:2016, illustrates that risk management as defined in ISO 14971:2019 – for each constituent part and the combination product as a whole – is a lifecycle process with inputs and outputs occurring throughout the product life, including disposal of the product. Activities are not performed at just a point in time but continue throughout the development, manufacturing, distribution, and disposal of the combination product. The US FDA assesses the results of the development phase risk management process, summarized in the Risk Management Report, as they review safety, efficacy and usability of each constituent part and combination product as a whole in the Marketing Application (Figure 6.15).

A sample Risk Management Report is shown in Figure 6.16.

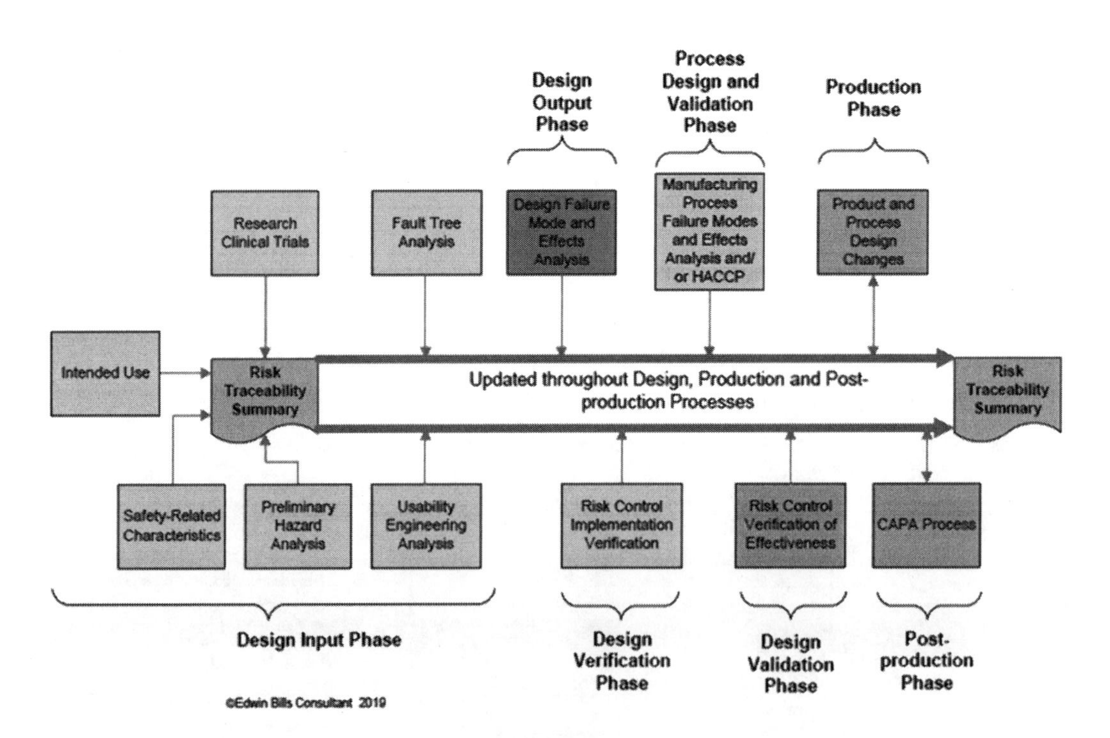

FIGURE 6.14 Typical information flow into risk traceability summary during combination product lifecycle.

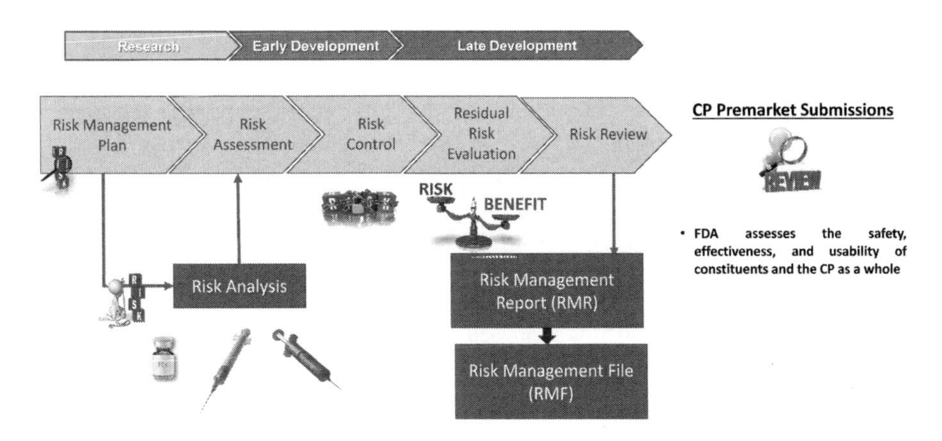

FIGURE 6.15 Risk management submission (©2022 Combination Products Consulting Services, LLC. All Rights Reserved).

Risk Management Report

Document #	
DHF #	

PRODUCT NAME

Instructions: Based on the final review of the Risk Management File prior to commercial distribution, confirm the following:

Has the risk management plan has been appropriately implemented?	☐ Yes ☐ No
Is the overall residual risk is acceptable?	☐ Yes ☐ No
Are appropriate methods are in place to obtain relevant production and post-production information?	☐ Yes ☐ No

APPROVALS			
Function	Meaning of signature	Print Name	Approval Signature/Date
Design Owner (or designee)	Content of the report is complete and accurate.		
Clinical / Medical	Conclusions are supported by documentation		
Quality	Conclusions are supported by documentation.		
☐ N/A			

REFERENCE DOCUMENT(S):

Risk Management Plan Document Number:	
Risk Traceability Summary Document Number:	
Risk Management Tools Used:	
Risk Management Usability Tools Used:	
Risk Management Design FMEA Tools Used:	
Risk Management Process FMEA Tools Used:	
Other Risk Management Tools:	
Other Risk Management Tools:	
Other Risk Management Tools:	

Page 1 of 1

(current revision)

FORM NUMBER:
FORM ISSUE DATE:
CHANGE NUMBER:

FIGURE 6.16 Example risk management report.

Production and Post-Production

As seen in Table 6.4, ICH Q9, ISO 14971, and AAMI TIR 105 risk management process models each require activities to monitor and act upon product risk information after the manufacturing and release of the product to market – the drug constituent part, the device constituent part, and the combination product as a whole.

After the combination product is released, the manufacturer will receive information on product experience through processes such as market/patient surveys, post-market studies, complaints, and adverse events. The product may also undergo a change in its design or manufacturing. During the entire product lifecycle, the RMF must be updated as new information is obtained. The RMF must also be reviewed each time a change is proposed to the product or the manufacturing process, to determine if the product risk, product benefit, or risk controls may be impacted.

Change will occur to the product and the manufacturing process over time in response to improvement initiatives or preventive actions and as a result of corrective actions in the face of newly discovered issues. It is important that change management processes include a review of the RMF as a part of the change evaluation prior to the implementation of any changes. This review should not simply be a "checkbox" that is marked to indicate that the review step has been completed. Often this approach results in an incomplete or inadequate review and can lead to additional corrective actions at a later time when additional issues are discovered. ISO 14971:2019 Clause 10 Production and post-production activities were extensively revised to indicate that manufacturers should *actively* monitor product performance and safety after release, taking action where necessary when information suggests that there is a need for improvement. Formerly, it was accepted that manufacturers would act on complaints and that active monitoring was not required. Some manufacturers currently actively monitor product performance, especially new products, after release to determine if data show that the product is performing at or above the expected safety level forecast in the RMF. The new accepted process is to actively monitor product performance, and not just waiting for complaints.

The RMF does provide an important reference document in the investigation of product performance. ISO 14971:2019 Clause 10.3 provides questions to guide this investigation. The first question is to determine if the event being reported is identified in the RMF. If not, then the risk management process should be started to identify any new hazard and the hazardous situation identified via the information and to evaluate the risk. If the risk is shown to be unacceptable, then the ISO 14971:2019 Clause 10.4 Actions process should be implemented. But manufacturers should also monitor publicly available information on their product performance, such as social media and clinical literature to evaluate these indicators of product safety performance. An extensive list of data sources for monitoring is shown in ISO TR 24971:2020 which also discusses *information collection, information review*, and *actions* as a result of this review. This list was developed from an even more extensive list of data sources and data elements in GHTF SG3/N18 Quality Management System – Medical Devices – Guidance on Corrective Action and Preventive Action and related QMS processes Annex B.[43]

If the information from an investigation or other source, determines that the hazard has been previously identified per the RMF, then the second question to be asked is to determine if the event is occurring with the severity and/or the probability that was forecast in the RMF. If the event is occurring at or below the frequency and severity that was predicted and was identified in the product submissions to the regulatory authorities, then a basis for no action may exist and that conclusion may be documented. The event must also be reported in accordance with the requirements of the regulatory authorities.

If either the severity of harm or the probability of occurrence of harm in the event being investigated is different than predicted in the RMF, then the risk management process must again be invoked to update the information in the RMF, including evaluation of the risk against the manufacturer's risk acceptability criteria. If the risk is no longer acceptable or the overall residual risk is no longer acceptable, then the risk management process demands that additional risk controls be developed to reduce the risk. Again, in these cases, the manufacturer usually invokes the ISO 14971:2019 Clause 10.4 Actions process. There are also cases where the benefit may have changed, for example, due to a new product release. This information may appear during the Information Collection phase and also needs to be evaluated to determine if the BRA is still acceptable.

In the event of product changes, then regulatory requirements may require that regulatory authorities be notified prior to the implementation of the proposed change. This is especially true if the changes are due to changes in the product risk being discovered, as this directly impacts the safety of the product.

The RMF is required to be open for review by those responsible for post-market surveillance through the entire life of the product, which means until the last unit is taken out of service in the field and destroyed. In the case of combination products, some tend to be single-use disposable products, such as prefilled syringes, that are used once and then destroyed. Until the last unit of the last lot produced and distributed is taken out of service and destroyed, the risk management process is a living process. The manufacturer is responsible to provide information on the proper disposal of the combination product, including the possible impact on people, property, or the environment. One example is the use of hazardous material disposal boxes for syringes.

Some combination products, such as drug-eluting stents, are implanted and in use for many years. The RMF for those products is open for review and update until the last unit has been explanted or the product is no longer in use. Perhaps the drug is no longer viable, but the stent may still be in place, thus the product is still in use and the RMF should still be kept open.

Additional issues occur with the use of kits assembled from individually manufactured constituents and identified in labeling. Who is responsible for issues with these types of products? It is important that any supplier agreements are developed in advance of product release to identify responsibilities and authorities for actions and notifications of regulatory authorities and customers of potential issues as required by regulations and good business practices. These risks need to be identified in the Risk Traceability Summary (RTS) and connections to appropriate documents identified so that at any necessary investigation quick location of appropriate documents can be made to expedite actions.

The use of the Risk Traceability Summary document aids in the management of the knowledge of product and process risk by providing both a central index as well as basic information regarding the risks. The traceability summary lists all of the decisions that were made about the product, including those hazards that were identified in the product and the manufacturing process. The traceability summary also lists the hazardous situation that may lead to harm, the severity of the harm, and the probability that harm may occur. The document identifies the documents providing the information on the hazard and the hazardous situation as well as any harm that may occur. The traceability summary documents the risk estimates and the risk evaluation. For those risks that are identified as unacceptable, the traceability summary points to the risk controls that were developed, documents showing where they were implemented, and those documents providing verification of implementation and verification of effectiveness objective evidence. Finally, the traceability summary documents the residual risk though the estimates of the severity of harm and the probability of harm following the implementation of the risk controls. From these two estimates, the residual risk level for each hazard is determined.

Since the traceability summary contains the basic information regarding the risk management for the product, in our case both the device and the drug/biologic, it is the first place that those conducting a change review or an investigational review should search. Since the traceability summary also contains links to the documentation of the various processes used to perform the risk management activities, it provides ease in accessing all of the risk management information.

There may be some cases where no events were identified from complaint processes, or no risk management trends were identified in the quality data review processes. In those cases, the periodic review should document that no new risk information has been identified. It is recommended that an annual review of the complaint and quality data for the product and manufacturing process should be conducted and documented. The results of this review should form part of the quality system management review agenda and can also be included in annual product reviews. This documentation may be useful in product liability documentation demands.

The Closed Loop Lifecycle of Combination Products Risk Management

Figure 6.17 and Table 6.10 illustrate the closed loop lifecycle of risk management. Table 6.10 aligns this activity with Combination Product Design Controls (see chapter on Integrated Development in this book).

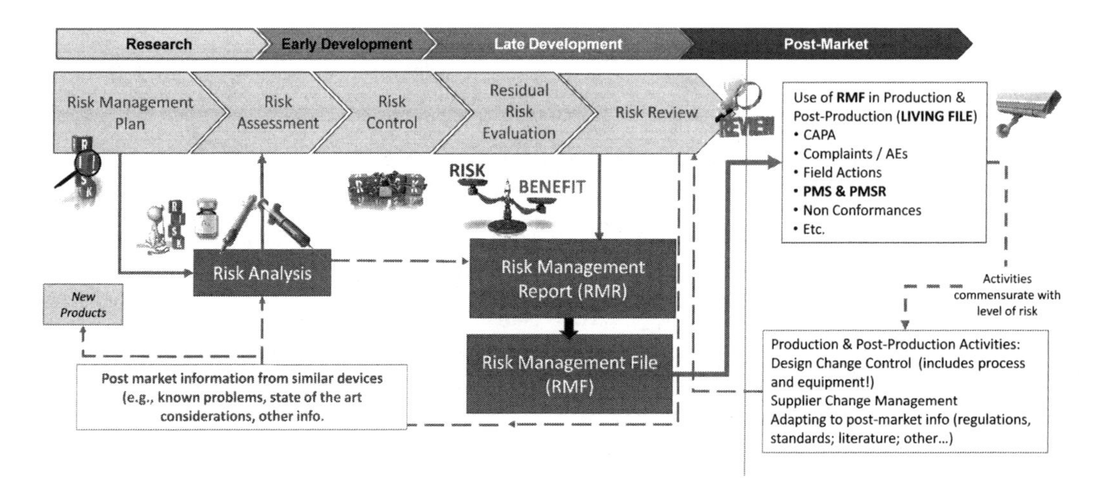

FIGURE 6.17 Combination products closed loop lifecycle (©2022 Combination Products Consulting Services, LLC. All Rights Reserved).

TABLE 6.10
Risk Management and Design Controls (© 2022 Combination Products Consulting Services, LLC)

Risk Management Phase	Focus		Design Control Element
Risk Management Plan	• Establish Context of Use • Risk Acceptability Criteria	⇨	Design & Development Plan
Risk Analysis	• Functional, Performance, & Safety Requirements • Safety Information from same/similar designs • ID hazards, harms, risk estimation/evaluation for each constituent part and their combined use • Requirements for risk control measures	⇨	Design Inputs
	• Risk assessment acceptable?	◊	Design Review
Risk Control	• Risk control measures feasible? • Design risk controls, including device, drug, combination product, & process risk controls, if needed	⇨	Design Outputs

(Continued)

TABLE 6.10 (CONTINUED)
Risk Management and Design Controls (© 2022 Combination Products Consulting Services, LLC)

Residual Risk Evaluation	• Individual risks meet Risk Acceptability Criteria?	◊	Design Review
Risk Control	• Determine individual residual risks after the application of risk controls	⇒	Design Verification
Residual Risk Evaluation	• Individual residual risks meet Acceptability Criteria?	◊	Design Review
Risk Control	• User needs met?	⇒	Design Validation, including Human Factors
Residual Risk Evaluation	• Residual risks meet Acceptance Criteria? • Benefits of the overall combination product outweigh risks?	◊	Design Review (acceptable risks, redesign, or cancel)
Risk Review	• Translate product and process for commercialization to combination product and process risk control specifications • Process Validation/Continued Process Verification	⇒ ◊	Design Transfer Design Review

Use of Third Parties

The design and production processes for the constituent parts of the final combination product often take place at different locations and perhaps different companies. These will impact the specifics of the risk management process. A separate Risk Traceability Summary may exist for each of the constituent components and/or constituent parts, and a composite Risk Traceability Summary may have to be created by the final combination product manufacturer.

Since the FDA typically states the manufacturer under whose name the product is sold is responsible for everything in the combination product as a whole, this party becomes the ultimate responsible party for the risk management of the combination product and thus should be the owner of the Risk Management File. A Sponsor *must* proactively evaluate the unique elements and considerations for the specific combination product application. Even the selection of a pre-existing constituent part for purchase should trigger integrated thinking about joint-use risks and approaches. A Sponsor can leverage the risk management data/information from a third-party supplier, the Sponsor should not over-rely on that data, which was very likely established independent of any drug-device interaction considerations. The questions raised by the joint use of drug and device should be proactively addressed throughout the development process.

The Sponsor needs to have access to all of the necessary information for the combination product risks. It will be necessary to combine the separate traceability documents from each of the various constituents of the combination product as well as any risk information from the supply chain into a single traceability summary. This would facilitate the identification of all risks for the entire combination product system. It is important to include the hazards created by one constituent part on the other, e.g., a chemical action of the drug on the device. ***The user and patient see risks of the entire combination product system and not just the device or the drug or the biologic. So, the final manufacturer needs to take a system view of the risks of the entire system in the risk management of the combination product being placed on the market.***

Importantly, contract manufacturers and component suppliers should not shirk their responsibilities to support the combination product risk management activities. This is an important consideration in establishing Quality Agreements (see Chapter 10 on Supplier Quality in this book). In preamble comment 15 (78 FR 4313-4314) to 21 CFR Part 4, US FDA states:

> **We reject the proposal that the CGMP requirements applicable to a constituent part come into effect only after that constituent part has been formed.** Such an approach would be inconsistent with the application of the underlying CGMP regulations listed in §4.3. The trigger is whether the facility is conducting manufacturing operations that would be subject to the underlying CGMP requirements.
>
> For example, if a facility is manufacturing only device components, it might not be subject to CGMP requirements under the QS regulation. However, **a facility that is manufacturing a finished device from such components is subject to the QS regulation. Therefore, for example, if a facility is manufacturing a finished combination product, a prefilled syringe for instance, from device components and drug components, that facility is subject to both the QS regulation and drug CGMPs.**

Further. the US, FDA has issued Compliance Program 7356.000 "Inspections of CDER-led or CDRH-led Combination Products" (June 4, 2020).[44] This compliance program defines a combination product manufacturer as

> An entity (facility) engaged in activities for a combination product that are considered within the scope of manufacturing for drugs, devices, biological products, and HCT/Ps. Such manufacturing activities include, but are not limited to, designing, fabricating, assembling, filling, processing, sterilizing, testing, labeling, packaging, repackaging, holding, and storage, **including a contract manufacturing facility**.

With these statements, FDA is clearly indicating that component suppliers that are also serving as contract manufacturing facilities have an obligation to augment their own quality management systems with the called-out provisions under 21 CFR Part 4 that are applicable to the services they provide (see Chapter 4 on Combination Products cGMPs in this book). In the case of several drug-led combination products, many component suppliers that have historically served as contract manufacturing organizations – supporting pharmaceutical companies in the design of the device constituent part, injection molding components, and even doing filling and assembly – have considered themselves as simply container-closure (packaging) manufacturers, with 21 CFR 211-based quality systems.

Table 6.11 summarizes expectations for a contract manufacturing organization that meets the definition above of "Combination Product Manufacturer" with an established 21 CFR 211-based quality system. Such Contract Manufacturing Organizations (CMOs) and Contract Design & Manufacturing Organizations (CDMOs) are expected to have established design controls – which includes mandatory risk management – for the components/device constituent part whose combination product manufacturing (design, fabrication, assembly, filling, processing, sterilizing, testing, labeling, packaging etc.) they support.

TABLE 6.11
Current Good Manufacturing Practice Requirements for a Drug- or Biologic-Led Co-Packaged or Single-Entity Combination Product

Application Type for Pre-Approval Inspection	Base cGMPs	Base Compliance Program	Additional Coverage of Called-Out Provisions applicable to Combination Product Manufacturer CMO
NDA/BLA/ANDA	Drug cGMP (21 CFR Part 211)	7346.832 (NDA/ANDA) 7356.002M 7356.002A (BLA)	Attachment B of Compliance Program 7356.000* 1) Management Controls (21 CFR 820.20) 2) Design Controls (21 CFR 820.30) 3) Purchasing Controls (21 CFR 820.50) 4) Corrective & Preventive Action (21 CFR 820.100)

*Installation (21 CFR 820.170) and Servicing (21 CFR 820.200) requirements under 21 CFR are also among called-out provisions; however, they typically do not apply to most combination product manufacturers of drug-led combination products

The device may be a stand-alone product that is combined with other stand-alones like the drug or biologic. The combined product then becomes a system to deliver treatment, such as in a case where a syringe is combined with a drug and is provided to the healthcare provider or the patient for ease in delivering the needed treatment. Each of the components, the device, and the drug or biologic each has a healthcare objective, but when put together as a combination or a system, provides the needed therapy with fewer steps or improved processes to deliver the care. The provider and the patient see the product as a system or combination and not as individual components, so it becomes important to provide a complete risk management result for the entire combination product and not just the individual component. It has also been shown that there can be interaction risks between the components when they are combined into a single combination product. So, the end result for the user is a single combined product with a single set of risks.

Proper risk management relies on the determination and understanding of the intended use of the product including the various user population(s) and environment(s) in which the combination product will be used. This information is critical to assure proper identification of risks and the determination of appropriate risk mitigation measures.

Table 6.12 summarizes typical responsibilities and considerations of a supplier versus combination product applicant in the combined-use context.

TABLE 6.12
Purchasing Control Considerations

Supplier, e.g., of Device, Component, Platform[a]	Combination Product Applicant
• Identification and evaluation of inherent component/device or process performance failures associated with the device design, manufacturing process and use, including foreseeable misuse, of the device/component • Potential harms associated with the device platform failures, *AGNOSTIC OF THE SPECIFIC MEDICINAL PRODUCT* • Risk controls independent of specific drug therapy • Partial verification of risk controls • Lifecycle management: Supply, Change Management, Investigations, PMSR e.g., - Loss of supply availability - Changes to component and constituent part sourcing - Changes to materials of construction - Distribution concerns - Could lead to product quality changes resulting in constituent part malfunction or missed dose.	• Unique interactions of the component or constituent part with the finished product: • Evaluation of device delivery system performance failures/risks for a specific medicinal product, intended use, intended users, and use environment, linking failures to hazardous situations and harms. • Final verification of risk controls and their effectiveness. • Residual risk assessment including benefit/risk analysis. • Post-Market Surveillance information needs to be fed back into Risk Management File. • Align and link Constituent Part/Supplier Risks to Specific Combination Product.

[a] Whether leveraging a platform approach or starting from scratch with a device constituent part design, siloes of activities for device versus drug are not a great idea. Integrated development and proactive risk-based consideration of each of the constituent parts throughout the development process will help eschew unwanted surprises late in development or- worse-post market.

RISK MANAGEMENT FOR COMPONENTS/CONSTITUENT PARTS *NOT* DEVELOPED UNDER ISO 14971

Combination products manufacturers may come to realize that their device constituent parts or components were not designed and developed under ISO 14971:2019, e.g., for legacy products, or because their supplier has not historically viewed themselves as a device constituent part supplier. The Risk Management File may be insufficient to demonstrate the required conformance. How should one handle this precarious situation? (See ISO 24971:2020, Annex G, for a further discussion on this topic.)

A successful practice starts with assessing historical information:

- Actual use of the combination product (legacy data)
- Products on the market with the same intended use
- Technology used in other same or similar medical device constituent parts

Manufacturers should build the risk management file for use throughout the remainder of the product lifecycle.

Risk Management Plan

Start with a forward-looking Risk Management Plan that includes all the elements required under ISO 14971:

- Risk management activities for the remaining phases of the lifecycle of the device and combination product
- Activities and responsibilities
- Requirements for review of risk management activities
- Criteria for risk acceptability for the device constituent part and combination product as a whole, based on the manufacturer's policy for determining acceptable risk
- Criteria for acceptability of overall residual risk
- Verification activities, both for existing risk control measures and for new risk control measures that may be required
- Activities for the collection and review of production and post-production information

Build the Risk Management File

With the forward-looking Risk Management Plan in place, build the risk management file:

1. *Identify intended use and reasonably foreseeable misuse and characteristics related to safety.* Regulatory files, production and post-production feedback/information, instructions for use, and design and development documentation are helpful resources.
2. *Identify all the technical solutions within the device that serve as risk controls.* Review information from device changes, the Device Master Record/Medical Device File, Master Batch Record for the combination product, drawings, and Design History File.
3. *Identify hazards, hazardous situations, and harms – with a focus on each constituent part and the combination product as a whole.* Great sources of data for this effort include complaint files, Instructions for Use (warnings, precautions, limitations), Post-Marketing Safety Reports (e.g., Medical Device Reports (MDR) and Adverse Event (AE) Reports), existing FMEA, Fault Tree Analysis, etc. Harms, including severities, can be identified using MDR and AE reports, regulatory agency medical events databases, clinical evaluations, and any health hazard evaluations.
4. *Determine if there are hazards or hazardous situations with no risk control, and follow the risk management process. Production and post-production information are helpful here. Data should be reviewed against the now pre-determined acceptance criteria from the Risk Management Plan.*

5. *Document Traceability. Use the Device Master Record/Medical Device File or Master Batch Record (one that includes device constituent part information) and Design History File to help identify the required information.*
6. *Evaluate overall residual risks. Post-production feedback, expert medical input (both internally gathered and external independent feedback), in addition to clinical evaluation, to vet the overall residual risks of the combination product and its constituent parts.*

Drug-Agnostic Risk Assessments

One of the challenges faced by combination product manufacturers can be the competing interests of bringing products to the market quickly while executing an effective and robust risk management process. This situation has been exacerbated in recent years by the speed at which certain therapies have been commercialized in order to address urgent global needs. An expedited process used for one product is now an expectation for all future projects. However, while increasing the rate at which products are developed is important, from both patient and business perspectives, it is vital to ensure that such products meet the requirements in terms of safety and effectiveness.

A major factor which may impede quick development can be the repetition of risk management activities that have been already conducted on similar marketed products. The larger the manufacturer, the more likely it is that similar products exist across the rest of the portfolio. These products will have already been demonstrated to be safe and effective and will have been approved accordingly by regulatory authorities in those territories into which the products have been released.

In addition to the above, initiating risk management activities from first principles in instances where, for example, the delivery device is the same as an already marketed product, but the drug product differs, can lead to inconsistencies across the portfolio in the estimation of the level of individual risks. The determination of hazardous situations and, hence, the associated harms and their severities, is commonly the task of clinical and medical subject matter experts. However, within the pharmaceutical industry, these subject matter experts are normally assigned to a combination product development program on the basis of their knowledge of a specific type of therapy (e.g., inflammation, vaccines, oncology etc.) and may not be adequately trained to assess the level of harm associated with non-dosing-related hazardous situations arising from the device constituent part of the combination product. The net result of this can be that what may seem to be a relatively innocuous risk on one product can be described as leading to a greater and more severe spectrum of harms on another. This represents an undesirable situation.

An example of this is a simple cut sustained by the user from a broken glass flange on a prefilled syringe presentation being documented as leading to minor injuries on one product, but more serious lacerations, possibly requiring clinical intervention, on another. The only difference between these two presentations is the drug product contained therein, the device constituent part is identical. What has happened is that clinical subject matter experts, whose primary knowledge is based on one or other of the two drug products, have arrived at a slightly different answer for the same hazardous situation. Neither are necessarily wrong, but this causes an inconsistency within the portfolio which may be subsequently identified during the regulatory approvals process itself or an audit. Consider also that this situation is not only limited to within an individual manufacturer but is also present across the industry at large. The final consequence of this is that we are left with multiple different estimations of the level of risk associated with the same hazardous situation. This might be manageable if it was only restricted to one or two instances but, depending on complexity, there can be many non-dosing-related hazardous situations associated with delivery devices. The potential for inconsistency and confusion is therefore high if the industry continues to approach risk assessments in this manner.

What can we do to address these issues? A possible remedy is the development of drug-agnostic risk assessments whereby the common hazards, hazardous situations and harms associated with a specific delivery device platform are defined and agreed across an organization, leaving an individual project team with the task of determining the level of harms associated with dosing of the specific drug product in question. In time this approach could also be extended to the industry at large and could even be broadened to define the language used for all dosing-related hazardous situations leaving the project team only with the task of determining the physiological harms associated with such instances.

Similarly, risk management activities within the drug product realm are concerned with the effect of the manufacturing process on the critical quality attributes of that drug product. Using the same rationale as above, it could also be possible to predict the critical process parameters in advance, especially when knowledge of these has been acquired from previous products, leaving project teams the task of determining what the effects would be on the specific drug product itself rather than establishing how these effects manifest themselves. By taking this approach, the twin possibilities of reducing the time associated with generating robust risk assessments and enhancing intra-company and cross-industry consistency could be achieved.

The rest of this section is concerned with how such drug-agnostic risk assessments might be constructed and then deployed across a portfolio of different drug products that are contained within similar delivery devices.

Identification of Characteristics Related to Safety and Associated Hazards

Aside from determining the intended use (and reasonably foreseeable misuse) of a combination product, one of the first activities mandated by ISO 14971:2019, when it comes to analyzing risk, is the identification of the characteristics related to the safety of the product. A simple but effective way of achieving this aim is to utilize the questions detailed in ISO/TR 24971:2020, Annex A.

Within any given organization, multiple sources of inputs, both internal and external, will already exist to enable these questions to be answered comprehensively and hazardous situations to be identified accordingly. These inputs are described in AAMI TIR 105:2020 and include the following:

- Complaints and adverse events
- Nonconformances and corrective and preventive actions
- Clinical evaluations, studies, and reports
- Audits and management review
- Information from predicate combination products
- Databases, literature, media
- Standards, regulations, guidance

Using these inputs, it is possible to not only identify the characteristics of the product which are related to safety but also to extrapolate them to determine the associated hazards, hazardous situations, and harms.

It is important to note that the questions contained in Annex A are designed to encompass a wide variety of both medical devices and drug delivery devices, ranging from a simple dosing cup to complex electromechanical devices incorporating software, for example. With that in mind, it is permissible to exclude questions which may not relate to the combination product under consideration. It is also worth noting that this is not an exhaustive list, and it is also permissible to add additional questions which may also be more relevant. The addition of extra questions will be discussed in further detail later, as this can be an effective method of identifying specific hazards associated with drug product manufacturing within combination product risk assessments.

Before we can start describing how these questions can be used to build drug-agnostic risk assessments, it is worth taking some time to see how they would be utilized by means of a real-world example. Consider a prefilled syringe containing a drug product used for the treatment of Crohn's disease. The first question encountered in the list relates to the intended use of the product and could be answered as shown in Table 6.13.

TABLE 6.13
Identification of the characteristics That Could Impact Safety, and the Associated Hazards, When Considering the Intended Use of a Prefilled Syringe Containing a Therapy for Crohn's Disease

Question	Applicability (Y/N)	Answer	Characteristic(s) That Could Impact on Safety	Hazards
What is the intended use and how is the medical device to be used? Factors that should be considered include: what is the medical device's role relative to: • diagnosis, prevention, monitoring, treatment, or alleviation of disease • diagnosis, monitoring, treatment, or alleviation of or compensation for an injury • investigation, replacement, modification, or support of anatomy or a physiological process, or • control of conception? • what are the indications for use (e.g., patient population, user profile, use environment)? • what are the contraindications? • does the medical device sustain or support life? • is special intervention necessary in the case of failure of the medical device? • can the performance of the medical device be impacted in the event of a security breach (performance degradation or loss of availability)? • can unauthorized access, unauthorized activities, or loss of data affect the medical device safety?	Y	The delivery device consists of a 1.0 mL prefilled syringe with a fixed needle, a plunger for closure, a finger flange, and a plunger rod The prefilled syringe (PFS) is intended to deliver a subcutaneous dose of the (product name) drug product This dose may be self-administered or given by a third party into the patient's thigh or lower abdomen The PFS is intended to treat the following patient condition: • Crohn's disease Indicated in treatment of moderately to severely active Crohn's disease in adults and pediatric patients >6 years of age The frequency of use is defined as an initial dose, with maintenance frequency of use every two weeks for all indications Each PFS dispenses a single dose and is for single use only The PFS should be disposed of after use. It is not intended to be refilled or re-used	• Single use combination product • Dose delivery • Route of administration • Biological contamination • Use errors	Biological agents • Bacteria, fungi, parasites, viruses • Prions • Toxins Delivery • Quantity • Route Functionality • Critical performance Miscellaneous • Small parts • Handling

The first step is to determine the applicability of the individual question to the combination product. As mentioned earlier, not all the questions may be relevant and can therefore be excluded. In our example of a single-use prefilled syringe containing medicine to alleviate Crohn's disease, questions relating to implantation and software are not applicable and therefore do not need to be answered.

Having determined applicability, the next step is to answer the question as completely as possible, using the factors described therein as prompts to determine the answer. In this case, we are concerned with the intended use of the combination product and are focused on attributes such as what it is supposed to do, the indications for use, who is supposed to use it, how it is supposed to be used, what is the environment of use, and the frequency of use. Once these parameters have been established, it is then possible to determine how these constraints may impact on the safe and effective use of the device. In this instance, those characteristics will be primarily concerned with how the dose is delivered (e.g., quantity, route of administration, number of times the delivery device is expected to be used, etc.).

Finally, once the characteristics have been documented it is then possible to predict the associated hazards. One possible and simple way of doing this is to use the list of hazards presented in ISO 14971:2019 Table C1. Note that this is not an exhaustive list; it is merely a list of examples, and it is possible to augment it with additional hazards should the need arise. With that in mind, consider the miscellaneous hazards documented in Table 6.13 (small parts and handling). These two hazards are not represented within the table shown in ISO 14971:2019 but are regularly encountered when considering the risks associated with combination products (and perhaps other medical devices), i.e., the ingestion of a small part (e.g., a rigid needle shield holder) or a handling issue (e.g., the prefilled syringe is dropped, rendering it unusable).

Once the hazards associated with one of the safety characteristics have been defined, it is necessary to repeat the steps described earlier for all the other questions in the list deemed applicable. Given the number that may be relevant, this could become an arduous and time-consuming process if repeated from first principles across an entire portfolio of similar delivery devices (e.g., prefilled syringes, prefilled pens etc.). A possible way to simplify and expedite this activity is to prepopulate a drug-agnostic risk assessment with either suggested text or guidance to show project teams how this assessment should be conducted.

Using the same or similar prefilled syringe as is designated for the Crohn's disease indication described in Table 6.13, the determination of the characteristics relating to safety and, hence, the hazards can be documented in a drug-agnostic risk assessment. This is achieved by simply replacing the drug product-specific text with comprehensive guidance for the risk management team developing a different drug product in the same delivery device to follow. This is shown in Table 6.14. A further example, using generic text to define how contact with the patient/user affects the safety characteristics and hazards of the prefilled syringe, is shown in Table 6.15.

The final point to make in this section relates to the addition of supplementary questions concerned with combination products or, more specifically, the drug product constituent part, and how this approach could be used to determine hazards associated with both the properties of the drug product and the process required for its production.

One of the challenges faced by pharmaceutical companies in recent years has been to devise an effective yet pragmatic method of estimating the level of the risk holistically for the combination product, as opposed to looking at the two constituent parts separately. AAMI TIR 105:2020 describes how concurrent risk assessments could be integrated and fed into a master risk management traceability matrix. The first step in this process is to use questions similar to the characteristics related to safety to determine the potential hazards associated with the drug product and its production. Table 6.16 shows two questions, and their resolution/answer, as a possible means of achieving this aim. Such an approach could be incorporated into an agnostic drug product risk assessment as a prompt to further investigation on specific projects.

One of the main considerations that must be made relates to how the manufacturing process can impact the critical quality attributes of the combination product, specifically the drug product

TABLE 6.14
Guidance to Identify the Characteristics That Could Impact on Safety, and the Associated Hazards, When Considering the Intended Use of a Prefilled Syringe in a Drug Agnostic Risk Assessment

Question	Applicability (Y/N)	Answer	Characteristic(s) That Could Impact on Safety	Hazards
What is the intended use and how is the medical device to be used? Factors that should be considered include: what is the medical device's role relative to: • diagnosis, prevention, monitoring, treatment, or alleviation of disease, • diagnosis, monitoring, treatment, or alleviation of or compensation for an injury, • investigation, replacement, modification, or support of anatomy or a physiological process, or • control of conception? • what are the indications for use (e.g., patient population, user profile, use environment)? • what are the contraindications? • does the medical device sustain or support life? • is special intervention necessary in the case of failure of the medical device? • can the performance of the medical device be impacted in the event of a security breach (performance degradation or loss of availability)? • can unauthorized access, unauthorized activities, or loss of data affect the medical device safety?	Y	This should be completed with product-specific information such as: 1. High-level delivery device/medical device description (e.g., single-use glass prefilled syringe with calibrated graduation marks in 0.5 g increments containing 3 ml of 0.5 g/ml <drug name>). 2. Intended use statement of the delivery device/medical device 3. Clinical application • Therapeutic intent • Administration route • Frequency of use/administration • Duration of use 4. Patient population: age, gender, pre-existing conditions, handicaps 5. User and user population: Same as patient, caregiver, HCP 6. Use environment: Home, ER, clinical setting, surgical setting	• Single-use combination product • Dose delivery • Route of administration • Biological contamination • Use errors	Biological agents. • Bacteria, fungi, parasites, viruses • Prions • Toxins Delivery • Quantity • Route functionality • Critical performance Miscellaneous • Small parts • Handling

TABLE 6.15
Guidance to Identify the Characteristics That Could Impact on Safety, and the Associated Hazards, When Considering the Patient/User Contact of a Prefilled Syringe in a Drug Agnostic Risk Assessment

Question	Applicability (Y/N)	Answer	Characteristic(s) That Could Impact on Safety	Hazards
Is the medical device intended to be in contact with the patient or other persons? Factors that should be considered include the nature of the intended contact, i.e., surface contact, invasive contact, or implantation and, for each, the period and frequency of contact.	Y	The (product name) PFS is intended to be handheld. The (product name) PFS is categorized per ISO 10993-1:2020 as being: • Category – Surface medical device • Contact – Intact skin (syringe barrel, plunger, cap) • Contact Duration – Long term (C) >30 days • Contact – Invasive – Specify nature of contact (SubQ, IM, IV, ID, etc.)	• Surface and invasive contact • Material selection	Immunological agents • Allergenic • Immunosuppressive • Irritants • Sensitizing

constituent part and how it can be integrated into the wider risk assessments. Traditionally, the way drug product manufacturing risk assessments are conducted focuses on the parts of the process which may affect the critical quality attributes of the drug product, namely the critical process parameters. This work is also performed for the device constituent part but can be addressed using the framework as defined by ISO 14971:2019 and other applicable standards.

When it comes to the drug product constituent, a possible method of linking the separate risk assessments is by equating the critical quality attributes to one of the hazards shown in ISO 14971: 2019 Table C1. Typically, the drug product critical quality attributes will have been defined prior to dosing decisions, such as the desired delivery device, having been taken. It is therefore possible to equate these attributes to a hazard. For example, the critical quality attribute of sterility becomes a biological hazard, visible particles become a chemical hazard, concentration becomes a delivery quantity hazard, and so on. Once this conversion has been made it is then possible to use the same hazardous situations and harms that would also be used for the device constituent part of the combination product.

Note that drug product critical quality attributes are likely to have been identified long before the delivery method has been determined. The delivery device is often only selected after the proof-of-concept phase 2 clinical study and, in some cases, selection can occur as late as just before the commencement of phase 3 clinical studies. Given that this information is available early in the development process, it is possible to determine the hazards, hazardous situations, and some of the harms prior to identifying those attributes in the device constituent part of the combination product. This information can also be utilized in drug-agnostic risk assessments by virtue of the same critical quality attributes having been defined over multiple products. For example, the hazardous situation and harms associated with biological contamination will be the same irrespective of the type of drug product under consideration. Where this will differ is when considering the physiological harms resulting from effects of the manufacturing process on some of the other critical quality attributes, namely factors such as drug concentration.

TABLE 6.16
Guidance to Identify the Characteristics That Could Impact on Safety, and the Associated Hazards, When Considering the Factors Related to the Drug Product Constituent in a Drug Agnostic Risk Assessment

Question	Applicability (Y/N)	Answer	Characteristic(s) That Could Impact on Safety	Hazards
Questions specific to combination products. Factors that should be considered include: • are there potential delivery method errors which may result due to the characteristics of the drug product (e.g., crystallization of drug product in the needle)? • is there any physical or biological property which may prevent the drug from being administered properly? • are there any patient conditions that could affect the delivery of the medicine?	Y	Question relevant dependent upon the route of administration, the specific drug properties (e.g., inadequate resuspension of drug product), use errors which may be encountered, and the effect the drug product may have on the device constituent part itself	• Reconstitution of drug product • Use errors	Chemical agents • Particulates (including micro- and nanoparticles) • Extractables & leachables Delivery • Quantity • Route Functionality • Critical performance
Questions relating to combination product manufacturing. • are any processes employed which may lead to degradation of or damage to the drug product or device components?	Y	Consider processes used as part of production which may affect the critical quality attributes of the combination product	• Effect of the manufacturing process on both the drug product and the delivery device	Biological agents • Bacteria, fungi, parasites, viruses • Prions • Toxins Chemical agents • Manufacturing & cleaning residues Immunological agents • Allergenic substances Delivery • Quantity Functionality • Critical performance

Once the characteristics relating to safety and the associated hazards have been established, the next stage is to identify the hazardous situations and harms which may occur. The simplest way of achieving this aim is by the means of a preliminary hazard analysis.

Preliminary Hazard Analysis

Identification of hazardous situations and harms can be one of the more time-consuming and contentious parts of the risk management process. A common issue can be that the estimation of the severity of harm(s) associated with any given hazardous situation is overly conservative, attaching a higher level of risk than may be warranted. While it is important to ensure all possible harms are considered, it is also imperative that they are realistic and reflect the actual harms that could occur. Moreover, as previously discussed, differences can also arise in risk assessments across multiple products for the same hazardous situation.

One method of preventing these issues is to develop a library of the common hazards, hazardous situations and harms which may be encountered when using the combination product. As described in the previous section and in AAMI TIR 105:2020, several inputs can be used to determine the common hazardous situations. This also includes those hazardous situations identified in preliminary hazard analyses for predicate products already marketed by the manufacturer. Note that in developing such libraries it is important to engage the relevant clinical and medical subject matter experts as early in the process as possible to ensure agreement around the estimation of the level of harm.

Table 6.17 shows a possible drug-agnostic preliminary hazard analysis, detailing the common hazards, hazardous situations, and harms, that could be generated for a prefilled syringe platform using information from the inputs described earlier and from predicate products. At this stage, and for simplicity, the hazards are described in broad terms. Extra granularity can be added in supplementary and more comprehensive risk assessments. Following engagement with clinical and medical subject matter experts, it is possible to define the hazardous situations, harms, severity of harms and, if considered in the manufacturer's risk management process, an estimation of the probability that the hazardous situations will result in their associated harms (i.e., the P2 value).

It is important to note that not all manufacturers use the P1 and P2 probability scales. Decomposition into P1 and P2 are not required by ISO 14971 but merely discussed in more detail in ISO TR 24971:2020 as a way to explain why the probability of occurrence of a hazard is not the same as the probability of harm occurrence. Care must also be taken when estimating the P2 value as it is challenging, without clinical or post-market data, to quantify the likelihood that when a hazardous situation occurs a certain harm will result. Some manufacturers automatically assume that this is the case and estimate the P2 value as being almost inevitable. Others will make a qualitative judgment based on previous experience and various data sources. In any case, neither of these approaches is incorrect, but for the example shown in Table 6.17, we will consider the latter.

It is also important to note that the probability of the hazard leading to the hazardous situation (P1) is not considered in this preliminary hazard analysis. Preliminary hazard analysis is just that – preliminary – and it must be followed with more investigation to get sufficient detail for the risk analysis. It is also important to understand that we are not just analyzing the hazard in the development of combination products, but rather, analyzing the risks associated with the products, and hazard analysis is just a small component of the complete risk analysis. These values are defined in the more comprehensive supplementary risk assessments, into which this preliminary hazard analysis feeds, where the sequence(s) of events leading from the hazard to the hazardous situation are defined and estimated accordingly and are based on how the specific hazard originated. Inclusion of these sequence(s) of events within a document such as the preliminary hazard analysis would lead to a much larger document which may end up being difficult to manage in the long term.

Despite the preliminary hazard analysis being focused on drug-agnostic elements, it is also possible to predict some of the hazards which could lead to the hazardous situations associated with dosing. Obviously, in a drug-agnostic risk assessment, it is impossible to list the specific harms

TABLE 6.17
Drug Agnostic Preliminary Hazard Analysis for a Prefilled Syringe Presentation

Hazard	Hazardous Situation	Harm	Severity of Harm	P2
Biological & Chemical Hazards				
Biological Agents (Bacteria, Viruses, Fungi, & Parasites)	Administration of biological agent into the patient	Death	Catastrophic	Extremely Unlikely
		Systemic infection (sepsis), permanent impairment, or life-threatening/irreversible injury	Critical	Extremely Unlikely
		Localized infection, abscess, systemic effects, requiring medical/surgical intervention	Serious	Unlikely
		Localized infection, not requiring medical intervention	Minor	Possible
	User/patient exposed to microbial agents on packaging or external surfaces of the prefilled syringe	Localized infection, abscess, systemic effects, requiring medical/surgical intervention	Serious	Unlikely
		Localized infection, not requiring medical intervention	Minor	Possible
	Subcutaneous injection performed with unclean hands	No harm	Negligible	Extremely Unlikely
	Subcutaneous injection given through unclean skin	Minor local irritation or inflammation	Minor	Possible
Biological Agents (Toxins)	Administration of endotoxins into the patient	Endotoxic fever reaction. Death.	Catastrophic	Extremely Unlikely
		Endotoxic fever reaction. Permanent impairment or life-threatening/irreversible injury.	Critical	Extremely Unlikely
		Endotoxic fever reaction, requiring medical/surgical intervention	Serious	Unlikely
		Endotoxic fever reaction, not requiring medical intervention	Minor	Possible
Biological Agents (Prions)	User/patient exposed to prions present from materials of animal origin	Spongiform encephalopathy. Death.	Catastrophic	Extremely Unlikely
Biological Agents (Virus)	User exposed to contaminated needle which has contacted with blood, tissue, or other body fluid containing blood-borne pathogen (such as HIV, HBV, or HCV)	Blood-borne pathogen infection	Critical	Possible
Chemical Agents – Particles (including Micro- and Nanoparticles)	IV Administration of particulate matter	Occlusion of small blood vessel leading to tissue necrosis.	Critical	Extremely Unlikely
	IM or SubQ Administration of particulate matter	Tissue granuloma (including necrosis), requiring medical attention.	Serious	Unlikely
		Granuloma, foreign body reaction, limited to mild symptoms, not requiring medical intervention	Minor	Unlikely

(Continued)

TABLE 6.17 (CONTINUED)
Drug Agnostic Preliminary Hazard Analysis for a Prefilled Syringe Presentation

Hazard	Hazardous Situation	Harm	Severity of Harm	P2
Chemical Agents (Extractables & Leachables)	Patient exposed to toxic extractables and leachables	Toxicological injury – permanent impairment or life-threatening/irreversible injury	Critical	Extremely Unlikely
		Cytotoxicity, localized necrosis requiring medical intervention.	Serious	Extremely Unlikely
		Localized irritation including edema, rash, not requiring medical intervention	Minor	Extremely Unlikely
Chemical Agents (Manufacturing & Cleaning Residues)	Patient exposed to toxic or caustic chemical residues	Toxicological injury – permanent impairment or life-threatening/irreversible injury	Critical	Extremely Unlikely
		Cytotoxicity, localized necrosis requiring medical intervention	Serious	Extremely Unlikely
		Localized irritation including edema, rash, not requiring medical intervention	Minor	Extremely Unlikely
Immunological Agents – Allergenic Substances	Skin contact (non-invasive contact) allergenic substance (device-related substance e.g., plasticizer, latex)	Generalized allergic reaction – injury or impairment requiring medical or surgical intervention	Serious	Unlikely
Immunological Agents – Biocompatibility	User exposed to a non-biocompatible substance (device-related materials of construction e.g., plasticizer, unbound monomers, latex)	Toxicological injury – permanent impairment or life-threatening/irreversible injury	Critical	Extremely Unlikely
		Cytotoxicity, localized necrosis requiring medical intervention	Serious	Extremely Unlikely
		Localized irritation including edema; rash, not requiring medical intervention	Minor	Extremely Unlikely
Performance-Related Hazards				
Critical Performance	Administration of air – adults & adolescents (1–2 ml air IV)	Air embolism, tissue ischemia, requiring medical intervention	Serious	Possible
	Administration of air – adults & adolescents (1–2 ml air IM or SubQ)	Air embolism, tissue ischemia, requiring medical intervention	Serious	Extremely Unlikely
		Localized injury, subcutaneous or intramuscular air	Minor	Unlikely
	Administration of air – children 2–12 years (1–2 ml air IV)	Death	Catastrophic	Extremely Unlikely
		Life-threatening pulmonary air embolism	Critical	Unlikely
		Air embolism, tissue ischemia, requiring medical intervention	Serious	Possible
	Administration of air – children 2–12 years (injection of 1–2 ml air IM or SubQ)	Air embolism, tissue ischemia, requiring medical intervention	Serious	Extremely Unlikely
		Localized injury, subcutaneous or intramuscular air	Minor	Possible

(*Continued*)

TABLE 6.17 (CONTINUED)
Drug Agnostic Preliminary Hazard Analysis for a Prefilled Syringe Presentation

Hazard	Hazardous Situation	Harm	Severity of Harm	P2
	Administration of air (IV administration of air 1–2 ml) – neonate (0–4 weeks), infant (5 weeks–12 m), and toddler (12–24 m) populations	Death	Catastrophic	Extremely Unlikely
		Pulmonary air embolism, life threatening	Critical	Unlikely
		Air embolism, tissue ischemia, requiring medical intervention	Serious	Extremely Unlikely
		Localized injury, subcutaneous or intramuscular air	Minor	Possible
	Administration of biological agent into the patient	Death	Catastrophic	Extremely Unlikely
		Systemic infection (sepsis), permanent impairment or life-threatening/irreversible injury	Critical	Extremely Unlikely
		Localized infection, abscess, systemic effects, requiring medical/surgical intervention	Serious	Unlikely
		Localized infection, not requiring medical intervention	Minor	Possible
	User exposed to contaminated needle which has contacted with blood, tissue, or other body fluid containing blood-borne pathogen (such as HIV, HBV, or HCV)	Blood-borne pathogen infection	Critical	Possible
Miscellaneous Hazards				
Small Parts	Ingestion of a small part	Ingestion of a small part – injury or impairment requiring medical or surgical intervention	Serious	Unlikely
	Inhalation of a small part	Choking – inhalation of a small part – permanent impairment or life-threatening/irreversible injury	Critical	Extremely Unlikely
		Results in injury or impairment requiring medical or surgical intervention	Serious	Unlikely
Handling	Error necessitating a preparation of a replacement PFS	Temporary inconvenience	Negligible	Likely
	Device damage, exposure to sharp edges	Laceration, puncture, injury, or impairment requiring medical or surgical intervention	Serious	Unlikely
		Contusion, abrasion, minor pinch injury, not requiring medical intervention	Minor	Likely
		Temporary discomfort	Negligible	Extremely Likely

(Continued)

TABLE 6.17 (CONTINUED)
Drug Agnostic Preliminary Hazard Analysis for a Prefilled Syringe Presentation

Hazard	Hazardous Situation	Harm	Severity of Harm	P2
	User exposed to pinch point	Contusion, abrasion, minor pinch injury, not requiring medical intervention	Minor	Likely
		Temporary discomfort	Negligible	Extremely Likely
	Excessive force used to deliver dose	Contusion, abrasion, minor pinch injury, not requiring medical intervention	Minor	Likely
	Excessive force used to deliver dose	Temporary discomfort	Negligible	Extremely Likely
	Exposed needle	Sterile needle stick injury	Minor	Likely
	Damaged or broken needle	Deep skin laceration or cut	Minor	Extremely Unlikely
		Minor skin laceration or scratches	Negligible	Unlikely
	Excessive delivery rate (IM/SubQ)	Temporary discomfort	Negligible	Likely
	Excessive delivery rate (IM/SubQ)	Localized tissue damage leading to mild pain; edema, not requiring medical intervention	Minor	Likely
	Slow delivery rate	Temporary discomfort	Negligible	Possible

associated with various drug products but, notwithstanding this, manufacturers can predict how dosing-related hazardous situations can occur.

Table 6.18 shows some of the common hazards, again at a broad level, which may result in dosing-related hazardous situations. Note that the overall harm, severity, and P2 values have been estimated in these instances. Also, note that this is not an exhaustive list and could be augmented with other drug product-related hazards as they become apparent.

TABLE 6.18
Dosing-Related Hazards and Hazardous Situations to Use in Drug Agnostic Preliminary Hazard Analysis for a Prefilled Syringe Presentation

Hazard	Hazardous Situation	Harm
Delivery – Quantity	No dose	Drug specific – Seek clinical and medical subject matter expert input
	Overdose	
Delivery – Route	Injection delivered to wrong injection site	
Critical Performance	No dose	
	Overdose	
	Injection too deep	
	Injection too shallow	
	Effect of drug being exposed to a higher temperature than specified	
	Effect of drug being exposed to a lower temperature than specified	
	Injection of degraded drug product	
Handling	Ophthalmic exposure of drug to user/patient	
	Topical exposure of drug to user/patient	

Once the preliminary hazard analysis has been defined, it can now be used to inform the supplementary risk assessments. The next section will be concerned with how this can be achieved.

Supplementary Risk Assessments

As stated in both ISO 14971:2019 and AAMI TIR 105:2020, there are many different tools to provide inputs into risk assessments that can be employed to determine the risks that could be faced by the patient/user of any given combination product or medical device. These tools include, but are not limited to:

- Failure Mode Effects Analysis (FMEA)
- Failure Mode, Effects and Criticality Analysis (FMECA)
- Fault Tree Analysis (FTA)
- Hazard Analysis and Critical Control Points (HACCP)
- Hazard Operability Analysis (HAZOP)

The main purpose of these risk assessment tools is to define the failure modes and use errors which cause the potential hazards, the sequence(s) of events which lead from the hazardous situations and, hence, calculate the P1 value utilized to determine the overall level of risk when combined with the already established severity and P2 values. Also captured within these risk assessment tools are the various risk controls which are implemented and verified to reduce the level of risk to an acceptable level.

It is also important to understand that each of the tools has benefits and limitations in its use, and no one tool by itself can provide complete coverage of the risk analysis. For example, FMEA is a single-fault tool that requires design outputs to be defined before it can be used. A complete risk analysis covers all faults, not just single-fault. Also, Design Input requires (in ISO 13485:2016 7.3.3 c) that the outputs of risk management be Design Inputs, and FMEA occurs too late in the process to meet this requirement. But FMEA provides a check on the final design to assure no hazards have been overlooked.

It should be noted that the purpose of this section it not to describe each of these tools in turn. As stated earlier, manufacturers can employ multiple methods to assess the level of risk, the main point is that this information is gathered from assessments made on, not only the drug and device constituent parts but the interactions between them. This is then used to populate a holistic risk assessment for the combination product, where all the risks are considered from a patient/user harms point of view, which can also be used as a traceability matrix to the other risk analyses. As mentioned earlier, it is possible (using information gathered from previously marketed products) to generate drug-agnostic risk assessments. Taking this approach can reduce the amount of time taken to generate specific combination product risk management files without compromising either quality or safety.

Table 6.19 shows an example of how information generated both within the preliminary hazard analysis and supplementary risk assessments could be used to populate an overall risk assessment which can also be used as a traceability matrix. Note that the table only shows the initial estimation of risk; it is recommended that the table is expanded to show the risk control measures, the verification of their implementation, the verification of their effectiveness, an assessment of the level of risk post-implementation and, finally, an assessment of whether this risk is acceptable or not based on the criteria detailed in the risk management plan. This information is not shown for brevity but could be leveraged from predicate products where the identified hazardous situations are similar.

As can be seen in Table 6.19, and in the above sections, it is possible to use information leveraged from already marketed products and other inputs, to populate a risk management traceability matrix which is agnostic of the drug product.

TABLE 6.19

Drug Agnostic Risk Management Traceability Matrix for a Prefilled Syringe Presentation

Failure Mode	Source Document(s)	Hazard	Hazardous Situation	Harm	Severity	P1	P2	POH	Risk Level
Plunger rod moves or deforms the plunger prior to activation	Design FMEA Preliminary Hazard Analysis	Biological Agents	Administration of biological agent into the patient	Death	Catastrophic	High	Extremely Unlikely	Improbable	Marginal
				Systemic infection (sepsis), permanent impairment, or life-threatening/irreversible injury	Critical	High	Extremely Unlikely	Improbable	Marginal
				Localized infection, abscess, systemic effects, requiring medical/surgical intervention	Serious	High	Unlikely	Remote	Marginal
				Localized infection, not requiring medical intervention	Minor	High	Possible	Occasional	Marginal
Syringe barrel damaged during de-nesting operation	Process FMEA Preliminary Hazard Analysis	Biological Agents	Administration of biological agent into the patient	Death	Catastrophic	Moderate	Extremely Unlikely	Improbable	Marginal
				Systemic infection (sepsis), permanent impairment, or life-threatening/irreversible injury	Critical	Moderate	Extremely Unlikely	Improbable	Marginal
				Localized infection, abscess, systemic effects, requiring medical/surgical intervention	Serious	Moderate	Unlikely	Remote	Marginal
				Localized infection, not requiring medical intervention	Minor	Moderate	Possible	Remote	Low
User selects expired/ non-sterile/ previously used needle	Use FMEA Preliminary Hazard Analysis	Biological Agents	Administration of biological agent into the patient	Death	Catastrophic	Low	Extremely Unlikely	Improbable	Marginal
				Systemic infection (sepsis), permanent impairment, or life-threatening/irreversible injury	Critical	Low	Extremely Unlikely	Improbable	Marginal
				Localized infection, abscess, systemic effects, requiring medical/surgical intervention	Serious	Low	Unlikely	Improbable	Low
				Localized infection, not requiring medical intervention	Minor	Low	Possible	Remote	Low
Product incorrectly stored at temperatures greater than 8°C	Use FMEA Preliminary Hazard Analysis	Critical Performance	Patient does not receive efficacious dose	Drug specific – Seek clinical and medical subject matter expert input	Drug specific – Seek clinical and medical subject matter expert input	Refer to specific combination product risk assessment	Drug specific – Seek clinical and medical subject matter expert input	Refer to specific combination product risk assessment	Refer to specific combination product risk assessment

Conclusion

In this section, we have discussed the possibility of preparing drug-agnostic risk assessments with a view to reducing the time it takes to develop risk management files for combination products without compromising on safety or quality. The process involves using multiple sources to achieve this aim and can be expanded to include a variety of device delivery platforms. It can also be used across a multitude of risk assessments if the requisite information is available to populate them. Finally, such templates can and should be iterated in the event of any new information becoming available, via production and post-production information. This information can be fed into the model and used in the risk management activities of subsequent combination products.

While this section has given a mere snapshot of a proposed process, it is hoped that the reader has seen the utility in such an approach. It is also hoped that, in time, this method will be adopted across the industry with a view to improving consistency and maintaining quality while making risk management activities both faster and easier.

SUMMARY

For drug-/biologic-device combination products, a single risk management process covering the entire lifecycle of each constituent part of the product can be developed following the elements of the processes identified in ISO 14971, the device risk management standard, and ICH Q9, the drug/biologic risk management guidance. Those two documents provide information on a four-step process that is described by FDA in the 21 CFR 820 Preamble. This process has been interpreted by ISO 14971 and ICH Q9 in the processes they describe to address risk management.

The Risk Traceability Summary is a method for documenting information discovered in the risk management process and identifying the risks of the product. The risk traceability summary document was developed by expanding the Global Harmonization Task Force (GHTF) SG3/N15R8 document and providing additional information beyond the GHTF document. The information contained in this document and its references are valuable in managing product- and manufacturing process-introduced risks throughout the lifecycle of the product.

Once the product is released for sale, the monitoring phase of risk management begins as part of post-market product surveillance activities. The manufacturer released the product based on knowledge gained during the development process and based largely on estimates of risk. Following design transfer and during the product post-market lifecycle, information on actual risk performance is received and may be different from the estimates in the RMF. If the data reveal different risk levels for the product, the manufacturer needs to evaluate the new information to determine if the product needs additional risk controls to return risks to acceptable levels. In the release phase, product and manufacturing process changes occur to improve the product, to add new features, and to improve the manufacturing processes. These changes must also be evaluated for impact on the risk profile of the product.

The manufacturer who assumes responsibility for product complaints needs the risk management documentation to identify the issue and its source. If the risk management process has been well done, the documentation should identify all potential issues with the product in the form of an estimated probability of occurrence for each harm, and the severity of each harm. The documents should identify the hazardous situation associated with each of the potential harms, providing information for the complaint investigator to determine if the events are occurring with the predicted severity and probability. If this is not the case, then the risk management documentation needs to be updated with the information developed from the complaint investigation.

Manufacturers must have complete risk management lifecycle processes in place to assure customers that the combination products are and continue to be safe for use (Figure 6.18). By monitoring the combination product and constituent part events against the risk management file, the manufacturer can monitor product safety and provide action to correct issues earlier.

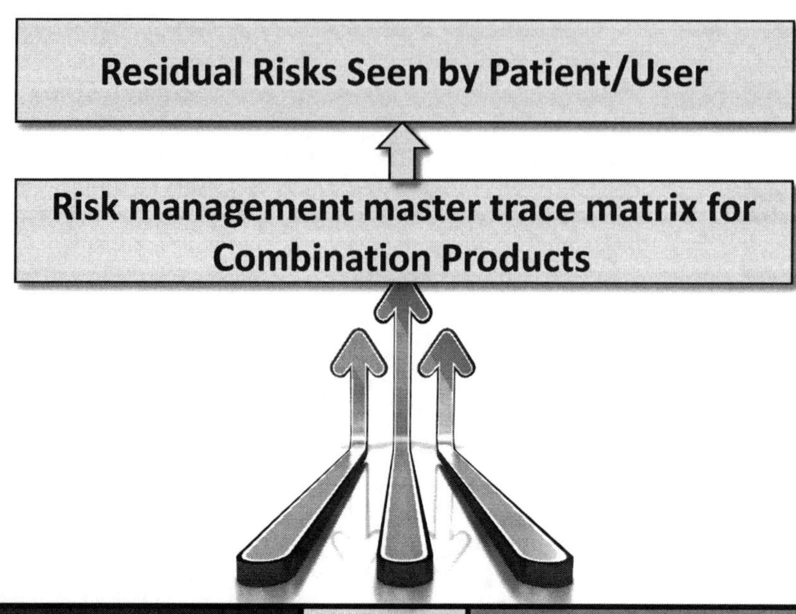

	ICH Q9(R1) (Medicinal Product Quality Risk Management)		AAMI TIR 105: 2020 Combined Use Risk Management		ISO 14971:2019/ ISO 24971:2020 (Medical Device Risk Management)
Risk Communication	**Risk Assessment** • Risk Identification • Risk Analysis • Risk Evaluation	Risk Management Tools	Risk Identification & Assessment	Risk Management Plan	**Risk Assessment** • Risk Analysis o Intended Use o Identify Hazards o Estimate Risk • Evaluate Risk
	Risk Control • Risk Reduction • Risk Acceptance		Risk Control		**Risk Control** • Option Analysis • Risk Control Implementation • Residual Risk Evaluation • Benefit-Risk Analysis • Manage Risks Arising from Risk Controls • Assure Completeness of Risk Contro
	Process Output/ Result of the Quality Risk Management Process		Residual Risk Evaluation		**Evaluation of Overall Residual Risk** **Risk Management Review**
	Risk Review • Review Events		Risk Review		**Production and Post Production Activitie** • General • Information Collection • Information Review • Actions

FIGURE 6.18 The resulting risk management view as seen by user/patient (©2022 Combination Products Consulting Services, LLC. All Rights Reserved).

NOTES

1. International Conference of Harmonization (ICH) Q9 was developed in 2005 from the first edition of International Standards Organization (ISO) 14971, which was released in 2000. The ISO committee participated in the development of ICH Q9 by providing expertise to the ICH committee, through a liaison that had drug expertise. That may explain why ICH Q9 uses many of the same terms and concepts as in ISO 14971.

2. ISO 14971:2019 (and its European-only amendment EN ISO 14971:2019+Amd11:2021), at the time of this writing, is currently state-of-the-art for medical devices risk management. Currently, ICH Q9(R1) drug "Quality Risk Management" (https://www.ema.europa.eu/en/documents/scientific-guideline/draft-international-conference-harmonisation-technical-requirements-registration-pharmaceuticals_en-1.pdf) and United States (https://www.fda.gov/regulatory-information/search-fda-guidance-documents/q9r1-quality-risk-management).
3. The content of this chapter is not necessarily sufficient to demonstrate compliance with any legal/regulatory requirements. The chapter is intended to inform combination product risk management activities.
4. ICH Q9(R1) and ISO 14971:2019.
5. In most cases, severity can only be reduced by redesigning the product.
6. See Chapter 5, Integrated Product Development, in this book.
7. 21 CFR 820.30 (d) *Design output procedures shall contain or make reference to acceptance criteria and shall ensure that those design outputs that are essential for the proper functioning of the device are identified. Design output shall be documented, reviewed, and approved before release.*
8. Notably, the combination product risks and the need to manage those risks don't necessarily end once a patient has finished their use of a product. A manufacturer should assess those risks across the product's lifecycle, i.e., even including disposal and ultimately the waste stream. For example, consider the example of a transdermal patch designed to treat opioid-tolerant patients who need daily, round-the-clock, long-term medicine by releasing the opioid through the skin over the course of the treatment. The patch is generally replaced every three days. If the patch is not disposed of carefully, young children can overdose on new and used patches, thinking the patch is something fun to explore by putting it in their mouths, or thinking the patch is a sticker, tattoo, or bandage, and sticking the patches – with full or residual doses – on their skin. Overdose can cause death by slowing the child's breathing and decreasing levels of oxygen in their blood. US FDA warns patients, caregivers, and healthcare professionals about the dangers of accidental exposure to such patches, and the need to properly store and dispose of the product. What controls might a combination product manufacturer proactively design into such a product to prevent such catastrophic harms? Application of a robust transparent adhesive film placed over the patch to make sure the patch doesn't prematurely detach from the patient's body, coupled with Instructions for Use that clearly communicate risks, appropriate storage, handling, and disposal are among potential mitigations. (https://www.fda.gov/consumers/consumer-updates/accidental-exposures-fentanyl-patches-continue-be-deadly-children#:~:text=The%20FDA%20has%20included%20fentanyl%20patches%20on%20a,Do%20if%20a%20Child%20Is%20Exposed%20to%20Fentanyl).
9. Harm is defined as "a physical or mental injury or damage to the health of people, or damage to property of the environment." A lot of time is spent focusing on harm to people's health, but there is not typically a great deal of discussion relating to harm to property and/or the environment. Given the advent of increasingly complex electromechanical drug delivery systems and increased awareness of environmental sustainability, impacts to property and environment are increasingly points for consideration. Property includes, for example, accessing personal information without permission, or improper operation of a medical product without the user's knowledge, something that could be the subject of a cybersecurity breach from a connected health perspective. From an environmental sustainability perspective, harm to the environment could include the effect of plastics in waste and even climate. Those property or environmental damage harm impacts would depend on the requisite SMEs to assess severity, e.g., a software security SME or an environmental specialist, respectively. Harms to peoples' health are in the realm of the subject matter expertise of a clinician/medical professional. In the combination products space, benefit-risk assessments likely would contend that the benefit to the patient outweighs the other risks, but that is an opportunity for consideration and discussion in any risk assessment.
10. GHTF.SG3.N15-R8: Implementation of Risk Management Principles and Activities Within a Quality Management System (2005) accessed August 3, 2022 at https://www.imdrf.org/sites/default/files/docs/ghtf/final/sg3/technical-docs/ghtf-sg3-n15r8-risk-management-principles-qms-050520.pdf
11. Helpful US FDA Guidance: "Factors to Consider When Making Benefit-Risk Determinations in Medical Device Premarket Approval and De Novo Classifications" (August 30, 2019) at https://www.fda.gov/media/99769/download, and "Benefit-Risk Assessment for New Drug and Biological Products" (September 26, 2022 at https://www.fda.gov/regulatory-information/search-fda-guidance-documents/benefit-risk-assessment-new-drug-and-biological-products).
12. ISO 14971 Clause 7.4 says "unacceptable residual risks." The EU MDR (and IVDR) says for "all residual risks." ISO 14971 Clause 8 also requires a Benefit-Risk Analysis for the Overall Residual Risk for the entire device – and for combination products, for the entire combination product.

13. US FDA Draft Guidance "Technical Considerations for Demonstrating Reliability of Emergency-Use Injectors Submitted under a BLA, NDA, or ANDA" (April 2020, accessed August 1, 2022 at https://www.fda.gov/regulatory-information/search-fda-guidance-documents/technical-considerations-demonstrating-reliability-emergency-use-injectors-submitted-under-bla-nda).
14. Neadle, Susan (editors: Bills, E. and Mastrangelo, S.) (2016). "Risk Management Considerations and Strategies in Product Development" in *Lifecycle Risk Management for Healthcare Products: From Research Through Disposal*. Davis Healthcare International Publishers.
15. International Ergonomics Association (IEA).
16. FDA Draft Guidance for Industry: Comparative Analyses and Related Comparative Use Human Factors Studies for a Drug-Device Combination Product Submitted in an ANDA (2017) accessed August 1, 2022 at https://www.fda.gov/regulatory-information/search-fda-guidance-documents/comparative-analyses-and-related-comparative-use-human-factors-studies-drug-device-combination
17. IEC 62366-1; FDA, Applying Human Factors and Usability Engineering to Medical Devices (2016).
18. US FDA CDRH Draft Human Factors Engineering Guidance, "Applying Human Factors and Usability Engineering to Medical Devices" (February 2016), accessed July 31, 2022 at https://www.fda.gov/media/80481/download
19. Note that CDER does not include the word "serious" in their definition of critical use task, so the scope of tasks that are considered critical is broader than in the definition from CDRH. US FDA CDER Draft Human Factors Engineering Guidance, "Human Factors Studies and Related Clinical Study Considerations in Combination Product Design and Development"(February 2016), accessed July 31, 2022 at https://www.fda.gov/regulatory-information/search-fda-guidance-documents/human-factors-studies-and-related-clinical-study-considerations-combination-product-design-and
20. This definition is adapted from IEC 60601-1-11:2015: Medical electrical equipment – Part 1-11: General requirements for basic safe and essential performance – Collateral standard: Requirements for medical electrical equipment and medical electrical systems used in the home healthcare environment," which defines essential performance as performance of a clinical function, other than that related to basic safety, where loss or degradation beyond the limits specified by the manufacturer results in unacceptable performance.
21. ICH Q12 (Essential Conditions) [ICH Q12: Implementation Considerations for FDA-Regulated Products Guidance for Industry (May 2021 Draft) at https://www.fda.gov/media/148947/download].
22. See also Chapter 8 on "Process Controls" and Established Conditions.
23. Neadle, S.; Riter, J.; McAndrew T.P.; (Pharmaceutical Online, Nov. 30, 2020) "*Considerations in Combination Product Risk Management*"; https://www.pharmaceuticalonline.com/doc/considerations-in-combination-product-risk-management-0001
24. See FDA Guidance titled "*Guidance for the Content of Premarket Submissions for Software Contained in Medical Devices*" and "*Content of Premarket Submissions for Management of Cybersecurity in Medical Devices*".
25. Preliminary Hazard Analysis, or PHA, is useful as it can be used as early as Research Phase to identify possible hazards even before a design is solidified, unlike FMEA which cannot be performed until Design Outputs are formed. PHA can be used to form safety requirements which are Design Inputs.
26. GHTF.SG3.N15-R8: Implementation of Risk Management Principles and Activities Within a Quality Management System (2005) accessed August 3, 2022 at https://www.imdrf.org/sites/default/files/docs/ghtf/final/sg3/technical-docs/ghtf-sg3-n15r8-risk-management-principles-qms-050520.pdf
27. As a plaintiff's expert, one author has noted that issues in Risk Management Files may lead to large awards to a plaintiff.
28. For example, US FDA has issued product-specific guidance and draft guidance, documents based on their own observation of industry challenges; these guidance serve as great resources to proactively inform product development risk analysis – and broader risk management – activities. One such FDA guidance is "Technical Considerations for Pen, Jet, and Related Injectors Intended for Use with Drugs and Biological Products"(2013) (accessible at https://www.fda.gov/regulatory-information/search-fda-guidance-documents/technical-considerations-pen-jet-and-related-injectors-intended-use-drugs-and-biological-products). The reader is encouraged to research what pertinent guidance might be available for your products at www.fda.gov. Chapter 5 on Integrated Product Development in this book also includes a reference list of a range of international standards, regulations, and guidance that may be helpful to the reader.
29. Reliability and reliability block diagrams are discussed in detail in IEC 61078:2016 – Reliability Block Diagrams. Combination product reliability is the probability (likelihood of occurrence) that a combination product will perform without failure for a given time interval under specified conditions. In FDA

guidance "Technical Considerations for Pen, Jet, and Related Injectors Intended for Use with Drugs and Biological Products" (2013) (https://www.fda.gov/media/76403/download), FDA identifies the need for demonstrating the reliability of the mechanism to deliver the drug/biological product. Manufacturers need to guard against systematic failures, i.e., design flaws that will be hazardous under certain scenarios (e.g., software failures that have been the source of a Class I recall for Medfusion Syringe Infusion 3500 and 4000 Syringe Infusion Pumps (https://www.fda.gov/medical-devices/medical-device-recalls/smiths-medical-recalls-certain-medfusion-3500-and-4000-syringe-infusion-pumps-software-issues-may) distributed October 2004 to February 2022).

Reliability issues that are the result of random failures (e.g., failure of a component/hardware; failure due to use; software failure in combination with a hardware failure or use interaction) may be a bit more difficult to detect and address. Design and process FMEAs are helpful to vet such hazards during design.

30. An extensive list of tools and their uses is identified in IEC 31010:2019 Risk management – Risk assessment techniques.
31. Carlson, C. (2012). *Effective FMEAs. John Wiley & Sons, Inc.*
32. FMEAs are developed using a different definition of "risk," which is "probability of failure occurring and severity of the effect of the failure." This is much different from the definitions in ISO Guide 63. Additionally, FMEA requires Design Outputs to be available to perform FMEA, while ISO 13485 7.3.3 c) Quality System standard (future 21 CFR 820) requires ***"outputs of risk management are design inputs"***. FMEA is a good check tool to make sure you have not missed anything but has less value than tools such as Preliminary Hazard Analysis (PHA), which can provide that Design Input much earlier in the process, thus lowering cost and reducing time caused by redesign when problems are found in the design later in the process.
33. The author is quite purposefully repeating the definitions of hazard, hazardous situation, and harm, because if one frames out the risk analysis in terms of those definitions, one hopes it becomes less confusing.
34. A medical professional on the risk management team would be the judge of harm severity that may vary based on specific considerations of intended use, intended users, and use environment and sequence of events.
35. *State-of-the-art* is defined in ISO 14971:2019 3.28 as *"developed stage of technical capability at a given time as regards products processes, and services based on relevant consolidated findings of science, technology, and experience."* See ISO 14971:2019 for additional information.
36. EU MDR (2017/745) Article 1 Chapter 1 Item 2.
37. However, if cost makes a product unavailable, there may be a different – non-monetary- cost consideration. This has been a discussion point in EU.
38. Also see GSPR 4 under EU MDR, which is well aligned with ISO 14971:2019. It states that "in selecting the most appropriate solutions, manufacturers shall, in the following order of priority: (a) eliminate or reduce risks as far as possible through safe design and manufacture; (b) where appropriate, take adequate protection measures, including alarms, if necessary, in relation to risks that cannot be eliminated; and (c) provide information for safety (warnings/precautions/ contra-indications) and where appropriate, trainings to users. Manufacturers shall inform users of any residual risks."
39. It is up to the manufacturer to determine what is significant.
40. US FDA "Principles of Premarket Pathways for Combination Products" Guidance for Industry and Staff (January 2022) accessed July 31, 2022 at https://www.fda.gov/media/119958/download
41. FDA Guidance "Factors to Consider Regarding Benefit/Risk in Medical Device Product Availability, Compliance, and Enforcement Decisions"(December 2016) at https://www.fda.gov/files/medical%20devices/published/Factors-to-Consider-Regarding-Benefit-Risk-in-Medical-Device-Product-Availability--Compliance--and-Enforcement-Decisions---Guidance-for-Industry-and-Food-and-Drug-Administration-Staff.pdf
42. FDA Draft Guidance "Benefit-Risk Assessment for New Drug and Biological Products" (September 2021) at https://www.fda.gov/regulatory-information/search-fda-guidance-documents/benefit-risk-assessment-new-drug-and-biological-products
43. *"GHTF SG3/N18 Quality management system – Medical Devices – Guidance on corrective action and preventive action and related QMS processes",* accessed 8/13/2022 at https://www.imdrf.org/sites/default/files/docs/ghtf/final/sg3/technical-docs/ghtf-sg3-n18-2010-qms-guidance-on-corrective-preventative-action-101104.pdf
44. US FDA Compliance Program 7356.000 (June 4, 2020) "Inspections of CDER-led or CDRH-led Combination Products" accessed August 3, 2022 at https://www.fda.gov/media/138592/download

REFERENCES

AAMI TIR 105: 2020 Combination Product Risk Management.

Edwin Bills, Stan Mastrangelo, and Fubin Wu (Spring 2015), Documenting Medical Device Risk Management through the Risk Traceability Summary, *Horizons*, AAMI.

Federal Register, Vol. 78, No. 14, January 22, 2013, Department of Health and Human Resources, Food and Drug Administration, Docket No FDA-2009-N-0435, Current Good Manufacturing Practice Requirements for Combination Products, Final Rule.

GHTF SG3/N15 (2005) Implementing Risk Management within a Quality Management System.

GHTF SG3/N18 (2010) Quality Management System –Medical Devices – Guidance on Corrective Action and Preventive Action and Related QMS Processes.

ICH (2005) ICH Q9: 2006 Quality Risk Management.

ICH (2009) ICH Q10: 2009 Pharmaceutical Quality System.

ICH (2009) ICH Q8: 2009 Pharmaceutical Development.

IEC 60812:2018 Analysis of system reliability-failure modes and effects analysis (FMEA).

IEC 61025 Fault Tree Analysis (FTA) (2006).

ISO (1998) ISO 14971-1:1998 Medical Devices – Risk Management – Part 1: Application of Risk Analysis (Withdrawn in 2000-No Longer Available).

ISO (2009) Guide 73:2009 Risk management – Vocabulary.

ISO (2018) ISO 31000:2018 Risk Management – Principles and Guidelines.

ISO (2019a) ISO 14971:2019 Medical Devices-Application of Risk Management to Medical Devices.

ISO (2019b) ISO TR 24971:2020 Guidance on Application of ISO 14971.

ISO (2019c) ISO/IEC 31010:2019 Risk Management – Risk Assessment Techniques.

ISO/IEC (2014) Guide 51:2014 Safety Aspects – Guidelines for their Inclusion in Standards.

Mike W. Schmidt, (March 2004), The Use and Misuse of FMEA in Risk Analysis. *Medical Device and Diagnostic Industry.*

Nancy G. Leveson (2011) *Engineering a Safer World: Systems Thinking Applied to Safety*, The MIT Press, Cambridge, MA.

The differences and similarities between ISO 9001:2015 and ISO 13485:2016, Mark Swanson, President and Lead Consultant, H&M Consulting Group, BSI White Papers, http://www.bsigroup.com/en-GB/our-services/medical-device-services/BSI-Medical-Devices-Whitepapers/

Tony Chan and Edwin Bills (May 2004), Exploring Post development Risk Management, *Medical Device and Diagnostic Industry.*

Tuesday, January 29, 2013, http://www.justice.gov/iso/opa/civil/speeches/2013/civ-speech-130129.html?goback=%2Egde_1977384_member_214363625 Accessed on March 1, 2013.

US Department of Justice Speeches (2013) *Deputy Assistant Attorney General Maame Ewusi-Mensah Frimpong Speaks at the 2013 CBI Pharmaceutical Compliance Congress.*

US FDA 21 CFR 210 Current Good Manufacturing Practice in Manufacturing, Processing, Packing, or Holding of Drugs; General.

US FDA 21 CFR 211 Current Good Manufacturing Practice for Finished Pharmaceuticals (1978).

US FDA 21 CFR 4, Current Good Manufacturing Practice Requirements for Combination Products.

US FDA 21 CFR 803 Medical Device Reporting.

US FDA 21 CFR 820 Quality System Regulation (1996).

US FDA Draft Guidance tor Industry and FDA Staff: Human Factors Studies and Related Clinical Study Considerations in Combination Product Design and Development.

US FDA Guidance for Industry and FDA Staff: Current Good Manufacturing Practice Requirements for Combination Products.

US FDA Guidance for Industry and FDA Staff: Factors to Consider Regarding Benefit-Risk in Medical Device Product Availability, Compliance, and Enforcement Decisions.

7 Human Factors Engineering in the Design, Development, and Lifecycle of Combination Products

Shannon Hoste, Stephanie Canfield, Susan W. B. Neadle, Bjorg Hunter, and Theresa Scheuble

CONTENTS

DOI: 10.1201/9781003300298-8

HUMAN FACTORS ENGINEERING/USABILITY ENGINEERING

The human factors engineering/usability engineering (HFE/UE) process is used to enhance and demonstrate the safe and effective use of the combination product (or product). Within this regulated space, human factors engineering and usability engineering are considered synonymous. The term human factors engineering (HFE) will be used throughout this chapter. The HFE process provides a systematic approach to identify use-related risks, develop the product user interface (UI), and support a conclusion that the product can be used safely and effectively by the intended users. When applied throughout the design development process of the product, HFE can benefit end users by optimizing product safety, efficacy, and usability, while also driving development efficiencies and cost-effectiveness.

The HFE process is one performed throughout the lifecycle for medical devices, drugs, biologics, and combination products and is designed to provide a structured methodology to identify, assess, and determine controls/mitigations for use-related risk.

The HFE process is defined in the International Electrotechnical Commission (IEC) standard IEC 62366-1:2015/AMD-1:2020 titled "Medical Devices – Part 1: Application of usability engineering to medical devices – Amendment 1" (henceforth referred to as IEC 62366-1) [Ref 2]. At the time of this writing, IEC 62366-1 is a United States (US) Food and Drug Administration (FDA)–recognized consensus standard and is cross-referenced by several international standards, including ISO 14971:2019 titled, "Medical devices – Application of risk management to devices" [Ref 4]. In addition to IEC 62366-1, there exist other standards (i.e., IEC TIR 62366-2:2016 [Ref 3], AAMI HE75:2009 [Ref 1]) which are helpful in the support of a successful HFE program. The FDA has issued several drafts and final guidance documents describing the FDA's expectation for HFE program execution and data within regulatory submissions. While not explicitly reviewed in this chapter, Refs. [5,6,7,9,10,11] are included for further information.

The goal of this chapter is to explore the framework, the type of data generated, regulatory review, and case studies for HFE in the combination product space, specifically products that include both medical devices and one or more drug(s) or biologic(s). This chapter will focus on use-related risk and the human factors engineering process around it. Following the HFE process allows a manufacturer to not only reduce the overall combination product risk but design a user interface that supports safe and effective use, along with generating the evidence to support such conclusions in regulatory submissions.

This will be an overview, as further details on the process can be found in international standards and in various guidance documents referenced in the Further Information section at the end of this chapter, to be used for continued learning.

GENERAL TERMS

When comparing IEC-62366-1, FDA guidance, and other sources for HFE programs, the practices are similar; however, there are some differences in terms. As already stated, some sources refer to human factors engineering, while others use usability engineering. Before proceeding with the chapter's content, here are some definitions and the terms used throughout.

Critical Task – The FDA draft guidance titled "Human Factors Studies and Related Clinical Study Considerations in Combination Product Design and Development" defines critical tasks as those "user tasks that, if performed incorrectly, or not performed at all, would or could cause harm to the patient or user, where harm is defined to include *compromised medical care*" ("Human Factors Studies and Related Clinical Study Considerations in Combination Product Design and Development" [Ref 11], henceforth referred to as Human Factors (HF) Considerations in Combination Product Design). Critical tasks are determined by the severity of harm and potential to compromise medical care; these can be identified at a threshold established by manufacturers, which should be defined in relevant Standard Operating Procedures (SOPs) and/or other internal instructions, and be consistent among all internal products. For example, any task that has a severity at or above this threshold and any task where a use error could impact effectiveness (e.g., underdose) is a critical task.

Hazard-Related Use Scenario – IEC 62366-1 defines hazard-related use scenarios as a use scenario that could lead to hazardous situations or harms. It is noted that these can often be linked to a potential use error and are not related to a failure of the medical device, unless the medical device failure was caused by a use error. Unless all hazard-related use scenarios are tested in the summative validation evaluation, it is part of the HFE process to determine which of the hazard-related use scenarios are to be evaluated in the summative validation evaluation, based on the severity of potential harm. This selection of hazard-related use scenarios can be analogous to the FDA-defined critical tasks.

Summative Validation Evaluation – IEC 62366-1 refers to the final UE study as the summative evaluation. FDA guidance refers to it as human factors validation testing [Ref 5]. This chapter will refer to it as a summative validation evaluation.

Use-Related Risk Analysis – URRA – a method for identifying use-related risks by tasks and use errors, and their related potential hazards, hazardous situations, harms, and severity of harms [Ref 9].

Use Error – The definition of use error differs slightly between IEC 62366-1 and FDA. IEC 62366-1 defines use error as "user action or lack of user action while using the medical device that leads to a different result than that intended by the manufacturer or expected by the user" [Ref 2].

FDA guidance defines use error as a user action or lack of action that was different from that expected by the manufacturer and caused a result that (1) was different from the result expected by the user, (2) was not caused solely by device failure, and (3) did or could result in harm [Ref 5].

User Interface (UI) – User interface is defined similarly between IEC 62366-1 and FDA guidance. Both define the user interface to include all points of interaction between the user and the product. This includes physical aspects of the device/product, visual, auditory, tactile displays, packaging, labels, instructions for use (IFU), training material, etc.

FRAMEWORK OF THE HUMAN FACTORS ENGINEERING PROCESS

Before exploring the HFE process in further detail, it is important to emphasize that HFE is indeed a process. The summative validation evaluation is not a stand-alone safety test – it is a step in the process. When thinking of the standards associated with product design, there are many standards and test methods to assess different aspects of product safety, such as biocompatibility and electromagnetic compatibility. However, there are also product areas which do not lend themselves to standardized testing. One example of the latter is software development, where lifecycle processes are focused on identifying, assessing, and managing software risks and security vulnerabilities. Similarly, when considering use-related risk, there is no standardized pass/fail test to demonstrate that the user interface supports safe and effective use.

Thus, human factors engineering is a product lifecycle process. To better understand this, Figure 7.1 provides an example of the inputs and outputs from the HFE process. This figure is based on the framework from IEC 62366-1 [Ref 19].

The process diagram below (Figure 7.1) is a mapping of the human factors engineering process based on the framework from IEC 62366-1 [Ref 19]. The purpose of this example is to demonstrate how the HFE process informs and is informed by the product development process and the risk management process. While product development terminology can vary from company to company, the building blocks are similar.

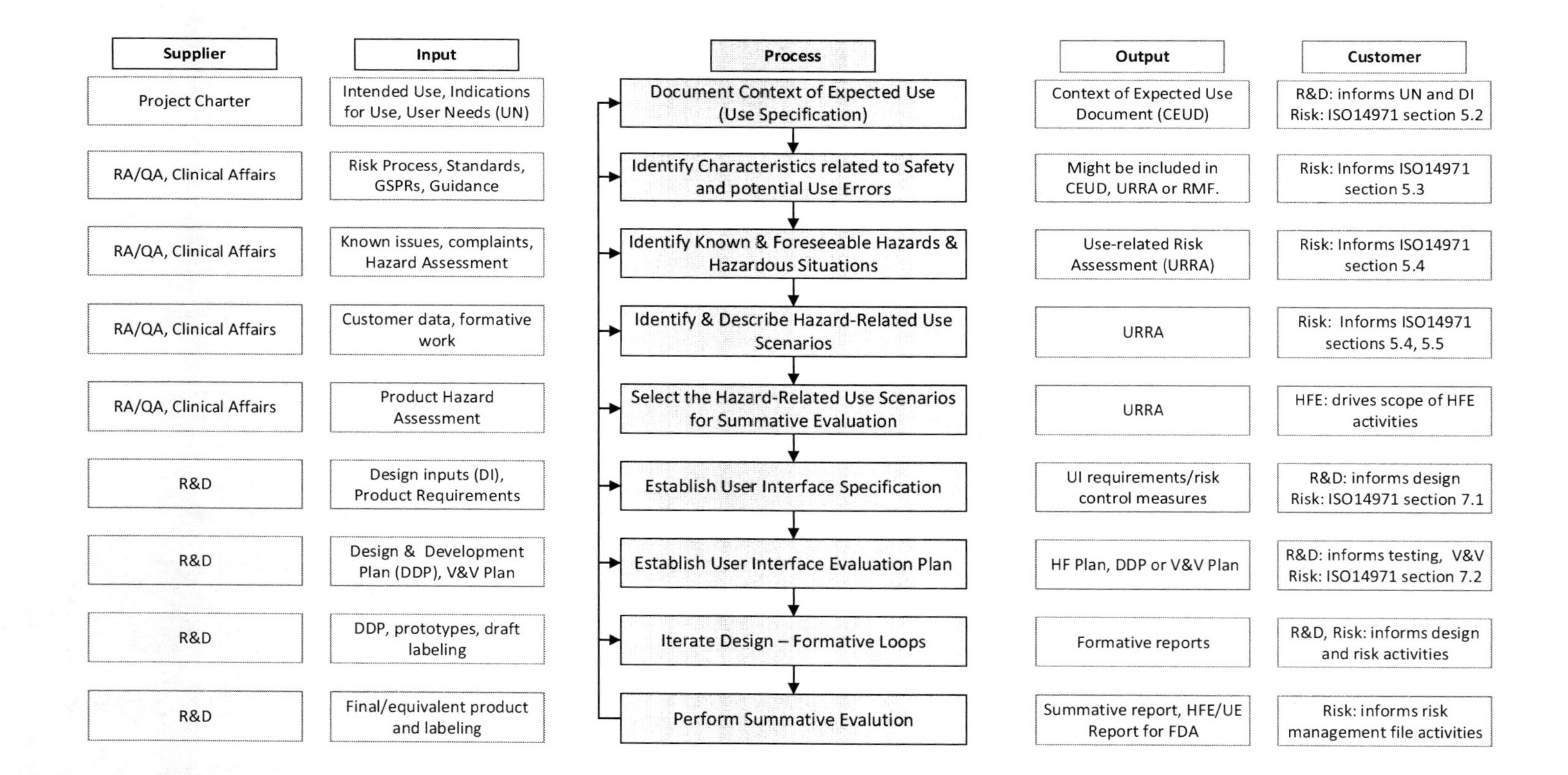

FIGURE 7.1 Example Human Factors process and Inputs-Outputs from the process steps [Ref 20]. © 2022 Agilis Consulting Group LLC. Used with Permission. All Rights Reserved. Also, to see a flowchart of how IEC 62366-1 directly relates to ISO-14791:2019, see Figure A.5 in IEC 62366 [Ref 2].

It is important to begin the combination product HFE process as early as possible within the development process to ensure learning and knowledge from the HFE process are incorporated in the early stages of the product design and are used to inform risk assessments and subsequent risk controls. If device constituent part development is started before a specific drug is identified, a representative product (e.g., saline or a placebo) can be utilized for early HFE activities. Formative HFE activities can be started early with concepts and mockups. The fidelity of the prototypes and user interfaces evaluated can increase as the product is further defined and refined.

By utilizing learning from HFE activities throughout the development process, the development team can optimize the device constituent part's design and therefore reduce the risk of complaints and adverse events in relation to the user interface. The HFE strategy should be planned as part of the design and development process. If the HFE process is left until the end of the product development, to satisfy regulatory requirements as part of design validation, blind spots in the UI design and failures in these validation-level studies can result in significant re-work and costly delays. It should be noted that the HFE process, specifically the summative validation evaluation, may be part of the design validation activities, however they are by no means all-encompassing of the design validation process.

HFE AND RISK MANAGEMENT

When working to identify, analyze, and control risks in combination products, each constituent part brings its own respective risk. For example, a medical device may be assessed for design failure modes that can contribute to risk. When bringing together the constituent parts, the various sources of risk need to then be assessed based on the combination product indications for use. It is within the context of this evaluation of overall combination product risks that the impact of use-related risk needs to be assessed. This is the analysis of use errors which could occur in the use of the combination product that could lead to harm or compromised medical care. The HFE process provides a structured method to understand, analyze, and manage these use-related risks. The documentation of this risk assessment is often called a use-related risk analysis (URRA), although Failure Modes & Effects Analysis (FMEA) tools are often used as well, such as use-FMEA (uFMEA) or applications FMEA (aFMEA).

USE-RELATED RISK AND COMBINATION PRODUCTS

The human factors engineering process is designed to inform product development and to provide a framework to evaluate use-related risk throughout the product lifecycle. To consider the contribution of this process from another perspective, consider a common post-market field issue scenario. A field issue, upon investigation, was attributed to a "user error." With this conclusion what can or should be done to address the issue? "User error" is not an attributable root cause as it does not identify the design factors that lead to the issue. Contributing factors could be one or a combination of issues, such as environmental, cognitive workload of the user, incorrect affordances of the design, incomplete procedures, etc. Having explored the context of use during product development provides a framework to support post-market issues and events, in order to more quickly identify and address use-related events.

The focus of human factors engineering in the medical and healthcare space is to understand what contributes to potential use errors and what can be done to prevent them. The goal is not to *ad nauseam* explore every error, mistake, slip, and lapse that can be made with a product. This is a risk-based process; therefore, the analysis is scoped to those steps of use (or tasks) where an error could impact the safety or efficacy of the product.

USE-RELATED RISK ANALYSIS

Below is an example of a URRA, adapted from Appendix B in the FDA guidance Contents of a Complete Submission for Threshold Analysis and Human Factors Submissions to Drug and Biologic

Applications [Ref 9]. This is a selection of tasks for a URRA for a single-use, single-dose-prefilled syringe for subcutaneous administration of a maintenance therapy. In practice this analysis is conducted in phases. The first step in this analysis involves identifying the sequential tasks of use for each use scenario. Once this is complete, then potential use errors can be identified. Subsequently, the evaluation of hazards and harms that can result from the use error can be assessed.

TABLE 7.1
Example of Use-Related Risk Analysis (URRA)

Task #	Task	Potential Use Error	Potential Hazards/ Harm and Severity	Critical Task?	Risk Mitigation	Evaluation Method
3	Remove PFS from blister pack	User drops the blister pack	PFS is damaged and unusable, leading to no dose, which leads to [clinical impact of no dose]	Yes	Blister pack designed with finger slots	Evaluated in formative round C and summative validation evaluation – Scenario X, Task 3
3	Remove PFS from blister pack	User cannot open the blister pack	PFS cannot be used, leading to no dose, which leads to [clinical impact of no dose]	Yes	Blister pack complies with the standards for accessible design	Evaluated in formative round C and summative validation evaluation – Scenario X, Task 3
6	Select injection site	User selects incorrect injection site	Intradermal injection causing injury and pain	No	Injection site information displayed on packaging, IFU, and training materials	Evaluated in formative round C and summative validation evaluation – Scenario X, Task 6
6	Select injection site	User selects incorrect injection site	Intradermal injection leading to delay in medication efficacy which leads to [clinical impact of delayed dose]	Yes	Injection site information displayed on packaging, IFU, and training materials	Evaluated in formative round C and summative validation evaluation – Scenario X, Task 6

(Adapted from Reference 9)

Combination Product Discussion Points

It is important to note that the previously provided definition of critical tasks in the HF Considerations in Combination Product Design guidance [Ref 11] is slightly different from that provided by FDA's Center for Devices and Radiological Health (CDRH) for stand-alone devices, which restricts critical tasks to those that may lead to serious harm [Ref 5]. As this text is on combination products, the remaining sections of this chapter will use the version of critical tasks as defined in the terms section above.

When it comes to combination products, based on the difference in definitions of critical tasks, there is the potential that more tasks will be classified as critical than they would if following the definition for a device alone. For example, consider the task of removing a needle shield from a prefilled syringe (PFS). This particular PFS is intended to allow a patient to self-administer a maintenance treatment medication (e.g., medication for rheumatoid arthritis). What would be the impact if that patient is unable to remove the needle shield when they need to administer the drug? Missing a dose or delaying therapy of a maintenance drug? Is removal of the needle shield considered a critical task? The answer to that question is "Yes." While the medication in the PFS might not be emergency-use, the intended use of that combination product is to be a product that allows patients to successfully self-administer their medication. If that patient cannot remove the needle shield, they cannot use the product as intended; therefore, they would experience *compromised medical care*. Whether such an event is a one-time occurrence, or even if that event does not lead to patient injury or a need for medical treatment, a missed dose or incorrect dose could still be of concern and a critical task.

FDA's HF Considerations in Combination Product Design guidance [Ref 11], as well as related "Safety Considerations for Product Design to Minimize Medication Errors" guidance [Ref 15], provides detailed, albeit not exhaustive, lists of examples of critical tasks and user interfaces, possible task failures, and possible hazards, harms, and results from failures and use errors. These are useful in the identification of hazards, hazardous situations, harms, severity of harms, use errors, and critical tasks.

With the inclusion of a drug and/or biologic in a combination product, a major category of potential hazards and hazardous situations surrounds medication error. A medication error is any preventable event that may cause or lead to inappropriate medication use or patient harm while the medication is in the control of the healthcare professional, patient, or consumer [Ref 28]. Medication errors can occur throughout the medication-use system (Figure 7.2), whether when prescribing a drug, when entering information into a computer system, when preparing or dispensing the drug, or when administering (or self-administering) the drug to (or by) a patient. A multitude of adverse events, sometimes with serious outcomes – including deaths, life-threatening situations, hospitalizations, disabilities, and birth defects – can result from medication errors.

When conceptualizing medication error, the Taxonomy of Medication Errors outlined by the National Coordinating Council for Medication Error Reporting and Prevention (NCC MERP) can be a useful reference to assist in the identification of use errors [Ref 23]. Within this taxonomy, error types are defined in Section 70 and include dose omission, overdose, underdose, extra dose, wrong drug, wrong route, wrong time, etc. This can be helpful when considering potential hazards and hazardous situations associated with a given product.

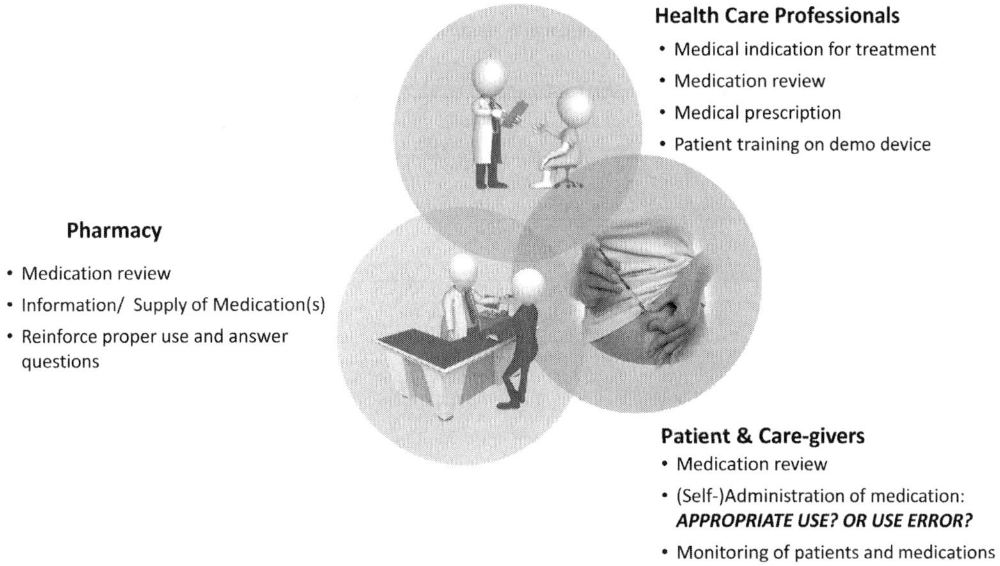

FIGURE 7.2 Medication-Use System. ©2022 Combination Products Consulting Services LLC. All Rights Reserved.

Beyond Medication Error

In addition to medication error, there are other hazards associated with products which can be associated with the device (or constituent parts), container closure, and packaging systems. These can include sharps hazards, electrical safety hazards for electromechanical devices, breach of sterile barrier, etc. Device-specific examples can be found in IEC 62366-1, IEC TIR 62366-2, and ISO 14971:2019 [Ref 2,3,4]. Another source is the considerations and expectations identified in The General Safety and Performance Requirements (GSPR) within European Union (EU) Regulation 2017/745 (Medical Device Regulation [MDR]) [Ref 17]and EU Regulation 2017/746 (In Vitro Diagnostics Regulation [IVDR]) [Ref 20].

HUMAN FACTORS DATA

The execution of a robust HFE process will result in HFE data and records, as with product development and risk management. While some of the deliverables are reports, such as an HFE formative report, other deliverables, such as risk management files, are lifecycle data, meaning they provide a framework that will aid in post-market surveillance activities and design changes and will need to be maintained through the product lifecycle.

While aligning with the more detailed process diagram in Figure 7.1, another way to think about the human factors engineering process is these four fundamental steps:

1. Establish and document the information necessary for a complete assessment of use-related risk;
2. Identify and assess use-related risk;
3. Evaluate the product–user interface to confirm the management of use-related risks; and
4. Monitor, assess, and mitigate use-related risk throughout the product lifecycle.

Breaking this down, for example:

1. The Context of Expected Use Document (Use Specification) is a living document initiating with preliminary HFE research to understand the intended use, users, use environment, and other details and limitations around the context of use. This information along with the characteristics related to safety, known use errors, and hazards informs the assessment of use-related risk. This process is typically iterative and supported by various research methods including heuristics and formative evaluations.
2. The Context of Expected Use Document is iterated along with the use-related risk analysis and User Interface Specification. This is typically initiated with a detailed breakdown of the tasks associated with use for each use scenario (task analysis). A use error assessment can be performed to identify the potential "failure modes" at each task. From this point, hazards, hazardous situations, and harms can be identified and evaluated based on the severity of potential harm, and risk mitigations can be identified. Use-related risk mitigations should inform User Interface Specifications to be evaluated through the design process.
3. When the design and research loops are complete (*see Combination Product Development and Risk Management chapters in this book*), the final user interface is evaluated through a summative validation evaluation, which evaluates each critical task of use with representative users and use scenarios. The output of this study informs the overall product risk assessment and supports a determination if the risks have been reduced as far as possible, within the accepted benefit–risk paradigm.
4. Finally, the HFE data and documentation of support product lifecycle management and are updated to reflect assessments of post-market surveillance data and changes made in post-market.

With this model of the HFE process, no stand-alone deliverable can support a conclusion of safe and effective use. It cannot be clear that the URRA is complete if the Context of Expected Use is not complete. Additionally, it cannot be determined if the summative validation evaluation scope is complete without a robust URRA. This is why IEC 62366-1 identifies the HFE process elements be stored within a usability engineering file (UEF) and why the FDA includes these elements in their requested HFE/UE report structure.

Regulatory Review of Human Factors Data

Throughout the HFE process, different records and data are captured. When it comes to compiling this information for a regulatory submission, different compilations of this data may be necessary for submissions to different regulatory authorities globally.

For jurisdictions accepting declarations of conformity to harmonized standards, conformance to IEC 62366-1 can be assessed by a third party per that standard. As an international standard, it can also be supportive of state-of-the-art process standard for use-related risks. For example, the requirements are outlined in the GSPRs in the MDR [Ref 22] and IVDR [Ref 20]. Where in Annex 1, Chapter 1 General Requirements, the following is defined:

> 5. In eliminating or reducing risks related to use error, the manufacturer shall:
>   (a) reduce as far as possible the risks related to the ergonomic features of the device and the environment in which the device is intended to be used (design for patient safety), and
>   (b) give consideration to the technical knowledge, experience, education, training and use environment, where applicable, and the medical and physical conditions of intended users (design for lay, professional, disabled or other users).

Noting that per the Council of the International Ergonomics Association (IEA), ergonomics (or human factors) is the scientific discipline concerned with the understanding of interactions among humans and other elements of a system and the profession that applies theory, principles, data, and methods to design in order to optimize human well-being and overall system performance [Ref 21].

In addition to this requirement, Chapters 2 and 3 of this Annex also identify other items that will need to be considered within a human factors engineering process.

In the United States, there is a pre-review process, wherein the applicable FDA Center provides feedback on the HFE approach, e.g., on the summative study protocol and IFU, prior to summative study execution. In the EU and other countries, the HF summative study report is submitted as part of Marketing Authorization Application (MAA) review, but typically there is no such interaction with health authorities on the HFE approach prior to submission.

For the US FDA, expectations around HFE activities are outlined in several guidance documents [Ref 5,7,9,11]. These also provide guidance and pathways for communications during the development process. One consideration when working with the FDA is that human factors engineering data is typically reviewed by specific specialized groups, such as the Division of Medication Error Prevention and Analysis (DMEPA) in the Center for Drug Evaluation and Research (CDER). This review is best supported by a compiled HFE package containing the information from the activities conducted in the HFE process, as outlined in the guidance document [Ref 5].

With the FDA, typically the review process is coordinated through the lead center, which is determined based on the Primary Mode of Action (PMOA) for the product [Ref 10,14]. Generally, if it is a drug-led combination product, the lead center would be CDER; if biologic-led, it would be the Center for Biologics Evaluation and Research (CBER); or CDRH if the device provides the PMOA. These primary review centers would then drive the review through the appropriate division or office and typically human factors would be consulted out to CDER/DMEPA and/or the CDRH HF team. For some combination products, HFE review is done by both CDER/DMEPA and the CDRH HF team. When these inter-center consults are necessary, the teams work to provide a single set of recommendations back to the Sponsor. Examples of device constituent parts that could need inter-center review include: gas delivery device, hyperbaric chamber, nebulizer, implantable infusion device, body-worn infusion device, external infusion device, software-driven medication dispenser, reusable autoinjector for multiple patient use or with external electrical components or mobile technology, and inhaler with external electronic components or mobile technology [Ref 25].

Globally, as evidenced by the range of international standards and guidance available, regulatory authorities recognize the value brought by the application of HFE in combination product development. The principles and application of HFE are relatively consistent across jurisdictions globally.

Compiling Human Factors Engineering Data

Even with an HFE process in place, when performing regulatory submissions, HFE data need to be compiled specifically for each submission type or regulatory authority. When compiling these data into reports, it is important to remember the purpose and the audience of the report.

Typically, report authors and internal reviewers have access to significant background data including knowledge of project constraints and design decisions. They also have access to other project records and information. In contrast, when this report is being reviewed by a regulatory or third-party reviewer, they come to it with knowledge on similar product reviews and current reported events and complaints on similar products. As for product- and project-specific data, they only have what is provided in the report. This review is focused on the regulatory question at hand. For example, an FDA 510(k) submission must provide data to support substantial equivalence to a predicate. This is different for a Prem-Market Approval Application (PMA), different for a 505(b)(1) New Drug Application (NDA), different for a 505(j) Abbreviated New Drug Application (ANDA), etc. It is important that the HFE package supports the regulatory question at hand and provides sufficient information and context for a complete review. When considering the HFE process, each step builds to the next, thus a gap in any one step propagates throughout. It is essential to provide details as requested in the FDA guidance documents.

For the FDA, Drug Approval Packages reviews are posted as public records on the FDA website [Ref 29]. DMEPA human factors reviews can be found in "Other Review(s)." These provide some visibility into reviewers' perspectives.

For example, in the HFE reviews, there is typically some version of Table 7.2.

TABLE 7.2
DMEPA Review Product Summary

Relevant Product Information	
Initial Approval Date	N/A
Active Ingredient (Drug or Biologic)	___umab-xxxx
Indication	Treatment of ____ in adults
Route of Administration	Subcutaneous
Dosage Form	Injection solution
Strength	120 mg/mL
Dose and Frequency	120 mg by subcutaneous injection at weeks 0, 2, and 4 and every 12 weeks thereafter
How Supplied	120 mg/mL prefilled autoinjector
Storage	Refrigerate at 2°C to 8°C (36°F to 46°F) in original carton to protect from light. Do not freeze
Container Closure/Device Constituent	Carton containing one autoinjector and alcohol pad
Intended Users	HCP, caregivers, patients
Intended Use Environment	Home, clinical

This table provides a succinct summary of the product under review. Before even walking through the human factors data, a list of questions can be compiled based on the product and application. These questions relate to the use scenarios indicated in the product review table:

- Are the user groups subcutaneous injection experienced, or is this a new therapy and route of administration for some, or all, user groups?
- Are there other treatment options in this space that patients could be familiar with or may be prescribed as well?
- Have known use errors with lay-user injections been considered, such as premature removal resulting in an underdose, unaware of storage requirements, confusion with use of autoinjector, and confusion around the dosing schedule.

These and other product-specific considerations should be identified in the URRA and evaluated through a summative validation evaluation as appropriate.

Within this review, there is typically a section containing the "Reason for review" and regulatory history. This typically outlines and defines the question to be answered by the review. Additionally, if the summative validation evaluation protocol is submitted for DMEPA review, this will be referenced here.

Historically, the bulk of the review is a table containing an analysis of the data from the summative validation evaluation. This table includes use errors, close calls, and difficulties identified in the study, Sponsor root cause analysis, Sponsor discussion of mitigation strategies, and DMEPA's analysis and recommendation.

These reviews can provide some insight into the perspective of a regulatory or third-party reviewer as they work to evaluate HFE data and other supporting documentation to determine if there is sufficient evidence to support the regulatory question at hand.

COMMON MISTAKES THAT COST TIME DURING REGULATORY REVIEW

The following are common mistakes made by development teams that impact timelines during regulatory review.

1. Human factors engineering process is not considered during development.
   a. One variation of this is if HFE is not conducted at all. While there are products for which a summative validation evaluation is not necessary with a regulatory submission, this conclusion is based on an assessment of use-related risk and the absence of critical tasks. Therefore, this does not negate the need for the HFE process in development, and furthermore, it is the URRA output that supports this determination.
   b. Another variation is where HFE is only considered during validation activities. To run a summative validation evaluation without the process building blocks behind it (i.e., Context of Expected Use Document, URRA, etc.) will yield an incomplete study at best. At worst it will identify issues that need to be corrected, which could have been identified earlier in development.
2. If working on a CBER- or CDER-led combination product, not submitting the summative validation protocol for review. Given the regulatory perspective, mentioned above, DMEPA will often have feedback that will need to be incorporated into the study protocol. Some common items for feedback are user groups, training and training decay, and completeness and representativeness of the study. These are items that, if identified after study execution, would likely require a new study.
   a. Similarly, project risk can be incurred when the project schedule does not allow sufficient time for this review and the subsequent labeling updates, protocol updates, and potentially recruiting changes based on DMEPA feedback.

3. If working on a CBER or CDER led combination product, not submitting patient-oriented labeling for Division of Medical Policy Programs (DMPP) review during submission of the summative validation evaluation protocol for review. Per the FDA Manual of Policies and Procedures, procedures for DMEPA Intra-Center Consult to DMPP on Patient-Oriented Labeling Submitted with Human Factors Validation Study Protocols, were introduced with the goal to ensure efficient, effective, and consistent combination product development and review, as it relates to patient-oriented labeling, including instructions for use materials, for those drug-device and biologic-device combination products regulated by CDER and CBER [Ref 26].

COMBINATION PRODUCT HUMAN FACTORS CASE STUDIES

This section contains hypothetical case studies to describe some best practices in the application of the HFE process for combination products. The HFE process is used to demonstrate that the hazards and risks related to the use of combination products and their constituent parts have been reduced as far as possible, providing evidence of safe and effective use. It is also part of design validation which demonstrates the product design meets the user needs. HFE summative validation evaluation provides information about those user needs for the product and associated labeling and/or training to support safe and effective use. Further, HFE activities can be a helpful source of use-related design insights for the development of not just the current product but also future product generations and iterations.

The case studies below are representative of common challenges faced within the industry. When possible, a third-party reviewer's perspective will be considered, as examples of questions someone may initially have when walking through the data presented without context on the product beyond an HFE submission package. This list is NOT all-inclusive but is intended to demonstrate how one might look at the data from a fresh perspective.

The case studies are:

- Case Study 1: Human Factors Engineering in Late-Stage Development: Device Changes and Platform Strategies
- Case Study 2: Development of a Novel Drug-Device Combination Product
- Case Study 3: The Co-packaged Combination Product
- Case Study 4: Lifecycle Management Change
- Case Study 5: Emergency-Use Combination Product
- Case Study 6: New Delivery Modality
- Case Study 7: Device-Led Combination Product
- Case Study 8: Digital Health and SaMD Combination Product

CASE STUDY 1: HUMAN FACTORS ENGINEERING IN LATE-STAGE DEVELOPMENT: DEVICE CHANGES AND PLATFORM STRATEGIES

A manufacturer is developing a combination product utilizing a prefilled syringe. Formative studies have been completed for this product configuration. The manufacturer makes a decision late in the development to commercialize the product as a single-use autoinjector (AI), using the already developed PFS as the primary container for the drug.

In preparation for the Phase 3 clinical study, the development team needs to consider what additional HFE activities, including usability testing, are needed to ensure the autoinjector can be used safely and effectively in the clinical study. The primary container closure in direct contact with the drug (i.e., barrel, stopper, and needle) remains the same. The assembly process used during manufacture to assemble the final autoinjector requires additional steps after filling. The drug formulation remains the same. The route of administration is the same (subcutaneous).

The user interface is different (i.e., PFS vs. AI) and therefore requires the manufacturer to consider which additional HFE activities are needed to evaluate the use-related risk for the autoinjector.

Manufacturer Questions

Question 1: *Can the summative validation evaluation protocol review with the FDA happen in parallel with the start of the clinical study?*

For the start of the clinical study, it is necessary to demonstrate the safety of the autoinjector for use in the clinical study. Use-related risks should be considered in the context of an Investigational Device Exemption/Investigational New Drug (IDE/IND) submission. From an HFE perspective, this would be at a minimum the completion of a use-related risk analysis (URRA) to determine if, within the clinical study, any identified use-related risks are controlled. This assessment may indicate that an HFE study is necessary to demonstrate that the autoinjector user interface supports safe during the clinical study. If it is determined that the HFE summative validation evaluation can be conducted after the start of the clinical study, then the protocol review can happen in parallel, prior to the start of the HFE study.

Question 2: *The IFU needs to be changed to accommodate the new autoinjector. The team wants to prepare the IFU to be as close as possible to what the FDA is expecting to minimize project risk. Is there guidance that can be consulted to ensure that IFU best practices can be leveraged?*

It is important that the IFU is developed as much as possible before the summative validation evaluation is executed, to ensure that the product is tested with the user interface developed as close to the commercial configuration as possible. It should be acknowledged that the study material may not have the final product name and will have other markings as necessitated by the nature of the study, such as "for investigational use only." These should be identified and rationales for their impact on the study results be provided. It is also important to ensure learnings from formative studies are addressed. The following specifics should be considered:

- Description and pictation – include an overview of key parts of the product and packaging;
- Description of storage, preparation, use, and disposal;
- Font size and type – it must be clear and easy for users to read;
- Process description – each step must be clearly identified;
- Language used – it should be plain and easy to understand; and
- Country-specific requirements, for example, for disposal.

The FDA offers IFU review as part of the summative validation evaluation protocol review and has also recently published draft guidance related to IFUs for combination products: Instructions for Use — Patient Labeling for Human Prescription Drug and Biological Products and Drug-Device and Biologic-Device Combination Products [Ref 12]. This guidance includes recommendations on both format and content of an IFU for a combination product.

The project team should, in this case, leverage this guidance to ensure the IFU is developed in line with FDA expectations. It is important to note that some of the recommendations in the FDA guidance may be specific to the US market, and the IFU may have to be modified to align with other specific country guidance/regulations depending in which markets the autoinjector is intended to be marketed.

Question 3: *The development team was recently made aware that the autoinjector on their product is already on the market for a different drug product within the company. Can the manufacturer leverage the human factors data from the already marketed product for this new submission?*

The summative validation evaluation protocol review is recommended by the FDA [Ref 11]. If there are sufficient data to support safe and effective use based on the previously acquired data of the users, uses, and use environment, then it may be sufficient to support the submission. This can also be described as bridging.

The team should consult FDA draft guidance "*Bridging for Drug-Device and Biologic-Device Combination Products*" [Ref 6], and "*Contents of a Complete Submission for Threshold Analyses and Human Factors Submissions to Drug and Biologic Applications*" [Ref 9], and the European Medicines Agency's (EMA) "*Guideline on quality documentation for medicinal products when used with a medical device*" [Ref 19]. The bridging guidance provides examples of the type of data needed to justify a bridging strategy while a threshold analysis can be used as a more detailed tool to identify differences and similarities between the previous and new combination product using the same device. EMA's guidelines provide an overview of their expected documentation for an EU submission.

The project team should conduct the threshold analysis and decide if the products, users, uses, use environments, use errors, and harms are deemed similar enough that the data from the summative validation evaluation of the previous combination can be used to mitigate the risks associated with the use of the new combination. The threshold analysis should include:

- The user population and groups;
- The user interface, including the device and packaging;
- The use environment; and
- The user task, their criticality, as well as the instruction for use.

If there are no impacted or new critical tasks identified through the URRA in the threshold analysis, then a submission can be made to the FDA via the summative validation evaluation protocol submission process [Ref 9]. This will provide the team with feedback from the FDA, as well as ensure the FDA agrees in the conclusions drawn by the team. This exercise should be done as early as possible in the process, as review times are approximately 90 days. If there is disagreement, the team must plan for running further HFE evaluations or summative validation evaluations which can take time to plan and implement. For other regions, a similar approach to the bridging documentation and justification can be applied.

Question 4: *The team realizes that this autoinjector is intended to be platform technology for other products within the company. The team would like to streamline the development process to improve the quality and consistency of products coming to market for the company and want to ensure patients can receive valuable medicines more quickly. As a result, the team would like to develop the autoinjector following a platform approach. What considerations should the team explore to document their efforts and enable leveraging for future products?*

The team should make considerations when planning to conduct formative studies that can be leveraged across the platform to evaluate features that are independent of the drug product utilized, potential indications, and user groups.

- A platform-based risk analysis as well as a URRA should be generated to ensure the approach identifies the appropriate testing methodology to support risk reduction across various formulations, disease states, and patient populations.
- Consider the integration of varying injection locations – arm, leg, stomach, etc., if in scope for the portfolio.
- The range of disease areas for which the device could be used should be taken into consideration, e.g., diabetes, psoriasis, or inflammatory bowel disease patients.
- The test populations should be representative in that they can cover the various disease states, with provided rationale to be able to be leveraged independently once drug-specific populations are tested.
- Take into consideration the potential different formulations of multiple viscosity/glide force variations of the medications, ensuring a range of solutions are tested by users.

When it comes to utilizing the data generated for a different product within the portfolio, it will be important to create a gap analysis to support the rationale for leveraging the data. It is also recommended when applying a platform approach, to review your strategy with the FDA.

Question 4

Third-party perspective: *Taking this scenario to a subsequent submission for a new indication using the platform technology.* "The team has compiled the information and it is being submitted to a regulatory reviewer or other third-party reviewer. Using the example Review Product Summary table discussed earlier, what questions might his reviewer have?" (See Table 7.3).

TABLE 7.3
Case Study 1: DMEPA Review Product Summary

Relevant Product Information	
Initial Approval Date	N/A
Active Ingredient (Drug or Biologic)	____xxxx
Indication	Treatment of ____ in adults and pediatrics 2–18
Route of Administration	Subcutaneous
Dosage Form	Injection solution
Strength	120 mg/mL
Dose and Frequency	120 mg by subcutaneous injection at weeks 0, 2, and 4 and every 12 weeks thereafter
How Supplied	120 mg/mL prefilled autoinjector
Storage	Refrigerate at 2°C–8°C (36°F–46°F) in original carton to protect from light. Do not freeze
Container Closure/Device Constituent	Carton containing one single-use autoinjector and alcohol pad
Intended Users	HCP, caregivers, patients
Intended Use Environment	Home, clinical

Some questions that would need to be addressed include:

- For this indication, are the user groups subcutaneous injection experienced or is this a new therapy and route of administration for some, or all, user groups? Does data exist for this or an equivalent user group?
- Are there other treatment options in this space that patients could be familiar with or may be prescribed as well? Are there other treatments in this space that may cause differentiation or negative transfer issues that were not previously evaluated?
- Known use errors with lay-user injections – premature removal resulting in an underdose, unaware of storage requirements, confusion with the use of autoinjector, and confusion around dosing schedule. Is there anything unique about the dosing schedule for this indication that necessitates new data to demonstrate safe and effective use?

Case Study 2: Development of a Novel Combination Product

A manufacturer is developing a novel device system to deliver a new drug which is administered subcutaneously on a monthly basis. The concentration of the dose is such that the formulation is likely to be 3 mL per injection. The manufacturer therefore decides to develop a novel on-body injection device which contains a cartridge with the drug for injection. The device is a single-use disposable device.

The drug has been developed for the treatment of a disease which is considered very rare, with only 50,000 patients worldwide. Out of these patients, about 10,000 are living in the United States.

Question 1: *As the patient population is rare, what can the company do to ensure recruitment for the study?*

When conducting HFE testing on combination products, it is important to ensure the simulated tests are as close to the real-use situation as possible. This also includes the user population, and therefore, the patient group is key. However, when it is difficult to recruit due to the patient population being rare, other options can be explored.

Options for recruitment could include the following:

- Recruit from the actual patient population; and
- Recruit from surrogate patient populations with similar characteristics that may impact use (e.g., similar dexterity profile as the actual patient population).

The priority should always be to recruit from the actual patient population. As it is likely that many iterative rounds of HFE activities will be needed with a novel device, it is important to ensure that it will be possible to continue to have access to these patients, especially for the summative validation evaluation. It is therefore an advantage to identify surrogate patient population who may be larger, and could be used in formative studies, or who could be recruited if it is not possible to recruit the full number of actual patients for the summative validation evaluation.

To identify potential surrogate patient populations, the project team should consult with physicians who are experts in the actual disease state and can help characterize the specifics of the patient population which will impact the user interface between the device and the user. This could include, but is not limited to, the following:

- The patient's perception – Does this patient population have affected hearing, sight, touch, smell, taste, and proprioception?
- The patient's cognition – Does the patient population have impacted cognitive abilities, including memory?
- The patient's mobility – Does the patient population have limited mobility, are they likely to be bedbound? Do they have any inability to use their extremities?
- The patient's strength – Does the patient population have any reduction in strength, stamina, or dexterity? Examples include grip strength, tremor, and ability to sustain an activity for several minutes.

Once the patient-specific characteristics have been identified, the experts can list other patient populations that may be able to function as surrogates.

For example, a surrogate population has been identified with rheumatoid arthritis (RA) which has a much larger patient population. The project team therefore decides to take the following strategy for recruitment:

> For formative studies, they aim to recruit at least 20% of the actual disease patient population and supplement the rest of the study with RA patients. For the final summative validation evaluation, they aim to recruit as many actual patients as possible with a minimum of 50% actual patient population and supplement with RA patients for the remaining participants.

This strategy ensures that enough patients are tested to support the study results while at the same time ensures that actual patients are available to use the product naively at each state of the design process.

It is strongly recommended, prior to conducting the summative validation evaluation, to review surrogate strategy with DMEPA through a protocol review submission.

Question 2: *The company is working with an external consultant who will be running the HFE studies. What should the company consider when working with the consultant?*

To facilitate the incorporation of HFE within design processes, some companies have chosen to bring HFE expertise in-house rather than outsourcing this skill set. Guidance and HFE services, ranging from study facilitation to coordinating the entire HFE process, are also available from design houses and independent HFE consultants and facilitators. One consideration when using an external group is who will hold the intellectual property (IP) if new design considerations are established as a result of the study and the implications of that decision. It is recommended that IP ownership be established during the contracting process with external groups. (Note: As part of Purchasing Controls, under 21 CFR 820.50, via the quality agreement and/or contract, the Sponsor should also ensure that such third-party consultants and facilitators are assessed and reflected on the Sponsor's Approved Supplier List, as they are performing activities that support design validation.)

When planning HFE activities, it would be an advantage for the project team to have a planning document (project plan, HFE plan, etc.). This is especially important when working with an external facilitator to ensure that both expectations and responsibilities are aligned. In this document all the elements of planning for HFE activities, including studies, can be captured as well as identify responsible parties, timing, and communication expectations. The document can also be used to clarify nomenclature and expectations. This will allow for running activities smoothly as well as functioning as a communication document.

Question 2, Third-party perspective: "*Taking this scenario further, the team has compiled the HFE data and submitted it within their NDA package. In the HFE data submitted, the report captures all the background and context information, reviews formative learning, and provides detailed results of potential user errors, close calls and difficulties identified in the study. These were not root caused, and some recommendations were made. What questions might a regulatory or third-party reviewer have?*"

This is an incomplete report. It is not clear how the findings play into the overall risk and if based on this what changes may be necessary. What recommendations were implemented, with what rationale for why or why not?

In an HFE submission, the data for the HFE process need to be documented. While much of this can be conducted by a consultant or contractor, the integration back into the risk process and discussion of risk acceptability and benefit–risk considerations needs to be identified within the larger product context. It is important that the submission does not only contain an HFE report, but the full analysis and rationale are necessary.

Question 3: *The project team discovers a problem with a formulation of the drug and makes a late-stage change to one of the excipients. How does the manufacturer decide if the HFE work is impacted by this change and what specific actions could they take to mitigate the risk?*

A change to the formulation may not seem like a change that could impact HFE activities, and manufacturers may not be aware that the impact of such a change should be considered on the combination product as a whole. However, it would be a mistake not to assess the impact, as there may be several implications of the change related to the handling of the product, even when the change does not alter the device constituent design of the immediate user interface.

The project team should consider a number of things including:

- Viscosity: The change in the formulation may have an impact on the drug viscosity. In this case, the following should be considered in relation to the HFE activities. If the formulation is mechanically expelled from the product, then a change in viscosity may change the delivery time. A change in delivery time will change the user interface for the user.
- Light sensitivity: The change in the formulation may increase the light sensitivity of the product which could mean that the IFU would have to be updated with this new information.

In this case, it is important to ensure through the HFE activities that users can understand this when reading the labeling.
- Storage: The change in the formulation may require a change in how the user stores the combination product. Again, it should be tested through the HFE activities that the users can understand this important information with reading the labeling.

It is important to highlight that this example only considers potential HFE impacts of the change, other evaluations should be conducted in relation to such a change.

Case Study 3: The Co-packaged Combination Product

Manufacturer is developing a solution for injection which will be marketed in a vial and will be administered either by a lay-user caregiver in a home setting or by a healthcare professional (HCP) in a clinical setting. To aid the injection, the manufacturer intends to co-package an off-the-shelf syringe with the vial.

Question 1: *The project team is planning to conduct HFE studies to ensure that caregivers can use the co-packaged product. However, they are uncertain if they should extend the HFE studies to the HCP population as well as if the product is clinically administered by trained professionals.*

It can be difficult to decide if an HCP group should be included in an HFE evaluation, as they are trained professionals and should have knowledge of injection. However, there may be several factors which impact if the product needs to be considered for further investigation, and there is little guidance.

The project team must consider if they are introducing any new risks related to the tasks and potential use errors with their co-package versus the standard of care. To do this, the team must first characterize the uses, users, and use scenarios. It may be, for HCPs, that the introduction of this co-packaged product does not produce new tasks or new scenarios of use due to their training and standard practice. However, for the lay-user caregiver, this would not be the case. Within the process defined earlier in this chapter, this research could be documented in the Context of Expected Use Document.

The next step is to work through the process steps to generate the URRA. This detailed assessment can help identify if there are any tasks specific to the use of this product that are not standard practice for the HCP user group. Additionally, this analysis will identify those tasks that are critical for the lay-user caregiver group.

Question 2: *The company sends the summative validation evaluation protocol to the FDA for feedback and asks to change the moderator script as it is suggested this is guiding the participant too much. How may the company change the script to ensure that the test participants have the appropriate context?*

There is little guidance available on how to construct moderator scripts when running an HFE study. The challenge is to ensure participants in the study get the information they need to be able to understand why they are there and what the purpose of the study is, while not leading them to conclusions or introducing instructions which may not be present in a real-life situation. However, when untrained user groups are tested without being provided with enough context, study artifacts may arise which are also not reflecting a real-life situation. Therefore, the team must be careful when updating their moderator script.

When composing the script, the team should focus on avoiding the wording that it may be leading. For example, an introduction could be changed from:

> I would like you to open the pack, take out the information leaflet, and read it carefully. Then you can prepare for the injection and let me know when you are ready.

to

> I would like to you do to whatever you would normally do with the product before you would use it for the first time as you would if you had been given this by your doctor. When you are ready to use the product let me know.

The difference between the two options above is that the first statement asks the participant to take specific actions, i.e., reading the information leaflet. In a real-world situation, the users may not do that naturally, nor would there be someone instructing them to review the instructions before every use; therefore, it would be leading to instruct the users to do so. It is more appropriate to ask the participants to do whatever they would normally do if it was their first time with the product in a real-use situation.

Question 2, Third-party perspective: *The team has submitted the protocol and moderator script to include lay-user caregiver training and a one-hour training decay. Using the example "Review Product Summary" table discussed earlier, what questions might his reviewer have?* (See Table 7.4).

TABLE 7.4
Case Study 3: DMEPA Review Product Summary

Relevant Product Information	
Initial Approval Date	N/A
Active Ingredient (Drug or Biologic)	____ xxxx
Indication	Treatment of ____ in adults and pediatrics 2–18
Route of Administration	Subcutaneous
Dosage Form	Injection solution
Strength	120 mg/mL
Dose and Frequency	120 mg by subcutaneous injection at weeks 0 and 4 and every 12 weeks thereafter
How Supplied	120 mg/mL prefilled autoinjector
Storage	Refrigerate at 2°C–8°C (36°F–46°F) in original carton to protect from light. Do not freeze
Container Closure/Device Constituent	Carton containing one co-packaged vial and syringe
Intended Users	HCP, caregivers
Intended Use Environment	Home, clinical

The goal of a summative validation evaluation is to represent the usage, user, and use environment. With just these elements identified, there may be a few questions:

- How will training be provided to every lay-user caregiver? If the expectation is that the prescribing clinician will train this user group or that the pharmacist will train if asked, then this does not ensure training. It would be acknowledged that, in practice, training may or may not happen. In this case, untrained users are a user group.
- However, perhaps the Sponsor (manufacturer) plans to put a training program in place that will deliver training tied to specialty pharmacy distribution in the US. The materials and modality for this program should be identified as part of the expected use scenarios and discussed in how this will be implemented in the summative validation evaluation to demonstrate representative use.
- With the delivery schedule for this indication, a few questions arise around the training. In this scenario, the initial doses are 4 weeks apart and subsequent doses are 12 weeks apart. If training is required, will users remember the training for these durations? Will a one-hour

training decay represent these durations? There is unfortunately little research in this area, and this is one of the areas where the FDA is funding ongoing research to inform the regulatory science. In the meantime, within the discussion around the establishment of representative use, any training and training decay should be discussed and rationale provided.

Case Study 4: Lifecycle Management Change

A manufacturer has a dry powder inhaler on the market for the treatment of acute asthma. As part of post-market surveillance (PMS), the manufacturer receives information on several medication errors, the rates of which have increased after several years of low error reporting. Patients and healthcare professionals (HCPs) report that the device is not working once the first dose has been taken, leading to the patients needing medical intervention. The manufacturer initiates investigations on the devices and finds that the devices work as intended, despite the report of the error. The only change they have been able to identify, which had been introduced just before the incidents of the medication errors, was a post-market change to the labeling. As part of an environmental review of the packaging, the IFU was changed from a one-sided leaflet to a two-sided leaflet to use less paper. The manufacturer determines that while this change was expected to have no impact on use, it is a potential reason for the increase in medication errors.

Question 1: *What should be the first priority for the manufacturer?*

First the manufacturer must find out why the error occurs by running an HFE study to understand what the impact of changing the format of the IFU has on the use. The manufacturer runs two separate studies: (1) participants are given devices and asked to follow the instructions to administer and (2) the participants are asked to first review the IFU and then asked to administer. The manufacturer discovers that a use error occurs when the device is used at the same time as reading the IFU for the first time. A critical last step – closing the device correctly, which resets the next dose – is on the back side of the new IFU. When using the inhaler, the participants did not have a free hand to turn the leaflet over, which resulted in them missing the step during the use scenario. This use error did not happen in the group which had the opportunity to read the IFU before using the inhaler. The manufacturer therefore concludes that the root cause of the use error is the new IFU layout, which had only been tested as part of a readability study before being placed into production.

Question 2: *What can the manufacturer do to rectify the error?*

The manufacturer has found the root cause and updates the IFU to ensure all user steps are on the same page of the IFU. To confirm they have resolved the problem, they repeat the previously performed study with the updated IFU to ensure the mitigation has fixed the problem.

A Corrective and Preventive Action (CAPA) needs to be opened in order to address this issue, and also to address their labeling design change control process, as fundamentally, this HFE activity should have been performed prospectively, not reactively. The manufacturer needs to address these points to avoid similar incidents of labeling changes with inadvertent usability impacts. Human factors engineering and regulatory review and approval need to be added as an element of the change control process. As part of the corrective action, a review of any other labeling updates has been made needs to be performed to determine if they had human factors and regulatory review as part of the label change control. Any additional discoveries should have the same investigations performed as were done with this specific instance.

Case Study 5: Emergency-Use Combination Product

A manufacturer has been designing injection devices for many years for maintenance medication. A new growth opportunity for the manufacturer has been initiated, and the company will be developing medication to treat anaphylactic shock, i.e., an acute-care medication. The development team decides they can base the new combination product on their existing injector platform. However, what the team does not realize is that the users, uses, and use environment for this new indication are significantly different. One of the issues that is discovered late in the development is that, in the actual use situation, it is not practical for users to consult the IFU. It is therefore important that the injector is as intuitive as possible. The manufacturer explores options of how to best provide the information to the user, considering the user may be a complete stranger who knows nothing about using the injector or a teacher in a school class with children around.

Question 1: *What aspects should the manufacturer focus on in the HFE activities to develop the most intuitive injector?*

The manufacturer may be able to base the design of this new injector on their existing platforms. However, changes may be necessary. The manufacturer should execute the full HFE process to evaluate this new indication. This starts with defining the users, use scenarios, and use environments for this indication. For example, these patients may have the injector for months or years before they need it. How do they need to store it until they need it? How do they store their current devices? Does this match? Who are the different users of this injector – trained users, untrained users, HCPs? This activity can help the manufacturer identify the user needs and product requirements for this indication. In conjunction with this, the manufacturer needs to use the learned information to update the use-related risk analysis for this indication. This work could lead to design activities as risk mitigations such as looking at the number of tasks needed to operate the injector and aiming to minimize these. For example, if the injector has a button which users must actively press to start the injection, this could be designed out in favor of press-on-skin activation mechanism to start the injection. Different colors and other indicators can be used to ensure it is clear which end of the injector has the needle. Graphics with use steps could be incorporated in the design or as part of the injector label to better guide the user. Different versions should be tested through a series of formative studies, taking into consideration the identified user groups, including untrained lay people, HCPs, and users with injection experience. It is often those users who think they know what they are doing based on prior knowledge who are most likely to have use errors.

Question 2: *A URRA is developed, and the injector design is optimized – is it ready for summative validation evaluation?*

It is important to remember that the user interface of the injector encompasses the injector itself, labels, instructions for use, training, packaging, etc. In the design optimization, all steps of use must be evaluated, and all user interface elements are evaluated to determine if they support safe and effective use. This includes the information available to the patient before and during the use scenario. In the case of an emergency-use product, it is not possible to rely on the user having access to any of the information (i.e., paper leaflets) which has been supplied with the injector in its packaging, as the patient may have removed the injector from its packaging to carry it around. Therefore, one solution is to have the instructions incorporated in the label on the injector, as stated above. This would aid in reducing use errors if the injector was used by someone without injection experience, or in a situation of panic where the users may forget their prior training or review of the instructions. The instructions must be carefully designed to ensure the information is adequate and understandable without leaving any critical tasks out. The instructions need to be appropriately sized to ensure it is possible for the user to interpret it. Text may not be ideal, whereas simple graphics or even incorporating technology (e.g., voice-guided use) may present a better interface. These interface designs should be evaluated in formative studies with different user groups in different environments to ensure that as many potential use errors as possible are mitigated.

Question 2, Third-party perspective: *The Sponsor executed the HFE process up to the point of creating a summative validation evaluation and submitted this data to the FDA for protocol review. What are some initial questions that a reviewer might have?* (See Table 7.5)

TABLE 7.5
Case Study 5: DMEPA Review Product Summary

Relevant Product Information	
Initial Approval Date	N/A
Active Ingredient (Drug or Biologic)	____ xxxx
Indication	Emergency Treatment of anaphylactic shock in adults and pediatrics 2–18
Route of Administration	Intramuscular
Dosage Form	Injection solution
Strength	0.15 mg/0.15 mL
Dose and Frequency	• Weight 15 to <30 kg: 0.15 mg; may repeat dose q5–15 min • Weight ≥30 kg: Two injections of 0.15 mg; may repeat dose q5–15 min
How Supplied	0.15 mL autoinjector
Storage	Refrigerate at 2°C–8°C (36°F–46°F) in original carton to protect from light. Do not freeze
Container Closure/Device Constituent	Carton containing two 0.15 mL autoinjectors
Intended Users	HCP, caregivers, patients, lay-users
Intended Use Environment	Home, clinical, community

Some questions that would need to be addressed in the HFE deliverables:

- Does the HFE data and subsequent URRA consider all user groups and use environments in the identification of critical tasks?
- There are user groups that have intramuscular (IM) injection experience and those that are IM injection naïve. Does this impact safe and effective use? If so, how are these user group characteristics accounted for?
- Are there other treatment options in this space that patients could be familiar with or may be prescribed as well? In Table 7.5, it indicates that for patients weighing ≥30 kg, two injections will need to be delivered. Is this different from other therapies on the market? Conversely if two products are marketed, a 0.15 mL and a 0.3 mL, are there any concerns about differentiation to users?
- Known use errors with lay-user injections – are these identified in the URRA and, if critical, accounted for in the summative validation evaluation study? For example, premature removal resulting in an underdose, unaware of storage requirements, unaware of acceptable injection sites?

Case Study 6: New Delivery Modality

A manufacturer is developing a new delivery modality for the treatment of insulin-dependent diabetics. Insulin administration is integral to the care and management of patients living with diabetes. The condition is chronic, with treatments daily up to multiple times a day. Through the years, there have been innovations to improve and modernize syringes, but ultimately the fear of injection is still a key detractor from patient compliance. Addressing this fear is a key user need.

The manufacturer decides to develop a product that would use inhaled insulin, encouraging patient compliance by addressing needle phobia.

Question 1: *How should the manufacturer ensure that the efficacy of the product was not compromised by use errors associated with a new delivery modality?*

When introducing a new delivery modality, the users perform different tasks with an inhaled medication compared to injecting. It is essential that preclinical HFE testing be executed to evaluate that users can perform the appropriate delivery technique to deliver the appropriate dose.

Question 2: *What HFE activities need to be taken into account when assessing patient needs when considering the design?*

Although it appears that an inhalation approach may address the patient's desire to avoid injections, it is important to plan HFE activities to define the new use scenario. These activities should be performed early in the development lifecycle to best understand the user needs and user groups who would be using this product. This research can then be iterated throughout development. For example, imagine the initial prototypes are large, awkward, and not discreet for patients to handle. What new tasks and potential use errors are associated with this new user interface configuration, and can they be addressed with further design modifications? Do they overall reduce or increase the risk associated with use?

It is important when pursuing a new delivery approach to consider previous results and known risks to (1) implement known learnings and (2) avoid making the same mistakes. A literature search should be conducted to ensure all known use-related risks have been assessed. These can also be found on some agencies databases (e.g., FDA MAUDE database [Ref 30]). Any previous products/studies and known use problem for the type of device should be addressed. For example, if a potential use issue (e.g., difficulty pushing a button) is identified, the product should address potential risks associated with this use issue and identify how this use issue will be examined during the HFE activities.

Performance considerations should be evaluated for the user groups, user interface, and use environment. Per FDA guidance on human factors [Ref 5], specific attention should be paid to:

- Physical size and strength;
- Physical dexterity, flexibility, and coordination;
- Sensory abilities (i.e., vision, hearing, tactile sensitivity);
- Cognitive abilities, including memory;
- Medical condition for which the device is being used;
- Comorbidities (i.e., multiple conditions or diseases);
- Literacy and language skills;
- General health status;
- Mental and emotional state;
- Level of education and health literacy relative to the medical condition involved;
- General knowledge of similar types of devices;
- Knowledge of and experience with the device;
- Ability to learn and adapt to a new device; and
- Willingness and motivation to learn to use a new device.

Case Study 7: Device-Led Combination Product

A manufacturer is developing a drug-coated balloon which will be marketed in a single-use kit that will be administered by trained cardiologists. The development team plans to perform a set of clinical studies.

Question 1: *As the user group of the combination product will have advanced training as cardiologists and will also be trained on the product prior to use, will HFE activities be necessary?*

It is recommended to still perform human factors activities. Considering the HFE process, the foundation of this analysis should be a robust assessment of use-related risk. For example, if the product utilized a new sizing technique that is critical because mis-sizing could lead to tissue damage. The use of this new sizing technique may necessitate further evaluations.

These activities would be in addition to the clinical studies. However, it would be advantageous to perform a round of formative testing prior to the clinical studies to ensure there are no parts of the UI which require further development. Finding UI design flaws during clinical studies are expensive discoveries, as they can cause the study to need to be repeated or put on hold, depending on the issue.

Additionally, based on the learning from the assessment of use-related risk, the training programs may need to be assessed through a summative validation evaluation. In such a case, the training program tested should be the same as the one expected for the commercial product, with representative trainers, training environment, training decay, and subsequent product usage evaluation.

Case Study 8: Digital Health and SaMD Combination Product

A manufacturer is developing a medication delivery system integrated with an implantable sensor which tracks blood pressure and a receiver that captures the information and processes the results through an artificial intelligence and machine learning (AI/ML) algorithm to produce and track medication recommendations. The connected on-body drug delivery device administers the recommendations automatically, once acknowledged on the integrated graphical user interface. The primary user of the system is the patient, under the prescription of their clinician.

Question 1: *The team has identified components of this system into an architecture where they are proposing the following categorizations:*

- The AI/ML algorithm as software as a medical device (SaMD);
- A monitoring display that captures the status of the product and displays it for third-party review; the display is not considered to be a medical device;
- The implantable sensor is a Class III medical device; and
- The on-body device is a revision of their current marketed combination product, adding the software and interface necessary to integrate to this system.

What are the human factors questions associated with the SaMD component, if any? How about the implanted sensor?

The development team may feel that neither the SaMD portion of this system nor the implantable sensor has a user interface and thereby feel that HFE is not necessary for the lay-user patient. However, in working through the HFE process, these components need to be considered through the framework of use-related risk. First consider the sensor – the output is information. The applicable hazards and hazardous situations are associated with this information being incorrect. Now consider the AI/ML algorithm – similarly the primary output of these systems is information. The HFE challenge with this system is interpreting the output, i.e., recommendations from the system. In walking this through the HFE process: the user group is lay-users, and the task is acting on the information presented. The potential use errors associated with this task are (1) not understanding the information and therefore taking incorrect action or (2) overreliance on the information when there is an indication that the results are inaccurate (i.e., the patient's symptoms do not match the information presented). Another way to state this is the task is deciding to trust the system output and knowing when to not trust the system output. This requires that the user has a proper mental model about how the recommendations are made and why. This can be fostered by an appropriate level of transparency in how the system has arrived at the recommendation so that the lay-user can make informed decisions when necessary.

This is a very complex system with many layers of questions around safety and effectiveness; however, the intent of this example is to illustrate that regardless of the system complexity, the HFE process scales accordingly.

CONCLUSION

This chapter has provided an overview of combination products human factors engineering, along with a number of case studies. These case studies highlight some of the considerations and challenges manufacturers of drug/biologic-device combination products may face when executing the HFE process at different stages in the product lifecycle, which in turn aids in the demonstration of the minimization of use-related risk as far as possible. Using the HFE process to supplement the development process, including risk management, manufacturers can investigate not only the device constituent but also the product labeling and the use environment of the combination product. It is possible to predict and eliminate a number of potential use errors within the design of the product. HFE is therefore a core process and technical skillset any manufacturer of combination products should ensure they have access to during the development of a new combination product and throughout the product's lifecycle, as seen in Case Study 4, when introducing seemingly minor changes to any of the interfaces between the products and the users.

Globally, regulatory bodies have identified the need for data around the safe and effective use of products, and HFE data review has increased accordingly. HFE has been identified as essential in ensuring patients are provided with safe and effective combination products and needs to be a part of the lifecycle process from the beginning of the combination product development program. By using HFE programs, manufacturers can develop product user interfaces that promote safe and effective use. The FDA highlights the importance of anticipating use errors through investigational design in their 2016 guidance on "Safety Considerations for Product Design to Minimize Medication Errors" where they write:

> Drug product design features that predispose end users to medication errors may not always be overcome by product labeling or health care provider or patient education. It is therefore preferable to eliminate, or minimize to the extent possible, these hazards from the product design. It is not possible to predict all medication errors; however, medication errors can be minimized by conducting premarketing risk assessments to evaluate how users will interact with the drug product within various environments of use within the medication use system, using well-established human factors engineering analytical methods. [Ref 15]

Though regulations and guidelines are being developed to help manufactures conduct HFE activities, there are still areas of discussion needed to ensure the HFE process works as intended. It is important to stay current with ongoing research, some of which is funded by the FDA, solicited from industry, academia, and other government agencies, to better understand the breadth of innovative scientific and technical solutions available to solve difficult regulatory science problems [Ref 27]. The FDA is continuing to update draft guidance, along with providing new guidance, such as the December 2022 released draft guidance titled "Content of Human Factors Information in Medical Device Marketing Submissions" [Ref 32]. It is recommended to regularly check the FDA's guidance website to stay up to date on the current guidance [Ref 33].

It is crucial that combination product developers use the HFE process integrated into the design process. This will ensure that products are optimized at the design level before they reach the market. The goal of this chapter has not been to provide all the details involved with the HFE process, as such information is readily available through guidance documents and international standards (listed in the Further Information section). The intent, however, is to provide the reader with an overview of the framework and purpose of the HFE process, the regulatory framework around the submission of HFE material to the FDA, as well as the types of considerations that should be considered throughout the development and lifecycle management.

FURTHER INFORMATION

International Standards

1. ANSI/AAMI HE75:2009/(R)2018: Human Factors Engineering – Design of Medical Devices.
2. IEC 62366-1:2015/AMD-1:2020 Medical Devices – Part 1: Application of Usability Engineering to Medical Devices.
3. IEC TIR 62366-2:2016 Medical Devices – Part 2: Guidance on the Application of Usability Engineering to Medical Devices.
4. ISO 14971:2019 Medical Devices – Application of Risk Management to Medical Devices.

FDA Guidance

5. Applying Human Factors and Usability Engineering to Medical Devices (Final, February 2016).
6. Bridging for Drug-Device and Biologic-Device Combination Products (Draft, December 2019).
7. Comparative Analyses and Related Comparative Use Human Factors Studies for a Drug-Device Combination Product Submitted in an ANDA (Draft, January 2017).
8. Considerations in Demonstrating Interchangeability with a Reference Product (Final, May 2019).
9. Contents of a Complete Submission for Threshold Analyses and Human Factors Submissions to Drug and Biologic Applications (Draft, September 2018).
10. Current Good Manufacturing Practice Requirements for Combination Products (Final, January 2017).
11. Human Factors Studies and Related Clinical Study Considerations in Combination Product Design and Development (Draft, February 2016).
12. Instructions for Use – Patient Labeling for Human Prescription Drug and Biological Products and Drug-Device and Biologic-Device Combination Products – Content and Format (Draft, July 2019).
13. Labeling for Biosimilar Products (Final, July 2018).
14. Requesting FDA Feedback on Combination Products (Final, December 2020).
15. Safety Considerations for Container Labels and Carton Labeling Design to Minimize Medication Errors (Draft, April 2013).
16. Safety Considerations for Product Design to Minimize Medication Errors (Final, April 2016).
17. Technical Considerations for Demonstrating Reliability of Emergency-Use Injectors Submitted under a BLA, NDA or ANDA (Draft, April 2020).
18. Technical Considerations for Pen, Jet, and Related Injectors Intended for Use with Drugs and Biological Products (Final, June 2013).

Other References

19. European Medicines Agency. Guideline on Quality Documentation for Medicinal Products When Used with a Medical Device. 2021.
20. Hoste, Shannon. Back to Basics – Human Factors Engineering Process for Medical Device Products. 2022. https://www.agilisconsulting.com/agilis-blog/2022/jan/human-factors-engineering-process.
21. In Vitro Diagnostic Regulation. Regulation (EU) 2017/746.
22. International Ergonomics Association. What is Ergonomics? https://iea.cc/what-is-ergonomics/.
23. Medical Device Regulation. Regulation (EU) 2017/745.
24. National Coordinating Council for Medication Error Reporting and Prevention. NCC MERP Taxonomy of Medication Errors. 2001. https://www.nccmerp.org/sites/default/files/taxonomy2001-07-31.pdf.
25. Nguyen, QuynhNhu (FDA). Product Quality Research Institute. 4th FDA/PQRI Conference on Advancing Product Quality. "The Role of Human Factors Engineering in Combination Product Post Approval Changes" Presentation. 2019. https://pqri.org/wp-content/uploads/2019/04/1-PQRI-April-2019_CDER-HF-Presentation_QNguyen.pdf.
26. U.S. Food & Drug Administration. FDA Staff Manual Guides, Volume IV – Agency Program Directions. Combination Products. Inter-center Consults for Review of Human Factors Information. 2019. https://www.fda.gov/media/120204/download.
27. U.S. Food & Drug Administration. Manual of Policies and Procedures. Center for Drug Evaluation and Research. Office of Surveillance and Epidemiology. Procedures for DMEPA Intra-Center Consult to DMPP on Patient-Oriented Labeling Submitted with Human Factors Validation Study Protocols. 2019. https://www.fda.gov/media/131008/download.
28. U.S. Food & Drug Administration. Regulatory Science Extramural Research and Development Projects. 2021. https://www.fda.gov/science-research/advancing-regulatory-science/regulatory-science-extramural-research-and-development-projects.
29. U.S. Food & Drug Administration. Working to Reduce Medication Errors. 2019. https://www.fda.gov/drugs/drug-information-consumers/working-reduce-medication-errors.
30. U.S. Food & Drug Administration. Drugs@FDA: FDA-Approved Drugs. https://www.accessdata.fda.gov/scripts/cder/daf/index.cfm.
31. U.S. Food & Drug Administration. MAUDE – Manufacturer and User Facility Device Experience. https://www.accessdata.fda.gov/scripts/cdrh/cfdocs/cfmaude/search.cfm.
32. Content of Human Factors Information in Medical Device Marketing Submissions (Draft, December 2022).
33. U.S. Food & Drug Administration. Search for FDA Guidance Documents. https://www.fda.gov/regulatory-information/search-fda-guidance-documents.

8 Combination Products Post-market Lifecycle Management

Susan W. B. Neadle and Khaudeja Bano

CONTENTS

INTRODUCTION: LIFECYCLE MANAGEMENT

Earlier chapters of this book have focused heavily on best practices throughout design and development to proactively ensure safety, efficacy, and usability of combination products. This chapter turns our attention beyond product development considerations to the commercialization and post-market phase of the combination product lifecycle, referred to here as "Lifecycle Management." We review production and post-production activities and controls to ensure the ongoing safety and effectiveness of the combination product once it is placed on the market (Figure 8.1). Key aspects and best practices for combination product lifecycle management that we'll review in this chapter include:

- Design Transfer, Process Validation/Continued Process Verification;
- Change Control for Post-market Modifications;
- Design History File Maintenance; and
- Post-market Vigilance, Surveillance, and Reporting.

DOI: 10.1201/9781003300298-9

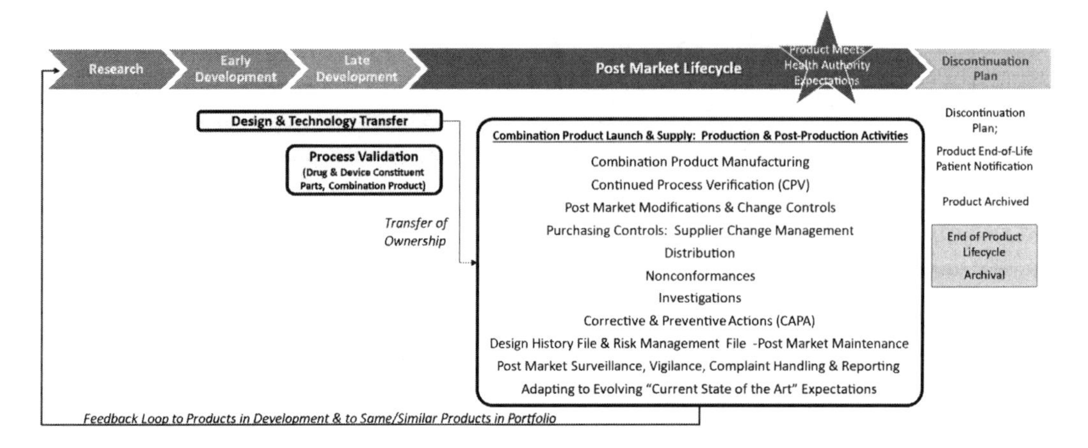

FIGURE 8.1 Combination products lifecycle management (©2022 Combination Products Consulting Services, LLC. All Rights Reserved.)

DESIGN TRANSFER (820.30(H)/ISO 13485:2016 CLAUSE 7.3.8)

The Sponsor has successfully worked its way through the product development process, proactively applying a risk-based systems approach to vet and address potential interactions of the drug/device constituent parts for the specific combined use application and proactively ensuring the safety, effectiveness, and quality of the combination product. The team is finally ready for commercialization!

The design and development process[1] transitions to commercial manufacturing via Design Transfer,[2] ensuring that the constituent parts and combination product design are correctly translated into production specifications (Figure 8.1). Design Transfer must be initiated so that production, or production-equivalent, units are the basis of Design Validation.

Per Preamble Comment #81 to the Quality System (QS) Regulation, when equivalent device constituent parts are used in the final Design Validation, the manufacturer needs to document how the device was manufactured, and how the manufacturing is similar to, and possibly different from, initial production. Where there are differences, the manufacturer must justify why Design Validation results are valid for production units, lots, or batches. A detailed justification considers all aspects of the Device Master Record/Tech File (see Chapter 4), including:

- Describing whether each aspect was in place and used to produce the device constituent part/combination product:
  - Components/raw materials specifications and suppliers;
  - Acceptance activities (incoming and final);
  - Inspection/measurement/test equipment;
  - Assembly/manufacturing process specifications/procedures;
  - Manufacturing equipment;
  - Device History Record(s);
  - Production, inspection/test, and installation personnel;
  - Storage and handling (electrostatic discharge (ESD), frozen/refrigerated, etc.);
  - Installation process instructions (if applicable).
- For each difference, explain the impact on the validity of the results.

During Design Transfer, all design and process specifications released to production get reviewed, approved, verified, and validated against pre-established acceptance criteria before they are implemented as part of the production process. Design Transfer is not a point in time. It is a transition period spanning across much of development, marked by a final stage of development intended to ensure all design outputs are adequately transferred, closing out the Design Transfer process (Figure 8.2).

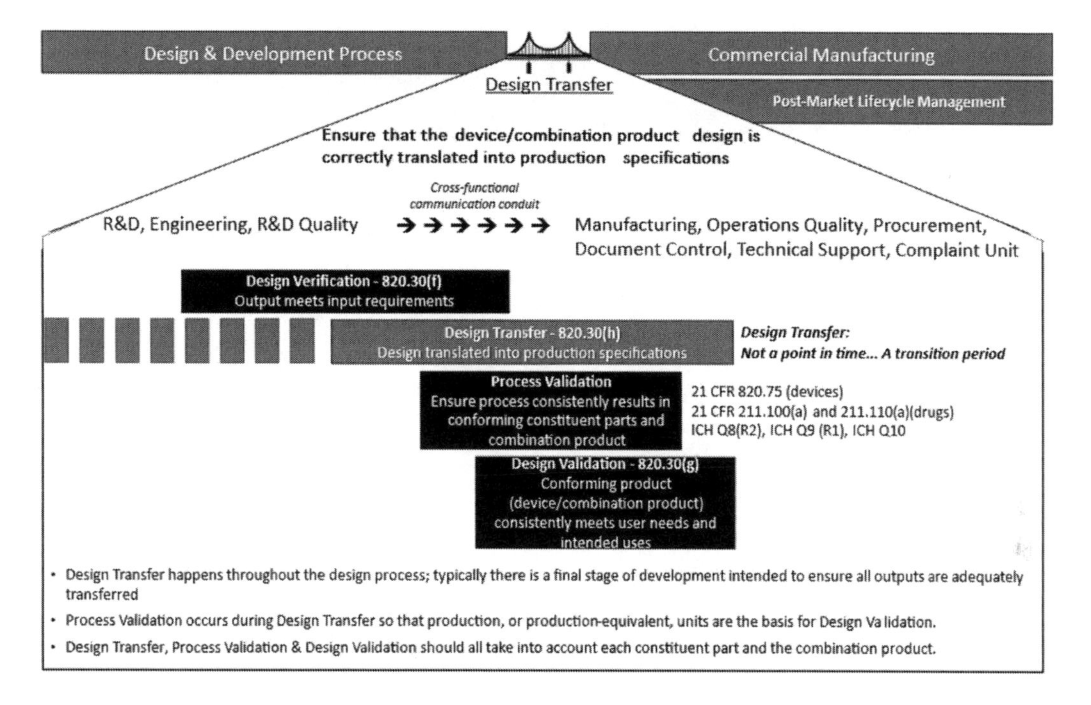

FIGURE 8.2 Design Transfer – a transition period (©2022 Combination Products Consulting Services, LLC. All Rights Reserved.)

Design Transfer serves as a conduit for cross-functional communication, including transfer activities reflected in the overall Design and Development Plan, or else following an explicit Design Transfer Plan, checklist, or protocol (an example of a Design Transfer Checklist is included in Chapter 4). A Design Transfer Plan generally includes the purpose, scope, site, roles and responsibilities, and specific deliverables related to the product. Additional deliverables during Design Transfer include:

- Finalized documents related to manufacturing, specifications, testing, and release;
- Bill of Materials (BOM);
- Device Master Record that includes the required drug constituent part information or a Master Batch Record that includes the required device constituent part information;
- A Trace Matrix and/or collection of documents needed to prove that all design activities up to Design Transfer were conducted as described in the Design and Development Plan;
- Agreement between the Design Team and the Manufacturing Teams on all activities that need to be completed to close Design Transfer;
- A Design Transfer Report, summarizing the results of all Design Transfer activities, with a conclusion stating if the Design Transfer was (or was not) successful;
- At the conclusion of Design Transfer, Design Validation will be able to assess if the design meets user needs and intended uses.[3]

R&D, Engineering, Clinicians, Quality Assurance, Manufacturing, Purchasing/Procurement, Technical Support, Logistics, Document Controls, and Post-market Complaint and Vigilance Units (among others) come together to ensure clarity and successful transition into full production. Design Controls begin with a focus on the customer and their needs, and through Design Transfer activities,

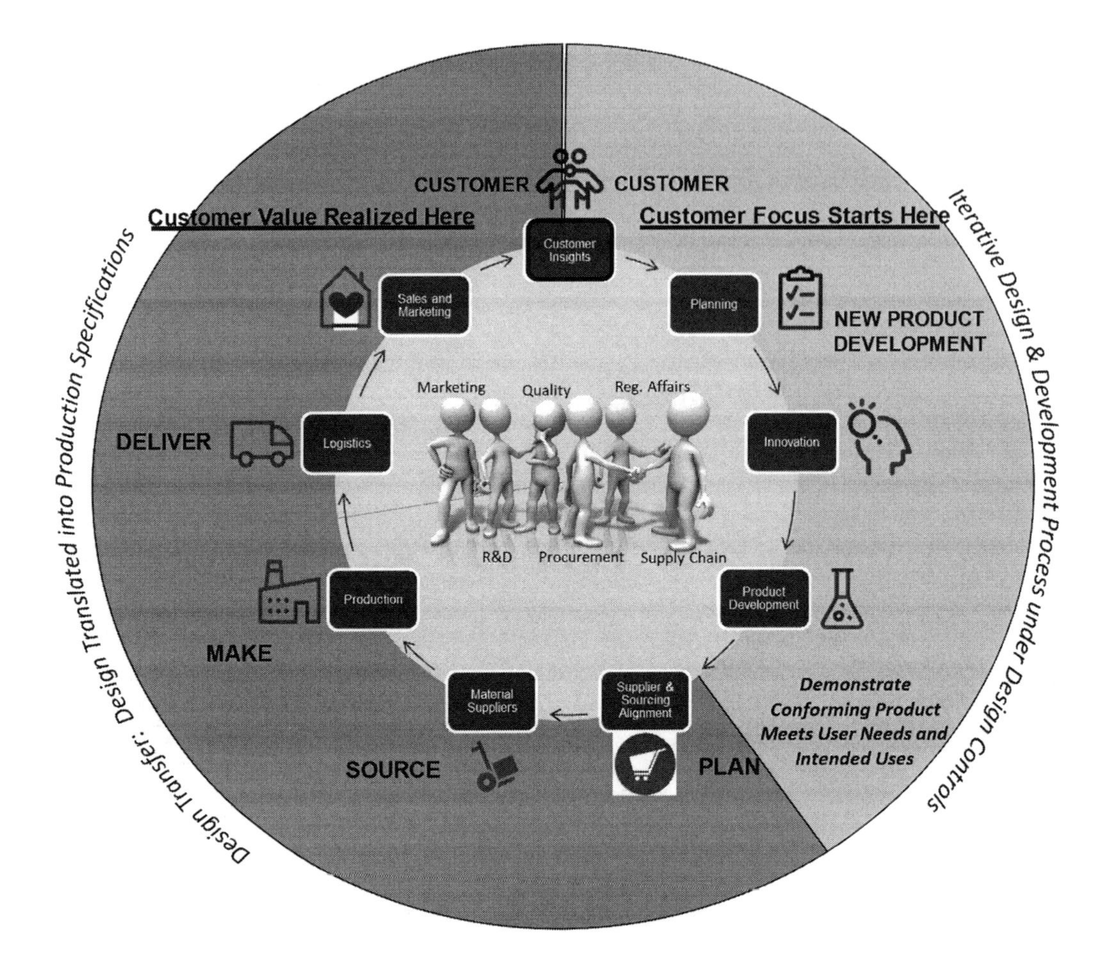

FIGURE 8.3 Interdependent activities delivering specific value to a customer (© 2022 Combination Products Consulting Services, LLC. All Rights Reserved.)

we come full circle, ensuring customer/end-user needs have been well understood and will be satisfied (Figure 8.3).

The Combination Product Sponsor[4] has ultimate accountability for design controls, including Design Transfer for the constituent parts and the combination product. This is the case even if a third-party designs or manufactures them. [It does not, however, absolve a Contract Manufacturing Organization (CMO) from their responsibilities as a combination product manufacturer, though. CMOs who meet the definition of combination product manufacturer are obligated to demonstrate compliance to 21 CFR Part 4 for the cGMP activities they perform, which may include Design Transfer (see Chapter 4).] A Quality Agreement (Purchasing Controls) should be established to define/refine the responsibilities and accountabilities between entities.

Design Transfer, Design Verification, Design Validation, and Process Validation

Design Verification (21 CFR 820.30(f)) confirms that the results of the design process, i.e., Design Outputs, meet design input requirements. While some Design Verification activities may be performed during the iterative design process, final Design Verification is generally performed on conforming product (i.e., initial production units/lots/batches), demonstrating that they conform to pre-defined specifications. Design Validation (21 CFR 820.30(g)) provides objective evidence that when users are provided conforming (or production equivalent) constituent parts and combination product, the product functions as intended to meet user needs/intended uses in the intended use environment. Per Comment 81[5] under the Preamble to the QS Regulations, Design Validation follows successful Design Verification.

"Certain aspects of design validation can be accomplished during the design verification, but design verification is not a substitute for design validation. Design validation should be performed under defined operating conditions and on the initial production units, lots, or batches, or their equivalents, to ensure proper overall design control and proper design transfer."

In order to successfully perform Design Validation, the Sponsor needs to ensure that the product being evaluated conforms with production specifications. Design Transfer supports this, as it ensures that the constituent parts and combination product from the development process are correctly translated into production specifications. Process Validation, an element of Design Transfer, applies the production specifications and demonstrates that production processes reliably and repeatedly produce conforming product. Process Validation is generally done at scale and is needed prior to distribution. The interdependence of Design Transfer, Process Validation, Design Verification, and Design Validation is clear (see Figure 8.2 and Table 8.1).[6]

TABLE 8.1
Key Questions Addressed: Design Transfer, Process Validation, Design Verification, and Design Validation

Design Transfer	Process Validation	Design Verification[a]	Design Validation
Have the constituent parts and combination product been correctly translated into production specifications?	Will the constituent parts and combination product production process (processes) ***reliably and repeatedly produce conforming products***?	Do the results of the design process (Design Outputs) meet Design Input requirements?	Will the constituent parts and combination product function as intended to meet user needs/intended uses in the intended use environment, ***when users are provided conforming product***?

[a] Design Validation may be conducted in parallel with final Design Verification and Process Validation. This is considered an "at risk" approach. If the Process Validation or Design Verification activities do not satisfy predetermined acceptance criteria, the Design Validation will not have been conducted using conforming product. This may necessitate repetition of a range of Process Validation, Design Verification, and Design Validation activities.

Notably, when "production-equivalent" product is used in the final Design Validation, the manufacturer needs to conduct a gap assessment, documenting in detail how the constituent parts and combination product are manufactured and how the manufacturing is similar to, or possibly different from, initial production. If differences exist, the manufacturer has to document a sound rationale for why the Design Validation results are valid for production units, lots, or batches. For each difference, the Sponsor should explain the impact on the validity of their results.

A combination product manufacturer Sponsor is not required to have both a Master Batch Record (MBR) (21 CFR 211.188) and Device Master Record (DMR)[7] (21 CFR 820.181), provided the relevant content of each constituent part and combined use system is reflected in the document used (DMR or MBR). A detailed justification considers all the aspects of the Device Master Record (including the drug constituent part aspects) or Master Batch Record (including the device constituent part aspects). A detailed justification would include whether each aspect was in place and was used to produce the constituent part(s)/combination product, for example:

- Bill of Materials (BOM), components/raw materials specifications, drawings, composition, formulation, software specifications, and suppliers;
- Quality assurance procedures and specifications, including acceptance activities (incoming and final), including inspection/measurement/test equipment to be used;

- Production process specifications, including the appropriate manufacturing equipment, equipment specifications, production methods, production procedures, and production environment specifications;
- Assembly/manufacturing process specifications/procedures;
- Packaging and labeling specifications, including methods and processes used;
- Device History Record(s)[8] or Batch Record(s);
- Production, inspection/test, and installation personnel;
- Storage and handling (electrostatic discharge, frozen/refrigerated, etc.); and
- Installation process instructions, maintenance, and servicing procedures and methods (if applicable).

Process Validation

As we've discussed, Process Validation ensures the process consistently results in conforming constituent part(s) and combination product. Under 21 CFR 211.100(a) and 211.110(a), aligned with ICH Q8(R2), ICH Q9 (R1), and ICH Q10, it describes expectations for process validation with respect to drugs. 21 CFR 820.75[9] describes process validation requirements for medical devices. For drug- or biological product-based cGMP operating systems, 21 CFR 820.75 is not a device called-out provision under 21 CFR Part 4. Similarly, for a device-based cGMP operating system, 21 CFR 211.100 and 211.110 are not called-out provisions. This is because the interpretation of the drug and device process validation expectations is similarly interpreted. When it comes to a combination product manufacturer, the Sponsor (and any associated CMO serving as a combination product manufacturer) needs to take into account process validation considerations for each of the constituent parts and their combination into the finished combination product. Table 8.2 contrasts example materials and process considerations for drugs, biological products, and medical device/ combination products.

At a high level, biological product manufacturing generally starts with establishing a cell bank of a specific protein which is introduced to a bioreactor containing a biological growth medium or "broth." The broth includes the required nutrients to brew in optimal conditions of temperature, pH, and oxygen for cell growth. The master cell bank is divided and put into deep freeze in disposable containers. The manufacturer thaws a vial from the master cell bank and "expands it," allowing it to multiply, and then through a series of filtration/harvest/capture, chromatography, and viral inactivation, and concentration steps, creates and stores the drug substance. When the manufacturer is ready to create the drug product, the drug substance is thawed, pooled, mixed, and filtered in preparation for filling in a container closure (which may also be considered a device constituent part).

The starting point for drug manufacturing is quite different from that of biological products. Generally, the starting materials include precursors, solvents, and reactors. The drug substance or active pharmaceutical ingredient (API) is synthesized through chemical synthesis steps and stored. When the manufacturer is ready to create the drug product, they will blend, granulate, dry, add excipients or fillers to the formulation. Subsequently, they may tablet, vial, or pre-fill the drug product into a container closure (which may also be considered a device constituent part).

Contrast biological product and drug product manufacturing with that of medical devices and combination products. With the wide gamut of medical device types, the range of starting materials can vary dramatically – e.g., from metals to resins or even glass. Plasticizers, coatings, or other processing aids may be used in their manufacture. Use of molds may be included for injection molding, or machining may be required, or a combination of approaches. From this foundation, components or devices are created. Where components are used, they may be assembled into sub-assemblies and, then through finishing processes, are assembled to create a medical device. Where the device is being used as a container closure for a drug or biological product, the drug or biological product is introduced through a fill-finish process. If the drug is ancillary (e.g., a coating or impregnated into the device material), the finishing process can take on numerous alternative approaches from what

TABLE 8.2
Example Critical Materials and Process Parameters (CPPs) for Biological Product, Drug, Medical Device, and Combination Product Manufacturing

Biological Product Manufacturing Process		Drug Manufacturing Process		Device Manufacture and Combination Product Assembly Process
Starting materials	• Master cell bank • Growth medium • Disposables	**Starting materials**	• Precursors • Solvents • Reactors	• Resin materials (or glass or metals) • Plasticizers, coatings • Mold tooling • Injection molding or machining
Drug substance	• Preculture→ expansion→ harvest/ capture→ viral inactivation→ chromatography→ concentration→ storage	**Drug substance/ API**	• Chemical synthesis → storage	⇩ • Components • Sub-assemblies • Fill/finish assemblies integral (or not) with the drug and/or biological product
Drug product	• Thawing→ pooling/ mixing→ filtration→ aseptic filling→ storage	**Drug product**	• Blending→ granulating→ drying→ excipients/ fillers→ tableting or vials→ sealing	⇩ • Finished combination product • Potential post-processing steps (e.g., sterilization, packaging)
Combination Products Manufacturing Process Development and Validation ⟷				

is described here. Post-processing, such as sterilization and/or packaging, is also a consideration in the device/combination product manufacturing process.

The production and process controls are different for each of these product types given how distinct from one another biological product, drug, and device/combination product manufacturing and assembly processes are. It only makes sense that establishing objective evidence that the manufacturing process consistently produces a result or product that meets its predetermined specifications, i.e., Process Validation, looks a bit different for each of these constituent part types and for their integration into a combination product.

Let's review the framework for process validation for drugs and biological products compared to that for medical devices, recognizing that for combination products, one needs to consider and apply applicable process validation and controls for each of the constituent parts and their combination.

Lifecycle Process Validation for Drugs and Biological Products

ICH Q8(R2) Quality by Design (Figure 8.4) is founded on building a robust understanding of both the product and the process for manufacturing that product. Robust product understanding is established through risk analysis and iterative development activities (see Chapter 5). The product requirements are translated into product specifications, including the identification of critical product quality attributes (CQAs) in the Target Product Profile (TPP). The process is designed based on the specific dosage form, stability expectations, etc., generally aligned with Table 8.2. Risk analysis (e.g., Failure Modes Effects and Criticality Analysis) is done in conjunction with the

FIGURE 8.4 General illustration of quality by design (ICH Q8(R2))

development and characterization activities to define critical process parameters (CPPs), and controls are implemented to ensure the process consistently produces a result or the product meets its predetermined specifications. Under QbD, a lifecycle concept of process validation links product and process development, qualification of the commercial manufacturing process, and maintenance of the process in a state of control during routine manufacturing. Within this concept, process validation involves activities that take place over the lifecycle of the product and process, including continual improvements. Figure 8.5 illustrates the stages of the Lifecycle Approach to Process Validation used for drug and biological products.

The Lifecycle Approach to Process Validation applied for drug and biological products is founded on scientific evidence across the product lifecycle. US Food and Drug Administration (FDA) Guidance (January 2011) "Process Validation: General Principles and Practices"[10] and EudraLex "EU Guidelines for Good Manufacturing Practice for Medicinal Products for Human and Veterinary Use Annex 15: Qualification and Validation" (March 2015)[11] similarly describe the state-of-the-art principles and practices for the lifecycle approach for drug and biological product process validation.

During Stage 1, "Process Design," the commercial process is defined based on the knowledge gained through development and scale up. During this stage, the manufacturer builds and captures process knowledge and understanding, and identifies process variables and control strategy. Deliverables from this stage include the Quality Target Product Profile (QTPP), Critical Quality Attributes (CQAs), process flow, design of experiments (DOEs), criticality analysis (CA or FMECA), and presumed critical process parameters (CPPs) and Critical Material Attributes (CMAs).[12]

During Stage 2, "Commercial Process Qualification," the manufacturer evaluates the process to demonstrate reproducible commercial manufacturing. This stage includes design of

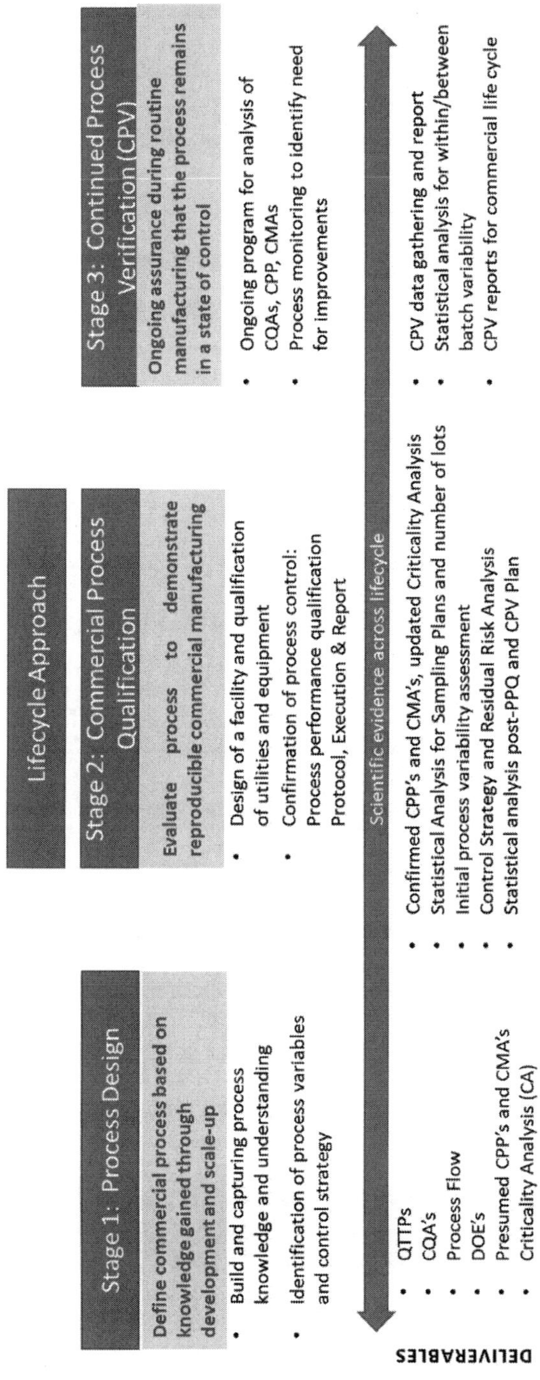

Lifecycle Approach

Stage 1: Process Design

Define commercial process based on knowledge gained through development and scale-up

- Build and capturing process knowledge and understanding
- Identification of process variables and control strategy

Stage 2: Commercial Process Qualification

Evaluate process to demonstrate reproducible commercial manufacturing

- Design of a facility and qualification of utilities and equipment
- Confirmation of process control: Process performance qualification Protocol, Execution & Report

Stage 3: Continued Process Verification (CPV)

Ongoing assurance during routine manufacturing that the process remains in a state of control

- Ongoing program for analysis of CQAs, CPP, CMAs
- Process monitoring to identify need for improvements

Scientific evidence across lifecycle

DELIVERABLES

- QTTPs
- CQA's
- Process Flow
- DOE's
- Presumed CPP's and CMA's
- Criticality Analysis (CA)

- Confirmed CPP's and CMA's, updated Criticality Analysis
- Statistical Analysis for Sampling Plans and number of lots
- Initial process variability assessment
- Control Strategy and Residual Risk Analysis
- Statistical analysis post-PPQ and CPV Plan

- CPV data gathering and report
- Statistical analysis for within/between batch variability
- CPV reports for commercial life cycle

FIGURE 8.5 Lifecycle process validation approach

a facility and qualification of utilities and equipment. The manufacturer also confirms the process is in control through Process Performance Qualification (PPQ) protocol, execution, and report. CPPs and CMAs are confirmed, and the criticality analysis is updated. Other deliverables include statistical analysis for sampling plans and number of lots, initial process variability assessment, establishing a control strategy and residual risk analysis (see Chapter 6), and statistical analysis post PPQ. A Continued Process Verification (CPV) Plan is also created during this stage.

During Stage 3, the manufacturer is in "Continued Process Verification" (CPV) phase of the lifecycle validation approach, i.e., ongoing assurance during routine manufacturing that the process remains in a state of control. CPV entails an ongoing program for analysis of CQAs, CPPs, and CMAs, as well as process monitoring to identify the need for improvements by gathering, statistically analyzing batches for within- and between-batch variability, and reporting on the analysis. CPV reports are generated across the commercial lifecycle of the product, with a focus on increasing knowledge and understanding of product performance, to ensure it will remain in a state of control throughout the product lifecycle. Table 8.3 summarizes the shift from a traditional approach to process validation to the current state-of-the-art lifecycle approach.

TABLE 8.3
Traditional vs Lifecycle Approach to Process Validation (PV) for Drug and Biological Products

Considerations	Traditional Approach	Lifecycle Approach
PV execution activity	Discreet	Ongoing
Understanding of CQAs, CMAs, and CPPs	Limited	Deep
Use of development data	Limited	Increased use/increased robustness
Assessment of analytical and process behavior post validation	Ad hoc	Regularly assessed
Continued process learning and understanding	Limited	Increased
Troubleshooting	Reactive	Proactive and reliable
Process improvement	Limited	Targeted and more robust

Process Validation for Medical Devices

Process validation for medical devices is conducted within the context of design controls – *the manufacturing process for the device constituent part is considered to be part of the product design, and so should be addressed as part of design controls.* Process specifications, equipment specifications, sampling, test methods, acceptance criteria, etc. are considered design outputs.

The device manufacturing process should be capable and stable to assure the products produced are safe and effective for their intended use/users/use environment, conforming to the product requirements. Robust product design (see Chapter 5) should enable the product to withstand variations in manufacturing materials and processes.

Global Harmonization Task Force (GHTF) SG3/N99-10:2004 process validation guidance for medical devices[13] includes a decision tree that is useful in determining whether or not a medical device/device constituent part process should be validated. Also included are examples of processes which should be validated, may be satisfactorily covered by verification, and those processes that may be verifiable, but for which a business might choose to do validation (Figure 8.6 and Table 8.4).

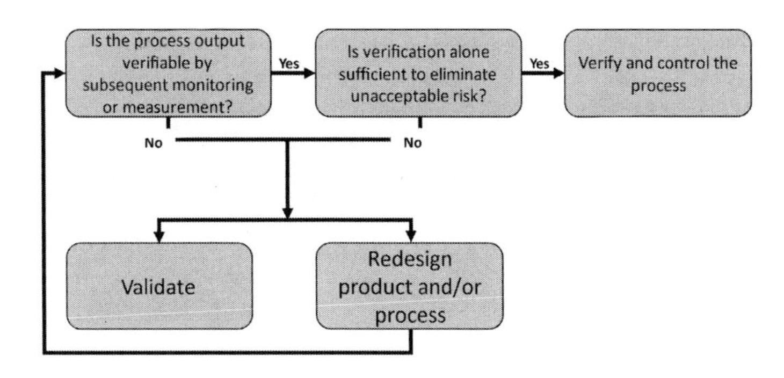

FIGURE 8.6 Process validation decision tree (adapted from GHTF SG3/N99-10:2004)

TABLE 8.4
Examples of Processes and Verification or Validation Approaches[a]

Processes Which Should Be Validated	Processes May Satisfactorily Be Covered by Verification	Processes That May Be Verifiable, But That a Business Might Choose to Do Validation
Sterilization Clean room ambient conditions Aseptic filling processes Sterile packaging sealing processes Lyophilization processes Heat treating processes Plating processes Plastic injection molding processes Assembly processes Software that is used in a process whose output may be verifiable	Manual cutting processes Testing for color, turbidity, total pH for solutions Visual inspection of printed circuit boards Manufacturing and testing of wiring harnesses	Certain cleaning processes Certain human assembly processes Numerical control cutting processes Certain filling processes

[a] adapted from GHTF SG3/N99-10:2004.

As with design and development and risk management, device process validation starts out with a plan, in this case a Validation Master Plan (VMP), which is a summary document for a particular project or process undergoing validation. The VMP defines the process, strategy, key steps, and roles and responsibilities of those engaged in the process validation activities. It generally follows a staged approach (Figure 8.7).

Process validation is generally done in three phases: Installation Qualification (IQ), Operational Qualification (OQ), and Performance Qualification (PQ):

IQ: Establishes objective evidence that all key aspects of the process equipment and ancillary system installation adhere to the manufacturer's approved specification as well as recommendations of the supplier of the equipment are appropriately considered. Software documentation is needed if the equipment or tool includes software.

OQ: Operation is challenged at worst-case process settings (limits of acceptability criteria) for specific parameters, demonstrating that the output of the process is still within the defined specifications, even at the process limits. OQ therefore establishes objective evidence that the process control limits and action levels consistently result in a product that meets all predetermined requirements/specifications.

PQ: Demonstrates that the process, under standard operating conditions, consistently produces a product that meets all predetermined requirements. Table 8.5 summarizes key considerations for a PQ Protocol.

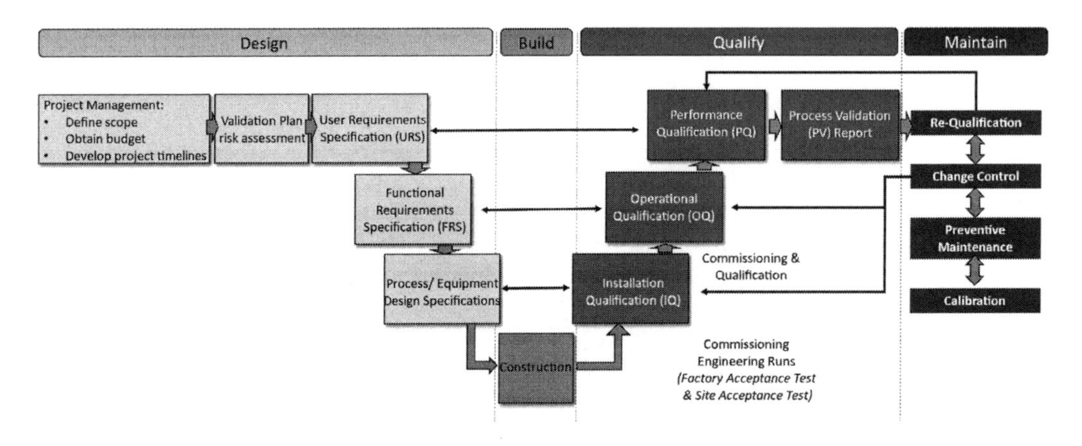

FIGURE 8.7 Process validation stage approach

TABLE 8.5
Process Validation: PQ Protocol Considerations and Contents

1	Identify and describe the process(es) to be validated.
2	Identify the medical device(s) to be manufactured using the process. Validation efforts should be proportional to the associated risks/potential safety issues for the patient, user, or other persons who have the potential to be impacted by the device use.
3	Ensure that all inspection, measuring, and test equipment, including mechanical, automated, or electronic inspection and test equipment, are suitable (precision, accuracy, repeatability, reproducibility) for its intended purposes and are capable of producing valid results.[a] Ensure that the equipment is calibrated, inspected, checked, and maintained.
4	Objective and measurable criteria for a successful validation should be prospectively defined (e.g., specifications that relate to product, components, or product characteristics to monitor).
5	Statistical methods for data collection and analysis should be prospectively defined.[b]
6	Length and duration of the validation should be defined using a risk-based approach. Justification and rationale should be documented on the length and duration of validation activities chosen.
7	Operators to be used in the process should be identified (operators should be representative of the personnel who will conduct the commercial manufacturing activities).
8	Operators should be trained, including awareness training of device defects which could result from the improper performance of their specific jobs. The training should be documented.
9	Equipment to be used in the process should be identified and calibrations performed and documented.

[a] 21 CFR 820.72 (related drug cGMP provisions include 21 CFR 211.165, 211.84, and 211.194(a)(2)). Particular attention should be paid to the relationships between requirements. If a combination product manufacturer is operating under a streamlined approach, citations are given for provisions based on the base cGMPs and the relevant called out provisions from the other cGMP system. So, for a drug-cGMP based quality system, if test method validation is deemed inadequate according to 820.72, the related 211 provision(s) might be cited.

[b] 21 CFR 820.250, ISO 13485 Clauses 8.1 and 8.4 or related 21 CFR 211.110 (b), 211.165, and 21 CFR 211.84

Once process validation is complete, a Process Validation Report (PV Report) documents conclusions on the overall results of the validation activities. Daily measuring and monitoring are then conducted post market according to the process control plan, which is typically largely developed during process validation.

Process Validation for Combination Products

The backdrop of process validation expectations for drugs/biological products and medical devices are an important foundation for combination products process validation expectations. According to US FDA Combination Products Compliance Program 7356.000:

> For both CDER-led and CDRH-led combination product post-approval/postmarket inspections, **inspectional coverage should include assessment of the combination product** and **not just a constituent part**, as appropriate. For example, **if process validation for the specific combination product is covered during the inspection, process validation for any device, drug, or combination product manufacturing processes occurring at the facility should be assessed**.

If you are a Combination Product Sponsor, and/or your facility is conducting the activities of a combination product manufacturer, for example:

- A CMO that is bringing components together to manufacture more than one constituent part,
- A CMO bringing the drug and device constituent parts together to create the combination product, or
- A facility conducting other related combination product processing activities, e.g., sterilization or testing, for the combination product (as opposed to only one type of constituent part)

then your facility is responsible for the process validations for each of the cGMP processes your facility performs.

The following CMO Warning Letter excerpt reflects this:

> You and [your customer] have a quality agreement regarding the manufacture of [combination] products. You are responsible for the quality of combination products you produce as a contract facility, regardless of agreements in place with {Sponsor} or with any of your suppliers. You are required to ensure that your combination products are compliant with the CGMP requirements applicable to each manufacturing process that occurs at your facility.[14]

A Sponsor should be cautious of over-reliance on suppliers or CMOs of the purchased components, constituent part, or combination product. Purchasing Controls activities, such as supplier qualification, ensuring a supplier is suitable and capable of providing the product(s) or service(s) the Sponsor purchases from them, are critical, particularly for combination products, where most Sponsors purchase at least one constituent part.[15] Purchasing and Supplier Controls include, for example, review and acceptance of the supplier's quality system, or arrangements for a supplier to use the Sponsor's quality system. Even with such Supplier Controls in place, the Sponsor should not just assume that the supplier/CMO has appropriately conducted required process validation activities.

While the supplier/CMO *is* indeed responsible for the cGMP activities they perform, ultimately, the Sponsor holds responsibility for the (design and process) validation of each constituent part and combination product. The Sponsor *must* ensure the quality of their combination products, including appropriate process validation activities for the constituent parts and combination product, even if those products are purchased or produced at a contract facility, regardless of agreements in place with any suppliers.

Requirements for process validation for the secondary constituent part and combination product may be a blind spot for some combination product manufacturers, mistakenly believing their process validation focuses just on the constituent part that is the primary mode of action (PMOA). The Validation Master Plan needs to address each constituent part and the combination product. Now that you have read this section of the chapter, please do not make that mistake! Figure 8.8 summarizes the integration of drug and device process validation activities.

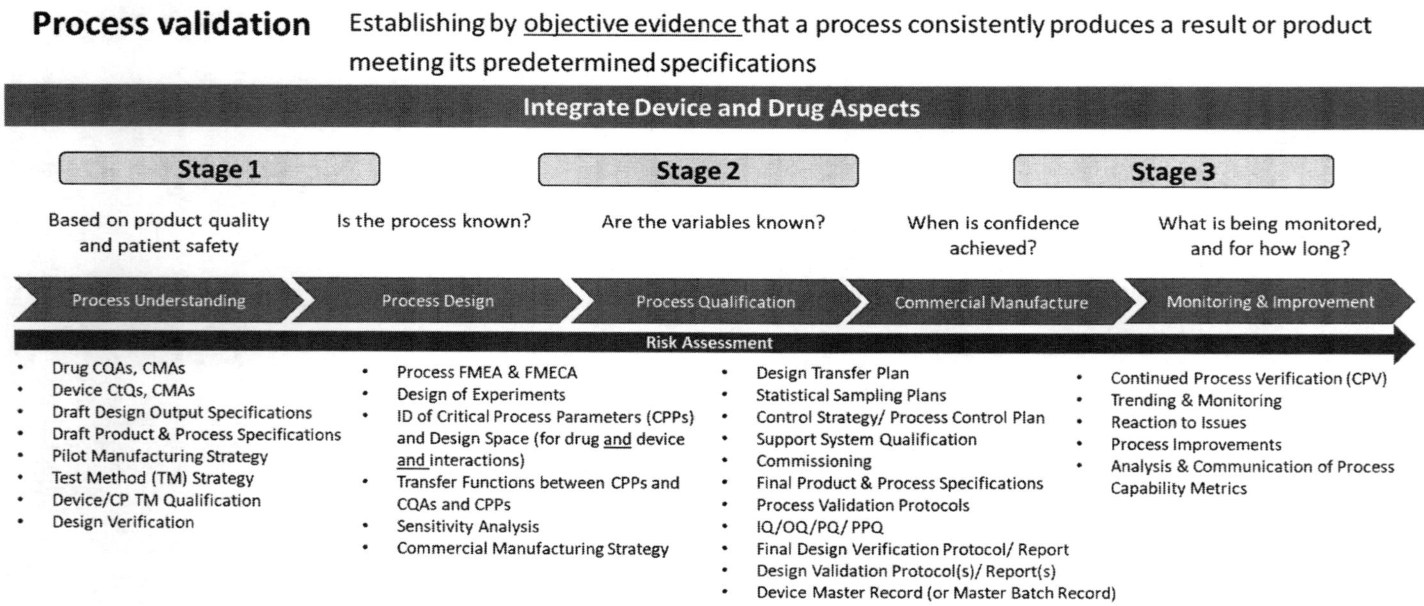

FIGURE 8.8 Integration of drug and device process validation activities for combination products. CQA: Critical Quality Attribute; CMA: Critical Material Attribute; CtQ: Critical to Quality; CP: Combination Product; FMEA: Failure Modes Effects Analysis; FMECA: Failure Mode Effects and Criticality Analysis; CPP: Critical Process Parameter; IQ: Installation Qualification; OQ: Operational Qualification; PQ: Performance Qualification; PPQ: Product Performance Qualification; CPV: Continued Process Verification (©2022 Combination Products Consulting Services, LLC. All Rights Reserved.)

Process Controls

"Process control" is the active changing of a process based on the results of process monitoring. There are varied approaches that may be taken for implementing process controls. They may be, for example, 100% in-line, including real-time controls in automated processes, or they could involve manual sampling and measurement outside the process. Once process monitoring tools have detected an out-of-control situation, the person responsible for the process, or automated controls, makes a change to bring the process back into control (Figure 8.9).

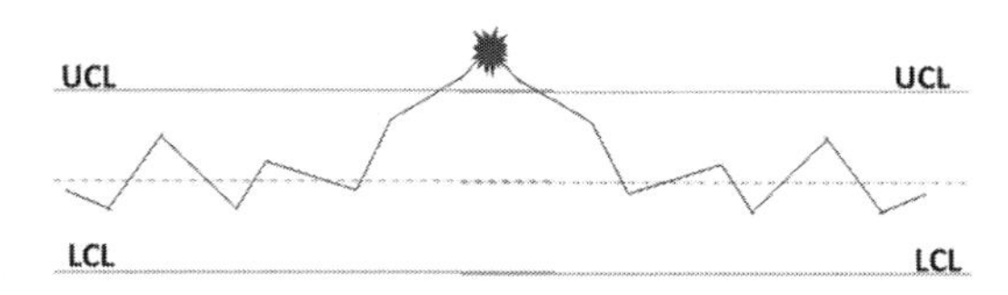

FIGURE 8.9 Process control

Consider the example of a prefilled syringe. In process closed-loop controls could include automated in-line check weigh stations for 100% assessments, enabling real-time adjustment to volume dosed if required. Alternatively, in-process controls may be done to isolate reject parts, sorting them from the process if control limits are exceeded. This may be done, for example, for attribute testing, e.g., for missing labels or missing parts. Alternatively, manual offline testing may be conducted based on samples drawn from the line, with appropriate reactions. (This is a higher risk as it typically requires longer test result lead times and can even result in rejection of portions of a batch, e.g., due to bracketing; one would have to toss part of the lot that lies between successful tests if there is a failed test.)

A post-market process control plan is developed through design controls,[16] risk management,[17] and, largely, through process validation activities. This control plan defines daily (or other periodic) measuring and monitoring to be conducted post market.

ICH Q12 defines "established conditions" (ECs) as legally binding information considered necessary to assure product quality, as helps clarify which elements of the product, manufacturing process, facilities, equipment and control strategy are considered to be ECs.[18] ICH Q12 and FDA Guidance "ICH Q12: Implementation Considerations for FDA-Regulated Products"[19] apply to drug substances, drug products, biological products, and combination products with device constituent parts. The ICH Q12 Implementation Considerations Guidance provides a framework to support the management of post-approval Chemistry, Manufacturing, and Controls (CMC) changes. It also includes tools, enablers, and guiding principles that can help combination product developers/manufacturers identify Essential Performance Requirements, their associated critical processes, and needed controls. Specifically, FDA suggests an approach to identifying ECs for device constituent parts of a combination product in a New Drug Application (NDA), Biologic License Application (BLA), or Abbreviated New Drug Application (ANDA). The combination product as a whole, including the roles and interactions of the constituent parts, should be considered in identifying combination product ECs. Figure 8.10 is adapted based on Appendix B of the ICH Q12 Implementation Considerations Guidance.

Risk-based considerations for the identification of Essential Performance Requirements are discussed in Chapter 6, Risk Management. The Essential Performance Requirements (EPRs) are therefore related to the Essential Conditions for the manufacturing process.[20] The questions in Figure 8.10 help identify "primary characteristics" of the combination product that are essential to its safe and proper use. The design features identified as critical to ensure product safety and efficacy/use are generally EPRs. The associated manufacturing process elements that need to be controlled to ensure the combination product produced conforms to requirements, so that its use

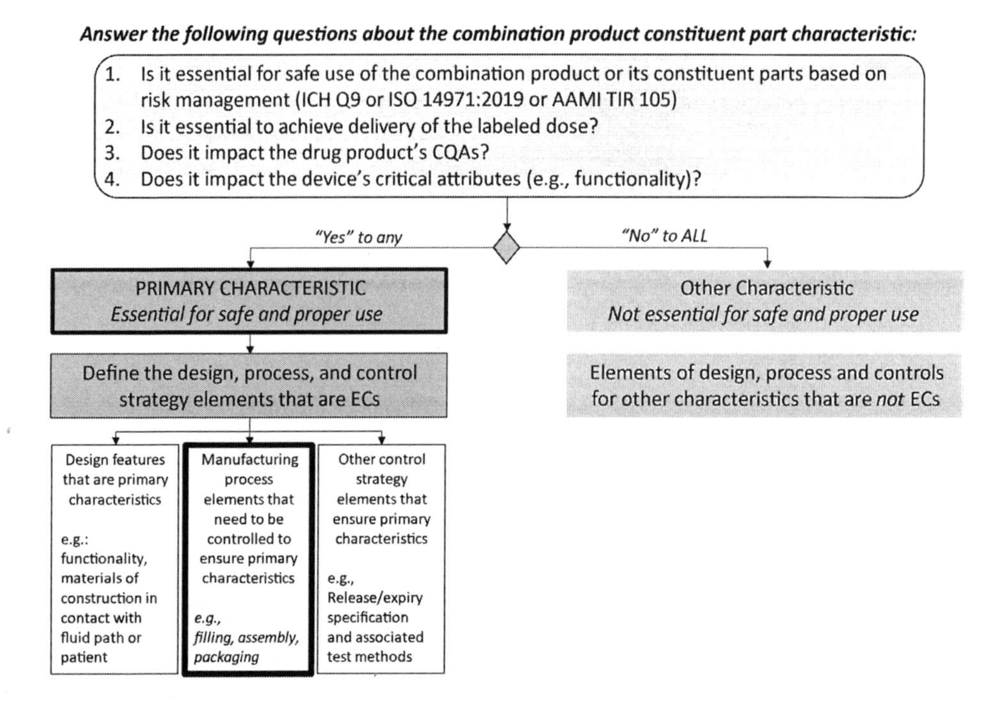

FIGURE 8.10 Established conditions (ECs) for the Device Constituent Part of an NDA, BLA, or ANDA Combination Product (adapted from US FDA ICH Q12 Implementation Considerations Guidance, Appendix B)

is safe and effective, are ECs. Some examples of manufacturing processes that would typically be considered ECs include:

- Filling: In-process controls might include filtration checks to avoid unwanted microbial growth, a check weigh station to ensure appropriate fill volume for a syringe (see "Typical Filling Process" later in this chapter).
- Sequence of the assembly process.
- Manufacturing process parameters such as temperature at blister-sealing and time of exposure for a temperature-sensitive drug prefilled in a syringe going through the blister-sealing process.
- Storage temperatures and device constituent part handling: Temperature-sensitive drugs or biological products may need to be stored under refrigerated conditions. A device constituent part may become brittle with extended exposure to cold temperatures. Carefully monitored handling controls, e.g., transition time to ambient temperatures prior to subsequent handling/assembly processes, may be needed.
- Storage temperatures and hold times: Typically, sterile drug products are stored under refrigerated conditions (2–8°C). Device assembly and packaging are conducted at ambient conditions. Given the sensitivity of drug products to temperature during any processing, careful monitoring and control of time outside refrigeration are needed to maintain drug stability.

Application of risk analysis (coupled with Figure 8.10 and related Guidance) can help a constituent part and/or combination product manufacturer identify those unit operations, sequence of manufacturing process, manufacturing process parameters, and material attributes that may impact product safety and performance and are thus considered as ***essential***. Early risk management activities typically focus directly on risks related to the device constituent part and combination product design. When designing, qualifying, and validating *production processing*, risks to users and patients that could arise based on the production process also need to be considered. Analyzing risks during

Design Transfer may help to identify key process steps or entire processes that could impact the ability of a product to perform and to be used safely and effectively. Analysis of processes might uncover missing or incorrect assembly steps; invalid/unreliable acceptance tests; unqualified equipment; or un-validated processes – all of which could lead to the production of defective devices, leading to unsafe, ineffective product performance.

In-process controls (IPCs) are established end-to-end throughout manufacturing and assembly, ensuring the product conforms to specifications. In-process controls can be implemented, particularly where variability impacts primary characteristics. ECs should be verified and, where appropriate, validated (Figure 8.6 should be helpful in this determination).

Following are pertinent excerpts from an FDA Warning Letter that reflects the importance of adequately analyzing combination product processes and work operations, implementing appropriate in-process controls, as well as ensuring appropriate process capability analysis.

FDA WARNING LETTER EXCERPT

Your firm does not distinguish between the different failure modes of rejected components/units that are collected in reject bins on the (combination product) manufacturing assembly line. For example, the *Packaging and Inspection Master Specification for [combination product]'s Automated Assembly, Labeling, and Packaging* instructs [manufacturing process] which leads to commingling of different types of rejected components. Your firm does not assess the types or causes of rejects, and instead only records the total number of rejects. Therefore, ***your firm does not adequately analyze processes*** to identify the existing and potential causes of nonconformities related to products or other quality problems.

Your firm plans to develop a procedure to assess performance variability and to require routine [manufacturing process] trending of reject levels at [manufacturing process] which will determine action limits based upon process capability. Your firm's response is not adequate because you have not provided this updated procedure with the aforementioned action limits, shown how you plan to use process capability in your analysis of processes, or indicated how this data will feed into your firm's corrective and preventive action system.

FDA WARNING LETTER EXCERPT

Your firm does not use appropriate statistical methodology for process capability in order to analyze the quality of production machinery output at critical process steps and to detect recurring quality problems. Your firm's Process Capability Report for [combination] products states that you performed capability analysis on [manufacturing process] test results to determine the process capability of the manufacturing operations involved in production. However, various specifications were only analyzed at the finished product attribute level. Since capability is not determined at the [manufacturing process step], and since the capability calculations were performed using final batch data collected after some defective units were removed, this analysis does not adequately demonstrate the ability to detect recurring quality problems.

Your firm stated that it will incorporate routine machine capability studies and periodic reviews into ongoing trend analysis. However, you have not provided information regarding how you intend to monitor these studies, how the information will be used, and how it will feed into your corrective and preventive action system to detect and prevent recurring quality problems. Further, your response does not address **the need to assess whether this capability analysis reveals other potential problems with the product and the need to review the capability of other processes.**

Combination Product Manufacturing Challenges: Drug PMOA[21]

Drug PMOA Example

Our discussion of combination product process validation and in-process controls would be incomplete if we did not address combination product manufacturing challenges. For purposes of this discussion, we'll review a typical drug PMOA development pathway (Figure 8.11). In this case, combination product manufacturing process development and validation begins with the drug substance and extends through to the drug product for clinical and commercialization phases. Let's focus on the drug product manufacturing at a typical fill-finish site. The drug product serves as a drug constituent part of the combination product. From a design controls perspective, for a drug PMOA, the drug product, and its characteristics (e.g., intended users/user needs/intended uses/intended use environment, risk analysis, critical quality attributes, critical material attributes, and critical process parameters) serve as design inputs to the combination product design controls process.

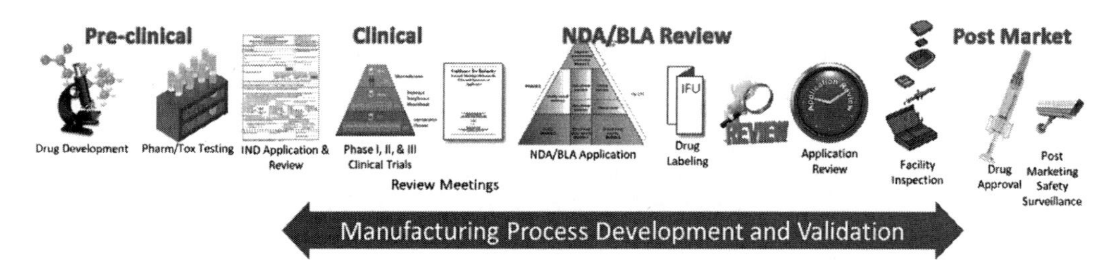

FIGURE 8.11 Typical drug PMOA development pathway (©2022 Combination Products Consulting Services, LLC. All Rights Reserved.)

Prior to fill-finish, drug substance is shipped and stored frozen to avoid microbial growth, e.g., at −40°C in plastic 10-L bottles (Figure 8.12). Sterile equipment is used throughout the process. The formulated bulk drug substance is thawed and pooled in preparation for fill-finish activities

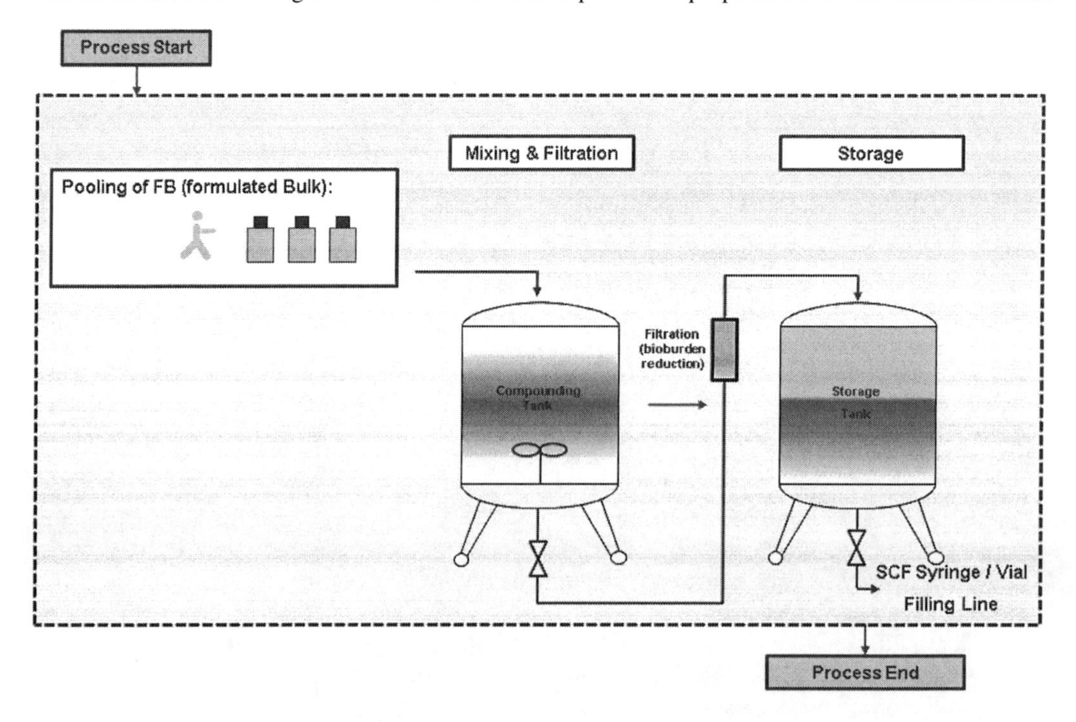

FIGURE 8.12 Example preparation of formulated bulk for filling

in a Class 10,000 cleanroom environment and mixed to achieve homogeneity. Ensuring bioburden thresholds are met, an in-process test ("in-process control" or IPC) is conducted pre-filtration. Media is transferred from the pooling and mixing vessel through a filter into a storage vessel.[22] The media is now ready to use to fill syringes.

Prior to filling, syringes and stoppers are prepared. Syringes are delivered sterile from the supplier, in a ready-to-fill state. Stoppers are supplied pre-washed, in ready-to-sterilize bags. Stoppers are sterilized offline in an autoclave. The tub that will be used to hold the syringes is decontaminated using e-beam. The production line is also decontaminated, e.g., with hydrogen peroxide, prior to any use. At this point, we are ready to start the filling process. As an extra precaution, in order to ensure bioburden is kept at an absolute minimum, the media is sterile-filtered using a 0.22-micron filter and back-up filter. A typical filling process is illustrated in Figure 8.13.

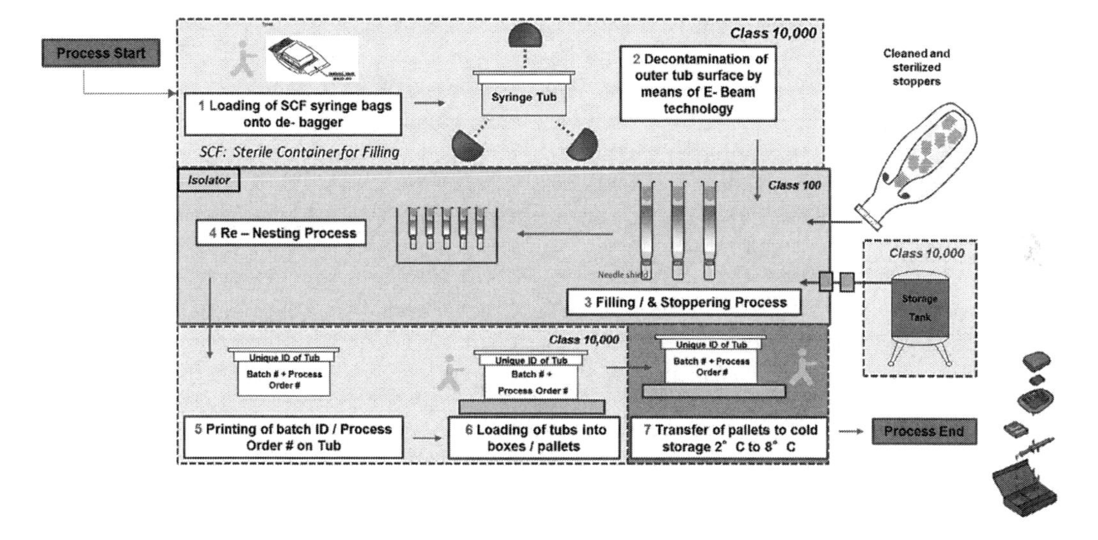

FIGURE 8.13 Typical filling process

Filling begins in a Class 100 clean room environment. An in-line weight check is used as another IPC to ensure fill volume accuracy during syringe filling. Following syringe fill, the syringes are stoppered. Stoppering is controlled with an offline check.

Filled syringes are then placed in tubs and transferred from the Class 100 clean room, back into a Class 10,000 clean room environment. Tubs are marked with unique identification codes for traceability and then transported to cold storage in a cold storage chamber. Syringes are subsequently given unique identifier labeling and optically inspected for defects, e.g., cracks and particles. The prefilled syringe is now ready for any required subsequent assembly (e.g., into an autoinjector) and/or packaging.

Combination product manufacturing challenges will now be discussed in the context of this typical filling process. There are five main areas of challenge for drug-led combination products:

1. Drug substance formulation;
2. Flexible facility designs;
3. Drug product manufacturing;
4. Stability considerations; and
5. Manufacturing logistical considerations.

Drug Substance Formulation

The combination product manufacturer has to assess and control potential interaction effects between the drug formulation, its excipients, and the device components/constituent part. The design control and risk management should take this into account, ensuring the drug's CQAs are maintained (e.g., identity, strength, quality, and purity) as well as the device's and combination product's safe and effective performance. A few common examples of drug-device interactions include:[23]

- Biocompatibility;
- Surface interactions (e.g., adverse chemical reactions or particle formation);
- Contaminants (e.g., silicone oil);
- Absorption or adsorption of API by the device components or constituent part; and
- Extractables and leachables.

Flexible Facility Designs

Shared facilities are often the standard for fill-finish operations. This requires diligence to preclude cross-contamination. The changeover from one product to another needs to be planned and carefully yet efficiently executed, in order to minimize downtime in production. The isolator (in our "Typical Filling Process" example) must be completely cleared of the previous batch, including any supporting materials (e.g., stopper bags, tools, or other equipment). The entire isolator needs to be decontaminated, and tanks and small parts have to be cleaned and sterilized. At that point, all the preparations for the next batch have to be addressed. Procedures and checklists are very helpful in ensuring robust changeover.

Drug Product Manufacturing

There are multiple challenges with respect to drug product manufacturing. These include clean room requirements; equipment sterilization; quality of utilities; in-process monitoring and controls; managing hold times; assembly, labeling, and packaging; and unique product identification.

- **Clean room requirements:** Clean room requirements must be maintained according to industry guidelines.
- **Equipment sterilization:** Effective filling equipment sterilization is critical.
- **Quality of utilities:** The quality of utilities, e.g., for water-for-injection, gas for filtration, and air particulates (both viable and non-viable) must be maintained. Placement of active or passive monitoring settle plates in defined locations is essential, as is making sure there are enough of them. Risk-based methods should be used to determine the particulate monitoring strategy. (It depends on the type of aseptic process – is it close to filling, where it would be susceptible? By the isolator?) It is not possible to 100% check your product for sterility, as it is destructive testing. In the case of prefilled syringes, we would have no syringes left to market!
- **In-process monitoring and controls:** As we've discussed earlier in this chapter, it is essential that critical process parameters (essential conditions) be monitored throughout manufacturing. In-process testing is used to evaluate the product throughout the process, end-to-end, identify undesirable trends relative to established control limits and make appropriate adjustments to ensure the process remains within validated conditions.

  Aseptic Processing Simulation (APS) is used to help ensure controls developed will provide adequate process control in-use. Simulation is done using a nutrient medium in the process. As part of the assessment, prefilled syringe units are filled with a nutrient medium, incubated, and tested for microbial growth. This simulation approach ensures that no colony-forming units (CFUs), indicating lack of sterility, are formed. Examples of routine and simulated interventions that are used as part of the simulation process include:

  - Removal of stopper blockage;
  - Removal of vials/syringe blockage;

- Removal of blockage during syringe filling;
- Bleed off from filter/distributor;
- Execution of active air sampling;
- Stopper refill;
- API charging (aseptic compounding);
- Exchange of filling needle;
- Exchange of a pump;
- Exchange of the reservoir;
- Exchange of all product contact parts;
- Exchange of the filter;
- Exchange of the stopper container;
- Disconnecting the product vessel from the isolator, emptying the product tubing, and afterward re-connecting the product vessel and the connecting pipes; and
- Cleaning and functional control of the scale.

APS is used to support process validation as well as periodic monitoring:

- For a new filling process, the process simulation challenge must be repeated multiple times (typically three), with representative batch sizes, demonstrating effective outcomes to support validation.
- Periodically, simulations are repeated to ensure the process remains in control.
- One should repeat validation and/or APS prior to significant changes to the actual process, such as installation of new or modified critical components (e.g., a dosing system, process setup, filling unit, or line speed adjustments).

- **Managing hold times:** Based on the nature of the drug constituent part, holding times and conditions are critical factors impacting product quality. Holding times during the manufacturing process must be well understood, controlled, validated, and maintained during drug product filling and device assembly and packaging. Typically, sterile drug product is stored under refrigerated conditions (2–8°C). Device assembly and packaging are conducted at ambient conditions (15–25°C). Given the sensitivity of drug product to temperature during any processing, careful monitoring and control of time outside refrigeration are needed. Assembly of larger batch sizes must be planned in consideration of the total holding time; production planning must consistently account for hold times given varied batch sizes. For example, in manual assembly processes, only partial lots are processed at a time, to ensure cold chain requirements are met. Holding times are based on available stability data. Adherence to holding time is essential for batch release decisions.
- **Assembly, labeling, and packaging**: Earlier in the chapter we also discussed the criticality of the Validation Master Plan including each constituent part and the combination product (Figure 8.7). Either separately or as part of validation activities, the combination product manufacturer should provide evidence that the manufacturing process(es) associated with the device assembly, labeling, and packaging do not adversely impact drug product quality attributes. Temperature, light, and vibration sensitivity are key considerations. Heat exposure during blister packaging, light exposure during secondary processing, or even processes like eject/reject cycles might introduce vibration and could impact the drug.

  The complexity and execution time associated with assembly, labeling, and packaging processes are typically a function of material handling and automation. Device partners have the design component and constituent part design specifications, but perhaps not the experience of device/combination product final assembly. A Sponsor and/or combination product manufacturer should be aware of these considerations.

For device suppliers, mold tooling, assembly equipment, test equipment and methods, and a Test Plan with prospectively defined acceptance criteria are typically part of a Validation Master Plan. Mold tool cavitation, e.g., for injection molding of components and/or device constituent parts, is typically a big lifecycle management "watch-out." Mold tooling cavitation directly relates to component complexity, capital investment, cost of goods, and forecasted volumes. In an effort to reduce the cost of goods and/or to meet increasing volume forecasts, component or device constituent part suppliers may increase mold tooling cavitation. Generally, increased cavitation brings increased variability. In the author's experience, such increased variability may result in assembly and/or product functionality issues if appropriate tolerance stack-up analysis has not been done during product design. If purchasing components from multiple suppliers, this becomes an even more challenging hurdle to overcome. Inattention to such design and processing details can have significant deleterious impacts on product assembly or performance issues, so please stay attuned to this challenge!

The assembly process and tertiary packaging (e.g., folding box, tray, Instructions for Use, label) must consider container closure integrity requirements to maintain product sterility. Design of the assembly process, whether it be executed manually, semi-automated, or fully-automated, is critical to ensure product quality. Human interfaces are the most impactful aspects of the processing particularly for manual processes. Operators require clear working instructions and specifications. In the semi-automated assembly process, the process consists of both manual and automated steps, with manual loading and removal of product from equipment. Increasing automation requires robust validation to ensure proper functioning of the system. Regardless of level of automation, critical process parameters must be characterized and qualified.

For the combination product manufacturing site, the VMP could also include shipping, labeling, and secondary/tertiary packaging, cleaning, and Continued Process Verification (CPV). Assembly, packaging, and labeling process robustness and stability assessments, typically with a minimum of three batches, generally include:

- Inter- and intra-batch variability;
- Batches including low- and high-process operating ranges (if applicable);
- Key performance indicators (e.g., yield, eject rate);
- Commercially representative material and handling steps;
- Commercially representative processes (e.g., production breaks, shift changes, line clearance); and
- Application of previously identified control strategy and Acceptable Quality Limits (AQLs) (critical, major, and minor) to determine sampling and testing requirements (variable and attribute).

- **Unique product identification: National Drug Code (NDC) and Unique Device Identification (UDI) requirements:**

The National Drug Code (NDC) is the US FDA standard for uniquely identifying drugs marketed in the United States.[24] Similarly, under 21 CFR 801.20, "the label of every medical device shall bear a unique device identifier (UDI)."[25] Each of these codes is a globally unique and unambiguous (numeric or alphanumeric) product identifier, intended to support product traceability and, if needed, post-market actions (e.g., field actions/recalls). 21 CFR 801.30(a)(11)[26] and 21 CFR 801.30(b)(1), (2), and (3) clarify the expectations of UDI for combination products. The requirements for the application of UDI and NDC codes vary based on the combination product category, i.e., single entity, co-packaged or cross-labeled. Table 8.6[27] summarizes the high-level UDI and NDC requirements for combination products in the United States.[28]

TABLE 8.6
US FDA Combination Product NDC and UDI Requirements Based on PMOA and Category

Combination Product PMOA	Combination Product Category	Combination Product Identification Code	Constituent Part ID Requirement		Example
			Drug	Device	
Drug (CDER/ CBER)	Single Entity	NDC	*No NDC*	*No UDI*	Metered dose inhaler
	Co-packaged	NDC	NDC	UDI	Drug vial co-pack with an empty syringe
	Cross-labeled		NDC	UDI	Photosensitizing drug and activating laser
Device (CDRH)	Single Entity	UDI	*No NDC*	*No UDI*	*Drug-eluting stent*
	Co-packaged	UDI	NDC	No UDI	First aid kit
	Cross-labeled		NDC	No UDI	?

Stability Considerations

Stability is another manufacturing challenge. Normal storage and excursions must both be considered. For biological products, refrigeration is typically required (e.g., 2°C–8°C). Shelf life of the product must be determined through stability studies of the drug product, the device constituent, and interactions between the two (see discussions in Chapters 4 and 11). Drug degradation may be a factor, and device constituent functionality must also be assessed. Impacts can be due to interaction effects or storage conditions. For example, cold storage may make the device constituent more brittle or could impact its functionality, while the same cold storage prolongs drug product expiration dating. The shelf life of the product is the shorter of the shelf lives of each of the constituent parts. (In the event that there are interactions between the constituent parts, the shelf life could be even shorter than that of the stand-alone constituent parts.) Temperature excursions must also be characterized during stability studies to determine what might be acceptable if temperature ranges are exceeded during transport. During transport, the storage conditions are monitored (best practice: each package has a unique temperature recording device to ensure storage conditions were acceptable).

Manufacturing Logistical Considerations

On a day-to-day basis, there are additional challenges, in part due to the global nature of supply chains, and differing product configuration or labeling requirements country by country. This results in a large number of packaging material combinations. For example, configurations of materials at the packaging stage may be different, due to variations in Instructions for Use (IFU), folding boxes, labels, Unique Product Identification Codes, etc. Proposed changes must be carefully evaluated and implementation effectively managed, to be consistent with local country requirements. Planning and execution systems must be set up effectively to enable proper identification of material and to perform required processes.

Combination Product Manufacturing Challenges: Device PMOA

This chapter has focused extensively on drug PMOA considerations for combination product manufacturing. Those for device PMOA are more aligned to standard cGMP practices, with a few notable exceptions. Production and process control cGMPs are similar for drugs and medical devices, but for a combination product manufacturer, the attention to detail to considerations of each constituent part and the combination product is more pronounced. Consider, for example, cross-contamination and microbial contamination. It is essential that the CQAs of the drug be preserved when introduced to the device manufacturing process. Unwanted material may spread from one area, equipment, surface, or product to another. Such undesired material transfer needs to be carefully monitored and controlled as it can adversely affect the quality of the combination product. Further, cleaning methods applied need to not only ensure machinery, parts, or equipment do not contribute to contamination but also that approaches used do not interact with the drug when it is time for its appropriate use. Cleaning validation is a critical activity. Facility controls, ensuring the physical building in which work is performed, including the surrounding area, are critical to avoid mix-ups and contamination issues as well. Additional air handling capabilities might need to be considered, depending on the required handling of the API or drug product being introduced to the medical device. Labeling basics, such as missing, incorrect, and illegible labeling on the product are cGMP activities for which controls should be in place even for stand-alone device manufacture. Additional considerations of labeling due to UDI and NDC code requirements (mentioned earlier in the chapter) likewise need to be addressed for device PMOA combination products.

The two areas where device manufacturers seem to experience more distinct challenges during Design Transfer and manufacturing for device PMOA combination products include laboratory controls and calculation of yield. In terms of laboratory controls, the prescriptive expectations of testing for approval or rejection of components (21 CFR 211.84), testing and release for distribution (21 CFR 211.165), and reserve sample expectations (21 CFR 211.170) with respect to the drug constituent part are new and require procedural updates/training. Calculation of yield (21 CFR 211.103) also presents challenges. Actual yields and percentages of theoretical yield are to be performed at the conclusion of each appropriate phase of manufacturing, processing, packaging, or holding of the combination product. For manufacturers whose devices are produced using continuous manufacturing processes, yields for stand-alone devices are typically looked upon as opportunities to increase efficiencies and reduce the cost of goods. Now those processes are scrutinized differently under the magnifying glass of drug calculation of yield expectations.

With the wide range of device technologies and associated gamut of risks and complexities, the challenges of device PMOA combination product manufacture are highly correlated to the specific product type.

Design Changes: During Design Transfer and Post Market

Stating the obvious: Change happens. There are a variety of triggers for changes in a combination product, its constituent parts/components, and associated manufacturing activities. These changes can happen before, during, and after Design Transfer. There are a variety of triggers for change(s) to occur, including, for example:

Pre-Design Transfer:

- Critical task analysis;
- Risk (and residual risk) analysis results;
- New information/insights on user needs, intended uses, or use environment;
- Design review results;
- Manufacturability or installation issues;
- Supplier raw material/component/constituent part issues;
- Design Verification or validation results; and
- Shifting regulatory requirements, e.g., updates to needle-based injection system standards (2022 updates to ISO 11608 series of standards).

Post-Design Transfer:

- Improve product performance, safety, and/or usability;
- Improve manufacturability or production yields, or other continuous improvement initiatives;
- Add new features or functionality;
- Improve or discontinue materials or components;
- Risk management decisions in response to field performance (e.g., complaints or adverse events);

- Response to customer feedback;
- New regulatory requirements, e.g., shifts in expectations or market expansion;
- Expanding indications for use;
- CAPA-related/non-conformances, e.g., correct design deficiencies discovered post transfer;
- Manufacturability issues;
- Capacity constraints/site expansion;
- Adding second-source suppliers; and
- Supplier requests for change.

According to the QS Regulation Preamble Comment # 87, all design changes made after the design review that approves the design inputs for incorporation into the design and those changes made to correct design deficiencies once the design has been released to production must be documented. Change Control [21 CFR 820.30(i) or ISO 13485:2016, clause 7.3.9, 820.40, 820.70(b)] (see also Chapter 4) includes requirements for the identification, documentation, validation, or, where appropriate, verification, review, and approval of design (which includes process) change *before* their implementation, to avoid unintended consequences of a proposed change. At a minimum this includes: (1) establishing a design change control procedure and post-market change controls; (2) determining the risk presented by each potential change; (3) implementing controls *commensurate with the risk* presented by the proposed change; and (4) conducting an impact analysis on design output activities.

Change control applies to any requirement with which a product, component, process, service, or other activity must conform. Included in the scope for change control are:

- Packaging and labeling specifications (user interface);
- Device specifications (each individual part, as well as each sub-assembly requires a unique specification, and is under change control);
- Drug specifications (and components);
- Raw material specifications;
- Production equipment and process specifications;
- Device software (source code, build instructions, libraries, and executables);
- Quality assurance procedures, test methods, and specifications; and
- Installation, maintenance, and servicing procedures and methods.

As discussed in Chapter 6, risk management is an essential aspect of managing changes. Risk analysis tools such as URRA, dFMEA, pFMEA, and Hazard Analysis documents should be reviewed and updated, and risk acceptability determined. Each change should be evaluated against overall residual risk acceptability and the Risk Management File should be updated.

For combination products, change control demands a systems approach: Risk-based assessment and application of controls relative to the combination product, *each of its* constituent parts, *and* their associated components to protect against any potential negative interactions or unanticipated/undesirable outcomes. It is critical that the team assessing proposed changes works cross-functionally to ensure holistic consideration of risks and controls for any proposed change.

Consider the example of changes to an autoinjector. The change assessment begins with the intended outcome of safe and effective use of the combination product. Some of the initial considerations[29] include:

Intended users:	Education, age, impairment, training (e.g., trained healthcare provider, caregiver, or patient), experience;
Use environment:	Frequency of use, home care versus clinical setting; emergency care versus routine;
User interface:	Instructions for Use (e.g., text versus images), outer shell (e.g., grip, color); notifications (e.g., visual indicators and/or audible alarms).

Then there are considerations of the autoinjector underlying mechanism of action, components, and fluid path contacting materials that will influence the ability of the product to perform safely and effectively to meet the user needs. Each element requires vetting to ensure there are no unintended consequences of change:

Delivery mechanism: Spring (force and travel), needle shield safety function; syringe siliconization (break-loose/glide force);

Fluid path: Hub/connector, needle (length, material, diameter), glue.

When systems thinking is not applied for combination product change control, consequences could be disastrous. Table 8.7 includes a few illustrative examples of combination product changes gone wrong.

TABLE 8.7
Illustrative Examples of Combination Product Changes

What Changed	Impact(s)
Epoetin case study[a] (≈ concurrent changes) 1. A shift from intravenous administration to subcutaneous administration using a PFS[b] 2. Manufacturing process change to freeze drying 3. The biological product formulation was changed to incorporate a different biosurfactant 4. Silicone oil was used as a lubricant in prefilled syringes	1. Problems in storage and handling; induced antibody formation 2. Facilitated oxidation or aggregation of protein, enhancing immunogenicity 3. Interaction of new biosurfactant with/solubilizing organic leachables from rubber plungers 4. Increased immunogenicity ↓ Increased incidence of epoetin-associated pure red-cell aplasia/deaths
Change in adhesive for transdermal patch[c]	1. Potential interactions with drugs impacting E&L, drug strength, and efficacy 2. Inadequate adhesive strength for labeled conditions of use, leading to: • Partial detachment of patch, uncertainty about the resulting drug delivery profile, uncertainty about the rate and extent of drug absorption from the transdermal patch • Complete detachment of patch; uncertainty about the extent of drug absorption achieved; potential overdose due to predictable misuse (application of a second patch); unintentional exposure of unintended recipient of drug, who handles the medicated patch
Change in autoinjector for the administration of drug formulation[d]	Incompatible PFS/autoinjector system

[a] Charles Bennet, M.D., Ph.D., et al., "Pure Red-Cell Aplasia and Epoetin Therapy" New England Journal of Medicine 351 (September 30, 2004)

[b] Prefilled Syringe

[c] "Assessing Adhesion with Transdermal Delivery Systems and Topical Patches for ANDAs Draft Guidance for Industry" (October 2018) accessed September 24, 2022 at https://www.fda.gov/regulatory-information/search-fda-guidance-documents/assessing-adhesion-transdermal-delivery-systems-and-topical-patches-andas-draft-guidance-industry; "Accidental Exposures to Fentanyl Patches Continue to Be Deadly to Children" accessed September 24, 2022 at https://www.fda.gov/consumers/consumer-updates/accidental-exposures-fentanyl-patches-continue-be-deadly-children

[d] "FDA alerts patients, caregivers, and health care providers of cross-compatibility issues with autoinjector devices that are optional for use with glatiramer acetate injection" (8/18/2022) at https://www.fda.gov/drugs/drug-safety-and-availability/fda-alerts-patients-caregivers-and-health-care-providers-cross-compatibility-issues-autoinjector

The scope and impact of a proposed change should be assessed against the entirety of the design controls process (Figure 8.14).[30] Depending on the scope and impact of a proposed change, it may send you back into the design controls process. If over the course of development, Design Transfer or post market, new or modified user needs, intended uses or use environments are identified, that could trigger the need to repeat Design Validation (and/or human factors) activities. An existing device that is intended for a new user population, new indication for use, or a new use environment would be such a significant change.[31] Changes to design input requirements, perhaps due to shifting state-of-the-art standards, would trigger an assessment of needed updates to design outputs, which in turn could lead to repetition of applicable Design Verification and possibly process validation.

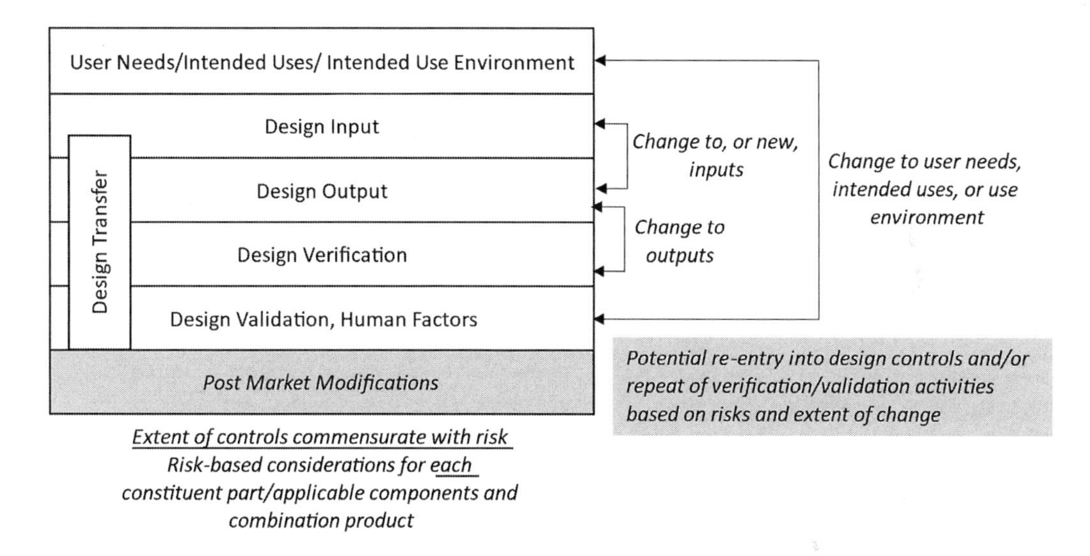

FIGURE 8.14 Change control risks and associated design control activities

Below is a list of some risk-based considerations to evaluate when assessing a potential change. Other questions may apply based on the specific combination product/combined use system you are addressing. For each question, consider *each* constituent part, *each* component, *and* the combination product/combined use system.

1. What is the impact on the risk profile?[32]
   a. Are new hazards, hazardous situations, or harms introduced? Are new controls required?
   b. Is the probability of occurrence of harm or severity of that harm increasing? Does it require modifications to risk controls?
2. Are there changes in "State-of-the-Art" *standards, regulations, competitive products, and treatments and in clinical practice that impact product use?*
3. Does the change impact other devices/accessories or interfaces to other products?
4. Does change impact Essential Performance Requirements (EPRs) and/or Established Conditions (ECs)?
5. Does the scope imply a need to conduct a Design Review?
6. Does the change impact the overall Quality System, e.g., Production and Process Controls? Design Verification? Design Validation? Human Factors? Process Validation?
7. Does the proposed change impact regulatory submissions?[33,34]

As indicated in "question 7," depending on the scope and impact of a change, it may require a new premarket submission, supplement, or study. The FDA has limited guidance on submissions for post-approval modifications to a combination product. According to their 2013 draft guidance on "Postapproval Modifications to combination products,"[35] FDA states:

> a combination product is comprised of different constituent parts. These constituent parts retain their regulatory identity as a drug, device or biological product. Therefore, if a change is made to any constituent part of the combination product that would have required a postmarket submission to FDA if the constituent part were a standalone product, then a postmarket submission is required for the combination product. In addition, a postmarket submission would also be required for the combination product if a change to any of the constituent parts would otherwise trigger the requirements associated with the application type used for approval of the combination product. In cases where the regulatory identity of the constituent part differs from the approved application type for the combination product, and a change is made that would require a postmarket submission to FDA, the requirement for submitting information about the change to the agency is generally satisfied with one postmarket submission to the original application. The type of submission to provide for the change will depend on the type of application used to obtain approval of the combination product. For example, a change to the device constituent part of a combination product approved under an NDA should be reflected in the appropriate postmarket NDA submission and be submitted to that NDA. In some cases, it may be easier to first identify the type of submission typically associated with the constituent part before determining what type of submission is required to the original application that was used for approval of the combination product.

Building off of Figure 8.10, "ICH Q12 Implementation Considerations" Draft Guidance (May 2021) and "Submissions for Post-approval Modifications to a Combination Product" Draft Guidance (January 2013), some broad guidelines for changes and associated submission types, are summarized in Figure 8.15. FDA's 2013 draft guidance includes tables that provide a bit more prescriptive information on submission expectations. These include:

- Draft Guidance Table 1: Type of NDA/BLA Submission for a Change in a Device Constituent Part of a Combination Product Approved under an NDA/BLA; and
- Draft Guidance Table 2: Type of Pre-Market Approval (PMA) Submission for a Change in a Biological Product/Drug Constituent Part of a Combination Product Approved under a PMA.

Those with questions on submission requirements for combination product post-approval changes are encouraged to reach out to the US FDA Office of Combination Products at combination@fda.gov and/or FDA review staff throughout the lifecycle[36] of a combination product.[37] *Note that regulatory filing requirements associated with such changes are different and evolving across jurisdictions globally, so the reader is encouraged to familiarize themselves with the applicable country/regional requirements (see also Chapters 3 and 14).*

Design changes must always be controlled. Changes during development may occur frequently as part of the iterative design process. The frequency of changes slows once you are at Design Transfer and post market. The level of effort and formality of documentation for change control increases as the development stages approach commercialization and are at the maximum at the point of final Design Transfer and commercial manufacturing. Significant changes during transfer can stall the Design Transfer process. Post-market manufacturing changes require ***careful*** review prior to implementation. Dependent on the scope and nature of a change, Design Verification and Design Validation may need to be repeated, and prior approval may be needed before marketing.[38]

Following is a phased approach for managing post-market changes. Risk management activities (plan, assessment, control, evaluation, and review) should be conducted as a foundation for the change control effort. Such analysis helps to determine what information/testing/data are needed given a specific change.

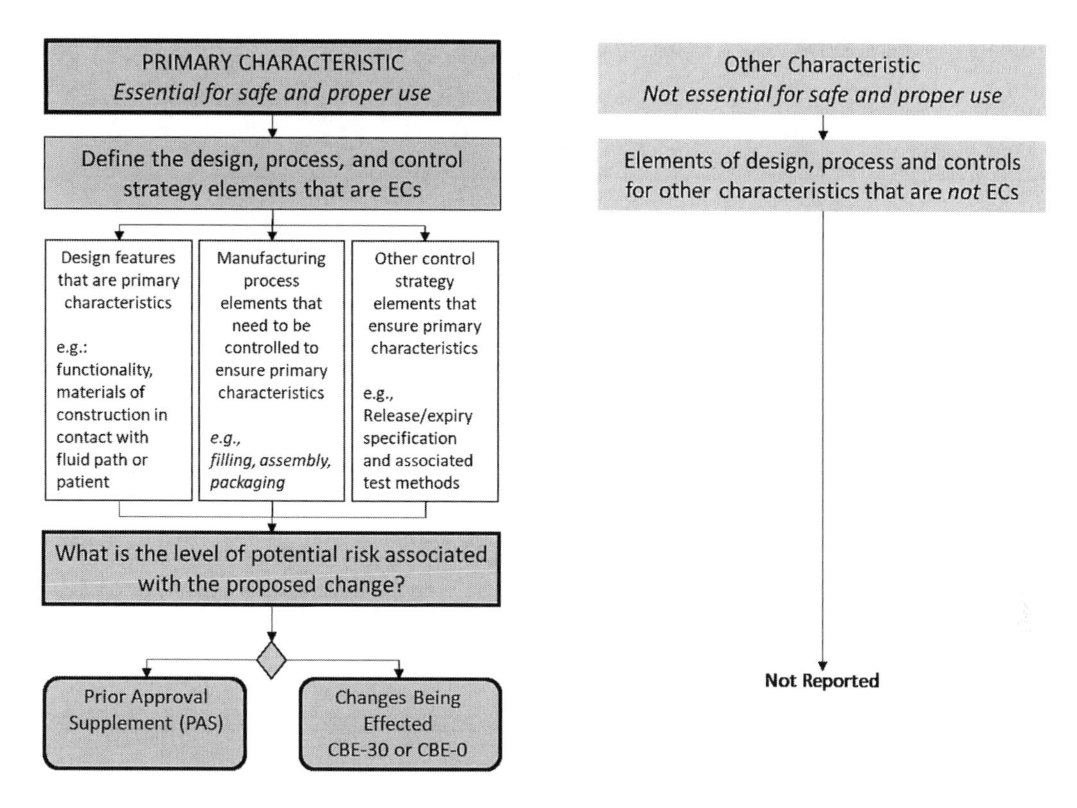

FIGURE 8.15 US FDA Regulatory submission requirements for a combination product post-market change

I. **Propose and plan design change**:
   a. Define the change and change objective;
   b. Identify change type, e.g., **finished product** (material, construction, form, fit, function, software, drug product), **process**, and **user interface**;
   c. Establish risk management plan, and assess risks (**intended use**, **reasonably foreseeable misuse**, identification of characteristics related to **safety**, identification of **hazards and hazardous situations**, **risk** estimation);
   d. Assess the impact on verification and validation;
   e. **Agreement to develop the proposed change plan**?
   f. **Evaluate risk control options and define risk control measures as appropriate.** Establish plan of execution commensurate with impacts. Considerations include, e.g., compatibility; performance; stability; shelf life or use life; validation activities; human factors; clinical evaluation.

II. **Verification and validation of design change**:
   a. **Implement risk control measures**;
   b. **Evaluate risks arising from risk control measures**;
   c. Execute verification testing, e.g., performance, stability, shelf life, use life, compatibility, etc.;
   d. Execute validation studies, e.g., process validation, human factors, software validation, and clinical evaluation.

III. **Final assessment**:
   a. **Evaluate residual risks, controls, and risk acceptability/completeness of risk controls**;
   b. Assess impact to Essential Performance (EPRs, ECs); If Essential Performance is impacted, assess clinical bridging and/or threshold analysis;
   c. Decision: **Proceed with change**?

IV. **Evaluate effectiveness of change implementation:**
   a. Information collection;
   b. Information review;
   c. Actions, if required.

Bridging for Drug-Device and Biologic-Device Combination Products[39]

Bridging refers to the process of establishing the scientific relevance of information developed in an earlier phase of a development program or in another development program to support the combination product for which an applicant is seeking approval. Bridging is relevant to post-market modifications to combination product where, for example, a drug may have been developed or even marketed in one dosing configuration, like a prefilled syringe, and there is a desire to change the administration to an autoinjector. FDA Draft Combination Product Bridging Guidance (2019) guides Sponsors in steps for clinical bridging, where appropriate. The process is founded upon conducting a gap analysis between the approved and proposed product configurations. All differences should be identified, considering the potential effect of the individual and aggregate differences on the safety and effectiveness profile of the proposed combination product configuration. The safety and effectiveness profile should include a clear and comprehensive listing of the differences in the device constituent part, the drug constituent part, and the combination product as a whole. Example considerations include:

- Changes to the local injection adverse reaction profile associated with changes in drug concentration, viscosity, formulation, or injection rate;
- Changes to the dose accuracy of the same device constituent part when the drug formulation is changed;
- Changes in manufacturing process and/or device constituent part that could impact drug quality;
- Changes to product usability by intended users when the device constituent user interface is changed;
- Changes to bioavailability of the drug and/or its metabolic profile associated with changes in the device, formulation, or route of administration (e.g., changes in needle depth, tissue plane, or rate of infusion; changes in drug formulation that lead to differential lung depositions, even with the same device); and
- Changes in drug formulation that can affect the extractable and leachable profiles of the combination product.

See Chapter 3, Combination Products Regulatory Strategies, for more discussion on this important topic.

Threshold Analyses and Human Factors for Combination Products[40]

As part of evaluating combination products and post-market changes for safety and effectiveness, one should consider the product user interface (see Chapter 7). Use-Related Risk Analysis and Human Factors studies[41] are key components of an appropriate evaluation. Threshold analyses are generally used in comparing two drug products. Similar to the bridging discussion above, Sponsors should include comparison of the product configurations – in this case, with a focus on the user interface. Considerations include:

- Labeling comparisons;
- Comparative task analysis of the proposed product and the product it references;
- Physical comparison of the device constituent part(s), e.g., detailed assessment of the physical features of the proposed product and the product it references;
- Assessment of the extent of design differences (characterized as, e.g., major or minor) that may exist between the two products, and a rationale for each characterization; and
- Comparative use human factors study may be required.

Demonstrating the product's relative safety and effectiveness and applying any relevant special controls are necessary steps to provide a reasonable assurance of safety and effectiveness for the combination product.

FDA published Final Guidance in January 2022 entitled "Principles of Premarket Pathways for Combination Products."[42] While this chapter is focused on post-market activities, that premarket guidance document is helpful in that it discusses "substantial equivalence." Understanding the principles of establishing substantial equivalence can help the Sponsor and/or combination product manufacturer who is conducting gap analyses and bridging activities. This guidance also provides helpful insights into considerations for submissions (see also Chapter 3). The following excerpt is drawn from the guidance:

> The standard for a determination of substantial equivalence in a 510(k) review is set out in section 513(i) of the FD&C Act. A product is substantially equivalent to a predicate product if it:
>
> - has the **same intended use as the predicate product**; and
> - has the **same technological characteristics** as the predicate product;
>
> **or**
>
> - has the **same intended use as the predicate product**;
> - has **different technological characteristics**; and
> - the information submitted to FDA, including **appropriate clinical or scientific data** if deemed necessary, demonstrates that **the product:**
>   - **does not raise different questions of safety and effectiveness than the predicate product; and**
>   - **demonstrates that the product is as safe and effective as the predicate product**.
>
> FDA considers the product's relative safety and effectiveness in the substantial equivalence determination, and safety and effectiveness considerations are also critical to the Agency's evaluation of compliance with any applicable special controls, all of which FDA has determined to be necessary to provide a reasonable assurance of safety and effectiveness for the product type.

Change Management and Purchasing/Supplier Controls

Changes associated with purchased constituent parts are a critical aspect of post-market lifecycle management and manufacturing controls. Arrangements (documented) with the manufacturers of constituent parts are important to ensure sufficient awareness of manufacturing changes to constituent parts that may occur – premarket or post market. These arrangements, e.g., in quality agreements, help ensure safety and effectiveness of the combination product. The combination product manufacturer needs enough time to assess the impact(s) of a proposed change commensurate with the risk and complexity of change and given the particular stage of combination product lifecycle management. Table 8.8 summarizes suggested considerations, expectations, and division of responsibilities for a supplier of a device constituent part and a drug-led combination products Sponsor.[43] Be wary that what one entity may historically have interpreted as simply a packaging 'component' could be interpreted as a 'device constituent part' in the combination products space. Table 8.8 should be interpreted cautiously.

Completion of Design Transfer

Design Transfer can happen in phases over time, with aspects of the Master Batch Record/Device Master Record released early (e.g., materials, components). Design Transfer is complete when: (1) the MBR (including the DMR elements) or DMR (including the MBR elements) is released; (2) adequate controls are in place to consistently produce constituent parts and combination product that conforms to specifications; (3) operations, quality, pharmacovigilance, and other related commercialization personnel are trained and qualified; and (4) the Design and Development Plan and Design Transfer Plan deliverables are completed.

TABLE 8.8
Division of Responsibilities between Device Supplier and Drug PMOA Sponsor

Component Supplier	Device Supplier	Combination Product Sponsor
• Identifies and evaluates inherent failures associated with the component manufacturing process, and use, including foreseeable mis-use of the component • Identifies potential harms, agnostic of the device or drug • Applies risk controls and conducts verification independent of the specific device or drug therapy • Assures controls for lifecycle management activities, e.g., change control, supply availability • Maintains a risk file for the component	• Identifies and evaluates inherent failures associated with the device design, manufacturing process, and use, including foreseeable mis-use of the device constituent part • Identifies potential harms, agnostic of the drug • Applies risk controls and conducts verification independent of the specific drug therapy • Assures controls for lifecycle management activities, e.g., change control, supply availability, investigations, and, if applicable, Postmarketing Safety Reporting (PMSR) • Maintains a risk file for the device constituent part	• Assess and address unique interactions of the component or constituent part with the finished product. If using the device as a platform technology, conduct such assessment for each specific medicinal product, intended use, intended users and use environment • Link (potential) failures to hazardous situations and harms • Final verification of risk controls and their effectiveness • Residual risk assessment including benefit/risk analysis • Postmarketing Safety Reporting on the combination product and its constituents • Maintain risk file for combination product and its constituent parts

In summary, Design Transfer is the systematic approach applied to pass the documented knowledge and experience gained during development and/or commercialization to an appropriate, responsible, and authorized party. It typically spans product development to commercial production but is also applicable when transferring a product or process between production sites. The Design Transfer Plan describes all the activities, roles, responsibilities, and expected outcomes for the commercialization process. Design Transfer is highly cross-functional! One of the first elements typically transferred are analytical methods, i.e., for comparability assessments of product between pilot/development and commercial scale product. Foundational to the entire Design Transfer risk-based effort is building a robust understanding of your process and process capabilities and ensuring controls are in place. As Design Transfer is completed, extensive documentation is established to ensure that the constituent parts and combination product design are correctly translated into production specifications, so that resulting product use is consistently safe and effective.

DESIGN HISTORY FILE

The Design History File (DHF) is a compilation of records that describe the design history of the finished combination product. The DHF contains or references records[44] that demonstrate the design was developed in accordance with the approved Design and Development Plan (see Chapter 4), company-specific design control procedures, and the requirements of 21 CFR 820.30(j) or ISO 13485:2016 clause 7.3.10. The DHF should have an accurate, clear index/table of contents to ensure the design's history is evident. A DHF is required for each combination product. If using a technology platform approach, you still need a DHF for each combination product to address its unique considerations based on the context of use (users, uses, use environment, drug interactions, etc.). Chapter 4 includes an example table of contents for a combination product DHF.

The DHF is necessary so that manufacturers can demonstrate and be accountable for control over the design process. According to QS Regulation Comment 90, the DHF maximizes the probability that the finished design conforms to the design specifications. QS Regulation Comment 92 states that "manufacturers who do not document all their design efforts may lose the information and experience of those efforts, thereby possibly requiring activities to be duplicated."

Procedures for the Design History File should define who:

- Maintains the DHF during the design process (e.g., indexing, adding documents, tracking pending documents, correlating electronic files with DHF – identifying and saving raw Design Verification data);
- Maintains the DHF after Design Transfer is complete, handling post-transfer DHF updates; and
- Review or audits the DHF.

Design Transfer/preliminary DMR (MBR) records may not be controlled when they are first established. The Design History File may contain early, uncontrolled versions. Design changes after Design Transfer affect the DHF. Industry practice is generally that post-transfer DHF records are maintained separately within the change control system. Alternatively, some create a separate "Chapter" of the DHF for post-market maintenance.

POST-MARKETING SURVEILLANCE AND SAFETY REPORTING

Once the combination product is released on the market, the monitoring phase begins (see Figure 8.16). The Sponsor monitors the safety and performance of the constituent parts and combination product in the marketplace and reports safety and efficacy issues identified through monitoring to health authorities.

The post-market safety of combination products is based on the safety, safe use, and efficacy/effectiveness of its multiple constituent parts, each with their own mode of action and safety risk profile coming together to meet the combination products' intended therapeutic or diagnostic use. The combination products are intended to enhance the benefit–risk profile of the product with benefits like ease of use, enhanced compliance, and broader population use across diverse environmental settings. Often the combination product may have additional safety risks that are introduced due to the physical, mechanical, or chemical interactions that may occur between the constituent parts that require additional mitigations and monitoring after distribution (see Chapter 6).

It is not the intent of this section of the chapter to exhaustively review the details of Post-market Surveillance and Reporting. Rather, the authors are introducing you to combination products' Post-market Surveillance (PMS) and Vigilance and to share best practices. If you are responsible for PMS, Vigilance, and Reporting for combination products in your organization, the authors recommend you familiarize yourself with the requirements and associated guidance for the specific countries in which you seek to market your combination product(s), as these requirements are not harmonized.

POST MARKET SURVEILLANCE PLAN

A Sponsor should establish a Post Market Surveillance (PMS) Plan prior to market launch[45] (Figure 8.17). Post Market Surveillance (PMS) can further refine, confirm, or deny the safety of a drug, device, or combination product after it is used in the general population. As illustrated in Figures 8.17 and 8.18, the PMS Plan is actively reviewed and updated and may drive risk-based actions through the Quality Management System.

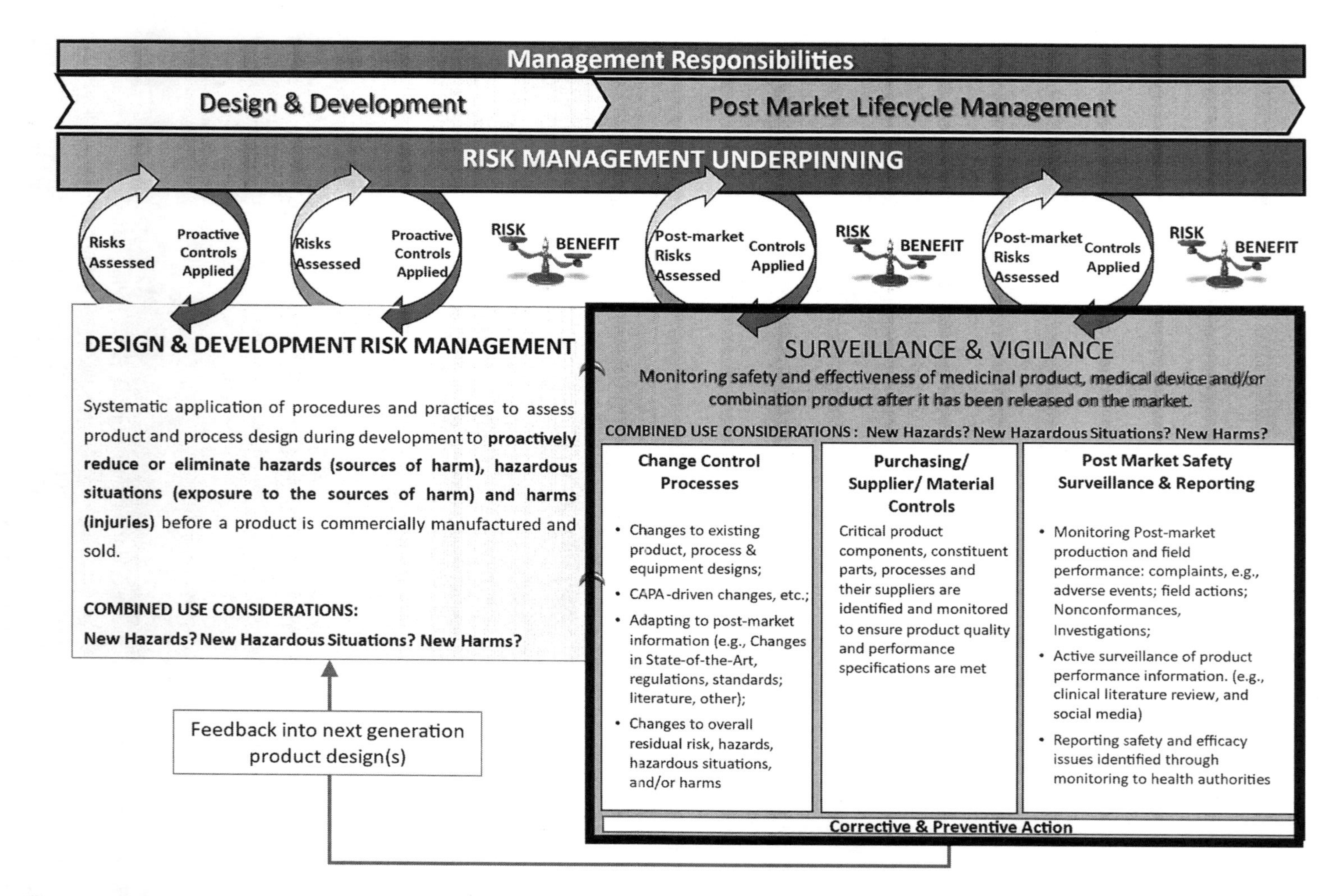

FIGURE 8.16 Combination product lifecycle (©2022 Combination Products Consulting Services, LLC. All Rights Reserved.)

Active monitoring of medical product safety is done through a variety of means, e.g.:

- Spontaneous reporting databases;
- Prescription event monitoring;
- Electronic health records;
- Patient registries;
- Record linkages between health databases;
- Social media;
- Market research; etc.

Data is reviewed to highlight potential safety concerns.

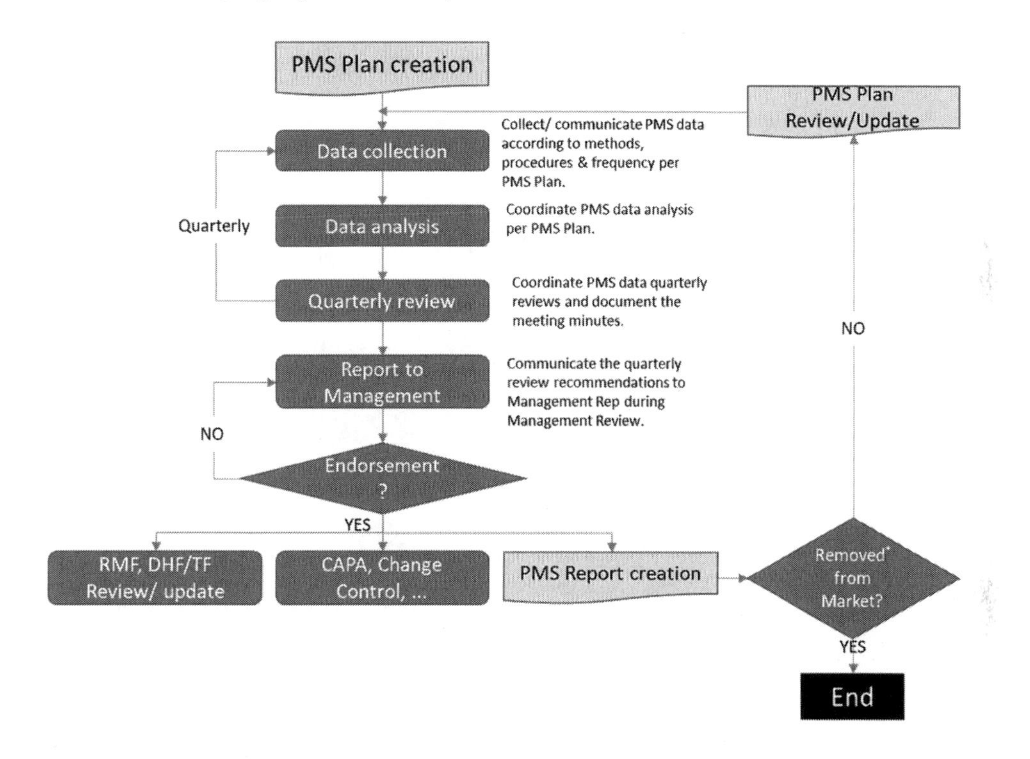

FIGURE 8.17 Post-market Surveillance (PMS) Plan

Postmarketing Safety Reporting (PMSR)

Sponsors are expected to report certain adverse events and other post-market issues associated with their products to health authorities through Postmarketing Safety Reporting (PMSR). In this section, we:

- Review the PMSR evolving global regulatory landscape;
- Dive into US PMSR expectations;
- Tease out global PMSR challenges, considerations, and lessons learned; and
- Summarize some key reporting best practices.

PMSR Evolving Global Regulatory Landscape

PMSR requirements are evolving globally, with varied regulations across geographies. There are multiple geographies that currently have some form of acknowledgment of "combination product," e.g.:

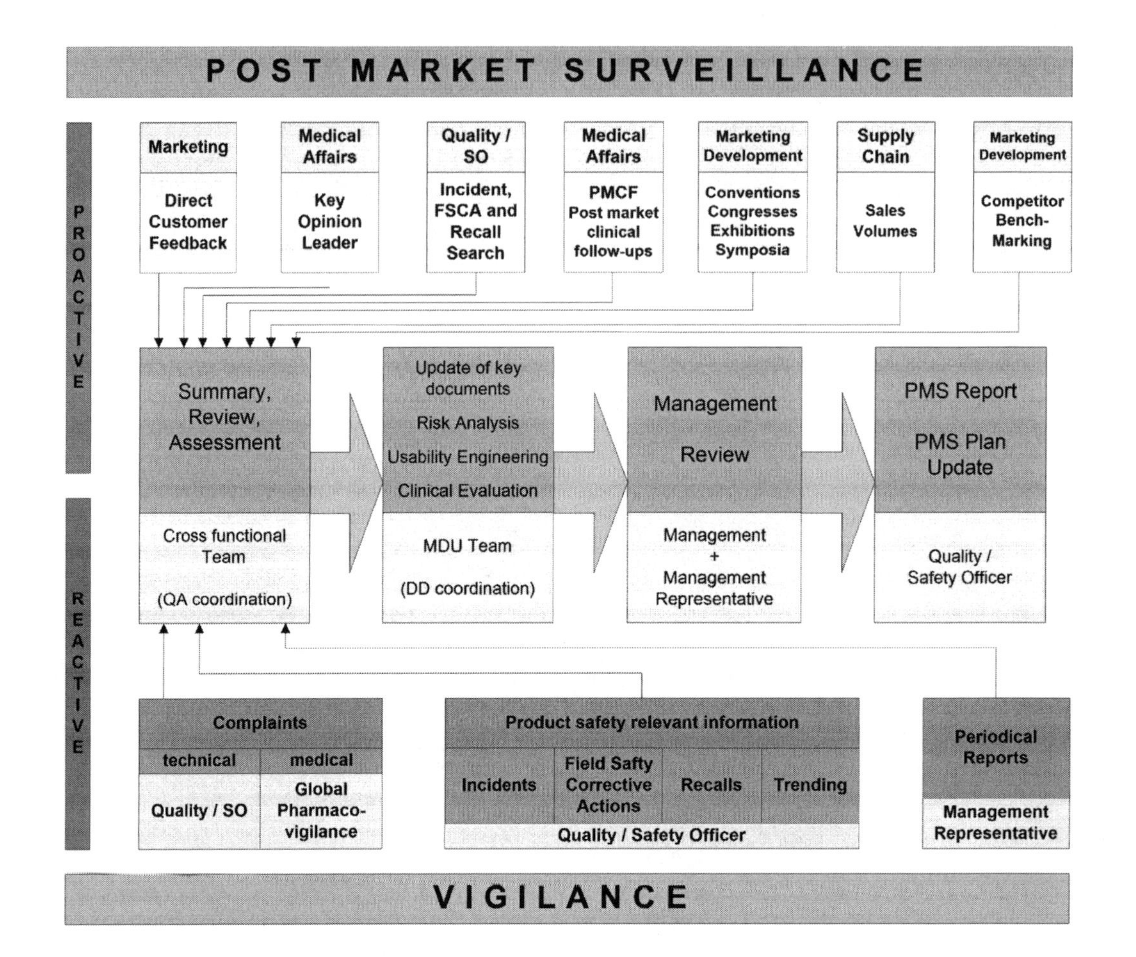

FIGURE 8.18 Post-market Surveillance

Europe: EU, Switzerland, Turkey
Asia-Pacific: Australia, China, Japan, Malaysia, Saudi Arabia, South Korea;
Americas: Brazil, Canada, Mexico, United States.

Looking at Combination Product Postmarketing Safety Reporting, and looking across these countries, the diversity in PMSR expectations is vast. There are three distinct models that are emerging:

1. One is the US model, where the intention is to reduce redundant/duplicate reporting. Only time will tell how effective that has been. Under 21 CFR Part 4B, there are streamlining mechanisms, where multiple reports can be combined and submitted through a common channel based on application type.
2. On the other end of the spectrum is the Asian emerging model, where there is a clear ask for industry to do duplicate reporting where we cannot pinpoint the causality;[46] if we cannot say for sure it was the drug or the device that was the source of the event, then the expectation is duplicate reporting. (Japan was actually among the first countries to issue formal combination product-specific PMSR requirements,[47] effective November 25, 2016.)
3. The third is the EU model, which is right in the middle. For some things the approach is streamlined, and for others the approach is duplicate reporting. For a single integral, non-reusable product (see Chapters 4 and 14), there is one set of expectations, but for co-packaged, single integral re-usable, or referenced products, there is a separate set of expectations.

PMSR requirements specific to combination products are the most expansively articulated under the US FDA regulations and guidance, so our review in this chapter and is admittedly rather US-centric, focusing on the US expectations. We will also review more globally applicable best practices for Combination Products PMSR.

Combination Product PMSR under US FDA

US FDA requirements for Combination Products PMSR are delineated in 21 CFR Part 4B and 81 FR 92603. FDA issued Final Guidance[48] and a helpful webpage[49] to clarify expectations for Sponsors on PMSR for combination products. Similar to the Part 4A approach to streamlined cGMPs, 21 CFR Part 4B recognizes that PMSR regulations for drugs, devices, and biological products share many similarities, but each also has some distinct reporting requirements, including triggers and timelines for reporting (Figure 8.19). 21 CFR Part 4B "addresses PMSR requirements for combination products that have received FDA marketing authorization, to ensure consistent and complete reporting while avoiding duplication."[50]

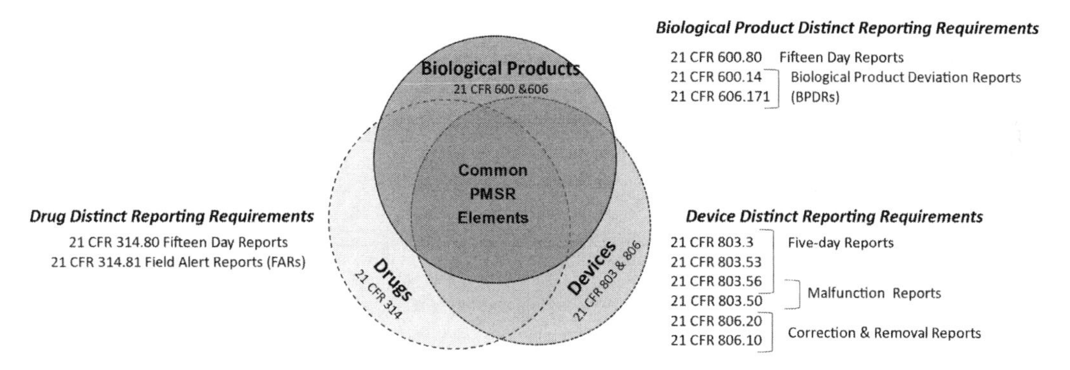

FIGURE 8.19 US Combination Product PMSR: common and distinctly interpreted reporting requirements

Key Terminology, Report Types, and Reporting

Table 8.9a is a handy summary of key terminology and examples associated with Combination Products PMSR. Please take the time to read each definition, interpretation, and examples, as they are important for effective understanding and execution of PMSR.

TABLE 8.9A
Key Terminology

Key Terminology (Constituent Part Type) (21 CFR, etc.)	Definition/Interpretation	Examples
Malfunction (device) [803.3(k)]	The failure of a device constituent part or of the product as a whole.	• Bent needle • Plunger broken • Label damage • Seal integrity compromised
Reportable malfunction (device) (806)	A malfunction is reportable when the applicant receives information that reasonably suggests that the product has malfunctioned ***and***	• Out-of-the box device malfunction where the autoinjector prematurely actuated; although no injury occurred, if the malfunction were to recur, a serious needle stick injury might result.

(Continued)

TABLE 8.9A (CONTINUED)
Key Terminology

Key Terminology (Constituent Part Type) (21 CFR, etc.)	Definition/Interpretation	Examples
	the product, or a similar product marketed by the applicant, "**would be likely*** to cause or to contribute to death or serious injury if the malfunction were to recur." For combination products, malfunctions may be reportable even if there is no patient adverse event.	• Inadequate adhesive strength for on-body injector (OBI) which led to premature detachment and inadequate delivery of life-saving drug, requiring patient hospitalization. • Premature deployment of drug-eluting stent out-of-the box. • OBI occlusion causing inability to deliver life-saving dose of insulin.
Serious injury (SI) (device) [803.3(w)]	An injury or illness that: Is life-threatening; Results in permanent impairment of a body function or permanent damage to a body structure; or Necessitates medical or surgical intervention to preclude permanent impairment of a body function or permanent damage to a body structure. Permanent means irreversible impairment or damage to a body structure or function, excluding trivial impairment or damage.	• Patient suffers a stroke and paralysis after the implantation of a stent. • Patient receives stitches due to explant of broken needle. • Someone chokes on parts of the product. • Label damage on a laser, leading to improper use of device, leading to blindness.
Adverse experience (AE) [314.80(a)] [600.80(a)]	Adverse experiences include any failure of expected pharmacological action and adverse events occurring: in the course of the use of the product in professional practice, from product overdose, whether accidental or intentional, from product abuse, or from product withdrawal. Drug: Any adverse event associated with the use of a drug in humans, whether or not considered drug related, including the following: An adverse event occurring in the course of the use of a drug product in professional practice; an adverse event occurring from drug overdose whether accidental or intentional; an adverse event occurring from drug abuse; an adverse event occurring from drug withdrawal; and any failure of expected pharmacological action. Biological Product:	• Nausea • Vomiting • Headaches • Dizziness • Cough • Sore throat • Hyperactivity • Agitation • Anxiety • Lack of effect • Allergic reaction to vaccine

(Continued)

TABLE 8.9A (CONTINUED)
Key Terminology

Key Terminology (Constituent Part Type) (21 CFR, etc.)	Definition/Interpretation	Examples
	Any **adverse event** (AE) associated with the use of a biological product in humans, whether or not considered product related, including the following: An **AE** occurring in the course of the use of a biological product in professional practice; an **AE** occurring from overdose of the product whether accidental or intentional; an **AE** occurring from abuse of the product; an **AE** occurring from the withdrawal of the product; and any failure of expected pharmacological action.	
Life-threatening adverse experience [314.80(a)] [600.80(a)]	Any adverse drug experience that places the patient, in the view of the initial reporter, at immediate risk of death from the adverse drug experience as it occurred, i.e., it does not include an adverse drug experience that, had it occurred in a more severe form, might have caused death.	Anaphylactic shock Miscarriage
Unexpected adverse experience (***Unexpectedness***) [314.80(a)] [600.80(a)]	Any adverse drug experience that is not listed in the current labeling for the drug product. This includes events that may be symptomatically and pathophysiologically related to an event listed in the labeling but differ from the event because of greater severity or specificity. "Unexpected," as used in this definition, refers to an adverse drug experience that has not been previously observed (i.e., included in the labeling) rather than from the perspective of such experience not being anticipated from the pharmacological properties of the pharmaceutical product.[a]	• Hepatic necrosis would be unexpected (by virtue of greater severity) if the labeling only referred to elevated hepatic enzymes or hepatitis. • Cerebral thromboembolism and cerebral vasculitis would be unexpected (by virtue of greater specificity) if the labeling only listed cerebral vascular accidents.
Serious adverse experience (SAE) [314.80(a)] [600.80(a)]	Any adverse experience occurring at any dose that results in any of the following outcomes: Death, a life-threatening adverse experience, inpatient hospitalization or prolongation of existing hospitalization, a persistent or significant disability/incapacity, or a congenital anomaly/birth defect. Important medical events that may not result in death, be life-threatening, or require hospitalization may be considered a serious adverse experience when, based upon appropriate medical judgment, they may jeopardize the patient or subject and may require medical or surgical intervention to prevent one of the outcomes listed in this definition.	• Hypoglycemic shock due to overdose of insulin • Allergic bronchospasm requiring intensive treatment in an emergency room or at home • Blood dyscrasias or convulsions that do not result in inpatient hospitalization • Development of drug dependency or drug abuse • Anaphylactic shock due to latex allergies • Hospitalization for transfusion • Broken embedded needle • Contaminated product causing sepsis • Stroke due to procedure implant drug-eluting stent

(*Continued*)

TABLE 8.9A (CONTINUED)
Key Terminology

Key Terminology (Constituent Part Type) (21 CFR, etc.)	Definition/Interpretation	Examples
Causality assessment	Causality assessment is a common procedure in pharmacovigilance. It generally includes physicians, investigators, professionals working in medical safety and national health authorities which can assist in taking regulatory decisions. Causality is the key factor for the identification of new signals, measuring the strength of evidence, and in evaluating benefit–risk profile of pharmaceutical medicinal products.[b]	

[a] For a cross-labeled CP, if the event is listed in the labeling accompanying either of the constituent parts, the event is considered expected.

[b] "Public Safety & Vigilance" (August 24,2020) accessed September 29, 2022 at https://www.linkedin.com/pulse/causality-assessment-pharmacovigilance-concept-and-vigilance/

Note: Historically, those working in the premarket clinical trial space use the terminology "adverse experience", while those working in the post market space use the terminology "adverse event". These terms are now generally used interchangeably.

Certain key concepts associated with the underlying requirements to be aware of when the combination products have a device and drug:

(1) For combination products and medical devices, reportability is not restricted to a serious injury or death that HAS occurred. The authors would like to specifically draw attention to the "reportable malfunction" terminology in the keywords in Table 8.9a. Malfunction reportability is not solely based upon a death or serious injury *having already occurred.* Rather, reportability is based upon whether, even if there was no such death or serious injury, if there is a ***likelihood*** that that malfunction would result in a death or serious injury ***if that malfunction were to recur***. Out-of-box failures that have the potential to cause death or serious injury if they were to recur may fall under this category. The concept is that "you are lucky it didn't happen when in-use; it was a near-miss." This contrasts with the interpretation of reportability for drugs and biologics, where reportability is driven by adverse events that HAVE occurred.

Medical professionals who are accustomed to working, e.g., with Class III pacemaker devices, have little difficulty determining which malfunctions would meet such likelihood criteria. Medical professionals in an organization that is not familiar with this device likelihood concept sometimes struggle, particularly when they are more accustomed to reporting on events that have already occurred, or those that are for chronic, low-risk, preventive therapy combination products. The likelihood assessment – whether qualitative or quantitative – requires fair and balanced thinking.

(2) For drugs, reportability takes into consideration the concept of "expectedness" based on the product label, accounting for the event in question (see Table 8.9a, "Unexpectedness"). If the label specifies that such a serious event could be anticipated, then it is deemed "expected." This includes events that may be symptomatically and pathophysiologically

related to an event listed in the labeling but that differ from the event because of greater severity or specificity (see 21 CFR 314.80(a) and 600.80(a)). The labeling for a combination product includes any labeling accompanying individual constituent parts. This contrasts with the interpretation of reportability for device/device-led products, under 803 for malfunctions, death, and serious injuries "expectedness" is not applicable and death, serious injury, and malfunction events are deemed reportable regardless of labeling expectedness.

Table 8.9b is a high-level overview of key report types, descriptions with respect to combination products, and reporting cadence expectations.[51] Table 8.9c summarizes where each of the report types should be submitted. The information in Tables 8.9a, 8.9b, and 8.9c is intended to serve as a useful backdrop for the rest of the discussion on Combination Products PMSR. Again, please take the time to read through the tables, as they will help you understand the reporting obligations discussed throughout this section.

TABLE 8.9B
Report Types, Descriptions, and Timing Requirements[a]

Case Safety Report Type	Report Type	General Description	Timeline for Reporting for Combination Products (CPs) Based on Application Type
ICSR	**Fifteen-Day Report** (21 CFR 314.80, 21 CFR 600.80) (21 CFR 4.102(c))	Report of adverse experience that is both serious and unexpected	**ANDA, NDA, BLA CPs**: No later than 15 calendar days from the initial receipt of information by the applicant **Device Application CPs**: No later than 30 days from the initial receipt of information by the applicant
ICSR	**Follow-Ups to Fifteen-Day Reports** (21 CFR 314.80(c) and 600.80(c)) (21 CFR 4.102(c))	The requirements for fifteen-day reporting for combination products include requirements for follow-up reports. Follow-up reports are required when the ICSR submitter becomes aware of reportable new information related to the event that was not available at the time of the initial report (see 21 CFR 314.80, 600.80, and 803.56). Under US Combination Products PMSR remit, there is flexibility to submit alternate constituent part-related report types as they become known.	**NDA, ANDA, BLA CPs**: Within 15 calendar days of receipt of new information **Device Application CPs**: No later than 30 days from the initial receipt of new information by the applicant
ICSR	**"Non-expedited ICSRs"**	ICSRs for adverse experiences that are either serious and expected or nonserious	Multiple report types and associate timelines may apply for the event,[b] e.g., 314.80 (15 days), 600.80 (15 days) 803.50 (malfunction)
ICSR	**Five-Day Report** (21 CFR 803.53)	Report of an event that necessitates remedial action to prevent an unreasonable risk of substantial harm to the public health	No later than five work days after the day that the applicant becomes aware[c] that remedial action is necessary

(Continued)

TABLE 8.9B (CONTINUED)
Report Types, Descriptions, and Timing Requirements[a]

Case Safety Report Type	Report Type	General Description	Timeline for Reporting for Combination Products (CPs) Based on Application Type
ICSR	**Death/Serious Injury/ Malfunction Report** (21 CFR 803.50)	Report of an event when the applicant receives information that reasonably suggests that the product has malfunctioned ***and*** the product, or a similar marketed product marketed by the applicant, "would be likely to cause or to contribute to death or serious injury if the malfunction were to recur."	No later than 30 calendar days from the day that the applicant receives information or otherwise becomes aware of the event
ICSR	**Supplemental/ Follow-Up Report to Five-Day/Death/ Serious Injury/ Malfunction Report[d]** (21 CFR 803.56) (21 CFR 4.102(b) (1))	The requirements for five-day and malfunction reporting for combination products include requirements for follow-up reports. Follow-up reporting requirements also apply to death and serious injury reports submitted by Combination Product Applicants for combination products that receive marketing authorization under a Device Application (see 21 CFR 4.102(b) (1) and 803.56). Under US Combination Products PMSR remit, there is flexibility to submit alternate constituent part-related report types as they become known.[e]	Within 30 calendar days of the day that the applicant receives information
ICSR	**"Same-Similar" Reporting** (21 CFR Part 4B)	There is an expectation of "Same-Similar Reporting" for events or experiences (all ICSRs) that meet US reportability criteria for an applicant's marketed same-similar constituent part(s) or combination product(s).	The reporting requirements for foreign events for combination products align with the underlying regulatory requirements for drugs, devices, and biological products for such events
Non-ICSR	**Field Alert Report (FAR)[f]** (21 CFR 314.81 (b) (1); 21 CFR 4.102(a); 4.102(b)(2); and 4.102(c)(2)(i))	Required the following issues could have resulted from the manufacturing process for any of the constituent parts of the combination product or for the combination product, even if due to material supplied to the applicant by a third party: "[A]ny incident that causes the [product] or its labeling to be mistaken for, or applied to, another article," or "concerning any bacteriological contamination, or any significant chemical, physical, or other change or deterioration in the distributed [product]" or "[A]ny failure of one or more distributed batches of the [product] to meet the specification established for it in the application" (21 CFR 314.81)	Within three working days of receipt of the information by the applicant
Non-ICSR	**Biological Product Deviation Report (BPDR)** (21 CFR 600.14(c), 21 CFR 606.171(c), 4.102(b)(3), and 4.102(c)(3)(i)	"Any event, and information relevant to the event, associated with the manufacturing, to include testing, processing, packing, labeling, or storage, or with the holding or distribution," which suggests that for a distributed product, there was either:	As soon as possible, but not to exceed 45 calendar days from the date of acquiring information reasonably, suggesting that a reportable event has occurred

(*Continued*)

TABLE 8.9B (CONTINUED)
Report Types, Descriptions, and Timing Requirements[a]

Case Safety Report Type	Report Type	General Description	Timeline for Reporting for Combination Products (CPs) Based on Application Type
		A deviation from cGMPs, applicable regulations, applicable standards or specifications that may affect the safety, purity, or potency of that product; or An unexpected or unforeseeable event that may affect the safety, purity, or potency of that product; and it occurs in the applicant's facility or another facility under contract with the applicant.	
Non-ICSR	**Correction and Removal Report** (21 CFR 806.10(b)), 4.102(b)(1), and 4.102(c)(1)(iii) (Also known as "10-day report")	Product is corrected or removed[g] from distribution due to a risk to health posed by the product or to remedy a violation of the FD&C Act which may present a risk to health.	Within ten working days of initiating correction or removal
Periodic Safety Reports[h] 21 CFR 314.80(c)(2) and 600.80(c)(2) Also recognized globally under ICH E2C(R2)	**Periodic Adverse Experience Report (PAER)** **Periodic Adverse Drug Experience Reports (PADERs)** **Periodic Safety Update Reports (PSURs)** ICH E2C (R1) **Periodic Benefit Risk Evaluation Reports (PBRERs)** ICH E2C(R2)	Applicants are required to submit post-marketing periodic safety reports for each approved application. Under the Combination Product PMSR rule, periodic reporting is required for combination products marketed under an NDA, ANDA, or BLA (21 CFR 4.102(d)(1), see 21 CFR 314.80(c)(2) and 600.80(c)(2)). If such a combination product includes a device constituent part, these periodic reports must include a summary and analysis of the Five-day and Malfunction reports submitted during the reporting interval for the periodic safety report (see 21 CFR 4.102(d)(1)). Periodic reporting is not routinely required for device applications, but there is a provision that allows FDA to request periodic reporting for device-led combination products (such requests are rarely used) [21 CFR 4.102(d)(2)].	The reports must be submitted quarterly for the first three years following the US approval date and annually thereafter. FDA accepts all three formats, the PADER/PAER, PSUR, and PBRER, to fulfill the postmarketing periodic safety reporting requirements under 21 CFR 314.80(c)(2) and 600.80(c)(2). Each format must be submitted according to the content and timelines specified in the regulations (PADER/PAER) or by ICH (PSUR and PBRER). See the Periodic Safety Reports (November 2016) Guidance for additional details.

[a] Adapted from US FDA Guidance "Post-Marketing Safety Reporting for Combination Products" (July 2019), Table 1, p. 28

[b] (See also "Streamlined Reporting" section later in this chapter.)

[c] Footnote 23 from US FDA Guidance on Combination Products PMSR (2019): "Applicants are considered to have become aware when any employee with management or supervisory responsibilities over persons with regulatory, scientific, or technical responsibilities, or whose duties relate to the collection and reporting of adverse events, becomes aware, from any information, including any trend analysis, that a reportable event or events necessitates remedial action to prevent an unreasonable risk of substantial harm to the public health (see 21 CFR 803.3(b)). If FDA has requested Five-day reports for certain events in accordance with 21 CFR 803.53(b), applicants are also considered to have become aware when any employee becomes aware of such an event occurring (see 21 CFR 803.3(b))"

[d] Ibid.

[e] see section on "Streamlining in this chapter.

[f] FDA Guidance "Field Alert Report Submission: Questions and Answers (July 2021) at https://www.fda.gov/drugs/surveillance/field-alert-report-form-questions-and-answers and https://www.fda.gov/vaccines-blood-biologics/guidance-compliance-regulatory-information-biologics/submitting-field-alert-reports-fars-cber.

[g] 21 CFR 806.2(d) defines "correction" as any repair, modification, adjustment, relabeling, destruction, or inspection (including patient monitoring) of a product without its physical removal from its point of use to some other location. 21 CFR 806.2(j) defines "removal" as the physical removal of a product from its point of use to some other location for repair, modification, adjustment, relabeling, destruction or inspection.

[h] US FDA Guidance "Providing Postmarketing Periodic Safety Reports in the ICH E2C(R2) Format (Periodic Benefit-Risk Evaluation Report)" (November 2016), accessed September 27, 2022 at https://www.fda.gov/media/85520/download

TABLE 8.9C
Where to Submit Reports

Report Type	Where to Submit Reports	Example
ICSRs[a]	Combination Product Applicants must submit ***all*** ICSRs in accordance with the procedural requirements in the regulations associated with the **application type** (see 21 CFR 4.104(b)) and should follow relevant **policies and procedures of the lead Center**. Device Application CPs: Submit per 21 CFR 803.12(a) and associated guidance. NDA/ANDA Application CPs: Submit per 21 CFR 314.80(g) and associated guidance. BLA Application CPs: Submit per 21 CFR 600.80(h) and associated guidance.	Combination product approved under a device (e.g., PMA) application. Fifteen-Day and Follow-up Reports to Fifteen-Day Reports are submitted in accordance with 21 CFR 803.12(a) for implementation specifications per CDRH eMDR.[b]
Non-ICSRs[c]	FARs, BPDRs, and Correction and Removal Reports should be submitted in accordance with the procedural requirements and policies associated with the **report type.**	

[a] FDA Adverse Event Reporting System (FAERS) Electronic Submissions (https://www.fda.gov/drugs/fda-adverse-event-reporting-system-faers/fda-adverse-event-reporting-system-faers-electronic-submissions); Electronic Medical Device Reporting (eMDR) (https://www.fda.gov/industry/fda-esubmitter/electronic-medical-device-reporting-emdr); Implementation specifications for CDRH eMDR (https://www.fda.gov/medical-devices/mandatory-reporting-requirements-manufacturers-importers-and-device-user-facilities/health-level-seven-hl7-individual-case-safety-reporting-icsr-files); Technical specifications for CBER/CDER electronic ICSRs, contained in Specifications for Preparing and Submitting Electronic ICSRs and ICSR Attachments: https://www.fda.gov/media/111763/download); CBER Vaccine ICSR Implementation (includes information on the Vaccine Adverse Event Reporting System or VAERS) (https://www.fda.gov/industry/about-esg/cber-vaccine-icsr-implementation)

[b] "Health Level Seven (HL7) Individual Case Safety Reporting (ICSR) Files" accessed September 29, 2022 at https://www.fda.gov/medical-devices/mandatory-reporting-requirements-manufacturers-importers-and-device-user-facilities/health-level-seven-hl7-individual-case-safety-reporting-icsr-files

[c] Field Alert Reports: https://www.fda.gov/drugs/surveillance/field-alert-reports and https://www.fda.gov/vaccines-blood-biologics/guidance-compliance-regulatory-information-biologics/submitting-field-alert-reports-fars-cber Recalls, Corrections and Removals: https://www.fda.gov/medical-devices/postmarket-requirements-devices/recalls-corrections-and-removals-devices; Biological Product Deviation Reports: https://www.fda.gov/vaccines-blood-biologics/report-problem-center-biologics-evaluation-research/general-instructions-completing-biological-product-deviation-report-bpdr-form-fda-3486

Report types are generally categorized as "Individual Case Study Report (ICSR)," Non-ICSR," and Periodic Safety Reports.[52] An ICSR is required for an event experienced by an individual user of a combination product, including adverse events (injury, death) and malfunctions.[53] (US FDA, EU, and WHO commonly refer to ICSRs as a mechanism for capturing information needed to support the reporting of adverse events, product problems, and consumer complaints.)

To Whom Does Combination Product PMSR Apply?

Combination Product PMSR applies to (1) **Constituent Part Applicants** and (2) **Combination Product Applicants**. A third group is also addressed in the FDA Guidance, i.e., **"Entities That Are Not 'Applicants'"** of the Combination Product (here referred to as "non-combination product applicant" entities).

(1) A ***Constituent Part Applicant*** holds an application for one constituent part, while another applicant holds the application for the other constituent part(s). This would be the case,

e.g., for a cross-labeled combination product, where one party is the MAH (Sponsor) for the drug, and a separate party is the Legal Manufacturer (Sponsor) for the medical device, but the constituent parts are intended for use with one another aligned to 21 CFR 3.2(e)(3) and/or 21 CFR 3.2(e)(4).

(2) A ***Combination Product Applicant*** holds the only, or all, applications for a combination product. A single entity generally has one Sponsor. This is similarly the case for a co-packaged combination product, which is approved under a single marketing application in the United States.

In the event that (a) a Sponsor has a cross-labeled combination product, and the Sponsor is both the Legal Manufacturer for the medical device constituent part and the Market Authorization holder for the drug or else (b) the Sponsor has elected to gain approval for each of the cross-labeled constituent parts together under one Market Authorization that Sponsor would also be considered a *Combination Product Applicant.*

We'll discuss the reporting obligations of Constituent Part and Combination Product Applicants in the sections below.

(3) **Entities that are not considered applicants**. There are reporting considerations for entities that are not considered applicants, reflected in Appendix 3 of the Combination Product PMSR Guidance (July 2019) (pages 43–44). Table 8.10 summarizes these "non-combination product applicant" entities and their respective reporting responsibilities:

TABLE 8.10
Combination Product PMSR: Considerations for Entities That Are Not 'Applicants'"

Entity	Reporting Requirements
Manufacturers, packers, and distributors, whose names appear on the label of over-the-counter (OTC) combination products that are not subject to premarket review and include a drug constituent part	Comply with the reporting and recordkeeping requirements described in section 760 of the FD&C Act (21 USC 379aa) for the combination product
Non-applicants listed as a manufacturer, packer, or distributor on the label of a combination product that contains a drug or biological product constituent part	Comply with reporting requirements as described in 21 CFR 314.80 and 600.80 for the product, as applicable
Manufacturers, packers, and distributors of unapproved prescription combination products that include a drug constituent part	• Report and maintain records as described in 21 CFR 310.305 for the combination product • Packers and distributors may meet these requirements by reporting to the combination product manufacturer within five days of receiving the information and maintaining records of these reports as described in 21 CFR 310.305(c)(3)
Manufacturers, importers, and user facilities (21 CFR 803.3) for combination products that include a device constituent part, whether or not subject to premarket review	• Comply with the requirements applicable to them in 21 CFR Part 803 for the combination product • May seek exemptions, variances, or alternatives to these requirements as described in 21 CFR 803.19(b) (e.g., where there is a Combination Product Applicant for the product) • Other such entities subject to 21 CFR Part 803 may seek an exemption from reporting to FDA if they choose instead to report to the Combination Product Applicant
Manufacturers and importers (21 CFR 806.2) of combination product that have a device constituent part	Comply with the requirements described in 21 CFR Part 806 for the combination product

Readers should take note, as these reporting duties may be different for combination products than they would be for other medical product types. In some cases, the constituent part of a combination product may also be manufactured or marketed separately by a different entity as a non-combination product. PMSR reporting obligations may also apply to such entities. Consider, for example, a Combination Product Applicant who purchases syringes to include in its co-packaged combination product from a syringe manufacturer, wherein the syringe manufacturer holds a marketing authorization to market the syringes for general use. The syringe manufacturer has no PMSR duties for the combination product (that is the Combination Product Sponsor's duty). The syringe manufacturer *does* have PMSR duties with regard to its syringes under 21 CFR Part 803.

If you are such a "non-combination product applicant" entity, you should identify the product as a combination product (and you *must* identify the product as a combination product if required under the applicable regulations) and provide a complete discussion of the event with respect to the combination product, including each constituent part, as appropriate, based on the information available to you. FDA Office of Combination Products encourages those with questions to reach out to them.

Reporting Obligations

The PMSR regulations that apply to drugs, devices, and biological products are summarized in Table 8.11. Combination Products PMSR is intended to only add reporting requirements needed to ensure comprehensive safety reporting for the combination product.

TABLE 8.11
US FDA PMSR Regulations for Drugs, Biological Products, and Medical Devices

Medical Product Type	Applicable PMSR Regulations
Drugs	21 CFR 314
Biological Products	21 CFR 600 21 CFR 606
Medical Devices	21 CFR 803 21 CFR 806

Constituent Part Applicants ***Constituent Part Applicants*** are obligated to submit post-marketing safety reports for the application type of their constituent part. For example, a device constituent part application holder would submit reports aligned to 21 CFR 803 and 806 expectations. A drug constituent part application holder would submit reports aligned to 21 CFR 314.

Under 21 CFR Part 4B, each of these applicants, in addition to their normal constituent part reporting obligations, is also required to share information with one another for events that involve a death or serious injury, or another adverse experience. The intent of 21 CFRR 4.103 is to ensure sharing of adverse event information between entities who are collaborating to market products that are intended for use together, to help ensure complete and timely reporting to FDA. (FDA "encourages such entities to share such information with one another" even if it is not a requirement, or even if the products do not necessarily compose a combination product, likewise to encourage complete and timely reporting.)

Per 21 CFR 4.103, the Constituent Part Applicant must share *initial* information received with the other Constituent Part Applicant(s) for the same combination product on an event associated with the combination product that involves either a **death or serious injury** (21 CFR 803.3) or **an adverse experience** (314.80(a) and 600.80(a)) (*see* Table 8.9a). Information must be shared within five days of receipt, regardless of expectedness and whether the Constituent Part Applicant believes it involves only its constituent part.

For example, a photodynamic drug therapy is marketed under an NDA held by one Constituent Part Applicant. A laser that is designed to activate the drug is being marketed under a Device Application held by another Constituent Part Applicant. The Drug Constituent Part Applicant receives information that during the use of the combination product, a patient received a severe skin burn (a "serious injury" per 21 CFR 803.3). The Drug Constituent Part Applicant must forward the initial information it receives on the event to the laser Device Constituent Part Applicant. Each applicant must report the event to FDA if required under the reporting requirements applicable to its constituent part.

Note, in the example above, if the Device Constituent Applicant received information on a malfunction did *not* result in a death, serious injury, or other adverse experience, there is no requirement to share that information with the Drug Constituent Part Applicant. They would need to report the event to FDA per 21 CFR 803 Device Application-type requirements.

Combination Product Applicants

Combination Product Applicants are subject to ***application-based*** reporting requirements as well as ***constituent part-based*** reporting requirements. Table 8.12 summarizes these reporting obligations.

Let's walk through some examples:

1. Consider a combination product that is approved under an NDA, with drug and device constituent parts. Based on the **application-based reporting requirements**, because it is a drug application type, reporting under 21 CFR 314 applies (e.g., 21 CFR 314.81 (Field Alert Reports) and 314.80 (Fifteen-Day Reports)). Then we see specific **constituent part-based reporting requirements**. For the medical device constituent part, **Five-Day Reports** (803.3, 803.53, and 803.56), **Malfunction Reports** (803.50 and 803.56), and **Correction and Removal Reports** (806.10 and 806.20) apply. In addition, to ensure consistent and complete reporting, under "Other Duties," the Sponsor is expected to address Five-Day and Malfunction Reports in periodic safety reports (21 CFR 4.102(d)) (Figure 8.20).

Application Type	Applicant's Product Type (CP = Combination Product)	Application Type-Based Requirements: See 21 CFR Section(s) 314	600 606	803 806	Additional Constituent Part-Based Reporting Requirements[53, 54] (see 21 CFR Section(s)): Field Alert Reports 314.81	Fifteen-day Reports 314.80	Biological Product Dev. Reports 600.14 606.171	Fifteen-day Reports 600.80	Five-day Reports 803.3 803.53 803.56	Malfunction Reports 803.50 803.56	Correction and Removal 806.10 806.20	Other Duties
NDA ANDA	Drug Constituent Part	X										Share information with other Constituent Part Applicant(s) (21 CFR 4.103)
	Drug-Device CP	X			Covered by Application Type-Based Requirements				X	X	X	**Address Five-day and Malfunction reports in periodic safety reports (21 CFR 4.102(d))**
	Drug-Biologic CP	X					X	X				
	Drug-Device-Biologic CP	X					X	X	X	X	X	Address Five-day and Malfunction reports in periodic safety reports (21 CFR 4.102(d))

FIGURE 8.20 PMSR requirements for a drug-device combination product under NDA

2. Consider a combination product approved under a PMA, with biological product and device constituent parts, e.g., a biological product-coated stent for implantation. Based on the application type (PMA), the Device Application-Type Reports apply per 21 CFR 803 and 806 as well as the biological product constituent part reporting obligations, i.e., Biological Product Deviation Reports (BPDRs), and Fifteen-Day Reports (Figure 8.21).

TABLE 8.12

Combination Product PMSR Requirements by Application Type and Constituent Part Type[a]

Application Type	Applicant's Product Type (CP = Combination Product)	Application Type-Based Requirements 21 CFR Section(s)			Additional Constituent Pat-Based Reporting Requirements (See 21 CFR Section(s))							Other Duties
		314	600 606	803 806	Field Alert Reports 314.81	Fifteen-day Reports 314.80	Biological Product Dev. Reports 600.14 606.171	Fifteen-day Reports 600.80	Five-day Reports 803.3 803.53 803.56	Malfunction Reports 803.50 803.56	Correction & Removal 806.10 806.20	
NDA ANDA	Drug Constituent Part	X			Covered by Application Type-Based Requirements							Share information with other Constituent Part Applicant(s) (21 CFR 4.103)
	Drug-Device CP	X							X	X	X	Address Five-Day and Malfunction Reports in periodic safety reports (21 CFR 4.102(d))
	Drug-Biologic CP	X					X	X				
	Drug-Device-Biologic CP	X					X	X	X	X	X	Address Five-Day and Malfunction Reports in periodic safety reports (21 CFR 4.102(d))
BLA	Biologic Constituent Part		X				Covered by Application Type-Based Requirements					Share information with other Constituent Part Applicant(s) (21 CFR 4.103)
	Biologic-Device CP		X						X	X	X	Address Five-Day and Malfunction Reports in periodic safety reports (21 CFR 4.102(d))
	Biologic-Drug CP		X		X	X						
	Biologic-Drug-Device-CP		X		X	X			X	X	X	Address Five-Day and Malfunction Reports in periodic safety reports (21 CFR 4.102 (d))
Device Application (PMA, 510(k), HDE, PDP, De Novo)	Device Constituent Part			X					Covered by Application Type-Based Requirements			Share information with other Constituent Part Applicant(s) (21 CFR 4.103)
	Device-Drug CP			X	X	X						Provide additional reports only if specified in writing by FDA (21 CFR 4.102(d))
	Device-Biologic CP			X			X	X				
	Device-Drug-Biologic CP			X	X	X	X	X				

[a] Appendix 1, US FDA Final Guidance "Postmarketing Safety Reporting for Combination Products" (July 2019), accessed September 27, 2022 at https://www.fda.gov/regulatory-information/search-fda-guidance-documents/postmarketing-safety-reporting-combination-products

Application Type	Applicant's Product Type (CP = Combination Product)	Application Type-Based Requirements — See 21 CFR Section(s) 314	600 606	803 806	Additional Constituent Part-Based Reporting Requirements[53, 54] (see 21 CFR Section(s)) — Field Alert Reports 314.81	Fifteen-day Reports 314.80	Biological Product Dev. Reports 600.14 606.171	Fifteen-day Reports 600.80	Five-day Reports 803.3 803.53 803.56	Malfunction Reports 803.50 803.56	Correction and Removal 806.10 806.20	Other Duties
Device Application (PMA, 510(k), H[illegible] D[illegible]	Device Constituent Part			X					Covered by Application Type-Based Requirements			Share information with other Constituent Part Applicant(s) (21 CFR 4.103)
	Device-Drug CP			X	X	X						Provide additional reports only if specified in writing by FDA (21 CFR 4.102(d))
	Device-Biologic CP			**X**			**X**	**X**				
	Device-Drug-Biologic CP			X	X	X	X	X				

FIGURE 8.21 PMSR reporting for a device-biological product combination product under a PMA

3. Consider a biological product-drug-device combination products under a BLA application (e.g., a co-pack containing a prefilled syringe (PFS) with water-for-injection co-packaged with a vial of a biological product). In this case, BLA-type PMSR reporting per 21 CFR 600 and 606 is required, as well as the drug and medical device constituent part report call-outs (Figure 8.22). Additionally, to ensure consistent and complete reporting, under "Other Duties," the Sponsor is expected to address Five-Day and Malfunction Reports in periodic safety reports (21 CFR 4.102(d)). (This is also the case for combination products approved under NDA/ANDA.)

Application Type	Applicant's Product Type (CP = Combination Product)	Application Type-Based Requirements — See 21 CFR Section(s) 314	600 606	803 806	Additional Constituent Part-Based Reporting Requirements[53, 54] (see 21 CFR Section(s)) — Field Alert Reports 314.81	Fifteen-day Reports 314.80	Biological Product Dev. Reports 600.14 606.171	Fifteen-day Reports 600.80	Five-day Reports 803.3 803.53 803.56	Malfunction Reports 803.50 803.56	Correction and Removal 806.10 806.20	Other Duties
NDA ANDA	Drug Constituent Part	X			Covered by Application Type-Based Requirements							Share information with other Constituent Part Applicant(s) (21 CFR 4.103)
	Drug-Device CP	X							X	X	X	Address Five-day and Malfunction reports in periodic safety reports (21 CFR 4.102(d))
	Drug-Biologic CP	X					X	X				
	Drug-Device-Biologic CP	**X**					**X**	**X**	**X**	**X**	**X**	**Address Five-day and Malfunction reports in periodic safety reports (21 CFR 4.102(d))**
BLA	Biologic Constituent Part		X				Covered by Application Type-Based Requirements					Share information with other Constituent Part Applicant(s) (21 CFR 4.103)
	Biologic-Device CP		X						X	X	X	Address Five-day and Malfunction reports in periodic safety reports (21 CFR 4.102(d))
	Biologic-Drug CP		X		X	X						
	Biologic-Drug-Device CP		**X**		**X**	**X**			**X**	**X**	**X**	**Address Five-day and Malfunction reports in periodic safety reports (21 CFR 4.102(d))**
Device Application (PMA, 510(k), HDE, PDP, De Novo)	Device Constituent Part			X					Covered by Application Type-Based Requirements			Share information with other Constituent Part Applicant(s) (21 CFR 4.103)
	Device-Drug CP			X	X	X						Provide additional reports only if specified in writing by FDA (21 CFR 4.102(d))
	Device-Biologic CP			X			X	X				
	Device-Drug-Biologic CP			**X**	**X**	**X**	**X**	**X**				

FIGURE 8.22 PMSR reporting for a drug-device-biological product combination product under BLA

Streamlined Reporting

Importantly, under the US model of PMSR, Combination Product Applicants can comply with *multiple* reporting requirements if they can be appropriately met *through* a *single ICSR for the same event.* Reporting would need to be done consistent with the shortest timeline of the multiple report types (see timelines listed in Table 8.9b), but including all the applicable information.

- Consider an applicant for an NDA applicant for a drug-device combination product (e.g., drug prefilled syringe). Per Table 8.12 and Figure 8.20, the NDA applicant must submit the following ICSR reports for an event:

  21 CFR 314.80 Fifteen-Day Report

21 CFR 803.50 Malfunction Report

According to 21 CFR 4.102(b) and (c), the applicant could actually satisfy ***both*** reporting requirements by submitting a ***single report*** within 15 days, provided that the Fifteen-Day Report includes *all the information required in both types of reports* for the event.

- Such flexibility also exists for a Combination Product Applicant whose product was approved under a Device Application. Per 21 CFR 4.102 and 806.10(f),[54] a correction or removal can be reported through:
  - A serious injury or death report; or
  - A fifteen-day report; or
  - A malfunction report; or
  - A five-day report.

  Consider a PMA-approved device-biological product (see Table 8.12 and Figure 8.21). The applicant is expected to submit:

  21 CFR 803.3 and 21 CFR 803.53 Five-Day Report;

  21 CFR 806.10 Correction or Removal Report.

  In lieu of submitting both reports, they can submit a single report that contains all the required information, no more than five days after determining that remedial action was needed.
- Consider an additional example: An applicant for a prefilled rescue inhaler approved under an NDA. The Combination Product Applicant receives a report of serious injury due to an inhaler malfunction. Upon investigation, it was determined that the product has a design defect that can result in actuator failure leading to the inability to use the inhaler in an emergency situation. This may pose an unreasonable risk of substantial harm to public health, as this design flaw exists in currently distributed products. Consequently, remedial action is needed (per 21 CFR §803.3(v)) to correct the design defect.
  - The applicant is required (Table 8.12 and Figure 8.20) to file three report types:

    One report for the serious injury (Fifteen-Day Report; 21 CFR §314.80) +

    One report for remedial action (Correction and Removal Report; 21 CFR § 806) +

    One report for significant risk of harm (Five-Day Report; 21 CFR § 803.53).

  Streamlined Reporting gives Industry flexibility to file these report types individually ***or*** combined as one report.
- Table 8.9b references "**non-expedited ICSRs**," i.e., ICSRs for adverse experiences that are either serious and expected or nonserious. If a Combination Product Applicant for a drug or biologic-led combination product is required to submit a malfunction report and a non-expedited ICSR for an event, following the Streamlined Reporting logic we've discussed, the applicant could submit a single report within 30 days that includes all of the information required in both types of reports for the event (see 21 CFR 4.102(b) and (c), 314.80, 600.80, and 803.50).
- Streamlined Reporting may also be applied for drug–biological product combination products. If an event for a distributed drug–biological product combination product demands submission of both a Field Alert Report (FAR) and a BPDR, the Combination Product Applicant may submit a single report for both the FAR and BPDR non-ICSRs provided that the report:
  - Contains sufficient information about the event and the combination product;
  - Is designated based on application type, i.e., if it is an NDA application, it should be designated as a FAR; if it is under a BLA, it should be designated as a BPDR;
  - Is submitted per the procedures of the lead Center at US FDA;
  - And (as with all the other Streamlined Reporting examples) is submitted within the shorter timeframe. This last one is important. Normally a BPDR (see Table 8.9b) report may be submitted within 45 days from the date of acquiring information reasonably suggesting that a reportable event has occurred. Under this Streamlined Reporting, the report would need to be submitted within three working days of receipt of the information by the applicant. (Alternatively, the applicant is still allowed to submit two separate reports per the normal timelines for each report type.)[55]

The FDA Guidance and webpage include several examples and flowcharts to support the applicant in clearly understanding reporting expectations.

Same-Similar Reporting

Among the Individual Case Study Report (ICSR) types identified in Table 8.9b is "Same-Similar Reporting."

A Combination Product Applicant's product/constituent part is considered to be a same-or-similar if it has:

a) The same general purpose and function; ***and***
b) The same type (e.g., would have the same device product code or drug product code); ***and***
c) The same basic design/performance characteristics, including slight modifications, as it relates to safety and effectiveness.
as the Combination Product Applicant's US-marketed product/constituent part.[56]

There is an expectation of "same-similar reporting" for *all* ICSRs events or experiences that meet US reportability criteria for an applicant's marketed same-similar constituent part(s) or combination product(s). Basically, a Combination Product Applicant is expected to report outside-the-US (OUS) events or experiences for their products that include *the same or similar device and/or the same active moiety*, whether that OUS product is a combination product or not. These reporting expectations apply to each constituent part (device, active moiety) or the combination product. The reporting requirements for foreign events for combination products align with the underlying regulatory requirements for drugs, devices, and biological products for such events or experiences.[57,58]

The same-similar reporting requirements reflect a desire to ensure complete reporting of events that raise an expedited reporting issue for the Combination Product Applicant's US product. Combination Product Applicants are not required to report on foreign events that did *not* involve their product (e.g., if they can confirm that the event was associated with another applicant's product, rather than their own). That said, the FDA recommends reporting foreign events for products marketed outside the United States *even if not a requirement*. Again, the intent is to enhance awareness and understanding of safety considerations that may be relevant to US combination products as for other medical products.

FDA Guidance does not provide details regarding whether an indication or intended use would fall within the same/similar "a)" and applicants may want to contact FDA/OCP (Office of Combination Products) with questions if in doubt. A best practice is for an organization to proactively determine same-similar products and document the rationale to support their decision. In the event that an event occurs OUS that meets US reportability criteria and that organization has multiple products registered in the US with that same-similar constituent part, an industry best practice is to submit an ICSR against the oldest approved combination product. It is the authors' recommendation to avoid non-value-added duplicate reporting and to include the list of all same-similar products in the narrative or constituent part investigation.

Table 8.13 is an example of a same-similar assessment form, where (a), (b), and (c) refer to the device constituent parts meeting same/similar criteria relative to the applicant's US-marketed medical device:

a) The same general purpose and function; ***and***
b) The same type (e.g., would have the same device product code); ***and***
c) The same basic design and performance characteristics, including slight modifications, as it relates to safety and effectiveness.

An organization should establish a cross-reference table to support same-similar reporting decisions. For example, when one concludes the investigation on a combination product and deems an event reportable, their investigation report should include reference to a list of all other products with the same-similar device constituent part (or the same active moiety).

TABLE 8.13
Same-Similar Assessment Examples

Device Constituent, Product 1, Marketed in US	Device Constituent, Product 2, Marketed OUS	Device Constituent Part Same-Similar or Not?	Why
Autoinjector "A" with biologic "B" for rheumatoid arthritis	Autoinjector "A" with same biologic "B" for rheumatoid arthritis	Same-Similar	a), b), and c) all hold true
	Autoinjector "A" with biologic "C" for rheumatoid arthritis	Same-Similar	a), b), and c) all hold true
	Autoinjector "A" with same biologic "B" for psoriatic arthritis	Same-Similar	a), b), and c) all hold true
	Autoinjector "A" with slight modification, e.g., color indicator for dose completion, for use with biologic "B" for rheumatoid arthritis	Same-Similar	a), b), and c) all hold true
	Autoinjector "A" with slight modification, e.g., label/carton/IFU update, for use with biologic "B" for rheumatoid arthritis	Same-Similar	a), b), and c) all hold true
Autoinjector "A" (3-step delivery) with biologic "B" for rheumatoid arthritis	Autoinjector "B" (single step delivery) for use with biologic "B" for rheumatoid arthritis	Not Same-Similar	"a)" and "b)" may be considered similar "c)" is different
Prefilled syringe "PFS A" for emergency use (e.g., anti-anaphylaxis)	Prefilled Syringe "PFS A" for cosmetic use (e.g., anti-wrinkle)	Not Same-Similar	"a)" is not similar "b)" and "c" might be considered similar
Cardiovascular stent "A" (CSA) coated with anticoagulant "AC"	Cardiovascular stent "CSA" coated with anticoagulant "AD"	Same-Similar	a), b), and c) all hold true
	Renal stent "CSB" (different material, different length, different design or mesh, different delivery mechanism than "CSA") coated with anticoagulant "AC"	Not Same-Similar	Neither a), b) nor c) is similar

The concept of Same-Similar Reporting has proven challenging, particularly for pharma and biologic organizations. Not only do you need to demonstrate that your procedures and processes are able to support Same-Similar Reporting, you also need to make sure that, based on the risk profile, you *are* able to file malfunction reports against all same-similar products in that report.

An industry best practice is that the report is filed against only one of the same-similar products, that with the oldest registration containing the device constituent part. In Table 8.14, for a reportable

TABLE 8.14
Proactive Same-Similar Assessment to Support Reporting

Reportable OUS Product SKU	Proactive Same-Similar Assessment vs OUS Product SKU (US Product SKUs)						Oldest Registered US Same-Similar Product	Report
OUS Product SKU	**PFS 0 (1960)**	**PFS 1 (1965)**	**PFS 2 (1973)**	**PFS 3 (2010)**	**PFS 4 (2020)**	**PFS 5 (1999)**	**PFS 0 (1960)**	**File ICSR against PFS 0 (1960), including reference to PFS 1 and PFS 4**
OUS PFS	Yes	Yes	No	No	Yes	No		

malfunction associated with "OUS PFS," there are three US products that meet the same-similar criteria for the device constituent part. Of these, PFS 0 (1960) is the oldest registered US product containing the same-similar device constituent part. Therefore, a single same-similar report should be filed against PFS 0 (1960), including reference to PFS 1 and PFS 4 as additional same-similar products potentially impacted by the malfunction.

Some Unique Reporting Considerations

Risk Management Across several chapters of this book, the authors stress that risk management underpins the combination product lifecycle, from development through post-market activities. Through the lens of PMSR, a helpful suggestion is that while you are conducting risk management Harm/Hazard Assessments (see Chapter 6), take the time to add an extra column to your risk management Harm/Hazard Trace Matrix. With the cross-functional collaboration of a medical professional, right then and there, as you are conducting your risk management assessment, based on harm, whether the severity of harm and/or likelihood of harm, make and document your reportability decisions. That way you are providing a tool right up front that enables efficient decision-making downstream during Post-market Surveillance, Vigilance, and Reporting.

Fifteen-Day "Malfunction-Only" Reporting: Unique Requirement for Combination Products with a Drug/Biologic Constituent Part of Application Type
A traditional drug or biologic Fifteen Day Report would require the following four elements to be valid:

1) An identifiable patient;
2) An identifiable reporter;
3) A suspect product;
4) An adverse experience.

A Malfunction Report may also be required when a Fifteen-Day Report is not, e.g., the malfunction occurs before use. Fifteen-Day "Malfunction-only" Reporting applies specifically to drug- or biologic-led PMSR types. A Fifteen-Day "Malfunction-only" is considered to be valid even in the absence of the above four elements. (As a best practice, in the authors' experience, organizations will need to establish default values for PMSR purposes that can be used for the four elements.)

For example, an NDA applicant receives a report that a healthcare provider (HCP) observed, before use, that the sterile barrier for a prefilled syringe (PFS) was compromised. The HCP discarded the PFS before using it on a patient. No Fifteen-Day Report is required because there was no adverse event; however, a Malfunction Report would be required if the breach in the sterile barrier would be likely to cause or contribute to a death or serious injury (e.g., an infection requiring hospitalization for treatment of the infection) if it were to recur (see 21 CFR 4.102(c)(1)(ii) and 803.50).

Similarly, if an NDA applicant receives a report that the needle of an autoinjector deployed early and either no dose or a partial dose was delivered, the NDA applicant must submit a Malfunction Report if the product would be likely to cause or contribute to a serious injury or death if the malfunction recurred, even if there was no patient injury or death associated with the reported malfunction event (see 21 CFR 4.102(c)(1)(ii) and 803.50).

The applicant has 30 days to report a malfunction. Malfunctions can be submitted concurrent with a Fifteen-Day Report for the same event, if both are submitted within 15 days. However, there is no need to rush the malfunction analysis to try to meet this shorter timeline. The applicant would still submit the report per application-type-based reporting through the lead center that approved the combination product. (A BLA-led combination product with a pure device malfunction would not use the 3500A form for its report; rather, it would report through Center for Biologics Evaluation and Research (CBER) on the E2b form, using the device-specific fields to reflect the malfunction information, against the lead-drug or lead-biologic product.)

Use Errors and Medication Errors Device constituent part use-error events and drug constituent part medication errors should be reported to the FDA in the same way as other reportable events, given the combination product type (see Chapter 7, Human Factors). The Final Rule on PMSR (21 CFR §4B) and associated Final Guidance do not explicitly comment on use error, so the authors' assumption is to follow the constituent part reporting requirements for use error or medication error, reporting through the lead center based on application type-reporting or constituent part applicant-type reporting.

Reporting Requirements for Investigational Combination Products Reporting responsibilities are for the approved combination product or constituent part and apply to the entity holding that approval/clearance. 21 CFR §4B, PMSR for combination products, does *not* apply to investigational combination products whose constituent parts have never received marketing authorization. It *does apply*, however, to those investigational combination products that (1) include at least one constituent part (a drug, device, or biological product) that has received marketing authorization or (2) include a combination product as a whole that has previously received marketing authorization, e.g., for a different indication, intended user, or strength. If the combination product or constituent part used in the clinical investigation is already legally marketed, entities must report adverse events that occur in the investigational setting as required by the applicable PMSR requirements for that entity based on the marketed combination product or constituent part.

For example, a Combination Product Applicant is legally marketing a BLA-led combination product comprised of an autoinjector with a biologic for a chronic condition. The Combination Product Applicant is also conducting a clinical investigation for an acute condition with the same marketed autoinjector constituent part, with an investigational biologic. In the event a reportable adverse event or malfunction occurs due to the autoinjector during the clinical trial, then the Combination Product Applicant must submit an ICSR to FDA (via FDA Adverse Event Reporting System (FAERS) or Vaccine Adverse Event Reporting System (VAERS), whichever system is used for reporting on the approved, against the marketed product BLA that includes the approved autoinjector constituent part). As a best practice, when submitting the ICSR, the causality must reference the device constituent part-related malfunction event and the specific clinical trial in the narrative.

Consider another example, in which a Combination Product Applicant is legally marketing a 510k-led combination product comprised of an implantable medical device coated with an anti-infective drug. The Combination Product Applicant is also conducting a clinical investigation with a different device with the same anti-infective drug. If there is a drug-related reportable adverse event, then the Combination Product Applicant must submit an Medical Device Reporting (MDR) against the approved, marketed medical device coated with the anti-infective drug. As a best practice, when submitting the MDR, the causality must reference the drug constituent part-related adverse events (AE) and the specific clinical trial in the narrative.

Recordkeeping Requirements

The FDA Guidance clarifies requirements on PMSR recordkeeping requirements for Combination Product and Constituent Part Applicants aligned to 21 CFR 4.105. These are summarized in Table 8.15.

Process Considerations

The FDA Guidance includes helpful flowcharts to help in the fulfillment of combination product ICSR requirements. Timing considerations based on those flowcharts are summarized in Table 8.9b, but the reader is strongly encouraged to review the flowcharts in Appendix 2 of the guidance (pp. 41–42). Some key points are summarized in Tables 8.16a and 8.16b.[59]

TABLE 8.15
Recordkeeping Requirements

Constituent Part Applicant Information Sharing	Example

• Retain reporting-related records for the time periods stipulated in the regulations applicable to the type of constituent part (see 21 CFR 4.102(b) and 4.105(a)(1)), ***and***
• Retain information-sharing records in accordance with 21 CFR 4.103, for the longest period required for any records under the PMSR requirements applicable to the Constituent Part Applicant who shared the information (21 CFR 4.105(a)(2)).
• **Information-sharing records must include:**
  • Copy of the information provided to the other Constituent Part Applicant(s);
  • Date information was received by the Constituent Part Applicant who shared the information;
  • Date information was shared;
  • Name and address of the other Constituent Part Applicant(s) with whom the information was shared (see 21 CFR 4.103(b)).
	PharmCo holds NDA for the drug constituent part of a combination product.
	MedDevCo holds a PMA for the device constituent part of that combination product.
	PharmCo receives a report of an AE and shares the information with MedDevCo.
	PharmCo must retain the required ***records of sharing the information*** with MedDevCo for ten years from the date PharmCo received the information, because the only PMSR recordkeeping period applicable under 21 CFR Part 314.80(j) and 21 CFR 4.105(a) for adverse drug experiences is ten years.
	MedDevCo must retain records associated with the event for the time period required under device PMSR regulations applicable with regard to the event. In accordance with 21 CFR Part 803, for malfunction, serious injury, or death reporting, MedDevCo would be required to keep records for ***the longer*** of two years from the date of the event or a period equivalent to the expected life of the device (21 CFR 803.3(f)), whichever is greater.

Combination Product Applicant	**Example**
• Per 21 CFR Part 4.105(b), retain records for the longest time period required for records under all PMSR requirements applicable to the combination product • Drug–biological product combination products: ten years (per 21 CFR 314.80(j); 600.80(k)) • Combination products with a device constituent part: the longer of either ten years or two years beyond life expectancy of the product (see 21 CFR 314.80(j), 600.80(k), 803.18, 806.20)	An implantable drug-device combination product has an expected life of 3 years. → The longest applicable recordkeeping requirement would be 10 years. An implantable drug-device combination product has an expected life of 9 years. → The longest applicable recordkeeping requirement would be the expected life (9 years) plus two years (equaling 11 years in this case).

Strong cross-functional collaboration is required to seamlessly meet the reporting obligations in a timely and robust way. Figure 8.23 illustrates a recommended process flow for drug/biological product organizations taking on Combination Product PMSR reporting obligations. Such pharma organizations sometimes struggle with the integration of intake, triage, vetting, and required coordination between product quality complaints (PQCs) and adverse events (AEs) to support timely decision-making, review, and reporting. Typically, the PQCs are managed through the Quality and Complaint Handling function, whereas the AEs are managed through the Pharmacovigilance Organization (PV). Creating a structured process flow is important for successful cross-functional coordination and collaboration. (Within a device organization, there is typically a single stream for handling complaints, adverse events and reporting, so such coordination has been less problematic.)

Global PMSR Challenges, Considerations, and Best Practices

With the context of the general global evolving PMSR regulatory landscape and understanding of US PMSR expectations, let's now turn our focus beyond. We'll tease out some of the global PMSR challenges, considerations, lessons learned, and best practices. There are a number of challenges. Some of these are cultural, some are associated with the complexities that come with manufacturing logistics, e.g., using suppliers, to support combination product manufacturing and reporting activities, and some are technical.

TABLE 8.16A
Combination Product PMSR Content Requirements

PMSR reports must contain all information required for the report type under the applicable regulations, including relevant information on the entire product (including each constituent part). Additional information to include: • Combination Product Identifier; • Report Types; • Patient Identifier; • Reporter Identifier; • Suspect Medical Device (procode, device common name, device brand name); • Malfunction reports for drug/biological product-led combination products should include the Device Common Name and Device Product Code *in all cases,* and the Device Brand name, if any; • Suspect Drug or Biological Product name(s) (e.g., trade name, active ingredient(s), dosage form, strength); • Adverse Event Coding (Patient Problem Code, MedDRA terms); and • Device Problem Code.
Constituent Part Elements in Reporting Systems • The goal is to improve the ability to look across reports, both within a center and across centers; • Include constituent part information for ICSRs, regardless of whether the constituent part was implicated in the event; • Reporters should refer to current technical specifications and other documents for detailed instructions on how to complete and submit electronic ICSRs;[a,b,c] • FAERS Device Constituent Part Fields: – When submitting information in FAERS, include as much information as is available for the constituent part; – Combination Product Identifier Procode: Three-letter code specific to a device type; – Device Brand Name; – Device Common Name; – Device Problem Code: FDA codes specific to device issues/failures; – Type of Reportable Event: Malfunction – Yes/No flag for reportable malfunction (see Table 8.9a and discussion on "likelihood"). • eMDR Drug/Biological Product Constituent Part Fields: – Combination Product Identifier; – Drug Product Name; – Drug Dose; – Drug Route.

[a] FDA Adverse Event Reporting System (FAERS) Electronic Submissions at https://www.fda.gov/drugs/questions-and-answers-fdas-adverse-event-reporting-system-faers/fda-adverse-event-reporting-system-faers-electronic-submissions

[b] Electronic Medical Device Reporting (eMDR) at https://www.fda.gov/industry/fda-esubmitter/electronic-medical-device-reporting-emdr#:~:text=Electronic%20Medical%20Device%20Reporting%20%28eMDR%29%20Collection%20of%20adverse,from%20manufacturers%2C%20user%20facilities%2C%20importers%20and%20voluntary%20reporters.

[c] Vaccine Adverse Event Reporting System (VAERS) at https://vaers.hhs.gov/

Our Biggest Challenge: Culture and Cross-Functional Collaboration

The complexities of Combination Product PMSR are challenging enough within a country, never mind the fact that there are different expectations between countries. To make things worse, the challenges industry faces are not just the logistical and regulatory expectations of PMSR. Culture and cross-functional collaboration together make up the biggest challenge (Figure 8.24).

Culture: People Skills and Competencies A key challenge that falls under the umbrella of culture change is talent. There are limited people with direct combination product expertise or experience. We need to invest in our talent pipeline. These resources have generally been grown organically. They are "pharma-born," "device-born," or "biological product-born," individuals

TABLE 8.16B
Combination Product PMSR Reporting Element Considerations

Reporting Element	Considerations[a]
FAERS ***Combination Product Identifier***	Select either "Yes" or "No" for the combination product identifier [Contact OCP with questions. "Combination Product" in our reporting is required, e.g., recognizing that an autoinjector or PFS is a combination product. This is a requirement, even if the product was approved as an NDA/BLA but has a delivery system with it.]
FAERS ***Device Brand Name***	Include the Common Device Name and Procode[b]
	If the device was purchased from another company and re-branded, include the brand name used by the other company
Common Device Name	Include a Common Device Name; *Do not* enter "unknown"
Procode	• Do not leave blank. Review the Procode website. Contact OCP for assistance in determining the product code that most closely aligns with the device constituent part or to discuss the need for a new Procode.
Device Problem Codes[c]	• Do not use inactive Device Problem Codes. Review "Coding Resources for Medical Device Reports"[d] *Examples*

Inactive Code	Inactive Device Problem Code Description	Replacement
1663	Device inoperable	Use more descriptive code
1059	Bent	2981 "Material Twisted/Bent"
2913	Device operates differently than expected	Use more descriptive code
1597	Sticking	4012 "Physical Resistance/Sticking"
1104	Detachment of device component	2907 "Detachment of Device or device component"
2379	Device issue	Use more descriptive code
2905	Delivery system failure	2906 "Activation, Positioning or Separation Problem"
1104	Detachment of device component	2907 "Detachment of Device or device component"

Event Summary	Inactive Device Problem Code (DPC) in Report	Better DPCs
A device used to deliver a drug was leaking and an incomplete dose was delivered to the patient		1250 – Fluid Leak 2339 – Inaccurate Delivery
The patient was unable to push the device button to administer the drug		2588 – Defective Device 4009 – Ejection Problem 2983 – Mechanism Jam 4012 – Physical Resistance/Sticking

(Continued)

TABLE 8.16B (CONTINUED)
Combination Product PMSR Reporting Element Considerations

Reporting Element	Considerations[a]
FAERS: ***Malfunction*** **Field**	• Do not leave blank. • When possible, select either "Yes" or "No" in the malfunction field; if the malfunction selection was incorrect in the original report, submit a follow-up report. • If steps were taken to determine there was – or was not – a malfunction, consider including this information in the narrative. • Malfunction reporting is a big challenge. Some organizations are significantly over-reporting, e.g., based on symptoms, versus likelihood of death or serious injury, were the malfunction to recur (see Table 8.9a and the section "Key Terminology, Report Types, and Reporting"). This seems to be related to concerns over the potential for late reporting. They would rather report and find out they did not need to, than to be found to, reporting late. The other side of the pendulum includes organizations only reporting if there is a confirmed malfunction, and no symptoms get reported. Fair and balanced reporting is essential. Industry is hoping for additional clarification from FDA to help level-set reporting expectations.
eMDR: Drug/ Biological Product Constituent Part ***Drug Product Name***	• Do not leave blank • Include at least the active pharmaceutical ingredient(s) (API) in the Drug Product Name field when available; where applicable (e.g., co-packaged drug constituent part), include the drug/biologic product name (brand/established)
eMDR: Drug/ Biological Product Constituent Part ***Drug Dose*** **and** ***Drug Route***	• Do not leave blank • Where applicable (e.g., co-packaged drug constituent part), include additional drug information (e.g., strength, route of administration, etc.)

[a] For any of these considerations, contact US FDA Office of Combination Products with questions

[b] Device Product Codes (procodes) for Device Constituent Parts of ANDA/NDA/BLA Combination Products at https://r.search.yahoo.com/_ylt=AwrhQATZFzdj4IkJim5XNyoA;_ylu=Y29sbwNiZjEEcG9zAzEEdnRpZAMEc2VjA3Ny/RV=2/RE=1664583770/RO=10/RU=https%3a%2f%2fwww.fda.gov%2fcombination-products%2fdevice-product-codes-procodes-device-constituent-parts-andandabla-combination-products/RK=2/RS=mjsO_.Jp6ISTHDlfDx_utko2CIg-

[c] FDA Device Problem Codes are mapped to International Medical Device Regulators Forum (IMDRF) terminology; the IMDRF site offers a search tool to find terms which may be helpful at https://www.imdrf.org/working-groups/adverse-event-terminology/annex-medical-device-problem

[d] "Coding Resources for Medical Device Reports at https://www.fda.gov/medical-devices/mdr-adverse-event-codes/coding-resources-medical-device-reports

with a keen interest in learning about the other "side." As leaders in this space, you are encouraged to find the people who have an open mind to learn and hire or cross-train them whether on-the-job or through formal education programs![60] The more we start doing this as an industrial community, the more we will stop struggling for qualified resources in this space. This actually applies more broadly than just to Combination Products PMSR! Quality functions, Operations, Safety, Procurement, Marketing, Regulatory Affairs, Development – in Chapter 5 we discuss language and cultural challenges, and no surprise, the underlying issue of talent development needs to be addressed.

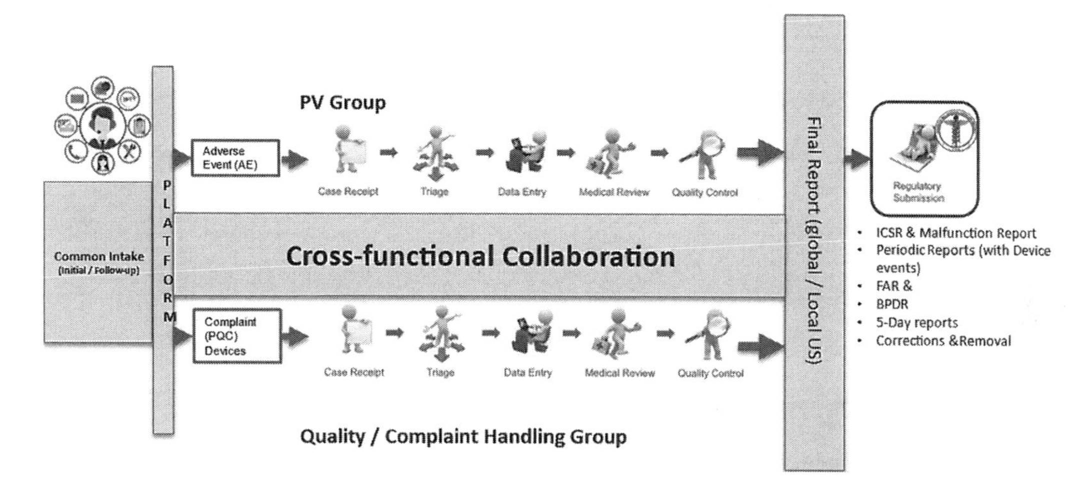

FIGURE 8.23 Cross-functional collaboration: drug/biological product organizational process flow

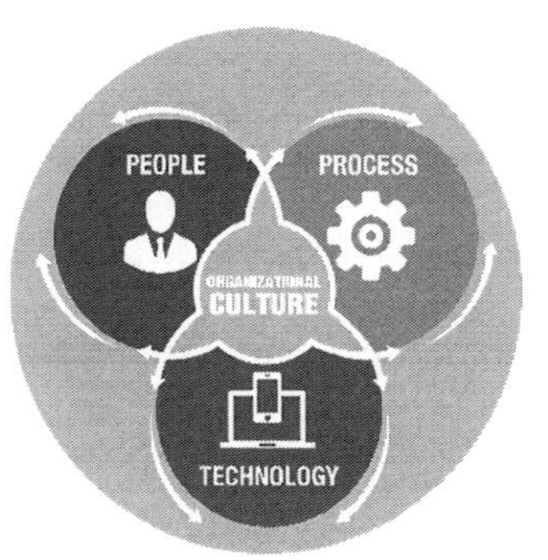

People: Skills & Competencies

Technology: Patient-centricity

Process: Who does what?

- True cross-functional collaboration
- Supplier engagement
- Global processes
- Automation (end-to-end Adverse Events and Product Complaint Linkage)
- Data/analytics
- Decision-making

FIGURE 8.24 Cultural change needed

Culture: Patient Centricity Cross-functional collaboration is essential in Combination Product PMSR. A key challenge in PMSR has been hesitancy to share data between Safety and Quality organizations within a company. "This is *my* safety data." "I have regulatory requirements, so I can't share it with you, 'Quality.'"

At one point, the authors were optimistic that Combination Product PMSR would help us break down organizational silos. Rather, there are areas where we see the silos deepening. What we are missing is the fact that if organizational functions like Manufacturing, Quality, and Commercial don't have the opportunity to see the safety data, we aren't looking at our problem holistically. Therefore, we aren't solving our problem holistically. This is especially the case with respect to usability-related/patient-centric topics.

A best practice recommendation is to get your Marketing and Safety peers on board with you to advocate for patient-centric combination product design, with a focus on enhancing the patient experience. It is all too easy to continue to promulgate products with legacy technology for speed-to-market under a mindset of "it's all about the drug therapy," while minimizing the importance of the device user interface and patient experience. Such a mindset has the potential to perpetuate known use errors and/or cause patients to live with outdated technology, subverting advancements in delivery technology that could genuinely improve patient physical, emotional, and mental well-being through the improved patient experience. To boot such technology improvements might even be more cost-efficient, as the health-challenged patient does not have to struggle with transportation to and from a clinical setting, nor cover the cost of a doctor's appointment when they are instead able to self-administer a product in the privacy and convenience of their own home. A patient is arguably more likely to comply with their therapeutic dosing regimen with a drug that is straightforward and convenient to administer (e.g., self-administering a high-volume on-body injector versus the limitations of having to regularly go to an infusion center; or a one-click autoinjector with an easy-to-view indication of dose completion versus one that requires five clicks or even, perhaps, requires a lyophilized vial and syringe with multiple reconstitution and assembly steps). Human Factors and Risk Management play central roles in developing such products. Supporting the investment of people resources in technology advancements takes leadership commitment and financial prioritization. Marketing and Safety need to see the value of these advancements to actively advocate and support them to fruition in the marketplace.

Culture: Information Gathering:

PMSR impacts are end-to-end – across the entire product lifecycle. The better design of clinical studies, the better the post-approval safety studies. Some of our biggest challenges right now are legacy studies that have been running, asking questions around the device, without knowing what to do with that data. Now, that data gets translated to malfunctions when you go through the data. This raises angst among Quality and Safety professionals.

A related aspect of this cultural challenge and the hesitancy to gather information is the power that digital solutions bring to our analytical capabilities. Digital solutions can get you consistent, real-time data that is so informative – especially around usability and medication errors – factors that you may be able to modify and address. But there is hesitancy about gathering that data because we do not know what to do with that data. This is particularly the case for data that comes from digital sources. The Quality or Safety groups may perceive that each bit of data is considered a complaint. Imagine taking terabytes of data and trying to weed through it to make complaints or to do safety reporting from it. We need to figure out that "right size" approach to digital data opportunities.

Challenges: Third-Party Suppliers and PMSR

In the global regulatory and manufacturing environment of today, many companies partner with third parties for sourcing components and constituent parts for their combination products or even for contract manufacturers serving as combination product manufacturing services to enable a Combination Product Applicant or Constituent Part Applicant to meet the increasing complexity and demands of our industry. A critical aspect of these industrial relationships is the establishment of robust quality agreements. These agreements need to reflect clearly delineated reporting responsibilities. Existing quality agreements need to be reviewed and potentially updated to reflect the additional reporting and information-sharing obligations related to Combination Products PMSR. Without such agreements, you may be caught off-guard, assuming the other party is doing the compliance work. Industry best practice is to create template language to be included in contracts for suppliers, vendors, and third-party service provider agreements making sure roles and responsibilities, information sharing, investigation timeliness, audit support, and PMSR-related requirements are clearly outlined.

Challenges: Outsourcing PMSR

A company might choose to outsource their PMSR obligations. It is important when identifying partners to support your combination product reporting requirements that the partners be carefully vetted to ensure they have the appropriate broad-based expertise in the regulatory reporting for the relevant constituent parts. Training on the additional constituent parts and their related requirements may be required. If a company has already established relationships with a third party for PMSR reporting, it is essential that an assessment be done to ensure their competence with respect to the evolving PMSR reporting needs.

Reporting Best Practice Key Takeaways

In the face of all these challenges and considerations, we can still end this section on a positive note: There are some best practices that apply wherever Combination Product PMSR expectations need be fulfilled. These are summarized here:

(1) Tell the whole story about all the constituent parts and treat the "combination product" as a whole;
(2) Reconsider how to conduct analysis for the product as a whole;
(3) Where known, identify the constituent part that is responsible for the event;
(4) Where possible, minimize duplicate/redundant reporting;
(5) Continue to utilize the existing systems, processes, and mechanisms for reporting e-MDR/ICSR based on the application type;
(6) Incorporate changes needed for the additional constituent part of the combination product;
(7) Remember that ***every*** combination product-related report is impacted!

CONCLUSIONS

Figures 8.25 and 8.26 bring this chapter to conclusion. Figure 8.25 summarizes the entire closed-loop, risk-based approach to combination products development through post-market lifecycle management. Figure 8.26 summarizes the integrated combination product development deliverables for a typical drug (or biological product) PMOA combination product. The deliverables for a device PMOA combination product are quite similar, adding an increased emphasis on ensuring the drug constituent part expectations are addressed. Some key takeaways:

- Not unique to combination products: User-centric/patient-centric design is fundamental to the medical product lifecycle. Everything starts and ends with the intended users fulfilling their needs and experience.
- Risk Management is the end-to-end underpinning of the combination product lifecycle!
- QbD and Design Controls together focus on improving the chances of product development success by proactively addressing risks throughout requirements definition, development, and lifecycle management.
- Often constituents, components, or services are outsourced for combination products. Outsourcing design or utilizing an existing device platform changes the "Design" process, necessitating the robust application of Purchasing Controls to assure rigor during technology evaluation, selection, and lifecycle management.
- Manufacturing combination products has its unique challenges. Manufacturing challenges that must be addressed span the gamut of drug formulation, facility designs, drug product manufacturing, stability considerations, and logistical challenges. Understanding and controlling interactions of drugs and devices throughout the process end-to-end is important in addressing these challenges.
- Change controls, Purchasing controls, CAPA, Post-market Surveillance, Vigilance, and Reporting all work interdependently during the post-market lifecycle to ensure sustained

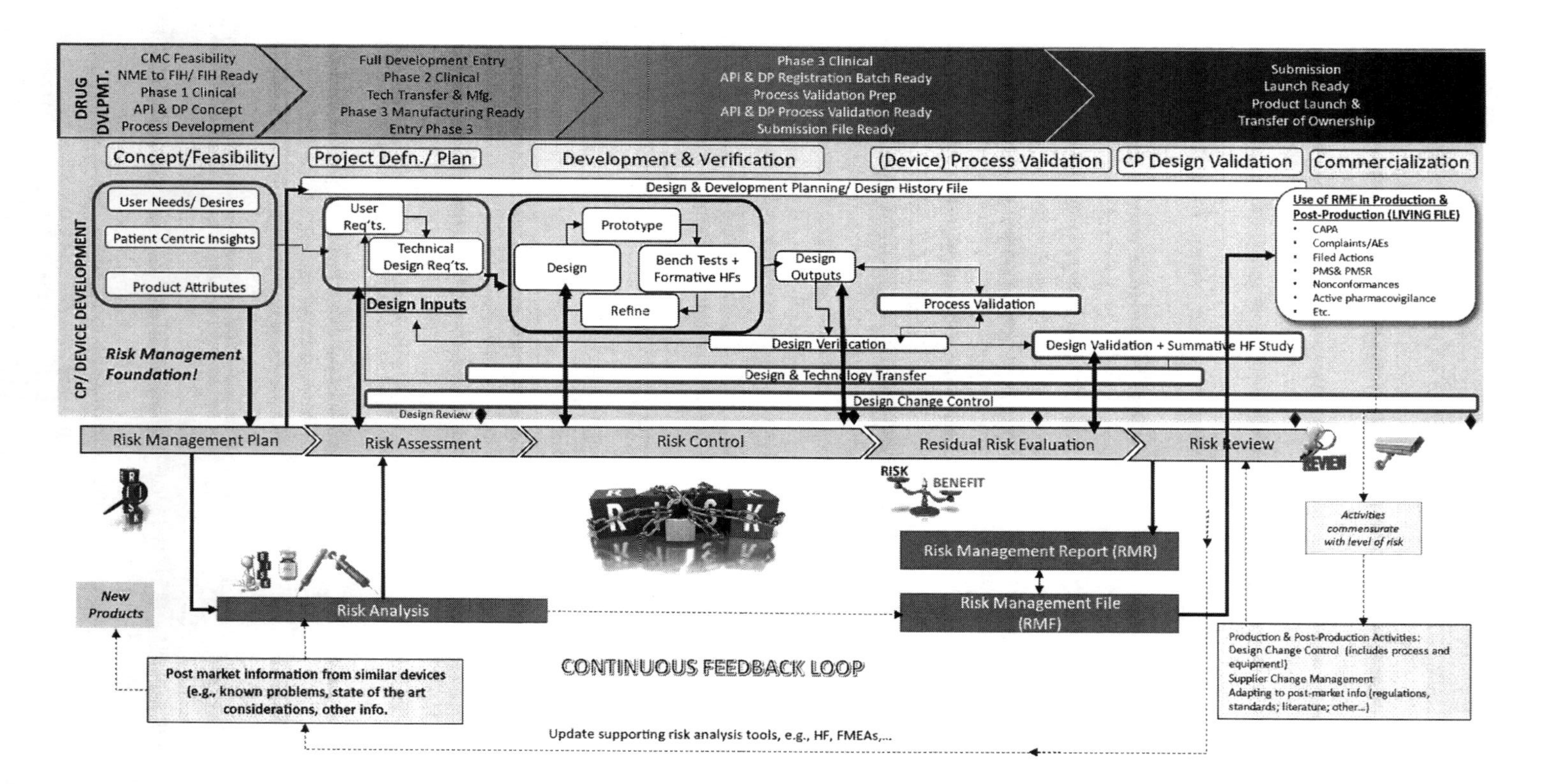

FIGURE 8.25 Combination products risk-based approach: continuous feedback loop (©2022 Combination Products Consulting Services, LLC. All Rights Reserved.)

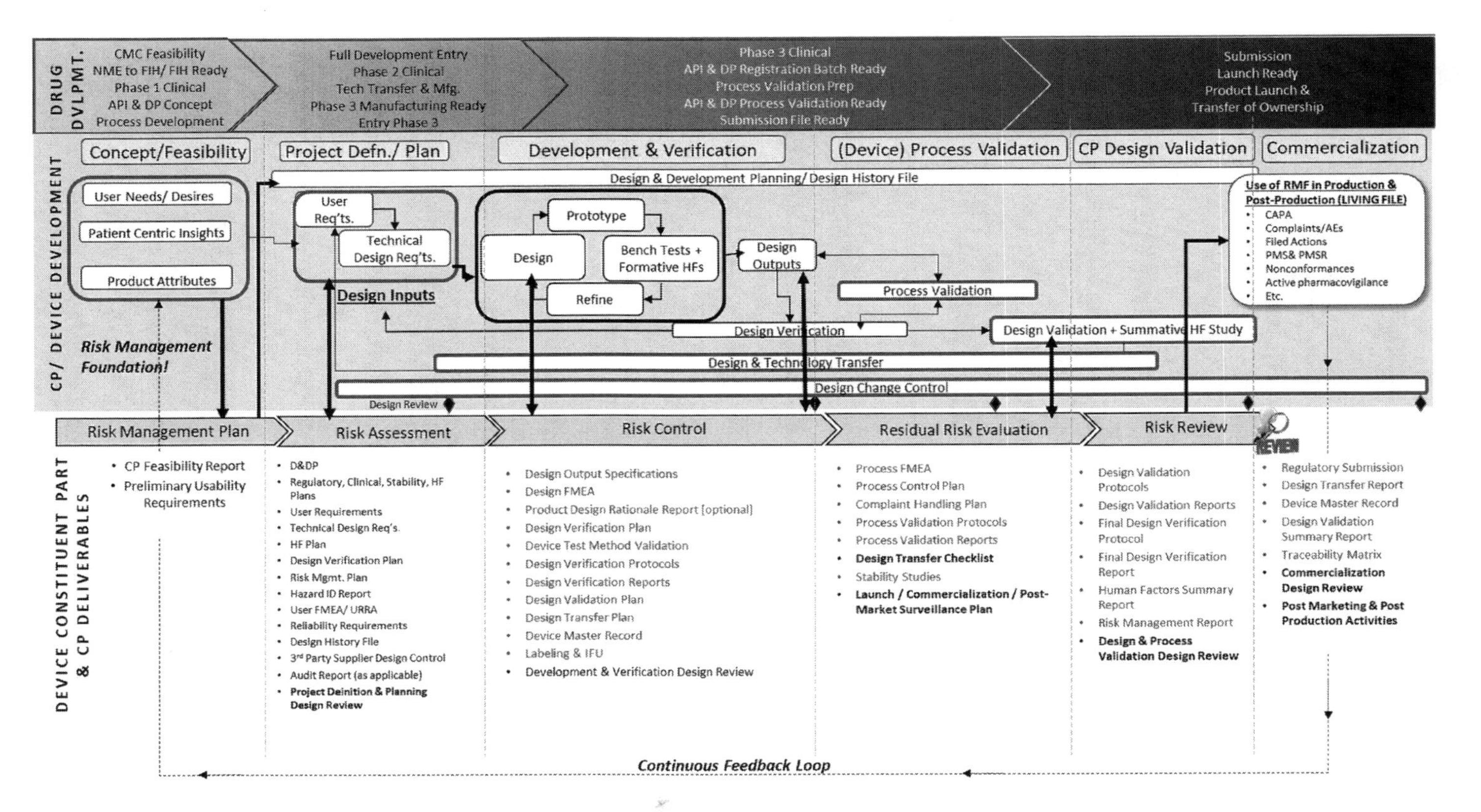

FIGURE 8.26 Integrated combination products lifecycle deliverables (©2022 Combination Products Consulting Services, LLC. All Rights Reserved.)

availability of safe and effective combination products, while also serving as a continuous feedback loop into products in the development pipeline, and informing currently marketed same-similar products.

Our combination product lifecycle management chapter has helped us complete the development journey and move to post-market. In Chapter 5, we began the journey of integrated development, stressing the importance of "language" and culture in development and their impacts throughout the product lifecycle. As we round out this chapter, it seems only appropriate to revisit these. The multilingual fluency with respect to drug, biological product, and device constituent parts clearly has an impact on expectations during Design Transfer, including ongoing risk management, process validation, design changes, CAPA, Supplier Controls, and into Post-market Surveillance and Reporting. There is a need for open-mindedness when working in the combination products space – a desire to expand our own personal design space to incorporate an understanding of the nuances surrounding each constituent part and the integration of those constituent parts for combined use. Those educated and/or experienced in the end-to-end combination product lifecycle management process will serve our patients (and businesses) well by educating our colleagues, building an understanding of what it takes to bring safe, effective, and usable combination products to market.

There is sometimes a mindset of "we've been on the market for 20 years with this exact same product, and nothing has changed." "We used to treat the autoinjector as a container closure; no one has been cited on this, so why can't I do … (fill in the blank), because that is the way I have always done it?" "It's been working the way we've done it, so why invest in improving the user interface?" We have to educate and reinforce across our organizations that "times have changed" and so have the regulatory expectations and understanding of the increased risks that combined use products bring. More importantly, we need to be the change agents that help better, and more consistently, meet the needs of the patients we serve.

NOTES

1. See Chapters 4–7.
2. 21 CFR 820.30(h) or ISO 13485:2016 Clause 7.3.8 (results and conclusions are documented per Clause 4.2.5).
3. Clarification: Final Verification and Validation occurs before Design Transfer is complete, as part of the premarket review process. Validation can include clinical studies, for example. But Design Transfer can occur over a period of Time (as illustrated in Figure 8.2). It is just that there is often a final stage of Design Transfer that we think of in which we know Design Transfer cannot be complete until all Validation is complete.
4. Drug or biological product-led Sponsor is generally referred to as the Market Authorization Holder (MAH) or Market Authorization Applicant (MAA); or medical device-led Sponsor is typically referred to as the Legal Manufacturer. In this chapter, the author generally refers to all of these as "Sponsor" or sometimes reverts to just saying MAH.
5. Preamble comment #81, accessed September 21, 2022, at https://www.fda.gov/medical-devices/quality-system-qs-regulationmedical-device-good-manufacturing-practices/medical-devices-current-good-manufacturing-practice-cgmp-final-rule-quality-system-regulation
6. A potential source of confusion is that some people simply refer to "Validation" or "V&V" (Verification & Validation). It is important to distinguish between Design Verification, Design Validation, and Process Validation. Process Characterization activities may also be used as a precursor to Process Validation, final Design Verification, and Design Validation. As illustrated in Table 8.1, they have different purposes.
7. Referred to as a Medical Device File (MDF) under Clause 4.2.3 of ISO 13485:2016. The MDF contents are similar to a DMR. The MDF includes a general description of the medical device, intended use/purpose, and labeling, including any instructions for use; specifications for product; specifications or procedures for manufacturing, packaging, storage, handling, and distribution; procedures for measuring and monitoring; as appropriate, requirements for installation; and as appropriate, procedures for servicing.
8. Device History Record (21 CFR 820.184), Medical Device Record(s) (ISO 13485:2016, clauses 7.1, 7.5.1, and 7.5.8) or Batch Record(s) (21 CFR 211.188).

9. ISO 13485:2016 Clause 7.5.5 similarly describes medical device process validation, in some respects, e.g., sterilization.
10. US FDA Guidance for Industry "Process Validation: General Principles and Practices" (January 2011) accessed September 22, 2022, at https://www.fda.gov/regulatory-information/search-fda-guidance-documents/process-validation-general-principles-and-practices
11. EudraLex Volume 4 "EU Guidelines for Good Manufacturing Practice for Medicinal Products for Human and Veterinary Use Annex 15: Qualification and Validation" (March 2015) accessed September 22, 2022, at https://health.ec.europa.eu/system/files/2016-11/2015-10_annex15_0.pdf
12. Neadle, S.W. (2016)"Risk Management Considerations and Strategies in Product Development." Ed Bills and Stan Mastrangelo. *Lifecycle Risk Management for Healthcare Products from Research through Disposal*. Davis Healthcare International (pp. 141–193).
13. GHTF SG3/ N99-10:2004 "Quality Management Systems- Process Validation Guidance" (January 2004) https://www.imdrf.org/sites/default/files/docs/ghtf/final/sg3/technical-docs/ghtf-sg3-n99-10-2004-qms-process-guidance-04010.pdf
14. Recall from Chapter 4, a CMO that is a combination product manufacturer is able to leverage the Streamlined Approach under 21 CFR Part 4. The Streamlined Approach obligates them to demonstrate compliance with their base cGMP-quality system as well as each of the called-out provisions under 21 CFR Part 4 for the cGMP activities that they perform, even though they are not the Sponsor.
15. Helpful resources: US FDA Guidance "Contract Manufacturing Arrangements for Drugs: Quality Agreements" (November 2016) accessed September 22, 2022, at https://www.fda.gov/media/86193/download; GHTF SG3 "Quality Management System – Medical Devices- Guidance on the Control of Products and Services Obtained from Suppliers" (December 2008) accessed September 23, 2022, at https://www.imdrf.org/sites/default/files/docs/ghtf/final/sg3/technical-docs/ghtf-sg3-n17-guidance-on-quality-management-system-081211.pdf and 21 CFR 820.50.
16. ISO 13485:2016: Medical devices – Quality management systems – Requirements for regulatory purposes, Clause 7.3.4 Product realization – Design and Development – Design and development outputs (also see Chapters 4 and 5).
17. ISO 14971:2019 Medical devices – Application of risk management to medical devices; ICH Q9(R1) Quality Risk Management (November 2021) and AAMI TIR 105:2020 Combination products risk management; see also Chapter 6.
18. Plans for inclusion of specific ECs in an application or post-approval supplement (PAS) may be considered an appropriate topic for pre-submission meetings with FDA.
19. ICH Q12: "Implementation Considerations for FDA-Regulated Products" Guidance for Industry (May 2021) accessed September 23, 2022, at https://www.fda.gov/media/148947/download
20. Some may refer to these Critical Process Parameters (CPPs) and Critical Material Attributes (CMAs).
21. Neadle, S. and Varadi, I. (February 2018). "Product Development and Manufacturing Challenges for Combination Products" in American Pharmaceutical Review, accessed September 23, 2022, at https://www.americanpharmaceuticalreview.com/Featured-Articles/347328-Product-Development-and-Manufacturing-Challenges-for-Combination-Products/
22. A filter pressure test is done to ensure the filter is suitable for its purpose. Generally, to hold back microorganisms, a 0.22-micron filter is commonly used.
23. See also Chapter 5, Table 5.10 and Neadle, S.; Riter, J.; McAndrew T.P.; (Pharmaceutical Online, Nov. 30, 2020) *"Considerations in Combination Product Risk Management"*; https://www.pharmaceuticalonline.com/doc/considerations-in-combination-product-risk-management-0001
24. US FDA Proposed Rule on Revising the National Drug Code Format (July 25, 2022), accessed September 23, 2022 at https://www.federalregister.gov/documents/2022/07/25/2022-15414/revising-the-national-drug-code-format-and-drug-label-barcode-requirements
25. US FDA "Understanding the UDI Format" at https://www.fda.gov/medical-devices/unique-device-identification-system-udi-system/udi-basics
26. 21 CFR 801.30(11) accessed September 23, 2022 at https://www.federalregister.gov/documents/2013/09/24/2013-23059/unique-device-identification-system#p-501
27. Adapted from Reed Tech "Drug & Biologic Product Submissions, Unique Device Identification" (August 10, 2022) accessed on September 23, 2022 at https://www.reedtech.com/knowledge-center/combination-products-regulatory-requirements-and-how-to-comply/
28. Notably, such unique identification requirements exist across regions globally, with different rules and code formats. The reader should be attuned to differences for the country in which they intend to market/distribute their product.
29. See Chapter 7, Human Factors.

30. This section of the chapter assumes that changes made are *not* ones that affect the PMOA designation, lead center assignment, or the underlying type of marketing application for the combination product.
31. Consider, for example, what questions may arise for a combination product that was designed for administration in a clinical setting by a trained healthcare provider, now being re-directed for self-administration by patients. Use-Related Risk Analysis (URRA) might trigger the need for updates to the user interface, e.g., Instructions for Use, in turn potentially necessitating new human factors and/or design validation activities.
32. See also Chapter 6, Risk Management.
33. US FDA Draft Guidance "Submissions for Post-approval Modifications to a Combination Product Approved Under a BLA, NDA, or PMA" (January 2013) at https://www.fda.gov/media/85267/download; *This guidance does not address changes to combination products that are not approved under a BLA, NDA, or PMA (e.g., those cleared solely under a device premarket notification submission or those marketed under an over-the-counter drug monograph). Nor does this guidance address changes to combination products that were approved under more than one marketing application. Further, while this guidance does address the type of submission to provide when making a change to a constituent part of a combination product approved under one marketing application, it does not address the scientific or technical content to provide in any such submission.*
34. "ICH Q12 Implementation Considerations for FDA Regulated Products" Draft Guidance (May 2021) at https://www.fda.gov/media/148947/download
35. US FDA Draft Guidance "Submissions of Post-approval Modifications to a Combination Product Approved Under a BLA, NDA or PMA (January 2013), accessed September 24, 2022 at https://www.fda.gov/regulatory-information/search-fda-guidance-documents/submissions-postapproval-modifications-combination-product-approved-under-bla-nda-or-pma
36. Assignment, development, premarket review, and post-market regulation.
37. Webpages with information on requesting meetings with US FDA: BLA Therapeutic Biologic Applications: http://www.fda.gov/Drugs/DevelopmentApprovalProcess/HowDrugsareDevelopedandApproved/ApprovalApplications/TherapeuticBiologicApplications/default.htm;NDA webpage: http://www.fda.gov/Drugs/DevelopmentApprovalProcess/HowDrugsareDevelopedandApproved/ApprovalApplications/NewDrugApplicationNDA/default.htm; PMA webpage: http://www.fda.gov/MedicalDevices/DeviceRegulationandGuidance/HowtoMarketYourDevice/PremarketSubmissions/PremarketApprovalPMA/default.htm; BLA webpage: http://www.fda.gov/BiologicsBloodVaccines/DevelopmentApprovalProcess/BiologicsLicenseApplicationsBLAProcess/default.htm
38. Ibid.
39. US FDA Draft Guidance "Bridging for Drug-Device and Biologic-Device Combination Products" (December 2019), accessed September 24, 2022 at https://www.fda.gov/regulatory-information/search-fda-guidance-documents/bridging-drug-device-and-biologic-device-combination-products. See also Chapter 3 of this book.
40. US FDA Draft Guidance "Contents of a Complete Submission for Threshold Analyses and Human Factors Submissions to Drug and Biologic Applications" (September 2018), accessed September 24, 2022 at https://www.fda.gov/media/122971/download. See also Chapters 3 and 7 of this book.
41. US FDA Draft Guidance "Human Factors Studies and Related Clinical Study Considerations in Combination Product Design and Development" (February 2016), accessed September 24, 2022 at https://www.fda.gov/regulatory-information/search-fda-guidance-documents/human-factors-studies-and-related-clinical-study-considerations-combination-product-design-and
42. https://www.fda.gov/regulatory-information/search-fda-guidance-documents/principles-premarket-pathways-combination-products
43. Under the Streamlined Approach, a combination product manufacturer must demonstrate compliance with their base cGMP-quality system as well as each of the called-out provisions under 21 CFR Part 4 for the cGMP activities that they perform, whether or not they are the Sponsor.
44. The DHF is not required to have all the documents in one place. As long as they are readily accessible, they can be included by reference.
45. Under EU MDR (2017/745) Article 10,9.(i), the PMS Plan is an expected element of the Quality Management System. The PMS Plan is potentially complex and may involve multiple stakeholders and, for manufacturers with more than one marketed device, the process will need to be repeated for every device or family of similar devices. See Article 83 of EU MDR for additional details on the Post Market Surveillance System (https://eumdr.com/post-market-surveillance-system/)
46. *See definition in Table 9a.*

47. MHLW/PMDA Notification (PFSB/ELD Notification No. 1024-2, PFSB/ELD/OMDE Notification No. 1024-1, PFSB/SD Notification No. 1024-9, PFSB/CND Notification No.1024-15)(October 24, 2014) "Handling of Market Applications of Combination Products."
48. FDA Final Guidance "Postmarketing Safety Reporting for Combination Products" (July 2019), accessed September 25, 2022 at https://www.fda.gov/regulatory-information/search-fda-guidance-documents/postmarketing-safety-reporting-combination-products
49. https://www.fda.gov/combination-products/guidance-regulatory-information/postmarketing-safety-reporting-combination-products
50. Ibid.
51. Several nuances apply to the timing of reports. If you are the responsible member of your organization for such reporting, it is strongly recommended that you review the guidance in additional detail.
52. FDA webpage "Individual Case Safety Reports" at https://www.fda.gov/industry/fda-data-standards-advisory-board/individual-case-safety-reports
53. An ICSR includes an identifiable patient, identifiable reporter, a suspect drug and/or device, and an adverse event or adverse drug reaction.
54. See also 81 FR at 92612-13 and 92615.
55. The Office of Combination Products encourages applicants of drug/biologic combination products who wish to create procedures to support the submission of combined FAR/BPDR reports to discuss their approach with them.
56. The considerations described here for the same or similar device or device constituent part are aligned with related content in the FDA Guidance for Industry and FDA Staff, "Medical Device Reporting for Manufacturers" (November 2016) at https://www.fda.gov/regulatory-information/search-fda-guidance-documents/medical-device-reporting-manufacturers. Importantly, though, for combination products, the same or similar concept applies not just to the device constituent part but also to the active moiety.
57. For additional information see FDA Draft Guidance for Industry, "Postmarketing Safety Reporting for Human Drug and Biological Products Including Vaccines" (March 2001) at https://www.fda.gov/regulatory-information/search-fda-guidance-documents/postmarketing-safety-reporting-human-drug-and-biological-products-including-vaccines
58. See Guidance for Industry and FDA Staff, *Medical Device Reporting for Manufacturers* (https://www.fda.gov/regulatory-information/search-fda-guidance-documents/medical-device-reporting-manufacturers)
59. Lauren Bateman, US FDA, "FDA Observations on PMSR Report Content Since Enforcement of the Combination Product Requirements" presentation at 2021 Xavier Combination Product Summit.
60. There are some formal educational programs and training available (e.g., through Association of Advancement in Medical Instrumentation (AAMI) or the Master's curriculum in Combination Products at University of Maryland Baltimore Campus (UMBC), part of the ISPE "Workforce-of-the-Future" Program).

9A Combination Products Inspection Readiness
Best Practices and Considerations

Susan W. B. Neadle

CONTENTS

Health authority inspections are performed by regulatory agencies and competent authorities to determine an organization's level of compliance with regulations for market authorization, post-market surveillance activities, and Quality Management System requirements. "Inspections." The word alone can be stressful for all involved. The consequences of day-to-day manufacturing activities within an organization can have a profound impact on public health, and accordingly, health authority findings during an inspection can likewise have a profound impact. Manufacturers subject to those inspections need to help give the regulatory authority confidence to support access/continued access to that manufacturer's products and/or services by demonstrating the organization's quality system is in control and suitable to ensure the safety and effectiveness of the products and services they seek to market/continue to market. In the United States, the inspection of manufacturers is risk-based, prioritizing facilities that pose the greatest potential risk to public health. (Table 9.1 summarizes the inspection types of US Food and Drug Administration (FDA).)

TABLE 9.1
US FDA Inspection Types

Inspection Type	Description
Bioresearch monitoring (BIMO)	Evaluate clinical investigators, contract research organizations (CROs), and institutional review boards (IRBs) to ensure the integrity of evidence collected during clinical trials, as well as the safety of research subjects
Pre-market *[**Pre-market approval inspections** (PMAs) for devices, **Pre-approval inspections** (PAIs) for drugs, and **Pre-license inspections** (PLIs) for biological products]*	Assess an organization's readiness for commercial manufacturing, verifying compliance to applicable cGMP/ quality system regulations and the accuracy and integrity of information submitted with a particular submission in regard to manufacturing
Post-market inspections and administrative/ enforcement activities	**Routine surveillance** – Ensure the continued quality of a product that is marketed **For-cause inspections**– Investigate known or suspected issues **Post-approval inspections** (PoAIs) – Assess compliance subsequent to changes in manufacturing

DOI: 10.1201/9781003300298-10

This chapter, building upon Chapter 4 "Combination Products Current Good Manufacturing Practices (cGMPs)," turns our focus to what's unique for inspections with respect to combination products. Currently, the United States is the only country globally with an inspection program specifically defined for combination products. In other countries, these products are generally inspected under the authority who serves as lead legislation for that product (*see Chapter 1 "Introduction," Chapter 14 "General Overview of Global Combination Product Regulatory Landscape," and the Appendix, "Comparative Overview of Global Combination Product Regulatory Landscape").* This chapter is divided into two parts:

1. A discussion of best practices for combination products inspection readiness (Chapter 9A); and
2. Given the fact that the United States is currently the only jurisdiction with a combination product-specific Compliance Program, a review of the US FDA Combination Products Compliance Program (Chapter 9B).

BEST PRACTICES AND CONSIDERATIONS FOR COMBINATION PRODUCTS INSPECTION READINESS

On a day-to-day basis, building and maintaining quality systems that are compliant with regulatory expectations and standards can take the angst out of inspection readiness. Understanding the applicable cGMP expectations based on the cGMP activities your organization performs clearly helps too. Beyond ensuring one's organization is meeting cGMP expectations (*see Chapter 4 on cGMPs*), there are other helpful considerations for a manufacturer's successful combination products inspection readiness.

In particular, these include challenges in **vocabulary** and **supplier collaboration**. *(See also discussion in Chapter 5, "Integrated Products Integrated Development," on "Keys to Combination Product Development Success.")*

VOCABULARY

i.e., the different dictionary definitions for device and medicinal product professionals, presents a challenge for industry and health authorities. Cross-functional teams and health authorities need to become keenly aware that there is inconsistent interpretations of terminologies by those whose dictionary is customarily based on "pharma" or "biotech" versus "device" (see Chapter 5, Table 5.1, "A Sampling of the Combination Products Dictionary"). The confusion in language and interpretation can lead to significant confusion during audits and health authority inspections. Being aware of, and attuned to, the language differences is an important factor for combination products inspection readiness. ***Communication with a focus on ensuring consistency of, and alignment on, interpretation is key throughout an inspection.*** During a combination products inspection, personnel within a company participating in the investigation need to be able to clearly articulate how their objective evidence satisfies the cGMP expectations applicable based on the activities performed at their facility. This includes understanding, and maybe helping the investigator(s) to understand and overcome, differences in interpretation and terminology.

If a Center for Device and Radiological Health (CDRH) investigator comes to a drug cGMP-based manufacturing facility for a combination products inspection, that investigator may be more familiar with the "device Quality System dictionary." Likewise, if a Center for Drug Evaluation and Research (CDER) investigator comes to a device cGMP-based manufacturing facility to conduct a combination products inspection, that investigator may be more familiar with the "drug cGMP dictionary." There is nothing wrong with you using the terminology you are accustomed to, but it may be very helpful to create a "translation document" that correlates the terminology applied internally to your organization to the terminology used by the particular FDA investigator to ensure mutual alignment and understanding. Table 9.2 illustrates an example of such translation to support a combination products inspection, in this case, focused on design controls.[1]

TABLE 9.2
Example "Translation Document" ***(In This Case, Correlating Terminology between "Drug" and "Device" Development)***

Device Development Design Controls Terminology	→Translation/Interpretation←	Pharma Development QbD Terminology
Design Controls (Mandatory)	Framework for product development	Quality by Design (Best Practices)
User Needs & Intended Uses	Who are the intended users and what are their needs? What is the intended use environment?	User Needs & Intended Uses
Design & Development Plan	Project Management Planning	Project Plan
Design Inputs	Translate the user needs to technical design requirements	Quality Target Product Profile (QTPP)
Human Factors/Usability Engineering	Ensure the product is safe, effective, and usable; avoid predictable misuse	Safety Considerations for Medication Error Reduction
Design Outputs	Translate product requirements to specifications	Product Specifications
Design Reviews	Reviews at development milestones; address problems	CMC Stage Gate Reviews
Design Verification	Ensure the product meets design specifications	Product Characterization
Design Validation	Ensure the product meets user needs and intended uses	Clinical Studies
Design Transfer	Scale up and commercialize the product	Technical Transfer Studies
Design History File (DHF)	Document the development data and activities	Product Dossier and Development Reports
Device Master Record (DMR)	Recipe for repeatable and reproducible product	Master Batch Record (MBR)
Design Change Control	Manage change to maintain performance	Change Controls
Risk Management (ISO 14971:2019/ISO 24971:2020)	Risk management and control strategies[2]	Risk Management (ICH Q9)

The terminology interpretation confusion is not restricted to design controls or risk management. For a drug cGMP-based Quality Management System (QMS) that is augmented by device called-out provisions for corrective actions and preventive actions (CAPA) (21 Code of Federal Regulations (CFR) § 820.100) or purchasing controls (21 CFR § 820.50), there are points of interpretation confusion. Medical device purchasing or supplier controls requirements under 21 CFR § 820.50 as well as corrective actions and preventive actions (CAPA) requirements under 21 CFR § 820.100 sound very familiar to activities and requirements within the pharmaceutical requirements in 21 CFR § 211. However, while the same terms may be utilized, regulatory expectations between the two regulated constituent parts and the combination product are very different on these two topics.

Inspection Logistics

Typically, particularly in larger or more complex organizations, there is a cross-functional team engaged to support an inspection, split into a "Front Room," directly interfacing with the investigator, and a "Back Room," supporting objective evidence requests. An example of such a configuration of team members and their responsibilities is summarized in Table 9.3.

TABLE 9.3
Inspection Support: An Example of Front Room and Back Room Participants

	Role	Description
FRONT ROOM	Facilitator	• Directly interfaces with Investigator • Hosts the inspection • Manages the Front Room • Defines Requests
	Subject Matter Expert(s)	• Represents the topic in the Front Room • Directly interfaces with Investigator
	Scribe(s)	• In the Front Room • Takes notes on the Front Room dialogue, serving as a communication conduit with the Back Room
↕	Runner(s)	• Communicates requests from Front Room to Back Room, e.g., for objective evidence based on Front Room dialogue
BACK ROOM	Back Room Manager	• Stays attuned to Front Room dialogue and drives Back Room activities accordingly • Manages the documents that enter the Front Room • Prioritizes requests • Coaches Subject Matter Expert(s) (SMEs) before they enter the Front Room • Responsible for logistics, schedule, and any last-minute or ongoing changes to the schedule
	Subject Matter Expert(s)	• Prepares documents for the Front Room • Supports the Front Room SME(s)
	Content Reviewer	• Consults with SME(s) • Reviews documents before they are released to the Front Room to ensure they are consistent with what was requested • Coordinates with the operators gathering documents
	Operators	• Gather documents • Document reconciliation: track those that have been shared and outstanding document requests

Those participating in *any* of these roles should be "multilingual." Consider the earlier example where a CDRH investigator, who might not be familiar with the "language" of pharm, comes to a drug cGMP-based facility. The investigator makes a request to review the traceability of specific design verification activities to design input requirements. A Design Traceability Matrix is a typical tool generated during device constituent part and combination product development; it summarizes such traceability and could easily be reviewed to identify the specific documents appropriate for sharing in order to satisfy the investigator's request. If the "language" comfort zone of the facilitator or subject matter expert (SME) engaged in the inspection does not recognize this or that of the scribe taking notes is solely based on drug cGMPs, they may struggle to effectively articulate the request to the Back Room, which then becomes frenetic with activity trying to figure out what documents they need to bring forward. When the wrong documents – or no documents – show up in the Front Room, the investigator (and those interfacing directly with the investigator) may get frustrated, the inspection gets drawn out, and there is even a risk that this inability to bring in the appropriate objective evidence might be construed as lack of compliance.

Those engaged in the inspection need to be able to both understand the investigator's questions and requests and clearly articulate responses in such a way as to demonstrate compliance with their quality system. In the event that an organization does not have a person fluent in both "drug" and "device" for any of the particular cross-functional team roles, then having both drug and device experts engaged in the Front Room and Back Room roles concurrently may help mitigate the potential for misunderstanding. This is obviously, and not ideally, a bit more resource-intensive approach, but it can work. Proactively training the broader cross-functional team who will support a combination products inspection will likely reduce the level of stress for all involved, enable more streamlined use of resources, and serve as one of the building blocks of a successful inspection outcome.

Supplier Collaboration

As has been discussed in previous chapters, suppliers play a significant role in most combination product organizations.[3] This makes purchasing controls and effective supplier collaboration an essential success factor across the lifecycle of combination products. Contract manufacturing organizations (CMOs) and suppliers of critical components and/or constituent parts may very well be the subject of inspection associated with a combination product sponsor's product. Pre-market approval inspection, post-market surveillance, or for-cause inspections could lead a health authority to scrutinize a supplier based on component/constituent part criticality for a to-be-marketed product or performance post-market. Generally, the FDA applies a risk-based approach in deciding to inspect a given supplier of a combination product sponsor. In evaluating CAPAs, change control, or complaints at a combination product manufacturer, for example, if there are issues that are pointers to a supplier, supplier qualification, or change controls, then purchasing controls may be called into question and an inspection of that supplier may be triggered. If that supplier has, for example, been operating as a "container-closure" component supplier under 21 CFR § 211, but whose scope of activities satisfies that of a combination product manufacturer[4] and is subject to inspection under the Combination Products Compliance Program, that presents additional challenges. The same "multilingual" drug/device/combination product skills that the combination product sponsor's internal resources have had to address are also an important supplier skill-set. Ensuring the success of the supplier's ability to demonstrate compliance to applicable cGMP requirements and its ability to articulate that compliance with an FDA investigator is an essential factor in inspection readiness.

SUMMARY

Health authority inspections verify the accuracy and integrity of information in a pre-market submission as well as an organization's compliance with applicable cGMP and quality system requirements. These inspections are a check and balance to protect public health. Given the unique considerations for the combined use of drug, device, and biological product constituent parts for combination products, FDA issued requirements for combination products in 21 CFR Part 4. The FDA subsequently issued a Combination Products Compliance Program (Chapter 9B) to provide more clarity of expectations and consistency in the execution of combination products inspections by the FDA. Building and maintaining quality systems that are compliant with combination product regulatory expectations and standards can take the angst out of inspection readiness, but may not be enough.

Beyond the helpful backdrop of ensuring one's organization is meeting combination product cGMP expectations, successful combination products inspection readiness requires overcoming barriers in communication. Differences in the interpretation of terminologies between those more familiar with medical devices versus drugs and/or biological products can lead to breakdowns during combination product inspections, so communication with a focus on ensuring consistency of, and alignment on, interpretation is key throughout. Building "multilingual skills" in the cross-functional inspection support team can support a successful inspection outcome.

The success of one's combination product is often heavily dependent on the supplier(s) of critical component manufacturers, constituent part manufacturers, and contract manufacturing organizations. Given that, gaining market approval, or the ability to continue to market a product, is reliant on an organization's critical suppliers to successfully demonstrate compliance to the applicable cGMP activities for their products and/or services. Overcoming suppliers' barriers to understanding combination product cGMP expectations and partnering with them to ensure communication/terminology hurdles are addressed is likewise an important aspect of the combination product sponsor's success.

Ultimately, this is all about the patients we serve. Inspection success enables businesses to stay focused on continued innovation in combination products therapeutic solutions and allows us to promote advances in public health.

9B US Combination Product Inspections

Kim Trautman

CONTENTS

Combination products have been in existence and inspected by US FDA investigators for over two decades even before the Office of Combination Products (OCP) was officially created on December 24, 2002, following the publication of the Medical Device User Fee and Modernization Act of 2002. Prior to 2002, staff within the US FDA Office of the Commissioner assigned to combination products would assist with combination product classifications, designations, coordination between different FDA centers, and coordination between the centers and the FDA Office of Regulatory Affairs (ORA).

ORA is the lead office for all FDA field activities. It serves as the eyes and ears of the agency through:

- Inspections of firms and plants producing FDA-regulated products;
- Investigations of consumer complaints, emergencies, and criminal activity;
- Enforcement of FDA regulations;
- Sample collection and analysis; and
- Review of imported products.[5]

FDA's Compliance Programs provide instructions to FDA personnel for conducting activities to evaluate industry compliance with the Federal Food, Drug, and Cosmetic Act and other laws administered by FDA. Compliance Programs for all FDA program areas may be accessed at Compliance Program Guidance Manual (CPGM). There are many different Compliance Programs across different centers.

The following is a list of biological Center for Biologics Evaluation and Research (CBER) Compliance Programs:

BIOLOGICS

- 7341.002: Inspection of Human Cells, Tissues, and Cellular and Tissue-Based Products (HCT/Ps)
- 7341.002A: Inspection of Tissue Establishments
- 7342.001: Inspection of Licensed and Unlicensed Blood Banks, Brokers, Reference Laboratories, and Contractors
- 7342.002: Inspection of Source Plasma Establishments, Brokers, Testing Laboratories, and Contractors

- 7342.007: Imported CBER-Regulated Products
- 7342.007 Addendum: Imported Human Cells, Tissues, and Cellular and Tissue-based Products (HCT/Ps)
- 7342.008: Inspection of Licensed In-Vitro Diagnostic (IVD) Devices
- 7345.848: Inspection of Biological Drug Products

The following is a list of medical device Center for Device and Radiological Health (CDRH) Compliance Programs:

Medical Devices

- 7382.845 Inspection of Medical Device Manufacturers
- 7383.001 Medical Device Premarket Approval and Postmarket Inspections
- 7385.014 Mammography Facility Inspections
- 7386.001 Inspection and Field Testing of Radiation-Emitting Electronic Products
- 7386.003 Field Compliance Testing of Diagnostic Medical X-Ray Equipment
- 7386.003a Inspection of Domestic and Foreign Manufacturers of Diagnostic X-Ray Equipment

The following is a list of drug Center for Drug Evaluation and Research (CDER) Compliance Programs:

Drugs

- 7348.001 In Vivo Bioequivalence
- 7348.809A Radioactive Drug Research Committee
- 7346.832 Pre-Approval Inspections/Investigations
- 7352.002 Unapproved New Drugs (Marketed, Human, Prescription Drugs only)
- 7353.001 Postmarketing Adverse Drug Experience (PADE) Reporting Inspections
- 7356.002 Drug Manufacturing Inspections
- 7356.002A Sterile Drug Process Inspections
- 7356.002B Drug Repackers and Relabelers
- 7356.002C Radioactive Drugs
- 7356.002E Compressed Medical Gases
- 7356.002F Active Pharmaceutical Ingredients
- 7356.002M Inspections of Licensed Biological Therapeutic Drug Products
- 7356.002P Positron Emission Tomography
- 7356.008 Drug Quality Sampling and Testing
- 7356.021 Drug Quality Reporting System (DQRS) (MedWatch Reports) NDA Field Alert Reporting (FARs)
- 7356.022 Enforcement of the Drug Sample Distribution Requirements of the Prescription Drug Marketing Act (PDMA)
- 7361.003 OTC Drug Monograph Implementation
- 7363.001 Fraudulent Drugs

So for more than 20 years, combination products in the United States were inspected under one of these many Compliance Programs assigned by whichever center was designated as the "lead center." This led to some inconsistencies as to how combination products were inspected and how Part 4 Good Manufacturing Practice (GMP) requirements were enforced. Sometimes two investigators would be assigned to a combination product new market application inspection – one from each of the commodities or constituent parts. This often was not the case, however, due to resource constraints. Therefore, as an example, a drug investigator trained to perform drug inspections under one of the many drug Compliance Programs and trained in 21 CFR § 210/211 would be trying to also assess medical device requirements under 21 CFR § 820 or human cell, tissue, and cellular and tissue-based products (HCT/Ps) under section 361 of the Public Health Service Act.

For years, CBER had a designated investigator cadre called "Team Biologics." As a result of CDER's "GMPs for the 21st Century Initiative" back in 2000, a designated investigator cadre for drugs was also formed called the "Pharmaceutical Inspectorate." Often these designated investigator cadre members did not have the exposure or opportunity to cross-train on different commodities or different regulatory requirements for constituent parts. Further adding to the complexities for inspections of combination products was the 2017–2019 two-year re-organization of FDA's ORA structure called "Program Realignment." ORA aligned by program stating:

> Creating distinct product-based and vertically integrated regulatory programs enables the agency to best achieve its objectives, to optimize the coordination and efficiency of the work performed between all FDA centers, directorates and ORA, to strengthen accountability and to reduce duplication.[6]

Combination Products

Due to these complexities, there was no particular Compliance Program for Combination Products when the final rule on current good manufacturing practice (cGMP) requirements for combination products was issued on January 22, 2013 (21 Code of Federal Regulations (CFR) Part 4). It was not until June 4, 2020, that a Compliance Program for Combination Products was issued by FDA for the ORA investigators (7356.000 Inspections of CDER-Led or CDRH-Led Combination Products).

This new Compliance Program discusses how CDER-led or CDRH-led combination product inspections will be handled. Note that this new Compliance Program for the Inspections of Combination Products does not include CBER-led inspections due to the unique nature of the biologic products and the composition of "Team Biologics" which includes both ORA and CBER personnel. For combination products that include biological products, all applicable cGMP requirements for biological products (including standards) that are found within 21 CFR Parts 600 through 680 (21 CFR § 4.3(c)) must be demonstrated. For a combination product that includes any human cells, tissues, and cellular and tissue-based products (HCT/Ps), the manufacturer must demonstrate compliance with all applicable regulations in 21 CFR § 1271. At this time, a specific Combination Product Compliance Program for the inspection of CBER-led products has yet to be published.

So for combination products that are under CBER as the lead center, Compliance Programs 7345.848 Inspection of Biological Drug Products and 7352.002 Unapproved New Drugs (Marketed, Human, Prescription Drugs only) will be utilized and not 7356.000: Inspections of CDER-Led or CDRH-Led combination products.

Let's explore the Combination Product Compliance Program 7356.000: Inspections of CDER-Led or CDRH-Led combination products. The lead center is based on which constituent part provides the primary mode of action for the combination product. While the FDA application type is aligned with the lead center, the combination product manufacturer can choose which base cGMP system to utilize for the combination product streamline approach, regardless of lead center or application type. The Compliance Program utilizes the lead center's base Compliance Programs for pre-approval inspections, post-approval inspections, surveillance inspections, for-cause inspections and other risk-based inspections. However, the Compliance Program emphasizes that the combination product manufacturer can choose either 21 CFR § 211 pharma based or 21 CFR § 820 medical device-based cGMPs. Remember, CBER-led inspections are not governed by this Compliance Program but instead utilize FDA Staff Manual Guide (SMG) 4101 "Inter-Center Consult Request (ICCR) Process" dated June 11, 2018.

The Compliance Program for CDER-led and CDRH-led combination products clearly lays out the instructions for FDA investigators to write and issue a single Establishment Inspection Report (EIR) and a single FDA 483 for any nonconformances. While the program is focused on combination products that are single entity or co-packaged, it discusses cross-labeled combination products. The Compliance Program notes for cross-labeled manufacturers that if both commodities are manufactured at the same manufacturing facility, the manufacturer can utilize the combination product Streamlined Approach.

Of note, the 21st Century Cures Act of 2016 directed FDA to develop a list of mechanisms for complying with the agency's cGMPs for pharmaceuticals under 21 CFR § 211 and the quality system regulation for medical devices under 21 CFR § 820. On September 13th, FDA published the Final Notice of "Alternative or Streamlined Mechanisms for Complying with the Current Good Manufacturing Practice Requirements for Combination Products."[7] The Notice discusses how to practically approach some specific requirements called out in the Streamlined Approach in a fashion that makes good scientific sense without undue burden or unneeded expensive testing of the entire combination product in certain circumstances. Also, the Notice says that additional approaches may be permissible in order to meet the Part 4 requirements and highly encourages manufacturers to discuss those approached with the agency. It will be important for manufacturers to document any such discussions and arrangements under the Streamlined Approach to show the FDA investigators during a combination product inspection. It should not be assumed that discussions with the FDA centers will automatically be relied to the FDA ORA investigators.

The Combination Product Compliance Program also discusses convenience kit manufacturers and reminds FDA investigators that the cGMPs related to production for the kit manufacturer should be limited to assembly, packaging, labeling, sterilization, and further processing of the kit. It also discusses the importance for kit manufacturers to also ensure compliance with purchasing controls, CAPA, complaints, and other cGMP requirements outside of production process as well.

The Compliance Program specifically states that if only one constituent part is manufactured at a specific facility and that facility is not the combination product manufacturer, then the commodity-specific Compliance Program should be utilized. However, any facility participating in design and development, including record keeping, for combination products is considered a combination product manufacturer and is subject to the Combination Product Compliance Program.

Component manufacturers are not regulated under either 21 CFR § 211 or 21 CFR § 820 cGMPs, but facilities that assemble components into combination products are subject to Part 4 and therefore subject to this Combination Product Compliance Program.

The Compliance Program discusses that the pre-market and post-market inspections, as well as the surveillance inspections, will primarily be governed by the lead center, but the risk and complexity of the manufacturer and each manufacturing facility will determine how each commodity's program will incorporate the other commodity's program. It also mentions that third-party inspection programs such as the Medical Device Single Audit Program (MDSAP) or the US-EU Mutual Recognition Agreement for Pharmaceutical GMP inspections may impact the implementation of the Compliance Program and advise the ORA investigators to contact the lead center for any questions or coordination.

FDA ORA investigators are encouraged to request pre-inspection meetings with the lead center. Investigators are instructed to determine ahead of time:

1. The facilities' cGMP operating system, reminding the investigator that combination product manufacturers can choose regardless of application type; and
2. The relationship between entities if multiple facilities are involved with the combination product and its constituent parts.

When a facility chooses to use both cGMP systems, the inspection may be conducted with dual program staffing or with an investigator with training or experience in combination products. Typical combination product inspections will include the base cGMP plus the call-out provisions from the other constituent part/commodity. Pre-approval and post-approval inspections are expected to be comprehensive or full inspections of both the base cGMP and the call-out provisions. Comprehensive inspections must be an assessment of the entire combination product and not just a constituent part. Comprehensive or full inspections must be utilized for the following inspections:

- Initial inspection of the combination product manufacturer;
- Initial inspections of the "call-out" provisions;
- Directed inspections; and
- CDRH-led foreign inspections.

Surveillance or "abbreviated" inspections can only be utilized when the combination product manufacturer has a satisfactory history of compliance and has been inspected against the full call-out provisions. Surveillance or abbreviated inspections are described as:

1. Abbreviated base plus full call-out provisions; or
2. Abbreviated base plus abbreviated call-out provisions.

Citations or FDA 483 items may be written against the base cGMPs and the relevant call-out provisions. When FDA 483 is issued, the inspection will be classified as Official Action Indicated (OAI), Voluntary Action Indicated (VAI), or No Action Indicated (NAI) based on the lead center's base Compliance Program.

The Compliance Program also has two valuable attachments that manufacturers should utilize as good guidance in understanding the FDA's intentions and expectations.

- Attachment A – Combination Product Inspectional Considerations for Call-out Provisions of 21 CFR Part 211; and
- Attachment B – Combination Product Inspectional Considerations for Call-out Provisions of 21 CFR Part 820.

There are several important references within the Combination Product Compliance Program and the attachments worthy of some discussion. There are two key guidance documents produced by the International Conference on Harmonisation of Technical Requirements for Registration of Pharmaceuticals for Human Use (ICH) that will be discussed here.

- ICH Guidance for Industry: Pharmaceutical Development Q8(R2) issued in June 2009; and
- ICH Guidance for Industry: Quality Risk Management Q9 issued in November 2005.

In November 2011, FDA Issued "Guidance for Industry Q8, Q9, and Q10 – Questions and Answers (R4)."[8]

> Since the Q8, Q9, and Q10 guidances were made final, experiences implementing the guidances in the ICH regions have given rise to requests for clarification … FDA's guidance documents, including this guidance, do not establish legally enforceable responsibilities. Instead, guidances describe the Agency's current thinking on a topic and should be viewed only as recommendations, unless specific regulatory or statutory requirements are cited.

One of the first issues discussed in this FDA Guidance on the use of ICH guidance documents addresses the use of ICH Q8(R2) Pharmaceutical Development within the context of the FDA regulatory framework.

Q1: Is the minimal approach accepted by regulators?

A1: Yes. The minimal approach as defined in Q8(R2) (sometimes also called "baseline" or "traditional" approach) is the expectation that is to be achieved for a fully acceptable submission. However, the "enhanced" approach as described in ICH Q8(R2) is ***encouraged*** [emphasis added] (Ref. Q8(R2) Annex, appendix 1). (Approved June 2009)

"Alternative or Streamlined Mechanisms for Complying with the Current Good Manufacturing Practice Requirements for Combination Products,"[9] issued by FDA on June 13, 2018, states:

> Using existing pharmaceutical development practices and documentation that align with the design control principles and requirements of § 820.30.

This is also confirmed in the September 13, 2022, Final Notice of "Alternative or Streamlined Mechanisms for Complying with the Current Good Manufacturing Practice Requirements for Combination Products," which states:

> Robust pharmaceutical development practices would address many design control requirements to assure compliance with § 820.30, where applicable. CP manufacturers need to demonstrate how development processes and terminology align with design control principles and requirements in § 820.30, when required, including, where necessary, developing additional design control elements. When evaluating the adequacy of existing pharmaceutical development processes, particular attention should be given to postmarket management of design changes to the combination product and the alignment of change control practices with the principles and requirements of § 820.30, as applicable.

So is the ICH Q8(R2) Pharmaceutical Development approaches accepted by US FDA regulators for Part 4 compliance? – It depends!

Design control principles and requirements of 21 CFR § 820.30 would ONLY align with the "enhanced" approach as described in ICH Q8(R2) and "encouraged" by FDA. Combination product manufacturers wishing to utilize ICH Q8(R2) under the FDA alternative or streamlined mechanism would have to align with the "enhanced" approach and not the "minimum" approach discussed in that ICH guidance document. The "enhanced" approach as described in ICH Q8(R2) utilizes a more systematic approach to include the use of quality risk management and the use of knowledge management throughout the lifecycle of the product, aligning it with the requirements of 21 CFR § 820.30 and the medical device intended for design controls as described in the regulation's preamble.[10]

Let's also look at ICH Q9 Quality Risk Management, where the introduction to that ICH guidance document states:

> Although there are some examples of the use of quality risk management in the pharmaceutical industry today, they are limited and do not represent the full contributions that risk management has to offer … it is becoming evident that quality risk management is a valuable component of an effective quality system.[11]

"Annex II: Potential Applications for Quality Risk Management" states:

> This Annex is intended to identify potential uses of quality risk management principles and tools by industry and regulators. However, the selection of particular risk management tools is completely dependent upon specific facts and circumstances.
>
> These examples are provided for illustrative purposes and only suggest potential uses of quality risk management. ***This Annex is not intended to create any new expectations beyond the current regulatory requirements***. [emphasis added]
>
> II.1 Quality Risk Management as Part of Integrated Quality Management

In contrast to the medical device quality system regulation preamble, which discusses risk analysis or risk management by saying:

> manufacturers are expected to identify possible hazards associated with the design in both normal and fault conditions. The risks associated with the hazards, including those resulting from user error, should then be calculated in both normal and fault conditions. If any risk is judged unacceptable, it should be reduced to acceptable levels by the appropriate means, for example, by redesign or warnings. An important part of risk analysis is ensuring that changes made to eliminate or minimize hazards do not introduce new hazards. Tools for conducting such analyses include Failure Mode Effect Analysis and Fault Tree Analysis, among others.
>
> **[Comment 83]**

In addition, the 1996 preamble to the quality system regulation discusses integrating risk decisions or risk management activities through the quality system in areas outside of design controls as well. See preamble comments 4, 31, 33, 159, and 161 for discussion on making decisions commensurate to the risks which are originally established in the design control process but carried through the entire quality system. Also, FDA participated in the Global Harmonization Task Force (GHTF)[12] where Working Group 3 on quality systems produced a guidance document entitled "Implementation of risk management principles and activities within a Quality Management System"[13] issued in May 2005. This GHTF guidance document has been utilized by FDA along with the ISO 13485:2016 "Medical devices – Quality management systems – Requirements for regulatory purposes" and ISO 14971:2019 "Medical devices – Application of risk management to medical devices" to establish what is "current" for medical device cGMPs with respect to risk management activities.

For medical devices, ISO 14971 "Medical devices – Application of risk management to medical devices" was recognized by FDA CDRH's standard recognition process. The most recent ISO 14971 version that was released in December of 2019 was immediately recognized by FDA on December 23, 2019.[14] For medical device manufacturers, ISO 14971 became a global regulatory expectation where global regulatory auditors have assessed manufacturers to the requirements in this comprehensive risk management system standard.

Pharmaceutical manufacturers and combination product manufacturers that use the pharmaceutical GMPs as the base cGMPs have not been inspected by FDA or other global regulatory auditors to ICH Q9 or the rigors of being audited against a full risk management standard like ISO 14971. Therefore, when FDA medical device investigators inspect the 21 CFR § 820.30 call-out requirements which include risk management procedures and records within the design and development process as well as in the design history file (where the cGMPs and expectations align with ISO 14971), the rigor of the design control and risk management processes often expected catch combination product manufacturers more familiar with CDER-led inspections by surprise with numerous medical device-related FDA 483 items.

It is important to remember that these two ICH guidance documents (Q8 and Q9), while of value, have not been utilized for inspections or global regulatory audits. Therefore, no regulatory investigators or auditors have used these documents to write up nonconformances or FDA 483 items. Without assessment or audits against these ICH documents, many combination product manufacturers that utilize pharmaceutical GMPs as the base cGMPs find themselves at a disadvantage when the medical device call-out requirements for design controls which include risk management are assessed.

Two additional key guidance documents related to the cGMPs for medical device "call-outs" relevant to combination products are:

- GHTF/SG3/N17:2008 – Quality Management System – Medical Devices – Guidance on the Control of Products and Services Obtained from Suppliers[15] [specifically called out in the FDA Combination Product Compliance Program]; and
- GHTF/SG3/N18:2010 – Quality Management System – Medical Devices – Guidance on Corrective Action and Preventive Action and Related QMS Processes[16] [used to train medical device FDA investigators].

Again, medical device purchasing or supplier controls requirements under 21 CFR § 820.50 as well as corrective actions and preventive actions (CAPA) requirements under 21 CFR § 820.100 sound very familiar to activities and requirements within the pharmaceutical requirements in 21 CFR § 211. However, while the same terms may be utilized, regulatory expectations between the two regulated commodities or sectors are very different on these two topics similar to the discussions above for design controls and risk management.

Medical device manufacturers and combination product manufacturers that use the medical device GMPs as the base cGMPs do not appear to have the same challenges on expectations with CDER-led inspections with the pharmaceutical call-out requirements. In part this may be because many of the pharmaceutical call-out requirements are more test-based versus quality system process-based requirements. Or maybe it is also attributed to the existence of very specific FDA or industry documents that have established detailed methods of successfully meeting the pharmaceutical call-out regulatory expectations over the years.

NOTES

1. See also Chapter 4 on combination product cGMPs. Importantly, while the general translation of terminology to support mutual understanding is important, the specific objective evidence required demonstrating compliance to each of the terms highlighted in Table 9.2 may be different. One must ensure that there is understanding of both the terminology applied and the data/objective evidence required to support demonstration that quality system expectations have been met.
2. See also Chapter 6.
3. See Chapters 4, 5, 6, 8, and 10 on cGMPs, integrated development, risk management, lifecycle management, and supplier quality considerations, respectively.
4. Combination Product Manufacturer: "An entity (facility) engaged in activities for a combination product that are considered within the scope of manufacturing for drugs, devices, biological products, and HCT/Ps. Such manufacturing activities include, but are not limited to, designing, fabricating, assembling, filling, processing, sterilizing, testing, labeling, packaging, repackaging, holding, and storage, including a contract manufacturing facility."
5. https://www.fda.gov/about-fda/office-regulatory-affairs/ora-overview
6. https://www.fda.gov/about-fda/office-regulatory-affairs/program-alignment-and-ora
7. https://www.federalregister.gov/documents/2018/06/13/2018-12634/alternative-or-streamlined-mechanisms-for-complying-with-the-current-good-manufacturing-practice
8. https://www.fda.gov/media/78668/download
9. https://www.federalregister.gov/documents/2018/06/13/2018-12634/alternative-or-streamlined-mechanisms-for-complying-with-the-current-good-manufacturing-practice
10. https://www.fda.gov/medical-devices/quality-system-qs-regulationmedical-device-good-manufacturing-practices/medical-devices-current-good-manufacturing-practice-cgmp-final-rule-quality-system-regulation
11. https://www.fda.gov/regulatory-information/search-fda-guidance-documents/q9-quality-risk-management
12. The Global Harmonization Task Force (GHTF) was a voluntary group of representatives from national medical device regulatory authorities and the regulated industry. Since its inception, the GHTF was comprised of representatives from five founding members grouped into three geographical areas: Europe, Asia-Pacific, and North America, each of which actively regulates medical devices using their own unique regulatory framework. http://www.imdrf.org/ghtf/ghtf-mission.asp
13. http://www.imdrf.org/docs/ghtf/final/sg3/technical-docs/ghtf-sg3-n15r8-risk-management-principles-qms-050520.pdf
14. https://www.accessdata.fda.gov/scripts/cdrh/cfdocs/cfStandards/detail.cfm?standard__identification_no=41349
15. http://www.imdrf.org/docs/ghtf/final/sg3/technical-docs/ghtf-sg3-n17-guidance-on-quality-management-system-081211.pdf
16. http://www.imdrf.org/docs/ghtf/final/sg3/technical-docs/ghtf-sg3-n18-2010-qms-guidance-on-corrective-preventative-action-101104.pdf

Part 2

Special Topics

10 Considerations for Supplier Quality: Raw Materials, Components, and Constituent parts

Fran DeGrazio, Meera Raghuram

CONTENTS

INTRODUCTION

The quality, safety, and performance of a combination drug-device delivery product (hereby referred to as combination product (CP)) is dependent on the materials, device components, constituent parts and finished drug. Development of combination products (products that combine a medical device with an active pharmaceutical ingredient (API)) has been a growth area since it was first demonstrated that drugs could be delivered with controlled release kinetics from polymers. In this chapter, we bring together the world of drugs and devices and present practical steps that a combination products manufacturer and supplier of materials, components and constituent parts can take to ensure compliance with regulatory standards and, ultimately, assuring patient safety.

DOI: 10.1201/9781003300298-12

The focus of this chapter is not on drug development or considerations related to starting materials, drug substances and inactive ingredients used in a drug product. The drug approval process is well established globally with well-defined regulatory requirements and processes. The drug constituent part can be treated as a design input into the development of the combination product, and development of the finished drug itself is outside of the scope of this chapter. The US Food and Drug Administration (FDA) has recognized the advances and role of materials sciences in spurring innovation in medical devices and the importance of material quality and biocompatibility in assuring patient safety. FDA has clearly communicated that they continue to evaluate and monitor scientific evidence as well as adverse event reports to learn more about risks from devices including those that may be related to the materials used in devices.[1] FDA does not clear or approve individual materials that are used in the fabrication of medical devices. The agency recommends that the risk assessment of the finished device includes materials, processing of the materials, manufacturing methods (including sterilization process) and any residuals from manufacturing aids used during the process. Combination products raise new quality and safety concerns for raw materials in device components and constituent parts which may come into contact with the drug constituent and potentially impact the quality, safety and performance of the finished product.

Combination product development involves purchasing materials, components or constituent parts from third-party suppliers. Though performance and material properties often drive the selection of materials in various industrial applications, using a trusted supplier with adequate quality procedures, well-characterized materials and biocompatibility is equally important when considering materials, device components and constituent parts. Recognizing the importance of supplied products and services is forcing the industry to interact with suppliers differently across the entire development, commercialization and lifecycle management of combination products. Much about the development of combination products is built around the concept of a risk-based approach: a 2019 survey reported in Pharma IQ, an online community that conducts market research and related support to the pharmaceutical industry, identifies the top risks associated with work processes and interactions between the developer/manufacturer and their suppliers along with recommendations on how to manage and mitigate risks. The most significant challenge with suppliers/supply chain is poor communication (69%), followed by logistics coordination (47%), lack of visibility (35%) and issues like lack of responsibility related to medical device/components (8%). This reinforces the need for improved communication and collaboration with critical suppliers.

The focus of this chapter is on the quality and safety of materials, device components and device constituent parts used in a combination product whether internally manufactured or purchased. In addition, the chapter covers best practices that a raw material, component or constituent part supplier and a combination product manufacturer can proactively adopt to assure patient safety. Because oftentimes, combination product applicants are purchasing at least one component, if not an entire constituent part, from a third party, the application of purchasing controls (21 CFR 820.50) becomes a crucial part of combination product quality, safety, and efficacy. Further, as articulated in an FDA guidance,[6] design controls (21 CFR part 820.30) apply to drug-device combination products, and design control activities should confirm that there are no negative interactions between constituent parts and assure that their combined use results in a combination product that is safe and effective and performs as expected. For example, in the case of a drug delivery system, the interaction of the device materials with the drug constituent part is important and needs to be carefully evaluated. FDA[2] has further explained in guidance documents that for drug-led combination products, sufficient information should be included to demonstrate that the non-lead constituent part is compatible for use with the final formulation of the drug constituent part. Similarly, for device-led combination products, the agency has stated that a device that is not combined with a drug or biologic constituent would not be successfully used as a predicate for a 510(k). This is because the addition of the drug or biologic constituent would likely raise different questions of safety and effectiveness as compared to the predicate. There is increasing recognition among both industry and regulators of the additional complexity related to quality and safety of raw material, components and constituent parts in combination products that may not necessarily be a consideration when dealing solely with medical devices or pharmaceutical products.

BACKGROUND

The devices and drugs used in combination products are selected to serve specific purposes. The intended use of the combination product and corresponding design requirements are important criteria in material selection. As the complexity and range of combination products increases, understanding the use environment and manufacturing process requirements is equally important in addition to other performance attributes in the selection of materials for device components and constituent parts in combination products. The examples below illustrate the range of combination products and the variables that may impact material selection.

- Prefilled syringes and injection devices are gaining strong acceptance as delivery systems for injectable drugs. In this case, the device may serve as a container closure, containing and protecting the drug, while also delivering measured doses of the drug. Examples of compatibility issues include leachables from uncoated syringe plungers or needles causing degradation of drug, silicone oil causing protein aggregation or sterility issues due to sealing integrity problems. There are many FDA guidance documents[3,4,5] including USP and ISO standards addressing container closure and related regulatory requirements. These are not further discussed in this chapter, but more details and guidance are provided in Chapter 11.
- Transdermal delivery systems (TDS) are designed to deliver an active ingredient (drug substance) across the skin and into the systemic circulation and are considered combination products. In the draft FDA guidance,[6] there is a discussion of the importance of excipients and components (permeation enhancers, rate-controlling or non-rate-controlling membranes, solubilizers, plasticizers/softeners, tackifiers, etc.) that can influence the quality and performance attributes of TDS.
- Implants intended for drug delivery as a primary mode of action (PMOA) need to be chemically resistant, biocompatible and stable for the intended purpose. Most often composed of polymers; solute diffusion, polymer swelling and polymer erosion or degradation are generally considered to be the main driving forces for drug transport from a polymeric matrix.[7] It is important for the materials to have consistent composition and be manufactured following good manufacturing practice (GMP) principles to ensure quality. In some cases, the device itself serves as a biodegradable delivery vehicle, delivering the drug as it degrades.
- Finally, the device itself may serve a primary mode of action in a combination product, leveraging a drug or biological product to enhance its performance, reduce inflammation or reduce the likelihood of infection. A drug-coated stent, in which the drug is applied to prevent restenosis of the artery, is one such example. In each of these scenarios, the raw materials for the construction of the device must meet the physical, mechanical and performance characteristics required for the intended use of the medical product and the specific drug- or biologic-device interactions.

RAW MATERIALS

Once a manufacturer has decided that creating a combination product is the most suitable approach for fulfilling a therapeutic need, they must consider the materials of construction for that product. Often, a drug manufacturer will turn to third-party suppliers for support with their device constituent part, that is, outside their typical core competencies. A device manufacturer will likewise often turn to a third-party supplier for the drug or biological product being used. It is important to understand the structure of the material, physical, chemical and mechanical properties and how it would impact the processing of the device component or constituent part. The key requirements for a drug delivery device are that the material in a device component or constituent part should not adversely impact the drug constituent part or the end user whether it is externally contacting (e.g., skin) or is intended for the implant. Historically, the most commonly used materials in devices have been limited to elastomers, metals, ceramics and plastics. Metals/alloys have included the use of titanium, tantalum, nitinol and stainless steel for durability, biocompatibility and corrosion resistance in joint replacements, surgical instruments, etc. Advances in material and manufacturing sciences continue to foster the development of new alloys that would improve the properties of metals for use in medical device manufacturing. Ceramics are neither metallic nor organic and include glass, clay, etc. and are chemically non-reactive. They are good insulators and can be molded in small sizes. Sensors made with piezoelectric ceramics are replacing metal sensors in many devices. Though the use of metals is still popular for many applications, polymers are increasingly being used in various device applications and specifically for drug delivery combination systems. Polymer selection is based on many factors including physicochemical (e.g., drug-polymer miscibility and stability) and pharmacokinetic properties (e.g., rate of release, absorption, etc.).

Traditional non-biodegradable polymers used in medical devices include thermoplastic polyurethane (TPU), ethylene-vinyl acetate (EVA), silicone, etc. Typical biodegradable polymer systems include polylactic acid (PLA), polyglycolide (PGA) and poly (lactic-co-glycolic acid) (PLGA). Other materials typically used in sterile injectable or orally inhaled and nasal drug product (OINDP) components or constituent parts are:

- Polyethylene terephthalate (PET)
- High-density polyethylene (HDPE)
- Polyvinyl chloride (PVC)
- Polycarbonate (PC)
- Low-density polyethylene (LDPE)
- Polypropylene (PP)

Although the above materials are common in the construction of combination product components, other less common materials include thermoset or thermoplastic elastomers, for metal containers, cannulas or other components designed for complex other types of combination products. In understanding the risk of these components, the two most significant considerations are long-term contact with the drug product or direct contact with the patient. Details around specifics pertaining to the primary packaging of the drug itself and associated testing approaches can be found in Chapter 11.

In general, raw materials used in device components and constituent parts in combination products come from different sources, and there is no one defined standard for suppliers to follow. At the most basic level, generally raw materials used in a combination drug delivery system may be categorized to fall under three distinct buckets: (a) industrial grade ingredients, (b) medical grade ingredients, and (c) excipients (pharmaceutical grade inactive ingredients).

Industrial Grade Ingredients

Polymers, metals, ceramics and other materials used in the construction of device components and constituent parts are a small part when compared to other industrial applications ranging from automotive, engineered materials and other commodity applications. The requirements related to composition, consistency, quality and safety for industrial applications are less stringent, and suppliers may not follow the quality processes in manufacturing including evaluation of biocompatibility that are a must for use in device components and parts to ensure patient safety. There are many examples of use of industrial grade materials that have compromised patient safety due to insufficient characterization for the presence of toxic impurities or other quality-related issues. The device component/constituent part manufacturer should conduct a thorough evaluation and qualification of both the raw material and supplier based on the evaluation of risk to the finished product. Additional controls may need to be implemented to assure acceptability and safety.

Medical Grade Ingredients

Currently, there is no specific regulatory definition for medical grade raw materials, which include metals, ceramics, polymers, plastics, etc. The raw material supplier will generally establish "applicable quality and performance specifications" for "medical grade" materials they supply into the medical devices market. Some best practices for consideration when establishing quality and performance specifications include, for example, biocompatibility and certain GMP aspects which would include quality in manufacturing processes. Generally, the biocompatibility of materials including polymer resins is determined by conducting select ISO 10993 and/or USP <1031> testing based on whether the material is intended for surface contact or in-body applications and for the expected duration of exposure. In addition to product performance and properties, consistency in composition, impurity profile, manufacturing process, management of change and agreements related to quality and supply are equally important for medical grade materials and are discussed in detail in this chapter.

Excipients

Excipients are defined as substances other than the active pharmaceutical ingredient (API), which have been appropriately evaluated for safety and are intentionally included in a drug delivery system.[8] According to 21 CFR 210.3(b)(8), an inactive ingredient (excipient) is any component of a drug product other than the active ingredient. The FDA Guidance for Industry Non-clinical Studies for the Safety Evaluation of Pharmaceutical Excipients[9] further clarifies that excipients mean any inactive ingredients that are intentionally added to therapeutic and diagnostic products and are not intended to exert therapeutic effects at the intended dosage, although they may act to improve product delivery (e.g., enhance absorption or control release of the drug substance). Examples of excipients include fillers, extenders, diluents, wetting agents, solvents, emulsifiers, preservatives, flavors, absorption enhancers, sustained-release matrices and coloring agents.

There is a strong case to be made that quality standards and associated processes for excipients supplied into traditional dosage forms also apply to the polymeric ingredients constituting device components and constituent parts that contact the finished drug formulation or influence drug delivery in a combination drug delivery device. The drugs may be either impregnated into the polymer matrix or contained in a reservoir system where a drug depot is surrounded by a rate-controlling membrane or surface coated on a substrate. Solute diffusion, polymer swelling and polymer erosion or degradation are generally considered to be the main driving forces for drug transport from a polymeric matrix,[10] resulting in sustained release of the drug product from the implant. Combination product examples include topical patches (e.g., hormone, contraceptives, etc.), mucosal inserts (e.g., vaginal rings with contraceptives or anti-retroviral and contact lenses) and in-body implants (arterial stents, subcutaneous contraceptive rods, ocular implants, etc.). These combination products often deliver drugs similar to immediate or controlled release applications in traditional oral dosage forms.

Additionally, prefilled syringe or prefilled cartridges are typically used as a portion of an injectable combination product. These systems can act as "excipients" by impacting or interacting with the drug product. For instance, it is very common to use polydimethylsiloxane fluid (PDMS), commonly known as silicone oil, as a lubricant to aid in plunger movement down the barrel of the syringe or cartridge. Not only can the material composition of the barrel potentially interact with the drug product, but other ingredients, such as the PDMS, can potentially interact with the drug product. This can lead to chemical or functional performance issues. Therefore, it is important that the drug product be evaluated in the primary package and delivery system that will be used for clinical and commercial use.

Excipient Functionality in Combination Products

Excipients function similarly whether used in a traditional dosage form, like tablets or capsules, or in drug delivery combination product application that may be implanted in the body. Excipients provide key functionality to support drug product processing, integrity maintenance and stability for the finished dosage form. On administration, the excipient may serve as a vehicle for drug delivery, identity and/or acceptability of the dosage form in the patient.

Figure 10.1 provides an illustration of how a polymer matrix could serve as an excipient. In the first example (Figure 10.1A, matrix design), the API is impregnated (by hotmelt extrusion, dissolution or casting) into a thermoplastic TPU polymer to yield a uniformly mixed or spatially localized ribbon. The ribbon can simply be cut, or it can be further processed (e.g., injection molding) to produce the finished device. The drug diffuses out of the polymer to adjacent areas of lower concentration. Variables potentially controlling the diffusion rate include drug loading, drug type, water absorption and polymer composition.

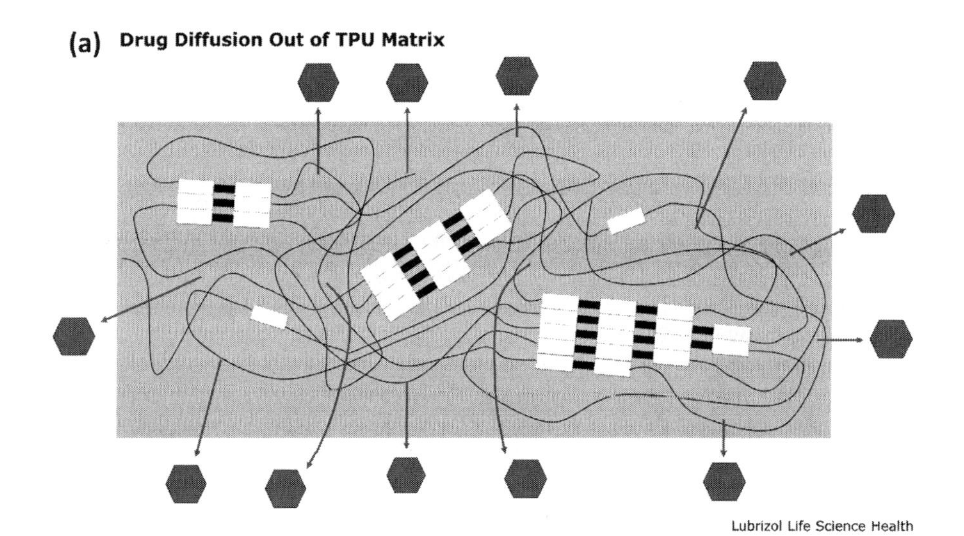

FIGURE 10.1 A) Matrix design drug delivery system utilizing thermoplastic polyurethane (TPU) polymer (*Reprinted by permission from Lubrizol Life Science Health, Lubrizol Advanced Materials Inc.*) B) Reservoir design drug delivery system utilizing thermoplastic polyurethane (TPU) polymer (*Reprinted by permission from Lubrizol Life Science Health, Lubrizol Advanced Materials Inc.*)

The second example (Figure 10.1B, reservoir design) shows a hollow, polymeric tube loaded with API (either neat or dispersed in a solvent or other suspension medium). A reservoir system can also be produced from a matrix device that is coated with a pure polymer by spraying, dipping or co-extrusion. In Figure 10.1B, the TPU tube acts as the rate-controlling membrane, and the rate of drug diffusion depends on the thickness and other membrane attributes. In other reservoir designs, there can be a thin coating of another polymer like EVA in Nuvaring® which serves as the rate-controlling membrane.[11]

Device Components

Components used as part of a combination product must be understood and characterized utilizing a risk-based approach as defined by the regulatory agencies. Components that come in direct contact with the drug product and are considered part of the primary package should meet both drug current good manufacturing practices (CGMPs) and device CGMPs. Per the *FDA Guidance on Current GMP Requirements for Combination Products published in January 2017*, if the article merely holds the drug, it is only subject to drug CGMPs as a container or closure. An article that holds or contains a drug, but also delivers it, may also be a device subject to the device quality systems (QS) regulation in addition to the requirements relating to drug containers and closures.

Drug product components, containers, and closures must be tested in accordance with 21 CFR 211.84.

An example of a component for a sterile injectable system would be a plunger (piston) for a prefillable syringe (Figure 10.2). The components of a syringe system are commonly sold separately, so that the combination product manufacturer chooses the combination that is best for their final specific CP application. It is important that the individual components, their composition and how they interface together be understood in the final application.

FIGURE 10.2 Example of prefillable syringe plungers (pistons). (*Photo courtesy of West Pharmaceutical Services, Inc.*)

Device Constituent Part

Examples of constituent parts are devices that may be used as part of a combination product kit or as part of a wearable injection or pump system used to deliver drug into a patient. The terms "constituent part" in part 4 and "component" in the CGMP regulations serve different regulatory purposes. The use of the term "constituent part" in part 4 refers to drugs, devices, and biological products included in combination products and does not alter the meaning of the term "component" or alter whether the regulations listed in 21 CFR 4.3 apply to component manufacturers.[12]

The device constituent part of a combination product may be purchased from a third-party supplier or may be based on a platform technology. Design controls start at the beginning of the process; therefore, it is important to clearly understand the selection criteria for each design. The developer/manufacturer of the device constituent part is responsible for maintaining a Design History File that can be referenced by the combination product applicant. The history must show that design controls were utilized from the beginning of the constituent part development. Because drug manufacturers that are using devices to deliver their drugs often purchase the device constituent part, Supplier Quality Agreements under purchasing controls, 21 CFR 820.50, are a critical aspect of ensuring the device constituent part will meet, and continue to meet, the performance requirements for that combination product.

Sharing of device risk management documentation is a critical input to the final combination product. It is important to have clarity between the supplier and combination product applicant for what information is needed, how it is to be shared and how it will be updated throughout the lifecycle of the products. There is an expectation that the development process for devices would follow the quality requirements and process outlined in 21 CFR 820 Quality System Regulations and the ISO 13485 Standard.

In February 2022, the FDA published a proposed rule to amend the CGMP requirements for medical devices under the quality systems (QS) regulations to harmonize with ISO 13485:2016. The intent of the proposed rule is to align the regulatory framework with that used by other global regulatory authorities. It is anticipated that harmonization will lead to more emphasis on risk management via the use of ISO 14971:2019 Medical Devices – Application of Risk Management to Medical Devices.

REGULATORY REQUIREMENTS

Under 21 CFR 820.50, "Each manufacturer shall establish and maintain procedures to ensure that all purchased or otherwise received product and services conform to specified requirements." This applies to purchased products, services and even sister companies providing services. **Purchasing controls are vital for combination products, as most include purchased materials/components/device assemblies**. Device companies may purchase the drug constituent part; pharma companies may purchase the device constituent part; both may purchase services (e.g., for human factors studies). **Outsourcing the design or utilizing an existing device platform changes the "design" process**. It doesn't start from rough drawings, but rather it starts with selection. ***Selecting materials/components/device assemblies is part of the design process***.

Purchasing controls assure rigor during technology evaluation, selection and lifecycle management and are critical for managing changes during the product lifecycle. Successful implementation of purchasing controls is essential to ensure a proactive vs a reactive regulatory strategy. Perhaps too strongly said, but **your design is only as good as your purchasing controls**. Without strong purchasing controls, the product design can change ***without you even knowing***. Customers and suppliers should agree on notification and approval of changes and include these terms in a Supplier Quality Agreement, and suppliers should ensure that **their** suppliers have adequate change control programs in place.

In general, there are no prescriptive regulatory requirements for excipients and raw materials (excluding the active pharmaceutical ingredient (API)) used in pharmaceutical drugs or devices. Components used as part of the primary drug package are required to meet CFR 210 and 211 drug CGMPs. For medical devices, the Quality System Regulations (21 CFR 820) explicitly requires that the finished device manufacturer assess the capability of suppliers, contractors and consultants to provide quality products pursuant to § 820.50 Purchasing Controls. These requirements supplement the acceptance requirements under § 820.80. Manufacturers must comply with both sections for any incoming component or subassembly or service, regardless of the finished device. Globally "*ISO 13485:2016 QMS, Medical Devices – Quality Management Systems – Requirements for Regulatory Purposes*" is an internationally agreed standard that sets out the requirements for a quality management system specific to the medical device industry. Additionally, AAMI TIR 102:2019 is the US FDA 21 CFR mapping to the applicable regulatory requirement references in ISO 13485:2016 Quality Management Systems.

There are some common elements between the CFR and the ISO standards. Several of these should be considered by the combination product applicant. These are sections such as:

Purchasing controls
Purchasing process
Evaluation of suppliers, contractors and consultants
Verification of purchased product
Purchasing information
Exceptions that may need to be considered

Beyond these, and various FDA guidance relating to combination products, most pharmaceutical and biotechnology companies are global in nature. Each country or region may have specific unique requirements that must be addressed. In addition to FDA requirements and ISO requirements from Europe, other game-changing regulations are being crafted for combination products. Under the EU Medical Device Regulation (MDR), combination products are ones in which there is a medical device and drug intended for use with each other in order to achieve the intended therapeutic effect. Based on a Q&A Guidance issued by European Medicines Agency (EMA) in October 2019, the regulatory framework for devices incorporating medicinal substances as an 'integral part' per Article 1(8) of MDR is driven by the primary mode of action (PMOA). If a single integral product is intended EXCLUSIVELY for use in the given combination and not reusable, it is considered to be a "drug-device combination" regulated as a medicinal substance. While under EU MDR it is not a requirement that medicinal substance-led drug-device combinations have ISO 13485 Certification, the device portion of the medicinal substance must go through an assessment of conformity by a notified body (NB). The device opinion of conformity must be included as part of the marketing authorization application to the European Medicines Agency (EMA) to comply with the new Medical Device Regulation. This could result in lengthier approval times in Europe since applicants must go through the 2-step process for regulatory approval. (See chapter on Evolving Global Regulations for more detailed discussion.)

A combination product applicant in the European Union could leverage a device supplier having a notified body certificate, however, they would still be required to provide evidence that the device works in their specific combination product application. If a combination product is considered integral in nature and the device supplier does not have a notified body certificate, it will be the responsibility of the combination product owner to assure that the notified body opinion or certificate is in place for any CPs to be marketed in the European region.

Under International Conference on Harmonization (ICH) Q10, pharmaceutical quality system vendor qualification likewise has explicit expectations for both outsourced activities and the quality of purchased materials. The quality risk management process is expected to include the following:

1. Assessing prior to outsourcing operations or selecting material suppliers, the suitability and competence of the other party to carry out the activity or provide the material using a defined supply chain (e.g., audits, material evaluations, qualification).
2. Defining the responsibilities and communication processes in a written agreement between the contract giver and contract acceptor.
3. Monitoring and review of the performance of the contract acceptor or the quality of the material from the provider, and the identification and implementation of any essential improvements.
4. Monitoring incoming ingredients and materials to ensure they are from approved sources using the agreed supply chain.

BEST PRACTICES FOR COMBINATION PRODUCT SUPPLIERS AND MANUFACTURERS

Excluding APIs, there is no distinct raw material (medical grade/excipients) or component-manufacturing industry focused solely on the needs of the drug and medical device sector. The lack of "industry best practices" or guidance presents a challenge with regard to consistency in quality and safety expectations for raw materials used in the medical device segment and in combination products. Most of the raw material manufacturers are chemical industry subsidiaries or divisions where a relatively small portion of the total production volume is sold to pharmaceutical and/or medical device uses. In general, the majority of the production volume for these materials goes into diverse industrial markets with different regulatory and quality standards and customer requirements.

It is typical for device component and constituent part manufacturers in regulated regions to be familiar with the regulatory requirements for those regions. Typically, these manufacturers segregate the production of products targeted for these market segments within their manufacturing facility (or facilities). Considering that the supply chain is global and fairly complex, as more manufacturers begin to supply ingredients and components to the pharmaceutical and medical device industry, especially from facility operations that have historically been less regulated, differences in good manufacturing practice operations and impact on quality of materials and components are very possible. It is therefore important for combination products manufacturers, suppliers and regulators to understand each other's needs and set clear expectations on best practices and to validate these are being met through sound auditing practices.

The International Pharmaceutical Excipient Council (IPEC) has developed best practices for inactive ingredients (excipients) that enable the use of safe and effective excipients in finished prescription, generic and over-the-counter (OTC) dosage forms. IPEC is a trade association whose members include excipient producers (makers), distributors and finished drug manufacturers (users). This unique partnership has enabled stakeholders to work together, and IPEC guides are recognized as a credible framework for setting industry best practices. These guides have gained wide acceptance from industry and regulators, globally. There are no such industry best practices or concise collection of standards for raw materials used in medical devices and combination products. Using the principles discussed in the IPEC Qualification Guide[13] and supply chain management concepts developed by Xavier University[30] as a resource, we have provided a framework that can be used by suppliers, users and regulators for assuring quality and supply chain security. This framework includes suppliers process, combination product manufacturer's process and communication process between the user and the supplier.

- **The supplier's process** – includes best practices a supplier should consider and implement for the material and/or device component/part. Manufacturers of raw materials and components have to ensure that adequate quality processes and procedures are in place from manufacturing through product disposition.
- **The combination product manufacturer (user's) process** – illustrates the path a combination product manufacturer may use in the evaluation of materials, parts, device and supplier qualification or use in a formulation. For users, it is important to source from high-quality suppliers with sound quality systems in place.
- **The communication process** – shows the process by which the supplier and user interact to reach a mutual agreement on quality requirements. Communication and agreement on specifications and supply chain considerations are equally important and covered below. The Xavier Health supply chain process and how its framework applies to supplier–user relationship (including outsourced activities like device components and constituent parts manufacturing) is discussed in this chapter.

Combination Product Framework: Key Considerations for Raw Material Supplier

The philosophy and concepts for "Total Excipient Control[14]" (TEC) articulated in various IPEC guides and position papers provide a road map for ensuring the quality and safety of raw materials used in combination products. Though all raw materials used in a combination product are not necessarily excipients, the IPEC guides can be used as a framework for developing best practices as the key principles of quality and safety are important for all raw materials. Each guideline fills a specific need related to an area of excipient control and illustrates best practices that can be applied to raw materials used in combination products. As shown in Figure 10.3, the key control elements are discussed below:

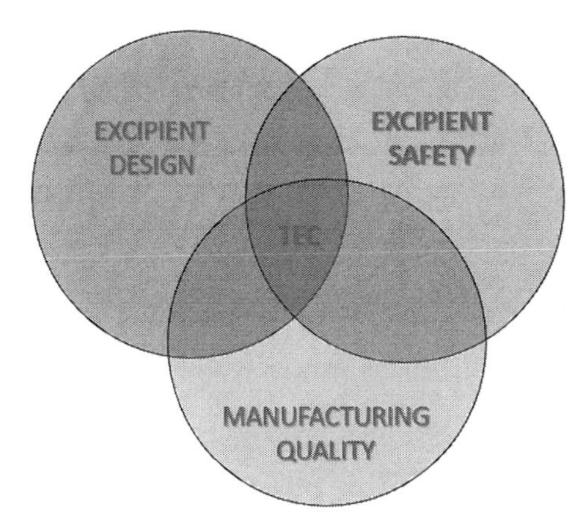

FIGURE 10.3 Total excipient control. (*Printed by permission from IPEC-Americas*)

The following sections cover key controls to ensure consistency in quality and safety for a raw material supplier to consider. It is envisioned that these will serve as industry best practices for all raw materials used in device components and constituent parts.

Raw Material Design Controls

Design controls include design criteria for the raw material to meet the requirements for the intended use. This section is intended to provide general guidance and is not meant to be prescriptive as specific requirements for a material may vary due to the diversity of combination product applications.

Composition of Raw Material

The IPEC Composition Guide[15] provides a well-thought-out approach for a material manufacturer to establish a composition profile for a pharmaceutical excipient. The concepts discussed would generally apply to medical grade ingredients and excipients supplied into drug delivery combination product applications. Evaluation of composition profile should be performed by the manufacturer using their knowledge and understanding of the manufacturing process and associated potential undesirable components. Raw material components (main/concomitant), additives, processing aids and undesirable components should be identified and quantified using suitable analytical techniques wherever possible. A composition profile may be used for regulatory purposes, quality consistency, manufacturing process monitoring and change control, product specification setting or for safety evaluation by the supplier and/or user. Presence of catalyst residues, decomposition or degradation products and process-related components are important as they may interact with the drug component(s) leading to undesirable results. The presence of additives and process aids in the raw material, such as biocides, anti-oxidants, or stabilizers, can lead to safety concerns for the combination product manufacturer, and their selection should be carefully reviewed and assessed for acceptability in the intended combination product application.

In the draft Guidance on Use of International Standard ISO 10993-1,[16] FDA has explained where chemical analysis of the materials used in a device in its final finished form can be informative. In certain cases, leachable and extractable testing may provide additional information on the extractable impurities under the test conditions. For example, in some circumstances, when evaluating a new raw material supplier or a change to a device component or part, a chemical analysis can demonstrate that the extractables and leachables in a biocompatibility extract have not changed, eliminating the need for additional biocompatibility testing using that type of solvent. In addition, chemical analyses of the raw material can be used to assess the toxicological risk of the chemicals that elute from device components and constituent parts. Extraction techniques could also be used to identify intermediate and final breakdown products in a raw material that is either synthesized *in vivo* (e.g., *in situ* polymerizing materials) or intended to be absorbable (e.g., degradable materials). However, chemical analysis is usually insufficient to identify all the risks of the device in its final finished form, because it will not consider aspects of the finished device component/part such as surface properties (e.g., rough versus polished surface) or device geometry that could affect the biological response in certain scenarios (e.g., thrombogenicity, implantation). In addition, the outcomes of chemical analyses are often sensitive to the parameters of the test. Extraction solvents should be selected to optimize compatibility with the device materials and provide information on the types of chemicals that are likely to be extracted in clinical use. For example, solvents that swell the polymer, cause the polymer to degrade or dissolve, or interfere with the detection of chemicals should be used with caution. These results should be carefully interpreted as the supplier cannot always simulate end-use conditions or the additional processing for the polymer.

Consideration should also be given to impact of sterilization processes on device materials. Some types of sterilization techniques may not be optimal for a material and may cause changes in the composition (residues) or performance of the material.

Certificate of Analysis (COA)

The COA is a legal document that certifies the quality of the raw material and demonstrates that the **batch** conforms to the defined **specifications**, has been manufactured under sound quality system utilizing GMP principles and is suitable for use in combination products. It is expected that a complete and accurate COA is provided to the user for each batch and/or delivery of raw material. The key required elements of a COA are discussed below, and the reader is referred to the IPEC COA guide[17] for additional details. The principles and concepts discussed here would generally apply to both excipients and medical grade ingredients.

- The **original manufacturer** and manufacturing **site** should be identified if different from the supplier and supplier location. The intent is to enable the user to assure that a change in manufacturing location has not occurred without their knowledge.[3] It is essential that the manufacturer be known to the user.
- The **identity** of the material should be definitively established by stating compendial (if applicable) and trade name, the grade of the material, etc.
- A batch number or other means of uniquely identifying the material quantity covered by the COA.
- The **date of manufacture** and, if applicable, the **expiration date**, recommended **retest date** or other relevant statement regarding the stability.
- The acceptance criteria and test results are preferably included for each characteristic (physical, chemical, mechanical, etc.) listed.
- The Certification and Compliance Statements Section is used to list various statements that may be required depending on the material and agreed user requirements. Any declaration

by the supplier as to compliance with compendial and/or other regulatory requirements is typically included in this section.

- The basis for COA approval should appear on the COA.

Stability of Raw Material

The supplier should have data to show stability in the commercial packaging and defined storage conditions that are intended to protect the raw material from degradation throughout the supply chain. If stability data is not available, the combination product manufacturer may have to generate it themselves. For a well-established material, the supplier will often know how it changes as a result of atypical processing conditions or due to thermal and related stresses to which it may be exposed in the supply chain. It is important to demonstrate that the excipient/medical grade ingredient will remain in conformance with the agreed to sales specification throughout the shelf life. Where formal stability studies have been conducted consistent with ICH,[18] IPEC Stability Guide[19] or other available protocols, such studies can provide similar information.

Regulatory Credentials

A regulatory assessment for raw materials should be conducted and credentials established prior to marketing the raw material. Examples of key considerations are presented below and by no means constitute an exhaustive list of all regulatory requirements that may apply. Applicable regulatory credentials should be established based on the analysis of national, regional and local regulations that may apply to the marketing of the raw materials and finished goods. If certain information is considered confidential by the supplier and critical to the qualification of the material, then the information may be presented under a non-disclosure agreement acceptable to the raw material supplier and combination product manufacturer.

- Conformance with compendial monograph (USP, EP, any country-specific) if applicable.
- Animal-derived materials lead to concern for issues related to BSE/TSE which can restrict acceptance. Processing information and risk management measures may be required.
- Elemental impurities.
- Presence of allergens.
- Susceptibility to microbial contamination.
- Information on additives, phthalates, solvents, processing aids and colorants.
- Presence of substances of very high concern (SVHC) under EU REACH regulations.
- Declaration of California Proposition 65 chemicals.
- Conformity with EU Restriction of Hazardous Substances Directive.
- Other federal, regional and local requirements as applicable.

Establishment of Drug Master File (DMF)/Master Access File (MAF)

Submission of an excipient Drug Master File (DMF) or a Master Access File (MAF) is not required by law or FDA regulations. It should be noted that DMFs and MAFs for excipients or raw materials used in devices are neither approved nor disapproved by FDA. DMFs and MAFs are only reviewed as part of a drug, device or combination product application review process when the DMF/MAF holder provides authorization to FDA to access the files. The sole purpose for establishing a DMF or an MAF is to make confidential or business proprietary information on the ingredient available for regulatory review without divulging information to the drug, device or combination product manufacturer. Alternatively, suppliers may prefer to share information directly with the combination product manufacturer for regulatory submissions under a non-disclosure agreement.

The DMF for an excipient (designated as a Type IV by FDA) is compiled in ICH Common Technical Document (CTD) format and may include information on Chemistry, Manufacturing and Controls (CMC), specifications and test methods for raw materials, in-process testing, safety data, packaging details, label content and safety data. For more details on what to include, refer to IPEC-Americas Drug Master File Guide for Pharmaceutical Excipients.[20] There is no required format or recommended content for an MAF for raw materials used in medical devices. An MAF may include information on composition, manufacturing and testing and material properties that is deemed important and confidential by the supplier. In cases where an assessment of biocompatibility is required for the material, the FDA guidance[14] (Appendix B) provides recommendations on information to be included in a master file.

Manufacturing Process and Distribution

As explained in the IPEC Qualification Guide, it is important that the excipient is produced using a manufacturing process that is in a state of control, often referred to as a capable process. This is important for both excipients and medical grade materials used in combination products as any variance in material processing can potentially impact the composition, performance and biocompatibility of the combination product. The process for manufacturing the excipient and medical grade ingredient should be capable of producing material that consistently meets the established specifications. According to 21 CFR 820.3 (z)(1), process validation means establishing by objective evidence that a process consistently produces a result or product meeting its predetermined specifications. There is no regulatory requirement to conduct formal validation studies for raw materials. As an alternative to a process validation study, statistical techniques can be used to demonstrate that the process capability is sufficient.

Whether the process is by batch or continuous, there should be a written set of manufacturing instructions listing the raw materials, operating equipment, operating conditions, in-process controls, sampling plan, packaging operations, packaging components and labeling materials with their content. While dedicated manufacturing equipment is preferred, multi-use equipment is acceptable provided there is no incidental carryover of **contaminants** from the manufacture of another chemical. Packaging operations include steps in the manufacturing process where the raw material may be exposed to potential environmental contamination. Suitable controls need to be put in place in this area to ensure raw material quality. Additional information is available on establishing good manufacturing practices for excipients in the IPEC-PQG GMP guide.[21] Though there is no regulatory expectation that medical grade ingredients be manufactured to excipient GMP standards, the principles regarding batch-to-batch consistency, ensuring quality and supply chain integrity applies to medical grade ingredients as well. In cases where an industrial grade material is being used, the user should ensure that appropriate quality criteria for their intended uses are met.

In addition to manufacturing, maintaining integrity of the raw material during storage and distribution is equally important. The supply chain can be complex with a material moving globally and being managed through distributors, brokers and traders. The IPEC GDP guide[22] as well as the Xavier supply chain management process are good references for suppliers.

Management of Change

Management of change and assessment of the impact of change is critical for excipients and medical grade ingredients used in device components and combination products. As presented in the FDA guidance,[14] a change in polymer resin supplier or a change in the processing of the raw material could alter the physicochemical characteristics which could impact the biocompatibility of the final device. For example, if the new resin supplier does not remove all processing solvents (some of which may be known toxic compounds, such as formaldehyde), the final manufactured device could cause unexpected toxicities (e.g., cytotoxicity, irritation, sensitization, genotoxicity) that were not seen with devices manufactured from the original resin. This illustrates the importance of the supplier having a change management process which includes notification of significant change to users. The FDA 510(K) guidance[23] includes discussion and examples of when a change impacting material may be significant and may require a new regulatory application which further illustrates the importance of suppliers having a change management process.

The IPEC Significant Change Guide[24] provides a framework for assessing and establishing a process for the management of change. The IPEC guide presents key principles that should be considered to determine the significance of the change with or without the use of a formalized risk assessment approach. The purpose of the evaluation is to consider the impact of the change on the raw material and to determine whether the raw material user and/or regulatory authority should be informed. These principles would apply to materials in combination products, and the supplier should consider developing a process along the following lines:

- Complexity of the change(s) (including possible cumulative effects)
- Level of understanding of historical norms
- The ability to fully characterize the impact of the change on the:
  a. Excipient properties (i.e., chemical, physical, microbiological, composition profile, etc.)
  b. Excipient performance in intended uses (**critical material attributes**)
  c. The equivalency of the composition profile comparing pre-change and post-change batches
- The ability to assess the change in trial batches and/or model products
- Level of understanding of the users' application(s) and use(s) of the product
- The potential for prediction of the impact on the users' application
- The content and requirements of any quality or technical agreements that are in place
- In the case of raw material changes, the level of knowledge, understanding, credibility and reliability of the raw material manufacturer, and the relationships that exist within the raw material supply chain
- Impact of change on the content of regulatory documents (**DMF**, **MAF**) or submitted to customers for their own regulatory applications (technical dossier).

Quality Agreements

Quality Agreements (QAs) enable customers and suppliers to create a partnership between the companies that ensures all quality requirements are defined. QAs are legally binding agreements that are negotiated between customers and suppliers of components, sub-assemblies, constituent parts, excipients and medical polymers to ensure that there is a clear delineation of responsibilities related to the quality of each constituent part and the combination product.

By clearly delineating responsibilities, costly product quality issues resulting from miscommunication can be reduced or eliminated as well as ensuring the customer meets their regulatory expectations and requirements. A sample template for a combination product manufacturer is attached to this chapter. IPEC has developed Quality Agreement templates[25] that are designed to provide excipient customers and suppliers with a common starting point to create mutually beneficial and regulatory compliant Quality Agreements. These can be used as a resource by both raw material and combination product manufacturers in developing Quality Agreement templates. Other resources include FDA final guidance entitled Contract Manufacturing Arrangements for Drugs: Quality Agreements (November 2016)[26] which provides key principles to develop a Quality Agreement.

It is recommended that quality requirements established by the manufacturer for raw materials should be discussed and agreed upon with the suppliers. Appropriate aspects of the production, and control, including handling, labeling, packaging and distribution requirements, complaints, recalls and rejection procedures should be documented in a Quality Agreement or specification.

By utilizing the IPEC QA template structure and level of detail, customers and suppliers will reduce the time and effort needed to complete QAs. Quality Agreements should be reviewed by the quality departments to verify all requirements are addressed and are achievable.

Material Safety/Biocompatibility

Biocompatibility is the ability of the device material to perform with an appropriate host response in a specific use situation. Though FDA requires testing to be conducted on the final device in the evaluation of biocompatibility risks for a device, the potential for an unacceptable adverse biological response is based on the evaluation of materials used in the device. A first step for the combination product manufacturer is therefore to assure that materials selected in device components and constituent parts are biocompatible for the intended use and duration of exposure. Chemical constituents and potential extractables should be identified and quantified for the overall safety assessment of the device. It is important to assure that the device materials either directly or through the release of leachable constituents will **NOT** produce adverse local or systemic effects in the body. Biocompatibility of the material is important in both direct contact (physical contact with body tissue) and indirect contact (delivery of fluid or gas through device component which comes into contact with body tissue) applications. New biocompatibility testing may not be needed if the device is made of materials that:

- Have been characterized chemically and physically in the public literature and have a long history of safe use; and
- Materials and manufacturing information support no new biocompatibility concerns.

An excipient/medical grade ingredient manufacturer cannot assess the risk for a finished device component or constituent part in the combination product but can evaluate the raw material as manufactured to assure that it does not have toxic chemical constituents. The manufacturer should select studies sufficient to screen the material for biocompatibility based on whether the material is being offered for external communication or implant device applications or both.

Biocompatibility testing is addressed by ISO Standard 10993 and related 10993 series of standards including an FDA guidance on how to apply and interpret the ISO Standard. ISO-10993-1 provides an overview of biocompatibility and an approach for risk assessment based on device design, material and manufacturing processes. The remaining chapters address sample preparation, toxicological study methodologies and risk assessment. In general, the suggested biological tests for determining the potentially adverse or toxic effects of medical devices include procedures designed to evaluate cytotoxicity; acute, sub-chronic, and chronic toxicity; irritation to skin, eyes, and mucosal surfaces; sensitization; hemocompatibility; short-term implantation effects; genotoxicity; carcinogenicity and effects on reproduction, including developmental effects. In 2021 revisions were made to the ISO 10933 standards. The more significant changes related to ISO 10993-12 Sample Preparation & Reference Materials and ISO 10993-18 Chemical Characterization of Device Materials. This also includes clarification on extraction procedures.

Many manufacturers use the USP Class VI testing as a criterion for offering medical grade polymers and excipients. These tests are designed to evaluate the biological reactivity of various types of plastic materials in vivo. The USP defines six plastics classes, from I to VI (VI remaining the strictest). Consequently, several plastic manufacturers find it beneficial to have their plastic resins certified as USP Class VI, particularly if the resin is a candidate to be used in medical devices. A plastic resin material that has passed Class VI certification is expected to be more likely to produce favorable biocompatibility results. For a product to pass USP Class VI standards, it must exhibit a very low level of toxicity by passing all of the test requirements. Compliance with USP Class VI is often requested by end users.

For in situ polymerizing and/or absorbable materials, biocompatibility may need to be evaluated over the course of polymerization and/or degradation to ensure that starting, intermediate and final degradation products are assessed. It is advisable to screen the candidate materials at an early stage to eliminate those that are toxic and select those that are sufficiently biocompatible or nontoxic for their intended use.

Key Considerations for Combination Products Manufacturer as They Interact with Component and Constituent Part Supplier

Many of the concepts captured for raw material suppliers also apply to component and constituent part suppliers. Design controls begin with building a product with the end in mind. This requires a clear definition of the goals of the development program from the beginning (see Chapter 5). This can apply to a drug or biological product component that may be developed based on drug CGMP using techniques such as Quality by Design (QbD), ICH Q8, Q9 and Q10 guidance's[27,28,29] or through device design controls as defined through CFR 820.30 regulations or compliance standards for devices as defined in ISO 13485 quality standards.

The important considerations in all these cases are to understand if a platform or "off the shelf" component is used as part of a combination product and if the process capabilities and specification of that product are adequate to meet the needs of the intended use, use environment and avoidable predictable misuses. Quite often because the combination product applicants may not truly understand their requirements at the component level sufficiently or they have not engaged early enough with their critical suppliers, they assume that standard components with standard specifications and tolerances are appropriate for their product's safe and effective use. This can lead to undesirable breakdowns and issues with unanticipated and unwanted variability over time.

Early engagement with suppliers to understand and collaborate on the choice and specifications of the right component(s) and constituent devices, and associated test methods, can help to avoid longer term potential problems. It is of much more long-term value to invest time working early with suppliers to understand these critical issues than be forced to retrospectively try to control a product that was not capable in the final application.

The expectations of component and constituent part/device suppliers are very consistent with those of material suppliers. These are things such as signing non-disclosure agreements depending on the type of information being exchanged. This typically can go two ways. For example, although the pharmaceutical customers may be sharing confidential information about a drug product or application with a supplier, the supplier is often sharing intellectual property (IP) or other information that is historically held as secure with the company. This level of collaboration is necessary to produce a product that is well understood and achieves its end goals. Quality Agreements are a standard expectation and define and outline expectations and ownership around various responsibilities relating to the component or constituent part. Suppliers that are approved for specific purposes under these Quality Agreements should be reflected on an Approved Supplier List within the combination product manufacturer's quality system.

A combination product manufacturer should not assume that a supplier is aware of their responsibility in meeting the applicable requirements for a product or service. The combination product manufacturer is ultimately responsible for meeting all aspects of their quality system and applicable regulations. A Quality Agreement is needed to ensure the supplier acknowledges and agrees with their defined responsibilities. This covers everything from management responsibility to how complaints will be handled.

Other types of information expected from suppliers are things like compliance with environmental and governmental standards, business continuity standards, control strategies for the items being purchased and an understanding of change control procedures. Certainly, if products are being developed specifically for a pharmaceutical company, for example, there should be a clear understanding of the product development plan, the regulatory requirements relating to the product, how communications will occur, specifics around handling, storage and distribution, and corrective and preventive actions.

Change Control

Change control/notification requirement is a requirement in CFR 820.50. If change notification requirements cannot be agreed upon, a CP manufacturer should put in supplemental controls. It is important to list the criteria in which change notification applies, as the definition of change varies between suppliers and industries. It is also important to consider including a timeframe to be notified prior to change, accommodate the time needed to assess and complete necessary testing/ revalidation if applicable.

Changes by a component or constituent part supplier should be categorized based on the evaluation of the change to impact the material of construction, design, form, fit, or function. Changes are typically categorized as:

- Major
- Moderate
- Minor

The major and moderate changes are communicated well in advance of change to the customers, so that customers have the opportunity to evaluate, conduct testing and provide prior approval of proposed changes relative to potential impacts on their product's safety and efficacy. Stipulations for communication in the event of an emergency need for change should also be clearly agreed upon in Supplier Quality Agreements, to prevent unintended consequences in product performance. Best practices are that even seemingly minor changes from a supplier perspective be shared with customers in advance, to avoid unintended consequences, where a supplier may not realize that a proposed change, they deemed minor actually has a large impact on combination product performance. Changes require evaluation of the need for repeating design verification or even validation activities. Examples of supplier changes are:

- Manufacture site location (one location to another)
- Process manufacturing changes/tooling changes (compression molding to injection molding)
- Changes to documentation (adjustments to Certificate of Analysis – COA)
- Changes to specifications (tightening or making broader)
- Raw material component no longer available, requiring supplier to change source or material
- Change in mold cavitation (e.g., to increase capacity), potentially increasing variation of critical part dimensions, impacting assembly of components
- Changes to component materials (e.g., changes implemented as cost improvement opportunities may lead to unintended consequences, such as interactions between drug and device)

It is critical that the combination product applicant owner and the supplier have full alignment on the process and categorization of changes, and this should be documented in the Quality Agreement.

Scope of Products and Services

Any procured product or service that can impact the quality of the finished product is in scope of the regulation. Services include any provider of processes, contracted work and consultants. Purchasing controls apply to all providers of products or services received from outside the organization, whether payment occurs or not. If a combination product manufacturer receives a product or service from its "sister facility" or some other corporate or financial affiliate, the sister facility must be included in the purchasing controls system.

The range of procured products and services for a combination product is wide and their use spans across the organization. Departments such as distribution, production, regulatory, quality, facilities, planning, engineering, labs, development and maintenance can use purchased products and services that can impact the quality of the finished product. It is important to have controls in place to ensure the suppliers of these products and services meet the quality needs of the organization.

Risk Management

A combination product manufacturer purchases a wide range of products and services which pose varying levels of risk to the quality of the finished combination product. These risks need to be identified and managed in the purchasing controls process. The purchasing controls procedures should include appropriate requirements for the management of suppliers based on the risks that the product, service and supplier pose to the finished product (3,4).

For instance, the purchasing control activities for a low-risk supplier of a low-risk product (i.e., cleaning agent) need not necessarily be subject to the same level of controls as compared to a high-risk supplier of a high-risk product (i.e., an excipient controlling drug release or a material interacting with the drug constituent part). Thus, the degree of supplier evaluation, selection, controls, monitoring and the respective purchasing data required by 820.50 should be commensurate with the risk of the product, service and supplier to the final product quality and integrated into the organizations' overall risk management process.

Pre-defining risk to purchased products or services provides the basis for subsequent supplier management activities. Defining risk to a product or service provider requires a team of experts, including experts from manufacturing, product, quality and engineering to adequately assess the impact of a product or service on the overall finished product quality and provides the needed structure for which the purchasing controls procedures can be based.

Purchasing Controls

A significant challenge for a combination product manufacture regarding purchasing controls is the sheer volume, diversity and risk level of the products and services, as well as suppliers used in the development and manufacture a combination product. If a supplier or service provider is providing materials or services in support of GMP activities for the combination product's or constituent part's design verification or validation activities in a Design History File, they must have a **Supplier Quality Agreement** and be approved to be included on the **Approved Supplier List**. The combination product manufacturer must evaluate the quality impact of the purchased product or service on the finished product and ensure appropriate controls are implemented. The supplier as well can impact the finished product quality, and this must also be assessed and controlled. It is important that the purchasing controls process for a combination product manufacturer can accommodate and appropriately manage these risks in order to assure compliance and quality for the finished product.

There are some considerations for establishing a robust combination product purchasing control process, which are mentioned here. One consideration is control of purchases of products or services that are subject to the purchasing control regulation. Purchasing controls do not apply to every purchase made in an organization. It is pertinent to think about how to control the purchasing of products and services to the regulations do apply. A common practice is to train the organization on purchasing controls, but this practice requires all individuals to make an assessment on whether the product or service impacts quality products, and this is not always effective. Additionally, failures occur when an assumption is made that a supplier can provide any product as long as they are on the Approved Supplier List, where in fact, they may only be approved for a certain product. Solving these issues may not be easy, especially in a large organization, with hundreds or thousands of individuals who make purchases of products or services. System controls are an option, as are limiting the individuals who can make purchases.

A second consideration for establishing a robust purchasing controls process is how to ensure the appropriate risk level is applied to a product or service. The activities associated with supplier management should be commensurate with (1) the risk that a product or service poses to the finished product quality and (2) the risk that the supplier poses to the quality and the business. The product or service risk should be assessed by individuals with the expertise to determine the potential product impact. An organization may have thousands of products or services that are subject to the purchasing controls requirements. If the risk assessment is performed on each product or service individually, the process will take significant effort and resources and is subject to inconsistencies over time. An industry practice is to establish risk levels or categories and assign products and services types to a risk category. The risk category of a product or service is predefined and can be used on the basis of the supplier management activities, including supplier selection and evaluation. Supplier management should focus on the risks associated with the supplier. This provides a standard framework for the organization to use for risk management of suppliers. The purchasing controls activities are based on the product or service risk, so it is important to ensure the risk assigned is accurate and consistent.

Supplier Controls and Monitoring

Supplier monitoring is necessary to demonstrate a supplier's continued conformance to the requirements. Controls include periodic evaluations of the supplier and the product or services provided. The requirements for the type and frequency of supplier evaluations should be defined based on the significance of the product or service provided and the supplier performance. Supplier controls typically include:

- Audit Program;
- Periodic review and update of the Quality Agreement;
- Incoming Acceptance Testing; and
- Supplier Performance Monitoring.

Periodic audits are used to determine if a supplier is maintaining quality system conformance and complying with the requirements agreed upon in the Quality Agreement. Business factors can impact a supplier over time, and an audit is an effective method of assessing the quality and overall operations of the supplier.

Quality Agreements should be reviewed periodically, to determine if requirements stated in the Quality Agreement need to be modified or if additional requirements should be added. New regulations may impact the requirements of the supplier. A periodic review of the Quality Agreement is important to assure compliance.

Acceptance Procedures

A combination product manufacturer must incorporate acceptance testing for all received products. This includes defining procedures for incoming acceptance of products including inspection or testing the product to specified requirements and documenting acceptance or rejection (21 CFR 21 820.80 (b)). Acceptance testing requirements that are explicit for certain products still apply. For example, 21 CFR 211.84 testing and approval or rejection of components, drug product containers and closures. Documentation of acceptance activities is required. The records must include:

(1) The acceptance activities performed;
(2) The dates acceptance activities are performed;
(3) The results;
(4) The signature of the individual(s) conducting the acceptance activities; and
(5) The equipment used, where appropriate.

These records shall be part of the DHR (21 CFR 21 820.80 (e)).

Performance

Monitoring a supplier's performance is a very effective method for identifying trends and issues with suppliers or services. If issues are identified, communication with the supplier should be done along with a request for improvements. Monitoring supplier performance requires procedures to be put in place for collecting performance data and trending. Performance indicators may include:

1. Regulatory inspection observations
2. Market actions
3. Customer complaints
4. Non-conformances
5. Response time for investigation
6. On-time completion of audit observations
7. On-time completion of corrective actions
8. Number of "Out of Specifications" or "Out of Trend"

Purchasing Data

Purchasing data include the record of the purchase of a product or service, accompanied by all the necessary documentation needed by a supplier to successfully manufacture a product or provide a service. A purchase order with appropriate information describing or referencing the product to be purchased include as appropriate:

- Product specifications
- Consulting or service contract
- Requirements for product acceptance
- Quality Agreement or at a minimum a written agreement that the supplier notifies the organization of changes in the purchased product prior to implementation of any changes that affect the ability of the purchased product to meet specified purchase requirements

As explained in the ISO 13485: 2016 Standard, "the degree of specificity of the purchasing information should be dependent upon the effect of the purchased product or service on the medical device, for example, as determined during risk management activities."

Communication Process and Supplier Relationship

A research initiative to better understand and provide guidance for industry–supplier holistic relationships was initiated by Xavier University in 2012. A cross-functional team of FDA, pharmaceutical and medical device professionals and suppliers worked together to uncover and present best practices for this interaction. In 2018, the culmination of this research led to a document entitled Good Supply Practices for the 21st Century. Critical to all of this is the foundation that the suppliers must become an integral part of the supply chain in all aspects. There were three themes that emerged from this holistic research.[30]

The first was that the Combination Product manufacturers need to understand the variability of their own products and processes in order to know what is truly needed from their suppliers. Quite often suppliers may not even be brought into the development process to better understand what specifications may be truly needed for the product being supplied. Very typically standard specifications are used instead of understanding the uniqueness of each combination product application and how this may have implications on a material, component or device. Assuring cross-functional involvement – including the supplier when appropriate – should be strongly considered.

The second theme was that combination products manufacturers may often circumvent their own processes which can eventually lead to poor decisions being made. Quite often it is found that supply agreements and Quality Agreements with a supplier can be in conflict. In addition to the criteria of capability, capacity and cost, key risk factors such as financial viability, geopolitical implications and culture are sometimes not considered. It is critical that the manufacturer has a cross-functional alignment when engaging with the supplier to avoid satisfying only a portion of the real requirements of a holistic program. A miss in this alignment can then lead to the need to "control" a supplier who may not have been the right partner, to begin with. This ultimately **leads to a lot of**

expended energy and angst that could have been avoided if the appropriate considerations were made early in development.

The third theme is the lack of developing a trustful relationship with suppliers. Working in a collaborative manner with both parties sharing concern for costs, disruptions, timelines, etc. is a critical foundation.

The Xavier University research found that it is extremely important to "recognize a supplier as a valued partner rather than one to be controlled." Changing this dynamic is critical in achieving the ultimate goal of high confidence in product and support through the supply chain and delivery to the patient.

A critical difference in working with devices and constituent parts of combination products is that due to purchasing controls, CFR 820.50 means that the outsourcing of the design or the use of an already existing device platform changes the typical "design" process as it now starts with this component selection. This selection of components and device constituents is part of the overall design.[31] These purchasing controls are critical for managing changes, not only during development, but throughout the lifecycle of a product. Strong purchasing controls are vital for these purchased products as without this a design change could occur without the manufacturer becoming aware.

In working with suppliers, it is also important to understand their change control procedures upstream in the supply chain. Anticipating and mitigating potential issues early can assure smoother lifecycle management downstream.

CONCLUDING REMARKS

The commentary above presents a strong case for a collaborative relationship between suppliers and combination product manufacturers. The combination product framework includes framework for suppliers, manufacturers and communication elements which are all very important. Suppliers of materials, components and device constituent parts are critical to the supply chain and quality and safety of the final combination product. It is critical for combination products manufacturers to source from reliable suppliers and establish requirements for quality, consistency and safety of raw materials commensurate with risk. A key aspect of the collaborative relationship is continuous improvement and lifecycle management in supporting the combination product owner in achieving compliance and patient safety. This is a great opportunity for industry and regulators to work together and develop best practices which will not only streamline industry efforts and make things more efficient, but the ultimate benefit will be to patients!!

NOTES

1. FDA statement on continued efforts to evaluate materials in medical devices to address potential safety questions. Director – CDRH Offices: Office of the Center Director. September 30, 2019.
2. Principles of Premarket Pathways for Combination Products. Guidance for Industry and FDA Staff. FDA Office of Combination Products. CBER. CDER. CDRH. January 2022.
3. Guidance for Industry: Container Closure Systems for Packaging Human Drugs and Biologics.
4. "Draft Guidance for Industry and FDA Staff: Technical Considerations for Pen, Jet and Related Injectors Intended for Use with Drugs and Biological Products" (when finalized).
5. Guidance on the Content of 510(k) Submissions for Piston Syringes.
6. Guidance for Industry. Guidance for Industry. Transdermal and Topical Delivery Systems – Product Development and Quality Considerations. November 2019.
7. Arifin, D.Y.; Lee, L.Y.; Wang, C. Mathematical modeling and simulation of drug release from microspheres: Implications to drug delivery systems. Adv. Drug Deliv. Rev. 2006, 58, 1274–1325.
8. General Glossary of Terms and Acronyms. The International Pharmaceutical Excipients Council (IPEC). IPEC-AMERICAS (ipecamericas.org)
9. Guidance for Industry Nonclinical Studies for the Safety Evaluation of Pharmaceutical Excipients. FDA CDER CBER. May 2005.

10. Lowinger Michael et al. Sustained Release Drug Delivery Applications of Polyurethanes. Pharmaceutics MDPI. May 2018.
11. Drug Approval Package. NuvaRing (Etonogestrel/Ethinyl Estradiol Vaginal Ring). FDA. 10/3/01 (https://www.accessdata.fda.gov/drugsatfda_docs/nda/2001/21-187_NuvaRing.cfm).
12. Guidance for Industry Current Good Manufacturing Practice Requirements for Combination Products, FDA, January 2017.
13. Qualification of Excipients for Use in Pharmaceuticals, International Pharmaceutical Excipients Council. IPEC-AMERICAS (ipecamericas.org).
14. Total Excipient Control: A Pathway to Increased Patient Safety. Pharmaceutical Technology, Volume 2011 Supplement, Issue 2.
15. The IPEC Excipient Composition Guide. The International Pharmaceutical Excipients Council. IPEC-AMERICAS (ipecamericas.org).
16. Use of International Standard ISO 10993-1, "Biological evaluation of medical devices – Part 1: Evaluation and testing within a risk management process" Guidance for Industry and Food and Drug Administration Staff. CDRH. September 2020.
17. The IPEC Certificate of Analysis Guide. The International Pharmaceutical Excipients Council. IPEC-AMERICAS (ipecamericas.org).
18. International Conference on Harmonization (ICH) guideline Q1A *Stability Testing of New Drug Substances and Products.* ICH Q1A(R2), Step 5.
19. The IPEC Excipient Stability Program Guide. The International Pharmaceutical Excipients Council. IPEC-AMERICAS (ipecamericas.org).
20. The IPEC U.S. Drug Master File Guide for Pharmaceutical Excipients. The International Pharmaceutical Excipients Council. IPEC-AMERICAS (ipecamericas.org).
21. The IPEC-PQG Joint Good Manufacturing Guide for Pharmaceutical Excipients. The Pharmaceutical Quality Group and The International Pharmaceutical Excipients Council. IPEC-AMERICAS (ipecamericas.org).
22. The IPEC Good Distribution Practices Guide for Pharmaceutical Excipients. The International Pharmaceutical Excipients Council. IPEC-AMERICAS (ipecamericas.org).
23. Deciding When to Submit a 510(k) for a Change to an Existing Device. Guidance for Industry and FDA Staff. CDRH. October 25, 2019.
24. The IPEC Significant Change Guide for Pharmaceutical Excipients. The International Pharmaceutical Excipients Council. IPEC-AMERICAS (ipecamericas.org).
25. The IPEC Quality Agreement Guide and Template. The International Pharmaceutical Excipients Council. IPEC-AMERICAS (ipecamericas.org).
26. http://www.fda.gov/downloads/drugs/guidancecomplianceregulatoryinformation/guidances/ucm353925.pdf
27. Guidance for Industry ICH Q8(R2) Pharmaceutical Development, FDA, 2009.
28. Guidance for Industry ICH Q9 Quality Risk Management, FDA, 2006.
29. Guidance for Industry ICH Q10 Pharmaceutical Quality System, FDA, 2009.
30. Xavier University, Good Supply Practices for the 21st Century.
31. Suzette Roan, Combination Product Development – Integrating Regulatory – Quality Requirements, Massachusetts Biotechnology Council, March 2019.

11 Analytical Testing Considerations for Combination Products

Jennifer Riter and Daniel Bantz

CONTENTS

INTRODUCTION

As more drug and device manufacturers are developing combination products, new and complex interactions among drug constituent parts, medical device constituent parts, and related components have arisen. Although a combination product may be comprised of an already approved drug and an already approved device, new scientific and technical issues may arise when the drug and device are combined for use together. It is critical to understand the compatibility and performance of both the drug constituent part and the device constituent part, and of the drug-device combined use as a system, in order to ensure successful development of the combination product. Developing and applying a strategy for component and system qualification is critical to ensure drug and medical device constituent part quality, safety, and efficacy. Beyond constituent parts, manufacturers also need to understand how chemical, physical, and functional aspects of drug and device all come together to function safely and effectively as a combination product system.

Consider, for example, a device-led combination product, such as a drug-eluting stent, in which the mechanical attributes of the polymer-coating system that contains the drug substance are important for stent deployment, drug release, biocompatibility, and stability. Alternatively, consider a drug-led combination product, for example, a drug-prefilled syringe, for which changes in the stability of the drug constituent may occur if it interacts with a coating on the syringe. The list of potential interactions goes on. Leachables/extractables of a device constituent's materials could impact a drug/biologic substance or final combination product. Drug adhesion/adsorption to device materials could change the delivered dose of a drug. Inactive breakdown products or manufacturing residues from device manufacture could impact safety. Perhaps device functionality could impact the drug performance at the time of use. Changes in stability or activity of a drug constituent part could occur when used together with an energy-emitting device.

DOI: 10.1201/9781003300298-13

Given the array of potential interactions, analytical testing is an important element in characterizing drug constituent(s), device constituent(s), and their potential interaction(s) in a combination product. For many combination products and their constituent parts, using consensus standards, such as test methods, may be appropriate. Health authorities, such as Food and Drug Administration's (FDA) Center for Drug Evaluation and Research (CDER) and Center for Devices and Radiological Health (CDRH), frequently recognize the use of existing consensus standards for the constituents of a combination product. Given the wide variety of innovations and associated development strategies in the evolving combination product space, developers are encouraged to engage with health authorities when exploring the application of standards or alternative methodologies for candidate constituent parts and combination products.

In this chapter we introduce the types of testing generally required, for drug or device companies entering into the combination product development space. We include a roadmap of the types of testing required, with an emphasis on the fact that where historically testing might have been thought of as drug OR device testing, now it needs to take into account the other constituent part and potential interactions between the two. We will highlight examples of combination product-specific considerations. (Note: While drug–biologic products are considered combination products in the US, this chapter does not address specialized questions relating to the combination of those two types of products.)

There are several key areas that need to be considered for analytical testing during combination product development, namely:

- Compendia;
- Biocompatibility;
- Extractables and leachables;
- Container closure integrity;
- Potential particle generation;
- Performance; and
- Stability.

Data integrity and documentation are also key to a successful testing strategy. Several techniques and approaches should be considered when assessing and qualifying the primary packaging system, the medical device, and the drug product with one another. Best practices for each of these key areas are discussed in this chapter. These best practices apply for drugs and/or biological products that leverage medical devices for delivery and for medical devices whose performance is enhanced by drugs and/or biologics.

COMPENDIAL TESTING OF COMPONENTS

Some of the most basic aspects of qualifying delivery systems for drug products and drug-led combination products are compendial compliance of the components of the systems. Typical testing includes analysis of the materials that will be utilized in the system, such as glass, plastic, and elastomers. There are several major compendial chapters that address testing of these materials for use with parenteral injectable products:

- United States Pharmacopeia (USP) <381> Elastomeric Closures for Injections[17]
- USP <660> Containers – Glass[18]
- USP <661> Plastic Packaging Systems and Their Materials of Construction[19]
- Ph. Eur. 3.2.9 Rubber Closures for Containers for Aqueous Parenteral Preparations, for Powders and for Freeze-Dried Powders[20]
- Ph. Eur 3.2.1 Glass Containers for Pharmaceutical Use[21]
- JP 7.03 Test for Rubber Closure for Aqueous Infusions[22]
- JP 7.01 Test for Glass Containers for Injections[23]

Primary packaging and medical device suppliers will normally provide compendial compliance information on their components and containers. In addition, a pharmaceutical manufacturer will further verify this compliance by completing the compendial testing on their particular system; this verification testing is normally completed after the selection of the medical device and related components.

Similarly, compendial and other testing standards (e.g., ISO, ASTM) may be applied to the qualification of device-led combination products, assessing both device constituent functionality, drug performance characteristics, and their interactions. Specific criteria will vary based on a particular drug and its combined use in the combination product, as well as specific functional performance requirements of the device constituent part. Manufacturers apply methods such as assay, residual solvent, content uniformity, etc. as guides, but a Market Authorization Holder (MAH) could use a proprietary method if appropriately vetted and validated. For example, a drug-coated stent MAH may apply compendial testing to assess the drug release characteristic, e.g., flow-through cell dissolution tests as described in United States Pharmacopeia (USP) <711>, Apparatus 4, and European Pharmacopoeia 2.9.3 as robust method for the determination of elution of an active drug substance. In a drug-coated stent, an *in vitro* profile over a day can mirror a month-long *in vivo* porcine profile, providing an *in vitro* release method that captures the entire release profile. Flow-through dissolution using special cell designs is also suitable for other medical devices such as drug-eluting beads, implants, and drug-eluting contact lenses.[1] Some examples of pertinent testing standards applicable for drug-eluting and drug-led devices are listed below:

- Biorelevant Dissolution Testing- Drug release/elution: USP <711>, Apparatus 4 and European Pharmacopoeia 2.9.3
- Uniformity of coating: ASTM F2743 Standard Guide for Coating Inspection and Acute Particulate Characterization of Coated Drug-Eluting Vascular Stent Systems
- Biocompatibility ISO 10993
- Prefilled syringes – ISO 11040 series
- Needle-based injection systems – ISO 11608 series

BIOCOMPATIBILITY TESTING OF COMPONENTS

Combination product biocompatibility is the analysis of a combined container closure system with a pharmaceutical drug product when in contact with, or in, the human body. The drug component of a combination product basically adds additional chemical variables to the medical device profile, and this requires risk assessment and planning. A range of systematic biological tests needs to be completed, as outlined in ISO 10993 Biological Evaluation of Medical Devices,[24] not only on the device constituent part of the combination product itself but also on other constituent parts as relevant, such as the primary containment components.

These tests include irritation, acute toxicity, chronic- and sub-chronic toxicity, *in vivo* cytotoxicity, sensitization, carcinogenicity, hemocompatibility, and implantation. Testing may include in-process materials and processing aids that come in contact with the drug or device constituent parts, or combination product as a whole, during production, to control against negative safety or performance impacts. Testing may also include reproduction and developmental effects. As there are several components of the combination product and its constituent parts that may have patient and/or drug product contact, it is important to define the appropriate testing that also considers exposure to the patient.

EXTRACTABLES AND LEACHABLES ANALYSIS

Extractables, as defined in USP <1663> Assessment of Extractables Associated with Pharmaceutical Packaging/Delivery Systems, are organic and inorganic chemical entities that are released from a pharmaceutical delivery system and into an extraction solvent under laboratory conditions.[1] In other words, extractables are substances in a device constituent part or a container closure system that can be extricated, or "drawn out," under stress conditions, e.g., application of strong solvents, elevated temperatures, and/or by increasing surface area exposure. Extractable substances are characterized, identified, and assessed to determine if the substance presents a potential hazard. Plasticizers, processing aids, surfactants, degradants, and container closure system materials are common extractables. Therefore, extractables analyses are focused on materials and systems.

Whereas extractables can be drawn out or pulled out of a device constituent part or a container closure system under stress condition, leachables are chemical entities that migrate out under normal conditions over the life of a product. Leachables, as defined in USP <1664> Assessment of Drug Product Leachables Associated with Pharmaceutical Packaging/Delivery Systems, are organic and inorganic chemical entities that are present in a packaged drug product because they have migrated into the packaged drug product from a delivery system, related component, or material of construction under normal conditions of storage and use, or during drug product stability studies.[2] In drug-led combination products, leachables analyses focus on what has migrated into the drug product, for example, in a drug-prefilled syringe. Notably, extractables themselves, or substances derived from extractables, have the potential to leach into the drug product under normal conditions and thus become leachables. They can migrate from the delivery system materials into a drug product and be delivered to a patient. They can be of little or of great consequence. There are many potential sources of leachables, including, for example, inks and adhesives from labels or packaging materials, intravenous bags, springs, and valves. Leachable species to consider as part of a study could include antioxidants and stabilizers, anti-static coatings, lubricants, slip agents, emulsifiers, dyes and colorants, residual monomer, polymer, and oligomeric species. The plethora of sources necessitates thorough evaluation.

Leachables analyses are also pertinent to the device constituent part of either drug-led combination products or device-led combination products. One must ensure that drug interactions with the device constituent do not impact its safety, efficacy (functionality), and usability. For example, consider implantable pumps, where one must assess and address drug adsorption or other chemical interactions that could lead to device damage and performance issues. Implanted pumps are medical devices surgically implanted underneath the skin, connected to an implanted catheter (FDA November 2018).[2] They are used to deliver fluids within the body, to treat pain, muscle spasms, and other diseases or conditions. Periodically, implanted pumps are refilled with medicines or fluids by healthcare professionals. Some medicines or fluids may contain preservatives or other characteristics that can leach out and damage the pump tubing or lead to corrosion of the pumping mechanism over a period of time. This may cause the implanted pump to perform unpredictably, including pump motor stalls, which may ultimately stop medication delivery.

Extractables and leachables analysis and monitoring are critical to ensure human exposure to these substances stays within appropriate safety limits. Extractables and leachables analysis is an essential part of qualifying the delivery system during drug, device, and combination product development. Regulatory agencies require each drug product filing to contain sufficient data and information to prove that the chosen delivery system and related components maintain the quality, sterility, and efficacy of the drug product. Likewise, one must have sufficient evidence and data to demonstrate that there are no adverse interactions between the drug and the device that may impair the device's performance. Where extractables and leachables analysis lies in the overall process of delivery system qualification is shown in Figure 11.1.

USP <1663> and <1664> are informational chapters that provide guidance on a framework for extractables and leachables studies. As these guidelines are not prescriptive, it is important that pharmaceutical and medical device manufacturers have a strategy and an approach for extractables and leachables studies for their products early in drug product, device, and combination product development. Recognizing how crucial it is to maintain safety, identity, strength, quality, or purity of a drug constituent part, and likewise the criticality of preserving the safety, functionality, and usability of the device constituent part, for combination products, an extractables and leachables study approach may include:

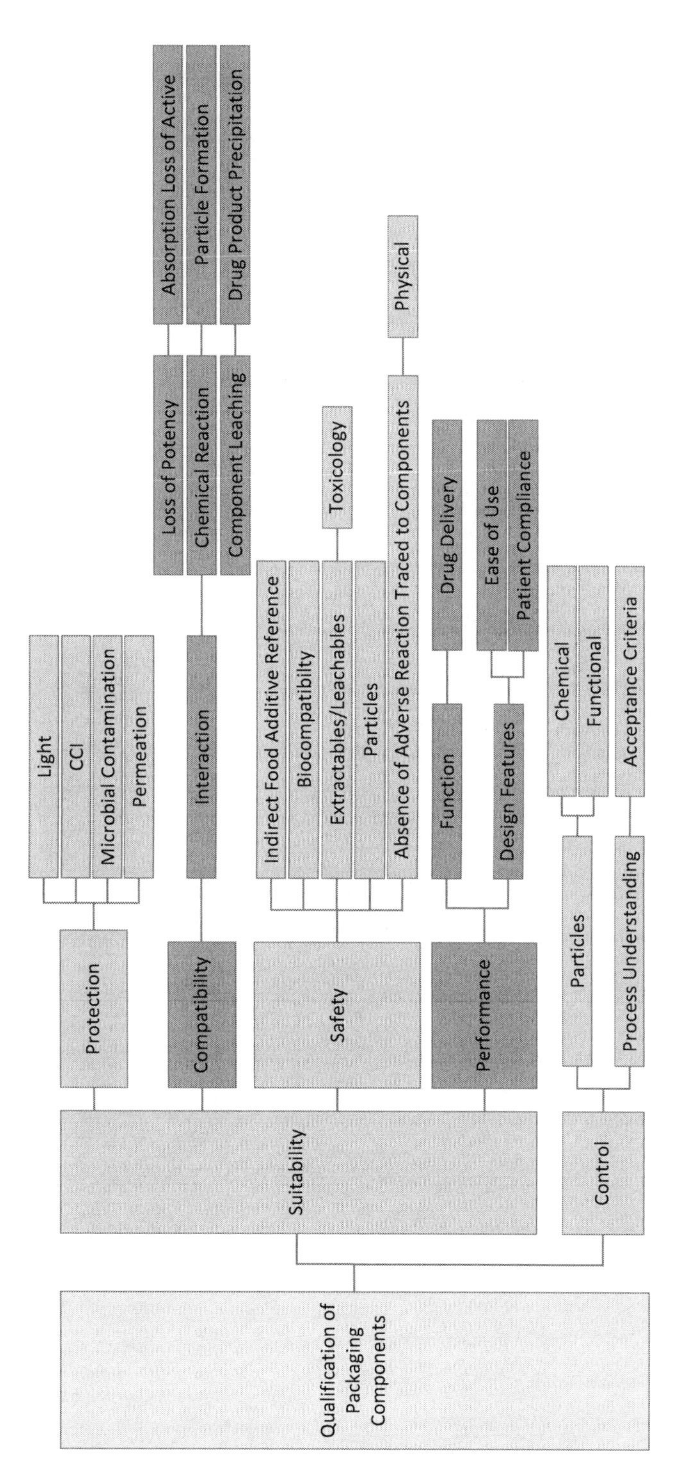

FIGURE 11.1 Flowchart of device constituent part and packaging component qualification

- Preplanning and component/system selection
  - Partner with a vendor and evaluate available vendor information and literature to select appropriate materials
- Extractables assessment and evaluation
  - Assess the extractables information and design "fit-for-purpose" studies
  - Develop extractables studies that involve:
    - Use of multiple solvents of various polarities and solvent properties
    - Use of multiple extraction techniques such as reflux, microwave digestion, etc.
    - Use of various instruments to detect extractables such as liquid chromatography-mass spectrometry, gas chromatography-mass spectrometry, quadrupole time of flight mass spectrometry, inductively coupled plasma mass spectrometry, etc.
    - Use of known and surrogate standards to obtain semi-quantitative data
- Risk assessment
  - Evaluate and organize extractable data based on risk ranking
  - Determine which extractables will be the focus for leachables analyses based on the risk assessment
- Leachables evaluation
  - Develop and validate leachables analyses methods and monitor leachables over the shelf life of the drug product, the device constituent part, and the combination product as a whole
  - Assess leachables safety and quality over the shelf life of the drug product, medical device, and combination product to ensure no drug or device safety, efficacy, or functionality impacts.

With this approach, a flowchart that a pharmaceutical manufacturer may use is shown in Figure 11.2.

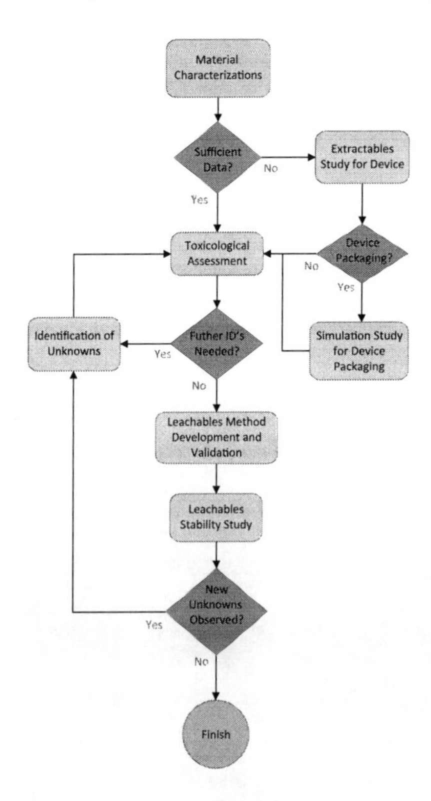

FIGURE 11.2 Typical extractables and leachables program flowchart

As mentioned, typical studies would start with understanding the information available from the supplier of the material and any characterization testing or extractables information they have. From there, the pharmaceutical manufacturer would need to determine if the information provided contains adequate extractables information and decide the need to perform additional extractables studies. A risk assessment would then be completed on critical extractables to determine which compounds will be a target for leachables analysis. Leachables method development and validation would then be performed using drug product. These methods would be utilized to monitor leachables throughout the stability study over drug product shelf life. Likewise, as previously indicated, such methods would be used to monitor impacts on drug-device interactions that could potentially impact device safety and functionality over its lifetime.

In addition to the delivery system materials, the secondary shipping packaging materials, interim storage containers, and processing equipment may need to be evaluated for compounds that could leach into the drug product. For example, an advanced polymer vial container, or prefilled syringe, system that utilizes a cyclic olefin polymer (COP) or cyclic olefin copolymer (COC), which is a semi-permeable material, may need to be evaluated when a label is applied. The adhesives and inks from the label may have compounds that can migrate through the polymer and leach into the drug product over time. Interim storage containers of active pharmaceutical ingredients (API), drug substances, or drug products also need to be considered for extractables and leachables analysis due to contact with products over a prolonged period.

For pharmaceutical manufacturers, it is important to partner with the manufacturer of the delivery system from the beginning, to assist with component/system and secondary shipping packaging selection, and in defining a strategy to address extractables and leachables. This will reduce the risk, later in development, that incompatibilities between the drug product and the delivery system are discovered.

When designing leachables and extractables studies for drug-coated devices, some important factors to consider are potential introduction of impurities during the coating process, the effect of sterilization on the device constituent materials, and their potential interaction with the drug constituent part of a combination product. Specific to the effects of sterilization on device materials, sterilization may generate leachables, and one would want to ensure studies of extractables and leachables incorporate effective evaluation of potential impacts of such production and post-production processes on the drug, device, and their interactions.[3] Associated toxic by-products resulting from sterilization should be considered when conducting biocompatibility tests as well. Testing should be conducted on the sterilized final combination product (see CDRH guidance document entitled "510(k) Sterility Review Guidance (2/12/1990) #K90-1 (blue book memo)" for more information). When conducting preparing samples for extractables and leachables testing, a best practice would be to sterilize all test articles using the same procedure that is to be actually used in the manufacturing and sterilization of the final combination product. Qualitative and quantitative description of all constituent materials in the device constituent part and drug constituent part before extraction should be provided, and the material specifications for the device constituent part and drug constituent part should be comprehensive. If any toxic leachables, by-products, or metabolites exist in the extracts from a sterilized combination product, the results of the toxicity tests on the extracts should represent the cumulative toxicities from the extracts. Extraction procedures should be rigorous to ensure that the extract toxicity results are representative of the toxicity of the combination product in actual human use. In order to create a safety window, extractions should be conducted under worst-case conditions as compared to those expected from the natural extraction in blood and other human tissues. If toxic responses are obtained from the extracts, then chemical analysis of the extract should be performed to address the identity of the toxic compound(s). If a device constituent part or its materials are found to be toxic, the sponsor should attempt to find an alternate material that is non-toxic.

Fundamentally, the inherent co-existence of a drug and a device gives rise to the possibility of unwanted migration of chemical entities between them, and as such leachables and extractables studies are an important element of combination product analytical characterization.

CONTAINER CLOSURE INTEGRITY (CCI) TESTING

Where the medical device constituent of a combination product is a container closure, prefilled syringe, or cartridge holding and protecting the drug in addition to delivering the drug, container closure integrity (CCI) is another key area that needs to be evaluated in the development phase of drug products. CCI is defined as the ability of device components to prevent product loss, to block the ingress of microorganisms, and to limit the entry of detrimental gases or other substances over the shelf life of the drug product – thus ensuring that the drug product meets all necessary safety and quality standards. This terminology is synonymous with device integrity. CCI failures for devices include leakage, loss of product, increased concentration of product, contamination, sterility failure, critical headspace loss, over-pressurization, or loss of pressure. USP <1207> Package Integrity Evaluation – Sterile Products addresses CCI and provides guidance on how to evaluate the integrity of devices intended for sterile products.[3] Divided into several sections, it includes an overview, general introduction, and glossary. It addresses the need for understanding the package system integrity throughout the life cycle of the drug product – from package development and validation to product manufacturing and then through stability. Where CCI lies in the overall process of device qualification is shown in Figure 11.1.

This first section of USP <1207> discusses test methods selection and validation. It states that to choose appropriate test methods, there is a need to understand the device design, materials of construction, and mechanics of operation. As part of that understanding, a pharmaceutical manufacturer must also understand and define the maximum allowable leakage limit (MALL) for the drug product, in order to properly evaluate the integrity of the device. Device and related components integrity is defined in terms of the MALL, and this is to be determined for each drug product. USP <1207> defines the MALL as the greatest leakage rate (or leak size) tolerable for the given devices that pose no risk to drug product safety and quality. Integrity requirements should not just be focused on sterility risks from microbiological contamination but also product quality risks. Table 11.1 outlines leakage rates versus leakage size.[3]

TABLE 11.1
Parameters of Deterministic Methods Per USP <1207.1>, Section 3.9[4]

Row	Detectable Leaks		Applicable Method
	Air Leakage Rate[4] (std cm^3/s)	Orifice Leak Size (μm)[4]	
1	$<1.4 \times 10^{-6}$	<0.1	Tracer gas leak detection Frequency-modulated spectroscopy
2	1.4×10^{-6} to 1.4×10^{-4}	0.1 to 1.0	Tracer gas leak detection Frequency-modulated spectroscopy
3	$>1.4 \times 10^{-4}$ to 3.6×10^{-3}	>1.0 to 5.0	High-voltage leak detection Frequency-modulated spectroscopy
4	$>3.6 \times 10^{-3}$ to 1.4×10^{-2}	>5.0 to 10.0	High-voltage leak detection Frequency-modulated spectroscopy Vacuum decay
5	$>1.4 \times 10^{-2}$ to 0.360	>10.0 to 50.0	High-voltage leak detection Vacuum decay Tracer liquid
6	>0.360	>50.0	tracer liquid

Once the MALL is defined for a product, the next step is choosing the appropriate technique for testing CCI. This is addressed in the second section of USP <1207>,[3] which outlines leak test methods and their proper use. When considering techniques, it is important to understand that there is no one test method that is appropriate for all devices and related components. Method selection and validation should be based on what is best suited for the application, and the methods should target quantitative results (with deterministic methods) versus qualitative results (with probabilistic methods). The second section of USP <1207>[3] discusses different types of probabilistic methods, such as blue dye testing and microbial challenge, which were very common in the industry in the past. However, as outlined in the section, expectations have since changed and deterministic methods should be utilized when possible. Examples of deterministic methods include tracer gas leak detection, high-voltage leak detection, vacuum decay, and frequency-modulated spectroscopy headspace analysis.

Based on the MALL, there may be a need to perform more than one deterministic test on a product. For example, it may be necessary to evaluate a vial system for microbial ingress with tracer gas leak detection. If drug product is sensitive to oxygen, it may be necessary to evaluate the system for oxygen ingress over time with frequency-modulated spectroscopy. Determining the appropriate technique based on the drug container system and the MALL is important when developing the CCI approach for the drug product within the intended device.

The tracer gas leak technique with helium is commonly used with vial systems; however, it can be used also with cartridge and prefilled syringe systems, with appropriate modifications to instrument fixtures. This technique uses a mass spectrometer that is tuned to only measure helium. The instrument applies vacuum to a sample; any helium escaping is measured, and a leak rate is calculated. It is highly sensitive, can quantify orifice leak size diameters as low as 0.2 μm, and can be employed at various temperatures. The sample preparation is extensive, and because the vial is typically flooded with helium through the stopper, it is considered a destructive test. A caution in using this technique is a false reading from orifice/leak-site blocking by drug product, particularly with protein-based drug products. A photo of the equipment is shown in Figure 11.3.

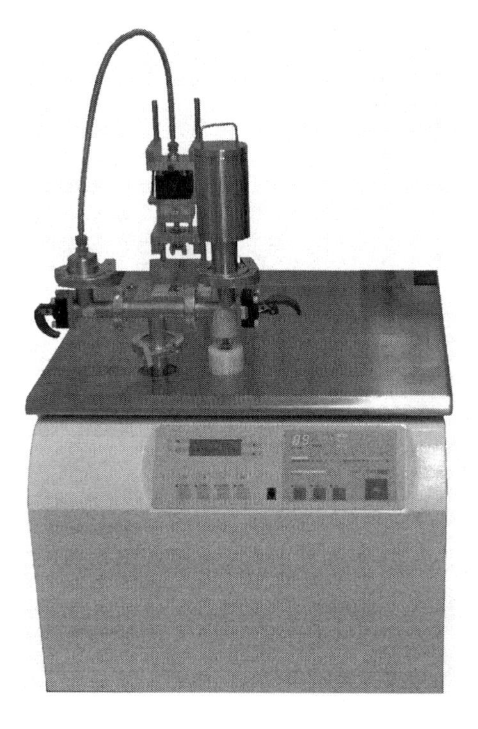

FIGURE 11.3 Tracer gas leak detection instrument

High-voltage leak detection (HVLD) commonly is used with prefilled syringe and cartridge systems. HVLD uses electrical current to detect leaks in a package and is also capable of detecting solidified product near a leak site. It is considered a non-destructive test, but the effect on product stability from exposure to high voltage should be evaluated. Specific points of this technique are that the container system must be non-conductive, and fixtures are required for every container type. Various adjustments must be made depending on the fill volume, viscosity, and conductivity of drug product; these factors need to be taken into consideration when developing and validating methods. A photo of the equipment is shown in Figure 11.4.

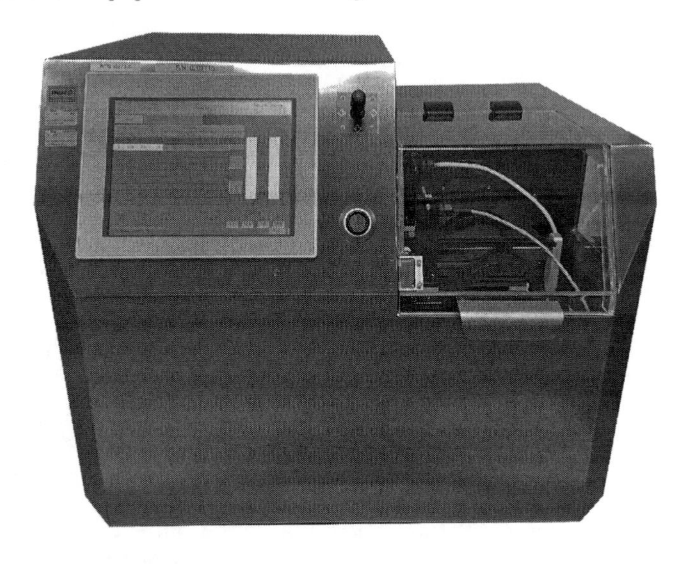

FIGURE 11.4 High-voltage leak-detection instrument

Vacuum decay can be used for various containment systems; however, it is of lesser sensitivity. Using specially designed fixtures, it applies a vacuum for a period of time to the exterior of a containment system at atmospheric pressure. Then the vacuum source is then isolated from the test system, and the pressure change (using transducers) is monitored for a length of time. A pressure rise above an established value is considered a failure. A photo of the equipment is shown in Figure 11.5.

FIGURE 11.5 Vacuum decay instrument

FIGURE 11.6 Laser-based headspace analysis instruments

Frequency-modulated spectroscopy headspace analysis (Figure 11.6) is another technique that is commonly used for vial systems, as it needs an appropriate amount of headspace to properly measure the concentration of gas (usually oxygen). This technique uses a laser light tuned to a frequency absorbed by the gas of interest. Light is absorbed as a function of gas concentration or pressure. Each instrument will be specific to the gas of interest.

In the third and final section, there is a discussion of package seal quality test methods, such as residual seal force. These tests are used to characterize and monitor the quality and consistency of parameters related to the package seal. They provide some assurance of the ability of the device and related components to maintain integrity. Seal quality tests and CCI tests work together to ensure this integrity.

As with extractables and leachables, it is important to consider CCI during development and include it as part of stability evaluation over the shelf life of the product and through life cycle management.

PARTICLE ANALYSIS

Another critical aspect of qualifying devices and related components is particles and the particle burden. Particles in a drug product are a very serious concern (e.g., obstruction of blood vessels, inducement of immunogenicity) for pharmaceutical and medical device manufacturers. Particles may be visible or subvisible and categorized as intrinsic, extrinsic, or inherent. Intrinsic particles may come from processing or device materials, including stainless steel components, gaskets, glass containers, rubber components, fluid transport tubing, and others. Extrinsic particles are foreign to the manufacturing process, such as hair, non-process-related fibers, and other foreign matter. Inherent particles are associated with specific drug product formulations such as suspensions, emulsions, aggregates, etc. Depending upon the drug product, there are different USP requirements and guidances that can be used:

- USP Injections and Implanted Drug Products (Parenterals) – Product Quality Tests[4]
  - Products designed to exclude particulate matter
- USP <790> Visible Particulates in Injections[5]
  - <1790> Guidance for Visual Inspection for Defects and Particles > 50μm[5]
- USP <787> Subvisible Particulate Matter in Therapeutic Protein Injections[6]
  - <1787> Measurement of Subvisible Particulate Matter in Therapeutic Protein Injections[6]
- USP <788> Particulate Matter in Injections[7]
  - <1788> Methods for Detection of Particulate Matter in Injections and Ophthalmic Solutions[7]
- USP <789> Particulate Matter in Ophthalmic Solutions[8]

Table 11.2 is a specification comparison for these compendial methods:

TABLE 11.2
USP <787>, <788>, and <789> Particle Specification Comparison[6–8]

USP Chapter	Parenteral Volume	Method 1 – Light Obscuration			Method 2 – Microscope		
		≥10 µm	≥25 µm	≥50 µm	≥10 µm	≥25 µm	≥50 µm
<787>	**Small volume injection 100 mL and lower**	6000 per container	600 per container	Not applicable	Not recommended – need to demonstrate inherent particles can be counted or should be used for extrinsic and intrinsic particles only		
	Large volume injection above 100 mL	25 per mL	3 per mL	Not applicable			
<788>	**Small volume injection 100 mL and lower**	6000 per container	600 per container	Not applicable	3000 per container	300 per container	Not applicable
	Large volume injection above 100 mL	25 per mL	3 per mL	Not applicable	12 per mL	2 per mL	Not applicable
<789>	**Any volume**	50 per mL	5 per mL	2 per mL	50 per mL	5 per mL	2 per mL

Delivery systems and related components can contribute to particles in drug products. The pharmaceutical manufacturer has to understand not only the particles in the drug product but also the particle burden coming from the delivery system. Contributing factors for particles on components and systems include materials of construction, processes, environment, and secondary shipping packaging. Another contributing factor is methodology, i.e., sample preparation and appropriate method. The techniques, methodologies, and instrumentation that are used for testing components and systems are different from those used for testing drug product solutions. ISO 8871-3:2003 Elastomeric Parts for Parenterals and for Devices for Pharmaceutical Use – Part 3: Determination of Released-Particle Count[9] is commonly used and referenced when testing components. However, with particle testing, there is still concern around test method variability. This method variability can be from the test samples, testing environment, filter preparation, and counting methodology (Figure 11.7).

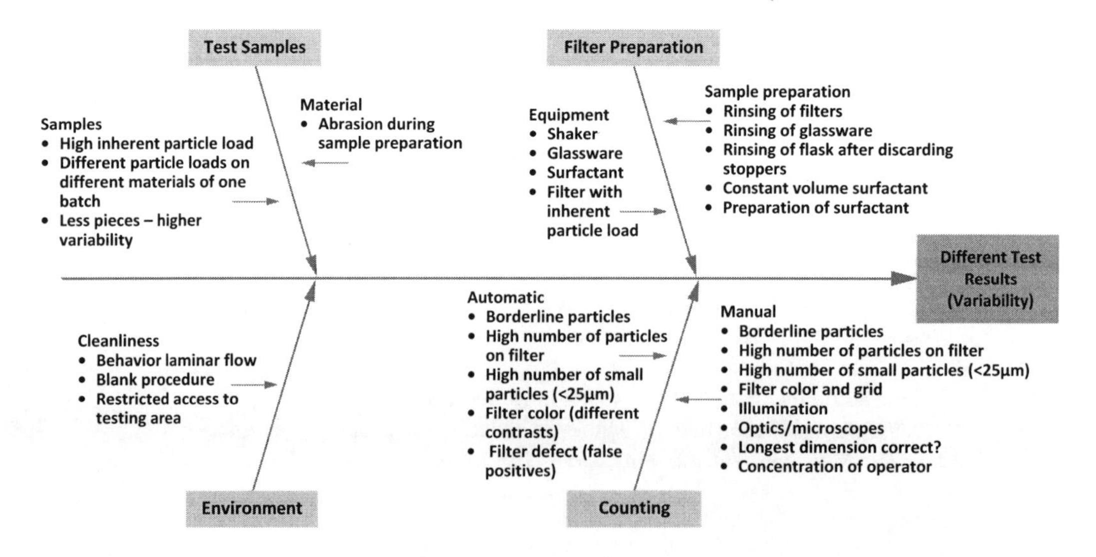

FIGURE 11.7 Particle testing variables © 2019 West Pharmaceutical Services, Inc. All rights reserved

All these factors need to be considered when developing and validating methods for particle analysis, in order to develop a robust and repeatable method. It is important for pharmaceutical manufacturers to engage with, and partner with, their suppliers to understand the particle burden and methodology for measuring particles on components and systems.

PERFORMANCE TESTING

The testing strategy will set the foundation for understanding whether the combination product will meet requirements and thus be safe and effective. This strategy is not just a best practice, but an expected practice, and it should have the full intent to understand risks and weaknesses, design limits, misuse, and environmental impacts. The objective of the testing strategy can in one sense be achieved by assuming there is an issue, doing that is necessary to identify it, resolve it, and then prove the said resolution. An all-encompassing assessment of the combination product design starts with the patient, and closely considers, *inter alia*: (a) interaction and compatibility among components, (b) drug-related compatibility, (c) total functional ability, (d) regulatory requirements, and (e) quality and cost targets.

Fundamental Performance – Drug-enhanced medical devices and related drug-filled containers, like syringes, vials, and cartridges, typically have multiple mechanical interfaces – often comprised of components developed independently. For any device and corresponding container, these interfaces must be understood thoroughly, and fundamental performance aspects substantiated and demonstrated – often through what may appear very rudimentary tests. An example is shown in Figure 11.8. There are various performance tests that can be applied to a prefilled syringe. Some are highly dependent on the intended use, history of combined components, and unintended use. Selection of tests is based upon assessment by subject matter experts.

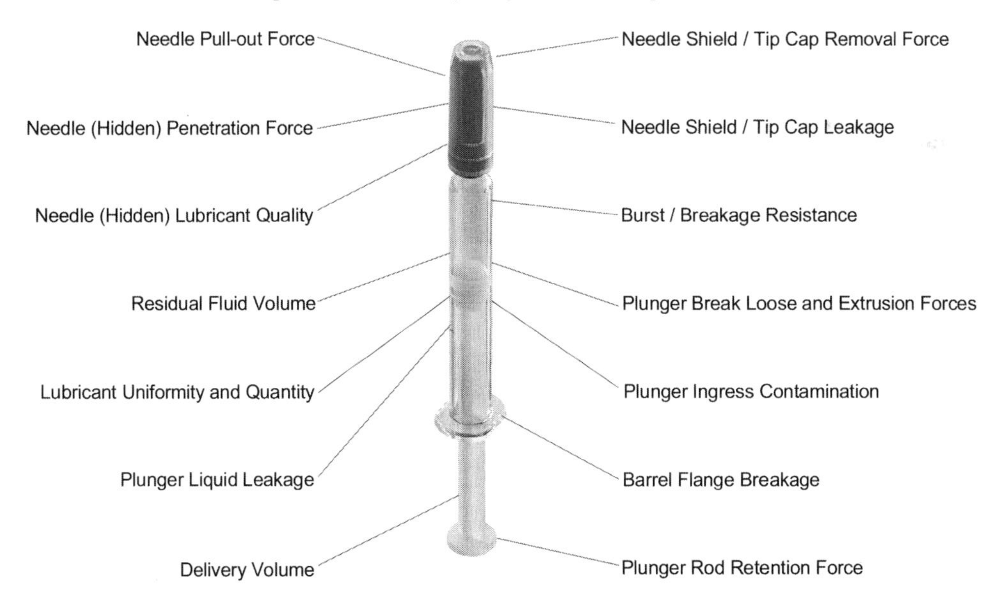

FIGURE 11.8 Syringe testing considerations

Of the tests to be performed, several will be considered more critical – these are tests that consider the essential performance requirements (EPRs). Based on risk assessment, product knowledge, and intended (or unintended) use, EPRs are of particular importance. It is expected they will be tested and understood not only through all stages of product development but through long-term environmental stability studies as well. Examples of tests for EPRs of a prefilled syringe are: (a) break loose and extrusion, (b) deliverable volume, and (c) leakage. This does not mean other tests should not be performed, but that these are aspects identified as most critical to understand.

Break Loose and Extrusion – This EPR test comprises evaluation of (a) maximum force (at a fixed speed) required to initiate plunger movement (i.e., break loose) and (b) force required (at same fixed speed) to maintain plunger movement (i.e., extrusion) to discharge all drug product. Figure 11.9 shows an example profile generated from an automated test machine. The data shows the syringe plunger movement starting force (Fs), the maximum force after the starting force (Fmax), and the average force (F) and is essential to quantitatively show force profiles associated with the fit of the plunger, plunger rod, and syringe barrel. The data also creates a thorough understanding of equally important factors such as drug product interactions, and the effects of surface tension, viscosity, and interactions among syringe components (e.g., plunger with lubrication on the syringe walls).

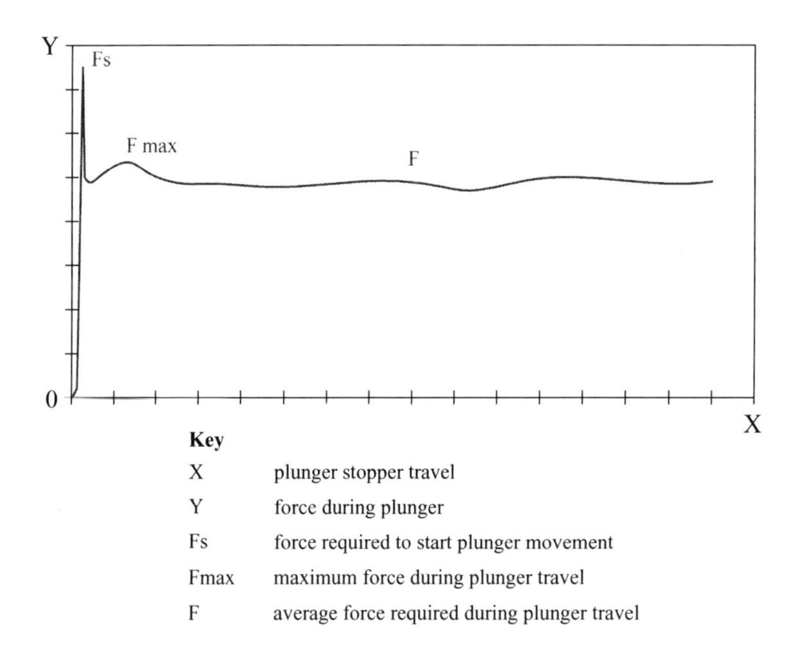

FIGURE 11.9 Break loose and extrusion force profile

Deliverable Volume – This EPR test is designed to provide assurance that the syringe will deliver the expected or declared prefilled drug volume. This test is typically performed manually utilizing a gravimetric system that converts weight to volume based on the specific gravity of the drug product and can be done in conjunction with other tests, for example, by utilizing the automated test instrument for break loose and extrusion testing. The data generated are essential in that they quantitatively confirm the dose volume delivered based on a target-filled volume. The fill volume will directly impact the deliverable volume as there is extra volume associated with the fill to accommodate any residual volume within the end of the syringe and within the needle. This test also indicates if there is any potential leakage that could occur during handling, shipping, and storage.

Leak Performance – This EPR test is designed to observe leakage of drug-filled containers that have several interfaces, e.g., syringe barrel to Luer needle, syringe barrel to plunger, and cartridge to seal, as shown in Figure 11.10. The sealing properties of physically contacting interfaces must be, as a standard practice, thoroughly understood, within a variety of environmental and in-use stress conditions. Each component may be designed and manufactured perfectly; however, it must be challenged to confirm it is compatible with its contacting component(s), in order to meet its design intent and support a combination product that is safe and effective. As mentioned previously, CCI of the system would also be performed as part of the evaluation in addition to leak performance.

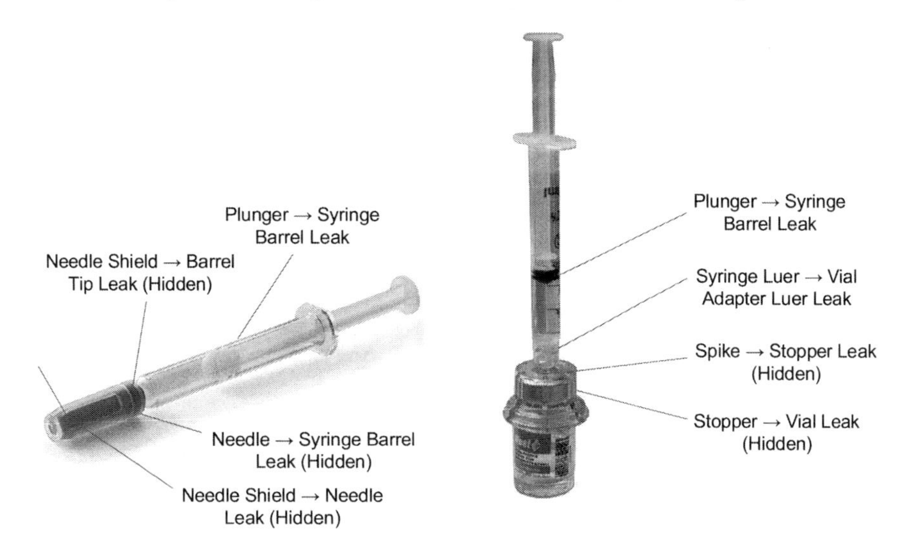

FIGURE 11.10 Potential interfaces leaks

Leakage from a device (i.e., egress and ingress) can result from intended (or unintended) use, environmental conditions, fill process, incompatible components, or components incompatible with the drug product. Depending on severity, leakage can cause safety and effectiveness issues, such as a low dose (adversely affecting therapeutic regimen). It can also cause drug exposure to those not intended, resulting in a significant health risk.

Standards and Guidances – Where does an organization start when it comes to developing an all-encompassing assessment of combination product performance? The answer is published standards and guidances. Many helpful standards and guidances related to combination product testing are available and can be used to establish frameworks for testing strategy. Good examples are from the International Organization for Standardization (ISO) and Food and Drug Administration (FDA) guidances. Two examples of ISO standards are the ISO 11040[10] and ISO 11608 series.[11]

ISO 11040 *Prefilled Syringes* consists of the following parts:

- Part 1: Glass cylinders for dental local anesthetic cartridges
- Part 2: Plungers stoppers for dental local anesthetic cartridges
- Part 3: Seals for dental local anesthetic cartridges
- Part 4: Glass barrels for injectables and sterilized sub-assembled syringes ready for filling
- Part 5: Plunger stoppers for injectables
- Part 6: Plastics barrels for injectables
- Part 7: Packaging systems for sterilized sub-assembled syringes ready for filling
- Part 8: Requirements and test methods for finished prefilled syringes

ISO 11608 *Needle-Based Injection Systems* consists of the following parts:

- Part 1: Needle-based injection systems
- Part 2: Needles
- Part 3: Finished containers
- Part 4: Requirements and test methods for electronic and electromechanical pen-injectors

- Part 5: Automated functions
- Part 6: On-body delivery systems
- Part 7: Accessibility for persons with visual impairment

Not all parts of these ISO standards are directly applicable to combination product testing; however, several have very close linkages and are very insightful when formulating the testing strategy. For example, ISO 11040-8 Requirements and Test Methods for Finished Prefilled Syringes provides guidance for testing drug- or placebo-filled syringes. It contains a wide range of recommendations, as well as testing annexes, which provide detailed guidance on how to perform functional tests. These performance-related tests may not meet all possible testing needs. Below is a list from ISO 11040 Part 8 and ISO 11608 Part 3:

ISO 11040 Part 8:

- Break loose and extrusion forces
- Burst resistance
- Break resistance
- Closure system forces and torques
- Connectivity with fluid path connectors
- Residual volume
- Needle penetration force
- Needle pull-out force
- Sharps injury protection requirements
- Liquid leakage beyond plunger
- Markings

ISO 11608 Part 3:

- Plunger force
- Leakage
- Dimensions
- Resealability
- Sealability
- Coring

The above-mentioned tests from ISO 11040-8 and 11608-3 can be performed at the combination product level based on the risk assessment and related product knowledge gaps associated with drug and materials interactions and other unknowns as components are selected and come together with the drug product. There are many other ISO standards referenced within the ISO 11040 and 11608 series to provide further guidance in developing the proper testing strategy; however, this requires critical thinking and a well-assembled risk assessment to understand knowledge gaps and performance-based risks. There are also many other ISO standards that are available for device-led or other drug-led combination products. One of the core benefits of having a standards-based testing strategy, such as from ISO standards, is that industry and regulatory bodies recognize ISO standards rather than new-to-world testing methodologies.

It is also important to note that standards don't necessarily cover all possible aspects needed to be considered to understand overall device performance relates to safety and effectiveness. For example, there are other factors not in typical standards that should be considered such as in the case of transdermal patches. In this example, it may be critical to understanding drug migration into the device adhesive intended as the primary patient interface for drug delivery. The fundamental physiochemical interactions between the drug, device adhesive, and the patient are critical to understanding from a safety and effectiveness perspective.

DRUG/BIOLOGIC CONSTITUENT AND STABILITY TESTING CONSIDERATIONS

Drug and biologic product considerations which are necessary to characterize the safety and effectiveness of the drug and biological product whether it is a new molecular entity (NME) or a drug/biologic constituent that is already approved for another use. Some of these considerations when devising a development plan for drug/biologic constituent include:[25]

- *In vivo* pharmacokinetic (PK) studies may be necessary to assess changes in formulation, strength, route of administration, dosing, population, or other factors that may alter the extent or time course of systemic exposure. These studies might be used to determine drug release kinetics such as release rate, local peak concentrations of the drug, local distribution, and systemic bioavailability (C_{max}, T_{max}, etc.).
- Dose-ranging or dose-finding studies[13] in humans may be appropriate to determine dose adjustments for safety/effectiveness when therapy is targeted to a local site.
- Acute and repeat dose toxicity studies using the new route of administration or method or delivery may be appropriate to determine the no observed adverse effect level (NOAEL) and toxicity profile of the combination product. Typically, these studies would evaluate the intended clinical formulation and dosing regimen/frequency that will closely approximate its use in clinical settings.
- Special safety studies may be appropriate for certain patient populations or risk profiles; e.g., hepatotoxicity, QT prolongation, and special populations.
- Specific safety monitoring in the clinical study may be appropriate to obtain data on the novel aspects presented by the combination product; e.g., local toxicity for a new route of administration.

There are several decisions that need to be determined for clinical studies which include trial design, sample size, statistical methods, clinical endpoints, appropriate number of studies, and appropriate indications/claims. It is recommended that you consider the science and technology of the combination product when determining sample size, use of statistical approaches, surrogate endpoints, techniques to measure drug levels in areas not typically accessible, or techniques to evaluate drug-device interactions. It is also important to evaluate the human factors of device use on the safety and effectiveness of the combination product. These evaluations should take place early in the combination product development process to determine any design features that may need modification before conducting the key studies to establish the safety and effectiveness of the combination product.[24]

The stability of a combination product as a whole may be different from that of the separate constituent parts. Certain drug or biological product constituent parts may be altered or destroyed by terminal sterilization techniques. Stability testing is done to confirm combination product performance through the expiration date when filled with the actual drug product and under actual storage conditions. The test data should prove that the EPRs, and all other performance targets, are met at all time points. Accelerated testing can be done at elevated temperatures; however, it is mandated by the FDA to complete real-time stability testing as well.

There are also guidances for establishing stability testing criteria as noted in Table 11.3. Per the World Health Organization International Council for Harmonization (ICH) Annex 2 WHO Technical Report Series, No. 953, 2009:[12] "At the accelerated storage condition, a minimum of three time points, including the initial and final time points (e.g., 0, 3, and 6 months), from a six-month

TABLE 11.3
General Case Accelerated Conditions and Durations Per ICH Annex 2

Study	Storage Condition	Minimum Time period Covered by Data at Submission
Long term[a]	25°C ± 2°C/60% RH ± 5% RH or 30°C ± 2°C/65% RH ± 5% RH or 30°C ± 2°C/75% RH ± 5% RH	12 months or 6 months
Intermediate[b]	30°C ± 2°C/65% RH ± 5% RH	6 months
Accelerated	40°C ± 2°C/75% RH ± 5% RH	6 months

[a] Whether long-term stability studies are performed at 25°C ± 2°C/60% RH ± 5% RH or 30°C ± 2°C/65% RH ± 5% RH or 30°C ± 2°C/75% RH ± 5% RH is determined by the climatic condition under which the API is intended to be stored (see Appendix 1). Testing at a more severe long-term condition can be an alternative to testing conditions, i.e., 25°C/60% RH or 30°C/65% RH.

[b] If 30°C ± 2°C/65% RH ± 5% RH or 30°C ± 2°C/75% RH ± 5% RH is the long-term condition, there is no intermediate condition.

study is recommended."[12] However, other approaches to confirm stability can be used, with proper scientific justification.

Additionally, the same ICH Annex 2 guidance recommends testing beyond just accelerated conditions:

The long-term testing should normally take place over a minimum of 12 months for the number of batches specified in section 2.1.3 at the time of submission and should be continued for a period of time sufficient to cover the proposed re-test period or shelf-life."[12]

Typically for combination product testing, the frequency with which components are removed from storage simulators (or pulls) is at three-month intervals during the first year, at six-month intervals during the second year, and annually during the third year and beyond. Obviously, EPR data is to be collected; however, in addition, general observations should also be collected, e.g., changes in stability, and experiences from following instructions for use (IFU), which enable real-life scenarios. These provide assurances that normal use performance can be achieved throughout the life cycle.

Lastly per ICH Annex 2: "reduced designs, i.e., matrixing or bracketing, where the testing frequency is reduced, or certain factor combinations are not tested at all, can be applied if justified."[12] The primary objective of this aspect is to ensure EPRs can be evaluated where a wide variety of factors must be considered, e.g., temperature, humidity, and time (accelerated and real-time). Use of this guidance is highly recommended for stability test program strategy creation.

Data Integrity Considerations. What is data integrity? According to the FDA, from the guidance Data Integrity and Compliance with Drug cGMP Questions and Answers (2018), data integrity "refers to the completeness, consistency, and accuracy of data. Complete, consistent, and accurate data should be attributable, legible, contemporaneously recorded, original or a true copy, and accurate (ALCOA)."[16] Data integrity is of utmost importance when generating results and conclusions from combination product testing following current good manufacturing practices (cGMP).

The reason the FDA has published such a guidance is due to recent increases in violations from data audits during cGMP inspections. Combination product (meaning the drug combined with the device) quality, efficacy, and safety are highly dependent on data integrity. Data-related violations have the potential for dire consequences, such as warning letters from the FDA, import alerts, and consent decrees. As part of the FDA regulations 21 CFR parts 210.1[13] and 212.2,[14] there are minimum requirements for meeting the Food, Drug and Cosmetic Act with regard to drug quality, potency, safety, identity, strength, quality, and purity. Regulations 21 CFR part 211[15] and part 212[14] provide several guidance aspects around data integrity. Some examples are:

- 21 CFR 211.68 – Requiring that "backup data are exact and complete" and "secure from alteration, inadvertent erasures, or loss" and that "output from the computer ... be checked for accuracy."[15]
- 21 CFR 212.110(b) – Requiring that data be "stored to prevent deterioration or loss."[14]
- 21 CFR 211.100 and 211.160 – Requiring that certain activities be "documented at the time of performance" and that laboratory controls be "scientifically sound."[15]
- 21 CFR 211.180 – Requiring that records be retained as "original records," or "true copies," or other "accurate reproductions of the original records."[15]
- 21 CFR 211.188, 211.194, and 212.60(g) –Requiring "complete information," "complete data derived from all tests," "complete record of all data," and "complete records of all tests performed."[15]

The device-side regulations under 21 CFR 820 Quality System Regulations for Medical Devices also need to be incorporated as part of the combination product and data integrity. Part 820 includes several subparts such as Quality System Requirements, Design Controls, Document Controls, Purchasing Controls, Production and Process Controls, etc.[26]

Completeness, consistency, and accuracy of data are equally important as the quality, thoroughness, design, and robustness of the combination product testing strategy. How the data are managed is crucial in establishing credibility for filing and audit purposes. Even more important, complete and accurate data will provide true and factual measures of performance that can withstand the test of time and future audits.

SUMMARY

Combination products are becoming ever more complex – resultant from advancements in drugs, technologies, and materials – all of these very often developed independently. Components are expected to merge and deliver a safe and effective therapy over long periods of time and varying environmental conditions – without a concern or even a second thought from the patient. Because the merging of device and drug is so complex, and requirements for performance are so high, demonstration of performance can only be accomplished with a thorough and multifaceted testing strategy that is highly effective in searching for, uncovering, and mitigating risks, while addressing regulatory expectations and properly managing data.

NOTES

1. https://www.sotax.com/dosage_form/medical_devices_stents_implants accessed on July 24, 2020.
2. https://www.fda.gov/medical-devices/safety-communications/use-caution-implanted-pumps-intrathecal-administration-medicines-pain-management-fda-safety "Use Caution with Implanted Pumps for Intrathecal Administration of Medicines for Pain Management: FDA Safety Communication (November 14, 2018); accessed July 24, 2020.
3. Guidance for the Submission of Research and Marketing Applications for Permanent Pacemaker Leads and for Pacemaker Lead Adaptor 510(k) Submissions Document issued on: November 1, 2000, pages 5–6.

REFERENCES

21 CFR Part 210.1 – Status of Current Good Manufacturing Practice Regulations.
21 CFR Part 211 – Current Good Manufacturing Practice for Finished Pharmaceuticals.
21 CFR Part 212.2 – Applicability of Current Good Manufacturing Practice Regulations.

21 CFR Part 820 – Quality System Regulation FDA, March 2008: Guidance for Industry Coronary Drug-Eluting Stents – Nonclinical and Clinical Studies Draft Guidance.
a. USP <1207> Package Integrity Evaluation – Sterile Products and Section. b. USP <1207.1> Package Integrity Testing in the Product Lifecycle – Test Method Selection and Validation.
a. USP <787> Subvisible Particulate Matter in Therapeutic Protein Injections, b. USP <1787> Measurement of Subvisible Particulate Matter in Therapeutic Protein Injections.
a. USP <788> Particulate Matter in Injections, b. USP <1788> Methods for Detection of Particulate Matter in Injections and Ophthalmic Solutions.
a. USP <790> Visible Particulates in Injections. b. USP <1790> Guidance for Visual Inspection for Defects and Particles >50µm.
Data Integrity and Compliance with Drug CGMP Questions and Answers Guidance for Industry Guidance for Industry December 2018.
Guidance for Industry and FDA Staff: Early Development Considerations for Innovative Combination Products, U.S. Department of Health and Human Services Food and Drug Administration Office of the Commissioner Office of Combination Products, September 2006.
ICH Annex 2 – Stability Testing of Active Pharmaceutical Ingredients and Finished Pharmaceutical Products.
ISO 10993 Biological Evaluation of Medical Devices.
ISO 8871-3:2003 Elastomeric Parts for Parenterals and for Devices for Pharmaceutical Use - Part 3: Determination of Released – Particle Count.
ISO Series 11040 – Pre-filled Syringes.
ISO Series 11608 – Package Integrity Evaluation – Sterile Products.
JP 7.01 Test for Glass Containers for Injections.
JP 7.03 Test for Rubber Closure for Aqueous Infusions.
Ph. Eur 3.2.1 Glass Containers for Pharmaceutical Use.
Ph. Eur. 3.2.9 Rubber Closures for Containers for Aqueous Parenteral Preparations, for Powders and for Freeze-Dried Powders.
USP Injections and Implanted Drug Products (Parenterals) – Product Quality Tests.
USP <1663> Assessment of Extractables Associated with Pharmaceutical Packaging/Delivery Systems.
USP <1664> Assessment of Drug Product Leachables Associated with Pharmaceutical Packaging/Delivery Systems.
USP <381> Elastomeric Closures for Injections.
USP <660> Containers - Glass.
USP <661> Plastic Packaging Systems and Their Materials of Construction.
USP <789> Particulate Matter in Ophthalmic Solutions.

12 Considerations for Development of Biological Products

Manfred Maeder

CONTENTS

CONSIDERATIONS FOR DEVELOPMENT

Combination products are an indispensable element of biological products.

Opposed to small molecules, which can be formulated in most cases in an oral pharmaceutical form, biological products need to be injected. Biological molecules, with very few exceptions, need to circumvent the complete digestive tract in order to avoid degradation.

This problem was identified at the early beginning of the 1900s. Soon after that, the first treatments were made with insulin. After having the capability to isolate insulin, the first treatments were made on children to treat diabetic ketoacidosis in 1922.

At this time the concept of vial/syringe (reusable) has been developed. Since these days it has been a long journey from the development of insulin cartridges and the use of reusable pens to the development of disposable pen systems (1985). Not only the primary containers, including closure systems saw significant improvements, but also injection needles. The gauge size increased remarkably to 33- or even 34-gauge needles for insulin injection systems. This was made possible with the steady improvement of the manufacturing process including improvements of steel quality.

Insulin could be considered one of the first and smallest (5808 Da) biological molecules for treatment. By now, biological treatments are typically in the >100 kDa range. During the last three decades, the development of biologics has increased significantly. Since a few years, the first biological products are reaching their end of patent life, and the first biosimilars are reaching the market as a consequence. During the last years, the application for medicinal products at the Food and Drug Administration (FDA) sees steadily increasing application numbers for Biologic License Application (BLA) compared to small molecules (New Drug Application (NDA)). Consequently, this means that almost all of these applications come as a combination product to the market. Again, this becomes necessary not only to enable application but also to improve patient convenience.

The development of pharmaceuticals follows a predefined pattern. Starting with preclinical activities, clinical phases I, II, and III follow with increasing numbers of patients involved in each phase. Then preparation for submission and commercial launch of these products takes place.

During the early days of biological products, there has been no consideration of drug-device combinations (i.e., combination products) during the clinical phases. This means during this time only the vial/syringe concept was used during all clinical phases. Typically, also the market entry was done with this setup. During recent years there had been a trend to find more patient convenient applications tools for the commercial presentation. Frequently, this happened a few weeks prior

DOI: 10.1201/9781003300298-14

to the launch of the commercial product, where marketing realized a competitive, patient-friendly presentation (combination product) is needed. During recent years, also based on increased expectations of some competent authorities, combination products are introduced earlier in the development process. Now, usually there is some testing during clinical phase II in order to gain some experience with a new medical device as an application tool. There is a target to have the final, commercial combination product evaluated and introduced in clinical phase III. This will enable the pharmaceutical company to already gather more information, including customer feedback to see less potential surprises after the launch of the product. As one further step in this combination product development, now products are entering clinical phase I or even preclinical stages. For example in the case of intrathecal or intratympanic injections or injection in the nL range over a longer period of time the device becomes the enabler for a treatment. An application of the drug product (biological or small molecule) enables a therapy to become possible with the correct device. Furthermore, connected devices could now be used to track usage and patient compliance during especially the clinical stage.

BIOLOGICAL PRODUCTS CONSIDERATIONS

Typically, all application tools like syringes, autoinjectors, pen systems or injection pump systems can be used for small as well as large molecules. For large molecules/biologics, some additional considerations need to be taken into account. For primary container systems like prefilled syringes or cartridges, siliconization will be necessary in order to be able to control the break-loose and gliding force (BLGF). This siliconization however might interact with the biological products to form agglomerates, which appear like particles deeming the product unacceptable. In other cases, the biological molecules are forming fibrils in the presence of silicone of up to 1 mm in length. This can be experienced if the product is exposed to silicone headspace and movement (e.g., during transport). Unfortunately, these unfavorable events only happen after sometime during stability studies rendering the formulation chosen unacceptable. Other interactions of the biologic and silicone have been observed. In some preparations the biological product managed to degrade the silicone layer on the inside wall of the container. The silicone layer within the closed container system was almost depleted after 18 months under controlled stability conditions leading to very high, unacceptable gliding forces, in several cases making an injection impossible. Today some container closure systems are in development without the usage of silicone or other gliding-enhancing liquids.

One of the current trends for injection needs is more to smaller sizes of needles (= higher gauge numbers). Smaller needle dimensions following the human factors principles clearly show less injection pain. Currently for injections with reusable or disposable pen systems needle gauge sizes of 32–34 are easily possible. These gauge sizes, however, are not usable for the vast majority of biological products, as the shear forces being developed within a pen system or an autoinjector would destroy the biological molecule during the injection. This is the reason why most products today are offered with a 27- or maybe a 29-gauge needle. Newer developments, such as thin wall needles, allow to move to slightly higher gauge sizes, as they are designed to have a larger inner diameter enabling a smooth injection. This is one more reason why the appropriate formulation of the drug product is of utmost importance to ensure the viscosity of the biological formulation is not getting too high, still enabling appropriate injection forces.

CONSIDERATIONS FOR BIOSIMILARS

The first biosimilar approved in the US was Zarzio, biosimilar for figrastim (2015). For the EU the first biosimilar approved (2006) was Omnitrope, a recombinant human growth hormone (rhGH). Other approvals followed quickly in the EU, whereas in the US there was a significant delay for biosimilar approvals. An important fact about biosimilars is the sufficient similarity to the originator product, which has to be proven in-depth applying multiple analytical methods. Based on this extensive work, a molecule may be declared biosimilar. This in turn means that the only differentiator of a biosimilar product is the device constituent part. As most older products reaching their end of patent protection are having less sophisticated device setups such as a vial/syringe/filtration-needle/injection-needle presentation. During this time, obviously, there have been no human factors considerations in the foreground during development. This example of a combination product presentation would need 12–15 handling steps, where two to three of them would be critical steps influencing the right dose for the patient. Unfortunately, following the US FDA guideline for interchangeability, it would be the easiest way to achieve the well-sought label for interchangeability in copying exactly the combination product setup of the originator (given the device patent landscape allows for that). The stony path considering the interchangeability claim, but by far more patient friendly, would be the development of a modern, less complex, easy-to-use combination product. For the above example, a two-step autoinjector could be a viable solution. There, all aspects of modern human factors considerations could be taken into account. Patients could benefit from the progress in the development of injection systems during the last 15 years. This would enable the pharmaceutical company to deliver a product with less painful injections and by far less handling steps to a patient compared to the older products. A significant improvement in patient compliance would be almost guaranteed, considering a system with multiple handling steps and self-injection with a syringe compared to a two-step, spring-driven autoinjector with a hidden needle during the complete process.

In more general terms there are a lot of interactions between the drug product or biologic, the device, and the patient. For patients, age and education have to be considered in designing a device properly. Therefore, the instructions for use (IFU) need to consider the reading capabilities of the user. This could also depend on the country the person is living in. Taking this into consideration, the IFUs will need a significant amount of pictures to guide the persons through the overall injection process. Designing a device will also be of utmost importance, if the device is intended to be used by patients or by caregivers only. In case of healthcare professionals (HCPs) being the only users, the designers can be sure that they are having quite some exposure to these types of devices and are familiar with most of the concepts. This familiarity however also bears some risk, because HCPs might assume the device works in a specific way, but maybe it has been designed in a slightly different way compared to currently marketed products. During the last few years, there had been a number of series of complaints demonstrating that.

If a device is designed to be used by a patient, any potential impairments considering the indication of the products need to be taken into account. For example, patients with arthritis or multiple sclerosis will not be able to handle very small devices. They might not be able to use devices where a healthy subject has no difficulties at all. Therefore, it is very important during development that the devices can be used not only by the healthy, young engineers designing them but also by persons being partially impaired by their disease. Size and shape of the device need to be adapted to the user needs.

Considerations like frequency of use and homecare vs. use in hospital are also very important considerations. In case a patient needs to apply medication multiple times a day, it would be counter-intuitive to develop a large bulky device which needs to be carried all the time. One of the reasons that the inhaled insulin was not a great breakthrough was the device was very large and difficult to set up and carry. If such a device would be used once a day or once a week, this would have been by far a smaller hurdle, because everything could have been planned to perform the treatment at home. It would have been cumbersome only during travel and vacation time. Likewise, for injection systems, the size matters less. There, the most important aspect is the ease of handling for the patients. Patients with impairments usually have less difficulties to handle a larger device that they can hold in a firm way. As almost all biological injection products have to be kept refrigerated, the storage space might matter. Also, there the frequency of treatments matters. As most of the products now are injected on a monthly or quarterly basis, the storage space will not be the most important decision factor.

One important aspect for the patients or users are the feedback signals of the device. For any injection system, there is an expectation to have visual and audial feedback for the start and end of injection. For the audial feedback, e.g., click at the beginning and end of injection, the environment of the usage has to be taken into consideration. In case the system is being used in a noisy environment, e.g., as an emergency medication after an accident, it might be very difficult to hear the clicks. For this reason, an additional feedback needs to be present to indicate the proper usage to the patient or HCP.

For the performance of the combination product, the interaction of the drug constituent part with the device constituent is relevant. Injection volume, needle size, gliding forces, needle diameters, and viscosity of the solution are some of the parameters influencing the injection time. With increasing injection time, the risk of premature removal of the injection system will increase leading to incomplete injection and incorrect dose. Therefore, these parameters need to be set and controlled properly to ensure the constant performance of the injection systems with very little variability. One important factor for injection time considerations is the aging of the product. During stability, this time will change. What typically can be observed is that the injection time will decrease with increasing temperature. This is not surprising as the viscosity will decrease too. Another factor to be considered is the increase of injection time due to aging of the drug-device combination.

Both factors can be seen in Figure 12.1, where both the decrease in injection time of a 1-month-old and a 24-month-old preparation with increasing temperature and also the increased injection time of the aged material for all temperature points can be observed.

One of the reasons for such a phenomenon could be the decrease of the silicone layer during storage as observed for multiple preparations.

In Figure 12.2, the level of silicone of the unfilled syringe, the syringe after 1, 24, and 33 months are shown. In the beginning the silicone layer is nicely distributed over the complete syringe and after storage time the top part of the syringe is almost depleted of silicone. In these cases, a smooth injection will not be possible.

Due to the large number of factors influencing the performance of the device and possible interaction of the drug/biologic with the device constituent parts and the change over time, the development of a combination product can be challenging.

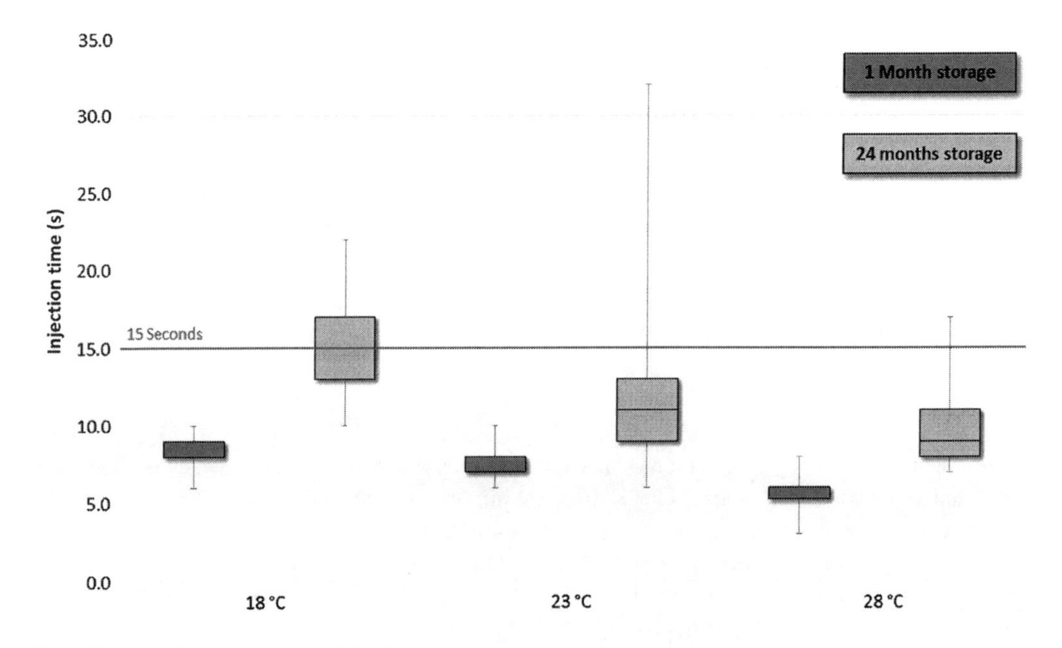

FIGURE 12.1 Impacts of material aging and temperature on injection time

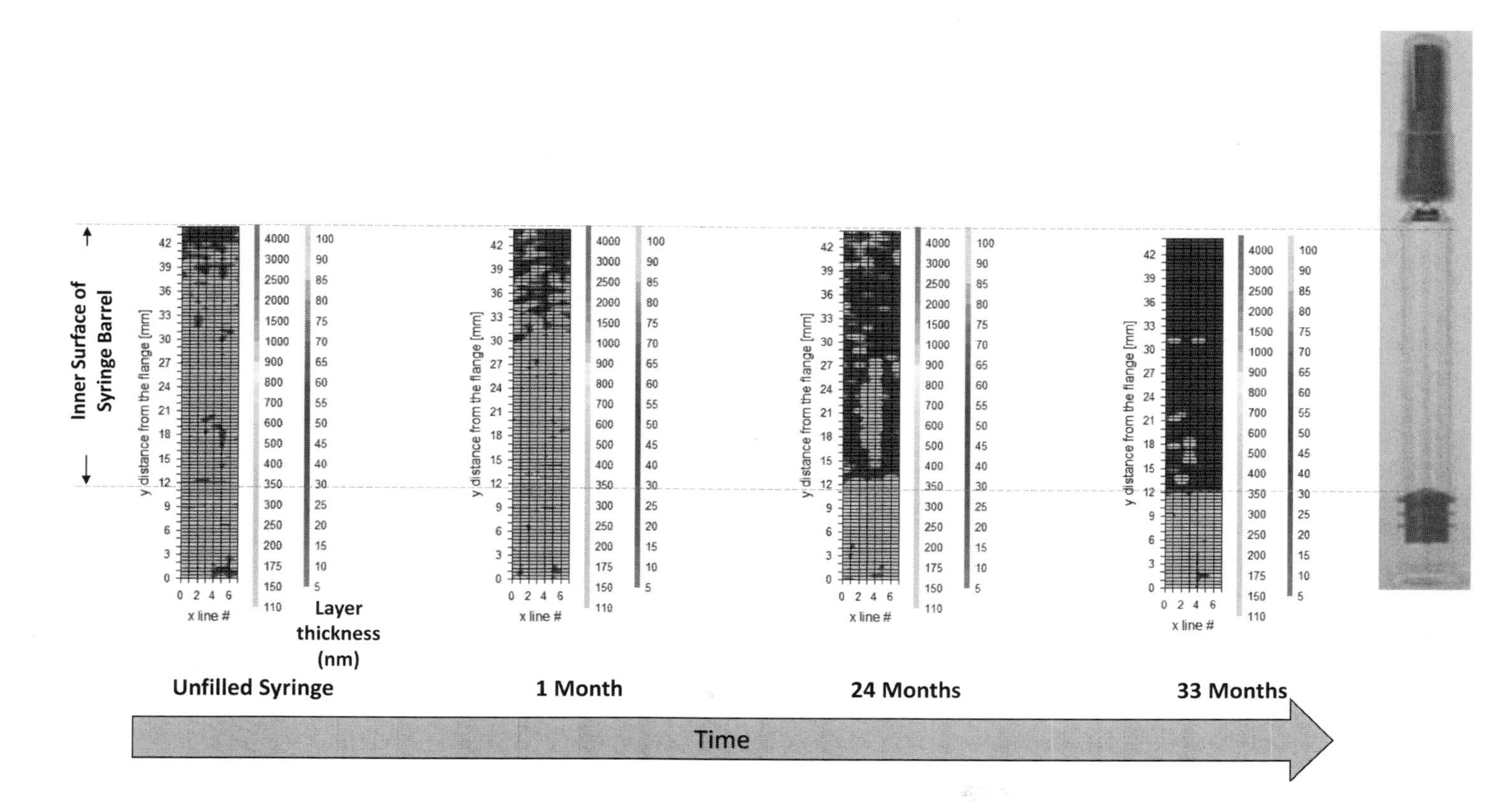

FIGURE 12.2 Impact of aging on level of silicone in an un-filled syringe

Specific Considerations on Interchangeability

Some biosimilars may be termed and treated as interchangeable products. If that should be achieved, they would need to follow specific requirements. The following regulation is defining the requirements: The BPCI Act (Biologics Price Competition and Innovation Act of 2009) created an abbreviated licensure pathway in section 351(k) of the Public Health Service Act (PHS Act) for biological products shown to be biosimilar to, or interchangeable with, an FDA-licensed biological reference product. FDA believes that guidance for industry that provides answers to commonly asked questions regarding FDA's interpretation of the BPCI Act will enhance transparency and facilitate the development and approval of biosimilar and interchangeable products. The BPCI Act was enacted as part of the ACA on March 23, 2010. The BPCI Act amended the PHS Act and other statutes to create an abbreviated licensure pathway for biological products shown to be biosimilar to, or interchangeable with, an FDA-licensed biological reference product. To meet the standard for "interchangeability," an applicant must provide sufficient information to demonstrate biosimilarity to the reference product and also to demonstrate that the biological product can be expected to produce the same clinical result as the reference product in any given patient, and if the biological product is administered more than once to an individual, the risk in terms of safety or diminished efficacy of alternating or switching between the use of the biological product and the reference product is not greater than the risk of using the reference product without such alternation or switch (see section 351(k)(4) of the PHS Act). Interchangeable products may be substituted for the reference product without the intervention of the prescribing healthcare provider.

Especially the very last statement of the guidance makes it very attractive for market authorization holders to strive for interchangeability, in spite of additional studies which need to be performed.

The interchangeability guidance is silent, however, with respect to application systems, e.g., those device constituent parts needed for injection of most biological drug constituent parts. Typical subcutaneous injection systems are prefilled syringes and autoinjectors. Now, for the device constituent part also some interchangeability should be expected. It would be hard to imagine that a vial kit (vial, syringe, needle, and filtration needle) with 12–15 handling steps should be comparable to a two-step autoinjector (AI). As the AI would be a very significant improvement for the handling of the patient, it would be tough to change back and forth. In these cases, a discussion with the relevant health authorities would be indicated, also in order not to prevent significant improvements for the patients (because originator systems are copied 1:1, which may have been designed 20 years ago).

13 Digital Combination Products and Software

Ryan McGowan

CONTENTS

This chapter discusses an emerging topic in the combination product market: the use of digital health technologies including software and connected systems. It presents a number of examples, provides advice on when digital products may be regulated as medical devices and therefore might be considered to comprise "combination products" with medicines, and offers regulatory strategies for successful marketing approval. The content of this chapter represents the personal views of the author and is intended only for educational purposes. It does not replace independent professional judgment. Regulatory requirements are discussed as needed to frame issues for the reader, but the reader should refer to relevant regulations, guidance, and standards regarding what requirements and policy expectations apply and seek the advice of legal counsel and other experts as appropriate. Requirements and policies may change and vary among health authorities.

THE PROMISE OF DIGITAL HEALTH

In recent years, the phrase "digital health" has enjoyed substantial buzz in the pharmaceutical, medical device, and combination product industries. Although the field is largely in its infancy, medical product manufacturers have placed considerable resources and realigned business goals around the promise of improved healthcare delivery through digital technologies.

DOI: 10.1201/9781003300298-15

Leaders in the digital health space advise that companies can no longer afford to think of themselves simply as manufacturers of a singular type of medical product. They must embrace digital technologies as support tools to better study, understand, market, and provide operational support for their products.

For professionals in product development, quality, regulatory, and supply chain roles to be successful within these companies, it is increasingly important for them to demonstrate digital literacy. Clinical development and sales and marketing teams are highly motivated to use digital technologies and product delivery and compliance functions now have the enviable task of developing and defending the relevance, safety, and efficacy of this new class of products to health authorities who may not be prepared, or resourced, for the challenge. This chapter aims to provide the reader with a basic knowledge of strategies needed to advise their company on creating approvable digital tools intended for use with combination products and medicines.

THE BIG DEAL WITH BIG DATA

To be successful supporting digital programs, one must understand the fundamental motivations behind the promise of digital health. At the heart of the matter is the seemingly endless pursuit of data. Health technology companies are now founded and valued purely on the number of users they can recruit and retain on their digital platforms. This quest for user data is so important that many digital health start-up companies run at a loss for years but maintain solvency through investors that believe that ultimately these companies will be successful. Venture capitalists and hedge fund managers investing resources in these companies believe that, eventually, big data will drive big insights and that is the commodity these stakeholders are interested in.

"Gathering insights" is admittedly a bit of a nebulous concept; however, most of us have at least a basic idea of what it implies. In truth, insight can mean different things for different industries. For social media companies, insights might be associated with better directing advertising to particular populations that are most likely to purchase goods and services. In the case of the medical product industry, insights are more focused on finding new health patterns, improving diagnostics, and targeting treatments to the patients who can best benefit. Ultimately these improvements hold promise for new and expanded claims for the drug, medical device, and combination product manufacturers who effectively wield these digital technologies.

Claims are at the heart of the medical product world. They are the basis of the labeling that is approved by health authorities and the fundamental language used in the marketing of drugs, medical devices, and combination products to healthcare professionals and patients. Claims are also important, albeit in a slightly different way, when it comes to convincing payers to reimburse for medical products. Thus, the collection and analysis of clinical and commercial digital data, if planned and executed correctly, can help companies make new claims for their products. This very concept is spurring the drive for the implementation of digital health technologies in the combination product space.

TAXONOMY OF DIGITAL COMBINATION PRODUCTS

Although digital health solutions may vary in their individual design, it is possible to broadly describe distinct categories of digital functions and intended uses that support the use of medical devices, drugs, and combination products. This section proposes a taxonomy for digital technologies that have been observed and proposed in the combination product space.

It should be noted that most examples presented in this section relate to drug primary mode of action combination products. The traditional medical device industry and associated health authorities have had experience with regulated software for a number of decades and tend to be more comfortable with the technology. Conversely, drug and drug-led combination product manufacturers and regulators tend to be new to the digital space.

Digital Components Embedded in Combination Products or Packaging

Historically, combination products and medicines have been released to market with expectations that safe and effective use will be facilitated through instructions and training materials. However, improvements in manufacturing technology have resulted in the miniaturization of digital sensors and interfaces allowing for the opportunity to sense aspects of combination product environments and use and to provide positive or negative feedback and other information to the user. For example, it may be possible to create digital labeling present on the package of a medicine to notify the user when the product has expired. The same type of technology might allow manufacturers to warn patients when their product refrigeration has been out of the acceptable temperature range. One can also imagine a future state where products or packaging could automatically prevent users from accessing medication when certain time, environmental, or use conditions apply.

Connected Digital Components Embedded in Combination Products or Packaging

Building on the digital sensor capabilities described above, it is possible to engineer embedded technologies with wireless antennae that allow the combination product to communicate some aspect of its environment or use to a remote receiver. Wireless communication increases the utility of the sensor as remote systems can offer more functionality on a larger interface and can integrate data coming from multiple users and products together to support a broader intended use. An example of the connected sensor approach would be onboard detection of injectable medication usage and communication of this information to a software application on the patient's or provider's mobile phone. The connection to a remote interface means that the patient can receive reminders or trend their medication usage over time, encouraging improved adherence to the dosing regimen. Such a feature might not be practical nor economically viable directly on the interface of a disposable combination product.

Connected Digital Components Added to Combination Products or Packaging

Similar to the concept of embedded sensors, so-called add-on sensor devices allow for the detection of combination product use and environmental data but are not enclosed within the combination product. These devices are often reusable, which allows for additional functions to be offered, such as a digital display or audible feedback. Like embedded sensors, these add-on devices may connect to ancillary software through the use of wireless antennae and share data. An example of an add-on sensor technology would be a clip-on sensor attached to an inhaler combination product. Such a sensor might detect when the patients have taken their medicines or count the remaining doses left in the inhaler through sensing the actuation of the inhaler or listening for inhalation sounds.

Software Applications for Use with Combination Products

Most combination product digital programs will offer some type of software application as part of the overall solution. The range of functions that might be offered by these applications is as wide as the therapeutic intended use of the combination products themselves. While it is not possible to enumerate the breadth of potential software uses in a single section, higher-level considerations and themes for software's intended use are provided below.

Interaction with Combination Product or Medicine

Receiving Information – a software application could be created to receive information from the connected combination product and apply further display, interpolation, or calculation to the parameters collected. For example, a software application might intake medication usage information (as collected and reported by the combination product sensor) to trend usage data and create a composite of "medication adherence." That same software might use a "lack of medication usage" signal to trigger a reminder for the patient to administer the medicine and prevent a missed dose.

Sending Information – a software application might be used not only to read information from the connected combination product but may also provide information back to the product. In the future, a software application might be used to transfer information to the combination product to control some aspects of delivery. For example, an application could provide instructions to the combination product to inject medication at a certain tissue depth or for a duration based on patient parameters (e.g., weight, body mass index) or patient preference. Another example is an artificial pancreas device system (NIH, October 2019), which automatically monitors and regulates blood glucose levels. This closed-loop control monitoring and delivery system tracks blood glucose levels using a continuous glucose monitor and automatically delivers the hormone insulin when needed using an insulin pump.

Influence Use of the Combination Product – a software application may not need to directly control a combination product in order to influence its use. Software may be used to provide instruction or feedback to the patients or to remind them of the proper use of their medicine. Software might display a digital version of patient instructions or play a video to remind the patients of how to use their medicine. It might also be more advanced and sense misuse from sensors onboard the device and attempt to intervene. For example, a syringe could be supplied with an onboard accelerometer that can sense if the product had been properly inverted to clear air bubbles prior to injection. If the software application does not receive a signal this step has occurred, it might alarm with a reminder and display the complementary instruction.

Further, a software application also may not require any communication with the combination product to influence use. Dose calculators, for example, could be used to suggest the next dose of medicine based on factors such as patient-reported exercise, caloric intake, and weight and instruct the patient to administer themselves with a particular amount of medication. (Note: Such a dose calculator would not be considered as a constituent part of a combination product. On the other hand, it may be considered as Software as a Medical Device (SaMD), subject to device requirements.) This type of software functionality does not require direct communication with the combination product but has the potential to significantly affect the use of the combination product. There may be interesting issues associated with labeling consistency when software is used to support prescription drug use, as reflected in the Food and Drug Administration (FDA) proposal as well as if the software is disseminated by or on behalf of the drug/combination product sponsor (Prescription Drug-Use-Related Software (PDURS)) [Docket 2018-25206; 83 FR 58574, November 2018)].

Nature of Use with Combination Product

When considering software applications to be used with combination products, it is fundamental to define how the software will be used with the product. Creating separation, both in a technical and in a regulatory sense, reduces risks, but it may also decrease the utility of the combined use.

Optional Use – most digital combination products will fall under this category. Ancillary digital functions such as patient support messaging, adherence management, symptom reporting, and even simple dose calculation are rarely required for the combination product to function or to be safe and effective. In this case, it might be possible to mention the use of optional software in the combination product label, but there would not be a clinical or regulatory requirement to use the digital product.

Required Use – although currently rare, required use of software and combination products may become more prevalent in the future as the area of "digital therapeutics" matures. Combination product manufacturers may consider the use of a software function to be required where there is some aspect of benefit, or indeed product safety, that requires digital support. An example of this may be a dosing schedule or calculation that is very difficult to understand without an accessory digital application or perhaps a highly addictive medicine that requires lockout controls to prevent unintended access between doses. In these cases, the combined use of the software and combination product would likely need to be studied and assessed in a pivotal clinical program.

Ancillary Digital Technologies for Use in Clinical Trials

Although the majority of development time and energy in the area of digital support for combination products is spent on new bespoke solutions for individual products, there is an entirely separate category of solutions that supports patients, investigators, and study sponsors within clinical trials for combination products. These products may help recruit and retain patients, monitor them within the study, measure and report endpoints, or reduce administrative burden associated with clinical study requirements. Examples include connected measurement tools such as spirometers and weight scales, medication tracking systems, e-diaries and/or electronic patient-reported outcomes (e-PROs), and appointment reminders. Although the goal of these technologies is not necessarily to advance the commercial use of the combination product, the proper selection and qualification of digital support products help to ensure a successful clinical study outcome supporting the eventual product approval.

REGULATORY CLASSIFICATION

In addition to the intended use of a product (indications, user population, use environment), a fundamental question that needs to be answered for digital health technologies is, "what will the system be classified as?" The question is relevant at the beginning of the product development cycle since the regulatory identity of the product (e.g., medical device, combination product, clinical measurement tool) can affect the general regulatory requirements that must be followed to deliver that product into clinical or commercial use. While regulatory expectations should be driven by risk analysis and what information is needed to demonstrate product safety and effectiveness, regulatory requirements may vary depending on product classification, such as the type and extent of quality management systems that must be followed, as well as the types of premarket evidence needed to obtain product approval (engineering performance testing, simulated use human factors testing, and/or clinical study data).

The first piece of the classification puzzle relates to how the digital product is associated with the combination product or medicinal product of interest. If the digital product is physically connected to the combination product or medicinal product, then the digital product is very likely to be regulated according to combination product or medicinal product requirements. For example, if a digital sensor is embedded within the plastic components of a prefilled syringe to detect the volume of drug delivered, that sensor would be considered a medical device constituent part of a combination product in the United States, or an integral device part of a medicinal product in Europe/United Kingdom. Clinical and other studies (e.g., bench data) to support any claims associated with the incorporation of the digital product would proceed according to the relevant investigational drug or medicinal product pathways, as would follow-on marketing applications. In these cases, the combination product would be reviewed under the drug or medicinal product pathway, with the digital device parts being assessed by health authority staff specializing in medical devices, or through a notified body opinion (EU).

If the digital product is labeled for use with a combination product or medicinal product but is not physically combined with the medicine, things get more complicated. In this case, the key question becomes, "Does the separate digital product meet the definition of a medical device on its own merit?" It is important to note that if the digital product meets the definition of a medical device on its own, then the combined use of a separate (i.e., non-integral) digital product with a combination product or medicinal device might satisfy the conditions of a "cross-labeled" combination product in the United States, as defined in 21 CFR Part 3.2 (e)(3) and (e)(4). In brief, according to these regulations, if both the digital product and the combination product or medicinal product are required for users to achieve an intended use or are studied together under the same clinical protocol, they may be considered as a new combination product system.

The constituent parts of cross-labeled combination products are generally manufactured at separate facilities subject to the same current good manufacturing practices (cGMPs) requirements ordinarily applicable to the product type (e.g., 21 CFR 210/211 for drugs and 21 CFR 820 for devices) (see Chapter 4, cGMPs). Accordingly, if the digital product is considered a device constituent part, and its use may influence the safety and efficacy of the medicinal product, then there are expectations that the digital product device sponsor will take into account its safe and effective use, with the specific drug as an input to design control (more general consideration of combined use with drugs as a design input may be appropriate where the digital product is intended for use with drugs in general or a class of drugs). The US FDA offers flexibility where cross-label constituent parts are being manufactured at the same facility – in which case one can apply a streamlined cGMP approach. Sponsors of a cross-label combination product may prefer and generally have the flexibility to choose either separate marketing applications for the constituent parts or a single application for the entire combination product. In most regions outside of the United States, there is no corollary to a "cross-labeled" combination product, although considerations related to the combined use of the digital medical device product and the medicinal product would still be evaluated through the independent medicinal product and medical device marketing approval pathways.

The process of determining if a digital product will be regulated as a medical device is theoretically straightforward: compare the intended use of the product to the functions of similar products already on the market. For physical devices which have 50 or more years of regulatory precedent, in most cases, the manufacturer can find similar products using regulatory and legal professionals or, if still uncertain, can contact the health authority directly for a classification ruling. For digital products, the same historical precedent often does not exist, which requires manufacturers and health authorities to consult the statutory definition of a medical device in the study or marketing region of interest. Unfortunately, these definitions have a reputation for being very broad and unhelpful. Most regions refer to a medical device as a system with features that achieve or aid in "diagnosis, prevention, monitoring, treatment or alleviation of disease." This is a definition that is so wide almost any tool used in the delivery of healthcare could be argued to be within its scope.

As broad as these definitions may be, a regulatory professional should study the regional regulations closely when considering if a digital technology might be regulated by the local health authority as a medical device and must be ready to defend why digital functions do not meet the letter or spirit of the law. Fortunately, in some cases, health authorities have provided guidance to better advise when commonly used digital health functions might meet the definition of a device. A list of relevant guidance documents is provided below. Note, this is not an exhaustive list of all regulatory guidance related to medical device software. It covers only the subset of guidance and health authority statements which may aid in determining if software functions meet the definition of a medical device in the region of interest. As regulatory guidance on digital topics is dynamic and evolving, and new policy statements are constantly being published that are important to this topic, health authority websites should be consulted to ensure new policy statements and guidance are accounted for.

Canada – Health Canada

Software as a Medical Device (SaMD) – Draft Guidance Document – January 2019

This draft guidance document describes software functions which Health Canada intends to regulate as medical devices, as well as those which are not considered to meet the definition of a device. Where possible, Heath Canada has aligned their approach with other health authorities.

European Union – European Commission

Guidelines on the qualification and classification of standalone software used in healthcare within the regulatory framework of medical devices – July 2016

This draft guidance document provides an interpretation of the definition of a medical device as found in the EU Medical Device Directive/EU Medical Device Regulation as it applies to software functions and includes examples of systems that fall within and outside the scope of the definition.

Guidance on Qualification and Classification of Software in Regulation (EU) 2017/745 – MDR and Regulation (EU) 2017/746 – IVDR – October 2019

Although not a European Commission document, this guidance was composed by the Medical Device Coordination Group (MDCG) which is composed of representatives of all Member States and chaired by a representative of the European Commission. The document provides an overview of common embedded and standalone software functions and compares those functions to the definition of a medical device and an in vitro diagnostic device under the latest EU regulations.

Infographic – Is Your Software a Medical Device? – March 2021

Another resource published by the MDCG is a two-page infographic intended to help manufacturers determine if their digital technology meets the definition of a medical device under the EU Medical Device or In Vitro Diagnostic Regulations.

Germany – Federal Institute for Drug and Medical Devices (BfArM)

Medical Apps – Guidance – June 2015

This draft guidance document builds on the European Commission guidelines regarding the definition of a medical device as applied to software and contains some additional regional interpretation and examples.

International Medical Device Regulators Forum

"Software as a Medical Device": Possible Framework for Risk Categorization and Corresponding Considerations – September 2004

This document, composed by a voluntary group of global medical device regulators from around the world, offers context for the definition of medical device software as well as a potential risk-based classification structure.

Japan – Ministry of Health, Labor and Welfare (MHLW)

Guideline of Determining Whether Software Is Classified as a Medical Device – March 2021

This document is intended to help digital health developers understand what types of software functions will cause the system to be regulated as a medical device and the specific regulatory requirements that apply to medical device software.

United Kingdom – Medicines and Healthcare Products Agency (MHRA)

Medical Devices: Software Applications (Apps) – Guidance – September 2022

This draft guidance document builds on the European Commission guidelines regarding the definition of a medical device as applied to software and contains some additional regional interpretation and examples.

United States – US Food and Drug Administration (FDA)

Proposed Regulatory Framework for Modifications to Artificial Intelligence/Machine Learning (AI/ML)-Based Software as a Medical Device (SaMD) – Discussion Paper and Request for Feedback – A Working Model – April 2019

Although not a formal Agency guidance, Center for Devices and Radiological Health (CDRH) released a proposed framework for the regulation of artificial intelligence/machine learning medical device software. The document largely focuses on the practical difficulties of regulating these technologies according to traditional medical device regulations, given the pace at which the systems may change and the fact that they may change autonomously without human direction or review. The document includes a proposed approach for post-approval change review and the potential for artificial intelligence/machine learning system to change without FDA review within certain boundaries.

Medical Device Data Systems, Medical Image Storage Devices, and Medical Image Communications Devices – Guidance – September 2019

This guidance originally provided information on certain types of software functionalities related to medical device data transfer, storage, display, and the extent to which those functions meet the definition of a medical device and/or will be practically regulated by the FDA. The guidance was later superseded by a change in the US Code under the 21st Century Cures Act which removed systems that simply transfer, store, and display medical device data from the formal definition of a medical device. The final guidance now accounts for this and incorporates elements of the 21st Century Cures Act within advice and examples.

General Wellness: Policy for Low-Risk Devices – Guidance – September 2019

This guidance provides information on products used in the general health support of patients. Although not all products covered by this guidance are software or related to software, many digital systems have a component of wellness functionality and may be covered.

Developing a Software Precertification Program – A Working Model – January 2019

Although not a formal Agency guidance, the Center for Devices and Radiological Health (CDRH) has proposed a framework for the regulation of medical device software known as "precertification." The precertification model would evaluate a manufacturer based on their quality practices and software development approach. If the Agency finds the company's quality policies and systems to be acceptable, it may reduce the amount of evidence required during premarket review. Note: In September 2022 the FDA published a document titled *The Software Precertification (Pre-Cert) Pilot Program: Tailored Total Product Lifecycle Approaches and Key Findings*, which suggested that the Agency required additional statutory authorities to implement the Pre-Cert program in the future.

Multiple Function Device Products: Policy and Considerations – Guidance – July 2020

This guidance interprets statutory changes to the definition of a medical device in the United States as enacted by the 21st Century Cures Act and specifically discusses the extent to which digital technologies and software can be considered to have multiple, separately architected, software functions, with some functions being regulated as a medical device and others not being so regulated by FDA.

Content of Premarket Submissions for Device Software Functions – Draft Guidance – April 2021

This draft guidance provides information regarding the recommended documentation sponsors should include in premarket submissions for FDA's evaluation of the safety and effectiveness of device software functions.

Digital Health Technologies for Remote Data Acquisition in Clinical Investigations – Draft Guidance – April 2021

This draft guidance outlines recommendations intended to facilitate the use of digital health technologies in a clinical investigation as appropriate for the evaluation of medical products, including some of the information that should be contained in an investigational new drug (IND) application or an investigational device exemption (IDE) application for a clinical investigation.

Medical Device Data Systems, Medical Image Storage Devices, and Medical Image Communications Devices – Guidance – September 2022

This guidance provides information on how FDA intends to regulate Medical Device Data Systems, Medical Image Storage Devices, and Medical Image Communications Devices in the context of section 3060(a) of the 21st Century Cures Act (Cures Act) amended section 520 of the Federal Food, Drug, and Cosmetic Act (FD&C Act), which removed certain software functions from the definition of device in section 201(h) of the FD&C Act.

Cybersecurity in Medical Devices: Quality System Considerations and Content of Premarket Submissions – Draft Guidance – September 2022

This draft guidance provides information on how cybersecurity practices should be considered and implemented within medical device quality management systems and how sponsors should submit cybersecurity information for premarket review.

Policy for Device Software Functions and Mobile Medical Applications – Guidance – September 2022

This guidance provides a classification of three software regulatory categories: those which meet the definition of a medical device and will be regulated as such, those which meet the definition of a medical device but will not be regulated by FDA (i.e. under enforcement description), and those which do not meet the definition of a medical device. Additional examples are provided for each category.

Clinical and Patient Decision Support Software – Guidance – September 2022

This draft guidance interprets recent statutory changes to the definition of a medical device in the United States as enacted by the 21st Century Cures Act and specifically discusses the extent to which digital technologies offering clinical decision support will be regulated as medical devices.

Regardless of if a digital technology meets the definition of a medical device, regulatory obligations must still be considered. Examples of such requirements are provided below. Note it is not the intention of this chapter to cover regulations outside of the medical product space, and thus the below list is not considered to be exhaustive:

- Privacy regulation such as the General Data Protection Regulation (GDPR) which is a data protection and privacy regulation applicable to all individual citizens of the European Union
- Privacy regulation for medical insurance and other healthcare information, such as US Health Insurance Portability and Accountability Act (HIPAA)
- Digital transmission and wireless system regulations as administered by authorities such as the US Federal Communications Commission
- Financial regulations such as US Sarbanes-Oxley Act for systems handling financial data or payments or the US Sunshine Act for benefits provided to investigators or clinical customers
- False consumer advertising and promotion regulations as enforced by the EU Free Trade Association or the US Federal Trade Commission

Additionally, since most combination product manufacturers will study or market a non-medical device digital asset in conjunction with an already regulated product (e.g., medicine), that digital asset may still face health authority scrutiny as part of its use as a clinical tool or through its marketing association with the regulated product; examples include:

- Good Clinical Practice (GCP) regulations which dictate the nature of clinical information that can be collected and how it must be controlled and review of patient and investigator facing digital materials.
- Post-market safety reporting (e.g., adverse event reporting) obligations for medical devices, drugs, or combination products which are used in conjunction with the digital asset (see Chapter 8). Manufacturers of these regulated products (i.e., medical devices, medicinal products) retain the obligation to receive information on, and report certain types of, malfunctions and adverse events to health authorities, which may include duties to report information received due to the use of a non-regulated digital system in conjunction with the regulated product.
- Regulations concerning digital signatures, data integrity, and storage of digital information for use in medical product studies and commercial use. For example, *21 CFR Part 11 – Electronic Records; Electronic Signatures – Scope and Application,* FDA's 2018 Guidance for Industry on *Data Integrity and Compliance with Drug CGMP Questions and Answers,* and the European Medicines Agency's (EMA) *Guidance on good manufacturing practice and good distribution practice: Questions and answers* which were updated in 2016 with 23 new data integrity topics.

PROMOTION OF DIGITAL PRODUCTS BY MEDICINAL PRODUCT MANUFACTURERS

The section above describes scenarios where a digital product used with a medicinal product might be classified as a medical device or a medical device constituent part of a combination product. Regardless of the medical device classification status of the digital product, it is important for pharmaceutical manufacturers to consider medicinal product promotional regulations when making statements about the use or benefits of digital technologies. Each region has particular requirements specifying the public statements the pharmaceutical company and its representatives can make concerning a regulated product. Referencing the use and benefits of a digital technology in combination with a medicinal product will likely be considered as promotion of the medicinal product itself.

In the United States, the FDA has released a Federal Register request for comment document which contains some of their current thinking on the promotion of prescription drugs and biological products in combination with digital technologies United States Federal Register (2018). This request for comment document provides an overview of the FDA's intent to regulate software output related to prescription drugs as required labeling or promotional labeling. For software outputs not considered as crossing the line into required drug labeling, the pharmaceutical manufacturer would follow typical promotional content submission pathways. At this time, the Federal Register comment is simply a request for public input, and so the Agency's position on this topic may change.

While other regions have not directly stated their expectations regarding the content of digital products as medicinal product labeling, it stands to reason that public statements concerning the use of a medicinal product would need to be considered under applicable labeling regulations.

DIGITAL COMBINATION PRODUCT DEVELOPMENT AND MANUFACTURING

In the scenario where a digital product will be regulated as a medical device or device constituent part of a combination product, the next major consideration is how the product will be developed and manufactured. Unlike other industries such as digital consumer electronics, the design, development, and manufacturing activities associated with medical devices and combination products are regulated processes. This means that any entity taking part in medical device or combination product development will be accountable to the relevant health authority in each marketing region. This section focuses on developmental and manufacturing considerations. While beyond the scope of this discussion, regulatory requirements may also include premarket approval requirements to ensure the safety and effectiveness of the product for its intended use(s) (see "Evidence Generation" section).

In the drug product and traditional combination product industries, the owner of the marketing application often takes on development and manufacturing responsibilities directly. This is because the company has years of experience and facilities dedicated to these activities. A combination product manufacturer may or may not be willing to take on digital development activities directly, given the breadth of additional activities required. Although it is important to acknowledge that the owner of the medical device or combination product cannot abdicate accountability for the compliant development of the product and its ongoing safety and effectiveness once on the market, certain aspects of development and manufacturing may be allocated to properly qualified contract organizations. If the company is relying on partners to deliver particular aspects of the design or manufacturing of the digital system, it is important that the quality management responsibilities between the companies be clearly established and documented consistent with supplier and purchasing controls.

If a company determines they will act as the manufacturer of a digital combination product or medical device, they will be accountable for performing those activities according to regulatory requirements, including updating their quality management systems to account for the new type of regulated device they will market.

The medical device quality management system regulations and standards applicable to medical device software are, for the most part, harmonized globally with a particular overlap in regions such as the EU, UK, Canada, and Australia. In these regions, conformance with the latest revision of ISO 13485 is often accepted as evidence of compliance with local regulations. The United States is somewhat different, as currently ISO 13485 compliance is not considered sufficient. FDA's Quality System Regulation, 21 CFR 820 is considered highly similar to ISO 13485; however, the reader is cautioned that some differences do exist between the FDA and ISO approaches. However, the FDA has recently released a draft rule to the Federal Register describing the processes that will be taken to further harmonize 21 CFR 820 and ISO 13485. See Chapter 4 for more information. Additionally, the Association for the Advancement of Medical Instrumentation (AAMI) recently published TIR 102:2019, which offers a detailed comparison between 21 CFR 820 and ISO 13485.

The extent of medical device quality management systems that must be followed for digital combination products depends on both the nature of the regulated activities taking place within a facility and the type of product being produced. In the case where responsibility for aspects of product design, development, and maintenance are conducted by multiple parties, the overall owner of the system must qualify sub-contractors to perform their tasks. For example, if a digital design firm is recruited to develop and release a piece of software, they must follow the appropriate regulations for design control/product realization and associated consensus standards. In this example, the owner would then retain overarching responsibility, including all follow-on production, maintenance, problem resolution, and complaint management activities.

As mentioned, the type of digital product being produced also affects the extent of the medical device quality management system that must be implemented. If the product is to be marketed in the EU, UK, and US, and the digital elements are physically integral to the medicine (e.g., embedded sensor to detect medicine administration), the digital product would be considered a medical device constituent part of a combination product (US) or a device part of a medicinal product (EU/UK). With regard to current good manufacturing requirements, the manufacturer would need to maintain drug and medicinal product quality management systems and current good manufacturing practices and would also need to incorporate certain additional medical device elements. Additional detail on this approach is provided in Chapter 5 (cGMPs).

If a digital device is not physically integrated or co-packaged with the medicine, in most cases it will be appropriate to design, develop, and maintain the device within a full medical device quality management system. Similarly, in the EU/UK, if a digital device is not integral with the medicine, it would not be considered a component of a medicinal product eligible for review under a Marketing Authorization Application, and the full EU/UK Medical Device Regulation(s) would apply.

There are differences in the quality system approaches used by drug and device manufacturers that must be approached carefully. Compliance with these additional development and manufacturing regulations for a software device takes knowledge and patience to implement, especially within organizations new to being accountable for traditional medical devices and software compliance activities.

In addition to quality system regulatory requirements, the use of international technical consensus standards is particularly relevant for the development and manufacturing of digital medical devices and combination products. Regulations tend to be written at a high level and are often agnostic of individual product types, and so both health authorities and manufacturers depend on the more detailed standards documents for guidance on how to create and manufacture products.

Table 13.1 contains a listing of regulations and international consensus standards relevant to the design, development, and manufacturing of digital medical devices and combination products. As regulatory guidance on digital topics is dynamic and evolving, health authority websites should be consulted to ensure new policy statements and guidance are accounted for.

EVIDENCE GENERATION

Conventional combination products such as prefilled syringes, inhalers, and drug-coated implants often have a well-understood product approval pathway. There are, of course, many details to be worked through around study design, statistical powering, and selection of endpoints; however, the basic tenants of reliance on well-controlled clinical studies continue to be the fundamental standard for these types of products.

The concept of evidence generation for digital technologies is only just emerging, and there is far less certainty around the types of data needed to convince patients, prescribers, health authorities, and payers of the utility of a digital asset.

TABLE 13.1
Summary of Applicable Regulations and Standards

Regulations for Medical Device Quality Management Systems	
Source	**Description**
US FDA Code of Federal Regulations 21 CFR Part 820 – Quality System Regulation 21 CFR Part 4a – Current Good Manufacturing Practice Requirements for Combination Products	These US and EU regulations cover the quality system requirements associated with medical device and software manufacturing. These resources cover aspects of management, controls, production controls, personnel, facilities, and other considerations. Although all sections are important for compliance, particular emphasis for software development and maintenance relates to sections 21 CFR 820.30, design controls and general safety and performance requirements within Annex I of the EU Medical Device Regulation.
EU European Commission Council Regulation 2017/745 concerning medical devices, OJ No L 117/1 Article 10 – General Obligations of Manufacturers (quality management system requirements found in section 9)	
Standards for Medical Device Quality Management Systems	
Source	**Description**
ISO 13485 – Medical devices – quality management systems – requirements for regulatory purposes	This standard contains detailed requirements for medical device quality management systems. For software design, development, and maintenance, particular attention should be paid to ISO13485 Section 7 – Product Realization.
Standards for Medical Device Software Development, Software Lifecycle, and Connected Systems	
Source	**Description**
IEC 62304 – Medical device software – Software life cycle processes	Provides detailed requirements for the design, development, and post-release maintenance of medical device software with a focus on requirements gathering through design verification stages.
IEC 82304-1 – Health software – Part 1: General requirements for product safety	Provides detailed requirements for medical device design validation process including clinical validation steps.
IEC 62366-1 – Medical devices – Part 1: Application of usability engineering to medical devices	Provides a process for human factors engineering and testing activities for medical devices. Not exclusively written for software, but applicable to software development.
ISO 14971 – Medical devices – Application of risk management to medical devices	Provides a process for identifying, analyzing, evaluating, and accepting technical and usability risks related to medical devices, including software
IEC 60601-1 – Medical electrical equipment – Part 1: General requirements for basic safety and essential performance	Provides requirements for electromechanical medical devices, including demonstration of basic safety and performance of software-controlled medical devices
IEC 60601-1-2 – Medical electrical equipment – Part 1-2: General requirements for basic safety and essential performance – Collateral standard: Electromagnetic disturbances – requirements and tests	Provides requirements for assessing electromagnetic emitting devices (e.g., connected combination products) for compatibility with environment.
IEC 60601-1-11 – Medical electrical equipment – Part 1-11: General requirements for basic safety and essential performance – Collateral standard: Requirements for medical electrical equipment and medical electrical systems used in the home healthcare environment	Provides requirements for use of electromechanical medical devices labeled for use in a home healthcare environment.
Guidance for Medical Device Software Development, Software Life Cycle, and Connected Systems	
Source	**Description**
AAMI TIR 45 – Guidance on the use of AGILE practices in the development of medical device software	Provides suggestions for developing software in an "agile" manner yet meeting regulatory documentation and process requirements which are typically associated with being "waterfall" in nature
FDA Software Validation Guidance	Describes regulatory expectations for the software development and life cycle, including connections to the conventional design control process
FDA Design Control Guidance	Provides guidance on the traditional design control process as specified in 21 CFR Part 820.30

Typically, the best approach to understanding the type of evidence customers and regulators will find convincing is to engage in early conversations with each stakeholder group. Digital combination products are no exception; however, in many cases, the manufacturer may find that the stakeholders themselves have not fully determined what the evidence requirements may be to gain a prospective claim. This may be frustrating initially, but it is also an exciting opportunity to innovate and create a "best fit" approach to digital health development. The sections below discuss types of evidence that may be appropriate to support digital combination product claims.

Evidence for Tool-Based Digital Products

The concept of a "tool-based claim" is often used in the traditional medical device arena. These products provide a clinical service to a patient or healthcare provider but do not make additional inferences or judgments of how the data generated should be used. For example, a combination product manufacturer may desire to offer a companion mobile application that allows the patient to record when they have taken their medicine and to remind them if a dose is missed, perhaps utilizing a sensor embedded into a drug delivery system such as an inhaler or injection device, without making any suggestions of how the function should be used in the treatment of an underlying disease. Health authorities have historically been willing to approve these simple digital device functions, particularly in the area of medication management without the need for clinical experience data.

In the case of tool-based claims, it may be possible to receive regulatory approval for complementary digital technologies simply by showing evidence that the technology accurately and reliably performs its function. Considerations for evidence generation include demonstrating that the device functions as intended across a range of bench-top conditions, proving the digital system does not adversely affect the medicine or combination product and that the user can consistently and safely use the digital system in a simulated use environment through human factors studies.

Evidence for Enhanced Tool-Based Digital Product

The current industry approach to digital combination product development often starts with the creation of a tool-based "minimum viable product" which offers expanded functionality over time. This approach allows users to become accustomed to the use of a digital support function and then introduces additional complexity in a stepwise manner. Examples of incremental enhancements include collection of patient symptoms, trending of medication usage and medication reminding, calculating medication dosages based on label instructions, and facilitating conversations between patients and healthcare practitioners.

Offering additional enhancements and features to combination product software increases the utility to and retention of users to the digital system; however, it may also mean additional and more convincing types of clinical evidence are needed to reach the market. This is particularly true where there is a perceived interdependence between the digital functions and the medicine or combination product. It is logical that a health authority reviewer might see a risk that the combined use of a respiratory symptom tracker and a respiratory medicine could possibly invite changes in the treatment regime, by either the healthcare practitioners or the patients themselves, even if that is not intended by the manufacturer.

It is reasonable to expect that some enhanced digital functions may require clinical experience to receive marketing approval, especially if the digital functions in some way relate to the dosing of a medicine or medical decision-making, even if no claims related to treatment or clinical benefit are being made. In cases where the enhanced digital functions require clinical experience to validate their intended use with an already approved drug, it may be possible to conduct relatively small observational studies or include the digital product within a long-term extension (Phase 4) study for the therapeutic product. The focus of these studies would likely be on demonstrating that the combined use of the digital product and the medicine or combination product does not introduce additional risk or inappropriately influence the care of the patient. It is not envisioned that these digital functions would require clinical evidence to demonstrate the efficacy of the digital solution, if the manufacturer is not making treatment or other clinical benefit-related claims (see below for evidence needed when making such claims).

Evidence Supporting Digital Therapeutic Claims

Digital therapeutics is an emerging area of health products. Although not currently a recognized term by health authorities, the concept of digital therapeutics has been defined by a key industry coalition as products that deliver

> evidence-based therapeutic interventions to patients that are driven by high quality software programs to prevent, manage, or treat a medical disorder or disease. They are used independently or in concert with medications, devices, or other therapies to optimize patient care and health outcomes.

Digital therapeutics have evolved from "digiceuticals" that focused more on consumer health claims such as exercise and weight loss rather than substantial clinical benefit claims.

In the context of combination products, a digital therapeutic would take the form of a digital support function that works together with the medical product to achieve a new treatment effect, indication, or other marketing claims. Consider for example the treatment of depression. A pharmaceutical manufacturer may currently offer treatment in the form of a pharmaceutical product which has been well characterized and shown to improve symptoms of depression. Knowing that patient counseling is also an accepted treatment for depression, the manufacturer may decide to develop a companion software application to provide pre-configured counseling exercises for patients. If the manufacturer can demonstrate that the combined use of the medicine and the software application realizes some additional benefit, in this example, consistently improved depression survey scores, they would be creating a new therapeutic claim. In the United States, while there is a limited official policy on what is, and what is not, a cross-labeled combination product, this example might be considered a combination product digital therapeutic.

There is currently no published guidance from health authorities or healthcare payer groups around the type of evidence required for digital therapeutic claims, especially when the product relies in part on a medicine to deliver effectiveness. For now, the industry has largely assumed that a drug level of evidence would be required if the digital therapeutic includes a medicine. This would include at least one, but generally two, adequate and well-controlled study demonstrating the agreed-upon clinical endpoints to claim superiority over the benefits of the use of the medicine or medical device in isolation. It is important to note, however, that discussion, and often negotiation, with health authorities is critical when designing methods to demonstrate safety and efficacy of digital therapeutic claims.

Real-World Evidence

Several health authorities have signaled that they are open to and interested in the concept of "real world evidence" to support medical product claims. The US FDA defines real-world evidence as "clinical evidence regarding the usage and potential benefits or risks of a medical product derived from analysis of real world data." Real-world data are sources such as electronic health records (EHRs), claims and billing activities, product and disease registries, and patient-generated data including in home-use settings. Importantly, this latter category (patient-generated data on home use) can be bolstered through the use of digital sources, such as connected medical devices, connected combination products, and software. Although some insights from real-world evidence can be planned for in advance, many will come up naturally through observing data collected using software and devices. In order to take advantage of insights that may present themselves, combination product manufacturers must take steps to collect data in a structured way that can provide convincing evidence that the findings are accurate and meaningful. While health authorities are very interested in real-world evidence to support future product claims, they will likely need to apply the traditional level of scrutiny to data integrity and trial design to grant those claims.

DELIVERING ON THE PROMISE OF DIGITAL HEALTH

As described within this chapter, the advent of digital technologies is already creating significant advances in the combination product space. Technologies such as connected devices, advanced data analytics, and machine learning hold promise not only to improve the way patients interact with their medical products but also to enable a more precise and personal approach to treatment. Although the development and use of these digital assets are in their infancy, and there are many unknowns around social and regulatory adoption, given the pace at which medicinal product and combination product manufacturers are moving to adopt these technologies, good collaboration between investigators, manufacturers, and health authorities will be essential. This will no doubt take innovative thinking and hard work on the parts of each of these stakeholders; however, the promise these technologies offer to patients and providers is well worth the effort.

REFERENCES

114th Congress. (2016). 21st Century Cures Act. H.R. 34. https://www.gpo.gov/fdsys/pkg/BILLS-114hr34enr/pdf/BILLS-114hr34enr.pdf. Accessed May 21, 2020.

Association for the Advancement of Medical Instrumentation. (2012). AAMI TIR45: 2012. *Guidance on the Use of AGILE Practices in the Development of Medical Device Software.*

Association for the Advancement of Medical Instrumentation. (2019). AAMI TIR102: 2019. *Technical Information Report U.S. FDA 21 CFR Mapping to the Applicable Regulatory Requirement References in ISO 13485:2016 Quality Management Systems.*

Digital Therapeutics Alliance. *What are Digital Therapeutics?* https://dtxalliance.org/dtx-solutions/. Accessed May 21, 2020.

European Commission. (2016). *Guidelines on the Qualification and Classification of Standalone Software Used in Healthcare Within the Regulatory Framework of Medical Devices.*

European Commission. (2017). *Council Regulation 2017/745 Concerning Medical Devices, OJ No L 117/1.* Article 10 – *General Obligations of Manufacturers.*

European Commission Medical Device Coordination Group. (2019). *Guidance on Qualification and Classification of Software in Regulation (EU) 2017/745 – MDR and Regulation (EU) 2017/746 – IVDR.*

European Commission Medical Device Coordination Group. (2021). *Infographic - Is Your Software a Medical Device?*

Federal Institute for Drug and Medical Devices. (2015). *Medical Apps – Guidance.*

Health Canada. (2019). *Software as a Medical Device (SaMD) – Draft Guidance Document.*

International Electrotechnical Commission. (2014). 60601-1-2:2014. *General Requirements for Basic Safety and Essential Performance – Collateral Standard: Electromagnetic Disturbances – Requirements and Tests.*

International Electrotechnical Commission. (2015a). 60601-1:2015. *Medical Electrical Equipment – Part 1: General Requirements for Basic Safety and Essential Performance.*

International Electrotechnical Commission. (2015b). 60601-1-11:2015. *General Requirements for Basic Safety and Essential Performance - Collateral Standard: Requirements for Medical Electrical Equipment and Medical Electrical Systems Used in the Home Healthcare Environment.*

International Electrotechnical Commission. (2015c). 62304-2006+A12015. *Medical Device Software – Software Life Cycle Processes.*

International Electrotechnical Commission. (2016). 82304-1-2016. *Health Software – Part 1: General Requirements for Product Safety.*

International Medical Device Regulators Forum. (2004). *"Software as a Medical Device": Possible Framework for Risk Categorization and Corresponding Considerations.*

International Standards Organization. (2015). 62366-1-2015. *Medical Devices – Part 1: Application of Usability Engineering to Medical Devices.*

International Standards Organization. (2016). 13485-2016. *Medical Devices – Quality Management Systems – Requirements for Regulatory Purposes.*

International Standards Organization. (2019). 14971-2019. *Medical Devices – Application of Risk Management to Medical Devices.*

Medicines and Healthcare Products Agency. (2022). *Medical Devices: Software Applications (Apps) – Guidance.*

Ministry of Health, Labor and Welfare. (2021). *Guideline of Determining Whether Software is Classified as a Medical Device.* (PSEHB/MDED Notification No.0331 1, PSEHB/CND Notification No. 033115).

NIH: Artificial Pancreas System Better Controls Blood Glucose Levels than Current Technology, October 16, 2019. https://www.nih.gov/news-events/news-releases/artificial-pancreas-system-better-controls-blood-glucose-levels-current-technology. Accessed May 22, 2020.

United States Federal Register. (2018). *Prescription Drug-Use-Related Software – Public Request for Comment.* Document Citation 83 FR 58574. Pages: 58574–58582. Docket No. FDA-2018-N-3017. Document Number: 2018-25206.

US Code of Federal Regulations. (2019a). 21 CFR Part 4a – Current. *Good Manufacturing Practice Requirements for Combination Products.*

US Code of Federal Regulations. (2019b). 21 CFR Part 820. *Quality System Regulation.*

US Food and Drug Administration. (1997). *Design Control Guidance for Medical Device Manufacturers – Guidance Document.*

US Food and Drug Administration. (2002). *General Principles of Software Validation – Guidance Document.*

US Food and Drug Administration. (2008). *FDA Update Transition to ISO 13485:2016.* https://www.fda.gov/media/123488/download. Accessed May 21, 2020.

US Food and Drug Administration. (2019a). *Clinical and Patient Decision Support Software – Guidance.*

US Food and Drug Administration. (2019b). *Developing a Software Precertification Program – A Working Model.* https://www.fda.gov/media/119722/download. Accessed May 21, 2020.

US Food and Drug Administration. (2019c). *General Wellness: Policy for Low Risk Devices– Guidance.*

US Food and Drug Administration. (2019d). *Medical Device Data Systems, Medical Image Storage Devices, and Medical Image Communications Devices – Guidance.*

US Food and Drug Administration. (2019e). *Policy for Device Software Functions and Mobile Medical Applications – Guidance.*

US Food and Drug Administration. (2019f). *Proposed Regulatory Framework for Modifications to Artificial Intelligence/Machine Learning (AI/ML) Based Software as a Medical Device (SaMD) - Discussion Paper and Request for Feedback – A Working Model.* https://www.fda.gov/medical-devices/software-medical-device-samd/artificial-intelligence-and-machine-learning-software-medical-device. Accessed May 21, 2020.

US Food and Drug Administration. (2020). *Multiple Function Device Products: Policy and Considerations – Guidance.*

US Food and Drug Administration. Real World Evidence. https://www.fda.gov/science-research/science-and-research-special-topics/real-world-evidence. Accessed May 22, 2020.

14 General Overview of Global Combination Product Regulatory Landscape

Stephanie Goebel, Vicky Verna, Cherry Marty, and Susan W. B. Neadle

CONTENTS

DOI: 10.1201/9781003300298-16

INTRODUCTION

In general, healthcare products are categorized as either medicinal products, to include drugs and biological products, or medical devices, including in vitro diagnostic devices (IVDs). Meanwhile, the evolution of science and technology has fostered the invention of innovative novel products using both categories simultaneously. However, innovations like these challenge the regulatory status quo and blur the regulatory lines. This has been the case for the technologies which gave birth to "combination products."

THE CONCEPT

The general concept of "combination product" has been adopted in many regions around the world to refer to medical products that consist of a combination of two or more regulated products such as a drug, biological product, or device. Classic examples of these products include drug-eluting stents, gene therapy systems, chemotherapeutic drugs combined with monoclonal antibodies, and even novel nanotechnology-based drug delivery systems. It should be noted, however, that the official regulatory definitions and terminologies can vary from one jurisdiction to another, resulting in some products being categorized differently in certain jurisdictions compared to others.

THE BENEFITS

In recent years, there is a marked public interest in combination products that is consistent with the observed market growth. This can be explained by the nature of combination products, which have greater therapeutic potential than their individual components. For instance, a drug-eluting stent prevents fibrosis that, together with clots, could otherwise lead to restenosis of the stented artery; or a condom (medical device) incorporated with a spermicidal drug will decrease the probability of contraception. Others, such as drugs combined with drug delivery devices can help control side effects through lower, consistent, and localized doses; support patient compliance; and improve access to difficult-to-reach target sites. In these cases, improved therapeutic effectiveness is anticipated in patient care. Moreover, when considering the aging and growing population globally which influences healthcare costs, these products have the potential to not only directly impact the quality of life, but also to decrease healthcare costs, by allowing for home-use or ease-of-use at point-of-care facilities. With such clinical potential and financial savings benefits, it is no wonder that the market is booming – projected at $177.7 billion by 2024 [1].

Need for Appropriate Controls

While the potential therapeutic advantages are evident, combination products raise a variety of regulatory and review challenges compared to individually categorized products, especially in regulatory jurisdictions outside of the leading regions of the US and EU. Though the regulatory controls under the main categories of drugs, devices, and biologics often share commonalities, the regulatory approach under each category can be very different due to the nature of the industries they oversee and issues raised by the different categories of products addressed. Therefore, when it comes to combination products, the separate regulatory worlds which differ in fundamental principles, approaches, and scientific bases need to be considered and reconciled. Regulatory expectations for each combination product, its constituent parts, and any issues that may be raised through their combined use need to be vetted through some application of regulatory requirements. And with the growing market, amendments and regulatory changes may be made as more jurisdictions consider how best to address the wide range of regulatory questions raised by this category of product.

GLOBALIZATION AND HARMONIZATION EFFORTS

In this section we will briefly discuss the global efforts currently taking place. It is important to note, while these efforts are addressing global regulatory challenges, they are often not specific to combination products. Instead, they cover requirements specific to medical devices (i.e., IMDRF) or medicinal products (i.e., ICH) which indirectly also influence and can help assure the safety and quality of combination products.

International Medical Device Regulators Forum (IMDRF): Medical Device-Related Considerations

A significant development to note is the establishment of the International Medical Device Regulators Forum (IMDRF) in October 2011, which built on the foundational work of the Global Harmonization Task Force on Medical Devices (GHTF) founded in 2007. With representatives from the medical device authorities of Australia, Brazil, Canada, China, European Union, Japan, Russia, Singapore, South Korea, and the United States, as well as the World Health Organization (WHO), the IMDRF provides guidance on strategies, policies, directions, membership, and activities of the Forum. GHTF/IMDRF materials may be helpful when addressing device constituent part considerations, albeit they do not go into how such issues should be addressed for the combination product as a whole, or at least not to combination product-level issues for drug/biologic-led combination products (see ICH, below, for drug/biologic constituent considerations).

Indeed, in the past, discussions over combination products were already recorded in 2007. These led to the creation of guidance documents which also reference combination products. The following GHTF/IMDRF guidelines are applicable for these types of products:

- GHTF/SG3/N17:2008 – Quality Management System – Medical Devices – Guidance on the Control of Products and Services Obtained from Suppliers
- IMDRF/MDSAP WG/N24 FINAL: 2015 – Medical Device Regulatory Audit Reports
- IMDRF/GRRP WG/N47 FINAL:2018 – Essential Principles of Safety and Performance of Medical Devices and IVD Medical Devices
  - Sections of note are the following:
    - Sections 5.13 and 6.1 define expectations for medical devices and in vitro diagnostics incorporating materials of biological origin, as well as administer medicinal products.
    - Section 6.5 states

> Where a medical device incorporates, as an integral part, a substance which, if used separately may be considered to be a medicinal product/drug as defined in the relevant legislation that applies in that Regulatory Authority and which is liable to act upon the body with action ancillary to that of the medical device, the safety and performance of the medical device as a whole should be verified, as well as the identity, safety, quality and efficacy of the substance in the specific combination product.

  - The document also indicates that the essential principles are not intended to provide definitions for combination products, since these definitions are not harmonized. Further the IMDRF reflects that how combination products are handled varies among different regulatory authorities.

Additionally, while the IMDRF's primary focus is on medical devices, it has also put together working groups (WG) to specifically work on combination product deliverables. With this aim, it assigned the task of developing a guideline on combination products to the "WG1 – Pre-Market" which resulted in the release of the "White Paper on summary of Combination Product Guideline" [2]. This activity, which seems to have stalled, was led by the Asian Harmonization Working Party (AHWP) in 2015 and was followed by the release of the draft "Guidance on Regulatory Practices for Combination Products" in 2016.

International Council for Harmonisation (ICH): Drug and Biological Products-Related Considerations

The same way that the IMDRF influences medical device requirements globally, the International Council for Harmonisation of Technical Requirements for Pharmaceuticals for Human Use (ICH) influences drug requirements, tracing back its origins to the 1980s. Since then, the ICH has gradually evolved, with the mission to achieve greater harmonization worldwide to ensure that safe, effective, and high-quality medicines are developed and registered in the most resource-efficient manner. The ICH now includes 20 members and 34 observers (status per January 2023).

The standards and guidelines released by the ICH have also greatly supported the setting of aligned quality and technical requirement expectation of drug/medicinal products. In the same direction as the IMDRF, this group is also working on projects which will improve the control over combination products especially drug-led combination products. In contrast to IMDRF/GHTF, ICH documents are finalized only if all member countries agree to implement them. The ICH group does not only continuously review the guidelines to ensure that they are up to date with latest technological requirements, but it also promotes the introduction of new tools which can help expedite the development and review of products. Therefore, future updates to better cater to combination product can be expected if sufficient consensus is obtained among all ICH members.

Other Harmonization Efforts

Another international standards organization is the ASTM International, formerly known as the American Society for Testing and Materials, which is an international standards organization that develops and publishes voluntary consensus technical standards for a wide range of materials, products, systems, and services. It has formed an ASTM E55 Committee to harmonize definitions for combination products.

Another entity which establishes foundations for global harmonization is the International Organization for Standardization (ISO), an independent, non-governmental international organization with a membership of 167 national standard bodies (status per January 2023). Though the ISO does not have developed standards especially for combination products, many drug and device manufacturers follow ISO standards for quality and safety for their systems, processes, and also products. There are a range of standards relevant to combination products; for example, some concerning delivery systems include:

- ISO 11040 – Prefilled syringes
- ISO 11608 – Needle-based injection systems for medical use
- ISO 20069 – Guidance for assessment and evaluation of changes to drug delivery systems
- ISO 20072 – Aerosol drug delivery device design verification – Requirements and test methods
- ISO 12417 – Cardiovascular implants and extracorporeal systems – Vascular device-drug combination products

Several countries or groups of countries are also finding reasons to align their regulatory control activities for mutual benefits. There are several joint programs initiated by the US and the EU. One noteworthy example is FDA's proposal to transition from the 21 CFR 820 Quality System Requirements (QSR) to align with ISO 13485:2016. Though a definitive statement of the timeline has not been made, this shift would be a significant harmonization initiative with widespread impact, including for combination products. It is worth noting, however, that the revisions to the ISO 13485:2016 are introducing several alignments with the FDA QSR 21 CFR Part 820. Thus, organizations which are already compliant with either set of regulations will have less burdensome regulatory adjustments once the transition takes place.

Meanwhile, another prime example of convergence is Health Canada's recent transition to the Medical Device Single Audit Program (MDSAP) on January 1, 2019, marking the success of such a jurisdiction-wide transition to the international regulatory auditing program. Furthermore, with Australia's recent adjustments toward aligning their device regulations with the European Union Medical Device Regulations (EU MDR), as well as acknowledging licenses and certificates from foreign authorities such as the United States Food and Drug Administration (US FDA) and MDSAP, it is evident that the Australian Therapeutic Goods Administration (TGA) is shifting toward global harmonization as well.

In the same line of thinking, FDA is also promoting "regulatory convergence" with its involvement in the Asia-Pacific Economic Cooperation Life Sciences Innovation Forum Regulatory Harmonization Steering Committee, The International Pharmaceutical Regulators Programme (IPRP), and the Pan American Network for Drug Regulatory Harmonization, to name a few. While these efforts are not combination product specific, it can be expected that they will influence the control of these products as well, especially but not limited to device-led combination products.

The US FDA also has two activities relating to post-market collaboration potential in the drug and device manufacturing inspectional space to work with foreign counterparts: the FDA-EU Mutual Reliance Initiative and the Medical Device Single Audit Program (MDSAP). FDA has spoken informally, e.g., at a RAPS/TOPRA virtual event on June 12, 2020, to the potential to reengage with counterparts on convergence possibilities for combination products [3]. FDA noted these two initiatives as examples of underlying medical product efforts that may signal that we are at a point where combination product convergence/coordination work might be tenable to start pursuing again. FDA has also spoken to the issue of figuring out what might be appropriate for regional/multilateral work on combination product issues, and the challenges of trying to converge on pre-market safety/efficacy data expectations as opposed to process, technical standards, and post-market issues such as inspection coordination [4]. Lastly, but not least, AAMI, via its Combination Products Committee, has done some useful work on CGMP (TIR 48) and, more recently, AAMI TIR 105 (Combination Products Risk Management, issued in October 2020). While national/US-centric, both speak to issues of potential value for global purposes and were seen to have value from that perspective among the member companies who participated in their development.

While the development of combination product regulatory frameworks on a global scale is not on the near-term horizon, there are globalization and harmonization efforts being undertaken which are shaping a positive outlook and that can help build a foundation for corresponding work with respect to combination products. Not many regulatory bodies classify combination products as a product category on its own; nevertheless, initiatives have been developing globally in order to address the growing need to clarify the application of the governing regulations which impact this category.

OBTAINING AND MAINTAINING MARKET ACCESS

For any regulatory jurisdiction, obtaining and maintaining market access defines the regulatory strategies that companies must develop throughout the lifecycle of their product from research, development, production, and pre- and post-market evaluation. The step-wise approach we illustrate below reflects the general steps that are similar across all regions, comprising milestones that will aid in defining the developmental and regulatory strategies throughout the product lifecycle. Figure 14.1 serves as a guide to the sections discussed under the jurisdictions covered in this chapter. However, definitions and classifications, especially for combination products, are among the factors that will determine how a company's application for one region will differ from another.

FIGURE 14.1 Step-wise approach to obtaining and maintaining market access.

The logical first step of classifying a combination product and defining the primary mode of action (PMOA) and secondary MOA (see section "United States of America") would generally determine the regulatory pathway to be considered and the regulatory authority (or authorities) to be approached by a manufacturer in the jurisdiction of application.

Generating the supporting evidence would then entail the design and development activities discussed in Figure 14.6, which will ensure that the combination product is safe, effective, and efficacious while also taking into consideration the manufacturing controls, labeling, and extent of clinical trials required in a particular jurisdiction.

Often, regulatory authorities allow for consultation meetings in order for developers to clarify their development plans, thus confirming the adequacy of the development strategy and its appropriate execution.

Marketing authorization applications may then require submission to the lead regulatory authority as defined by the PMOA. Only the US and China hold a designated coordination body. (Of note as an interesting example of the value and importance of engagement among regulators in this space, CFDA created this body after engaging with US FDA/OCP on the US approach to regulation of combination products [5]. AdvaMed and the Commerce Department coordinated this effort.) Additionally, for co-packaged combination products in kits or sets, the different components may sometimes require separate marketing approvals for certain jurisdictions.

Lastly, once a combination product is on the market, maintaining compliance usually requires adverse event monitoring and reporting, ensuring that manufacturers communicate systematically

TABLE 14.1

Overview of Regulation of Combination Products in International Jurisdictions. Adapted from AHWP (2016)

	Classifying a Combination Product		Generating Supporting Evidence		Confirming Adequacy & Executing the Development Strategy	Market Authorization Application Submission & Approval		Maintaining Compliance & Post-Market Requirements
	Formal Definition in Regulation	**Formal Status Determination Mechanism**	**Separate Coordination Body**	**Evaluation Process**	**Clinical Trials & Data Requirements**	**Manufacturing Controls**	**Labeling**	**Post-market Reporting**
US*	✓	✓	✓					
EU*	✗	✓	✗					
Australia	✗	✓	✗					
Canada	✓	✓	✗					
Japan	✗	✓	✗					
China	✓	✓	✓					
Singapore	✓	✓	✗					

(Continued)

TABLE 14.1 (CONTINUED)

Overview of Regulation of Combination Products in International Jurisdictions. Adapted from AHWP (2016)

	Classifying a Combination Product		Generating Supporting Evidence		Confirming Adequacy & Executing the Development Strategy	Market Authorization Application Submission & Approval		Maintaining Compliance & Post-Market Requirements
	Formal Definition in Regulation	**Formal Status Determination Mechanism**	**Separate Coordination Body**	**Evaluation Process**	**Clinical Trials & Data Requirements**	**Manufacturing Controls**	**Labeling**	**Post-market Reporting**
Hong Kong	✗	✗	✗	[PMOA]	[PMOA]	[PMOA]	[PMOA]	[PMOA]
South Korea	✓	✓	✗	[All components]	[PMOA]	[PMOA] [All components]	[All components] [Cross-labeling]	[PMOA]
Malaysia	✗	✓	✗	[Under development]	[PMOA]	[PMOA]	[Under development]	[PMOA]
Saudi Arabia	✓	✓	✓	[PMOA] [Under development]	[PMOA]	[PMOA]	[PMOA]	[PMOA]

✓ Yes

[PMOA] Regulations and/or guidance based on PMOA applied

✗ No

[All components] Regulations and/or guidance for all components applied

[Under development] Regulations and/or guidance currently under development

[Cross-labeling] Cross-labeling requirements for co-dependent products

**Applicable regulations and guidance are not all "based on PMOA." In some instances, regulations and guidance documents are specific to combination products (e.g., FDA cGMP for combination products). Some do specifically/expressly address combination products, but others that do not may still apply and be relevant to consider (e.g., guidance and regulations relating to drugs, devices, and biological products may be relevant and applicable to combination products regardless of their PMOA).*

with the regulatory bodies where the product is marketed, to confirm product safety and effectiveness throughout its lifecycle. These post-market requirements also include reporting significant changes which might affect the safety and/or effectiveness of the product resulting in the need to go through the flowchart steps with a new review by the agency: Post-market modifications might be a basis for going through the flowchart again!

Table 14.1 aligns our step-wise approach with data adapted from the AHWP "Guidance on Regulatory Practices for Combination Products" of 2016. Though several other countries and concepts listed are not discussed in detail in this chapter, this summary table sets the stage for our in-depth look into the regulatory developments under the jurisdictions of the US, EU, Australia, Canada, Japan, and China.

For each country or region presented in this chapter, we will introduce the jurisdictions by looking briefly into their respective regulatory histories, which often justify the similarities and differences of the regulatory approaches that exist between the regions. Comparative overviews of the evolving global requirements are included in Appendix 2 (8.2 "Global Regulatory Landscape Comparative Overview") as well as in Tables 14.6 and 14.7. There is more focus on the United States, as they have been leading in this space, since US Office of Combination Product was created in 2002, and they have produced multiple rules and guidance, unlike any other region.

COMPARATIVE OVERVIEW OF INDIVIDUAL REGIONS

United States of America (US)

Regulatory Evolution

Perhaps because the US can be considered the pioneer in establishing an extensive regulatory framework for combination products, it is no surprise that many regulatory regions are now following suit and thus also experiencing similar challenges that the US has faced throughout the last century.

In the US, the Food and Drug Administration (FDA)'s function began with the original Pure Food and Drug Act, also known as the Wiley Act of 1906. This act was then amended with the Federal Drug and Cosmetic Act (FDCA) of 1938 which established quality standards for food, drugs, medical devices, and cosmetics. In 1944, the Public Health Service Act, along with the FDCA Act, established regulatory requirements for biological products. In 1976 amendments made some devices subject to pre-market regulation. The Kefauver-Harris Amendments of 1962 added efficacy requirements (in addition to safety requirements of 1938) for drugs. Until 1976 amendments, FDA had no way to apply pre-market review to products classified as devices. As a result, products that met the device definition were classified as drugs where such review was considered appropriate. The FDCA has otherwise been amended numerous times to clarify and adjust regulatory requirements, programs, and duties for medical products.

Combination product classification was first established under the FDCA in 1990 with the goal of improving regulatory clarity, consistency, and efficiency. Associated classification regulations were finalized in 1991. Significant policy work on combination products also included regulatory agreements among the centers signed in 1991. Intercenter agreements came into play in an attempt to minimize repetitive and conflicting regulations between the centers when handling the applications for combination products [6]. During this time, in an attempt to increase monitoring and traceability, the FDA Medical Device Reporting (MDR) regulations and the 1990 Safe Medical Device Act (SMDA) federal legislation were enacted, which included provisions regarding center assignment for combination products based on PMOA and general expectations for regulation of combination products.

By the turn of the century, the growing number of increasingly innovative and complex combination products, time, and costly resources, along with the blurring responsibilities between the FDA centers, were continuously complicating the pre-market approval and post-market surveillance of combination products. It became apparent that the centers needed to begin standardizing their efforts to keep up with industry developments.

In 2002, the Medical Device User Fee and Modernization Act (MDUFMA) called for a formalization of the regulations for combination products [7] and a mandate for an office to ensure their timely, effective pre-market review and consistent and appropriate post-market regulation, which resulted in the establishment of the "Office of Combination Products" (OCP) to oversee and hence accelerate the process of reviewing, regulating, and releasing the combination products to patients and healthcare givers in a timely manner. A major wave of work began after the establishment of OCP (in accordance with further amendments made to the FDCA intended to help ensure effective coordination and oversight to support clarity, consistency, and efficiency goals), including issuance of guidance beginning in 2004 and establishment of new SOPs and systems to facilitate regulatory coordination.

The first combination product-specific rulemaking occurred in 2013, on CGMP, and the second in 2018 for PMSR. Over the period from 2005 to date, FDA has issued dozens of guidance (draft or final) expressly addressing combination products, many of which focus specifically on combination products. Activity has increased since 2016, as part of the Omnibus 21st Century Cures Act, to enhance efficiency and consistency of agency practices, supported by further statutory amendments to clarify regulatory expectations for combination products, largely intended to codify and ensure ongoing pursuit of policy and procedural approaches adopted by FDA to that date. These amendments essentially codified the current FDA policies and practices for the regulation of combination products, to ensure that the agency would stay on its current path of applying a risk-based approach to their regulation based on coordination between the center assigned the product and the other center(s) responsible for the regulation of the other type(s) of constituent parts included in the combination product. These amendments also clarified and/or expanded the duties of OCP to ensure that these statutory expectations would be implemented.

Along the way, and through the time of writing (December 2020), FDA has continued to release guidance documents for pre-market and for post-market regulation of combination products, including guidance on specific categories of products and on general regulatory principles, procedures, and expectations, illustrating the agency's continued efforts toward identifying and addressing industry's needs and concerns.

In the US alone, it has taken almost a century of continuous developments for FDA to establish and continue to develop the robust regulatory pathways for drugs, devices, and biologics that there are today. With the continued release of rules and guidance especially for combination products, manufacturers and stakeholders hope to better navigate through the necessary regulatory strategies. Pre-market guidance allows streamlining of efforts for pre-market applications, potentially shortening the FDA approval process and bringing innovative combination products to the clinics, caregivers, and patients faster than before. And post-market guidance documents further allow predictability for planning and further development of combination products after release to the market.

While Figures 14.2a and 14.2b give a general timeline of the history of regulatory development in the US for combination products, Figure 14.3 complements this timeline by listing additional examples of draft and final guidance documents that have been released. FDA has issued guidance and regulations on a wide range of pre-market and post-market considerations for combination products to support robust design and development; to address compilation of technical evidence for market authorization submissions; as well as to maintain compliance and assure safety, efficacy, and usability throughout the product lifecycle. There are various guidance documents that expressly address combination products but that are not exclusively focused on them. US FDA continues efforts to improve clarity and transparency related to combination product regulation and review. A more comprehensive list of combination product guidance documents can be found on the FDA website (https://www.fda.gov/combination-products/guidance-regulatory-information/combination-products-guidance-documents).

(a)

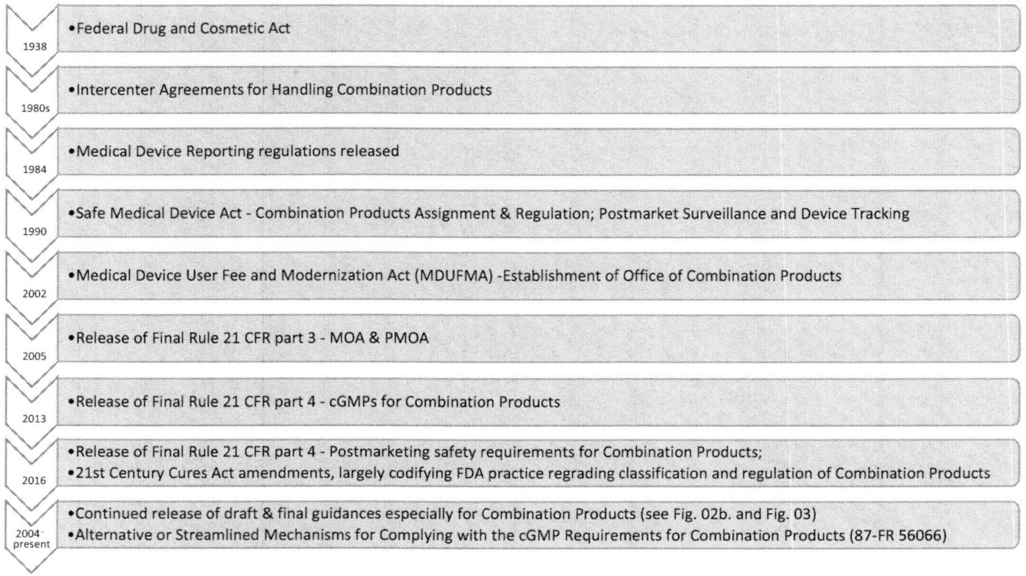

- 1938 • Federal Drug and Cosmetic Act
- 1980s • Intercenter Agreements for Handling Combination Products
- 1984 • Medical Device Reporting regulations released
- 1990 • Safe Medical Device Act - Combination Products Assignment & Regulation; Postmarket Surveillance and Device Tracking
- 2002 • Medical Device User Fee and Modernization Act (MDUFMA) -Establishment of Office of Combination Products
- 2005 • Release of Final Rule 21 CFR part 3 - MOA & PMOA
- 2013 • Release of Final Rule 21 CFR part 4 - cGMPs for Combination Products
- 2016 • Release of Final Rule 21 CFR part 4 - Postmarketing safety requirements for Combination Products; • 21st Century Cures Act amendments, largely codifying FDA practice regrading classification and regulation of Combination Products
- 2004-present • Continued release of draft & final guidances especially for Combination Products (see Fig. 02b. and Fig. 03) • Alternative or Streamlined Mechanisms for Complying with the cGMP Requirements for Combination Products (87-FR 56066)

(b)

2016	2017	2018	2019	2020	2021	2022
21st Century Cures Act (Section 3038) [8] Postmarketing Safety Reporting (Final Rule) [9] Human Factors Studies and Related Clinical Considerations (Draft Guidance) [10]	Current Good Manufacturing Practice Requirements (Final Guidance) [11]	Preparing a Pre-RFD (Final Guidance) [12] Inter-Center Consult Request Process (Staff Manual Guide) [13] Expectations and Procedures for Engagement on Regulations and Guidance (Staff Manual Guide) [14]	Principles of Premarket Pathways (Draft Guidance) [15] Postmarketing Safety Reporting (Final Guidance) [16] Inter-center Consults for Human Factors (Staff Manual Guide) [17]	Requesting FDA Feedback (Final Guidance) [18] Technical Considerations for Reliability of Emergency-Use Injectors (Draft Guidance) [19] Inspections of CDER-led or CDRH-led Combination Products (Compliance Program) [20]	Technical Considerations for Medical Devices with Physiologic Closed-Loop Control Technology: Draft Guidance for Industry and Food and Drug Administration Staff [21]	Principles of Premarket Pathways for Combination Products (Final Guidance) [23]

FIGURE 14.2 (a) Regulatory evolution of US combination product framework. (b) Combination product activities – Office of Combination Products.

Premarket
Provide additional clarity on use of regulatory pathways, efficiently interacting with FDA, and application review considerations
Availability/use of regulatory pathways for combination products • Premarket Pathways Draft Guidance (2019) • De Novo Classification Process Proposed Rule (2018) **Interacting with FDA on Combination products** • Requesting FDA Feedback Draft Guidance (2020) • Submission and Resolution of Formal Disputes Regarding the Timeliness of Premarket Review of a Combination Product (2004) **Challenging premarket regulatory areas** • Regulating Software for Use with Prescription Drug Products Federal Register Notice (2018) • Multi-function Device Products Draft Guidance (2020) **Human Factors** • Human Factors Intercenter Consults SMG (2019) • Contents of a Complete Submission Draft Guidance (2018) • Human Factors Draft Guidance (2016) • Safety Considerations for Product Design to Minimize Medication Errors Guidance for Industry Guidance (2016) **Product type-specific Guidance** • Emergency-use Injectors Draft Guidance(2020) • Transdermal and Topical Delivery Systems- Product Development and Quality Considerations (2019) • Evaluation of Devices Used with Regenerative Medicine Advanced Therapies (2019) • Glass Syringes for Delivering Drug and Biological Products: Technical Information to Supplement International Organization for Standardization (ISO) Standard 11040-4 (2013) • Technical Considerations for Pen, Jet, and Related Injectors Intended for Use with Drugs and Biological Products Guidance (2013) • New Contrast Imaging Indication Considerations for Devices and Approved Drug and Biological Products (2010) *[There are subsequent statutory amendments to the imaging guidance to enable addition of certain new uses to imaging device labels without having to add them to the drug label as well.]* • Evaluation of Devices Used With Regenerative Medicine Advanced Therapies – (2019) **Topic-specific Guidance** • Patient Labeling Draft Guidance (2019) • Selection of the Appropriate Package Type Terms and Recommendations for Labeling Injectable Medical Products Packaged in Multiple-Dose, Single-Dose, and Single-Patient-Use Containers for Human Use Guidance for Industry (2018) • Early Development Considerations for innovative Combination Products (2006) **Leveraging Data** • Bridging Draft Guidance (2019) • Comparative Analysis Draft Guidance (2017) **Application User Fees** • Application User Fees for Combination Products (2005)
Post-Market
Provide clarity on post-market expectations and how underlying requirements from drugs, devices, and biological products apply to combination products
• Postmarketing Safety Reporting (PMSR) Requirements - Final Guidance (2019) - Compliance Program (2019) - Final Rule (2016) - PMSR Webpage *Requirements in effect as of July 31, 2020 (vaccines January 2021)* • Current Good Manufacturing Practice Requirements (CGMPs) (2017) Compliance Program for CDER-led and CDRH-led Combination Products (2020) • Alternative or Streamlined Mechanisms for Complying With the Current Good Manufacturing Practice Requirements for Combination Products; List Under the 21st Century Cures Act (87 FR 56066) (2022) • Submissions for Postapproval Modifications to a Combination Product Approved Under a BLA, NDA, or PMA- Draft Guidance (2013)
Jurisdictional
Improvement and assessment of FDA internal processes related to combination products
• CDRH Planned Transition to ISO 13485 (Quality Management System) • PDUFA VI Assessment of Combination Product Review Practices in PDUFA VI (2020) [21] • Updates to inter-center consult process - Human Factors SMG (2019) - Inter-center Consult SMG (2018) - Coordination on Regulations and Guidance SMG (2018) • Combination product information in forms/checklists: Updates to 356h & 1571 forms, PMA/510(k) Acceptance/Filing Checklists • Preparing a Pre-RFD (Final Guidance)(2018) • Definition of the Term "Biological Product" (2020) [22] *(Clarifies definition of "protein" for purposes of biological product classification; helpful determination of drug v. device v. combination product.)* • FDA Regulation of Human Cells, Tissues, and Cellular and Tissue-Based Products (HCT/P's) Product List (2018) • Classification of Products as Drugs and Devices and Additional Product Classification Issues (2017) • How to Write a Request for Designation (RFD)(2011) Interpretation of the Term "Chemical Action" in the Definition of Device Under Section 201(h) of the FD&C Act (2011)

FIGURE 14.3 Examples of the wide range of pre- and post-market US FDA Draft and Final Guidance Documents and Regulations issued for Combination Products as of December 2020. [23]

Important Definitions

As indicated in Chapter 2 ("What is a combination product?"), the US FDA identified three categories of combination products: single entity, co-packaged, and cross-labeled. Table 14.2 reiterates definitions, respective regulation numbers, and examples for the different combination product configurations.

TABLE 14.2
Definition and Examples of Combination Products as per 21 CFR 3.2(e) (*See Also Chapter 2*).

Definition and Examples of Combination Products as per 21 CFR 3.2(e)		
Single-Entity CP **21 CFR 3.2(e)(1)**	**Co-Packaged CP** **21 CFR 3.2(e)(2)**	**Cross-labeled CP** **21 CFR 3.2(e)(3) or (4)**
Components are physically, chemically, or otherwise combined	*Components are packed together*	*Components are separately provided but specifically labeled for use together*
• Monoclonal antibody combined with a therapeutic drug • Device coated or impregnated with a drug or biologic (drug-eluting stent, pacing lead, transdermal patch) • Prefilled drug delivery system (syringes, insulin injector pen, metered dose inhaler)	• Drug or vaccine vial packaged with a delivery device • Surgical tray with surgical instruments, drapes, and anesthetic or antimicrobial wipes • First-aid kits containing devices (bandages, gauze), and drugs (antibiotic ointments, pain relievers)	• Photosensitizing drug and activating laser/ light source may comprise a cross-labeled combination product

As mentioned in Chapter 2, generally, the medical products that comprise a combination product are referred to as "constituent parts" (21 CFR §4.2). Of note, there is sometimes confusion over terminology: "constituent parts" versus "components" versus "accessories." Such terminology is currently the subject of harmonization efforts by the ASTM International E55 Committee developing standard definitions for combination products. Briefly, the following are working definitions, derived from a culmination of US FDA, World Health Organization (WHO), IMDRF, and EU definitions:

Constituent Part: The drug, device (including Software as a Medical Device) or biological product (including Human Cell Tissue/ Products) that is part of a combination product.
Component: Functional elements, formulations, and compositions, including raw material, substance, piece, part, software, firmware, labeling, or assembly intended to be included as part of the finished, packaged, and labeled device and/or ingredients intended for use in the manufacture of a drug product, including those that may not appear in such drug product (e.g., water, excipients). The individual components may be considered devices or drugs in and of themselves.
Accessory: An article which is intended by its manufacturer to be used together with one or several particular medical device(s) to specifically enable the medical device(s) to be used in accordance with its/their intended purpose(s) or to specifically and directly assist the medical functionality of the medical device(s) in terms of its/their intended purpose(s). Accessories intended specifically by manufacturers to be used together with a 'parent' medical device to enable that medical device to achieve its intended purpose should be subject to the same procedures as apply to the medical device itself. For example, an accessory will be classified as though it is a medical device in its own right. (This may result in the accessory having a different classification than the 'parent' device.) [24]

Generically, a mode of action (MOA) is the way a medical product achieves its intended effect(s) or action(s). As described in Chapter 2, a combination product is comprised of more than one medical product (constituent part), each of which has a mode of action. The primary mode of action (PMOA) is the single mode of action that makes the greatest contribution to the combination product's overall intended use(s). Correct PMOA designation is an important aspect of combination product development and product lifecycle. The correct designation of the PMOA is instrumental in defining a successful product development program as it will facilitate the management of applicable requirements and testing. For combination products, the PMOA is the basis for deciding the direction of the approval pathway. In many cases, researching precedents will help clarify the product jurisdiction. However, in some cases, especially with novel technologies, this can be challenging. The PMOA (21 CFR §3) and how it is determined is illustrated in the following flowchart (Figure 14.4). Once a PMOA is designated, an associated center is identified as the lead center and has primary jurisdiction for pre-market review and post-market regulation. It should be noted that while the provided chart

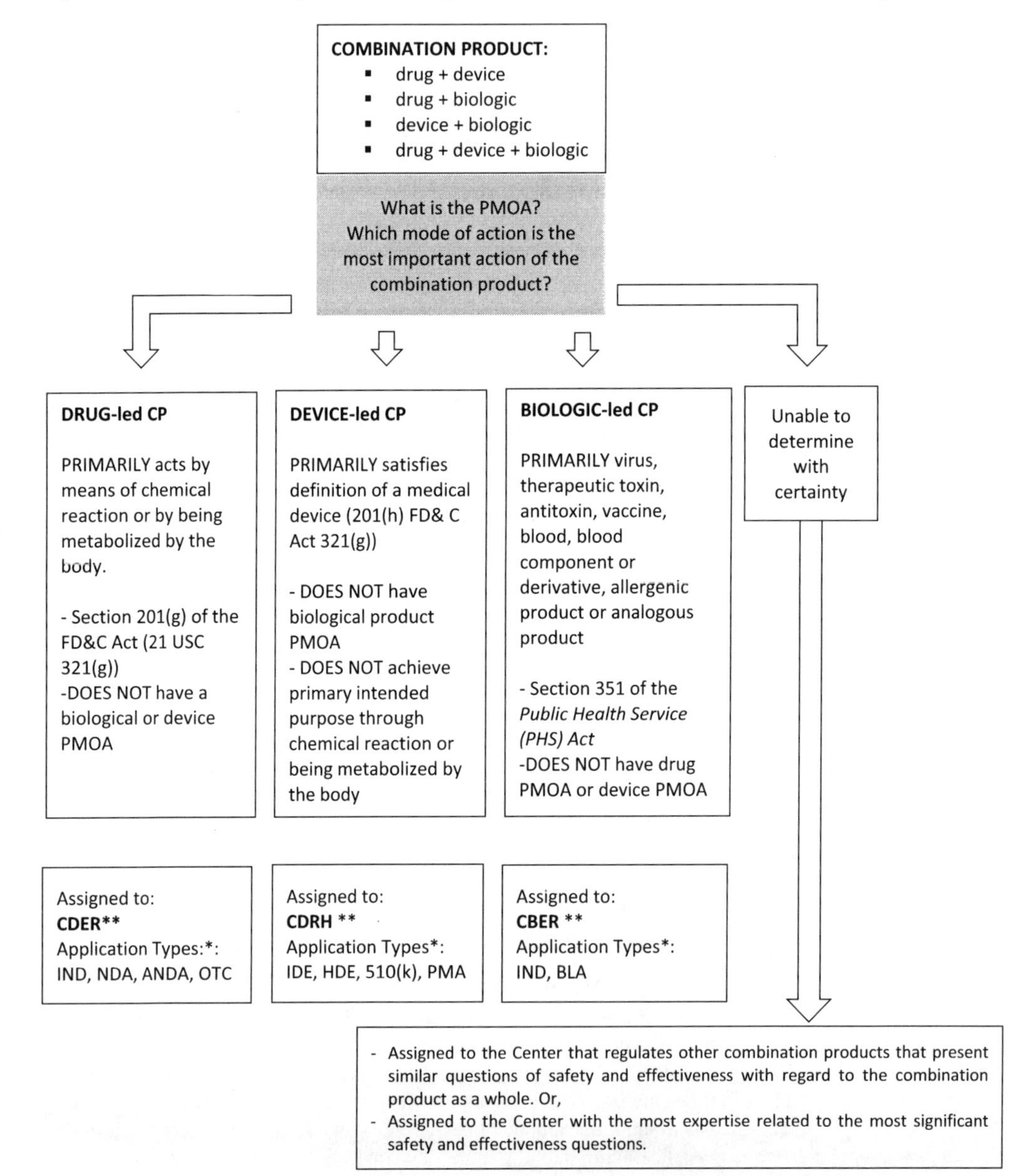

FIGURE 14.4 Flowchart for PMOA of CPs. *See acronyms at end of chapter and glossary for definitions. ***There are exceptions*. Not all drug PMOAs are assigned to CDER, nor all device PMOAs to CDRH, nor all biologic PMOAs to CBER (e.g., therapeutic proteins are assigned to CDER instead of CBER, so combination products made of a drug and such protein products would always be assigned to CDER regardless of PMOA; devices for use with blood and blood products are assigned to CBER, so a combination product comprised of such devices and biological products would always be assigned to CBER regardless of PMOA).

(Figure 14.4) describes the FDA's general approach for assigning combination products to FDA medical product centers, there are **exceptions** to the rule. Specifically, not all drug PMOAs are assigned to CDER, nor all device PMOAs to CDRH, nor all biologic PMOAs to CBER. For example, therapeutic proteins are assigned to CDER instead of CBER, so combination products made of a drug and such protein products would always be assigned to CDER regardless of PMOA. Similarly, devices for use with blood and blood products are assigned to CBER, so a combination product comprised of such devices and biological products would always be assigned to CBER regardless of PMOA.

Confirming the PMOA

In certain cases, determining the PMOA and identifying the applicable regulations influenced by the configuration can be clear based on the regulatory definition. However, in other cases such as those involving novel technologies, where data on relative contribution are limited, or the constituent parts have independent function rather than contributing to the same intended use, PMOA may not be possible to determine. In these situations, it may be in a firm's best interests to develop convincing scientific arguments to inform FDA's designation decision through the "Request for Designation" (RFD) process.

Submitting an RFD allows a company to obtain a formal FDA determination of a combination product's PMOA and subsequently be informed of the assigned lead center for the product's premarket review and regulation. In addition, companies and centers often informally request assistance from OCP in working out difficult jurisdictional issues not raised in an RFD submission. These are the situations where the importance of developing a collaborative relationship with OCP becomes instrumental as it is tasked to help answer any questions regarding combination product regulations, review, or product jurisdiction for FDA staff as well as industry.

The mission of OCP is to ensure the prompt assignment of combination products (drug-device, biologic-device, drug-biologic, or drug-device-biologic products) to FDA centers, the timely and effective pre-market review of such combination products, and consistent and appropriate postmarket regulation of these products.

Within the FDA structure, OCP resides in the Office of Clinical Policy & Programs (OCPP) in the Office of the Commissioner, which serves as the agency's focal point for special programs, developing and coordinating cross-center initiatives covering complex medical product policies [25].

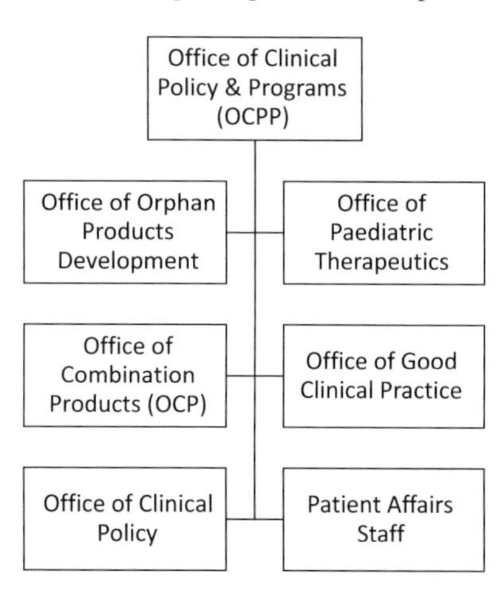

FIGURE 14.5 Current FDA organization chart under the new OCPP.

Defining What Is Effective and Efficacious

With the PMOA confirmed and the associated regulatory requirements identified, a firm will be in a better position to prepare a development strategy that will satisfy the agency's expectation on the type and level of evidence necessary to prove that the product is safe, effective, and efficacious.

A device's **effectiveness** as described in CFR 860.7(e)(1) is determined when "in a significant portion of the target population, the use of the device for its intended uses and conditions of use, when accompanied by adequate directions for use and warnings against unsafe use, will provide clinically significant results."

A drug's **efficacy** as first defined in the 1962 Drug Amendments refers to the "substantial evidence" that the drug "will have the effect it purports or is represented to have under the conditions of use prescribed, recommended, or suggested in the proposed labeling" [26].

The terms "effectiveness" and "efficacy" are often associated with the objectives of the evidential package for devices and drugs/biologics, respectively. However, as noted in the "Principles of Premarket Pathways for Combination Products" draft guidance document (February 2019) and consistent with the clarifications made by section 3038 of the 21st Century Cures Act,

> [r]egardless of which center may have the lead and which application type may be appropriate, consistent with section 503(g) of the FD&C Act, FDA is committed to applying a consistent, risk-based approach to address similar regulatory questions, including scientific questions, similarly, utilizing relevant expertise from the lead and consulted centers.

Essentially, FDA has said that similar data and information should be expected to address similar questions regardless of the center that has the lead for the combination product and that they apply a risk-based approach in accordance with the 21st Century Cures Act.

While there are fundamental differences in the steps involved in the process of developing a device in comparison to a drug, we align the concepts that should be considered for the complete combination product in Figure 14.6. Design control requirements apply to the design of the **entire combination product**. For drug/biologic-led, the drug/biologic may just be a design input since the goal generally is to make the device work with the already defined drug/biologic. For device-led combination products, the drug/biologic constituent part is likely being designed to some degree as part of the combination product design since no such drug/biologic product (same formulation, etc.) is already on the market, and combination product-specific considerations apply (e.g., about rate of release as a coating, risk of coating flaking off, etc.). These issues are discussed more fully in Chapters 4 (cGMPs) and 5 (Integrated Product Development). Regardless of whether the combination product is drug/biologics-led or device-led, design control and quality by design (QbD) principles, underpinned by robust risk management (including Human Factors), will generally provide the developmental framework.

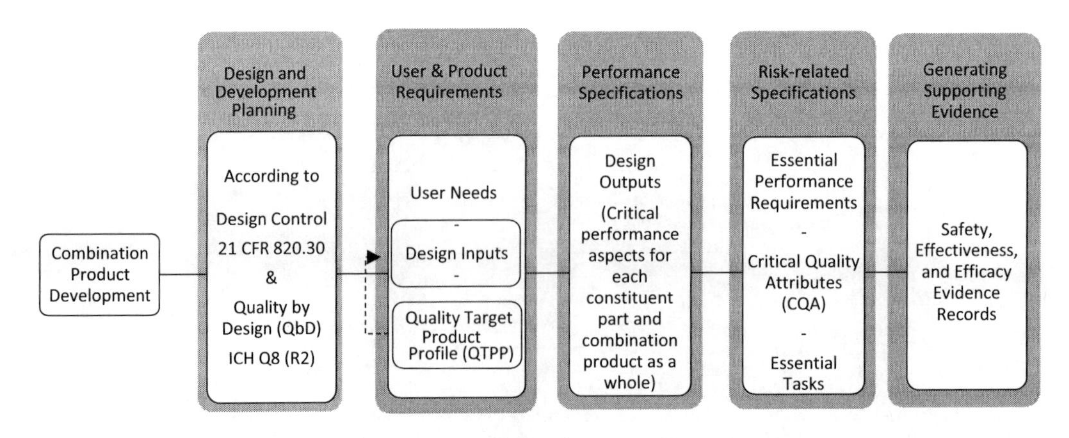

FIGURE 14.6 Aligning combination product component development.

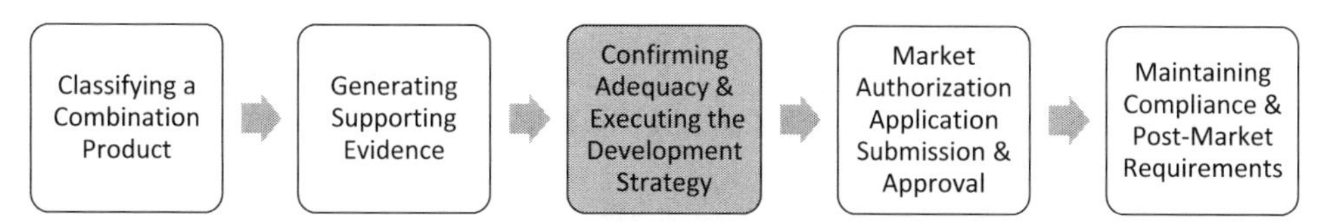

As the execution of development activities can require substantial resources and time, the importance of collaborating with the regulatory authorities at this stage cannot be overstated [see Chapter 4 (Combination Products Regulatory Strategy)]. To this end, the FDA has made available several means to open communication lines with product developers which include meetings. For combination products, these meetings are expected to be arranged with the lead centers designated according to the PMOA (see Figure 14.4). Taking advantage of these meetings is strongly recommended because they provide a great opportunity to confirm the adequacy of the overall approach to be undertaken to reduce the chances of failing to satisfy FDA's expectations.

Generating the necessary supporting evidence to prove the safety, efficacy, and effectiveness of a combination product requires an understanding of both the drug and development processes to cover the constituent parts. At a minimum, performing preliminary tests and animal studies is expected. However, for many cases, clinical studies involving human volunteers are required as well.

Pre/Non-clinical: Verification and Validation Tests

An important portion of the supporting evidence package consists of those data generated from the laboratory. Though the research and discovery phase of product development will be able to bench-test the safety and potential benefits of the product, animal studies may be necessary to better validate the developed acceptance criteria. There are also tests that can be performed at this stage to generate evidence of product efficacy and effectiveness without involving human subjects. In fact, there are possibilities to introduce medical devices and drugs to the US market through the 510(k) routes and abbreviated approval processes (i.e., ANDA), respectively, without having to perform rigorous clinical trials. These routes often rely on the ability to prove compliance with defined specifications or requirements set by the FDA (i.e., monograph, special controls) or equivalence with another approved product. The issue for product types subject to pre-market review is whether there is a sufficiently similar approved product already on the market to copy or bridge from. The more novel a combination product, the less likely that these opportunities will be available. The novelty may elevate the risk profile of the product or require additional evidence generation because the product is not sufficiently similar to an approved product already on the market to copy or bridge from. For example, bio-equivalence (BE) studies are often needed for generic drugs presented in a combination product form, and comparative human factors (HF) data in humans may also be needed if the device constituent used differs from the one used in the referenced innovator product; such differences may preclude the use of the ANDA pathway.

Clinical Trials

If bench tests are not sufficient to provide enough objective evidence of the efficacy and effectiveness of the combination product, clinical studies will be required. In these instances, the product is used on human subjects to confirm whether it is capable of reaching the treatment objectives while confirming the safety (see Chapter 6, Integrated CP Development). Combination Products human factors studies are also a related aspect to support design and development activities to prevent and minimize medication errors associated with the combination product's human interfaces (see Chapter 6.2, CP Human Factors).

A single application is generally appropriate for pre-market authorization, though separate applications for each constituent part may be acceptable for cross-labeled combination products, and use of master files and separate approval of platform technologies can be available and helpful. In the case of a single application, it is assigned to a lead center which becomes the sponsor's primary point of contact. The types of applications expected based on the PMOA are illustrated in Figure 14.4 and in Chapter 4, Regulatory Strategies.

Nonetheless, to assess combination products for approvability, FDA scrutinizes the information pertaining to each constituent part reviewed under separate sections which should be included in the application. The safety of the interaction between the separate constituent parts and the efficacy and effectiveness of their joint performance are evaluated as well, as part of the overall assessment. The reviews are performed by the centers responsible for the corresponding constituent part. The coordination between the activities of the involved parties is facilitated and managed as needed by OCP, though in most cases, combination product review does not demand OCP engagement. If the agency collectively determines that the benefits of the combination product outweigh the risks for the intended use, FDA will approve the application.

The regulatory obligations do not end with the FDA approval. Instead the combination product is expected to remain in compliance throughout its lifespan [see Chapter 4 (cGMPs) and the section "Inspection Readiness Considerations" in Chapter 9A)].

Pre-dating issuance of 21 CFR Part 4, separate cGMP requirements were in place specifically for medical devices (21 CFR 820 Quality System Regulations), drugs (21 CFR 210/211), biological products (21 CFR 600-680), and blood and HCT/Ps (21 CFR 1271). Prior to the issuance of 21 CFR Part 4 (Combination Product cGMPs), FDA's stated view was that the regulations applicable to each of a combination product's constituent parts needed to be addressed by combination product manufacturers. Recognizing that there are some commonalities between the product-type-specific cGMPs, FDA indicated streamlining could be utilized, while ensuring unique elements of the constituent part cGMPs were addressed. This perspective was reflected in draft guidance before the rulemaking.

Considering the need for guidelines specific to combination product cGMPs and the overlap that exists between those for drugs and devices, FDA released 21 CFR Part 4, subpart A. The rule codified FDA thinking on how to demonstrate compliance with both drug and device CGMP requirements, as well as those for biological products and human cellular and tissue products. The rule permits a "streamlined approach" to drug and device CGMP compliance, while calling out provisions applicable to each of the constituent parts that must be specifically addressed (see Chapter 5, cGMPs, for a detailed discussion on this topic).

When suppliers are involved, sponsors also need to be concerned with the compliance and quality of the constituent parts of the combination product and associated development and lifecycle services purchased. This means that adherence to 21 CFR Part 210 and 211 for the manufacturing of drug components and to 21 CFR Part 820 for device components needs to be assured (see Chapters 8, Lifecycle Management, and 10, Supplier Quality Considerations).

In addition to ensuring that products are manufactured in compliance with cGMP requirements, as part of the product's lifecycle management (Chapter 8), safety information is required to be monitored including from user feedback submitted to sponsors by the public. Specifically, a post-market monitoring report is expected to be provided by sponsors to the leading center which makes its review. While some aspects of the requirements are delineated in the guidance on cGMP for combination products, the final rule for Post-marketing Safety Reporting (PMSR) (21 CFR Part 4B) and guidance for the PMSR for combination products (released in December 2016 and July 2019, respectively) cover the expectations for adverse event reporting for combination products (see Chapter 8, Safety Assurance and Post-marketing Safety Reporting).

European Union (EU)

Regulatory Evolution

Whereas the current US regulatory framework has continued evolving throughout the last century, Europe's regulatory framework is comparably in its infancy stage. This is due to the history of the European Union, as it was only after the Maastricht Treaty of 1992 that a European Union – at this time still called European Economic Union – began existing in the geo-political and economic form it is today. Though the first Council Directive on medicinal products (Directive 65/65/EEC) was already released in 1965, the Council Directive concerning medical devices (MDD, Directive 93/42/EEC), on the other hand, was not released before 1993 in order to harmonize the applicable requirements for medical devices in what was then a grown EU (from six to twelve member states); however, the Directive on Active Implantable Medical Devices (AIMDD, Directive 90/385/EEC) had already been in force since 1990, while Directive on in vitro diagnostic medical devices (IVDD, Directive 98/79/EC) followed in 1998. After the MDD's release, based on the legal status of a Directive, member states of the EU had to implement MDD in national law and were by this enabled to define how to achieve the goals stated in MDD.[2] Legislation evolved for medicinal products as well as for medical devices, resulting in the currently applicable Directive relating to medicinal products for human use (MPD, Directive 2001/83/EC) as well as the Regulation (EU) on medical devices (MDR, Regulation 2017/745); MDR replaced MDD as well as AIMDD per May 26, 2021, and Regulation (EU) on in vitro diagnostic medical devices (IVDR, Regulation 2017/746) replaced IVDD per May 26 2022 with staggered transition periods[3]. A regulation in contrast to a directive must be applied as binding legislative act throughout the EU.

Despite the fact that there has never been and also still is no separate Directive or Regulation for combination products in the EU, the lawmakers have always strived to provide clarity under the existing legislative acts for medicinal products and for medical devices.

Accordingly, when MDD/AIMDD/IVDD and MPD were applicable, a combination product was either regulated as a medical device/active implantable medical device or as a medicinal product. To that effect, a so-called borderline document (Manual on Borderline and Classification in the Community Framework for Medical Devices) was developed to classify cases where it is unclear if the product is primarily a medical device, an in vitro diagnostic medical device, an active implantable medical device, a medicinal product, a biocidal product, or a cosmetic product. Though cases were also addressed in MEDDEV documents (MEDDEV = MEDical DEVIces Documents which state(d) guidance documents under MDD/AIMDD/IVDD[4]), MEDDEV 2.1/1 [27], MEDDEV 2.1/3 [28], and MEDDEV 2.14/1 (borderline and classification issue: A guide for manufacturers and notified bodies) [29], the guidance and approaches given, however, did not necessarily address the potential confusion brought about by (innovative) combination products that did not clearly fall in any regulated product category.

The MDD already included requirements on combination products: on the one hand, on combinations of medical devices that act ancillary to and are an integral part of the medicinal products – which were regulated as medicinal product. And on the other hand, on combinations of medicinal products/substances that act ancillary to the medical device – which were regulated as medical devices. Such combinations regulated as medical devices were assigned to the highest (risk) class (class III, according to rule 13, MDD) and a consultation process between the manufacturer's notified body and one of the National Competent Authorities for medicinal products or EMA was required [Ref MEDDEV 2.1/3]. However, with regard to devices ancillary and integral to medicinal products, the requirements stated were to ensure compliance with the "relevant essential requirements" of Annex I of the MDD related to the device's features on safety and performance (Article 1(3), 2nd sub-paragraph). Since there were neither in the MPD nor in any EMA guidelines details on compliance with Annex I, MDD, there were inconsistent approaches to compliance and check of compliance, both in industry and at medicinal product regulatory authorities.

And similar to the US, the constant developments in industry toward more software depending, connected and complex products as well as regulatory challenges on ensuring reliable and safe products like those involving but not limited to combination products, contributed to the need to adapt and amend the MDD in the EU. In 2010, a European company acquired EC certificate/CE mark for breast implants and was later found to be fraudulently using industrial-grade silicone gel that was more likely to rupture or leak than "state of the art" silicone used in such implants [27]. This situation caused a scandal which sparked criticisms toward the seemingly lax regulatory framework that allowed such a product to be marketed. In response, the European Commission initiated a movement to tighten the regulations to include among other measures heightened requirements for products, manufacturers and notified bodies, vigilance systems, post-market surveillance, and product traceability.

In 2017, this new Regulation (EU) 2017/745 (MDR) was introduced and is applicable since May 26, 2021 – the initially stated date of May 26, 2020, was replaced due to the COVID-19 situation and respective pandemic crisis.[5] The largest shift is seen as the new MDR adopts a lifecycle approach, in terms of increased requirements on the manufacturer's quality management system, e.g., to cover a proactive post-market surveillance system including post-market clinical follow-up as well as clinical evaluation. (Of note, at industry conferences, a topic of frequent discussion surrounds what is a sufficient amount of clinical evidence, i.e., sufficient clinical data and clinical evaluation results to demonstrate the device's safety and clinical benefits (Article 2(51), MDR), especially for innovative and legacy products. It is a question that often comes up for combination products now, too – as clinical evaluation and the respective clinical evaluation documentation (plan and report) are mandatory for manufacturers of medical devices (as stated in Article 10, 61 and Annex XIV, MDR) and are therefore not needed for (single) integral devices of medicinal products (e.g., not reusable, prefilled syringes) that only have to apply Annex I, MDR. However, as Annex I also covers clinical aspects, e.g., in General Safety and Performance Requirement (GSPR) 1, there are different views on the need for clinical evaluation reports – some notified bodies regard them as mandatory independent of any clinical related safety and/or performance claims of the device, while industry tendency is toward clinical data being only needed in case such claims exist. It is a topic that still needs to be addressed.) With the demographic changes in the EU since the implementation of the MDD in the 1990s, the lifecycle approach for medical devices caters well to the growing elderly European population, who would be using these products longer and potentially more frequently anyway. Not only is the lifecycle approach beneficial to consumers, it also promotes a more systematic post-market surveillance plan, if not more responsibility from the manufacturers, throughout the product's lifecycle. Another aspect of MDR is Article 117, which amends Annex I in the MPD, point 12 of section 3.2., and which provides more details on compliance to the GSPR (Annex I, MDR) than existing for the Annex I compliance under MDD. This Article 117 reflects and recognizes the need for combination products that are combinations of medical devices that act ancillary to and are an integral part of the medicinal products, to address the considerations for each of their parts, i.e., the medical device part as well as the medicinal product part, and the product as a whole. This will be discussed more detailed below (also see Chapter 2, "What Is a Combination Product?").

Following the transition to MDR, a new European database on medical devices (Eudamed) is scheduled to go fully online by May 2022,[6] a year after the EU MDR set in and aligned with the application date of the IVDR, providing a publicly accessible interoperable collaborative registration, notification, and dissemination system [28]. What used to be information restricted to the National Competent Authorities (NCA) is to be a database that is accessible to consumers (at least partly), regulators, and manufacturers alike. In this regard, the Medical Device Coordination Group (MDCG) has released guidance [29] to help all stakeholders adequately prepare for their data submissions once the platform goes online. An obvious means to increase traceability and transparency across the EU, this platform intends to reduce the probability of another scandal involving unsafe medical devices.

Since the introduction of the MDR, EMA and medicinal product's NCAs have new as well as updated responsibilities with regard to medical devices that form an integral product with a medicinal product/substance. Guidance and Questions & Answers (Q&A) documents were issued by regulatory authorities to aid in a smooth transition. In May of 2019 "Guidelines on the quality requirements for drug-device combinations" [30] was released having been developed by the EMA Committee of Human Medicinal Products for Human Use (CHMP) Quality and Biologics Working Parties, where definitions are laid out as well as requirements for the respective combination of medicinal products and medical devices (integral medicinal products, medicinal products with co-packaged medical devices, medicinal products with referenced medical devices); this Guideline came into effect per January 1, 2022 – the final name being "Guideline on quality documentation for medicinal products when used with a medical device" (EMA/CHMP/QWP/BWP/259165/2019). In October of 2019, the EMA and Coordination Group for Mutual Recognition and Decentralized Procedures – human (CMDh), representing national competent authorities, also released a first update to a Q&A document entitled "Questions and Answers on Implementation of the Medical Devices and In Vitro Diagnostic Medical Device Regulations ((EU) 2017/745 and (EU) 2017/746)" (EMA/37991/2019) which focused on so-called integral drug-device combinations, medicinal products that contain a co-packaged medical device, as well as the consultation procedure for ancillary medicinal substances in medical devices – Revision 2 is into effect since June 23, 2021. MDCG has issued MDCG 2021-24 on the classification of medical devices which supports the classification of medical devices that administer or incorporate medicinal substances.

In addition, industry working groups have been active in writing white papers in hopes of elucidating aspects of combination product requirements, such as definitions, co-packaged product expectations, and interpretation of requirements around significant changes. EMA hosted a virtual multi-stakeholder workshop to support the implementation of Article 117 of the MDR on November 27, 2020. The workshop included discussion and exchange of views and experience based on practical examples between European Union regulators, the European Commission, notified bodies, the pharmaceutical industry, and medical device manufacturers. During the workshop, notified bodies shared lessons learned per that date, and their perspectives (reflecting the medical device perspective) on lifecycle management and "substantial design changes." Subsequent to the workshop, Team NB's[7] working group on Article 117 issued a position paper on the interpretation of device-related changes in relation to a Notified Body opinion ("Position Paper for the Interpretation of Device Related Changes in Relation to a Notified Body Opinion as Required under Article 117 of Medical Device Regulation (EU)2017/745") [31]. Meanwhile, Team NB's working group on Article 117 has issued a further position paper on "Proposal for a Notified Body Opinion Template" to achieve a harmonized reporting approach.[8] Industry eagerly awaits further clarification on the requirements and authority/EU Commission expectations for drug-device combinations.

With all these current developments taking place in the EU, the regulatory landscape is slowly making headway toward a more robust regulatory framework. Below, we illustrate the discussed EU regulatory history and the important milestones reached since the Maastricht Treaty of 1992 (Figure 14.7).

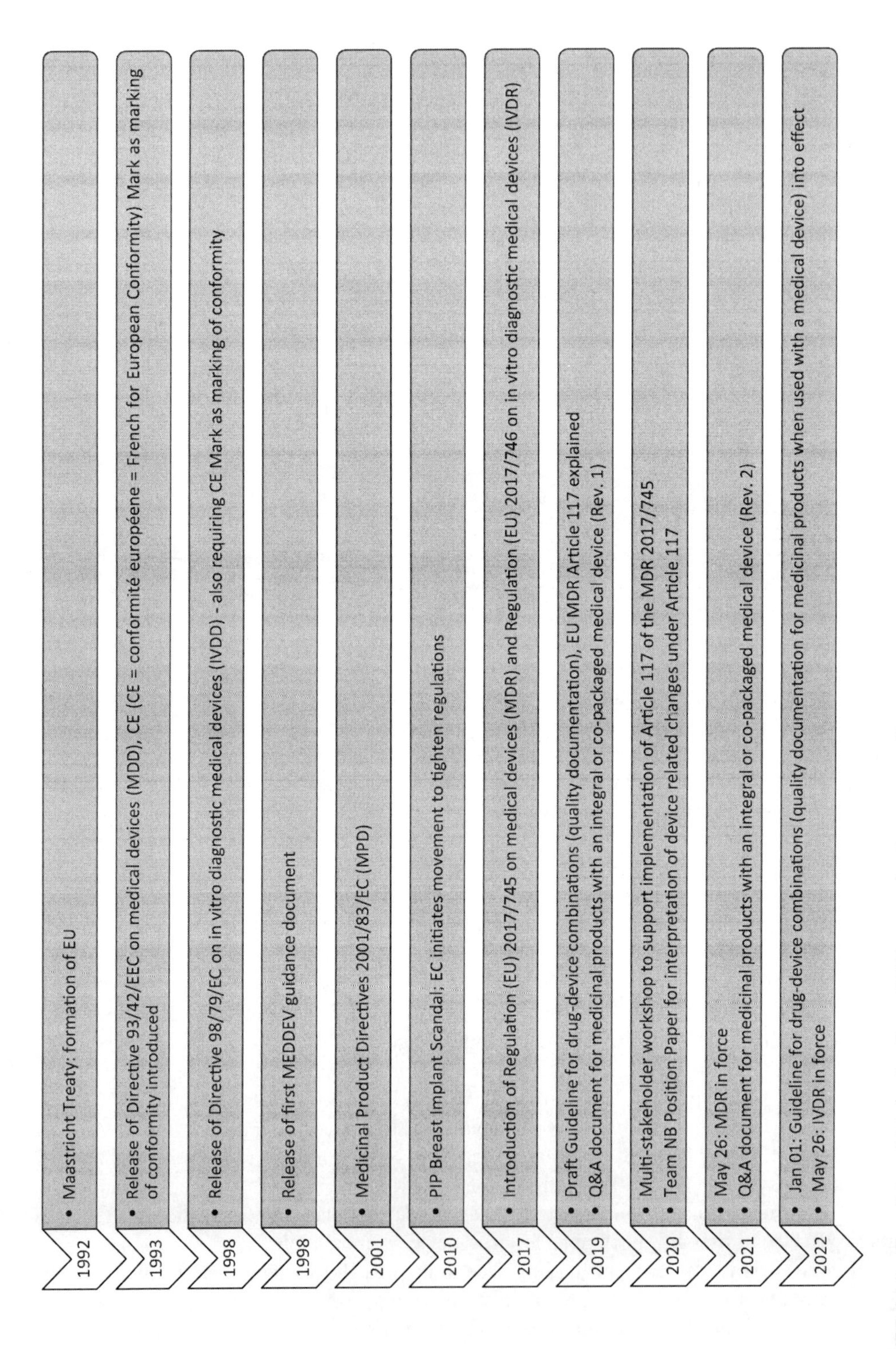

FIGURE 14.7 Regulatory history timeline of EU – covering important milestones.

EU's "Combination Product" Approach

Although the US-coined term "combination product" is officially defined neither in the MDR nor in the MPD, it is applied and in consequence interpreted differently, e.g., within EMA's Guideline (EMA/CHMP/QWP/BWP/259165/2019) (see Chapter 2, "What Is a Combination Product?"). It refers to (1) integral or (2) non-integral, i.e., co-packaged or referenced products (drug-device combinations (DDC)) or 3) a medical device with an ancillary medicinal product. All these terms refer to the concept of two categories of health products (category 1: medical devices, category 2: medicinal product) that are combined to provide treatment benefits to patients.

The control of such products – the leading legislative act – is also primarily governed by their principal mode of action (MOA) (of note – this term corresponds to the term "primary mode of action" in the US), which dictates whether a product falls under MPD or MDR (or formerly applicable MDD) and accordingly also which regulatory authorities are expected to be involved in the approval and oversight process [28] (see Table 14.3).

TABLE 14.3
Definition of Medicinal Product and Medical Device as Stated in Applicable EU Legislative Acts

Medicinal Product Directive (MPD) (Directive 2001/83/EC Relating to Medicinal Products for Human Use)	Medical Device Regulation (MDR) (Regulation (EU) 2017/745 on Medical Devices)
Article 1(2): Medicinal Product means (a) Any substance or combination of substances presented as having properties for treating or preventing disease in human beings; or (b) Any substance or combination of substances which may be used in or administered to human beings either with a view to restoring, correcting, or modifying physiological functions by exerting a pharmacological, immunological, or metabolic action or to making a medical diagnosis.	Article 2(1): Medical Device means Any instrument, apparatus, appliance, software, implant, reagent, material, or other article intended by the manufacturer to be used, alone or in combination, for human beings for one or more of the following specific medical purposes: – Diagnosis, prevention, monitoring, prediction, prognosis, treatment or alleviation of disease; – Diagnosis, monitoring, treatment, alleviation of, or compensation for, an injury or disability; – Investigation, replacement, or modification of the anatomy or of a physiological or pathological process or state; – Providing information by means of in vitro examination of specimens derived from the human body, including organ, blood, and tissue donations; and which **does not achieve its principal intended action by pharmacological, immunological, or metabolic means, in or on the human body,** but which may be assisted in its function by such means.

According to the following EU regulatory definitions, the intended principal MOA is determined based on that part of the combination product that is intended to provide the product's most decisive medical effects:

If the principal intended MOA of the combination product is achieved by the medicinal product, the entire product is regulated as a medicinal product under MPD and its market authorization is defined in Regulation (EC) No 726/2004 (Regulation on procedures for the authorization and supervision of medicinal products for human and veterinary use and establishing a European Medicines Agency (EMA)). In the case where the principal intended MOA is achieved by the medical device, the entire product will fall under the MDR.

The main focus related to combination products has so far been on the definition and regulation of **"integral medicinal products"** which refer to combination products often classified as "drug-led single-entity combination products" in the US (see Table 14.4). The terms used in MDR are (a) **"integral product"** and (b) **"single integral product"** and define the following combination products which are drug-led and therefore regulated as medicinal products:

> a) Any device which, when placed on the market or put into service, incorporates, as an integral part, a substance which, if used separately, would be considered to be a medicinal product … the action of that substance is principal and not ancillary to that of the device.
>
> **(MDR, Article 1(8), 2nd sub-paragraph)**

> b) Device intended to administer a medicinal product and the medicinal product are placed on the market in such a way that they form a single integral product which is intended exclusively for use in the given combination and which is not reusable.
>
> **(MDR, Article 1(9), 2nd sub-paragraph)**

Moreover, EMA's Guideline (EMA/CHMP/QWP/BWP/259165/2019) on quality documentation for medicinal products when used with a medical device has introduced the following definition of "integral product" which is aligned with MDR's terminology (see Table 14.4).

Additionally, this EMA Guideline defines "non-integral DDCs" as either co-packaged or referenced combination products for which MPD remains the leading legislative, but the medical devices co-packaged or referenced are regulated separately.

For cases where the distinction of the medicinal product from the device intended purpose is not clear and in consequence the above-mentioned principal intended MOA is not clear, the EU has developed guidance documents for these products termed as borderline products referring to "complex healthcare products for which there is uncertainty over which regulatory framework applies."

MDCG 2022-5[9] (Guidance on borderline between medical devices and medicinal products under Regulation (EU) 2017/745) was issued in April 2022 and contains clarifications on definitions used to classify the product, e.g., principal mode of action, as well as decision trees that support the product-per-product decision to decide if the respective product is a medical device or a medicinal product.

In addition, "Manual on borderline and classification for medical devices under Regulation (EU) 2017/745 on medical devices and Regulation (EU) 2017/746 on in vitro diagnostic medical devices"[10] was published in September 2022; it records the agreements reached by the member state members of the Borderline and Classification Working Group (BCWG).

If the applicable "borderline" guidance document does not cover a particular product, there is an option for applicants who are unclear on the correct classification to consult an NCA and provide information on the product's composition and constituents, a scientific explanation of the MOA, and its intended purpose. In addition, EMA's Innovation Task Force provides advice to medicine developers on eligibility to EMA procedures relating to the research and development of borderline products [28]. In case no conclusion and/or agreement between applicants and the NCA or further involved authorities is found, the binding decision can only be given by the European Court of Justice ("the Court").[11]

Classification with regard to Medicinal Product Legislation

With the entry into force of MDR, an amendment has also been made to MPD: Article 117 addresses the integral products (DDCs)) that fall under MDR's second sub-paragraph of Article 1(8) or the

TABLE 14.4
Integral and Non-integral (Co-packaged and Referenced) Products as Defined in the EU[1] [32]

Integral Products	Non-integral (Co-packaged) Products	Non-integral (Referenced) Products
A medical device (part) that falls under the second sub-paragraph of Article 1(8) or the second sub-paragraph of Article 1(9) of Regulation (EU) 2017/745, where the action of the medicinal product is principal: **(1)** Devices that when placed on the market **incorporate, as an integral part, a substance** that, if used separately, would be considered as a medicinal product and has an action that is principal and not ancillary to the action of the device (second sub-paragraph of Article 1(8)). Example: **Medicinal products with an embedded sensor** where the sensor is a medical device and its action is ancillary to the medicinal product. **(2)** Devices **intended to administer a medicinal product**, where the device and the medicinal product are placed on the market in such a way that they form a single integral product intended exclusively for use in the given combination and which is not reusable (second sub-paragraph of Article 1(9)). Typically, these devices have measuring or delivery functions. Examples: • Single-use prefilled syringes, single-use prefilled pens, and single-use prefilled injectors (including autoinjectors) used for the delivery of one or more doses of medicine and which are not intended to be re-used or refilled once the initial doses provided are exhausted. • Dry powder inhalers and pressurized metered dose inhalers that are preassembled with the medicinal product and ready for use with single or multiple doses but cannot be refilled when all doses are exhausted.	A medicinal product and a medical device **are packed together into a single pack** (e.g., carton), which **is placed on the market by the Marketing Authorization Holder** (MAH). Examples: • Oral administration devices (e.g., spoons, syringes). • Injection needles. • Refillable/reusable (e.g., using cartridges) pens and injectors (including autoinjectors). • Refillable/reusable dry powder inhalers and metered dose inhalers; spacers for inhalation sprays. • Nebulizers and vaporizers. • Single-use or reusable pumps for medicinal product delivery.	The product information (SmPC and/or package leaflet) of the medicinal product **refers to a specific medical device** to be used (e.g., identified by its brand name and/or specific description), **and the specified medical device is obtained separately by the user of the medicinal product**. Examples: • Oral administration devices (e.g., spoons, syringes). • Injection needles. • Refillable/reusable (e.g., using cartridges) pens and injectors (including autoinjectors). • Refillable/reusable dry powder inhalers and metered dose inhalers; spacers for inhalation sprays. • Nebulizers and vaporizers. • Single-use or reusable pumps for medicinal product delivery.

second sub-paragraph of Article 1(9), particularly devices used to administer medicinal products. These integral products (DDCs) will now require marketing authorization dossiers to include evidence on the device part's conformity with the relevant GSPR stated in MDR, Annex I, as either EU declaration of conformity (EU DoC), or EU Certificate, or opinion issued by NBs (so-called NB Opinion).

These three types of evidence require a closer look: only in case the device part is by itself a "stand-alone" medical device according to MDR (i.e., MDR is fully applicable), it is possible to provide either the manufacturer's EU DoC or the EU Certificate. In this context, the medical device classification is important: according to MDR's Recital (60) and Article 52 (in specific section 7), manufacturers of class I medical devices issue and possess an EU DoC, but *do not possess* a certificate of a notified body; a notified body is only involved related to medical devices offering a (risk) class higher than class I; that means being class I m (measuring function), or I s (sterile), or I r (reusable surgical instrument),[12] or IIa, or IIb, or III.

This MDR classification concept and in consequence the class-depending involvement of a notified body is also applicable for the NB Opinion. In case of device parts being no "stand-alone" medical device according to MDR, Article 117 defines the need for an NB Opinion "where for the conformity assessment of the device, if used separately, the involvement of a notified body is required in accordance with Regulation (EU) 2017/745"; this means an NB Opinion is only applicable for device parts classified as class I m, I s, IIa, IIb, or III.

Consequently, pharmaceutical manufacturers must apply the classification rules as stated in Annex VIII, MDR for the device part of their integral product to determine the need for notified body involvement; the intended purpose statement as well as the functional descriptions of the device parts should be considered to perform this "virtual" classification. As the device part is not treated as a "stand-alone" medical device discussed above, such a classification for the device part may be called a "virtual" one, as it is only needed to determine if a notified body needs to be involved (or not). Guidance on medical device classification is described in MDCG 2021-24.[13]

Regarding this demarcation for the need of a NB Opinion, device parts which are "virtually" classified as class I, i.e., device parts that are no "stand-alone" class I medical devices, require no NB Opinion and can possess neither an EU DoC nor an EU Certificate. Pharmaceutical manufacturers that offer such device parts in their integral product should generate and submit their documentation on the device part's conformity with the relevant GSPR stated in MDR, Annex I as part of their marketing authorization dossier, i.e., submit a "self-declaration" and related evidence of conformity to the relevant GSPRs.

In addition, pharmaceutical manufacturers that intend to market co-packaged non-integral products have to consider the MDR obligations stated for so-called distributors (of medical devices) who are defined in Article 2(34), MDR as "any natural or legal person in the supply chain, other than the manufacturer or the importer, that makes **a device available on the market**, up until the point of putting into service." These obligations as defined, e.g., in Article 14, MDR, describe the distributor's responsibilities related to vigilance, labeling, product traceability, and market activities related to complaints, non-conformities, withdrawals, and recalls.

Pharmaceutical manufacturers that intend to market referenced non-integral products have to consider MDR's transition timelines (transition from MDD to MDD as stated in Article 120, MDR) to ensure that medical devices referenced are still allowed to be placed on the market or continue to be available on the market.

Consequently, pharmaceutical manufacturers that provide medical devices as either integral or non-integral part of their medicinal products must also comply with the MDR where applicable or at least know and consider the requirements (see Figure 14.8).

Legacy integral products currently authorized for sale in the EU or integral products whose applications were submitted before the MDR entered into force, i.e., before May 26, 2021, were not impacted by Article 117 requirements. However, if a marketing authorization holder intends to perform changes to the device or to a device part "that may affect the safety and performance of the device part or the conditions prescribed for the intended use of the device part" or to introduce "and addition or full replacement of the device or device part" after May 26, 2021, Article 117 requirements are applicable as stated in EMA's Q&A document (EMA/37991/2019)[14] [33]. There are no further details provided in EMA's Q&A document on any change impact assessment rules; the recommendation in the document for situations where the type of change and the corresponding necessity or non-necessity for application of Article 117 is not clear is to consult the medicines NCA that issued the existing marketing authorization.

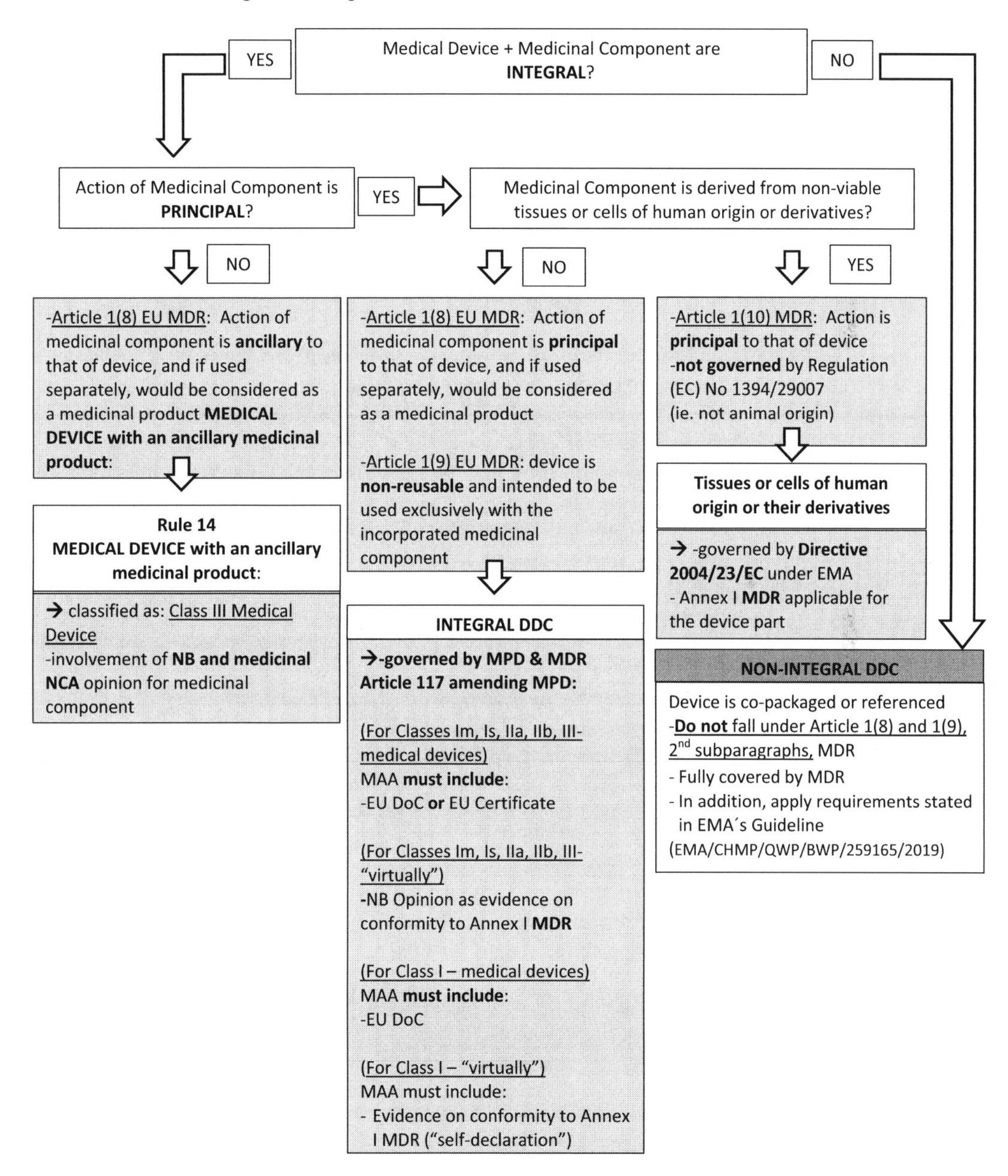

FIGURE 14.8 Flowchart for integral and non-integral products (DDCs) and other combination products under the EU MDR [30,34].

Classification with regard to Medical Device Legislation

Related to classification, the rules previously detailed in Annex IX of the MDD have been moved to and amended in Annex VIII of the MDR. Now what used to be Rule 13 under the MDD is defined in Rule 14 under the MDR, stating that devices with medicinal substances, considered as "medical devices with an ancillary medicinal product," fall under class III (highest risk class of medical devices).

Accordingly, the notified bodies (NB) which are conformity assessment bodies designated by the EU member states in accordance with MDR's requirements will assess the conformity of these products based on MDR conformity assessment procedures stated in Article 52 of the MDR. For Rule 14 products, the NB acts in addition as "communicator" for the device manufacturer and deals directly with a medicinal product NCA or EMA to seek their scientific opinion on specific aspects of safety and performance relevant to quality, safety, and usefulness of the ancillary medicinal substance (Annex IX, Chapter II, Section 5.2 of the MDR).

[Of note, Rule 21 under the MDR applies to devices that are composed of substances or combination of substances, but these substances not being medicinal products (e.g., many over-the-counter medical devices, such as salt water nasal sprays, wart removal gels, active coal for oral administration, vaginal lubricants, and dermal creams for eczema or psoriasis, see also MDCG 2022-04, section 3.3). Due to the market importance of these products, this rule and its implications are stated in this section: such substance-based medical devices can no longer be classified as class I, but at least in class IIa. In consequence, nasal and throat sprays or ear drops currently on the market as class I medical devices will face a conformity assessment procedure and notified body involvement. Manufacturers need to consider the transitional provisions stated in Article 120 (3) of MDR, and the pre-condition of these transitions periods is only applicable if "no significant changes in the design and intended purpose" take place; more information on what are regarded significant changes in the context of the transition is described in MDCG 2020-3.[15] However, post-market surveillance, market surveillance, vigilance, and registration of still MDD-compliant devices must already occur as stated in MDR – more details are provided in MDCG 2022-4.[16] Also, invasive substance-based devices currently often classified as class IIa will be up classified class IIb (e.g., vaginal lubricants) or even class III (e.g., anti-obesity capsules).]

With the classification of the combination product completed to better understand the corresponding requirements, developing a product development strategy to meet the expectations of the concerning regulatory authorities becomes more feasible. A summarized description of the DDC development process and other combination products is provided in Figure 14.6 in the US section of this chapter. While the development process as applicable for the US is in line with EU considerations, the application of Annex I of the MDR (General Safety and Performance Requirements) in addition to US-requested essential performance requirements is needed.

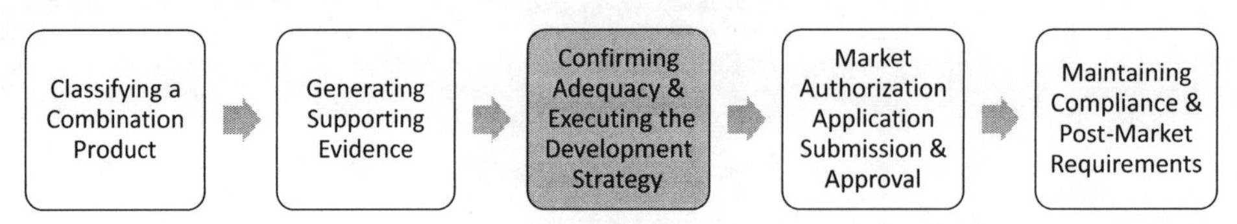

As also recommended for the US market, meeting with the concerned EU regulatory authorities is highly recommended before executing a development plan to assure the success of the approval process. There are meeting opportunities where the marketing authorization applicants or medical device manufacturer can consult with the respective NCA. While some of those can be free of charge (i.e., pre-submission meetings), others require a fee (i.e., scientific advice, protocol review). This diverges from the US approach where meeting requests don't require any fees. For innovative products where the current regulatory pathways are not clear, national regulatory agencies have innovation offices available to give advice, which can be coordinated across the EU-Innovation Office Network, with administrative support from the EMA. EMA also has its Innovation Task Force, which is increasingly dealing with combination products. These meetings are free of charge [39,40].

From the medicinal product side, the free pre-submission meetings with EMA are opportunities for companies to introduce their proposed development program and receive feedback (primarily non-scientific) which could also cover the administrative and procedural aspects. They are recommended for companies seeking general advice on specific types of medicinal products or therapies for which complex DDCs might be a good candidate.

Official approvability feedback is provided through the paid Scientific Advice and Protocol Assistance by Scientific Advice Working Party (SAWP) or Committee for Medicinal Products for Human Use (CHMP).

These opportunities are also provided by NCAs to cover both drug and device regulatory topics. For example, BfArM from Germany promotes its "Kick-off Meetings" through its "innovation office" to go over relevant aspects regarding mandatory regulatory requirements and essential procedural steps or available opportunities can be identified and discussed at a very early stage of the product development related to medicinal products and medical devices.

These kinds of meetings ensure that product developers perform the most appropriate robust tests and studies to ensure that no major objections regarding the design of the tests are likely to be raised during the evaluation of the marketing authorization application. This also helps avoid patients taking part in studies that will not produce useful evidence. There are even opportunities from EMA for these meetings to be held in parallel with the US FDA and/or Health Technology Assessment (HTA) bodies.

These processes are not developed specifically for combination products; however, they can be found to be instrumental for these cases. Therefore, discussing early on with the NCA who are expected to be involved in the review of the application is strongly recommended. Related to notified bodies, they are in a more restricted situation related to consulting; Annex VII, section 1.2.3 prohibits them to "offer or provide consultancy services to the manufacturer, its authorised representative, a supplier or a commercial competitor as regards the design, construction, marketing or maintenance of devices or processes *under assessment*," However, these restrictions still allow the exchange of "technical information and regulatory guidance between a notified body and a manufacturer applying for conformity assessment" (Annex VII, section 1.2.9).

Similar to the US, appropriate supporting evidence is expected to be generated to finally prove the quality, safety, and efficacy of medicinal product parts, and the safety and performance of medical device parts. These regulatory expectations are usually met first by executing pre/non-clinical verification and subsequent validation tests. In many cases, especially for products which are intended to enter the market for the first time (usually referred to as "innovator" or "brand name" products) and/or are class III medical devices and/or implantable medical devices, clinical investigations involving human subjects are necessary to provide evidence of the combination product's safety and benefit for the patient/users.

Market Access with regard to Medicinal Product Legislation

For the integral products regulated as medicinal products, the application process is based on ICH guidelines requiring all the quality, safety, and efficacy information to be presented in a common format (called Common Technical Document (CTD)). This approach has been adopted not only in Europe but in several other countries to promote global harmonization of market access documentation.

In June 2019, EMA released for consultation a draft guidance on DDCs which gave insight on the expectations for the quality information that should be contained in the application of medicinal product-led integral products. This draft guideline was entitled: "Quality requirements for drug-device combinations" and is called "Guideline on quality documentation for medicinal products when used with a medical device" in the final version that came into effect in January 2022. Additional clarifying information was issued in the form of a Q&A document in October 2019: "Questions & Answers on Implementation of the Medical Devices and In Vitro Diagnostic Medical Devices Regulations ((EU) 2017/745 and (EU) 2017/746)" which covered clarifications, e.g., for legacy integral DDCs and impact of changes.

Regardless of the EMA guideline or EMA's Q&A document, per Article 117 requirements medicinal product-led integral products, the MAA consisting of the CTD must include "the results of the assessment of the conformity of the device part with the relevant" GSPR of Annex I – this means an EU certificate or an EU DoC for the device part or, in certain cases as discussed above, an NB Opinion.

In Europe, for DDCs controlled under the Medicinal Product Legislation, marketing authorization applicants have the option to submit their application following the "centralized," "decentralized," or the "mutual recognition" procedures. For the first one, applicants are allowed to submit the MAA to EMA which then coordinates among member states, via the Committee for Medicinal Products for Human Use (CHMP), a scientific assessment of the application, and gives a recommendation to the Commission on whether to allow the product on the European market. The final market approval is given by the European Commission. Alternatively, the applicant can choose the "decentralized" procedure (remark: unless the central procedure is stated as mandatory for the respective medicinal product in line with EU Regulation 726/2004, e.g. for orphan drugs), which means approaching in parallel all the national CA of the EU member states where the medicinal product shall be marketed; the applicant can select the reference member state that will take care of the coordination among member states, etc. As soon as all affected Competent Authority (CA) have agreed, the medicinal product will be simultaneously approved in all these member states. After being approved in at least one EU member state, the "mutual recognition" procedure provides an option to expedite the approval process to enter additional EU member state markets. Specifically, the approval process is expedited as a new application submission is not required for the new targeted member state market. Instead, the MAA and the evaluation assessment performed by the CA of the member states where the product is already approved is shared with the CA of the targeted member state. This precludes the need to duplicate the review efforts.

Market Access with regard to Medical Device Legislation

As "medical devices with an ancillary medicinal product" (e.g., implants coated with antibiotics) are automatically classified as class III devices, they must undergo a conformity assessment according to Annex IX or Annex X coupled with Annex XI of the MDR to demonstrate that they meet legal requirements for the product as well as for the manufacturer's quality management system (QMS). To reach this objective, it is necessary to set up, document, and maintain the QMS with at least the procedures required by MDR (see e.g. Article 10, MDR) and to ensure that within this QMS the generation and lifecycle maintenance of the product-specific technical documentation and the technical documentation on post-market surveillance is existent as specified in Annexes II and III of the MDR.

For class III products clinical investigations are required (see e.g. Recital (63), MDR), which must consider MDR's requirements stated in Annex XV and which must be pre-approved by a NCA before they are conducted in the respective EU member state. Before market access is possible, the NB will audit and assess the manufacturer's QMS and the respective product's technical documentation. The medicinal product NCA or EMA reviews the medicinal information included in the technical documentation and issues the opinion known as Notified Body Consultation report. For consultations performed by EMA, Consultation Public Assessment Reports (CPARs) are made available following confirmation of the CE marking by the notified body, in the similar way to European Public Assessment Repots (EPARs) for medicinal products [41].

The NB is not allowed to issue the respective EU Certificate covering or including the respective product, if the scientific opinion is "unfavourable" – the final decision on certification must accordingly be taken by the medicinal products NCA consulted (Annex IX, Section 5.2 (e), MDR).

With the EU Certificate obtained after a successful audit and technical documentation review, the EU DoC issued, and the CE marking affixed to the product, the product is allowed to be placed on the market. In addition, any manufacturer with a registered business place outside the EU must have an Authorised Representative (see Article 2(32), MDR) located in the EU who is qualified and registered in the Eudamed database to handle regulatory issues as defined in Article 11 of the MDR; this representative (name and address) must also be stated on the product's label.

To date, the EU has yet to specify quality system (QS) and good manufacturing practice (GMP) requirements especially for medicinal product-led combination products and medical devices with ancillary medicinal products. However, as these products are currently subject to either the medicinal product or medical device legislations as leading legislative act, the corresponding requirements, specifically the GSPRs with quality system requirements (e.g. GSPR 3) and GMP per ICH requirements, could be enforced by the regulatory authorities. Until actual manufacturing controls are specified, it might be advisable to ensure compliance with applicable requirements from both device QS requirements (e.g., ISO 13485) and medicinal product cGMPs (e.g., ICH Q7, GMP for active substances). Furthermore, at the post-authorization phase, compliance with good distribution practice (GDP) and good pharmacovigilance practices (GVP) is enforced. Therefore, for specific cases which are unclear, discussions with the concerning regulatory authorities should be held. Additional insights on practical considerations relative to cGMPs, e.g., relative expectations for manufacturing controls as against US Part 4 approach, are considered in Chapter 4 (cGMPs).

Post-Market Activities with regard to Medical Device Legislation

Post-market oversight on medical devices according to MDR will now require a more proactive approach from manufacturers, regardless of medical device class. Logically, combination products that are "medical devices with an ancillary medicinal product" will also be affected. From a holistic approach, manufacturers must have in place a QMS that includes:

- Post-market surveillance system (PMS) including post-market clinical follow-up (PMCF);
- Risk management system;
- Management of corrective and preventive actions (CAPA);
- Monitoring and measurement of output, data analysis, and product improvement;
- System for reporting serious incidents and field safety corrective actions (vigilance) to the Competent Authorities.

These requirements are intended to promote clinical and safety-related data gathering by monitoring product performance to systematically identify risks from the product use. These activities are expected to be performed throughout the entire product lifecycle to continuously ensure that the benefit–risk assessment is up to date and that necessary measures are initiated when needed (vigilance).

For class III medical devices under rule 14, the manufacturer must prepare and submit Periodic Safety Update Reports (PSURs) containing the summarized results and conclusions of the analysis of the PMS data including the benefit-risk determination and post-market clinical (PMCF) follow-up information, among others. These PMCF information is expected to make up the post-market section of the Clinical Evaluation Report (CER); both requiring periodic updates as stated in Annex XIV of the MDR. Design changes and/or QMS changes performed as a result of the aforementioned activities or for other reasons shall be reported to the notified body whenever they are considered as change according to Annex VII, sections 4.9 and 4.11 and/or Annex IX (Chapter I, section 2.4; Chapter II, section 4.10/5.2/5.3.1) and/or Annex X (section 5) and/or Annex XI (section 9) of the MDR.

In the context of the medical device with an ancillary medicinal product, changes to the drug specifications or characteristics are also considered to be design changes which should be evaluated through the same design control procedures. Specifically, these changes should be evaluated to determine the "significance" of the change, i.e. the impact it may have on the overall medical device. Being a medical device, the primary regulatory authority to contact for these changes is the responsible NB who must consult with the respective medicinal product NCA or EMA to review the acceptability of the changes.

Post-Market Activities with regard to Medicinal Product Legislation

There are also requirements enforced on medicinal products including integral products after they have been approved. This phase is known as the "post-authorization stage" of the product lifecycle. The expectations for this stage are defined for areas such as pharmacovigilance, product profile changes requiring updates to the marketing authorization dossier, reporting of product defects or recalls, post-authorization safety studies (PASS), post-authorization measures (PAMs), PSURs, risk management plans (RMP), post-authorization efficacy studies (PAES), etc. (see Chapter 8, section on PMSR). Regarding changes of the integral product (DDC), according to EMA's Q&A (question 2.7), variation applications according to the variation guideline are needed if the registered medicinal product information are affected as well as for changes to the device part that have "an impact on the quality, safety and/or efficacy of the DDC." In addition, such changes will require a new EU DoC or EU Certificate or NB Opinion (see discussion above in the section on classification under Medicinal Product Legislation).

United Kingdom/Great Britain and Northern Ireland

The transition period for the United Kingdom leaving the European Union came to an end on December 31, 2020.

A new UK Conformity Assessment (UKCA) marking will be used for goods being placed on the market in Great Britain (England, Wales, and Scotland). It covers most goods which previously required the CE marking, and CE marking will be valid for a transition period until December 31, 2024.[17]

As the MDR and IVDR were not yet in force when the transition period with the EU has ended, they do not automatically apply in Great Britain. Consequently, the amendment to medicines legislation brought about by Article 117 of the MDR will not take effect in Great Britain. EU legislation will continue to apply in Northern Ireland and therefore the MDR (and in consequence also Article 117 requirements) is applicable since May 26, 2021.

After the EU MDR's date of application, the UK MHRA guidance on medical devices incorporating medicinal substances was withdrawn [43]. In effect, EU NBs are no longer able to consult with MHRA under the EU medical device legislation. Subsequently, a recent guidance document was released in November 2022 entitled "How the MHRA makes decisions on when a product is a medical device (borderline products), and which risk class should apply to a medical device"[18] which discusses the borderline products, known under the same moniker in the EU.

Based on this current guidance, MHRA decides whether a product is a medical device or medicinal product in case the manufacturer is unsure of their product. As in the EU and US, the decision about whether a product is a medical device is based on the intended purpose of the product and its primary/principal mode of action; however, further clarification on how the new UK system for combination products will function is awaited from the UK MHRA.

Australia

Regulatory Evolution

The Therapeutic Goods Administration also known as TGA is part of the Australian Government Department of Health and Ageing and is the responsible authority regulating therapeutic goods. The TGA administers the Therapeutic Goods Act (the "Act"), released in 1989 and later amended in 2002 by the Therapeutic Goods Amendment (Medical Devices) Bill and the Therapeutic Goods (Medical Devices) Regulations for legislative requirements [35].

In June 2017 and February 2018, Australia passed new legislations to improve its regulatory framework by streamlining and optimizing its regulatory processes. Namely, new conformity assessment body (CAB) designation criteria have been set by the TGA. These new criteria are consistent with those set for Auditing Organizations (AO) of the Medical Device Single Audit Program (MDSAP) by the IMDRF. Since then, the new legislation allows manufacturers to submit comparable overseas regulatory certificates to streamline their efforts. Certificates and licenses issued by foreign regulators like the Notified Bodies of the EU, FDA in the US, Health Canada, Japan, and the MDSAP will now be recognized by the TGA.

In September 2019 a consultation document was released listing the TGA's proposed changes to the Essential Principles of the Therapeutic Goods (Medical Devices) Regulations 2002 to (1) incorporate the IMDRF Essential Principles and EU GSPR details, (2) clarify existing requirements, (3) restructure the current presentation of the essential principles to improve clarity and readability, and (4) amend the inclusion criteria for the Australian Register of Therapeutic Goods (ARTG) number.

For combination products in particular, streamlining efforts are also evident. In particular, the repeal of regulation 4.1 in July 2021, referring to devices that contain medicines or materials of animal, microbial, recombinant, or human origin, and class 4 IVDs, will no longer require mandatory TGA conformity assessment certification for these devices. Now, TGA will accept conformity assessment evidence issued by the TGA, EU notified bodies, and Australian CABs.[19]

And most recently, a first version of a draft guidance document for "Boundary and combination products" was released in October 2022 and was opened to consultations until November of the same year.[20] This guidance document echoes very familiar principles on determining the appropriate regulatory requirements of combination products based on the PMOA and the product's intended use. The guidance also includes in Appendix 1 a practical list of specific boundary products and the applicable regulatory pathway.

Despite the substantial changes required for harmonization, the current amendments and ongoing changes in the Australian regulations and guidelines illustrate the TGA's commitment to work toward global harmonization.

Therapeutic goods are broadly defined as products for use in humans in connection with:

- Preventing, diagnosing, curing or alleviating a disease, ailment, defect, or injury;
- Influencing inhibiting or modifying a physiological process;
- Testing the susceptibility of persons to a disease or ailment;
- Influencing, controlling, or preventing conception;
- Testing for pregnancy;

This includes products:

- Used as an ingredient or component in the manufacture of therapeutic goods;
- Used to replace or modify parts of the anatomy;

Therapeutic goods are categorized as medicine, medical device, biological, or other therapeutic goods (Table 14.5).

TABLE 14.5
Therapeutic Goods in Australia

Medicines	Biologicals	Medical Devices	Other therapeutic goods
• prescription drugs • Over the Counter (OTC) • complementary medicines	• something made from or containing human cells or tissues, such as human stem cells or skin	• instruments • apparatuses • appliances	• not regulated specifically as medicines, biologicals, or medical devices • ex. sterilants, disinfectants, tampons, menstural cups

In addition, there are three definitions of combination products: a combination of (1) a medicine and medical device, (2) a biological with a medicine, or (3) a biological with a medical device. Determination of combination product type will in turn dictate which regulations must be followed. However, when the appropriate regulatory pathway is not immediately obvious or the therapeutic goods have some attributes of two or more categories of regulated goods, these cases are considered "boundary products" similar to the "borderline" products in the EU. For such products, the TGA will determine how these products are regulated based on determining the principle therapeutic effect and factors that influence this effect.[21]

Confirming the Classification

As a good starting point for determining if a product is regulated as a medicine, biological, or medical device, the TGA website has established an online assistant for small to medium enterprises (SMEs) [37]. The SME Assist is an interactive webpage that can guide inquiries about medical device classification, market authorization, GMP application requirements, among many others [38].

Identifying the Applicable Requirements

TGA enforces the applicable regulations for these different types on a risk-based paradigm, through pre-market assessments, post-market monitoring, and enforcement of standards, licensing of Australian manufacturers and verifying overseas manufacturers' compliance with the same standards as the Australian counterparts [38].

The Figures 14.9a illustrates the connection between the combination product classification and their respective governing regulatory framework. It is taken from the ARGMD of 2011 that is still currently under revision at the time of writing (December 2022); however, it compliments Figure 14.9b illustrating the definitions from the most recent draft guidance document on boundary and combination products (released in October 2022).

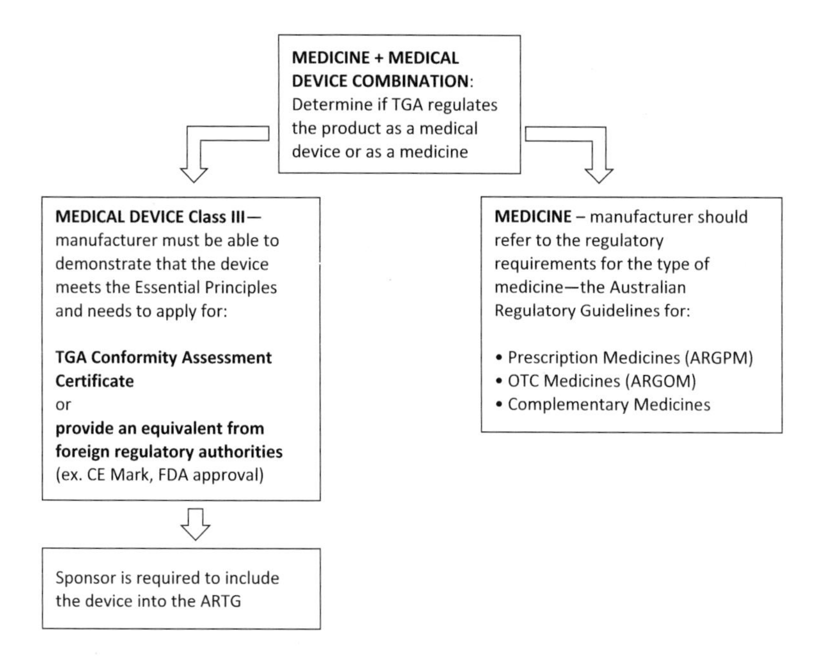

FIGURE 14.9 (a) ARGMD flowchart of determining regulatory pathway for medical device or medicine under TGA (under revision as of 2022) [38].

Combination Product: Medicine + Medical Device

Taken from most recent draft guidance for boundary and combination products (October 2022), Figure 14.9b is a graphic interpretation of the three types of medicine-device combination products defined in the guidance, illustrating how a manufacturer may determine the applicable regulatory requirements.

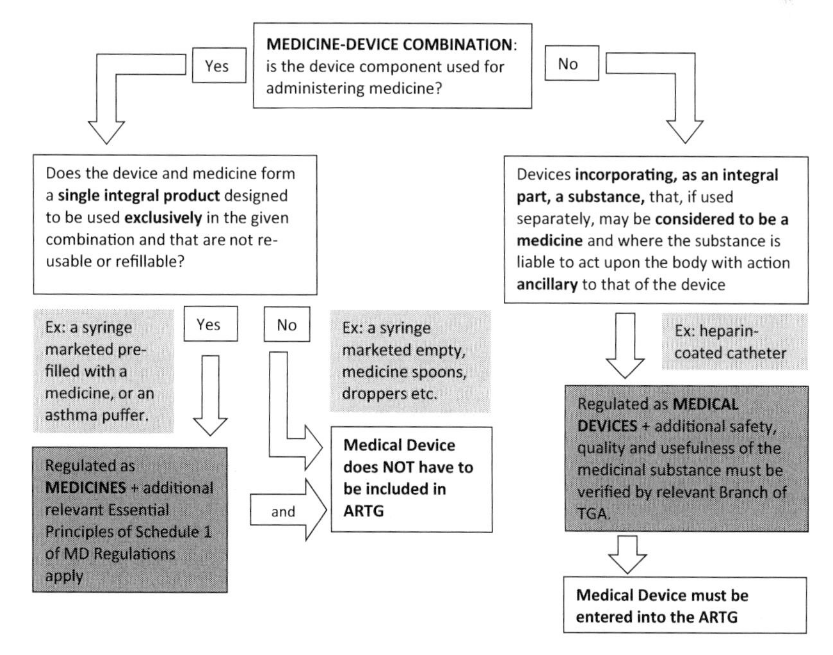

FIGURE 14.9 (b) Determining the type of medicine-device combination products according to the draft TGA guidance on combination products (TGA, Boundary and combination products – medicines, medical devices, and biologicals, Version 1.0, October 2022 (Accessed December, 2022)).

System or Procedure Packs (SOPPs)

In a separate document released in November 2021, more guidance is given regarding the regulation of system or procedure packs (SOPPs),[22] which are defined as medical devices that are intended to be used in a medical or surgical procedure, containing a combination of two or more goods where:

- At least one of the goods is a medical device (which may be an in vitro diagnostic (IVD) medical device); and
- All of the goods are packaged together or are to be interconnected or combined for use.

In this regard, these SOPPs may also include medicines, biologicals, other therapeutic goods, or other goods that are not considered to be therapeutic goods (see Figure 14.10).

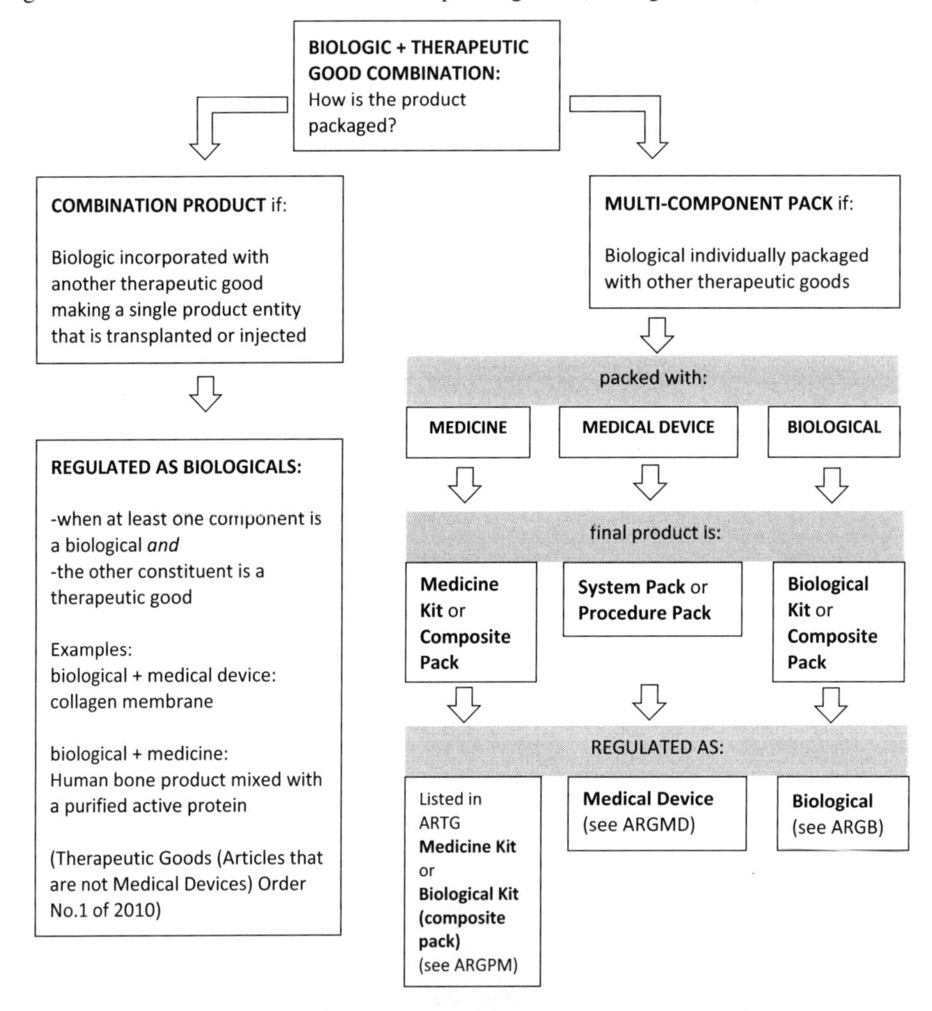

FIGURE 14.10 Flowchart as described in the ARGB for determining the regulations applicable to a biologic + therapeutic good combination product.

Combination Product: Biologic + (Medicine and/or Medical Device)

This last guideline on SOPPs also compliments the guideline released in 2018 as part of the Australian Regulatory Guidelines for Biologicals (ARGB) "Biologicals packaged or combined with another therapeutic good," which defines combination products that refer to a biological packaged or combined with another therapeutic good [36]. These combination products will be ruled as biologics, medicines, or medical devices depending on the packaging (see Figure 14.10).

As TGA accepts foreign approval decisions particularly those originating from the EU and the US and adopts the principles of international harmonization efforts, the approach to develop an adequate strategy can be based on the recommendations described in the previous EU and US sections of this chapter.

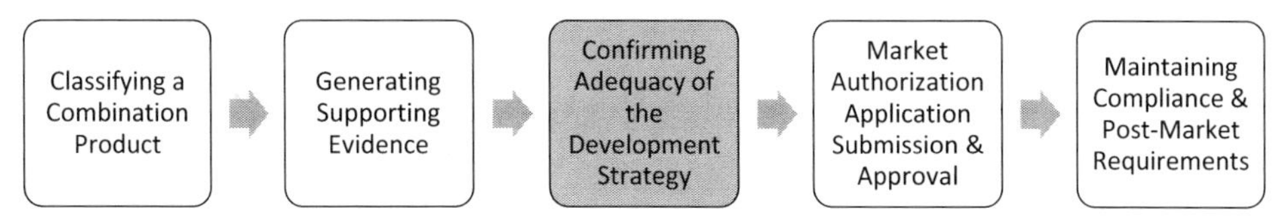

As for the US and the EU, there are also opportunities to meet with TGA to confirm the adequacy of a development strategy before executions. To this end, it has released a guidance document in March 2018 which provides instructions regarding pre-submission meetings related to applications [39]:

- *To enter therapeutic goods on the Australian Register of Therapeutic Goods (ARTG). This includes all medicines, biologicals (cell and tissue-based products), medical devices, and other listed and registered therapeutic goods;*
- *For designation of prescription medicines, e.g., orphan drugs, priority review;*
- *For TGA Conformity Assessment Certification (for the manufacture of medical devices).*

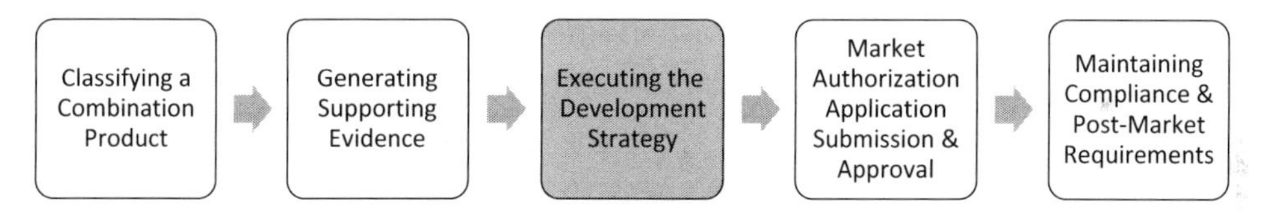

The quality, safety, efficacy, and effectiveness evidence requirements imposed by the US and the EU are generally applicable to Australia as well. These expectations are usually met first by executing pre/non-clinical drug verification and validation bench tests which are performed in laboratory settings. In general, for certain cases such as generics or "me-too" products where the risk profiles are understood or known to be low due to the regulatory authority's previous experiences with similar products, evidence from these bench tests may be sufficient to obtain market access.

Generally, the testing and study expectations for these cases are covered by internationally recognized and harmonized standards and guidelines such as the ISO or ICH. However, in many cases, especially for products which are accessing the market for the first time usually referred to as "innovator" or "brand name" products, clinical trials involving human subjects are necessary to provide evidence of the product's safety.

Any additional TGA-specific requirements should be addressed during the pre-sub-meeting.

To gain market access, products must apply to the ARTG, and depending on the posed risk of the therapeutic product, it will be listed or registered. This is realized first by submitting an application for market authorization via an online system. Once received, an administrative screening is performed to ensure that the necessary data and information have been provided. For all concerning therapeutic goods including combination products, supporting data should provide evidence of the quality, safety, efficacy, and/or performance of the combination product, kit, or pack. A complete application would be accepted and processed. If all regulation requirements are met, TGA will grant the market authorization and enter the therapeutic good to the ARTG.

In general, the approval pathway requires a TGA CA Certificate for medical devices. Evidence of the manufacturer's CA certification needs to be established and registered with the TGA for all medical devices except class I non-measuring and non-sterile medical devices. However, per the classification Rule 5.1 of Schedule 2 of the Regulations, medicine-device combinations products are classified. Therefore, these combination products are subject to the certification requirements described above.

It is important to note that TGA accepts equivalent certification (i.e., CE certificate) issued by a European NB. This lessens the burden of re-applying the medical device for a CA certificate. The sponsor must, nonetheless, register the device into the ARTG [40]. Additionally, approvals from the US FDA, Health Canada and Japan MHLW/PMDA, and certificates issued under the MDSAP are also accepted by TGA.

Respectively, for medicines and biologicals, a TGA license must be acquired by the manufacturer to illustrate compliance to the GMP for Medicines or GMP for Human Blood and Tissues [41]. Manufacturers are also required to follow the Australian Regulatory Guidelines for Medicines (Prescription, OTC, or Complementary) for medicines.

Therapeutic goods including combination products are bound to post-market requirements to ensure they continue to meet the requirements specified in the market authorization application. As a control measure, TGA randomly selects products for post-market reviews.

All medicines and biologicals, including those classified as combination products, are subject to the GMP requirements. Specifically, the expectations are described in the following:

- PIC/S Guide to Good Manufacturing Practice (GMP) [42];
- Australian code of good manufacturing practice for human blood and blood components, human tissues, and human cellular therapy products [43].

Once compliance with these principles is confirmed, a "Manufacturing License" is issued which must be maintained.

On the other hand, medical devices undergo conformity assessment to comply with the Essential Principles prescribed in Schedule 1 of the Therapeutic Goods (Medical Devices) Regulations 2002. The conformity assessment takes into account quality system requirements per the Medical Devices Essential Principles Checklist which are consistent with internationally recognized standards such as ISO 13485 and MDSAP requirements. Once compliance is confirmed, the TGA CA certification is issued but is also required to be maintained post-market authorization [44]. This procedure also holds true for device-classified combination products.

Like the lifecycle approach under EU MDR, TGA also monitors medical devices after they have entered the Australian market. These include conducting periodic inspections of manufacturers, assessing and reporting problems of approved medical devices, and checking that the medical devices continue to comply with the essential principles and that manufacturers regularly report adverse events within specific timeframes. Likewise, for medicines, post-market surveillance involves monitoring and evaluating the sponsors' and manufacturers' compliance with Australian pharmacovigilance guidelines, ensuring that medicines that are included in the ARTG maintain their standards of quality, safety, and efficacy. Sometimes, inspections are conducted unannounced by the TGA. Manufacturers need to have adequate processes routinely in place to support a surprise inspection since the opportunity to plan for announced inspections is not available.

With recent updates across the guidelines and legislations in Australia, it is apparent that harmonization with foreign regulatory bodies will complement TGA's local efforts to allow innovative products to gain market access in a timely manner at the same time ensuring that the products are safe and effective for the consumers. Given that Australia is geographically isolated from the medical development hubs in Europe and the US, recognizing foreign certificates and licenses will not only be favorable for the sponsors and manufacturers accessing the Australian market, it also brings these innovative products to the Australian population, advancing public health.

CANADA

Regulatory Evolution

Since the Food and Drugs Act of 1985, the drug/device combination products were only first recognized much later in 1998 with the release of the Policy on Drug/Medical Device Combination Products – Decisions. Under the policy, specific combination products were classified as drugs and others put under the medical device category. The policy also listed products that were difficult to classify as either drug or device although they were not necessarily always combination products. In this list, classic combination products such as prefilled syringes were placed under the drug category and those like drug-coated stents were treated as medical devices. However, other combination products that were not listed needed to comply with both the Food and Drug Regulations and the Medical Device Regulations.

It wasn't until 2006 when the requirement to classify a drug and device combination product based on its principal mechanism of action came into effect. Following one approval route lessened the burden on sponsors of drug/medical device combination products as well as the regulatory authorities involved in the review process. This relief is explained by the control over the entirety of a combination product being streamlined under either the Food and Drug Regulations or the Medical Devices Regulations, but no longer both simultaneously.

Health Canada (HC) announced its transition from the Canadian Medical Devices Conformity Assessment System (CMDCAS) to the MDSAP in 2012. Seven years later, as of January 1, 2019, HC has completed its transition. Ever since, manufacturers of Class II-IV medical devices must demonstrate compliance with ISO 13485 as well as compliance with the QMS under the Canadian Medical Device Regulations (CMDR). At the time of writing in October 2019, HC has reported 3000 transition submissions or MDSAP certificates (90% of medical device manufacturers), demonstrating the stakeholders' engagement in this important initiative toward harmonization.

According to the Food and Drug Act (RSC, 1985, c. F-27) [45]:

- A "therapeutic product" is defined as a drug or device, or any combination of drugs and devices, but does not include natural health products;
- A "therapeutic product authorization" refers to a license that is approved for the import, sale, advertisement, manufacture, preparation, preservation, packaging, labeling, storage, or testing of a therapeutic product.

Meanwhile, the Drug/Medical Device Combination Products policy of 2006 defines a combination product is

> a therapeutic product that combines a drug component and a device component (which by themselves would be classified as a drug or a device), such that the distinctive nature of the drug component and device component is integrated in a singular product.

Product Classification

Combination products are not classified in a particular risk class but are instead classified according to the 16 classification rules found in Schedule 1, Part 1 of the CMDR SOR/98-282. Nonetheless, according to the Policy on Drug/Medical Device Combination Products, Figure 14.11 illustrates the classification criteria by which HC is guided.

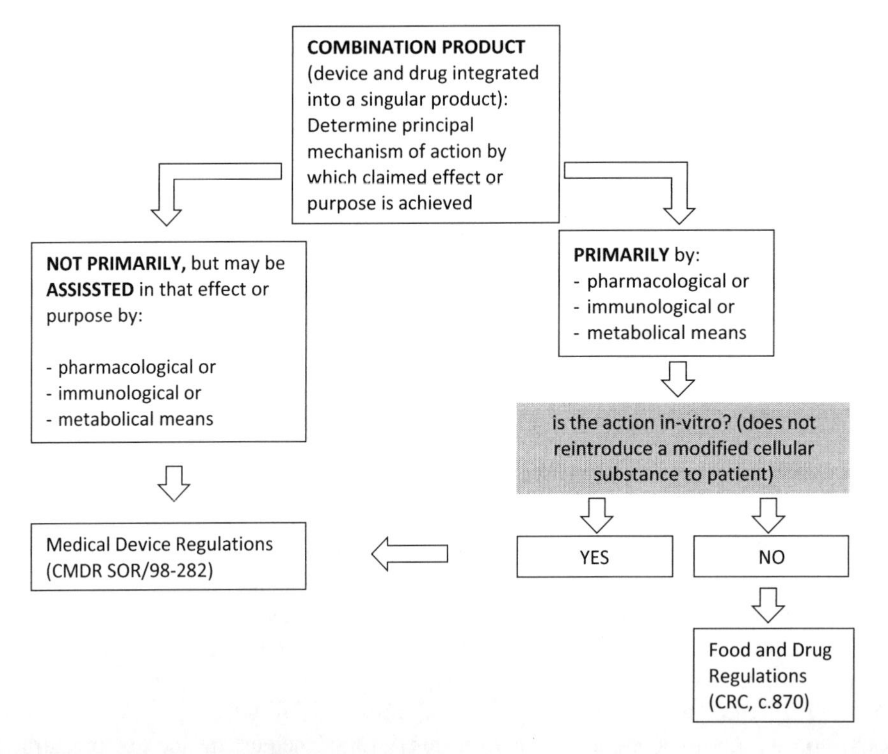

FIGURE 14.11 Criteria for classification of combination products according to the policy statements in the Drug/Device Combination Products Policy in Canada.

Regulatory Oversight

Canada's Health Products and Food Branch (HPFB) is their US FDA equivalent. It is the national authority that is responsible for regulating, evaluating, and monitoring the safety, efficacy, and quality of drugs, biologics, genetic therapies, and other health products available for the Canadian marketplace. The HPFB's mandate is to manage the health-related risks and benefits of health products and foods for Canadians.

Under HC, drugs and medical devices are regulated by the Therapeutic Products Directorate (TPD), while the Biologics Genetic Therapies Directorate (BGTD) has the regulatory authority over biological drugs (products made from living sources) and radiopharmaceuticals (drugs that have radioactivity) for human use.

Confirming the Classification

In cases when a product cannot be clearly classified as a drug or device, the Therapeutic Products Classification Committee (TPCC) may be consulted. Appointed by the Director General of TPD, the TPCC develops, maintains, evaluates, and recommends the approval of policies, procedures, and guidelines concerning the classification and review of therapeutic products as drugs, devices, or combination products. It also assesses submissions for combination products referred to it and determines an appropriate classification and review mechanism for the submission. Though a recent guidance has been published in February 2018 for the classification of products, entitled *Classification of Products at the (Medical) Device-Drug Interface*, this guidance specifically does not address combination products. Thus, manufacturers must continue to refer to the Drug/Medical Device Combination Products Policy of 2006.

When planning to enter the market, the sponsor of a combination product may make a presentation to TPD or BGTD (as appropriate) to classify the product by providing the following information:

- Name of the product and identification of the device/drug components;
- A synopsis of relevant data describing the mechanism of action of each component;
- The principal mechanism of action of the product, including composition, study design, measurements of efficacy in terms of structural, pharmacologic, metabolic, immunologic, and ADME studies conducted, toxicity studies, etc.

The submission will be referred to TPCC for a final decision within 30 days of accepting the submission.

The recommendations for developing an adequate strategy provided in Figure 14.6 are also applicable for Health Canada, as they aim to satisfy the globally recognized quality, safety, efficacy, and effectiveness evidence expectations. Nevertheless, the jurisdiction-specific expectations should be clarified with the concerning regulatory authorities before executing a development strategy.

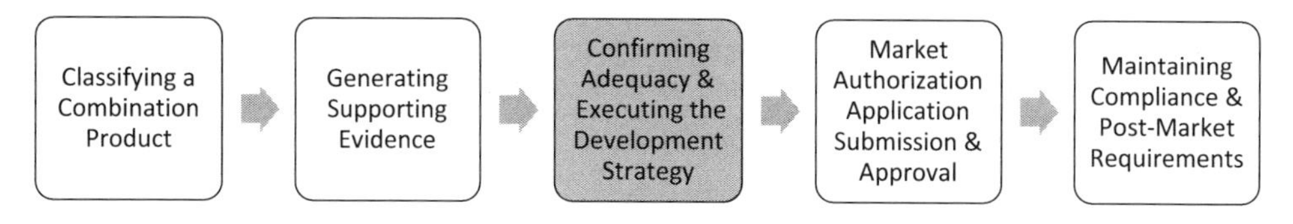

As for the US and the EU, there are also opportunities to meet with HC to confirm the adequacy of a prepared development strategy before executions. To this end, HC launched a new pilot project to provide regulatory advice to medical device manufacturers in September 2018 among other initiatives. These include the "Device Advice: Pre-Clinical Meetings" through which advice and recommendations on investigational testing protocols can be obtained early in the device development process.

There are "Pre-Submission" or "Pre-Application" meetings available through the drug review pathway, as well. Specifically, HC's website states that sponsors may request a meeting with the appropriate Directorate within the agency prior to filing the following among others:

- A drug submission/application (e.g., NDS, SNDS, ANDS, SANDS, DIN – see acronyms);
- A clinical trial application;
- A request for a combination product classification.

As described above, the Drug/Medical Device Combination Products Policy, (November 2005) dictates whether an application for combination products is processed under the Medical Devices Regulations or the Food and Drug Regulations. This is determined based on the principal mechanism of action by which the claimed effect or purpose is achieved. Regardless, the review of principal and ancillary components must meet acceptable standards of safety, efficacy, and quality.

Market Access Under the Canadian Food and Drug Regulations

In Canada, drugs are regulated under the Food and Drug Regulations (CRC, c 870). Accordingly, drugs including drug-led combination products are authorized to reach that market only once they have successfully gone through the relevant review process of the safety, efficacy, and quality of the drug. For these cases, a submission or an application should be sent to the Office of Submissions and Intellectual Property (OSIP).

Pre-Market Activities Under the CMDR

To market a device-led combination product in Canada, the appropriate licensing is required. Specifically, a Medical Device Establishment License (MDEL) is required for Class I products while a Medical Device License (MDL) for Class II-IV devices. To obtain and maintain the MDL or MDEL, an application that is fully compliant with the CMDR must be submitted for review to HC Medical Devices Bureau in conjunction with the appropriate documentation for the product's class level. For this purpose, the guidance document "Management of Applications for Medical Device Licenses" should be reviewed.

Under the Food and Drug Regulations

Even after a health product receives a favorable decision, monitoring of its safety and effectiveness continues. Lifecycle management activities (post-approval submissions to HC for new indications, new dosage forms, new strengths, manufacturing changes, etc.) are required to ensure the maintenance of the product license with its related improvements. In summary, sponsors need to ensure its continued compliance with the Food and Drug Regulations while their products are on the market. As these regulations are specific to drugs, it is not clear how medical device-related requirements will be applied to drug-led combination products.

Post Market Requirements under CMDR

Since the January 1, 2019 transition to MDSAP, medical devices of Class II-IV must comply with ISO 13485 as well as the QMS requirements of the CMDR. The compliance is confirmed through an audit performed by an MDSAP-accredited Auditing Organization (AO). These new expectations are also applicable to device-led combination products. However, it is not clear how the drug quality and safety aspects of a combination product will be controlled, as there are no regulatory notes found on the topic. The only reference found is in the "Guidance Document: Quality Management System - Medical Devices - Guidance on the Control of Products and Services Obtained from Suppliers" (2010) which states "This guidance document is also applicable to combination products which are regulated as medical devices. However, regulations may impose additional or differing requirements on suppliers and/or manufacturers of combination products (device/drug, device/tissue, device/biologic, etc.)." However, how these additional requirements will be imposed is not specified.

Post-market Activities under CMDR

In comparison to other jurisdictions, the post-market surveillance requirements appear to be less stringent. For example, while post-market surveillance studies are required for several cases in the EU, they are typically not required in Canada. HC does, however, enforce measures for the same purpose such as the Mandatory Problem Reporting (MPR). The MPR refers to the requirement for medical device manufacturers or importers to submit incident reports to HC per it's Food and Drugs Act covered by the "Medical Devices Regulations SOR-98-282 (CMDR)" and "Guidance Document for Mandatory Problem Reporting for Medical Devices."

Requirements for significant changes to be reported for review by the agency are also enforced per the "Guidance for the Interpretation of Significant Change of a Medical Device."

As part of its five-year initiative on the "regulatory review of drugs and devices" which includes the project, "Strengthening the use of real-world evidence and regulations for medical devices," HC intends to also strengthen the post-market surveillance and risk management of medical devices in Canada [46]. Once launched, the regulations will provide the authority to require "Analytical issue reports" and "Annual reports" to include significant changes among other information. They will also grant the authority to:

- *Compel manufacturers to conduct a reassessment (section 21.31 of the FDA) and the conditions under which this authority would be used;*
- *Require tests and studies (section 21.32 of the FDA) and the conditions under which this authority would be used;*
- *Require manufacturers to provide notification to Health Canada of any risk communications, changes to labeling, recalls, reassessments, and suspensions, or revocations of medical devices occurring in other countries.*

Post-Market Surveillance under the Canadian Food & Drug Regulations

On the drug side, post-market surveillance activities appear to be more established with several measures in place. Among the measures there are post-market drug benefit-risk assessment, [47] Periodic Safety Update Reports (PSUR), Periodic Benefit-Risk Evaluation Report (PBRER), RMPs, annual summary reports (ASRs), issue-related summary reports (IRSRs), and Post-Notice of Compliance (NOC) Changes [48].

JAPAN

Regulatory Evolution

The current Japan Pharmaceuticals and Medical Devices Agency (PMDA) regulations are laid out in the Pharmaceuticals and Medical Devices Act (PMD Act) [49,50]. The PMD Act came into force on November 25, 2014, and replaced the Pharmaceutical Affairs Law (PAL). With the implementation of Revised PAL, the following notifications have become effective:

- Handling of Marketing Application for Combination Products (released on October 24, 2014);
  (PFSB/ELD Notification No. 1024-2, PFSB/ELD/OMDE Notification No. 1024-1, PFSB/SD Notification No. 1024-9, PFSB/CND Notification No.1024-15)
- Q&A on Adverse Drug Reaction and Malfunction Reports of Combination Products (released on October 31, 2014).

Regulatory Oversight

Japan's Ministry of Health, Labor and Welfare (MHLW) is the regulatory body that oversees medical products in Japan, which includes creating and implementing safety standards for medical devices and drugs. In conjunction with the MHLW, the PMDA is an independent agency that is responsible for reviewing drug and medical device applications. The PMDA works with the MHLW to assess new product safety, develop comprehensive regulations, and monitor post-market safety.

The PMD Act is known as the Act on Securing Quality, Efficacy and Safety of Pharmaceuticals, Medical Devices, Regenerative and Cellular Therapy Products, Gene Therapy Products, and Cosmetics. It affects all aspects of Japanese medical product registration, including in-country representation, certification processes, licensing, and quality assurance systems. Accordingly, it also controls combination products.

Department groups which may get involved in the control of combination products include:

- PFSB: Pharmaceutical and Food Safety Bureau;
- ELD: Evaluation and Licensing Division;
- OMDE: Office of Medical Devices Evaluation.

Product Classification

Regulatory oversight on medical devices classify these products depending on risk level as a General Medical Device (Class I), Controlled Medical Device (Class II), or a Specially Controlled Device (Class III and Class IV). For General Medical Devices, only a notification/self-declaration is required, and the product does not need to undergo the approval process by the MHLW and PMDA. For Controlled Medical Devices, designation is certified by an authorized third-party certification party or reviewed by the PMDA. Only Specially Controlled Medical Devices must be reviewed and approved by the PMDA and MHLW.

On the other hand, combination products with a drug primary mode of action are expected to be controlled based on the drug regulatory pathways under the Japan PMDA regulations. The PMDA reviews new drugs, generic drugs, OTC drugs/behind-the-counter (BTC) drugs, and quasi-drugs and conducts re-evaluations of previously approved drugs.

A combination product can also be assigned to the regenerative medical product route. On this account, the PMDA regulations define cellular and tissue-based (regenerative medicine) products as processed human/animal cells or tissues that are intended to be used for:

- The reconstruction, repair, or formation of structures or functions of the human body;
- The treatment or prevention of diseases.

As of August, 2014, regenerative medicine products are regulated under the Japanese GCTP (Good Gene, Cellular, and Tissue-based Products Manufacturing Practice) for the manufacturing management and the quality control of regenerative medicine products.

Regulatory Definitions

Per the list of notifications released on October 24, 2014, combination products are referred to as

> products marketed as a single drug, medical device, or cellular and tissue-based product that combine two or more types of drug, device, processed cell, etc. (hereinafter referred to as "drugs etc.") that are expected to fall under the category of drugs, medical devices, or cellular and tissue-based products if marketed individually (hereinafter referred to as "combination products").

Confirming the Classification

A combination product is reviewed and regulated according to main function/purpose on a case-by-case basis. MHLW defines the PMOA based on rationales provided by the sponsor to decide whether the product falls under the category of drugs, medical devices, or cellular and tissue-based products.

Identifying the Applicable Requirements

A handling scheme is illustrated in the Figure 14.12 based on the notification entitled "Handling of Marketing Application of Combination Products" (October 2014). In addition, the following products are also handled as combination products per the notification:

- Products in which medical devices for puncture, such as catheters and injections, and external disinfectants used as drugs for disinfecting the skin at the puncture site are combined and then they are packaged together and sterilized (according to PFSB/ELD/OMDE Notification No. 0331002) if they include drugs;
- Products in which marketed drugs, medical devices, or cellular or tissue-based products are sold together by distributors (PMSB/IGD Notification No. 104);
- "Drugs Approved for Integral Marketing with Devices" specified in Article 98-2 and Article 228-20-3; "Medical Devices Approved for Integral Marketing with Drugs" specified in Article 114-60-2; and "Cellular and Tissue-based Products Approved for Integral Marketing with Devices etc." specified in Article 137-60 of the Ministerial Ordinance for Enforcement of the Act for Ensuring, etc. the Quality, Efficacy, and Safety of Drugs, Medical Devices, etc. (Ordinance of the Ministry of Health and Welfare No.1 of 1961).

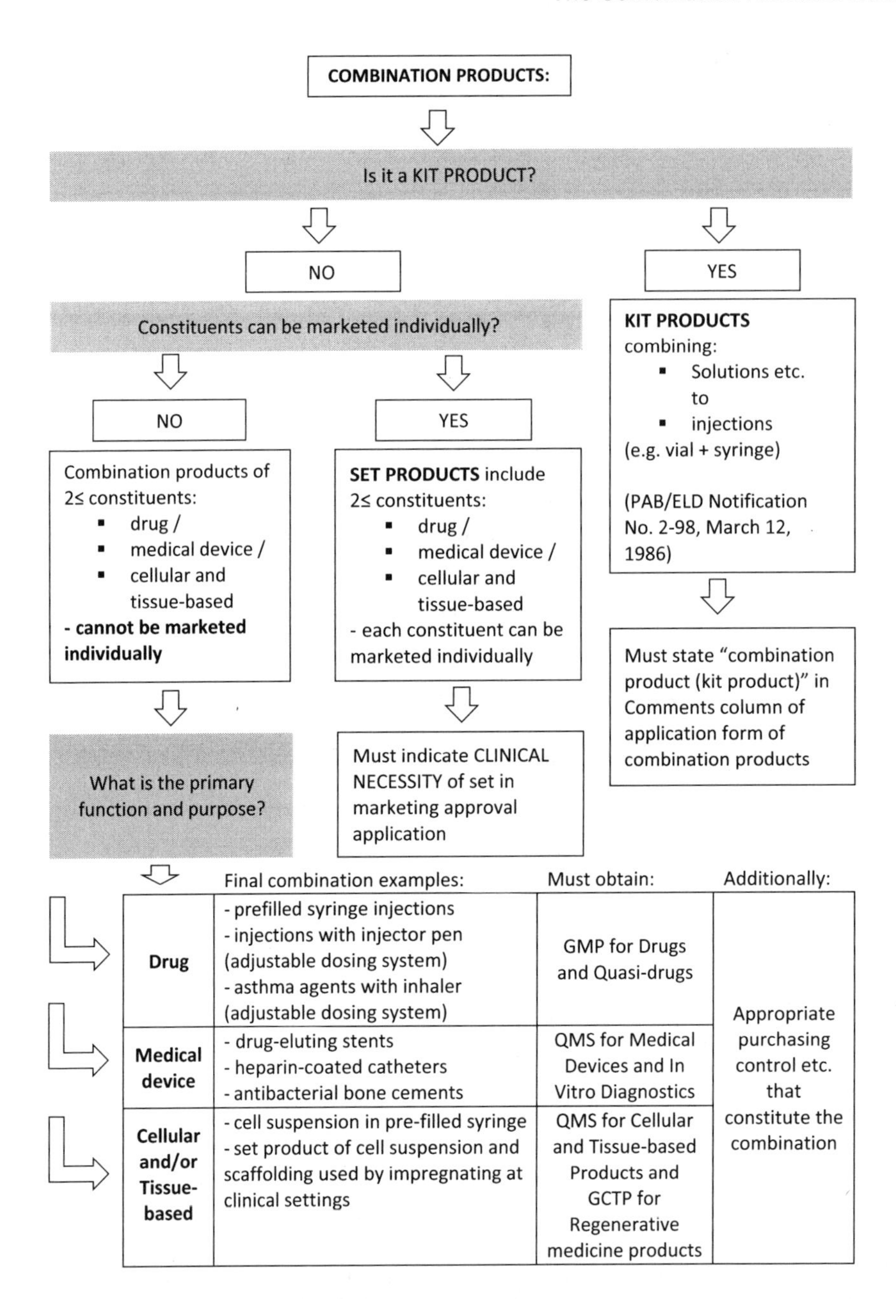

FIGURE 14.12 Combination product handling based on the notification entitled Handling of Marketing Application for Combination Products (2014).

The recommendations for developing an adequate strategy provided in Figure 14.6 (US Section) are also applicable for Japan as they aim to satisfy the globally recognized quality, safety, efficacy, and effectiveness evidence expectations (i.e., ICH, IMDRF, etc.). Nevertheless, the jurisdiction-specific expectations should be clarified with the concerning regulatory authorities before executing a development strategy.

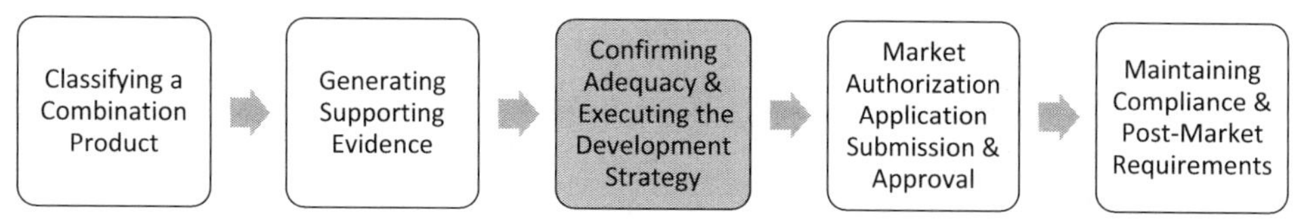

As for the primary jurisdictions, PMDA also offers opportunities to meet with sponsors to provide feedback not only on clinical trials but also on data for regulatory submissions of drugs, medical devices, and cellular and tissue-based products. One of the established meetings is also dedicated to innovative products where PMDA intends to give feedback on the development process.

In the case that a combination product is considered a medical device, the Office of Medical Device leads, while the drug gets reviewed by the Office of New Drug. The two offices work together in the approval of the medical product. In the opposite scenario where the drug is the lead, the same approach is adopted where the roles are switched, thereby ensuring that the review is handled within MHLW. Thus, combination products do not require individual marketing approval, certification, nor notification.

The GMP requirements enforced by PMDA according to the classification of the combination products are listed in Figure 14.12.

Similar to other regions, post-market surveillance guidelines in Japan involve reporting of adverse drug reactions and device malfunctions regardless of whether the combination product is primarily regulated as a drug or as a device. As per the "Attachment: Reports of Adverse Drug Reactions etc. of Pharmaceuticals" (October 2, 2014), guidelines and timelines are indicated as to what is defined as an adverse event and when these must be reported to the PMDA.

In addition to the common PMS controls, PMDA has introduced a couple of approaches which are different from those observed in other jurisdictions. For example, in addition to the RMPs, the following can be required as conditions of approval:

- EPPV: Early Post-marketing Phase Vigilance (6 months intensive monitoring)
- Re-EX: Re-examination (Pharmaceutical Affairs Act) (scheduled generally eight years after approval)
- Re-EVA: Re-evaluation (Pharmaceutical Affairs Act)

China

Regulatory Evolution

The State Food and Drug Administration (SFDA) No 16 Notification on Matters Concerning Registration of Drug and Medical Device Combination Products released in 2009 is one of the initial government communications regarding combination products. However, this notification was released in the midst of an array of changes to the overall regulatory system which started long before and are still occurring [51]. This notification has recently been abolished and replaced by the most recent NMPA Notice on Matters Concerning the Registration of Drug-Device Combination Products released in 2021.[23]

Many regulatory and administrative developments have taken place from the time of the SFDA notification in 2009 to the most recent NMPA Notice of 2021. One of the major reforms since the SFDA No. 16 publication consists of the conversion of the SFDA to the China Food and Drug Administration (CFDA) in 2013. Then in 2014, the entire medical device regulatory regime was revised, which was later amended and approved by the State Council in 2017. Likewise, on the drug side, a new system of registration pathways for small-molecule drugs was established in 2015. This pathway introduced the requirement for generic drugs to demonstrate therapeutic and quality equivalence with a fully evaluated reference or innovator drug.

These efforts led the way to the China's State Council new proposed regulations for drugs and devices to streamline the development and approval processes, and increase the enforcement of good practice guidelines and post-market obligations, particularly in the wake of a scandal related to the manufacturing of vaccines that emerged shortly before. Since May 2017, CFDA has published four new documents related to the clinical development and marketing of new drugs as document numbers 52–55:

- Encourage Drug/Medical Device Innovation and Accelerate Review/Approval (Document No.52) for the optimization of the review and approval process for product registrations by establishing a centralized dossier system, a panel review system, and more efficient communication channels;
- Encourage Drug/Medical Device Innovative Reform of Clinical Trials Management (Document No. 53) which introduced:
  - The option to recruit third parties to conduct trials instead of only allowing the used certified medical institutions; and
  - The acceptance of foreign clinical trial data for drugs and devices.
- Encourage Drugs/Medical Device Innovative Implementation of drugs/medical devices' lifecycle management (Document No. 54) which introduced the Marketing Authorization Holder (MAH) system, with the requirement for an entity to be appointed as holding the responsibility for pre- and post-approval activities;
- Protect Drug and medical devices innovators' rights and interests (Document No. 55) for the strengthening of intellectual property (IP) protection.

It should be noted that in June 2017, China also reached a milestone by becoming a full regulatory member of ICH, influencing the rapid issuance of guidelines aligned with global practice. Furthermore, following the Opinion on Deepening the Reform of the Review and Approval System and Encouraging the Innovation of Drugs and Medical Devices (Innovation Opinion) (No. 42 2017) in late 2017, a reorganization of government agencies was implemented as part of the activities launched to reach the core goal of the 13th Five Year Plan and related policy plans that the government put in place thereafter.

In 2018, the food function was removed from CFDA, and it was reorganized into the National Medical Products Administration (NMPA), which is now the primary regulator of drugs, medical devices, and cosmetics. The NMPA is subordinate to a new ministry, the State Administration for Market Regulation (SAMR). The NMPA also continued the regulatory framework reform by releasing drafts of the Drug Administration Law (DAL) and of the Regulations on the Supervision and Administration of Medical Devices (RSAMD) for comments in 2018.

Many of the activities continue to this day. A proposal to ease rules for the import of medicines, if they are approved in other countries, was made in August 2019 and was expected to go into effect in December of the same year. The Innovation Opinion is now final; however, the laws and regulations necessary for its effective implementation will continue being released. In 2021, the NMPA Notice was released addressing the registration of drug-device combination products, and most recently, in 2022, an announcement was made initiating the development of guidelines for drug-device combinations with a device PMOA.[24]

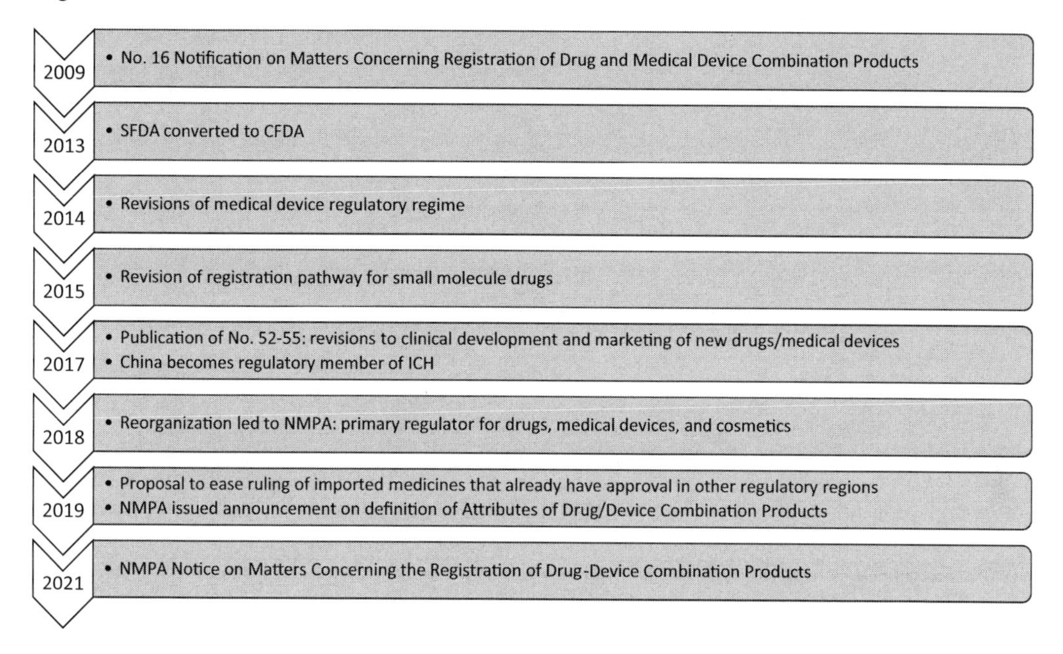

FIGURE 14.13 China's recent re-organizational timeline and regulatory revisions.

Regulatory Oversight

After reorganization of the CFDA, the NMPA is now the primary regulator of drugs, medical devices, and cosmetics. Table 14.6 lists the departments and functions under the NMPA.

Regulatory Definitions

Drug-device combination products refer to medical products composed of drugs and medical devices and produced as a single entity. This definition is drawn from the NMPA Notice on Matters Concerning Registration of Drug and Medical Device Combination Products (NMPA, 2021).[25]

Confirming the Classification and Identifying the Applicable Requirements

Accordingly, the MAH must evaluate all the attributes of the combination product to determine whether it is drug-led or device-led. In cases when this is unclear, an application must be applied for "attribute definition" to the Center for Medical Device Standards Management (CMDSM) prior to registration. If the drug plays the leading role, the MAH of the combination product shall apply for drug registration. If the device is device-led, it shall apply for medical device registration.

The supporting documentation is expected to contain information related to the structure and composition, intended use, action mode, usage and duration, development, manufacturing, and related products. Once the supporting material is submitted, the CMDSM will give an opinion within 20 business days; however, requiring supplementary dossiers and expert discussions may prolong this process.[26]

TABLE 14.6
China NMPA Departments and Functions

NMPA Departments	Function
Center for Medical Device Evaluation (CMDE)	– Dossier review during medical device registration process
General Administration of Quality Supervision, Inspection, and Quarantine (AQSIQ)	– Evaluation of food and cosmetics – Conducts mandatory safety registration, certification, and inspection of certain devices (e.g., radiation emitting)
Department of Medical Device Registration	– Registration for Class III devices – Optimizes control procedures – Organizes and implements classification administration – Supervises the implementation of best practices
Department of Medical Device Supervision	– Tracks and analyzes existing problems with medical devices – Recommends system modifications – Conducts medical device adverse events monitoring and reevaluation – Handles issues related to manufacturing and distribution of devices
Center for Drug Evaluation (CDE)	Assess safety and effectiveness data in clinical trial application justifies NMPA approval for: • Chemistry drugs • Traditional Chinese medicines • Biologic products
Departure of Drug and Cosmetics Registration	– Registration of pharmaceutical products – Optimizes registration and licensing procedures – Monitors drug non-clinical and clinical studies

China's efforts throughout the recent years to harmonize its regulatory framework with the rest of the world led to its introduction to the ICH. As being a member of this international harmonization group, it can be expected that China's requirements should be aligned with globally accepted quality, safety, efficacy, and effectiveness evidence. Nevertheless, the center-specific requirements should be clarified before executing a development strategy.

With the improvement of the regulatory pathways, China has introduced several processes for sponsors to meet with the agency. Particularly, NMPA introduced the Pre-IND/Pre-submission Consultation/Communication Meeting routes as well as the milestone meetings dedicated for development conversations with Center for Drug Evaluation (CDE) or CMDE [53]. This options are also open for products which would be considered to be combination products.

Pre-market Requirements

According to the NMPA Notice of 2021, where the drug or medical device contained in combination product has already been approved either in China or in the country of origin, the marketing approval document must be submitted along with the registration application.

In cases when the applicant disagrees with the attribute definition outcome, a request for re-review may be made within ten business days of the outcome. Thereafter, the opinions resulting from re-review would be considered final for the attributes definition.

With the determined attributes definition, the applicant may proceed with drug or medical device registration application to NMPA and indicate "drug-device combination products" in the application form.

The general approach to access the Chinese market with a combination product is summarized in Figure 14.14.

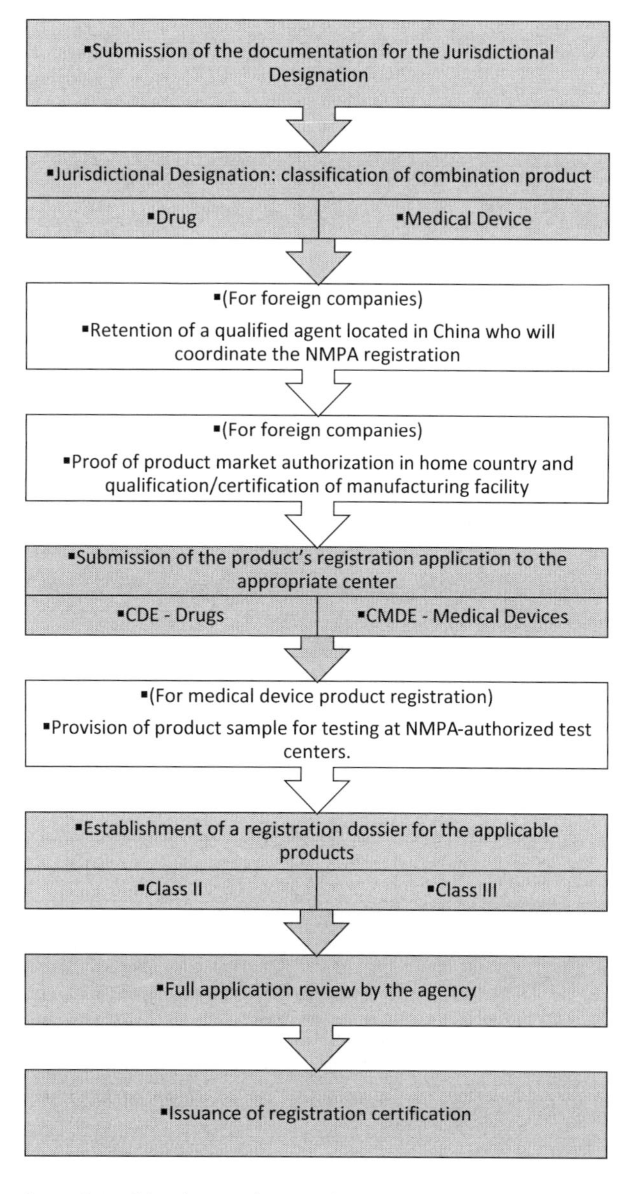

FIGURE 14.14 Flowchart of combination product market application process for Chinese market.

During the evaluation of a combination product, the experts from the relevant medical device departments work jointly with those from the drug department until the final approval.

Combination products with a device primary role follow the medical device regulatory pathways. Under these regulations, devices can be categorized into Class I, Class II, and Class III devices. Class I devices are simple devices that can be effectively monitored through regular administration, while Class II and Class III devices (i.e., complex implants or life-supporting devices) are held to higher regulatory standards. These medical device classification rules were issued in State Drug Administration's Order Number 15.

For market application, it is required that the MAH provide device samples to the NMPA for testing. In the case of registering Class II and Class III devices, manufacturers are obligated to send the appropriate documents showing that the device has been approved in its country of origin (i.e., CE Mark, 510(k) letter, ISO 13485 certification, approved Pre-market Approval Application). It may also be required to provide supportive clinical data along with the application. All product information on packaging and labeling must be translated into Simplified Chinese. And foreign manufacturers must also hire China-based agents that will represent their interests in China.

Combination products with a drug's primary role follow the drug regulatory pathways. Under the respective regulations, drugs can be classified into three types: Chemical Drugs, Biological Drugs, and Traditional Chinese Medicine/Natural Drugs. Within the Chemical Drugs class, drugs can be further categorized into Innovative New Drugs, Improved New Drugs, Generic Drugs, Domestic Generic Drugs, and Imported Drugs. Depending on the classification of the drug different types, levels of proof are imposed [54].

Among the improvement measures introduced by CFDA, enforcement of quality system requirements for medical devices were also considered through inspections carried out by the General Administration of Quality Supervision, Inspection and Quarantine (AQSIQ). The intensified requirements based on the new standards released in 2017 raised the bar on risk analysis and management, medical device supply chain and procurement, and post-market surveillance. There are also GMP certification requirements for pharmaceutical companies. These are expected to ensure the globally harmonized standards as China became a formal member of the ICH in 2017.

Post-market safety monitoring in China involves adverse event reporting from multiple stakeholders. In 2011, the Adverse Drug Reaction Reporting and Monitoring required drug regulatory authorities at national, provincial, and municipal levels to organize local collections systems that obligated the manufacturer, distributor, and healthcare centers to report adverse events within specified timeframes. For medical devices, 2018 saw the revisions made to Measures on the Administration of Medical Device Adverse Event Monitoring and Reevaluation which include "innovative devices." In summary, these documents indicate specified guidelines for adverse event reporting on all levels.

Regulatory Landscape Comparison for US, EU, Australia, Canada, Japan and China

Table 14.7 compares combination product terminology and guiding documents/ regulations, as well as regulatory pathways across the countries/ regions we have reviewed to this point in the chapter. See also Appendix: Global Combination Product Regulatory Framework Comparative Overview.

TABLE 14.7
Combination Product General Requirements Framework Comparison across Jurisdictions

<table>
<tr><td></td><td colspan="8">Combination Product Terminology and Guiding Documents/Regulations*</td></tr>
<tr><td rowspan="2">US</td><td colspan="4">Single-Entity Combination Product
---21 CFR 3.2(e)(1)---</td><td colspan="2" rowspan="2">Co-packaged CP
---21 CFR 3.2(e)(2)---</td><td colspan="2" rowspan="2">Cross-labeled CP
---21 CFR 3.2(e)(3) or (4)---</td></tr>
<tr><td>Device-Led CP</td><td colspan="2">Drug-Led CP</td><td>Biologic-Led CP</td></tr>
<tr><td>EU</td><td>Medical Device with Ancillary Medicinal Component
---MDR 2017/745---</td><td colspan="2">Integral DDC
--- MPD 2001/83/EC & Article 117 MDR 2017/745---</td><td>---MPD2004/23/EC---</td><td colspan="4">Non-integral DDC</td></tr>
<tr><td rowspan="2">Australia</td><td colspan="4">Combination Product</td><td colspan="4">Multi-component Pack</td></tr>
<tr><td colspan="2">Medicine-Medical Device
---ARGMD part 2 sect 14 (2011)[+]---
---ARGPM---</td><td colspan="2">Biological + Therapeutic Product
---ARGB---</td><td>Medicine Kit or Composite Pack
---ARGPM---</td><td colspan="2">System or Procedure Pack (SOPP)
---ARGMD---</td><td>Biologic Kit or Composite Pack
---ARGB---</td></tr>
<tr><td rowspan="2">Canada</td><td colspan="4">Combination Product
“drug and device combined in a singular product”</td><td colspan="4" rowspan="2">No clear equivalent
---CFDR--- & ---CMDR---</td></tr>
<tr><td>Medical Device PMOA
---CMDR SOR/98-282---</td><td colspan="3">Medicinal PMOA
---CRC, c.870---</td></tr>
<tr><td>Japan</td><td colspan="4">Combination Product
“cannot be marketed individually”
---Handling of Marketing Application for Combination Products (2014)---</td><td colspan="2">Set Product</td><td colspan="2">Combination Product
“Kit Product”
---PAB/ELD Notification No. 2-98, March 12, 1986---</td></tr>
<tr><td rowspan="2">China</td><td colspan="4">Combination Product defined by Attribute Definition
---NMPA Notice on Drug-Device Combination Products, 2021---</td><td colspan="4" rowspan="2">No clear equivalent</td></tr>
<tr><td>Medical device</td><td colspan="3">Drug</td></tr>
</table>

[+]Currently under revision

TABLE 14.8
Overview of CP Regulatory Pathways across Regions

Regional Comparison from Combination Product Classification to Post-market Surveillance*						
	US			**EU**		
Classification	Device-led CP	Drug-led CP	Biologic-led CP	Device classification		Medicinal product classification
Consultation meetings	Pre-submission			Pre-submission		
Application types	IDE, HDE, 510(k), PMA, IND, NDA, ANDA	IND, BLA, ANDA	NDA, IND	STED to NB		MAA to EMA/CA
Approval	FDA approval/clearance			CE mark	EMA/CA consultation (device-led) or NB opinion (drug-led)	EMA/CA authorized
Post-market	21 CFR Part 4 – cGMP & PMSR			PSUR, PMCF		PASS, PAMs, PSUR, RMP, PAES
	Australia			**Canada**		
Classification	Device-classified	Medicine or biologicals-classified		Device PMOA		Drug PMOA
Consultation meetings	Pre-submission			Device advice: Pre-clinical meetings		Pre-submission/pre-application
Application types	Market authorization process/application			MDL, MDEL application		NDS, SNDS, ANDS, SANDS, DINA
Approval	TGA CA Certificate	TGA License		MDL, MDEL		NOC, DIN
Post-market	Maintaining listing/registration on ARTG Randomly selected for post-market review			MPR		PSUR, RMP, PBRER, ASR, IRSR, NOC
	Japan			**China**		
Classification	Device-classified	Drug-classified		Device-classified		Drug-classified
Consultation meetings	Pre-submission			Pre-submission, communication meetings		
Application types	Application to Office of Medical Device	Application to Office of New Drug		Registration application to CMDE		Registration application to CDE
Approval	Approval/notification/certification			Registration certificate		
Post-market	RMP: EPPV, Re-EX, Re-EVA			Adverse event reporting and monitoring		

Rest of the World

In addition to the countries and regions described above, other jurisdictions have also begun to introduce the combination product concepts within their regulatory frameworks as well. Some of the notable jurisdictions rising from the horizon are India, Brazil, Saudi Arabia, Malaysia, and South Korea. The process of improving their basic regulatory framework for the mainstream products (drugs, biologics, devices) often takes precedence; thus, combination products may take the back seat. Increasingly, though, jurisdictions around the globe are recognizing the need to make regulatory amendments and changes as they consider how best to address the wide range of questions and considerations raised by this diverse category of products. Most jurisdictions do not have the benefit of a focal point, like US FDA Office of Combination Products, to support coordination between device and medicinal regulatory bodies. Perhaps as this space evolves, regions will evolve to similar models. At DIA Europe (July 2020 virtual conference), in a session co-led by EMA, it was suggested that an integrated evaluation pathway for drug-device combination products is needed to ensure timely access to patients of innovative treatments. Benefits of the FDA OCP, especially for small- to medium-sized enterprises, were explained by a company sharing their experiences in US and Europe. It is apparent that regulatory authorities worldwide are working toward improved convergence with respect to drugs, biologics, and devices. Combination products ultimately will present ample opportunities for harmonization as well. Through a variety of industry forums, health authorities are signaling a willingness to engage (e.g., June 2020 RAPS/TOPRA virtual conference and ASTM International efforts E55 Committee efforts). Clearly, there is more to come in this exciting space!

SUMMARY

Most countries do not have a regulatory definition for "combination products" as the US, but do adopt the concept and refer to similar terms. Accordingly, they do not always have regulations specific for combination product control. Most of the time, these products are regulated according to the part with the main function or purpose, usually referred to as the PMOA.

Currently, the FDA has a separate regulatory OCP although the regulatory oversight is led by the center responsible for the PMOA. Over the past 15 years, the FDA has released an accelerating number of regulations and guidance documents for pre-market and post-market expectations expressly and specifically for combination products. The EU has also started introducing regulations to better control these products as part of the new MDR roll out. In addition, guidelines are expected to be issued in the process. Many of the other jurisdictions such as Australia, Canada, Japan, and China have also taken major steps to assure adequate controls. We are seeing current drafts of regulatory documents being released and revised requirements being issued. And at the time of this writing, the latest communications indicate that either previously released documents are in revision or new regulatory guidelines or guidance documents are being released.

Global Challenges and Recommendations

The initiatives launched by the different countries and regions should be applauded as the need is evident. However, well-defined regulatory pathways are often missing and the regulatory discrepancies still exist not only geographically but also for several combination product specific situations. Consequently, companies find it difficult to develop appropriate and reliable regulatory strategies which can assure approvability, thereby discouraging innovation in this area to a certain extent.

Nevertheless, based on the regulations released, products regulated, and the directions being considered by the different jurisdictions thus far, manufacturers and developers of combination products should look to address the following challenges as part of their global regulatory strategy development early on in the design process.

Collaboration Misalignment

One of the inherent challenges that is introduced with the merger of two different worlds (device and drugs/biologics) is communication and collaboration effectiveness. Because combination products are a final combination of constituent parts, it is no surprise that successful development and subsequent compliance will rely on clear communication between the device and drug experts within a company but also within the country-specific authority setups. The collaboration is challenged by the fact that medical device experts often know very little about developing drugs, nor are pharmaceutical experts always versed in medical device regulations, since they traditionally work separately in siloes. This complicates development and later approval of products in which a device and a drug work together. Device experts would benefit from understanding concepts governing the drug properties, pharmacodynamics, pharmacokinetics, and the like. Likewise, the drug experts would benefit from understanding concepts such as requirements under design controls, management responsibility, purchasing controls, CAPAs, technical files, and Summary Technical Documentation (STED) documentation that are critical aspects for the development and regulation of medical devices [64]. Related to companies, collaboration should include the development of analytical methods, understanding user needs, stability tests, and sterilization among others. This situation can be complicated further if the teams involved are familiar with the regulations from some jurisdictions and not others. Therefore, it also becomes essential to establish the regulatory foundation and structure to foster synergies between these two groups.

In addition to the gaps in knowledge of each other's operation methods, the cultural, experience, and mindset differences between the drug and device world can also create complexities hindering the flow of information sharing. One simple example is the set of terminologies which can greatly differ sometimes even to refer to similar principles. The inability to speak the same regulatory language can rapidly result in discord in the development and manufacturing processes (see Chapters 5 and 8, Integrated Development and Lifecycle Management). Accordingly, cross-functional training for teams, understanding of common goals, and mentoring are crucial for developing a collaborative working environment.

Global Compliance Challenges

From a product development standpoint, what is often observed is that as drugs are developed, specific approaches are applied under the QbD which can be very different from the device design control methodologies. An ineffective development program can lead to a product failing to meet all of the required development regulatory requirements, i.e., generation of results which would not meet all necessary expectations. To control this risk, combination product manufacturers should ensure that the development strategies cover development requirements not only from drug and device requirements, but should also capture the differences in requirements from the different countries and regions.

When engaging with combination products, important medical device development requirements which could be novel to the drug world include human factors, software, cyber security, biocompatibility, mechanical properties, component material properties, to name a few. Procedural and conceptual differences can also introduce challenges such as the understanding of eCTD requirements and QTPP.

For the most part, many of the jurisdictions have similar levels of expectations. However, during the transfer of the knowledge between drug and device counterparts, the global aspect of the requirements should be considered. This is essential for the ability to develop and capture all strategies to convince the targeted regulatory authorities of the preferred PMOA as well as the acceptability of the level of evidence the manufacturer plans on providing. Once aligned internally, it is then highly recommended to engage in early conversations with the regulatory bodies of the targeted jurisdictions. This would further manage the potential regulatory risks that may not be the same across jurisdictions.

GMP Gap Assessments

Even with a well-developed product, there could still be a risk of not aligning the compliance of the established quality management systems against applicable drug and medical device regulations from all jurisdictions of interest. While FDA's guidance document "Current Good Manufacturing Practice Requirements for Combination Products" is intended for the US, companies can also follow the principles in preparation for the control of other jurisdictions, especially for those which have not yet established and communicated their post-marketing application authorization requirements and expectations.

To assure the readiness of the GMP to cover not only the quality and manufacturing aspects but also the product's technical dossier documentation, gap assessments should be performed against the applicable requirements from the different jurisdictions. In particular, these gap assessments should be performed on what are referred to legacy products, products that have been in the market prior to the release of new regulations they must comply to. Internationally accepted risk management approaches (i.e., ICH, ISO, etc.) should be followed in these processes to justify regulatory and quality decisions.

Future Outlook

When considering the projections on the market value for the near future, it is clear that combination products will continue to play a major role in the healthcare industry. With the rapid evolution of technologies, we may expect the development of even more complex combination product configurations. However, it seems we are only at the beginning of the regulatory development stage for these products. Combination products regulators may need to ramp up their communication and sharing of new policies to coincide not only with the pace of technology but also with the regulations specific to its constituent parts.

In terms of compliance requirements, there is a lack of combination product-specific policies. These (and other evolving issues) need to be taken into account by combination product manufacturers, including further engagement of relevant expertise. In addition to the need to ensure that regulatory frameworks are established and finalized, several specific regulatory topics will need to cover their impact on combination product oversight. In particular, the industry is lagging behind with unclear combination product-specific requirements on much of the following concepts of global harmonization, product traceability/counterfeit measures, generic and accelerated approval pathways, post-market/lifecycle requirements including post-market surveillance, software/digital health/cybersecurity, patient-specific/custom-made/3D printing, among others. Therefore, manufacturers need to stay abreast of the rapidly evolving regulations across multiple jurisdictions on all these topics, including those for specific constituent parts and combination products as a whole. As this can be challenging, in addition to staying up to date with the latest regulations and guidelines, it is strongly recommended to discuss these types of topics with the regulators of the jurisdictions of interest already in the beginning of any development program.

Opportunities and Challenges for Regulators (Harmonization)

On a regulatory framework level, the International Medical Device Regulators Forum (IMDRF) and the International Council for Harmonisation (ICH) are main players influencing the medical device requirements and the drug requirements, respectively. However, there exists no international organization specific to the regulations of combination products. Additionally, it is not unusual to come across different terminologies under these organizations as well as between jurisdictions that may be referring to one requirement under the context of combination products. This showcases the need for establishing globally aligned terminologies, which can be considered as one example of the array of harmonization needs particular to combination products.

Therefore, there might be an opportunity for these major organizations to merge their efforts or for new multinational organizations to be created, to develop global frameworks specific to combination products. This approach would help converge the divergence that is being observed despite the infancy of the development of regulatory frameworks in different jurisdictions. Furthermore, by having globally harmonized standards specified, new jurisdictions which are developing their frameworks would have principles to work from instead of duplicating effort, thereby saving time and financial resources.

CONCLUSION

The road to regulatory compliance for combination products has been a long one, and we have yet to arrive at our final destination. Nonetheless, ensuring that combination products are effectively managed throughout their product lifecycle is critical to long-term success in the market. However, navigating through the complex regulatory frameworks of different countries and regions is undoubtedly a costly investment in time and resources for the combination product industry and stakeholders.

While the necessary regulations ensure the safety and efficacy/effectiveness of products being introduced into the market, if not established correctly, the consequence is serious; it can be a delay in effective treatment for patients who may already have immediate needs for such products, in particular innovations designed along patient-centric principles and to enable efficient healthcare treatments. Therefore, close collaboration and transparency among regulators, combination product developers, and other stakeholders should be encouraged to help establish, improve, streamline, and harmonize global combination product regulatory oversight. This is essential to foster innovation and expedite market entry for the common goal of ensuring the availability of high-quality and safe new products to promote public health.

TABLE OF ABBREVIATIONS

ADME	Absorption, distribution, metabolism, and excretion
AIMD	Active Implantable Medical Device
AIMDD	Directive 90/385/EEC on active implantable medical devices
ANDA	Abbreviated New Drug Application
ANDS	Abbreviated New Drug Submission
AO	Auditing Organizations
AQSIQ	Administration of Quality Supervision, Inspection, and Quarantine
ARGB	Australian Regulatory Guidelines for Biologicals
ARGMD	Australian Regulatory Guidelines for Medical Devices
ARTG	Australian Register of Therapeutic Goods
ASEAN	Association of Southeast Asian Nations
ASR	Annual Summary Reports
ATMP	Advanced Therapy Medicinal Product
BGTD	Biologics Genetic Therapies Directorate
BTC	Behind-the-counter
CA	Conformity Assessment (Australia)
CA	Competent Authority (EU)
CAB	Conformity Assessment Body
CAPA	Corrective and Preventive Action
CDE	Center for Drug Evaluation

CDRH Center for Devices and Radiological Health
CE Conformité Européenne
CER Clinical Evaluation Report
CFDA China Food and Drug Administration
CGMP Current Good Manufacturing Practice
CHMP Committee for Medicinal Products for Human Use
CMDCAS Canadian Medical Devices Conformity Assessment System
CMDE Center for Medical Device Evaluation
CMDR Canadian Medical Device Regulations
CP Combination Product
CQA Critical Quality Attributes
CTD Common Technical Document
DAL Drug Administration Law
DDC Drug-device combinations
DHF Device History File
DIN Drug Identification Number submission
EMA European Medical Agency
EU European Union
Eudamed European Database on Medical Devices
FDA Food and Drug Administration
FDCA Federal Drug and Cosmetic Act
GDP Good distribution practice
GHTF Global Harmonization Task Force
GMP Good Manufacturing Practice
GSPR General Safety and Performance Requirements
GVP Good Pharmacovigilance Practices
HC Health Canada
HPFB Health Products and Food Branch
HTA Health Technology Assessment
ICH International Council for Harmonization
IMDRF International Medical Device Regulators Forum
IP Intellectual Property
IRSR Issue-Related Summary Reports
ISO Internal Organization for Standardization
IVD In Vitro Diagnostic
IVDD Directive 98/79/EC on in vitro diagnostic medical devices
IVDR Regulation (EU) 2017/746 on in vitro diagnostic medical devices
MAA Marketing Authorization Application
MAH Marketing Authorization Holder
MDCG Medical Device Coordination Group
MDD Directive 93/42/EEC on medical devices
MDEL Medical Device Establishment License
MDL Medical Device License
MDR Medical Device Reporting; in EU Regulation (EU) 2017/745 on medical devices
MDSAP Medical Device Single Audit Program
MDUFMA Medical Device User Fee and Modernization Act
MHLW Ministry of Health, Labor and Welfare
MOA Mode of action

MP	Medicinal Product
MPD	Directive 2001/83/EC on medicinal products for human use
MPR	Mandatory Problem Reporting
MRI	Magnetic Resonance Imaging
NB	Notified Body(ies)
NCA	National Competent Authority
NDS	New Drug Submission
NMPA	National Medical Products Administration
NOC	Notice of Compliance
OCP	Office of Combination Products
OCPP	Office of Clinical Policy & Programs
OSIP	Office of Submissions and Intellectual Property
PAES	Post-Authorization Efficacy Studies
PAL	Pharmaceutical Affairs Law
PAM	Post-Authorization Measures
PASS	Post-Authorization Safety Studies
PBRER	Periodic Benefit-Risk Evaluation Report
PMCF	Post-Market Clinical Follow-Up
PMD	Pharmaceuticals and Medical Devices
PMDA	Pharmaceuticals and Medical Devices Agency
PMOA	Primary mode of action
PMS	Post-market surveillance system
PMSR	Post-marketing safety reporting
PSUR	Periodic Safety Update Report
QbD	Quality by design
QMS	Quality Management System
QSR	Quality System Requirements
QTPP	Quality Target Product Profile
RFD	Request for Designation
RMP	Risk Management Plan
RSAMD	Regulations on the Supervision and Administration of Medical Devices
SAMR	State Administration for Market Regulation
SANDS	Supplemental Abbreviated New Drug Submission
SAWP	Scientific Advice Working Party
SFDA	State Food and Drug Administration
SMDA	Safe Medical Device Act
SME	Small to medium enterprises
SNDS	Supplemental New Drug Submission
STED	Summary Technical Documentation
TGA	Therapeutic Goods Administration
TPCC	Therapeutic Products Classification Committee
TPD	Therapeutic Products Directorate
WG	Working groups

NOTES

1. EMA/CHMP/QWP/BWP/259165/2019.
2. Source: https://european-union.europa.eu/institutions-law-budget/law/types-legislation_en
3. Updated – Joint Implementation Plan on actions considered necessary to ensure the sound functioning of the new framework for medical devices under the IVDR (europa.eu).
4. md_guidance_meddevs_0.pdf (europa.eu).

5. Regulation (EU) 2020/561 of April 23 2020 amending MDR as regards the dates of application of certain of its provisions.
6. Eudamed to Launch in 2022 for Both Devices and IVDs | RAPS.
7. Team NB = The European Association for Medical devices of Notified Bodies Home – Welcome to Team NB | Team NB (team-nb.org).
8. Team NB position paper on Template for NBOp_V4 (team-nb.org).
9. https://health.ec.europa.eu/system/files/2022-04/mdcg_2022-5_en_0.pdf
10. md_borderline_manual_09-2022_en.pdf (europa.eu).
11. Manual on Borderline and Classification in the Community Regulatory, Table 04. Integral and Non-integral (co-packaged and referenced) Products as defined in the EU [3232].

Framework for Medical Devices, Version 1.22 (05-2019).

12. Reusable surgical instruments are not possible to be device parts of integral products according to the definitions stated in MDR, Article 1(8) and 1 (9), 2nd sub-paragraph; class Ir is only mentioned to state all possible medical device classes.
13. mdcg_2021-24_en_0.pdf (europa.eu).
14. EMA/37991/2019, section 2.7.
15. MDCG 2020-3.
16. mdcg_2022-4_en.pdf (europa.eu).
17. Businesses to be given UK product marking flexibility – GOV.UK (www.gov.uk); information available as per November 14, 2022.
18. MHRA, Borderline products: how to tell if your product is a medical device and which risk class applies, published January 6, 2021, last updated November 16, 2022 (Accessed December 2022).
19. Australian regulatory guidelines for medical devices (ARGMD), Last updated: 19 August 2022 (Accessed December 2022).
20. TGA, Boundary and combination products – medicines, medical devices, and biologicals, Version 1.0, October 2022 (Accessed December 2022).
21. TGA, Boundary and combination products – medicines, medical devices, and biologicals, Version 1.0, October 2022 (Accessed December 2022).
22. TGA, System or procedure packs Guidance for sponsors, manufacturers and charities Version 1.0, November 2021, Accessed December 2022.
23. NMPA Notice on Matters Concerning the Registration of Drug-Device Combination Products (Accessed December 2022).
24. NMPA Announcement on Two Guidelines for Registration Review of Drug-Device Combination Products with Device Taking Primary Mode of Action (Accessed December 2022).
25. NMPA Notice on Matters Concerning the Registration of Drug-device Combination Products (Accessed December 2022).
26. NMPA issues the Announcement on Adjusting the Relevant Matters Concerning the Definition of the Attributes of Drug/Device Combination Products | govt.chinadaily.com.cn (Accessed December 2022).

REFERENCES

1. GlobeNewswire, "$177.7 Billion Drug Device Combination Products Market by Product & Region - Global Forecast to 2024," 30 April 2019. [Online]. Available: https://www.globenewswire.com/news-release/2019/04/30/1812682/0/en/177-7-Billion-Drug-Device-Combination-Products-Market-by-Product-Region-Global-Forecast-to-2024.html. [Accessed February 2020].
2. AHWP, "Update on AHWP Activity," in *IMDRF Stakeholders Forum*, 2015. https://www.imdrf.org/sites/default/files/docs/imdrf/final/meetings/imdrf-meet-150914-kyoto-presentation-ahwp-activity-update-01.pdf [Accessed 8 January 2023].
3. E. John Barlow Weiner, "Combination Products, Current Approaches, Challenges, Insights & Trends: The View from the United States," in *RAPS/TOPRA Inter-regulatory and Stakeholder Workshop on Alignment of Global Combination Products Regulations*, Virtual Conference, June 12, 2020.
4. E. John Barlow Weiner, "Combination Product Regulatory Development & Potential for Convergence," in *3rd Annual Lifecycle Management for Combination Products Summit*, Virtual Conference, November 17, 2020.
5. E. John Barlow Weiner, "Fireside Chat," in *Xavier/FDA Combination Products Summit*, Virtual Conference, October 2020.
6. D. C. Schillinger, "The Office of Combination Products: Its Roots, Its Creation, and Its Role," 2004. [Online]. Available: http://nrs.harvard.edu/urn-3:HUL.InstRepos:8852096.

7. MDUFMA, "Medical Device User Fee and Modernization Act of 2002," US FDA, 2002.
8. FDA, "21st Century Cures Act," Public Law 114–255 (42 USC 201), 13 December 2016. [Online]. Available: https://www.congress.gov/114/plaws/publ255/PLAW-114publ255.pdf. [Accessed 25 December 2020].
9. FDA, "81 FR 92603," Federal Register, 20 December 2016. [Online]. Available: https://www.federalregister.gov/documents/2016/12/20/2016-30485/postmarketing-safety-reporting-for-combination-products. [Accessed 25 December 2020].
10. FDA, "Human Factors Studies and Related Clinical Study Considerations in Combination Product Design & Development," FDA, 3 February 2016. [Online]. Available: https://www.fda.gov/regulatory-information/search-fda-guidance-documents/human-factors-studies-and-related-clinical-study-considerations-combination-product-design-and. [Accessed 25 December 2020].
11. FDA, "Current Good Manufacturing Practice Requirements for Combination Products - Guidance for Industry and Staff," FDA, 11 January 2017. [Online]. Available: https://www.fda.gov/regulatory-information/search-fda-guidance-documents/current-good-manufacturing-practice-requirements-combination-products. [Accessed 25 December 2020].
12. FDA, "How to Prepare a Pre-Request for Designation (Pre-RFD)- Guidance for Industry," FDA, February 2018. [Online]. Available: https://www.fda.gov/media/102706/download. [Accessed 25 December 2020].
13. FDA, "Combination Products Inter-center Consult Request Process (SMG 4101)," FDA, 11 June 2018. [Online]. Available: https://www.fda.gov/media/81927/download. [Accessed 25 December 2020].
14. FDA, "Expectations and Procedures for Engagement Among Medical Product Centers and OCP on Regulations and Guidance Pertaining to Combination Products – SMG 4103," FDA, 27 March 2018. [Online]. Available: https://www.fda.gov/media/112448/download. [Accessed 25 December 2020].
15. FDA, "Principles of Premarket Pathways for Combination Products-Draft Guidance for Industry," FDA, 6 February 2019. [Online]. Available: https://www.federalregister.gov/documents/2019/02/06/2019-01196/principles-of-premarket-pathways-for-combination-products-draft-guidance-for-industry-availability. [Accessed 25 December 2020].
16. FDA, "Postmarketing Safety Reporting for Combination Products- Final Guidance," FDA, July 2019. [Online]. Available: https://www.fda.gov/regulatory-information/search-fda-guidance-documents/postmarketing-safety-reporting-combination-products. [Accessed 25 December 2020].
17. FDA, "Intercenter Consults for Review of Human Factors Information - SMG 4104," FDA, 12 February 2019. [Online]. Available: https://www.fda.gov/media/120204/download. [Accessed 25 December 2020].
18. FDA, "Requesting FDA Feedback on Combination Products," FDA, 4 December 2020. [Online]. Available: https://www.fda.gov/media/133768/download. [Accessed 25 December 2020].
19. FDA, "Technical Considerations for Demonstrating Reliability of Emergency-Use Injectors Submitted under a BLA, NDA or ANDA- Draft Guidance," FDA, April 2020. [Online]. Available: https://www.fda.gov/regulatory-information/search-fda-guidance-documents/technical-considerations-demonstrating-reliability-emergency-use-injectors-submitted-under-bla-nda. [Accessed 25 December 2020].
20. FDA, "Inspections of CDER-led or CDRH-led Combination Products- Compliance Program 7356.000," FDA, 4 June 2020. [Online]. Available: https://www.fda.gov/media/138592/download. [Accessed 25 December 2020].
21. FDA, "Assessment of Combination Product Review Practices in PDUFA VI," 7 August 2020. [Online]. Available: https://www.fda.gov/media/141519/download. [Accessed 25 December 2020].
22. FDA, "Definition of the Term 'Biological Product'- 85 FR 10057–10063," FDA, 21 February 2020. [Online]. Available: https://www.federalregister.gov/documents/2020/02/21/2020-03505/definition-of-the-term-biological-product#:~:text=Under%20this%20final%20rule%2C%20the%20term%20protein%20means,statutory%20framework%20under%20which%20such%20products%20are%20regulated.. [Accessed 25 December 2020].
23. FDA, "Combination Products Guidance Documents," 23 December 2019. [Online]. Available: https://www.fda.gov/combination-products/guidance-regulatory-information/combination-products-guidance-documents.
24. FDA, "Medical Device Accessories - Describing Accessories and Classification Pathways," 2017.
25. FDA, "FDA Organization," 2020. [Online]. Available: www.fda.gov/about-fda/fda-organization.
26. S. W. Junod, "FDA and Clinical Drug Trials: A Short History," 2008.
27. NHS UK, "PIP Breast Implants," 21 January 2019. [Online]. Available: https://www.nhs.uk/conditions/pip-implants/. [Accessed June 2020].

28. European Commission, "European database on medical devices (EUDAMED)," February 2020. [Online]. Available: https://ec.europa.eu/growth/sectors/medical-devices/new-regulations/eudamed_en.
29. European Commission, "Guidance - MDCG Endorsed Documents," [Online]. Available: https://ec.europa.eu/health/md_sector/new_regulations/guidance_en. [Accessed 7 July 2020].
30. EMA, "Guideline on the Quality Requirements for Drug-Device Combinations," 29 May 2019. [Online]. Available: www.ema.europe.eu. [Accessed February 2020].
31. T. NB, "Position Paper for the Interpretation of Device Related Changes in Relation to a Notified Body Opinion as Required under Article 117 of Medical Device Regulation (EU)2017/745," 4 December 2020. [Online]. Available: www.team-nb.org. [Accessed 24 December 2020].
32. EMA, "Medical Devices," 2020. [Online]. Available: https://www.ema.europa.eu/en/human-regulatory/overview/medical-devices.
33. S. Eisenhard, "Europe's MDR and Combination Products: Clarification for Drug-Device Product Oversight," 21 May 2019. [Online]. Available: https://www.emergobyul.com/blog/2019/05/europes-mdr-and-combination-products-clarifications-drug-device-product-oversight.
34. EMA, "Q&A on Implementation of the EU MDR 2017/745 and IVDR 2017/746," 21 October 2019. [Online]. Available: www.ema.europe.eu/. [Accessed February 2020].
35. A. Government, "Therapeutic Goods (Medical Devices) Regulations 2002," [Online]. Available: https://www.legislation.gov.au/Details/F2017C00534.
36. TGA, "Biologicals Packaged or Combined with Another Therapeutic Good, ARGB," 2018.
37. TGA, "SME Assist," 2020. [Online]. Available: https://www.tga.gov.au/sme-assist.
38. TGA, "Australian Regulatory Guidelines for Medical Devices (ARGMD), V1.1," 2011. [Online]. Available: www.tga.gov.au.
39. TGA, "Pre-Submission Meetings with TGA," March 2018. [Online]. Available: https://www.tga.gov.au/sites/default/files/pre-submission-meetings-tga.pdf. [Accessed February 2020].
40. TGA, "Overview of Supplying the Therapeutic Goods in Australia," 2020. [Online]. Available: https://www.tga.gov.au/overview-supplying-therapeutic-goods-australia. [Accessed February 2020].
41. TGA, "Good Manufacturing Practice Overview," 2020. [Online]. Available: https://www.tga.gov.au/good-manufacturing-practice-overview.
42. TGA, "PE009-13, the PIC/S Guide to GMP for Medicinal Products," 2 January 2018. [Online]. Available: www.tga.gov.au/publication/pe009-13-pics-guide-gmp-medicinal-products.
43. TGA, "Australian Code of Good Manufacturing Practice for Human Blood and Blood Components, Human Tissues and Human Cellular Therapy Products," April 2013. [Online]. Available: https://www.tga.gov.au/publication/australian-code-good-manufacturing-practice-human-blood-and-blood-components-human-tissues-and-human-cellular-therapy-products.
44. TGA, "Essential Principles Checklist (Medical Devices)," 17 September 2019. [Online]. Available: https://www.tga.gov.au/sites/default/files/essential-principles-checklist-medical-devices.pdf.
45. HC, "Food and Drugs Act," 1985. [Online]. Available: https://laws-lois.justice.gc.ca/eng/acts/F-27/page-1.html.
46. HC, "Notice of Intent: Strengthening the Post-Market Surveillance and Risk Management of Medical Devices in Canada," April 2018. [Online]. Available: https://www.canada.ca/en/health-canada/services/drugs-health-products/public-involvement-consultations/medical-devices/noi-strengthening-post-market-surveillance-risk-management-medical-devices.html.
47. HC, "Format and Content for Post-Market Drug Benefit-Risk Assessment in Canada - Guidance Document," 8 February 2019. [Online]. Available: https://www.canada.ca/en/health-canada/services/publications/drugs-health-products/content-drug-benefit-risk-assessment/guidance-document.html.
48. HC, "Post-Notice of Compliance (NOC) Changes – Quality Guidance," 14 October 2010. [Online]. Available: https://www.canada.ca/en/health-canada/services/drugs-health-products/drug-products/applications-submissions/guidance-documents/post-notice-compliance-changes/quality-document.html.
49. NAMSA, "New Marketing Application Procedure for Combination Products in Japan," 11 July 2016. [Online]. Available: https://www.namsa.com/asian-market/new-marketing-application-procedure-for-combination-products-in-japan/.
50. P. B. Medical, "Japan MHLW & PMDA Medical Device and Pharmaceutical Regulations," 12 August 2018. [Online]. Available: https://www.pacificbridgemedical.com/regulation/japan-medical-device-pharmaceutical-regulations/.
51. J. Balzano, "China," in *The Life Sciences Law Review - Edition 7*, Covington & Burling LLP, 2019. [online]. https://thelawreviews.co.uk/title/the-life-sciences-law-review/china [Accessed 12 January 2023].

52. SFDA, "State Administration for Market Regulation," [Online]. Available: http://www.samr.gov.cn/. [Accessed February 2020].
53. Covington, "CFDA Releases Groundbreaking Drug and Device Policies for Public Comment," 23 May 2017. [Online]. Available: https://www.cov.com/-/media/files/corporate/publications/2017/05/cfda_releases_groundbreaking_drug_and_device_policies_for_public_comment.pdf. [Accessed February 2020].
54. ITA, "The GHTF Economics and Combination Products," 2019. [Online]. Available: https://www.trade.gov/td/ohit/assets/pdf/The%20GHTF%20Economis%20and%20Combination%20Products.pdf. [Accessed 2020].
55. 21 November 1991. [Online]. Available: 56 fed. reg. 58754–01, 1991 wl 242348.
56. vol. wl 242348, 1991.
57. A. Government, "Therapeautic Goods Administration," 2020. [Online]. Available: www.tga.gov.au/sme-assist.
58. Working Group 1, Pre-Market: General MD, "Draft Document: Guidance on Regulatory Practices for Combination Products," AHWP, 2016.

Appendix: Global Combination Product Regulatory Framework Comparative Overview

Dhiraj Behl and Susan W. B. Neadle

The contents of this Appendix are intended to be a high-level overview of the complex global regulatory combination products landscape. The information gathered and reflected under this Appendix is a culmination of public domain information, regulations, directives, guidelines, and standards coupled with author experience for the regions across global jurisdictions as of the time of this writing (January 2023). Given the very dynamic evolution of the global environment with respect to combination products, the reader is strongly encouraged to refer to the quick reference links in the table for respective health authority websites for the most current updates. (If any links are not working, please refer to the parent website and search under archival.)

Appendix Table 1: Regions/Individual Countries Covered in This Appendix

Association of Southeast Asian Nations (ASEAN) (Tables 2A–2D) • **Malaysia** • **Philippines** • **Singapore** • **Thailand**
Asia-Pacific (Tables 3A–3F) • **Australia** • **China** • **India** • **Japan** • **South Korea** • **Taiwan**
Europe (Tables 4A–4C) • **European Union** • **Switzerland** • **United Kingdom/Great Britain**
Latin America (Tables 5A and 5B) • **Brazil** • **Mexico**
Middle East North Africa (Table 6) • **Saudi Arabia**
North America (Table 7A and 7B) • **Canada** • **United States**
World Health Organization (Table 8) • **WHO**

Associations of Southeast Asian Nations (ASEAN)

TABLE 2A
Malaysia

Country	Malaysia
Is "Combination Product" Defined?	Yes, there is a definition of "combination product" available. https://portal.mda.gov.my/documents/guideline-documents/1771-guideline-for-registration-of-drug-medical-device-and-medical-device-drug-combination-products-4th-edition-1/file.html https://www.npra.gov.my/easyarticles/images/users/1047/drgd/APPENDIX-2---Medical-Device-Drug-Cosmetic-Interphase-MDDCI-and-Combination-Products.pdf
Scope	Single integral (drug/device, biological/device, or drug/device/biological) and co-packaged https://npra.gov.my/index.php/en/frequently-asked-questionscombo/faqs-general.html
Legal Framework	• Legal framework is available. • A product comprised of two or more regulated components, i.e., drug/device, biological/device, or drug/device/biological, which are physically, chemically, or otherwise combined or mixed and produced as a single entity. • Two or more separate products packaged together in a single package or as a unit and comprised of drug and device products, device, and biological products. • Regulated based on PMOA as either drug (incl. biologic) or as device. • Combination products regulated as medical device by Medical Device Authority in accordance with the requirements set forth in the Medical Device Act 2012 (Act 737) and its subsidiary legislations, and any other relevant documents published by MDA. • Combination products regulated as drug by drug Control Authority in accordance with the requirements set forth in the Control of Drugs and Cosmetics Regulations 1984, which is promulgated under Sale of Drugs Act 1952 and any other relevant documents published by NPRA. https://portal.mda.gov.my/industry/medical-device-registration/combination-product.html
Leading Authority/ Agency Responsible	1. **National Pharmaceutical Regulatory Agency (NPRA) – drug-led** 2. **Medical device Authority (MDA), Ministry of Health (MOH) – device-led** https://www.npra.gov.my/easyarticles/images/users/1047/drgd/APPENDIX-2---Medical-Device-Drug-Cosmetic-Interphase-MDDCI-and-Combination-Products.pdf
Contact Details	**Enquiry on registration of combination products** For submission or any enquiries, kindly contact Registration Unit: Email: combination.product@mda.gov.my Phone Number: Registration Unit +603 8230 0376 or Pn. Aidahwaty bt Ariffin +603 8230 0341 **Product & Cosmetic Regulatory Coordination Section, Centre for Coordination & Strategic Regulatory Planning,** National Pharmaceutical Regulatory Agency, Ministry of Health Malaysia, Lot 36, Jalan Universiti, 46200 Petaling Jaya, Selangor. Tel: +603-78835400 Fax :03-7956 2924 Email: helpdesk@npra.gov.my Website: https://www.npra.gov.my/ **Chief Executive Medical Device Authority (MDA),** Ministry of Health Malaysia, Level 6, Prima 9, Prima Avenue II, Block 3547, Persiaran APEC, 63000 Cyberjaya, Selangor, MALAYSIA Tel: +603-8230 0300 Fax : +603-8230 0200 Email: combination.product@mda.gov.my Website: http://www.mda.gov.my/

(*Continued*)

TABLE 2A (CONTINUED)
Malaysia

Country	Malaysia
Evaluation or Registration Process	CP registration is based on the primary mode of action/the principal mechanism of action by which the claimed effect or purpose of the product is achieved. The review process depends on the product category and follows. 1. Drug-medical device CP registration process (NPRA as primary agency) The registration process of drug-medical device combination product shall undergo the following three stages: Stage 1 – Obtaining Certification from Conformity Assessment Body (CAB) Stage 2 – Obtaining Endorsement from Medical Device Authority (MDA) Stage 3 – Application For Registration to National Pharmaceutical Regulatory Agency (NPRA) 2. Medical device-drug CP registration process (MDA as primary agency) The registration process of medical device-drug combination product shall undergo the following three stages: Stage 1 – Obtaining Endorsement from NPRA Stage 2 – Obtaining Certification from CAB Stage 3 – Application for Registration to MDA NOTE: Prior to registration, an applicant may apply classification application to NPRA through product classification form (BPFK 300-1) which is available at http://npra.moh.gov.my The review time required to perform product/ancillary dossier evaluation by each respective agency will be dependent on the CP evaluation efforts at three stages of the product (drug/device). More details in link below. https://portal.mda.gov.my/doc-list/guideline.html **Guideline for Drug-Medical Device and Medical Device-Drug Combination Products**, 5th Edition, 2023 https://www.npra.gov.my/index.php/en/guidelines-for-combination-products/425-english/announcement-main/announcement-2021/1527155-guideline-for-registration-of-drug-medical-device-and-medical-device-drug-combination-products-3.html
Fees	Fees applicable to all components (Fees imposed by: • NPRA under the CDCR 1984, Regulation 8(3) • MDA as specified to Fifth Schedule of Medical Device Regulation 2012 • CAB as per Circular Letter of Medical Device Authority No.2 Year 2014)
Master Files Allowed?	Yes
Biosimilar/Generics Guideline on Combination Products	No, but specific regulations/guidance are available for drug-device combinations.
Requirements	
GMP/Quality System	CP should meet essential principles of safety & performance (EPSP) as per Third Schedule, Conformity Assessment Procedure, Medical Device Regulation 2012 as below: 1. Manufacturer – **ISO 13485**, Medical devices, quality Management System certification required 2. Authorized representative – Good Distribution Practice for Medical Devices (GDPMD) 3. Importer/Distributor – Good Distribution Practice for Medical Devices (GDPMD) (https://mda.gov.my/documents/surat-pekeliling-pbpp-mda-s-circular-letter/mda-s-circular-letter/504-cl-22014-eng/file.html, https://asean.org/wp-content/uploads/Guidelines-on-OSH-Risk-Management.pdf)

(Continued)

TABLE 2A (CONTINUED)
Malaysia

Country	Malaysia
Packaging & Labeling	Yes, labeling requirement available, applicable as per PMOA Local language labeling required (Malay) Note: Specific labeling requirement as stated in drug registration guidance document (DRGD), Appendix 9: Labeling requirements, released by the NPRA, MDA/GD/0026: Requirements for Labeling of Medical Devices. (https://www.npra.gov.my/index.php/en/drug-registration-guidance-documents-drgd-e-book.html, https://pharmaboardroom.com/legal-articles/marketing-manufacturing-packaging-labeling-advertising-malaysia/, https://www.mda.gov.my/documents/guideline-documents/1771-guideline-for-registration-of-drug-medical-device-and-medical-device-drug-combination-products-4th-edition-1/file.html)
Unique Device Identification (UDI)	No clear requirement of UDI available for combination products
Postmarketing Safety Reporting (PMSR)	Yes (as per PMOA) (https://portal.mda.gov.my/announcement/561-medical-device-webinar-2020-new-medical-device-regulations-under-act-737-advertisement-and-post-market-requirements-august-17,-2020.html)
Clinical Trial/ Investigations	Yes (applicable on components for manufacture) Clinical trial data is required, which is applied as regulations to all components for manufacturer of the product component
Human Factors	No HF requirements/guideline available applicable to PMOA on CP
Lifecycle Management Guidelines	Yes, guidance available, but not in detail as CP. Combination products will have to meet requirements of both the Medical Device Authority (MDA) and the National Pharmaceutical Regulatory Agency (NPRA). (Links –https://www.npra.gov.my/index.php/en/component/sppagebuilder/925-drug-registration-guidance-document.html)
Quick Link	1. https://www.npra.gov.my/index.php/en/component/content/article/225-english/1527155-guideline-for-registration-of-drug-medical-device-and-medical-device-drug-combination-products-3.html?Itemid=1391 2. https://www.npra.gov.my/index.php/en/guidelines-for-combination-products.html
Comments on Evolution of Combination Product Expectations	**Drug Registration Guidance Document** (DRGD), 3rd Edition, Third Revision July 2022 – Appendix 2: Medical Device-Drug-Cosmetic Interphase (MDDCI) and Combination Products. https://www.npra.gov.my/index.php/en/component/sppagebuilder/925-drug-registration-guidance-document.html **Guideline for Drug-Medical Device and Medical Device-Drug Combination Products**, 5th Edition, 2023: https://www.npra.gov.my/index.php/en/guidelines-for-combination-products/425-english/announcement-main/announcement-2021/1527155-guideline-for-registration-of-drug-medical-device-and-medical-device-drug-combination-products-3.html *January 2023 Guidance updated*: • Name of the Guideline, Preamble, Glossary • 2.0 Registration Process of Combination Product • 4.0 Timeline for Registration of Combination Product: Evaluation Timeline by NPRA • 7.0 Adverse Drug Reaction and Incident Reporting • Appendix 1: Ancillary Medical Device Dossier Requirement for Drug-Medical Device Combination Product • Appendix 2: Ancillary Drug Dossier Requirement for Medical Device-Drug Combination Product • Appendix 3: Application Form for Endorsement Letter of Ancillary Component for the Registration of Combination Product • Appendix 6: Change to Ancillary Medical Device Components • Appendix 8: List of Relevant References • Appendix 7: Incident Reporting Form for Combination Product

TABLE 2B
Philippines

Country	Philippines
Is "Combination Product" Defined?	There is no formal legal definition of "combination product." Products are characterized as either medicinal products or medical devices.
Scope	No, legal framework available, a product comprised of two or more regulated components Single Integral/Entity (drug/device, biologic/device, drug/biologic or drug/device/biologic) **(Administrative Order No. 2020-0010 || Regulations on the Conduct of Clinical Trials for Investigational Products:** https://www.fda.gov.ph/administrative-order-no-2020-0010-regulations-on-the-conduct-of-clinical-trials-for-investigational-products/)
Legal Framework	No, legal framework available; however, as a general principle, a combination product is a therapeutic product which integrally combines features of two or more of medical device, medicine, biologic, or in vitro diagnostic device. Not clear. (https://www.fda.gov.ph/administrative-order-no-2020-0010-regulations-on-the-conduct-of-clinical-trials-for-investigational-products/, https://asean.org/wp-content/uploads/2016/06/22.-September-2015-ASEAN-Medical-Device-Directive.pdf)
Leading Authority/ Agency Responsible	No Committee, but CP applications assess by Food and drugs Administration – Product Services Division (FDA-PSD)
Contact Details	https://www.fda.gov.ph/contact-us/Landline: (02) 8857-1900 local 1000 (02) 8842-5635 Mobile: 09617709691 09616845994 09610574926 Email: fdac@fda.gov.ph
Registration Process	Combination products with primary action as A device or IVD mode of action is regulated as devices or IVDs, and classified according to the medical device classification Primary drug action are regulated as drugs Six categorizations (Details – Administrative Order No. 67 s. 1989: Revised Rules and Regulations on Registration of Pharmaceutical Products, March 15, 1989) The review time will be dependent on the category of the product (drug/device)
Fees	Fees applicable to PMOA
Master Files Allowed?	No
Biosimilar/Generics Guideline on Combination Products	No (but specific regulations/guidance available for drug/device)
Requirements	
GMP/Quality System	CP should meet essential principles of safety & performance (EPSP) as below: 1. Manufacturer – **ISO 13485, medical devices, Quality Management System** certification required pharmaceutical GMP or both 2. **Guidelines for Current Good Manufacturing Practices (cGMP) Clearance and Inspection of Foreign drug Manufacturers,** August 13, 2013 (https://www.fda.gov.ph/draft-for-comments-revised-guidelines-on-good-manufacturing-practice-gmp-clearance-for-foreign-drug-manufacturers/)
Packaging & Labeling	Labeling requirement available, applicable as per PMOA
Unique Device Identification (UDI)	No clear requirement of UDI available for combination products

(Continued)

TABLE 2B (CONTINUED)
Philippines

Country	Philippines
Postmarketing Safety Reporting (PMSR)	Yes (as per PMOA)
Clinical Trial/ Investigations	Applicable on components for manufacture (https://www.fda.gov.ph/draft-for-comments-guidelines-on-regulatory-reliance-on-the-conduct-of-clinical-trials-in-the-philippines/) Clinical trial data is required, which is applied as regulations to all components for manufacturer of the product component
Human Factors	No HF requirements/guideline available
Lifecycle Management Guidelines	Guidance available for pharmaceuticals and medical device separately, but not in detail for CP. (Links: 1. https://www.fda.gov.ph/administrative-order-no-2020-0010-regulations-on-the-conduct-of-clinical-trials-for-investigational-products/ 2. https://asean.org/wp-content/uploads/2016/06/22.-September-2015-ASEAN-Medical-Device-Directive.pdf 3. https://asean.org/wp-content/uploads/AVG-Revision-2-endorsed-31PPWG.pdf)
Quick Link	https://www.fda.gov.ph/
Comments on Evolution of Combination Product Expectations	In January 2022, PFDA circular on categorization of identified borderline health products under the jurisdiction of the food and drug administration. This provides 1. Preliminary list of borderline products and their designated product categories 2. Reiterating the considerations in determining the product classification 3. Producing the appropriate transitory period to allow the proper reclassification of borderline health products https://www.fda.gov.ph/wp-content/uploads/2021/12/Draft_FC-Illustrative-List-of-Borderline-Products.pdf

TABLE 2C
Singapore

Country	Singapore
Is "Combination Product" Defined?	There is no formal legal definition of "combination product." However, the term combination products is referenced in November 2020 public consultation. (https://www.hsa.gov.sg/docs/default-source/hprg-mdb/gn-15-r8-_guidance-on-medical-device-product-registration-(2022-jan)-pub.pdf https://www.hsa.gov.sg/docs/default-source/hprg-tpb/guidances/guidance-on-therapeutic-product-registration-in-singapore_jun22.pdf)
Scope	Single integral (chemical drug or biologic) combination product, medical devices incorporating registrable therapeutic/medicinal products are classified as Class D medical devices. It is mentioned under one public consultation, i.e., under the proposed regulation for cell, tissue, and gene therapy public consultation (dated on 6 November 2020), stating that risk-based regulatory approach for product registration and dealer licensing, based on the degree of manipulation (minimal or more than minimal), intended use (homologous or non-homologous use), and whether it is a combination product (whether combined with medical devices or therapeutic products), with Class 1 CTGTP being lower risk, and Class 2 CTGTP being higher risk. (https://www.hsa.gov.sg/docs/default-source/hprg-mdb/gn-15-r8-_guidance-on-medical-device-product-registration-(2022-jan)-pub.pdf, https://www.hsa.gov.sg/docs/default-source/default-document-library/pressrelease_public-consult-ctgtp-6nov20183b2975baff4ad98fe78893603ec290.pdf)
Legal Framework	According to Chapter 7 of **Guideline: GN-15 – Guidance on Medical Device Product Registration, Rev 8.0**, January 2022, a medical device may be incorporated with a therapeutic/medicinal product in an ancillary role (chemical drug or biologic), to achieve its intended purpose. (https://www.hsa.gov.sg/medical-devices)
Leading Authority/ Agency Responsible	According to Guideline: **GN-15 – Guidance on Medical Device Product Registration, Rev 8.0,** January 2022, medical devices incorporating registrable therapeutic/medicinal products are classified as Class D medical devices. Wherein the devices registration is evaluated by Medical Device Branch (MDB) together with Therapeutic Products Branch (TPB) of HSA (Health Sciences authority).
Contact Details	1. Enquiries should be submitted using the product enquiry form: https://form.gov.sg/60dfe2bcaaa4100012a01ad6 1. The applicant can enquire with HSA about the product classification for such products to determine the applicable regulatory control. Such enquiries should be submitted using the product enquiry webform found on the HSA website https://www.hsa.gov.sg/feedback There are two methods to contact HSA: 1. Pre-submission Enquiry via email 2. Pre-submission Meeting/Notification
Registration Process	Medical device assigned to Class D if it incorporates, as an integral part, a substance that is liable to act on a human body with an action ancillary to that of the medical device and the substance with: (i) a therapeutic product; or (ii) a medicinal product subject to the licensing requirements of section 5 or 6 of the Medicines Act. No specific guideline on process and timelines. (https://www.hsa.gov.sg/docs/default-source/hprg-tpb/guidances/guidance-on-therapeutic-product-registration-in-singapore_jun22.pdf, http://www.imdrf.org/docs/imdrf/final/consultations/imdrf-cons-rrar-cabc-mdrr.pdf)
Fees	Fee as applicable to components (https://www.hsa.gov.sg/medical-devices/fees, https://sso.agc.gov.sg/SL/HPA2007-S436-2010, https://www.hsa.gov.sg/therapeutic-products/fees)

(Continued)

TABLE 2C (CONTINUED)
Singapore

Country	Singapore
Master Files Allowed?	Yes (https://www.hsa.gov.sg/therapeutic-products/register/guides/drug-master-file)
Biosimilar/Generics Guideline on Combination Products	No (but specific regulations/guidance available for drug/device)
Requirements	
GMP/Quality System	CP should meet essential principles of safety & performance (EPSP) as below: 1. Manufacturer – **Good Distribution Practice for Medical Devices** (https://www.hsa.gov.sg/medical-devices/dealers-licence/good-distribution-practice) 2. **Guidance on Licensing of Manufacturers, Importers and Wholesalers of Medical Devices** (https://www.hsa.gov.sg/docs/default-source/hprg-mdb/gn-02-r4-2-guidance-on-licensing-of-manufacturers-importers-and-wholesalers-of-md(19apr-pub.pdf) 3. https://www.hsa.gov.sg/medical-devices/registration/risk-classification-rule 4. https://asean.org/wp-content/uploads/Guidelines-on-OSH-Risk-Management.pdf
Packaging & Labeling	Labeling requirement available, applicable as per PMOA (https://www.hsa.gov.sg/docs/default-source/hprg-mdb/gudiance-documents-for-medical-devices/gn-23-r1-1-guidance-on-labelling-for-medical-devices(20mar-pub).pdf)
Unique Device Identification (UDI)	UDI requirement for device is applicable and available. (https://www.hsa.gov.sg/docs/default-source/announcements/regulatory-updates/udi-implementation-for-singapore_19oct20.pdf, http://www.imdrf.org/docs/imdrf/final/technical/imdrf-tech-190321-udi-sag.pdf, https://www.imdrf.org/documents/unique-device-identification-system-udi-system-application-guide, https://www.hsa.gov.sg/docs/default-source/hprg-mdb/gudiance-documents-for-medical-devices/faq-(medical-device-udi-system)_updated-3-june-2022.pdf)
Postmarketing Safety Reporting (PMSR)	Requirements based on PMOA
Clinical Trial/ Investigations	Applicable on components for manufacture (http://www.imdrf.org/docs/imdrf/final/technical/imdrf-tech-191010-mdce-n57.pdf) Clinical trial data is required, which is applied as per medical device requirements. (https://www.hsa.gov.sg/medical-devices/clinical-trials, http://www.imdrf.org/docs/imdrf/final/technical/imdrf-tech-191010-mdce-n56.pdf)
Human Factors	No HF requirements/guideline available
Lifecycle Management Guidelines	Guidance available for pharmaceuticals and medical device separately, but not in detail for CP (Links: 1. https://www.hsa.gov.sg/docs/default-source/hprg/therapeutic-products/guidance-documents/guidance-on-therapeutic-product-registration-in-singapore_dec2020.pdf 2. https://www.hsa.gov.sg/medical-devices/guidance-documents 3. https://www.hsa.gov.sg/docs/default-source/hprg-tpb/guidances/guidance-on-therapeutic-product-registration-in-singapore_jun22.pdf)
Quick Links	1. https://www.hsa.gov.sg/docs/default-source/hprg-io-ctb/hsa_gn-ioctb-03_crm_1mar2021.pdf 2. https://www.hsa.gov.sg/docs/default-source/hprg-tpb/guidances/guidance-on-therapeutic-product-registration-in-singapore_aug22.pdf 3. https://www.hsa.gov.sg/docs/default-source/hprg-mdb/gn-15-r7-5-guidance-on-medical-device-product-registration-(aug21-pub).pdf
Comments on Evolution of Combination Product Expectations	**GN-15: Guidance on Medical Device Product Registration** – describes the procedures and general requirements for the submission of an application for a new Product Registration for medical devices, including medical devices incorporating medicinal products. https://www.hsa.gov.sg/docs/default-source/hprg-mdb/gn-15-r7-5-guidance-on-medical-device-product-registration-(aug21-pub).pdf

TABLE 2D
Thailand

Country	Thailand
Is "Combination Product" Defined?	No, definition on the CP is available in Thai FDA guideline or law.
Scope	N/A; however, established combinations as drug product has the primary action and the medical devices are administration devices only.
Legal Framework	No, legal framework available on combination products; however, Thai FDA had published below ordinances: • Evaluation of drug Eluting Stent, September 16, 2005 • Human Medical Devices that may Harm or Cause Risk, May 15, 2007 In April 2020, notification: Anticipated that CP soon classified as Class 4 when drug delivered or administered by implanted medical device with long-term duration or CP where drug exerts the performance of the medical device (https://www.qualtechs.com/index.php?route=newsblog/category&filter_name=%20Thai %20FDA)
Leading Authority/ Agency Responsible	No committee, but CP application is assessed by 1. CP as drug: Drug Division of the Thai FDA is the competent authority (CA) 2. CP as medical device: Drug Division and Medical Device Control Division are the CA (https://asean.org/wp-content/uploads/2016/06/22.-September-2015-ASEAN-Medical-Device -Directive.pdf, https://www.qualtechs.com/index.php?route=newsblog/category&filter_name=%20Thai %20FDA)
Contact Details	1. Importer preliminary consultation with the FDA official via electronic consultation system of the Thai FDA: https://www.fda.moph.go.th/sites/fda_en/Pages/Main.aspx 2. Consultation advice via letter: One Stop Service Center of the Office of FDA. Address: Office of the Thai Food and Drug Administration, Ministry of Public Health, Tiwanon Road, Nonthaburi, 11000, Thailand The application fee is around USD 30 per application.
Registration Process	Combination products classified as either drug or general medical device (Class I), details: 1. www.info.go.th. 2. https://www.fda.moph.go.th/sites/fda_en/SitePages/Medical.aspx?IDitem =LawsAndRegulations 3. https://www.fda.moph.go.th/sites/drug/EN/Pages/Main.aspx The review time will be dependent on the category of the product (drug/device) In the future, it is anticipated that the risk classification rules in accordance to ASEAN Medical Device Directive (AMDD) will be applied (means medical device risk analysis on all four classification of device)
Fees	Fee as applicable to PMOA
Master Files Allowed?	**No**
Biosimilar/Generics Guideline on Combination Products	No (but specific regulations/guidance available for drug/device)

(Continued)

TABLE 2D (CONTINUED)
Thailand

Country	Thailand
Requirements	
GMP/Quality System	CP should meet essential principles of safety & performance (EPSP) as below: 1. Good Manufacturing Practice (GMP) in accordance with the Pharmaceutical Inspection Co-operation Scheme 2. **ISO 13485 Medical Devices – Quality Management Systems** 3. **ISO 14971 Medical Devices – Application of Risk management** 4. **ISO 10993 Biological Evaluation of Medical Devices** The review time will be depend on the category of the product (drug/device) https://asean.org/wp-content/uploads/Guidelines-on-OSH-Risk-Management.pdf
Packaging & Labeling	Labeling requirement available, applicable as per PMOA
Unique Device Identification (UDI)	No clear requirement of UDI available for combination products
Postmarketing Safety Reporting (PMSR)	Yes, as per PMOA
Clinical Trial/ Investigations	Yes (applicable on components for manufacture, as per PMOA)
Human Factors	No HF requirements/guideline available applicable to PMOA on CP
Lifecycle Management Guidelines	**Yes, guidance available, but not in detail as CP.** 1. https://content.next.westlaw.com/practical-law/document/Id4af42fe1cb511e38578f7ccc38dcbee/Medicinal-product-regulation-and-product-liability-in-Thailand-overview?viewType=FullText&ppcid=755c2f0a3aaf471aa8e84760fb285833&originationContext=knowHow&transitionType=KnowHowItem&contextData=(sc.Default)&firstPage=true 2. https://asean.org/wp-content/uploads/2016/06/22.-September-2015-ASEAN-Medical-Device-Directive.pdf
Quick Link	1. https://www.fda.moph.go.th/sites/fda_en/SitePages/Medical.aspx?IDitem=LawsAndRegulations 2. https://www.fda.moph.go.th/sites/drug/EN/Pages/Main.aspx 3. https://content.next.westlaw.com/practical-law/document/Id4af42fe1cb511e38578f7ccc38dcbee/Medicinal-product-regulation-and-product-liability-in-Thailand-overview?viewType=FullText&ppcid=755c2f0a3aaf471aa8e84760fb285833&originationContext=knowHow&transitionType=KnowHowItem&contextData=(sc.Default)&firstPage=true
Comments on Evolution of Combination Product Expectations	1. After implementation of ASEAN Medical Device Directive (AMDD), combination product wherein the drug is integrated as part of the medical device is now regulated under the classification of "Licensed Medical Device" (class 4), whereas these were classified as Class 1 or as drug depending on the functionality of the device. 2. Draft MOPH Notification: **Quality Management System of Medical Device Manufacturing, Draft FDA Notification: Rules, Procedure, and Conditions for Medical Device Good Manufacturing Practices** (GMP). According to these drafts, it is anticipated that domestic medical device manufacturer must manage and control their manufacturing process to be in compliance with one of the quality management systems below: a) Medical Device Good Manufacturing Practices (GMP) – must align with the Thai Conformity Assessment Standard called "**Medical Devices – Quality Management Systems –Requirements for Regulatory Purposes No. Mor Tor Chor 13485-2562**" or the latest updated version. b) **ISO 13485:2016** or Medical Device Quality management system – Requirements for regulatory purposes – ISO 13485:2016. 3. Other quality management system standards certified by foreign government authorities that are recognized by the Thai FDA

Asia-Pacific

TABLE 3A
Australia

Country	Australia
Is "Combination Product" Defined?	Yes, there is a definition of "combination product" available. https://www.tga.gov.au/biologicals-packaged-or-combined-another-therapeutic-good
Scope	Single integral (medicines, medical devices, and biologicals/blood & tissue-based items) Your biological is considered to be a combination product if it is: 1. Combined or incorporated with another therapeutic good such that the other good may or may not act on the human body in addition to the biological. 2. The goods are combined and supplied for use as a single product entity that is transplanted or injected. NOTE: The following is not considered as CP: • Biologicals not packaged individually to the therapeutic good. Combination products are regulated as biologicals: • When at least one of the components is a biological. • The other constituent/s is a therapeutic good. https://www.tga.gov.au/biologicals-packaged-or-combined-another-therapeutic-good
Legal Framework	Yes, legal framework available, CPs are regulated, the TGA considers: • the primary intended purpose • the mode of action of the product as per to the definition of a medicine and a medical device (Therapeutic Goods Act 1989 as well as amendments – Section 3(1) meaning of a medicine & Section 41DB, meaning of a medical device). The proposed 2019 classification rule suggests that including at least one medical device and containing a medicine are regulated as medical devices. Additionally, as per Therapeutic Goods, Order No. 1 of 2010, in principle, covers similar groups of products and declares that the following articles are declared not to be medical devices: An article that is intended to administer a medicine in such a way that the medicine and the article form a single integral product which is intended exclusively for use in the given combination and which is not reusable (may be multidose). Currently there are no requirements for manufacturers of device components of such products to comply with requirements of the Australian MD Regulations. NOTE: It is proposed that a new classification rule be included in the Therapeutic Goods (Medical Devices) Regulations 2002 to align with Rule 20 of the EU Medical Devices Regulation. (https://www.tga.gov.au/sites/default/files/consultation-proposed-new-classification-rule-medical-devices-administer-medicines-or-biologicals-inhalation.pdf) (Additional Links – https://www.tga.gov.au/legislation-legislative-instruments, https://www.tga.gov.au/biologicals-packaged-or-combined-another-therapeutic-good)
Leading Authority/ Agency Responsible	No Committee, but CP application assessed by TGA (Therapeutic Goods Administration) 1. CP as drug: Drug Division of the Thai FDA is the competent authority (CA) 2. CP as medical device: Drug Division and Medical Device Control Division are the CA (https://www.tga.gov.au/biologicals-packaged-or-combined-another-therapeutic-good#:~:text=Multi%2Dcomponent%20packs-,Combination%20products,in%20addition%20to%20the%20biological)
Contact Details	1. TGA via email at info@tga.gov.au 2. Pre-submission meetings with TGA (https://www.tga.gov.au/publication/pre-submission-meetings-tga)
Registration Process	Combination products classified as either drug or medical device: 1. https://www.tga.gov.au/sites/default/files/ahmac-scheduling-policy-framework-medicines-and-chemicals.pdf 2. https://www.legislation.gov.au/Details/F2020C00822 The review time will be dependent on the category of the product (drug/device) NOTE: Australian Regulatory Guidelines for Medical Devices is currently under review by TGA (http://www.imdrf.org/docs/imdrf/final/consultations/imdrf-cons-rrar-cabc-mdrr.pdf)
Fees	Fee as applicable to PMOA

(*Continued*)

TABLE 3A (CONTINUED)
Australia

Country	Australia
Master Files Allowed?	Yes
Biosimilar/ Generics Guideline on Combination Products	No (but specific regulations/guidance available for drug/device)
Requirements	
GMP/Quality System	CP should meet essential principles of safety & performance (EPSP) as below: 1. Good Manufacturing Practice (GMP) in accordance with https://www.tga.gov.au/manufacturing-therapeutic-goods 2. **ISO 13485 Medical Devices – Quality Management Systems** 3. **ISO 9000** series QMS standards (http://www.imdrf.org/docs/imdrf/final/procedural/imdrf-proc-recognised-standards-australia-2015.pdf)
Packaging & Labeling	Labeling requirement available, applicable as per PMOA
Unique Device Identification (UDI)	UDI requirement for device are applicable and available. (https://www.tga.gov.au/medical-device-reforms-establishment-unique-device-identification-system, http://www.imdrf.org/docs/imdrf/final/technical/imdrf-tech-190321-udi-sag.pdf, https://www.imdrf.org/documents/unique-device-identification-system-udi-system-application-guide.)
Postmarketing Safety Reporting (PMSR)	As per PMOA, however, it is proposed that manufacturers of the device component of such products (i.e., devices prefilled with the medicine) should have evidence to demonstrate compliance of the device with the relevant essential principles/general safety and performance requirements. (https://www.tga.gov.au/sites/default/files/consultation-proposed-new-classification-rule-medical-devices-administer-medicines-or-biologicals-inhalation.pdf)
Clinical Trial/ Investigations	Applicable on components for manufacture (http://www.imdrf.org/docs/imdrf/final/technical/imdrf-tech-191010-mdce-n57.pdf) Clinical requirements as per PMOA (https://www.tga.gov.au/sites/default/files/clinical-evidence-guidelines-medical-devices.pdf, https://www.imdrf.org/sites/default/files/docs/imdrf/final/technical/imdrf-tech-191010-mdce
Human Factors	No HF requirements/guideline available applicable to PMOA on CP
Lifecycle Management Guidelines	Guidance available, but not in detail for CP. However, TGA adopts a number of EU guidelines and so the data requirements are broadly in line with to the requirements of the CP in EU. Information relating to the management of minor variations (post-approval changes) to prescription medicines, Appendix 2: Variation change types – biological medicines (v2.1; January 2018) is available: (https://www.tga.gov.au/publication/minor-variations-prescription-medicines-appendix-2-variation-change-types-biological-medicines, https://www.tga.gov.au/sites/default/files/substantial-changes-affecting-tga-conformity-assessment-certificate-and-transfers-certificates.pdf)
Quick Link	1. https://www.tga.gov.au/regulation-basics 2. **Australian Regulatory Guidelines for Medical Devices** (ARGMD) – https://www.tga.gov.au/sites/default/files/devices-argmd-01.pdf 3. https://www.tga.gov.au/publication/device-medicine-boundary-products 4. https://www.tga.gov.au/node/847489 5. https://www.tga.gov.au/sites/default/files/devices-guidelines-35.pdf 6. https://www.tga.gov.au/book-page/content-application-dossier
Comments on Evolution of Combination Product Expectations	New TGA draft guidance issued: "**Boundary and combination products guidance – medicines, medical devices, and biologicals**" (October 2022) https://consultations.tga.gov.au/tga/boundary_and_combination_products_guidance/; comments from this newest document will inform Final Guidance.

TABLE 3B
China

Country	China
Is "Combination Product" Defined?	Yes, there is a definition of "combination product" available. http://english.nmpa.gov.cn/2021-07/27/c_661024.htm
Scope	Single Integral (drugs and medical devices) (NOTE: The NMPA drug-device combination products guideline is under revision, January 2021, http://english.nmpa.gov.cn/2021-07/27/c_660305.htm)
Legal Framework	Yes, legal framework available, where the products consisting of drugs and medical devices and produced as a single product as per Announcement on **Issues Concerning Adjustment of Attribute Definition of drug-Device Combination Products** (NMPA Announcement [2019]). Few determined classifications by China Health authority: • Drug-coated stents, catheters with antimicrobial coating, drug condoms, and medicated contraceptive rings should be registered as medical devices for its more significant medical device function. • Adhesive bandages containing antibacterial and anti-inflammatory drugs and TCM sticking products for external application should be registered as drugs for its dominant drug function. (http://english.nmpa.gov.cn/2019-07/06/c_387749.htm, **NMPA Notification No. 2019/28: Issues on Adjusting Classification of Combination Products of Drugs and Medical Devices**, May 28, 2019)
Leading Authority/ Agency Responsible	No Committee, but CP application assessed by National Medical Products Administration (NMPA) (Previously CFDA) (http://english.nmpa.gov.cn/2019-07/06/c_387749.htm) Coordination mechanism between CDE and CMDE as deemed necessary by lead PMOA.
Contact Details	No specific contact for CP, however, can check on below: Address: No 1 Beiluyuan Zhanlan Road, Xicheng district, Beijing Postcode: 100037 Tel: 68311166 (http://english.nmpa.gov.cn/2019-07/18/c_380718.htm)
Registration Process	Combination products classified as PMOA: Drug-device combination products mainly used as drugs, they shall be registered as drugs; and for any drug-device combination products mainly used as medical devices, they shall be registered as medical devices. NMPA coordinates on this and leads. The applicant shall submit an application for registration of a drug or medical device to NMPA based on the attribute definition result and indicate "drug-device combination product" in the application form: 1. For drug-device combination products applied as drugs, CDE shall lead the evaluation. Where a joint evaluation is required, the registration application dossiers shall be transferred to CMDE for synchronized evaluation. 2. For the drug-device combination products applied as medical devices, CMDE shall lead the evaluation, where a joint evaluation is required, the registration application dossiers shall be transferred to CDE for synchronized evaluation. NOTE: The review time will be dependent on evaluation mechanism of drug-device combination products, such as the circulation process of registration application dossier between the CDE and CMDE. As per new September 2019 draft and discussions, CMDE will further study how to optimize the joint evaluation mechanism jointly with the CDE. In addition, for the requirements for application dossier of drug-device combination products, the Department of Medical Device Registration organizes the drafting of relevant guidelines; for relevant requirements for quality system verification for registration of drug-device combination products. (http://sfdachina.com/info/64-1.htm, http://www.imdrf.org/docs/imdrf/final/consultations/imdrf-cons-rrar-cabc-mdrr.pdf)

(Continued)

TABLE 3B (CONTINUED)
China

Country	China
	The specific requirements of the dossier formatting, product description, interaction between medical devices and drugs, drug dosage choice, chemical and physical properties, biological characteristics, animal testing research, stability research, product technical requirements, manufacturing information, clinical evaluation, etc., are discussed in detail in new guidance – NMPA Notification **No. 2022/03: Issuance of 2 Technical Guidelines for Registration Review including Technical Guidelines for Registration Review of Drug-Device Combination Products Based on the Function of Medical Devices**, dated on January 11, 2022)
Fees	Fee as applicable to all components
Master Files Allowed?	No
Biosimilar/ Generics Guideline on Combination Products	No (but specific regulations/guidance available for drug/device)
Requirements	
GMP/Quality System	CP should meet essential principles of safety & performance (EPSP) as below: 1. Good Manufacturing Practice (GMP) in accordance with http://english.nmpa.gov.cn/2019-07/12/c_398090.htm 2. http://english.nmpa.gov.cn/2022-10/11/c_819190.htm 3. **ISO 13485 Medical Devices – Quality Management Systems** (Links: a) IMDRF – Recognized Standards in China: http://www.imdrf.org/docs/imdrf/final/procedural/imdrf-proc-recognised-standards-china-2015.pdf b) **Provisions: SFDA Order No. 28: Regulations on Drug Registration Administration, Revision**, July 10, 2007: http://english.nmpa.gov.cn/2020-03/30/c_488526.htm and https://vdocument.in/provisions-for-drug-registration-cfda.html?page=1 c) **CFDA Order No. 4: Medical Device Registration Provision**, July 30, 2014: http://english.nmpa.gov.cn/2019-07/25/c_390617.htm d) **Provisions for Post-approval Changes of Drugs** (Interim): http://english.nmpa.gov.cn/2022-06/30/c_785633.htm) e) **FDA-SFDA China, Agreement on the Safety of Drugs and Medical Devices** at https://www.fda.gov/international-programs/cooperative-arrangements/fda-sfda-china-agreement-safety-drugs-and-medical-devices
Packaging & Labeling	Labeling requirement available where regulations applied for all components applied
Unique Device Identification (UDI)	UDI requirement for device is applicable and available. The registration requirements will apply from January 1, 2021: (http://www.imdrf.org/docs/imdrf/final/technical/imdrf-tech-190321-udi-sag.pdf, http://english.nmpa.gov.cn/2022-06/30/c_785636.htm, https://www.imdrf.org/documents/unique-device-identification-system-udi-system-application-guide)
Postmarketing Safety Reporting (PMSR)	As per PMOA; local language labeling required (Chinese)
Clinical Trial/ Investigations	Applicable on components for manufacture; clinical requirements as per PMOA (http://www.imdrf.org/docs/imdrf/final/technical/imdrf-tech-191010-mdce-n57.pdf) (http://www.imdrf.org/docs/imdrf/final/technical/imdrf-tech-191010-mdce-n56.pdf)

(Continued)

TABLE 3B (CONTINUED)
China

Country	China
Human Factors	Yes, HF requirements/guideline available applicable to PMOA on CP (https://www.qualtechs.com/en-gb/article/409#:~:text=Human%20factors%20design%20(also%20known,the%20usability%20of%20medical%20devices)
Lifecycle Management Guidelines	Guidance available, but not in detail as CP. Medical Device and Pharmaceutical Regulations 1. Very few applicable regulations for combination products 2. Primarily a negotiation between local operating company and agency
Quick Links	1. http://www.nmpa.gov.cn/WS04/CL2138/299892.html 2. http://www.nmpa.gov.cn/WS04/CL2439/353262.html 3. **Comments Soliciting on Announcement on Registration of drug-Device Combination Products (Exposure Draft for Draft Revision) NMPA [2021] No. 6**: http://english.nmpa.gov.cn/2021-07/27/c_660305.htm 4. J. Balzano, "The Life Sciences Law Review: China," Covington & Burling LLP, February 23, 2022. Accessed online January 12, 2023: https://thelawreviews.co.uk/title/the-life-sciences-law-review/china. 5. **NMPA Notification No. 2021/52: Matters Related to Registration of Drug-Devices Combination Products**: http://english.nmpa.gov.cn/2021-07/27/c_661024.htm
Comments on Evolution of Combination Product Expectations	1. In July 2021, NMPA published **NMPA Notification No. 2021/52: Matters Related to Registration of Combination Products of Drugs and Medical Devices**. 2. In January 2022, **NMPA Notification No. 2022/03: Issuance of 2 Technical Guidelines for Registration Review including Technical Guidelines for Registration Review of Drug-Device Combination Products Based on the Function of Medical Devices**, January 11, 2022.

TABLE 3C
India

Country	India
Is "Combination Product" Defined?	No, definition on the CP is available.
Scope	As per PMOA, however CP is classified as medical devices, i.e., 1. Medical devices incorporating medicinal products 2. Medical devices incorporating animal or human cells, tissues, or derivatives.
Legal Framework	No legal framework available, however, as per "The Drugs and Cosmetics Act," 1940 and Rules, 1945 and Medical Device Rules 2017 as well as amendment rules (2020) – According to classification of medical devices and in vitro diagnostic medical devices under First Schedule of the Medical Device Rules 2017, CPs classified as medical devices incorporating medicinal products or medical devices incorporating animal or human cells, tissues, or derivatives (https://cdsco.gov.in/opencms/opencms/en/Home/).
Leading Authority/ Agency Responsible	No Committee, but CP application assessed by Central Drugs Standard Control Organization (CDSCO) under the Ministry of Health and Family Welfare, Government of India, as competent authority.
Contact Details	N/A specific to combination products, however, below are contact details: Central Drugs Standard Control Organization, Ministry of Health and Family Welfare, Directorate General of Health Services, Government of India FDA Bhavan, ITO, Kotla Road, New Delhi – 110002 email: dci@nic.in, ithelpdeskcdscoMD@gmail.com Phone: 91-11-23236973, 91-11-23216367(CDSCO)/23236975 PRO Toll Free No.: 1800 11 1454 (https://cdscomdonline.gov.in/NewMedDev/Homepage, https://cdsco.gov.in/opencms/opencms/en/Search/index.html)
Registration Process	Based on medical devices incorporating medicinal products or medical devices incorporating animal or human cells, tissues, or derivatives. Combination products classified as medical device as medical device rules 2017: 1. https://cdsco.gov.in/opencms/opencms/system/modules/CDSCO.WEB/elements/download_file_division.jsp?num_id=MTAwOQ== 2. https://cdsco.gov.in/opencms/opencms/en/Notifications/Gazette-Notifications/
Fees	Fee as applicable to PMOA (https://cdscomdonline.gov.in/NewMedDev/resources/app_srv/NMD/global/helpfiles/nmd_fee.pdf)
Master Files Allowed?	No
Biosimilar/Generics Guideline on Combination Products	No (but specific regulations/guidance available for drug/device)
Requirements	
GMP/Quality System	CP should meet essential principles of safety & performance (EPSP) as below: 1. Good Manufacturing Practice (GMP) and QMS in accordance with https://cdsco.gov.in/opencms/opencms/system/modules/CDSCO.WEB/elements/download_file_division.jsp?num_id=MjAzNA== 2. https://cdsco.gov.in/opencms/opencms/en/Notifications/Gazette-Notifications/
Packaging & Labeling	As per PMOA

(Continued)

TABLE 3C (CONTINUED)
India

Unique Device Identification (UDI)	As of January 1, 2022, all medical devices shall bear the Unique Device Identifier (UDI). (https://www.fdanews.com/articles/180641-india-finalizes-new-regulations-specifically-for-devices#:~:text=Starting%20on%20Jan.,identifier%20and%20a%20production%20identifier.&text=A%20production%20identifier%20would%20include,manufacturing%20and%2For%20expiration%20date)
Postmarketing Safety Reporting (PMSR)	As per PMOA https://cdsco.gov.in/opencms/export/sites/CDSCO_WEB/Pdf-documents/medical-device/Essentialprinciples.pdf
Clinical Trial/ Investigations	Applicable on components for manufacture (https://cdsco.gov.in/opencms/opencms/en/Medical-Device-Diagnostics/Medical-Device-Diagnostics/) Clinical requirements as per PMOA (https://cdsco.gov.in/opencms/opencms/en/Medical-Device-Diagnostics/Medical-Device-Diagnostics/, https://cdsco.gov.in/opencms/opencms/en/Clinical-Trial/Global-Clinical-Trial/)
Human Factors	No HF requirements/guideline available
Lifecycle Management Guidelines	No guidance/regulation available on post-approval management for CP. (https://cdsco.gov.in/opencms/opencms/en/Medical-Device-Diagnostics/Medical-Device-Diagnostics/)
Quick Link	1. https://cdsco.gov.in/opencms/export/sites/CDSCO_WEB/Pdf-documents/acts_rules/2016drugsandCosmeticsAct1940Rules1945.pdf 2. https://cdsco.gov.in/opencms/export/sites/CDSCO_WEB/Pdf-documents/medical-device/Classificationg1.pdf 3. https://cdsco.gov.in/opencms/opencms/en/Home/ 4. https://cdscomdonline.gov.in/NewMedDev/viewChecklistReport 5. https://cdscomdonline.gov.in/NewMedDev/resources/app_srv/NMD/global/pdf/MD_User_Manual_Guidance.pdf 6. https://cdsco.gov.in/opencms/export/sites/CDSCO_WEB/Pdf-documents/IVD/FAQs/CDSCO-IVD-FAQ-03-2022-.pdf 7. https://cdsco.gov.in/opencms/opencms/en/Notifications/Gazette-Notifications/
Comments on Evolution of Combination Product Expectations	None

TABLE 3D
Japan

Country	Japan
Is "Combination Product" Defined?	Yes, there is a definition of "combination product" available. https://www.pmda.go.jp/files/000153158.pdf
Scope	Single integral (drug, medical device, or cellular and tissue-based product), co-packaged
Legal Framework	Yes, legal framework available, where CP as therapeutic products that combine active pharmaceutical ingredients (medicines), equipment (devices), and/or processed cell (regenerative products) as per: • Law No. 145/1960: Law on Securing Quality, Efficacy and Safety of Pharmaceuticals, Medical Devices, Regenerative and Cellular Therapy Products, Gene Therapy Products, and Cosmetics (Pharmaceutical and Medical Device Act), the amended version as of Law No. 50/2015 • Notification: PFSB/ELD No. 1024/2, PFSB/ELD/OMDE No. 1024/1, PFSB/SD No. 1024/9, PFSB/CND No. 1024/15: On Market Authorization Applications of Combination Products • Notification: PSEHB/PED, PSEHB/MED No. 0915/1, PSEHB/SD, PSEHB/CND No. 0915/3: Amendment On Market Authorization Applications of Combination Products
Leading Authority/ Agency Responsible	No Committee but CP application assessed by 1. PMDA (Pharmaceuticals and Medical Devices Agency) 2. MHLW (Ministry of Health, Labour and Welfare) (https://www.pmda.go.jp/files /000153158.pdf)
Contact Details	Not specific to combination products. PMDA (form): https://www.pmda.go.jp/english/0002.html PMDA FAQ: https://www.pmda.go.jp/english/about-pmda/0004.html
Registration Process	Combination products classified as drug or medical device or regenerative product as per PMOA: 1. Set products (i.e., independently distributed as drug, medical device, or regenerative product) 2. Kit product (i.e., drug and medical device) is marketed 3. Medical device, etc.: Combined medicinal components are inseparable from the medical device and each of them cannot be independently distributed 4. Products in which marketed drugs, medical devices, or cellular or tissue-based products are sold together by distributors 5. Drugs approved for Integral Marketing with Devices 6. Combination products shall be submitted for application as a single product corresponding to either drugs, medical devices, or cellular and tissue-based products The review time will be dependent on the category of the product (drug/device). NOTE: Combination products new concept in the Japanese regulatory system and lots of changes/amendment/harmonization in line with IMDRF (International Medical Device Regulators Forum) is expected. (https://www.pmda.go.jp/files/000153158.pdf)
Fees	Fee as applicable to PMOA
Master Files Allowed?	No
Biosimilar/Generics Guideline on Combination Products	No (but specific regulations/guidance available for drug/device)

(*Continued*)

TABLE 3D (CONTINUED)
Japan

Country	Japan
Requirements	
GMP/Quality System	No quality system requirements specific for combination products but similar requirement as medical device is applied. 1. https://www.pmda.go.jp/english/review-services/regulatory-info/0004.html 2. For pharmaceuticals as per GQP Ordinance. (http://www.imdrf.org/docs/imdrf/final/procedural/imdrf-proc-recognised-standards-japan-2015.pdf)
Packaging & Labeling	Labeling requirement available, applicable as per PMOA
Unique Device Identification (UDI)	UDI requirement for combination products, as per PMD act amendment 2019, anticipated enforcement date for traceability is December 4, 2023. Also, Japan is part of IMDRF management committee where UDI requirement is available. 1. https://namsa.com/japan-issues-amended-pmd-act/, 2. http://www.imdrf.org/docs/imdrf/final/technical/imdrf-tech-190321-udi-sag.pdf, 3. https://www.imdrf.org/documents/unique-device-identification-system-udi-system-application-guide
Postmarketing Safety Reporting (PMSR)	1. **Pharmaceuticals and Medical Devices Safety Information** (PMDSI) – https://www.pmda.go.jp/english/safety/index.html) 2. **Post-Market Change Process for Continuously Improved Device** (Enforcement Date – December 4, 2021, https://www.pmda.go.jp/files/000232939.pdf)
Clinical Trial/ Investigations	Applicable on components for manufacture (http://www.imdrf.org/docs/imdrf/final/technical/imdrf-tech-191010-mdce-n57.pdf) Clinical requirements (as per PMOA) (http://www.imdrf.org/docs/imdrf/final/technical/imdrf-tech-191010-mdce-n56.pdf)
Human Factors	No HF requirements/Guideline available applicable to PMOA on CP
Lifecycle Management Guidelines	Guidance available, but not in detail as CP. However, if the medical device part will be changed, sponsor should refer to PFSB/ELD/OMDE No. 1023001, PFSB/ELD No.0120-9, etc., which were notification of PCA for medical device and refer to below links. Pharmaceuticals and Medical Devices Agency (PMDA) 1. Key device specifications are registered in the commitments section of the dossier – dimensions with tolerances; conformity to standards; manufacturing/sterilization sites 2. Types: Partial change application (prior approval for change) [like PAS] minor change notification (within 30 days after implementation or shipping) [like CBE 30] (Non-approved matters) [like AR] (Links – https://www.pmda.go.jp/files/000153158.pdf, http://www.jpma.or.jp/english/parj/pdf/2019.pdf)
Quick Link	1. https://www.pmda.go.jp/files/000222459.pdf 2. **Handling of Marketing Applications for Combination Products**: https://www.pmda.go.jp/files/000153158.pdf 3. https://www.pmda.go.jp/english/rs-sb-std/rs/index.html 4. https://www.pmda.go.jp/files/000153158.pdf 5. https://www.pmda.go.jp/english/review-services/regulatory-info/0003.html 6. HOMEPAGE (English): http://www.pmda.go.jp/english/index.html 7. Regulatory Science Page: http://www.pmda.go.jp/regulatory/index.html
Comments on Evolution of Combination Product Expectations	Combination products were recognized as part of Japan's regulatory framework in 2014, and they were early adopters of combination products specific PMSR expectations (2017).

TABLE 3E
South Korea

Country	South Korea
Is "Combination Product" Defined?	Yes, there is a definition of "combination product" available. (https://www.mfds.go.kr/eng/brd/m_61/view.do?seq=94&srchFr=&srchTo=&srchWord=&srchTp=&itm_seq_1=0&itm_seq_2=0&multi_itm_seq=0&company_cd=&company_nm=&page=3)
Scope	Single integral (drugs, quasi-drugs, and medical devices) (https://www.mfds.go.kr/eng/brd/m_61/view.do?seq=94&srchFr=&srchTo=&srchWord=&srchTp=&itm_seq_1=0&itm_seq_2=0&multi_itm_seq=0&company_cd=&company_nm=&page=3)
Legal Framework	No, legal framework available, however CP is: • either drug combining or integrating with device or • device combining or integrating with drug or quasi-drug as per **Law No. 17248: Medical Device Act, Revised version**, April 7, 2020, regulations on market authorization for drug's Pharmaceutical Affairs Act (PAA) and device's Medical Devices Act apply to each portion of review. (https://www.mfds.go.kr/eng/brd/m_61/view.do?seq=41) **Rule to Handle Combination Products (MFDS No. 99**, July 27, 2017, http://www.kobia.kr/skin/bbs/downloads_e2/download.php?tbl=policy_report&no=407)
Leading Authority/ Agency Responsible	Yes, Item Adjustment Committee to review and adjust the following: 1. Decision of lead office and classification of CP, in case lead office is not clear 2. Administration standard of CP with no applicable or unclear regulation 3. Administration standard of items that are not CP but not classified 4. Complaints filed by the applicant against procedure or result of application
Contact Details	N/A, however for regulatory process: Ministry of Food & Drug Safety (MFDS) Osong Health Medical Administration Town 187 Osongsaengmyeong2(i)-ro Osong-eup, Heungdeok-gu, Cheongju-si, Chungcheongbuk-do, Korea 28159 Tel: +82-43-719-1564 Website: http://www.mfds.go.kr Consultation meeting request: https://www.mfds.go.kr/usr/reservation_38/list.do https://www.mfds.go.kr/usr/tCounsel_1024/list.do
Registration Process	Combination products classified as per PMOA. (Article 5 (Determination of the appropriate Regulatory Bureau/Division) and Article 6 (Handling product approval applications) of the MFDS Established Rule No. 31, https://www.mfds.go.kr/eng/wpge/m_39/denofile.do)
Fees	Fee as applicable to PMOA
Master Files Allowed?	No
Biosimilar/ Generics Guideline on Combination Products	No (but specific regulations/guidance available for drug/device)
Requirements	
GMP/Quality System	CP should meet essential principles of safety & performance (EPSP) as below: 1. Good Manufacturing Practice (GMP) in accordance with the Pharmaceutical Affairs Act and Regulations for Governing the Management of Medical Device 2. ISO 13485 Medical Devices – Quality Management Systems 3. ISO 14971 Medical Devices – Application of Risk management 4. ISO 10993 Biological evaluation of medical devices 5. https://www.mfds.go.kr/eng/brd/m_61/view.do?seq=94&srchFr=&srchTo=&srchWord=&srchTp=&itm_seq_1=0&itm_seq_2=0&multi_itm_seq=0&company_cd=&company_nm=&page=3

(*Continued*)

TABLE 3E (CONTINUED)
South Korea

Country	South Korea
Packaging & Labeling	Labeling requirement available, applicable as per PMOA
Unique Device Identification (UDI)	No UDI requirement for combination products; however, as per Medical Devices Act No. 14330 as well as the Regulation on KGMP No. 2016-156 – UDI enforcement would come into effect in: • 2019 for Class IV and high-risk devices • 2020 for Class III devices • 2021 for Class II devices • 2022 for low-risk Class I devices (https://www.imdrf.org/documents/unique-device-identification-system-udi-system-application-guide, https://www.emergobyul.com/blog/2017/03/south-korean-regulators-clarify-medical-device-udi-kgmp-rules)
Postmarketing Safety Reporting (PMSR)	Pharmaceuticals and Medical Devices Safety Information (PMDSI): (https://www.pmda.go.jp/english/safety/index.html) 2. Post-Market Change Process for Continuously Improved Device (Enforcement Date – December 4, 2021)
Clinical Trial/ Investigations	Applicable on components for manufacture (http://www.imdrf.org/docs/imdrf/final/technical/imdrf-tech-191010-mdce-n57.pdf) Clinical requirements as per PMOA (http://www.imdrf.org/docs/imdrf/final/technical/imdrf-tech-191010-mdce-n56.pdf)
Human Factors	No HF requirements/guideline available applicable to PMOA on CP. However, as per recent update, it is recommended to have human factors engineering (HFE) into medical device development in line with USFDA to create products that ensure good user experiences for patients and healthcare professionals. • **ANSI/AAMI HE 75:2009, human factors engineering – design of medical devices**. Provides detailed guidance on how to perform specific human factors analyses and provides a wealth of design principles • **ISO/IEC 60601-1-6, general requirements for basic safety and essential performance** – Collateral standard: Usability • **ISO/IEC 62366:2007, Medical devices – application of usability engineering to medical devices**: (https://www.fda.gov/regulatory-information/search-fda-guidance-documents/list-highest-priority-devices-human-factors-review, https://www.fda.gov/media/96018/download)
Lifecycle Management Guidelines	No guidance/regulation available on post-approval management for CP. Few details: 1. Treats device constituent part as a medical device; typically requires: General information, Photographs, Engineering drawings, Raw material information, Biocompatibility data, Performance data, Physical and chemical data 2. Manufacturing process related data, stability data 3. Changes to this information typically require submission (3–5 months) 4. License renewals can trigger significant additional device information (http://www.kobia.kr/skin/bbs/downloads_e2/download.php?tbl=policy_report&no=407)
Quick Link	https://www.mfds.go.kr/eng/index.do
Comments on Evolution of Combination Product Expectations	None

TABLE 3F
Taiwan

Country	Taiwan
Is "Combination Product" Defined?	Yes, there is a definition of "combination product" available. https://www.fda.gov.tw/upload/189/Content/2013011111564485285.pdf
Scope	Single integral (drug+medical device), co-packaged.
Legal Framework	Legal framework available, where CP refer to the medical products as a single product consist of drug + medical device or two/more drug and device products packed together as per the Ministry of Health and Welfare (MOHW) announced **MOHW Order No. 1041411228: Designation Guidelines of Combination Products**, January 21, 2016. (https://www.mohw.gov.tw/mp-2.html)
Leading Authority/Agency Responsible	CP application is assessed by Ministry's Food and Drug Administration (TFDA): 1. Medicinal CP Ad Hoc group (as per **MOHW Order No. 1041411228: Designation Guidelines of Combination Products**, January 21, 2016)
Contact Details	N/A
Registration Process	Combination products classified as per PMOA: 1. CP primary mode of action as pharmaceutical 2. CP primary mode of action as medical device The review time will be dependent on the category of the product (drug/device) (https://www.fda.gov.tw/upload/189/Content/2013011111564485285.pdf)
Fees	Fee as applicable to PMOA
Master Files Allowed?	No
Biosimilar/Generics Guideline on Combination Products	No (but specific regulations/guidance available for drug/device)
Requirements	
GMP/Quality System	CP should meet essential principles of safety & performance (EPSP) as below: 1. Good Manufacturing Practice (GMP) in accordance with the Pharmaceutical Affairs Act and Regulations for Governing the Management of Medical Device 2. ISO 13485 Medical Devices – Quality Management Systems 3. ISO 14971 Medical Devices – Application of Risk management 4. ISO 10993 Biological evaluation of medical devices
Packaging & Labeling	As per PMOA https://www.tga.gov.au/sites/default/files/poisons-standard-and-medical-devices.pdf
Unique Device Identification (UDI)	No UDI requirement for combination products. However, as per draft guidance by Taiwan Food and Drug Administration, the anticipated timelines for mandatory implementation of UDI requirements are as follows: • Class III implantable devices – June 1, 2021 • Cass III non-implantable devices – June 1, 2022 • Class II devices – June 1, 2023 (https://www.fda.gov.tw/TC/newsContent.aspx?cid=5072&id=26490, http://www.imdrf.org/docs/imdrf/final/technical/imdrf-tech-190321-udi-sag.pdf)
Postmarketing Safety Reporting (PMSR)	As per PMOA
Clinical Trial/Investigations	Applicable on components of manufacture; clinical requirements based on PMOA
Human Factors	Human factor requirements are available for CP as per guidance by TFDA (https://www.qualtechs.com/en-gb/article/416)
Lifecycle Management Guidelines	Guidance available, but not in detail as CP (https://www.cde.org.tw/eng/drugs/med_explain?id=41)

(*Continued*)

TABLE 3F (CONTINUED)
Taiwan

Country	Taiwan
Quick Link	1. **MOHW Order No. 1101603189: Regulations Governing the Classification of Medical Devices,** April 26, 2021 2. **MOHW Order No. 1101613379: Revision on Annex of Article 4 of Regulations Governing the Classification of Medical Devices**, December 9, 2021 3. **MOHW Order No. 1101602740: Charge Standards of Administrative Fees for Medical Devices**, April 28, 2021 4. **MOHW Announcement No. 1101609769: Soliciting Public Comment on Abolishment (Draft) of Regulations for Governing the Management of Medical Device, Regulation for Registration of Medical Devices and Charge Schedule of Review Fees for the Registration and Advertisements of Medical Devices,** September 30, 2021 https://www.mohw.gov.tw/mp-2.html
Comments on Evolution of Combination Product Expectations	None.

Europe

TABLE 4A
European Union

Country	EU
Is "Combination Product" Defined?	There is not a formal definition, but the term "drug-device combinations" is applied under EU MDR Article 117
Scope	Under EU MDR Articles 1(8) and (9), there are two types of "combination products" considered under the MDR: Integral and Co-packaged. The regulatory requirements for the medical device differ depending on whether or not it is integral. • *Integral*: The *medicinal product* and device form a single integrated product, e.g., prefilled syringes and pens, patches for transdermal drug delivery and prefilled inhalers. • *Co-packaged*: The *medicinal product* and the device are separate items contained in the same pack, e.g., reusable pen for insulin cartridges, tablet delivery system with controller for pain management. • "Referenced products" are also referred to as non-integral, but they are not treated as combination products.
Legal Framework	Legal framework under **(EU) 2017/745** on medical devices, in particular the requirements under Article 117. Applicable legal framework is applied under Medical Device Directive (MDD – only applicable during a transitional phase), EU MDR (2017/745), MEDDEV 2.1/3 rev. 3 (part: consultation procedure), EMA's Q&A on MDR's implementation. (Regulation EU 2017/745 – https://eur-lex.europa.eu/legal-content/EN/TXT/PDF/?uri=CELEX:32017R0745, https://www.ema.europa.eu/en/news/consultation-draft-guideline-quality-requirements-medical-devices-combination-products)
Leading Authority/ Agency Responsible	1. European Commission provides scientific, technical, and logistical support to coordinating national authorities of the member states, and ensures the regulatory compliance for devices with scientific evidence. 2. Notified Bodies (NBs) appointed by the Competent Authority of the Member State to carry out conformity assessment 3. Device Expert Group (working party) (DEG) on Borderline and Classification 4. Medical Device Coordination Group (MDCG)
Contact Details	As applicable per Competent Authority and Notified Body • List of notified bodies under Regulation (EU) 2017/745 on medical devices (https://ec.europa.eu/growth/tools-databases/nando/index.cfm?fuseaction=directive.notifiedbody&dir_id=34) • List of national competent authorities in the EEA (https://www.ema.europa.eu/en/partners-networks/eu-partners/eu-member-states/national-competent-authorities-human#list-of-national-competent-authorities-in-the-eea-section)
Registration Process	Drug-Device Combinations (MDR (Art.1(8),(9))) are either categorized/registered as medicinal product or as medical device and thus regulated either by the Directive on Medicinal Products (Directive 2001/83/EC) or the EU MDR. CP are classified as: 1. Drug-Device Combinations: Single integral, non-reusable, only for use in the given combination – EU MDR (2017/745) Article 1(9) Subparagraph 2 2. Devices which incorporate, as an integral part, ancillary medicinal products – EU MDR (2017/745) Article 1(8) Subparagraph 1 3. Medicinal products co-packaged with medical devices (Article 1(9) Subparagraph 1) 4. Medicinal product with ancillary medical device, single integral re-usable – Article 1(8) Subparagraph 2 (https://www.ema.europa.eu/en/human-regulatory/overview/medical-devices) The NB review of integral devices is usually between 2 and 6 months currently, whereas full review time (excluding clock-stops) for the MAA is 210 days. And for device incorporating medicinal substances, review time will vary at the NB, depending on workload and complexity of the device. Review of the medicinal component will be up to 210 days.

(Continued)

TABLE 4A (CONTINUED)
European Union

Country	EU
	NOTE: Changes expected in future: 1. Scope of MDR 2017/745 and article 117 2. The developing role of notified bodies 3. Clinical and regulatory environment 4. Falsified medicines 5. 5.EU MDCG Medical Device Coordination Group is working on this and revision of the classification document to update to MDR (http://www.imdrf.org/docs/imdrf/final/consultations/imdrf-cons-rrar-cabc-mdrr.pdf, https://www.ebe-biopharma.eu/mediaroom/ebe-refection-paper-medicinal-product-incorporating-a-drug-delivery-device-component-an-industry-perspective-on-developing-an-efficient-end-to-end-control-strategy/)
Fees	Fees as applicable to PMOA and need for NB
Master Files Allowed?	No
Biosimilar/ Generics Guideline on Combination Products	No (but specific regulations/guidance available for drug/device)
Requirements	
GMP/Quality System	The medical device part of drug-device combinations must at least fulfill the relevant General Safety and Performance Requirements (GSPRs (Annex I of the MDR)) despite the combined product being considered a medicinal product. Compliance to GSPRs now needs to be assessed by a Notified Body for Class Is, Im, IIa, IIb, or III device parts of single integral products. • These are medicinal products with a medical device part. • Conformity must be shown either by CE marking of the whole medical device part or by a Notified Body Opinion (NBOp) - as per Article 117. • NBOp is needed for all new marketing authorization application and submissions for substantial changes (component replacements, intended use extension). • No UDI or Eudamed registration/submission is required for these products. For all single integral products incorporating a device part of Class Is, Im, IIa, IIb or III, Notified Body involvement is required to confirm compliance of the device part of single-use integral products with relevant General Safety and Performance Requirements (GSPRs): 1. CE mark certificate and/or Declaration of Conformity (DoC) Or, if not available, a 2. Notified Body Opinion (NBOp) The term "Non-Integral Drug-Device Combinations" was defined in EMA's **Draft Guidance on Quality Requirements for Drug-Device Combinations** (DDCs) and adopted in EMA's Q&A on Article 117. These products are not per se subject to Notified Body assessment, and they are considered medicinal products subject to Directive 2001/83/EC. The devices within the medicinal product packaging are expected to comply in full with the MDR. In addition, the relevant output from the risk management, including the relevant safety information included by the manufacturer of the devices in the IFU, must be incorporated in the medicinal product labeling, unless it is in conflict with Directive 2001/83/EC. • **EMA, "Q&A on Implementation of the EU MDR 2017/745 and IVDR 2017/746,"** October 21, 2019. [Online]. Available: www.ema.europe.eu/. • http://www.imdrf.org/docs/imdrf/final/procedural/imdrf-proc-recognised-standards-eu-2015.pdf • https://www.efpia.eu/media/413395/ebe-efpia_-position-paper_pfp-case-study_final_20191211.pdf)

(Continued)

TABLE 4A (CONTINUED)
European Union

Country	EU
	The FDA has a Mutual Recognition Agreement (MRA) in place with the European Union that covers good manufacturing practice inspections of facilities making human drugs, but combination products have not been in scope, due to the differing approaches to inspections relative to the combination product device constituent part for medicinal-led products.
Packaging & Labeling	Labeling requirement available, applicable as per PMOA Local language labeling required (https://www.efpia.eu/media/413358/02-ebe-efpia_-position-paper_-industry-perspective-on-art-117-of-mdr_labelling-requiremts-final_20190808-1.pdf)
Unique Device Identification (UDI)	UDI requirements for device are applicable and available. Till date, it is applicable for standalone, CE-marked devices. For more details, refer below link (Links: https://ec.europa.eu/health/sites/health/files/md_sector/docs/md_mdcg_2019_2_gui_udi_dev_en.pdf, http://www.imdrf.org/docs/imdrf/final/technical/imdrf-tech-190321-udi-sag.pdf, https://ec.europa.eu/health/sites/default/files/md_sector/docs/md_mdcg_2019_2_gui_udi_dev_en.pdf) • UDI must be available on the label and all higher levels of packaging (excl. shipping containers) • UDI must be used when reporting AE and FSCA and provided to the patient for implanted devices • Basic UDI must be on the Declaration of Conformity (DoC) • UDI must be part of the Technical Documentation • UDI must be part of the traceability data kept by the economic operators
Postmarketing Safety Reporting (PMSR)	Yes, aligned to PMOA
Clinical Trial/ Investigations	Applicable on components for manufacture NOTE: IMPDs for integral, single-use combination products need to provide a statement as to conformity to MDD Annex 1. Effective March 2018: **EMA Guideline on the requirements for quality documentation concerning biological investigational medicinal products in clinical trials** (http://www.imdrf.org/docs/imdrf/final/technical/imdrf-tech-191010-mdce-n57.pdf, https://www.efpia.eu/media/413358/02-ebe-efpia_-position-paper_-industry-perspective-on-art-117-of-mdr_labelling-requiremts-final_20190808-1.pdf) Clinical Trial requirements per PMOA (http://www.imdrf.org/docs/imdrf/final/technical/imdrf-tech-191010-mdce-n56.pdf, https://www.ebe-biopharma.eu/mediaroom/ebe-efpia-position-paper-an-industry-perspective-on-article-117-of-the-eu-medical-device-regulation-clinical-requirements-for-prefilled-single-use-integral-drug-device-combination-products/, **EN ISO 14155-1:2003 – Clinical investigation of medical devices for human subjects** Part 1: General requirements and EN ISO 14155-2:2003 – Clinical investigation of medical devices for human subjects Part 2: Clinical investigation plans)
Human Factors	**Human factor requirements are available for CP as per MDR – Article 2** – Definitions, IEC 62366:2007 **Medical devices – Application of usability engineering to medical devices**, IEC 62366-1:2015, EC/TR 62366-2:2016 **Medical devices – Part 2: Guidance on the application of usability engineering to medical devices** (https://www.iso.org/standard/69126.html, **Human Factors and Usability Engineering – Guidance for Medical Devices Including drug-device Combination Products Version 1.0**, September 2017: https://www.medical-device-regulation.eu/mdr-guidance-documents/)

(*Continued*)

TABLE 4A (CONTINUED)
European Union

Country	EU
Lifecycle Management Guidelines	Guidance available, but not in detail for CP. The limitations of the EU variations guidance with respect to drug-device combination products are known and don't adequately manage typical types of post-approval changes. The guideline available is more on pharmaceutical changes but not on CP. More updates expected from EMA/Team NB in 2021. (Links: 1. https://www.ema.europa.eu/en/documents/scientific-guideline/draft-guideline-quality-requirements-drug-device-combinations_en.pdf 2. https://www.ema.europa.eu/en/human-regulatory/post-authorisation/classification-changes-questions-answers 3. https://www.ema.europa.eu/en/documents/scientific-guideline/concept-paper-developing-guideline-quality-requirements-medicinal-products-containing-device_en.pdf 4. https://www.ebe-biopharma.eu/mediaroom/ebe-reflection-paper-on-medicinal-product-incorporating-a-drug-delivery-device-component-an-industry-perspective-on-the-eu-marketing-application-technical-requirements-regulatory-review-process-and/ 5. https://www.efpia.eu/news-events/the-efpia-view/blog-articles/pragmatic-solutions-needed-for-lifecycle-management-of-drug-device-combinations-guest-blog/)
Quick Link	1. **Regulation EU 2017/745**: https://eur-lex.europa.eu/legal-content/EN/TXT/PDF/?uri=CELEX:32017R0745 2. **Commission Guideline MEDDEV. 2.1/3 Rev 3**: http://meddev.info/_documents/2_1_3_rev_3-12_2009_en.pdf 3. **Commission Guideline MEDDEV 2.4/1 Rev 9**: http://www.meddev.info/_documents/2_4_1_rev_9_classification_en.pdf 4. **Directive 93/42/EEC**: https://www.legislation.gov.uk/eudr/1993/42/2020-12-31 5. **Directive 2001/83/EC**: https://health.ec.europa.eu/publications/directive-200183ec_en 6. https://ec.europa.eu/health/md_sector/new_regulations/guidance_en 7. https://www.ema.europa.eu/en/human-regulatory/overview/medical-devices 8. **MDCG 2022-5: Guidance on Borderline between Medical Devices and Medicinal products under Regulation (EU) 2017/745 on Medical Devices** https://health.ec.europa.eu/latest-updates/mdcg-2022-5-guidance-borderline-between-medical-devices-and-medicinal-products-under-regulation-eu-2022-04-26_en 9. http://www.ema.europa.eu/ema/index.jsp?curl=pages/regulation/general/general_content_000083.jsp 10. http://www.ema.europa.eu/ema/index.jsp?curl=pages/regulation/general/general_content_000394.jsp#cad
Comments on Evolution of Combination Product Expectations	In April 2022, Guidance on borderline between medical devices and medicinal products under regulation (EU) 2017/745 on medical devices April 2022. This new guidance covers borderline products not easily categorized either as medical devices falling under MDR requirements or as medicinal products for human use falling under directive 2001/ 83/ EC (MPD). Separate chapters cover herbal products, substance-based devices, and medical device and medicinal product combinations. https://health.ec.europa.eu/system/files/2022-04/mdcg_2022-5_en_0.pdf

TABLE 4B
Switzerland

Country	Switzerland
Is "Combination Product" Defined?	Yes, there is a definition of "combination product" available. https://www.swissmedic.ch/swissmedic/en/home/news/mitteilungen/ch-bevollmaechtigter-kombiprodukte.html
Scope	Single integral (medicinal products with a medical device component), co-packaged
Legal Framework	Yes, legal framework available, where CPs are therapeutic or diagnostic products combine a medicinal product and a medical device. Combination products contain at least two components that if individually considered are licensed differently. An integral, non-separable combination, the Medical Device Ordinance does not apply to the combination product and the medical device component must satisfy the general safety and performance requirements set out in Annex I MDR (Art. 2 para. 1 let. f and g and para. 2 MedDO). For combination products that are placed on the market as co-packaged unit, the medical device component must satisfy the general safety and performance requirements set out in Annex I MDR and meet the conformity requirements of MedDO (CE mark) taking into account its intended use. (https://www.swissmedic.ch/swissmedic/en/home/news/mitteilungen/anforderungen-und-angaben-zu-kombinationsprodukten-arzneimittel-mit-einer-medizinproduktkomponente-im-formular-gesuch-zulassung-aenderung-fuer-humanarzneimittel.html, https://www.swissmedic.ch/swissmedic/en/home/news/mitteilungen/praxisauslegung-anforderungen-mepv-importeure-kombinationsprodukte.html **Regulation EU 2017/745**: https://eur-lex.europa.eu/legal-content/EN/TXT/PDF/?uri=CELEX:32017R0745)
Leading Authority/ Agency Responsible	No committee but CP application assessed by Swissmedic is the designating authority
Contact Details	Swissmedic is currently unable to receive telephone enquiries. Please submit your questions by email (anfragen@swissmedic.ch) FAQ: https://www.swissmedic.ch/swissmedic/en/home/input/overview_faq.html Switchboard/Reception desk +41 58 462 02 11
Registration Process	Combination products classified as PMOA: 1. Single-entity combination products (integrated combination products, medicinal product, and medical device) 2. Co-packaged combination products (non-integrated combination products, medical device is available separately) 3. Cross-labeled combination products (non-integrated combination products, where medical device is enclosed in the packaging or provided separately) The review time will be dependent on the category of the product (drug/device). NOTE: Swiss medical devices legislation updates inline to EU legislation, changed expected
Fees	Fees applicable to all components
Master Files Allowed?	No
Biosimilar/Generics Guideline on Combination Products	No (but specific regulations/guidance available for drug/device)
Requirements	
GMP/Quality System	CP should meet essential principles of safety & performance (EPSP) as below: 1. Good Manufacturing Practice (GMP) in accordance with the ICH GMP guidelines 2. ISO 13485 Medical Devices – Quality Management Systems

(*Continued*)

TABLE 4B (CONTINUED)
Switzerland

Country	Switzerland
Packaging & Labeling	Labeling requirement available, applicable as per PMOA -- Local language labeling required (German/French)
Unique Device Identification (UDI)	Yes, UDI requirement for device is applicable and available. (https://www.swissmedic.ch/swissmedic/en/home/medical-devices/market-access/produktidentifikation-udi.html)
Postmarketing Safety Reporting (PMSR)	As per PMOA
Clinical Trial/ Investigations	Applicable on components for manufacture and requirements based on PMOA
Human Factors	No, HF requirements/guideline available
Lifecycle Management Guidelines	Yes, guidance available, but not in detail as CP. However, as Article 117 of new MDR (2017/745) amends MPD 2001/83/EC in EU, similar changes expected by the federal council to carry out the adjustments in Swiss Law. (Links: **812.21 Federal Act of December 15, 2000 on Medicinal Products and Medical Devices** (Therapeutic Products Act, TPA) of December 15, 2000 (Status as of January 1, 2018): https://www.admin.ch/opc/en/classified-compilation/20002716/index.html **812.213 Medical Devices Ordinance**, October 17, 2001 (MedDO) of October 17, 2001 (Status as of November 26, 2017): https://www.admin.ch/opc/en/classified-compilation/19995459/index.html **Swissmedic Swiss Agency for Therapeutic Products**: https://www.swissmedic.ch/swissmedic/fr/home/dispositifs-medicaux.html **Swissmedic requirements and information relating to combination products**: https://www.swissmedic.ch/swissmedic/en/home/news/mitteilungen/anforderungen-und-angaben-zu-kombinationsprodukten-arzneimittel-mit-einer-medizinproduktkomponente-im-formular-gesuch-zulassung-aenderung-fuer-humanarzneimittel.html) https://www.swissmedic.ch/swissmedic/en/home/humanarzneimittel/authorisations/information/revidierte-anforderungen-kombinationsprodukte.html
Quick Link	1. https://www.swissmedic.ch/swissmedic/en/home/news/mitteilungen/anforderungen-und-angaben-zu-kombinationsprodukten-arzneimittel-mit-einer-medizinproduktkomponente-im-formular-gesuch-zulassung-aenderung-fuer-humanarzneimittel.html 2. https://www.swissmedic.ch/swissmedic/en/home/humanarzneimittel/authorisations/information/anforderungen_an_kombinationsprodukte.html 3. https://www.fedlex.admin.ch/eli/cc/2020/552/en 4. https://www.swissmedic.ch/swissmedic/en/home/medical-devices/market-access/ch-rep.html 5. https://www.swissmedic.ch/swissmedic/en/home/humanarzneimittel/authorisations/information/revidierte-anforderungen-kombinationsprodukte.html 6. https://www.swissmedic.ch/swissmedic/en/home/humanarzneimittel/authorisations/information/praezisierung_terminologie_bei_kombinationsprodukten.html
Comments on Evolution of Combination Product Expectations	On January 12, 2023, United States signed a Mutual Recognition Agreement (MRA) with Switzerland. The Swiss Confederation (Switzerland), the FDA, and the Swiss Agency for Therapeutic Products (Swissmedic) will be able to use each other's GMP inspections of pharmaceutical manufacturing facilities, avoiding the need for duplicate inspections. Excluded from the MRA scope are Advanced Therapy Medicinal Products (ATMPs), human blood, human plasma, human tissues and organs, cells (Swiss MRA), and veterinary immunologicals. There is no mention of combination products in the MRA. Historically, combination products have not been in scope of MRA (e.g., with Europe) given the differing approaches for review of cGMPs relative to the device constituent part of medicinal-led combination products.

TABLE 4C
United Kingdom/Great Britain

Country	United Kingdom (UK)
Is "Combination Product" Defined?	Definition is available.
Scope	Single integral (medicinal product and medical device) non-re-usable, intended exclusively for use in given combination. "Single-use combination product" means a product which comprises a medical device and medicinal product forming a single integral product which is intended exclusively for use in the given combination, and which is not reusable; and "system or procedure pack" has the same meaning as in article 12 of Directive 93/42 (UK MDR 2002) and Section 2 of Human Medicines Regulation 2012. https://www.legislation.gov.uk/uksi/2002/618/contents/made https://www.legislation.gov.uk/uksi/2012/1916/contents/made
Legal Framework	Products are regulated either by the Medical Devices regulations or by the Medicines Legislation. (https://assets.publishing.service.gov.uk/government/uploads/system/uploads/attachment_data/file/872742/GN8_FINAL_10_03_2020__combined_.pdf, https://www.gov.uk/guidance/borderline-products-how-to-tell-if-your-product-is-a-medicine, https://www.gov.uk/guidance/borderline-products-how-to-tell-if-your-product-is-a-medical-device, (MEDDEV 2.1/3 rev.3) https://ec.europa.eu/docsroom/documents/10328/attachments/1/translations)
Leading Authority/ Agency Responsible	CP application assessed by MHRA (Medicines and Healthcare products Regulatory Agency) involving the Commission on Human Medicines or as a medical device; the latter case involves the intervention of a Notified Body.
Contact Details	1. Borderline advice for medicine, sent form, and attachments to borderline_medicine@mhra.gov.uk. 2. Borderline advice for medical devices question on regulatory route: Devices.regulatory@mhra.gov.uk 3. Request for consultation: Tel 020 3080 6000 (General Inquiry) If specific to medical device: Tel: 020 3080 7386 Email: Area0-PLNumberAllocation@mhra.gov.uk 4. Submission queries (confirmation of receipt by MHRA): email: Area4-RENEW-PSUR-PIQ-BROMI-submission-queries@mhra.gov.uk 5. Contact the MHRA about borderlines with medical devices via email: devices.regulatory@mhra.gov.uk
Registration Process	Combination products as per a general rule according to MHRA, regulated by either 1. CP as the medical devices regulations or 2. CP as the medicines legislation The review time will be dependent on the category of the product (drug/device)
Fees	As applicable based on PMOA
Master Files Allowed?	No
Biosimilar/Generics Guideline on Combination Products	No (but specific regulations/guidance available for drug/device)
Requirements	
GMP/Quality System	CP should meet essential principles of safety & performance (EPSP) as below: 1. Good Manufacturing Practice (GMP) in accordance with the ICH GMP guidelines 2. ISO 13485 Medical Devices - Quality Management Systems https://www.gov.uk/government/news/uk-to-strengthen-regulation-of-medical-devices-to-protect-patients

(Continued)

TABLE 4C (CONTINUED)
United Kingdom/Great Britain

Country	United Kingdom (UK)
Packaging & Labeling	Labeling requirement available, applicable as per PMOA. Integral drug-device combination product manufacturer should follow the labeling requirements for a medicinal product
Unique Device Identification (UDI)	UDI requirement for device are available. The registration requirements apply from January 1, 2021. (https://www.imdrf.org/documents/unique-device-identification-system-udi-system-application-guide) https://www.gov.uk/government/consultations/consultation-on-the-future-regulation-of-medical-devices-in-the-united-kingdom/chapter-4-registration-and-udi NOTE: For integral drug-device combination product regulated as medicinal products, the labeling and UDI requirements of the MDR are not applicable.
Postmarketing Safety Reporting (PMSR)	As per PMOA
Clinical Trial/ Investigations	Applicable on components of manufacture; requirements per PMOA (https://www.hra.nhs.uk/planning-and-improving-research/policies-standards-legislation/clinical-trials-investigational-medicinal-products-ctimps/combined-ways-working-pilot/)
Human Factors	Human Requirements are available for combination products as per "**Guidance on applying human factors and usability engineering to medical devices including drug-device combination products Version 2.0**," January 2021 in Great Britain https://assets.publishing.service.gov.uk/government/uploads/system/uploads/attachment_data/file/970563/Human-Factors_Medical-Devices_v2.0.pdf
Lifecycle Management Guidelines	Guidance available, but not in detail as CP. The limitations of the UK/EU variations guidance with respect to drug-device combination products are known, and don't adequately manage typical types of post-approval changes. (https://www.gov.uk/topic/medicines-medical-devices-blood/marketing-authorisations-variations-licensing)
Quick Link	1. https://www.gov.uk/guidance/borderline-products-how-to-tell-if-your-product-is-a-medicine 2. http://www.legislation.gov.uk/ 3. https://www.gov.uk/guidance/borderline-products-how-to-tell-if-your-product-is-a-medical-device
Comments on Evolution of Combination Product Expectations	The amendment to medicines legislation brought about by EU MDR (2017/745) Article 117 is not applicable to Great Britain since BREXIT. EU legislation continues to apply to Northern Ireland. Guidance released November 2022 entitled "How the MHRA makes decisions on when a product is a medical device (borderline products), and which risk class should apply to a medical device" discusses borderline products. Based on this guidance, MHRA decides whether a product is a medical device or medicinal product in case the manufacturer is unsure of their product. As in the EU and US, the decision about whether a product is a medical device is based on the intended purpose of the product and its mode of action.

Latin America

TABLE 5A
Brazil

Country	Brazil
Is "Combination Product" Defined?	No definition available. In Brazil they are referred to as "combined products" defined in Technical Note No. 1/2021/SEI/COMEP/ANVISA as *"a product that comprises two or more components that are regulated as products subject to sanitary surveillance, such as medicine/medical device, vaccine/medical device, which combine physically, chemically or otherwise, produced as a single entity."*
Scope	N/A
Legal Framework	Brazil does not recognize combination products (medical devices and medicines or medicines and biologics) as a category of products. Each product must have its own registration with Anvisa. There are no additional regulations that apply specifically to combination products. Resolution RDC 185/2001 regulates the use of medical devices used in combination with another product. https://www.in.gov.br/en/web/dou/-/resolucao-rdc-n-751-de-15-de-setembro-de-2022-430797145 In case of doubts regarding the classification of a combined product, the product must follow the same flow of analysis defined for the so-called borderline products, given the absence of specific regulations in this regard. Borderline products are products that are difficult to distinguish such as medicine, medical device, cosmetic, food, among other categories, due to their technical characteristics that include composition, place of application/use, presentation, and mechanism of action. These products are referred to as borderline products until their classification is decided by Anvisa, resulting in the designation of a specific regulatory pathway. The technical criteria for framing borderline products is established in Technical Note No. 1/2021/SEI/COMEP/ANVISA. https://www.gov.br/anvisa/pt-br/centraisdeconteudo/publicacoes/medicamentos/pesquisa-clinica/notas-tecnicas/nota-tecnica-no-01-de-2021-copec.pdf https://www.gov.br/anvisa/pt-br/setorregulado/regularizacao/produtos-fronteira/arquivos/nota-tecnica_1_2021_definicoes-e-criterios-para-enquadramento.pdf https://www.gov.br/anvisa/pt-br/setorregulado/regularizacao/produtos-fronteira/arquivos/nota-tecnica_1_2021_definicoes-e-criterios-para-enquadramento.pdf, https://www.gov.br/anvisa/pt-br/canais_atendimento, https://www.emergogroup.com/sites/default/files/file/rdc_185_2001_classification_and_registration_requirements_of_medical_products_0.pdf https://uk.practicallaw.thomsonreuters.com/w-029-4257?contextData=(sc.Default)&transitionType=Default&firstPage=true
Leading Authority/ Agency Responsible	Anvisa's Collegiate Board Medical device application is assessed by Technical Chamber of Medical Devices Technologies (Câmara Técnica de Tecnologia de Produtos para a Saúde (CATEPS)), i.e., ANVISA's Advisory Committee (https://www.in.gov.br/en/web/dou/-/resolucao-rdc-n-751-de-15-de-setembro-de-2022-430797145)
Contact Details	http://portal.anvisa.gov.br/contato
Registration Process	The combined product stands out, which, as defined in Technical Note No. 1/2021/SEI/COMEP/ANVISA, as "a product that comprises two or more components that are regulated as products subject to sanitary surveillance, such as medicine/medical device, vaccine/medical device, which combine physically, chemically or otherwise, produced as a single entity."

(Continued)

TABLE 5A (CONTINUED)
Brazil

Country	Brazil
	"Application for registration of a combined product may be forwarded by the technical area of the Agency that initially received it for analysis by the Committee for the Qualification of Products Subject to Sanitary Surveillance (COMEP) in cases in which it is not entirely clear, according to the legislation, which category that fits and when different understandings can be considered when framing the combined product. In this case, the product will follow the workflow defined for the classification of borderline products, which must be registered according to the recognized categories. The workflow is concluded with the final decision of Anvisa's Collegiate Board (after analysis by the technical area and subsequent evaluation and manifestation of COMEP) and the decision is communicated by the technical area for which the product is intended to the company responsible for the product, which also receives any guidance in relation to the measures to be taken." [S. Bosenberg and S. Benvenides (September 23, 2022) "Combined Products–Brief Comments on ANVISA's Current Understanding" accessed January 14, 2023, at https://www.kasznarleonardos.com/en/combined-products-in-brazil-brief-comments-on-anvisas-brazilian-national-health-surveillance-agency-current-understanding/] https://www.lexology.com/library/detail.aspx?g=6108fe31-5114-4535-9080-9175091b6c34 Resolution RDC 185/2001, https://www.in.gov.br/en/web/dou/-/resolucao-rdc-n-751-de-15-de-setembro-de-2022-430797145 https://www.who.int/medical_devices/countries/regulations/bra.pdf?ua=1, http://www.imdrf.org/docs/imdrf/final/consultations/imdrf-cons-rrar-cabc-mdrr.pdf
Fees	Applicable to all components
Master Files Allowed?	Yes
Biosimilar/Generics Guideline on Combination Products	No (but specific regulations/guidance available for drug/device)
Requirements	
GMP/Quality System	Per ANVISA, Brazilian Good Manufacturing Practice (BGMP) certification will be required for the following manufacturing units for Class III and Class IV devices: 1. Manufacturing units producing final products either on their own behalf or for other companies 2. Manufacturing units performing final releases of final products and that are involved in at least one stage of production other than design, distribution, sterilization, packaging, or labeling stages 3. Manufacturing units for software as a medical device (SaMD) Affected manufacturers had until November 28, 2022, to come into compliance with ANVISA BGMP certification requirements. http://www.cvs.saude.sp.gov.br/zip/U_RS-MS-ANVISA-RDC-687_130522.pdf, Good Manufacturing Practices: https://www.gov.br/anvisa/pt-br/english/regulation-of-companies) CP should meet essential principles of safety & performance (EPSP) as below: 1. Good Manufacturing Practice (GMP) in accordance with the ICH GMP guidelines 2. ISO 13485 Medical Devices – Quality Management Systems 3. Certificate of Pharmaceutical Products (CoPP) for medicines 4. ANVISA's RDC 665/2022 at https://www.gov.br/anvisa/pt-br/assuntos/noticias-anvisa/2022/rdc-665-de-2022 5. Medical Device Single Audit Program (MDSAP) AU P0002.007 applicable. https://pesquisa.in.gov.br/imprensa/jsp/visualiza/index.jsp?data=31/03/2022&jornal=515&pagina=341&totalArquivos=444 http://www.imdrf.org/docs/imdrf/final/procedural/imdrf-proc-recognised-standards-brazil-2015.pdf)

(Continued)

TABLE 5A (CONTINUED)
Brazil

Country	Brazil
Packaging & Labeling	Labeling requirement available, applicable as per PMOA
Unique Device Identification (UDI)	UDI requirement for device are applicable and available. (Local legislation RDC 40/2015 and RDC 185/2001, https://www.imdrf.org/documents/unique-device-identification-system-udi-system-application-guide).
Postmarketing Safety Reporting (PMSR)	As per PMOA
Clinical Trial/ Investigations	Applicable on components for manufacture (http://www.imdrf.org/docs/imdrf/final/technical/imdrf-tech-191010-mdce-n57.pdf) Clinical requirements per PMOA http://antigo.anvisa.gov.br/legislacao#/visualizar/29313, http://antigo.anvisa.gov.br/legislacao#/, http://antigo.anvisa.gov.br/legislacao#/visualizar/29315, http://www.imdrf.org/docs/imdrf/final/technical/imdrf-tech-191010-mdce-n56.pdf)
Human Factors	No HF requirements/guideline available; however, it is advised to follow RDC 55/2010 for conducting clinical studies in human factors http://antigo.anvisa.gov.br/documents/10181/2718376/RDC_55_2010_COMP.pdf/bb86b1c8-d410-4a51-a9df-a61e165b9618
Lifecycle Management Guidelines	Guidance available, but not in detail as CP. Can request kitted device specifications – would be regulatory binding – requiring notification if there are changes.
Quick Link	https://www.gov.br/anvisa/pt-br
Comments on Evolution of Combination Product Expectations	None at this time.

TABLE 5B
Mexico

Country	Mexico
Is "Combination Product" Defined?	No definition of combination product
Scope	N/A
Legal Framework	Lack of regulatory framework, so authorization may become complex. Regulations as applicable for medicinal products and medical devices will be apply. For devices, classification is risk-based: • Class I: Those that are known in the medical practice with a proven safety and efficacy record • Class II: Those that are known in the medical practice with variations in their materials, and are not usually in the human body for more than 30 days • Class III: Those that are known in the medical practice and are usually in the human body for more than 30 days
Leading Authority/ Agency Responsible	*Comisión Federal para la Protección contra Riesgos Sanitarios (*COFEPRIS), under the authority of the Undersecretary for Prevention and Promotion of Health
Contact Details	A meeting for consultation can be requested in written to health authority. All documentation and communication are expected to be in Spanish. COFEPRIS stands for *Comisión Federal para la Protección contra Riesgos Sanitarios*: Federal Commission for Protection against Sanitary Risks controls and regulates drug products and medical devices in Mexico. https://www.gob.mx/cofepris Oklahoma 14 Col. Nápoles 03810 Ciudad de México Teléfono: 55 5080 5200 Atención Ciudadana: 800 033 50 50 Comisión Federal para la Protección contra Riesgos Sanitarios.
Registration Process	Sponsors should: 1. Identify the main therapeutic indication or intended use of their product, 2. List all components. 3. Confirm if the primary intended use of the product is mainly pharmaceutical, immunological, or metabolic (i.e., medicinal PMOA) *or not.* 4. Determine if your product or any component can be classified as a medical device. 5. Assess whether you can fulfill all the general requirements for registering the product as medicine or medical device. 6. Compare the product with other similar cases to help evaluate the process. Review time for registration is dependent on the product and category of the medical device. There are no specific regulations on medical software. Due to this gap, COFEPRIS has implemented a process to review these types of products on a case-by-case basis.
Fees	Fees applicable to all components
Master Files Allowed?	Yes
Biosimilar/Generics Guideline on Combination Products	No (but specific regulations/guidance available for medical devices and medicines)
Requirements	
GMP/Quality System	The main legislation for medical devices is the **General Health Law**, its regulations, and the NOM for good manufacturing practices (NOM-241-SSA1-2012). For medicinal products, NOM-059-SSA1-2015 and NOM-176-SSA1-1998 are applied. Product should meet essential principles of safety & performance (EPSP) as below: 1. Good Manufacturing Practice (GMP) in accordance with the ICH GMP guidelines 2. ISO 13485 Medical Devices – Quality Management Systems

(*Continued*)

TABLE 5B (CONTINUED)
Mexico

Country	Mexico
Packaging & Labeling	Labeling requirement applicable, for all components of the product. Regulated by the **General Health Law, Health Law Regulations**, and **NOM 072-SSA1-2012** relating to the labeling of medicinal products
Unique Device Identification (UDI)	No UDI requirements for devices
Postmarketing Safety Reporting (PMSR)	As per PMOA
Clinical Trial/ Investigations	Applicable on components for manufacture. Clinical requirements per PMOA. The main legislation for clinical trials is the **Health Law Regulations for Research** (*Reglamento de la Ley General de Salud en Materia de Investigación para la Salud*) (RLGSMIS) and **NOM for Health Research in Human Beings** (NOM-012-SSA3-2012). **The Guideline for Good Clinical Practice E6(R1)** is applied.
Human Factors	No HF requirements/guideline available.
Lifecycle Management Guidelines	Devices usually go through third-party evaluation; license renewals can request supplier Quality Certifications (e.g., ISO 13485) legalized, conformity declarations, and sometimes specific device information (needle materials)
Quick Link	1. http://www.gob.mx/cofepris/ 2. https://www.dof.gob.mx/nota_to_doc.php?codnota=5176019 3. https://www.paho.org/en/file/45093/download?token=raUHIhwr 4. https://data.miraquetemiro.org/sites/default/files/documentos/6criterios_clasif_riesgosan_DM_251108%201.pdf 5. https://uk.practicallaw.thomsonreuters.com/7-376-5204?transitionType=Default&contextData=(sc.Default)&firstPage=true
Comments on Evolution of Combination Product Expectations	1. New Agreement set the grounds for all digital applications. All medical device registrations may now be submitted online at the COFEPRIS submission website: https://tramiteselectronicos02.cofepris.gob.mx/Frontendnuevoportal/login.aspx 2. Good Manufacturing Process (GMP) Certification: New timelines, requirements, and forms for GMP certification and export authorization inspections. Timeline is extended from 15 business days to 90; video remote inspections are considered; these can only be submitted online now https://www.dof.gob.mx/nota_detalle.php?codigo=5641214&fecha=24/01/2022#gsc.tab=0

Middle East/North Africa

TABLE 6
Saudi Arabia

Country	Saudi Arabia
Is "Combination Product" Defined?	Yes, there is a definition of "combination product" available. https://sfda.gov.sa/sites/default/files/2022-04/Products-Classification-Guidance_0.pdf, https://sfda.gov.sa/sites/default/files/2020-10/FINAL_for_implementation.pdf
Scope	Single integral, co-packaged, cross-label
Legal Framework	Legal framework available, where product consists of two or more of items that subject to different SFDA's jurisdictions in terms of regulatory path, marketing, and/or manufacturing. In order to decide whether a product is regulated as a medical device or a drug product, the following points should be considered: • The intended purpose of the product taking into account the way the product is presented. • The method by which the intended purpose is achieved. Products that achieve their intended purpose by pharmacological, immunological, or metabolic action in/on the body are regulated by drug sector. Products that do *not* achieve their principal intended action in or on the body by pharmacological, immunological, or metabolic means, but that may be assisted in its intended function by such means, are regulated by medical device sector. https://sfda.gov.sa/sites/default/files/2020-10/FINAL_for_implementation.pdf, https://sfda.gov.sa/sites/default/files/2022-04/Products-Classification-Guidance_0.pdf
Leading Authority/ Agency Responsible	CP application is assessed by Saudi Food and Drug Authority (SFDA) Products Classification Department (PCD) (https://sfda.gov.sa/sites/default/files/2020-10/SFDAProductsClassificationGuidance-e.pdf)
Contact Details	Electronic-Products Classification System (ePCS): Submit a product classification request via the e-PCS. (https://pcs.sfda.gov.sa/Default.En.aspx)
Registration Process	Based on the following: • Statutory definitions • Proposed claim/indication • Classification of the product ("drug-medical device" combination product, or "medical device-drug" combination product) PMOA (detail below) Combination products classified by PMOA: 1. Single-entity combination products (integrated combination products) 2. Co-packaged combination products (non-integrated combination products) 3. Cross-labeled combination products (non-integrated combination products) Links: • "**Guidance for Combination Products Classification**" Saudi Food & Drug Authority Kingdom of Saudi Arabia (October 7, 2020), at https://www.sfda.gov.sa/sites/default/files/2020-10/FINAL_for_implementation.pdf • **MDS-G008 Guidance on Medical Devices Classification** (MDS-G-008-V2/221213), for more examples on combination products with medical device primary mode of action: https://sfda.gov.sa/en/regulations/65997 • **MDS-G7 Guidance on Criteria of Medical Devices Bundling/Grouping within one MDMA Application**: https://sfda.gov.sa/sites/default/files/2019-10/%28MDS-G7%29en_0.pdf • **MDS-G001 Regulatory Framework for Drugs Approval** (DS-G-001-V6.2/090705): https://www.sfda.gov.sa/sites/default/files/2021-10/RegulatoryFramework_1.pdf

(*Continued*)

TABLE 6 (CONTINUED)
Saudi Arabia

Country	Saudi Arabia
Fees	Fees applicable to all components
Master Files Allowed?	No
Biosimilar/Generics Guideline on Combination Products	No, but specific regulations/guidance available for drugs and devices.
Requirements	
GMP/Quality System	CP should meet essential principles of safety & performance (EPSP) as below: 1. Good Manufacturing Practice (GMP) in accordance with the ICH GMP guidelines 2. ISO 13485 Medical Devices – Quality Management Systems
Packaging & Labeling	Labeling requirement available, applicable as per PMOA – Local language labeling required
Unique Device Identification (UDI)	UDI requirements for device are applicable and available. (https://www.sfda.gov.sa/sites/default/files/2022-10/E-6.1.pdf)
Postmarketing Safety Reporting (PMSR)	As per PMOA (https://www.sfda.gov.sa/sites/default/files/2022-10/E-6.1.pdf)
Clinical Trial/ Investigations	Applicable on components of manufacture as per PMOA
Human Factors	No HF requirements/guideline available. Applicable based on PMOA of CP
Lifecycle Management Guidelines	Guidance available, but not detailed for CPs
Quick Link	1. https://www.sfda.gov.sa/sites/default/files/2020-10/FINAL_for_implementation_0.pdf 2. https://www.sfda.gov.sa/sites/default/files/2021-11/GuidanceBorderlineProductsClassificationE.pdf 3. https://www.sfda.gov.sa/sites/default/files/2021-10/RegulatoryFramework_1.pdf 4. https://sfda.gov.sa/sites/default/files/2020-10/FINAL_for_implementation.pdf
Comments on Evolution of Combination Product Expectations	Very similar to US FDA construct for the classification of combination products.

North America

TABLE 7A
Canada

Country	Canada
Is "Combination Product" Defined?	Definition of "combination product" available. https://www.canada.ca/content/dam/hc-sc/migration/hc-sc/dhp-mps/alt_formats/pdf/prodpharma/applic-demande/pol/combo_mixte_pol_2006-eng.pdf Also see Health Canada "**Issue Identification Paper: Drug-Device Combination Products (DDCPs) Draft for Consultation**" (May 10, 2021) at https://www.canada.ca/en/health-canada/programs/consultation-issue-identification-paper-drug-device-combination-products-draft/document.html
Scope	A therapeutic product that combines a drug component and a device component (which by themselves would be classified as a drug or a device), such that the distinctive nature of the drug component and device component is integrated in a **singular product**. The term "singular product" includes single integral and some co-packaged drugs and devices. Products classified as DDCPs include: • A drug delivery system may be combined at time of manufacture. • A drug delivery system that may be co-packaged and combined prior to administration of the drug. • Drug-enhanced devices. • Device-enhanced drugs (drug is the PMOA). Excluded: Kits, cross-labeled products, companion diagnostics, veterinary combination products, and equipment used for point-of-care manufacturing of drugs.
Legal Framework	Health Canada classifies each individual component as either a drug or a device PMOA. **Guidance Document: Classification of Products at the (Medical) Device-Drug Interface**. February 7, 2018. Health Canada. https://www.canada.ca/en/health-canada/services/drugs-health-products/classification-health-products-device-drug-interface/guidance-document-factors-influencing-classification-products-device-drug-interface.html Health products at the drug-medical device interface are those that do not readily fall within the definition of "drug" or "device" in section 2 of the Food and Drugs Act (F&DA). Under the F&DA, three sets of regulations may apply to these products: The Food and Drug Regulations; The Medical Devices Regulations; or The Natural Health Products Regulations. https://www.canada.ca/en/health-canada/services/drugs-health-products/drug-products/applications-submissions/policies/policy-drug-medical-device-combination-products-decisions.html
Leading Authority/ Agency Responsible	Classification is the first step in the review of health products by the Health Products and Food Branch (HPFB). When the classification of a health product is not evident, the *Office of Science of the Therapeutic Products Directorate* is consulted. The *Office of Science* provides recommendations on the classification of products as either drugs (i.e., pharmaceutical, biologic, or natural health product), medical devices, or drug-medical device combination products. In rare cases, the *Office of Science* may pursue additional consultation with the *Therapeutic Products Classification Committee* (TPCC). The TPCC includes representatives from various areas of Health Canada, within and outside of HPFB. The guidance document titled, "Guidance Document: Classification of Products at the Drug-Medical Device Interface" describes the factors considered by Health Canada in the classification of health products as either devices or drugs. https://www.canada.ca/en/health-canada/services/drugs-health-products/classification-health-products-device-drug-interface.html NOTE: For questions about the classification of products at the drug-medical device interface or the regulatory framework that may apply to a particular health product, please contact the Office of Science (Drug-Device.Classification.Drogue-Instrument@hc-sc.gc.ca).

(*Continued*)

TABLE 7A (CONTINUED)
Canada

Country	Canada
Contact Details	Request for designation: Email: drug-Device.Classification.Drogue-Instrument@hc-sc.gc.ca.
Registration Process	Single regulatory pathway based on PMOA of the DDCP; for drug-enhanced devices, there is more than one therapeutic effect and the component that provides the most significant effect may not be immediately apparent. Health Canada has established through practice that when a drug-enhanced device consists of a Class I device combined with a drug, the DDCP is authorized under the drug pathway. This is because Class I devices require no premarket license application. However, when a drug-enhanced device involves a device component of a higher risk class, additional clarity is needed for classifying the primary component. 1. Guidance Document: **Classification of Products at the Drug-Medical Device Interface**. February 7, 2018. Health Canada. https://www.canada.ca/en/health-canada/services/drugs-health-products/classification-health-products-device-drug-interface/guidance-document-factors-influencing-classification-products-device-drug-interface.html 2. https://www.canada.ca/en/health-canada/services/drugs-health-products/drug-products/applications-submissions/policies/drug-medical-device-combination-products.html, 3. http://www.imdrf.org/docs/imdrf/final/consultations/imdrf-cons-rrar-cabc-mdrr.pdf, 4. https://www.canada.ca/en/health-canada/programs/consultation-regulatory-review-drugs-devices.html)
Fees	Fees applicable based on primary component
Master Files Allowed?	Yes
Biosimilar/ Generics Guideline on Combination Products	No, but specific regulations/guidance are available for drugs/devices.
Requirements	
GMP/Quality System	A combination product is subject to either the *Food and Drug Regulations* or the *Medical Devices Regulations* (or in the case of herbal medicines, the *Natural Health Product Regulations).* The current framework for a DDCP requires that a sponsor comply with either the GMP or QMS standards when the primary component is considered to be, respectively, a drug or a device. https://www.canada.ca/en/health-canada/programs/consultation-draft-guidance-list-recognized-standards-medical-devices/document.html Evidence of GMP or QMS compliance for the ancillary component of a DDCP is not required. (This has created challenges for Health Canada when considering the classification of co-packaged drugs and devices. Specifically, packaging requirements under GMP apply only to the direct packaging of a drug and do not as a rule extend to the co-packaged device components that deliver a drug. This results in different GMP requirements for single-entity drug-delivery systems that are combined at time of manufacture, and co-packaged drug delivery systems that are combined prior to administration.) Most DDCPs regulated under the F&DR are also required to demonstrate compliance with Good Manufacturing Practices for the manufacture of Active Pharmaceutical Ingredients (APIs). This requirement is not currently applied to APIs in DDCPs where the primary component is a medical device and the product is licensed under the MDR.

(*Continued*)

TABLE 7A (CONTINUED)
Canada

Country	Canada
	CP should meet essential principles of safety & performance (EPSP) as below: 1. Good Manufacturing Practice (GMP) in accordance with the ICH GMP guidelines 2. ISO 13485 Medical Devices – Quality Management Systems 3. 21 CFR Part 211: Current Good Manufacturing Practice for Finished Pharmaceuticals 4. 21 CFR Part 820: Quality System Regulation https://www.canada.ca/en/health-canada/programs/consultation-draft-guidance-list-recognized-standards-medical-devices/document.html Both the principal and ancillary components need to meet acceptable standards of safety, efficacy, and quality. Risk Management Plans (RMPs) are not yet required by regulation but are currently requested by policy for drugs only. Health Canada is further assessing this policy with respect to DDCPs when the ancillary component is a drug. https://www.canada.ca/en/health-canada/programs/consultation-issue-identification-paper-drug-device-combination-products-draft/document.html
Packaging & Labeling	Health Canada currently applies labeling requirements to only the primary component of a DDCP, e.g., a DDCP authorized under the Food & Drug Regulation (F&DR) only needs to comply with the labeling requirements for drugs. There are differences in the labeling requirements for drugs, natural health products, and medical devices, though that are currently under review at time of this writing. 1. https://www.canada.ca/en/health-canada/programs/consultation-issue-identification-paper-drug-device-combination-products-draft/document.html 2. https://www.canada.ca/en/health-canada/services/drugs-health-products/drug-products/applications-submissions/guidance-documents/labelling-pharmaceutical-drugs-human-use-2014-guidance-document.html, 3. https://www.canada.ca/en/health-canada/services/drugs-health-products/medical-devices/application-information/guidance-documents/guidance-labelling-medical-devices-including-vitro-diagnostic-devices-appendices.html
Unique Device Identification (UDI)	June 2021 Health Canada introduced a proposal to introduce UDI for medical devices in Canada, following IMDRF guidelines. IMDRF requires UDI on medical devices. 1. http://www.imdrf.org/consultations/cons-udi-system-180712.asp 2. https://www.imdrf.org/documents/unique-device-identification-system-udi-system-application-guide, 3. https://www.canada.ca/en/health-canada/programs/consultation-unique-device-identification-system-medical-devices-canada/document.html
Postmarketing Safety Reporting (PMSR)	As per PMOA. Health Canada is grappling with the potential for insufficient data to evaluate the risk presented by the ancillary component as it evaluates policy updates.
Clinical Trial/ Investigations	Applicable to components for manufacture: http://www.imdrf.org/docs/imdrf/final/technical/imdrf-tech-191010-mdce-n57.pdf Clinical requirements per PMOA: http://www.imdrf.org/docs/imdrf/final/technical/imdrf-tech-191010-mdce-n56.pdf
Human Factors	HF requirements/guideline available and applicable to PMOA on DDCP: https://www.canada.ca/en/health-canada/services/consumer-product-safety/reports-publications/industry-professionals/guidance-document-application-human-factors-summary.html For further information, visit the resources in Appendix A of these guidelines or contact a Health Canada Consumer Product Safety Office via email hc.ccpsa-lcspc.sc@canada.ca

(*Continued*)

TABLE 7A (CONTINUED)
Canada

Country	Canada
Lifecycle Management Guidelines	Guidance available, but not in detail as CP. Officially recognizes DDCPs – emphasis is on ISO standards – Health Canada: Guidance Document Post-Notice of Compliance (NOC) Changes: Quality Document (2018). 1. Level I – Supplements (Major Quality Changes) 2. Level II – Notifiable Changes (Moderate Quality Changes) 3. Level III – Annual Notification (Minor Quality Changes) 4. Level IV changes – Record of Changes
Quick Link	1. https://www.canada.ca/en/health-canada/services/drugs-health-products/drug-products/applications-submissions/policies/policy-drug-medical-device-combination-products-decisions.html 2. https://www.canada.ca/en/health-canada/services/drugs-health-products/international-activities/drug-products/international-coalition-medicines-regulatory-authorities-icmra-fact-sheet-frequenlty-asked-questions.html 3. https://www.canada.ca/content/dam/hc-sc/migration/hc-sc/dhp-mps/alt_formats/hpfb-dgpsa/pdf/prodpharma/combo_mixte_dec_pol-eng.pdf 4. https://www.canada.ca/content/dam/hc-sc/documents/services/drugs-health-products/drug-products/drug-products/guidance-document-classification-products-device-drug-interface.pdf
Comments on Evolution of Combination Product Expectations	In May 2021, Health Canada published a Draft for **Consultation on Drug-Device Combination Products (DDCPs)**. This paper is intended to identify the key challenges associated with Health Canada's oversight of drug-device combination products (DDCPs) as per **Health Canada's Policy on Drug/Medical Device Combination Products** (the Policy). It emphasizes on 1. Initiative to update the policy on drug-device combination products 2. To identify any concerns they have with the current version of the policy. In 2021 Health Canada also started an initiative to engage industry and stakeholders for "Advanced Therapeutic Products" including combination products: "Regulating advanced therapeutic products" (https://www.canada.ca/en/health-canada/services/drug-health-product-review-approval/regulating-advanced-therapeutic-products.html)

TABLE 7B
United States of America

Country	United States (US)
Is "Combination Product" Defined?	Yes.
Scope	A product comprised of two or more regulated components, i.e., drug/device, biologic/device, drug/biologic, or drug/device/biologic (21 CFR 3.2(e)) at https://www.ecfr.gov/current/title-21/chapter-I/subchapter-A/part-3/subpart-A/section-3.2#p-3.2(e) **Single Entity** **Co-packaged** **Cross-Labeled** **Not included**: Drug/Drug or Biologic/Biologic combinations
Legal Framework	Comprehensive legal framework established. https://www.fda.gov/combination-products **Guidance & Regulatory Information**: https://www.fda.gov/combination-products/guidance-regulatory-information **Regulation of Combination Products: 21 CFR Part 4 (Subparts A & B)**: https://www.ecfr.gov/current/title-21/chapter-I/subchapter-A/part-4 **Classification and Jurisdictional Information:** https://www.fda.gov/combination-products/jurisdictional-information **Feedback on Combination Products**: https://www.fda.gov/combination-products/feedback-combination-products
Leading Authority/ Agency Responsible	**US FDA Office of Combination Products (OCP)** serves as primary point of contact for CP issues for FDA staff and industry: https://www.fda.gov/about-fda/office-clinical-policy-and-programs/office-combination-products **Center for Devices and Radiological Health (CDRH)** – device PMOA **Center for Drug Evaluation and Research (CDER)** – drug PMOA **Center for Biologics Evaluation and Research (CBER)** – biological product-PMOA
Contact Details	Email: combination@fda.gov Phone: 301-796-8930 Fax: 301-847-8619 Office of Combination Products Office of Clinical Policy and Programs Food and Drug Administration 10903 New Hampshire Ave. WO-32, Room 5129 Silver Spring, MD 20993 United States of America
Registration Process	US FDA CP regulation is built upon Primary Mode of Action, but also considers each constituent part that comprise the combination product. If classification is unclear and/or center assignment is unclear, a sponsor of the combination product makes a submission for **Pre-Request for Designation (Pre-RFD)** https://www.fda.gov/regulatory-information/search-fda-guidance-documents/how-prepare-pre-request-designation-pre-rfd or **Request for Designation (RFD)** https://www.fda.gov/regulatory-information/search-fda-guidance-documents/how-write-request-designation-rfd **FDA Guidance on Principles of Premarket Pathways** (January 2022): https://www.fda.gov/regulatory-information/search-fda-guidance-documents/principles-premarket-pathways-combination-products **FDA Frequently Asked Questions about Combination Products**: https://www.fda.gov/combination-products/about-combination-products/frequently-asked-questions-about-combination-products#process

(Continued)

TABLE 7B (CONTINUED)
United States of America

Country	United States (US)
Fees	Application User Fees for Combination Products are based on the fee associated with the type of application required for the product's premarket approval, clearance, or licensure. In some circumstances, a sponsor may choose to submit two applications covering the various components of a combination product when one application would suffice. In such cases, two application fees would be assessed, i.e., one fee for each application: https://www.fda.gov/regulatory-information/search-fda-guidance-documents/application-user-fees-combination-products
Master Files Allowed?	Yes See guidance for Drug Master Files and Drug Master File Binders, master files for biological products, and for master files for devices available on FDA's website.
Biosimilar/Generics Guideline on Combination Products	**US FDA Principles of Premarket Pathways for Combination Products Guidance for Industry and FDA Staff** (January 2022) https://www.fda.gov/media/119958/download **US FDA Comparative Analyses and Related Comparative Use Human Factors Studies for a Drug-Device Combination Product Submitted in an ANDA: Draft Guidance for Industry** (January 2017) http://www.fda.gov/downloads/Drugs/GuidanceComplianceRegulatoryInformation/Guidances/UCM536959.pdf
Requirements	
GMP/Quality System	**21 CFR Part 4** https://www.ecfr.gov/current/title-21/chapter-I/subchapter-A/part-4 **US FDA Current Good Manufacturing Practice Requirements for Combination Products Guidance for Industry and FDA Staf***f* (January 2017) at https://www.fda.gov/regulatory-information/search-fda-guidance-documents/current-good-manufacturing-practice-requirements-combination-products Offers either: (1) compliance with all the cGMP requirements that apply to each combination product constituent part; or (2) a "Streamlined Approach" that acknowledges commonalities between the quality systems associated with each of the constituent parts, supplemented by specific called-out provisions for quality system elements that are differently interpreted. For 21 CFR 211-based QMS, supplement with the following: • 820.20 Management Responsibility • 820.30 Design Controls • 820.50 Purchasing Controls • 820.100 CAPA • 820.170 Installation • 820.200 Servicing For 21 CFR 820-based QMS, supplement with the following: • 211.84 Testing and approval or rejection of components, drug product containers, and closures • 211.103 Calculation of yield • 211.132 Tamper-evident packaging requirements for the OTC human drug products • 211.137 Expiration dating • 211.165 Testing and Release for distribution • 211.166 Stability Testing • 211.167 Special Testing Requirements • 211.170 Reserve Samples For combination products including a biological product constituent part, in addition to either 21 CFR 211 or 21 CFR 820 quality system, you must supplement it with the applicable biological quality system elements (21 CFR 600-680, and for HCT/Ps, 21 CFR 1271).

(Continued)

TABLE 7B (CONTINUED)
United States of America

Country	United States (US)
	Alternative or Streamlined Mechanisms for complying with CGMP requirements for Combination Products (September 2022) https://www.federalregister.gov/documents/2022/09/13/2022-19713/alternative-or-streamlined-mechanisms-for-complying-with-the-current-good-manufacturing-practice Some products that have historically been regulated as drugs are now considered combination products, in particular ophthalmic dispensers packaged with drugs. Historically, under 21 CFR 200.50, dispensers packaged with ophthalmic drugs were regulated as drugs. FDA has declared 21 CFR 200.50 obsolete. The dispensers for ophthalmic preparations meet the "device" definition, and the ophthalmic preparations and dispensers are to be regulated as drug-led combination products. FDA is evaluating its approach to post-market manufacturing requirements for low-risk device constituents, e.g., an eye dropper bottle. **US FDA Certain Ophthalmic Products: Policy Regarding Compliance with 21 CFR Part 4 Guidance for Industry** (March 2022) at https://www.fda.gov/regulatory-information/search-fda-guidance-documents/certain-ophthalmic-products-policy-regarding-compliance-21-cfr-part-4-guidance-industry FDA has published a proposal to align its 21 CFR 820 device Quality Management System with International Consensus Standard ISO 13485:2016 (2/23/2022). The proposed updates are not intended to change the cGMP requirements established for combination products under 21 CFR Part 4. The final rule based on the proposal has not yet issued. https://www.federalregister.gov/documents/2022/02/23/2022-03227/medical-devices-quality-system-regulation-amendments See also AAMI TIR 48
Packaging & Labeling	Labeling requirement available, applicable as per PMOA
Unique Device Identification (UDI)	**21 CFR 801.30(a)(11)**, and **21 CFR 801.30(b)(1), (2), and (3)** clarify expectations of UDI for combination products. The requirements for application of UDI and NDC codes vary based on the combination product category, i.e., single entity, co-packaged, or cross-labeled. https://www.ecfr.gov/current/title-21/chapter-I/subchapter-H/part-801/subpart-B/section-801.30#p-801.30(a)(11) https://www.ecfr.gov/current/title-21/chapter-I/subchapter-H/part-801/subpart-B/section-801.30#p-801.30(b)
Postmarketing Safety Reporting (PMSR)	**21 CFR Part 4 Subpart B** https://www.ecfr.gov/current/title-21/chapter-I/subchapter-A/part-4/subpart-B FDA Final Guidance "**Postmarketing Safety Reporting for Combination Products**" (July 2019) https://www.fda.gov/regulatory-information/search-fda-guidance-documents/postmarketing-safety-reporting-combination-products
Clinical Trial/ Investigations	Applicable based on PMOA and constituent parts **US FDA Draft Guidance: Early Development Considerations for Innovative Combination Products** (September 2006) https://www.fda.gov/files/about%20fda/published/Guidance-for-Industry-and-FDA-Staff---Early-Development-Considerations-for-Innovative-Combination-Products-%28PDF%29.pdf **ISO 14155:2020 Clinical investigation of medical devices for human subjects – Good clinical practice** **IMDRF: Clinical Investigation (2019)** http://www.imdrf.org/docs/imdrf/final/technical/imdrf-tech-191010-mdce-n57.pdf **US FDA Combination Products Guidance Documents** https://www.fda.gov/regulatory-information/search-fda-guidance-documents/combination-products-guidance-documents) Product-specific guidance documents are available and may give further insights for considerations on specific products. https://www.fda.gov/regulatory-information/search-fda-guidance-documents/

(Continued)

TABLE 7B (CONTINUED)
United States of America

Country	United States (US)
Human Factors	**Human Factors Studies and Related Clinical Study Considerations in Combination Product Design and Development** *Draft Guidance for Industry and FDA Staff* (February 2016) https://www.fda.gov/regulatory-information/search-fda-guidance-documents/human-factors-studies-and-related-clinical-study-considerations-combination-product-design-and **Content of Human Factors Information in Medical Device Marketing Submissions** *Draft Guidance for Industry and Food and Drug Administration Staff* (December 2022) https://www.fda.gov/regulatory-information/search-fda-guidance-documents/content-human-factors-information-medical-device-marketing-submissions **Safety Considerations for Product Design to Minimize Medication Errors Guidance for Industry** (April 2016) https://www.fda.gov/regulatory-information/search-fda-guidance-documents/safety-considerations-product-design-minimize-medication-errors-guidance-industry **Comparative Analyses and Related Comparative Use Human Factors Studies for a Drug-Device Combination Product Submitted in an ANDA: Draft Guidance for Industry** (January 2017) https://www.fda.gov/regulatory-information/search-fda-guidance-documents/comparative-analyses-and-related-comparative-use-human-factors-studies-drug-device-combination **Contents of a Complete Submission for Threshold Analyses and Human Factors Submissions to Drug and Biologic Applications Guidance for Industry and FDA Staff Draft Guidance** (October 2018) https://www.fda.gov/regulatory-information/search-fda-guidance-documents/contents-complete-submission-threshold-analyses-and-human-factors-submissions-drug-and-biologic See also Human Factors Usability Engineering for Medical Devices ISO-IEC 62366-1:2015/AMD 1:2020; -2:2016 AAMI TIR 59:2017 Integrating Human Factors into Design Controls
Lifecycle Management Guidelines	1. **21 CFR Part 4: Post Market cGMPs** 2. **US FDA Draft Guidance "Submissions for Post-approval Modifications to a Combination Product Approved Under a BLA, NDA, or PMA"** (January 2013) at https://www.fda.gov/media/85267/download. *This guidance does not address changes to combination products that are not approved under a BLA, NDA or PMA (e.g., those cleared solely under a device premarket notification submission2 or those marketed under an over-the-counter drug monograph). Nor does this guidance address changes to combination products that were approved under more than one marketing application. Further, while this guidance does address the type of submission to provide when making a change to a constituent part of a combination product approved under one marketing application, it does not address the scientific or technical content to provide in any such submission.* 3. **"ICH Q12 Implementation Considerations for FDA Regulated Products" Draft Guidance** (May 2021) at https://www.fda.gov/media/148947/download 4. **US FDA "Bridging for Drug-Device and Biologic-Device Combination Products" Draft Guidance** (December 2019) https://www.fda.gov/regulatory-information/search-fda-guidance-documents/bridging-drug-device-and-biologic-device-combination-products
Quick Link	1. https://www.ecfr.gov/current/title-21/chapter-I/subchapter-A/part-3/subpart-A/section-3.2#p-3.2(e) 2. https://www.fda.gov/combination-products 3. https://www.fda.gov/combination-products/guidance-regulatory-information 4. https://www.ecfr.gov/current/title-21/chapter-I/subchapter-A/part-4
Comments on Evolution of Combination Product Expectations	The US FDA Combination Products Quality Management System continues to evolve. At time of this writing, FDA has communicated pending draft or final guidance for: Human Factors of Combination Products (final) Essential Performance Requirements (EPRs) Established Condition & Combination Products (draft) Technical Considerations for Demonstrating Reliability of Emergency-Use Injectors (final) Combined Use/Cross-Labeled Combination Products (draft) Post-Market Changes to Combination Products (draft) Guidance on Unique Device Identifiers & Combination Products (draft) Labeling Considerations for Insulin Pumps (draft)

World

TABLE 8
World Health Organization

Country	International/World Health Organization (WHO)
Is "Combination Product" Defined?	Yes WHO Global model regulatory framework for medical devices including in vitro diagnostic medical devices (GMRF) WHO/BS/2022.2425: https://cdn.who.int/media/docs/default-source/biologicals/bs-2022.2425_global-model-regulatory-framework-for-medical-devices-including-ivds_11-july-2022-dl_14-july.pdf?sfvrsn=28d107c5_1
Scope	"Two or more different types of medical products (i.e., a combination of a medicine, device, and/or biological product with one another), such that the distinctive nature of the drug component and device component is integrated in a singular product." (Note 42, WHO/BS/2022.2425)
Legal Framework	WHO proposes that in the interest of consistency, transparency and predictability, the national regulatory authority should adopt and publish guidance on how to: 1. determine what qualifies as a combination product; 2. determine an appropriate regulatory pathway; and 3. establish suitable pre- and post-authorization requirements. GMP requirements may be developed specifically for combination products, e.g., https://www.fda.gov/media/90425/download or should follow the regulatory requirements of the constituent parts of the combination product.
Leading Authority/ Agency Responsible	WHO (WHO/BS/2022.2425) recommends that the designation of a product that combines a medicine, a biological product, or a device as a combination product be decided by the national regulatory authority. Some combination products will be designated as primarily subject to the regulatory requirements for medicines; and some to the requirements for medical devices. This may require development of a single product-specific "hybrid" pathway, combining elements of both sets of requirements.
Contact Details	National health authorities may make decisions on a case-by-case basis, taking account of all the characteristics of the product. Department of Health Products Policy and Standards World Health Organization, 20 Avenue Appia, 1211 Geneva 27, Switzerland.
Registration Process	WHO proposes: To be predictable and transparent in their decision, the regulatory authority is best advised to use a streamlined regulatory pathway and develop criteria for determining the appropriate regulatory regime for combination products.
Fees	Per country and National authority involved
Master Files Allowed?	No details
Biosimilar/Generics Guideline on Combination Products	No specifics for combination products
Requirements	
GMP/Quality System	GMP requirements may be developed specifically for combination products, e.g., https://www.fda.gov/media/90425/download or should follow the regulatory requirements of the constituent parts of the combination product.
Packaging & Labeling	No specifics for combination products

(Continued)

TABLE 8 (CONTINUED)
World Health Organization

Country	International/World Health Organization (WHO)
Unique Device Identification (UDI)	**Unique Device Identification system (UDI system) Application Guide**," International Medical Device Regulators Forum, 2019, https://www.imdrf.org/documents/unique-device-identification-system-udi-system-application-guide.
Postmarketing Safety Reporting (PMSR)	As per PMOA
Clinical Trial/ Investigations	No specifics for combination products
Human Factors	No specifics for combination products
Lifecycle Management Guidelines	No specifics for combination products
Quick Link	1. https://cdn.who.int/media/docs/default-source/biologicals/bs-2022.2425_global-model-regulatory-framework-for-medical-devices-including-ivds_11-july-2022-dl_14-july.pdf?sfvrsn=28d107c5_1 2. https://www.federalregister.gov/documents/2005/08/25/05-16527/definition-of-primary-mode-of-action-of-a-combination-product 3. https://www.fda.gov/media/119958/download. 4. https://www.fda.gov/combination-products/guidance-regulatory-information.
Comments on Evolution of Combination Product Expectations	Encouraging the Health authorities to work towards the harmonization.

Glossary[1]

Term	Definition
A	
Abbreviated New Drug Application (ANDA)	(US FDA: https://www.fda.gov/drugs/types-applications/abbreviated-new-drug-application-anda) A submission to FDA for the review and potential approval of a generic drug product. Once approved, an applicant may manufacture and market the generic drug product to provide a safe, effective, and possibly lower cost, alternative to the brand-name drug it references.
Abbreviated New Drug Submission (ANDS)	(Canada, https://www.canada.ca/en/health-canada/services/drugs-health-products/drug-products/applications-submissions/guidance-documents/chemical-entity-products-quality/guidance-document-quality-chemistry-manufacturing-guidance-new-drug-submissions-ndss-abbreviated-new-drug-submissions.html) A written request to Health Canada to obtain marketing approval for a generic drug. An ANDS provides the necessary information for the government agency to evaluate the safety and efficacy of a generic drug compared with its brand name equivalent. The generic drug must be equally safe and effective to gain approval.
Abuse (of a medicinal product)	Persistent or sporadic, intentional excessive use of medicinal products which is accompanied by harmful physical or psychological effects.
Active Implantable Medical Devices (AIMD)	[EU 90/385/EEC Article 1B(c)] Any active medical device which is intended to be totally or partially introduced, surgically or medically, into the human body or by medical intervention into a natural orifice, and which is intended to remain after the procedure.
Administration of Quality Supervision, Inspection, and Quarantine	(China: https://www.aqsiq.net) A ministerial administrative organization directly under the State Council of the People's Republic of China in charge of national quality, metrology, entry-exit commodity inspection, entry-exit health quarantine, entry-exit animal and plant quarantine, import-export food safety, certification and accreditation, standardization, as well as administrative law-enforcement.
Advanced Therapy Medicinal Product (ATMP)	(EU, Article 17 of Regulation (EC) No 1394/2007) A medicine for human use that is based on genes, cells, or tissue engineering under the EMA of the EU.
Adverse Experience (AE) (biologic)	Any Adverse Event associated with the use of a biological product in humans, whether or not considered product related, including the following: An Adverse Event occurring in the course of the use of a biological product in professional practice; an Adverse Event occurring from overdose of the product whether accidental or intentional; an Adverse Event occurring from abuse of the product; an Adverse Event occurring from withdrawal of the product; and any failure of expected pharmacological action. Examples: Lack of effect; Allergic reaction to vaccine
Adverse Experience (AE) (drug)	[21 CFR §314.80(a)] Any adverse event associated with the use of a drug in humans, whether or not considered drug related, including the following: An adverse event occurring in the course of the use of a drug product in professional practice; an adverse event occurring from drug overdose whether accidental or intentional; an adverse event occurring from drug abuse; an adverse event occurring from drug withdrawal; and any failure of expected pharmacological action. Examples: Nausea; Vomiting; Headaches; Dizziness; Cough; Sore throat; Hyperactivity; Agitation; Anxiety

Term	Definition
Annual Summary Report (ASR)	(Canada, https://www.canada.ca/en/health-canada/services/drugs-health-products/public-involvement-consultations/medical-devices/proposed-amendment-medical-device-regulations/document.html) A comprehensive assessment of all known safety information for a licensed medical device. Sections 61.4 and 61.5 of the proposed regulatory changes to the Medical Devices Regulations (MDR) require medical device license holders to complete ASRs on the safety and effectiveness of their devices.
Applicant	A person holding an application under which a combination product or constituent part of a combination product has received marketing authorization (such as approval, licensure, or clearance). Examples: PharmCo is applicant for a drug prefilled syringe market authorization; MedCo is applicant for a drug-coated stent; PharmCo is an applicant for a photosensitive drug, labeled for activation with a specific laser; and MedCo is an applicant for the specific laser labeled for use with that drug.
Application	Submission type based on combination product Primary Mode of Action (PMOA), e.g., in the United States: Drug-led (CDER or CBER oversight); Biologic-led (CBER oversight); or Device-led (CDRH oversight) Examples: New Drug Applications (NDAs), Abbreviated New Drug Applications (ANDAs); Biologics License Applications (BLAs); and "Device Applications" (Pre-market Approval Applications (PMAs), Product Development Protocols (PDPs), Humanitarian Device Exemptions (HDEs), De Novo Classification Requests (De Novos), and Pre-market Notification Submissions (510(k)s)); Emergency Use Application (EUA)
Association of Southeast Asian Nations (ASEAN)	(International, https://asean.org/asean/about-asean/overview/) A regional intergovernmental organization comprising ten countries in Southeast Asia, which promotes intergovernmental cooperation and facilitates economic, political, security, military, educational, and sociocultural integration among its members and other countries in Asia.
Auditing Organizations (AO)	(International; IMDRF: IMDRFIMDSAP WG/N11FINAL:2014) An organization that audits a medical device manufacturer for conformity with quality management system requirements and other medical device regulatory requirements. Auditing Organizations may be an independent organization or a Regulatory Authority which perform regulatory audits.
Australian Register of Therapeutic Goods (ARTG)	(Australia, https://www.tga.gov.au/australian-register-therapeutic-goods) A register of therapeutic goods that can be lawfully supplied in Australia.
Australian Regulatory Guidelines for Biologicals (ARGB)	(Australia: https://www.tga.gov.au/publication/australian-regulatory-guidelines-biologicals-argb) A regulatory guideline that provides information on the supply and use of human cell and tissue-based therapeutic goods, and live animal cells, tissues, and organs and explains the legislative requirements outlined in the regulatory framework for biologicals, including specific biological standards.
Australian Regulatory Guidelines for Medical Devices (ARGMD)	(Australia, https://www.tga.gov.au/publication/australian-regulatory-guidelines-medical-devices-argmd) A regulatory guideline provides information on the import into, export from and supply of medical devices within Australia and Explains the legislative requirements that govern medical devices.
B	
Behind-the-Counter (BTC)	(US, https://www.ncbi.nlm.nih.gov/pmc/articles/PMC4115313/) Similar to over-the-counter (OTC) status, BTC allows a patient to access medications at the pharmacy without seeing a doctor. Unlike OTC, however, access would not be allowed without the intervention of a learned intermediary. But, unlike a prescription medication, BTC would allow a patient to access drugs after an assessment and decision by a pharmacist.

Term	Definition
Benefit	Positive impact or desirable outcome of the use of a medical product on the health of an individual, e.g., positive impacts to clinical outcome, quality of life, or diagnosis, or positive impact on patient management or public health.
Biological Product	§351(i) (as modified by the Patient Protection and Affordable Care Act) of the Public Health Service Act (PHS Act) (42 U.S.C. 262(i)) provides that the term "biological product" means: A virus, therapeutic serum, toxin, antitoxin, vaccine, blood, blood component or derivative, allergenic product, protein (except any chemically synthesized polypeptide), or analogous product, or arsphenamine or derivative of arsphenamine (or any other trivalent organic arsenic compound), applicable to the prevention, treatment, or cure of a disease or condition of human beings.
Biological Product Deviation Report (BPDR)	(21 CFR §600.14 and §606.171) "[A]ny event, and information relevant to the event, associated with the manufacturing, to include testing, processing, packing, labeling, or storage, or with the holding or distribution," of a product, if that event: Represents either: a deviation from current good manufacturing practice, applicable regulations, applicable standards, or established specifications that may affect the safety, purity, or potency of that product; or an unexpected or unforeseeable event that may affect the safety, purity, or potency of that product; Occurs in the applicant's facility or another facility under contract with the applicant; and Involves a distributed product.
Biologics Genetic Therapies Directorate (BGTD)	(Canada, https://www.canada.ca/en/health-canada/services/drugs-health-products/biologics-radiopharmaceuticals-genetic-therapies/regulatory-roadmap-for-biologic-drugs.html) Health Canada's directorate that reviews and provides market authorization of all drug submissions for biologic drugs for human use.
Blister	A primary package for a medical device (through lens of device manufacturer); A secondary or tertiary package for a medicinal product (through the lens of a drug manufacturer) (see Chapter 5)
Borderline Products	(ASTM International) In the combined use context, medical products that offer combined characteristics that are covered by at least two legislations (for example, both medical device and medicinal product[2]), whose lead legislation within a jurisdiction is unclear. (EMA) Borderline products are complex healthcare products for which there is uncertainty over which regulatory framework applies. Common borderlines are between medicinal products, medical devices, cosmetics, biocidal products, herbal medicines and food supplements. In Europe, National competent authorities classify borderline products either as medicinal products or, for example, as medical devices on a case-by-case basis. This determines the applicable regulatory framework. (accessed on August 17, 2020: https://www.ema.europa.eu/en/human-regulatory/overview/medical-devices#medicinal-products-that-include-a-medical-device-(%E2%80%98combination-products%E2%80%99)-section)
C	
Canadian Medical Devices Conformity Assessment System (CMDCAS)	(Canada, https://www.scc.ca/en/agl-cmdcas) A system designed to implement Canadian regulations requiring some medical devices be designed and manufactured under a registered quality management system (QMS). The system was developed by the SCC and Health Canada's Therapeutic Products Directorate (TPD). It came into effect January 1, 2003.

Term	Definition
Canadian Medical Device Regulations (CMDR)	(Canada, https://laws-lois.justice.gc.ca/eng/regulations/sor-98-282/page-1.html) The medical device regulation of Canada labeled SOR/98-282, registered in 1998 under the Food and Drugs Act.
Caused or contributed	[US, 21 CFR § 803.3(c)] A death or serious injury was or may have been attributed to a medical device, or that a medical device was or may have been a factor in a death or serious injury, including events occurring as a result of: (1) Failure; (2) Malfunction; (3) Improper or inadequate design; (4) Manufacture; (5) Labeling; or (6) User error. Examples: Infection of a newly implanted hip joint; A retroperitoneal bleed during vessel closure (has nothing to do with the actual device, but is contributed to by the procedure of implanting the device)
Center for Devices and Radiological Health (CDRH)	(US, https://www.fda.gov/about-fda/fda-organization/center-devices-and-radiological-health) The US FDA Center that assures that patients and providers have timely and continued access to safe, effective, and high-quality medical devices and safe radiation-emitting products.
Center for Drug Evaluation (CDE)	(China, http://subsites.chinadaily.com.cn/nmpa/2019-07/19/c_389169.htm) NMPA's Center for Drug Evaluation (CDE) is responsible for evaluating drug clinical trial applications, drug marketing authorization applications, supplementary applications, and overseas production drug re-registration applications.
Center for Drug Evaluation and Research (CDER)	(US, https://www.fda.gov/about-fda/fda-organization/center-drug-evaluation-and-research-cder) A division of the US FDA that performs an essential public health task by making sure that safe and effective drugs are available to improve the health of people in the United States. CDER regulates over-the-counter and prescription drugs, including biological therapeutics and generic drugs.
Center for Medical Device Evaluation (CMDE)	(China, http://english.nmpa.gov.cn/2019-07/19/c_389172.htm) A center under NMPA responsible for: (1)The acceptance and technical review of registration application of domestic Class III medical device products and imported medical device products; also responsible for the filing of imported Class I medical device products. (2)Participation in the drafting of relevant laws, regulations, and normative documents related to the registration administration of medical devices. Organize the formulation and implementation of relevant technical reviews norms and technical guidelines for medical devices. (3)Technical review of medical devices involved in emerging medical products such as regenerative medicine and tissue engineering. (4)Coordinating the inspection work related to the evaluation of medical devices review. (5)Research the theories, technologies, development trends, and legal issues related to medical device review. (6)Providing guidance and technical support for local departments in the technical review of medical devices. (7)Organizing relevant consulting services and academic exchange, and carrying out international (regional) exchange and cooperation related to medical device review.
Chemical	With respect to Modes of Action, an isolated chemical action that takes place in the body that can be replicated in vitro and which does not require an active physiological response to achieve its intended purpose. Example: Local change in pH

Term	Definition
China Food and Drug Administration (CFDA)	(see NMPA) CFDA is the former name of what is now called the National Medical Products Administration (NMPA) in China.
Clinical Evaluation Report (CER)	(EU, MEDDEV 2.7/1 revision 4: Guidelines on Medical Devices, Clinical Evaluation: A Guide for Manufacturers And Notified Bodies under Directives 93/42/EEC and 90/385/EEC) A document that summarizes and draws together the evaluation of all the relevant clinical data documented or referenced in other parts of the technical documentation. The clinical evaluation report and the relevant clinical data constitute the clinical evidence for conformity assessment.
Co-packaged Combination Product	Two or more separate products packaged together in a single package or as a unit and comprised of drug and device products, device and biological products, biological and drug products, or drug, device and biological products. [21CFR§3.2(e)]
Combination Product	A combination product is one that is composed of two or more differently regulated types of medical products (see Chapters 2 and 3 and Appendix for global interpretations and nuances).
Combination Product Applicant	The applicant that holds the only application or all applications for a combination product. (*Note: the Combination Product Applicant holds the application, but may not be the manufacturer, e.g., if the combination product is manufactured by a contract manufacturer.) Examples: PharmCo is combination product applicant for a drug prefilled syringe, where the syringe is general use; MedCo is combination product applicant for a drug-coated stent; PharmCo is combination product applicant for a co-pack of analgesic pain reliever and its delivery system; CellCo is a combination product applicant for a cell-gene therapy/IV bag delivery system.
Combination Therapy	Also referred to as fixed-dose combinations; combination therapies are treatments in which a patient is given two or more drugs, or two or more active pharmaceutical ingredients, or two or more biologics, for a single disease.
Combined Use Medical Products	(ASTM International) Two or more differently regulated medical products (regardless of configuration) that are to be used together, or are being studied for use together, to achieve the intended use, indication, or effect, for example, a medicinal product and a device.
Committee for Medicinal Products for Human Use (CHMP)	(EU, https://www.ema.europa.eu/en/committees/committee-medicinal-products-human-use-chmp) European Medicines Agency's (EMA) committee responsible for human medicines. the CHMP is responsible for: • Conducting the initial assessment of EU-wide marketing authorization applications; • Assessing modifications or extensions ("variations") to an existing marketing authorization; and • Considering recommendations of the Agency's Pharmacovigilance Risk Assessment Committee on the safety of medicines on the market and when necessary, recommending to the European Commission changes to a medicine's marketing authorization, or its suspension or withdrawal from the market. The CHMP also evaluates medicines authorized at national level referred to EMA for a harmonized position across the EU. In addition, the CHMP and its working parties contribute to the development of medicines and medicine regulation, by: • Providing scientific advice to companies researching and developing new medicines; • Preparing scientific guidelines and regulatory guidance to help pharmaceutical companies prepare marketing authorization applications for human medicines; and • Cooperating with international partners on the harmonization of regulatory requirements.
Common Technical Document (CTD)	(International Conference on Harmonization, https://www.ich.org/page/ctd) Under ICH, the CTD is a common format for assembling quality, safety, and efficacy information. The CTD is organized into five modules. Module 1 is region specific and Modules 2, 3, 4, and 5 are intended to be common for all regions. In July 2003, the CTD became the mandatory format for new drug applications in the EU and Japan, and the strongly recommended format of choice for NDAs submitted to United States FDA.

Term	Definition
Component/ Constituent Part	(US FDA) Under 21 CFR 820.3(c): **"**Any raw material, substance, piece, part, software, firmware, labeling or assembly which is **intended to be included as part of the finished, packaged and labeled device**." Under 21 CFR 210.3: "Any ingredient intended for use in manufacture of a drug product, **including those that may not appear in such drug product."** "Component" is used within FDA 21 CFR § 3.2 high-level CP definitions, however is then switched to "constituent" within more detailed guidance, to separate the major regulated parts of a product (drug, biologic, device constituents) from sub-parts e.g. springs, molded parts ("components"). "Combination product is a product comprised of any combination of a drug, a device, and/or a biological product. The drugs, devices, and biological products included in combination products are referred to as 'constituent parts' of the combination product." FDA eCTD guidance uses "constituent" language, e.g., 'Generally, drug or biological product information for combination drug and device product information and related engineering and manufacturing information should be located in the same eCTD sections that would provide similar information for the drug or biological product alone. This particularly applies to device constituent parts that also serve as the drug container closure system. For example, the M3 quality module should contain information on such device constituents in section 3.2.P.7. Supportive files for container closure device constituents should be located in section 3.2.R.' (EU) EMA Quality guideline (October 2019 draft) uses "constituent" and "component" somewhat interchangeably MDR 2017/745 generally uses "constituent" (e.g. Annex I 23.2r) to describe the major item, and "component" to describe sub-parts (e.g. Article 23) ISO 13485 uses "constituent parts" within 7.3.9, referring more to sub-parts. (Suppliers) Some suppliers of sub-assemblies call these "components," others call them "sub-assemblies" or "devices." Those that refer to the sub-assemblies as components, typically also refer to their sub-parts as components, making interpretation of documents more difficult.
Competent Authority (CA)	(EU, https://www.ema.europa.eu/en/glossary/competent-authority) Under the European Commission, a medicines regulatory authority in the European Union.
Concession	Permission to use or release a product that does not conform to specified requirements (GHTF/SG3/N18:2010)
Conformity Assessment (CA)	(Australia, https://www.tga.gov.au/publication/australian-regulatory-guidelines-medical-devices-argmd) Under Australian Regulatory Guidelines for Medical Devices (ARGMD), a Conformity Assessment is the systematic and ongoing examination of evidence and procedures to ensure that a medical device (including IVD medical devices) complies with the essential principles.
Conformity Assessment Body (CAB)	(EU, https://ec.europa.eu/growth/single-market/goods/building-blocks/accreditation_en) Laboratories, inspection or certification bodes, etc., who have the technical capacity to perform their duties in the EU. Used in regulated sectors and voluntary areas, accreditation increases trust in conformity assessment. It reinforces the mutual recognition of products, services, systems, and bodies across the EU.
Conformité Européenne (CE)	(EU, https://ec.europa.eu/growth/single-market/ce-marking) The letters "CE" appear on many products traded on the extended Single Market in the European Economic Area (EEA). They signify that products sold in the EEA have been assessed to meet high safety, health, and environmental protection requirements.
Constituent Part	A medicinal product or a medical device that is part of a combination product.

Term	Definition
Constituent Part Applicant	(US FDA) An applicant that holds an application for a constituent part of a combination product, the other constituent part(s) of which is marketed under an application held by a different applicant. Generally, there would be Constituent Part Applicants if the combination product is marketed by two different entities in a cross-labeled configuration. Example: PharmCo is a Constituent Part Applicant for a photosensitive drug, labeled for activation with a specific laser; and MedCo is a Constituent Part Applicant for the specific laser labeled for use with that drug.
Container Closure	Holds and protects the medicinal product; also known as "primary package"; if it delivers a metered dose of the medicinal product, it may also be considered a medical device.
Continued Process Verification (CPV)	Ongoing assurance during routine manufacturing that the process remains in a state of control. This is considered Phase 3 of the Lifecycle Process Validation Approach (ICH Q8(R2)).
Control Strategy	A planned set of controls based on product and process understanding that ensure a process performs as it should, to ensure product quality is maintained.
Convenience Kit	"Convenience Kit" is a term that can have a specialized meaning for purposes of regulation in the US. In the context of combination products, a kit is one that includes two or more different types of medical products (e.g., a device and a drug), and is considered a co-packaged combination product under Part 4. If the kit includes only products that are 1) legally marketed independently and 2) packaged in the kit as they are when marketed independently, including labeling used for independent marketing, it is considered a "convenience kit." [21 CFR 3.2(m)]
Co-packaged Combination Product	A combination product comprised of one or more differently regulated medical products that are packaged together as a unit.
Correction	Action to eliminate a detected nonconformity (GHTF/SG3/N18:2010)
Correction and Removal Report	(Adapted from 21 CFR 806.10 and 806.20) A report intended to reduce a risk to health posed by the product; or to remedy a violation of the [FD&C Act] caused by the product which may present a risk to health. (21 CFR 806.10). Correction means any repair, modification, adjustment, relabeling, destruction, or inspection (including patient monitoring) of a [product] without its physical removal from its point of use to some other location (21 CFR 806.2(d)). Removal means the physical removal of a [product] from its point of use to some other location for repair, modification, adjustment, relabeling, destruction, or inspection (21 CFR 806.2(j)
Corrective Action	Action to eliminate the cause of a detected nonconformity or other undesirable situation. There can be more than one cause for a nonconformity; correction action is taken to prevent recurrence whereas preventive action is taken to prevent occurrence. (GHTF/SG3/N18:2010)
Corrective and Preventive Action (CAPA)	(US, 21CFR§820.100/ ISO 13485:2016, Clause 8.2 and subclauses) Procedures a firm must establish for implementing corrective and preventive action to ensure conformance of product, processes, and quality systems.
Critical Material Attribute (CMA)	A physical, chemical, biological, or microbiological property or characteristic of an input material that should be within an appropriate limit, range, or distribution to ensure the desired quality of output material.
Critical Process Parameter (CPP)	A process parameter whose variability impacts a CQA or CtQ and therefore should be monitored or controlled to ensure the process produces the desired quality. (ICH Q8(R2))
Critical Quality Attributes (CQAs)	A physical, chemical, biological, or microbiological property or characteristic that should be within an appropriate limit, range, or distribution to ensure the desired product quality (e.g., identity, strength/potency, purity and safety). (ICH Q8(R2))
Critical to Quality (CtQ)	The quality of a product or service in the eyes of the voice of the customer; typically used to refer to the critical functionality/ features of a medical device.

Term	Definition
Critical Use Tasks (also called Critical Tasks)	CDRH: "A user task which, if performed incorrectly or not performed at all, would or could cause ***serious*** harm to the patient or user, where harm is defined to include compromised medical care."[3] CDER: "User tasks that, if performed incorrectly or not performed at all, would or could cause harm to the patient or user, where harm is defined to include compromised medical care."[4] 1. Consider harm to both the patient or user, such as a caregiver or healthcare professional. 2. Dosing, such as dialing a proper dose using a pen-injector. 3. Administration, e.g., the step of selecting the correct injection site. Urgency/time. In some cases, e.g., emergency use-administration of epinephrine to treat a severe allergic reaction, drugs must be administered as quickly as possible to be effective. Combination Products: User tasks that, if performed incorrectly or not performed at all, would or could cause harm to the patient or user, where harm is defined to include compromised medical care. Thus, categorizing a task as critical is dependent on the unique considerations *for each combination product*. The US FDA expects the risk analysis for the combination product to include an identification of all the critical tasks required for using the combination product, the consequences for failing to perform each critical task correctly, and the strategies that have been applied in the design of the user interface to eliminate or reduce risks to acceptable levels. Such an assessment should include considerations of the indication, the users, the environment, and other conditions that might influence the importance of a particular task. Some examples of critical tasks for a combination product: Ability of the patient to successfully self-administer a drug as described in its labeling; safe disposal of the used combination product; ability to distinguish a product from other similar products.
Cross-labeled Combination Product	Combination product category uniquely defined by the US FDA under 21 CFR 3.2(e)(3) and (4), including constituent parts that are separately packaged or but specified for use together, or are studied together, to achieve the intended use, indication, or effect.
Current Good Manufacturing Practices (cGMP)	A framework for the minimum requirements to assure product quality. cGMPs provide for systems that assure proper design, monitoring, and control of manufacturing process and facilities. *(See also definition of Combination Products Manufacturer)*
D	
Data Sources	The processes within a Quality Management System that provide quality information that could be used to identify nonconformities, or potential nonconformities (GHTF/SG3/N18:2010)
Design Change Control	Each manufacturer shall establish and maintain procedures for the identification, documentation, validation or where appropriate verification, review, and approval of design changes before their implementation. (21 CFR 820.30(i)/ ISO 13485:2016, Clause 7.3.9)
Design Controls	Framework to control the design process, made up of quality practices and procedures incorporated into design and development to ultimately assure the device (and combination product) meet user needs and intended use(s) and are safe, effective, and usable.
Design History File (DHF)	(US, 21 CFR §820.30/ ISO 13485:2016, Clause 7.3.10 and 4.2.3) The compilation of records which describes the ***design history*** of a finished device.
Design Inputs	Each manufacturer shall establish and maintain procedures to ensure that the design requirements relating to a device are appropriate and address the intended use of the device, including the needs of the user and patient. The procedures shall include a mechanism for addressing incomplete, ambiguous, or conflicting requirements. The design input requirements shall be documented and shall be reviewed and approved by a designated individual(s). The approval, including the date and signature of the individual(s) approving the requirements, shall be documented. [21 CFR 820.30(c)/ ISO 13485:2016 Clause 7.3.3]
Design Output	Each manufacturer shall establish and maintain procedures for defining and documenting design output, the results of the design process, in terms that allow an adequate evaluation of conformance to design input requirements. Design output procedures shall contain or make reference to acceptance criteria and shall ensure that those design outputs that are essential for the proper functioning of the device are identified. Design output shall be documented, reviewed, and approved before release. The approval, including the date and signature of the individual(s) approving the output, shall be documented. (21 CFR 820.30(d)/ ISO 13485:2016, Clause 7.3.4

Term	Definition
Design Review	Formal documented reviews of the design results that are planned and conducted at appropriate stages of the device's design development. The procedures shall ensure that participants at each design review include representatives of all functions concerned with the design stage being reviewed and an individual(s) who does not have direct responsibility for the design stage being reviewed, as well as any specialists needed. The results of a design review, including identification of the design, the date, and the individual(s) performing the review, shall be documented in the design history file (the DHF). (21 CFR 820.30(e)/ ISO 13485:2016, Clause 7.3.5)
Design Transfer	(21 CFR 820.30(h)/ ISO 13485:2016, Clause 7.3.8) The design and development process transitions to commercial manufacturing via Design Transfer,[5] ensuring that the constituent parts and combination product design are correctly translated into production specifications. Design Transfer must be initiated so that production, or production-equivalent, units are the basis of Design Validation.
Design Validation	Design Validation (21 CFR 820.30(g)/ ISO 13485:2016 Clauses 7.3.7/ 7.3.10) provides objective evidence that when users are provided conforming (or production-equivalent) constituent parts and combination product, the product functions as intended to meet user needs/intended uses in the intended use environment.
Design Verification	Each manufacturer shall establish and maintain procedures for verifying the device design. Design verification shall confirm that the design output meets the design input requirements. The results of the design verification, including identification of the design, method(s), the date, and the individual(s) performing the verification, shall be documented in the DHF. (21 CFR 820.30(f)/ ISO 13485:2016 Clauses 7.3.6 and 7.3.10)
Detectability	The ability to discover or determine the existence, presence, or fact of a hazard
Device	§201(h) of the FD&C Act (21 U.S.C. 321(h)) provides that the term "device" means: An instrument, apparatus, implement, machine, contrivance, implant, in vitro reagent, or other similar or related article, including any component, part, or accessory, which is – 1.Recognized in the official National Formulary, or the US Pharmacopeia, or any supplement to them, 2.Intended for use in the diagnosis of disease or other conditions, or in the cure, mitigation, treatment, or prevention of disease, in man or animals, or 3.Intended to affect the structure or any function of the body of man or other animals, and which does not achieve its primary intended purposes through chemical action within or on the body of man or other animals and which is not dependent upon being metabolized for the achievement of its primary intended purposes.
Device History Record (DHR)	(US, 21 CFR §820.30) A compilation of records containing the ***production history*** of a finished device.
Device Master Record (DMR)	(US, 21 CFR 820.30/ ISO 13485:2016, Clause 4.2.3 Medical Device File) A compilation of records containing the procedures and specifications for a finished device.
Drug	[US, § 201(g) of the FD&C Act (21 U.S.C. 321(g)] A) Articles recognized in the official United States Pharmacopeia, official Homeopathic Pharmacopeia of the United States, or official National Formulary, or any supplement to any of them (B) Articles intended for use in the diagnosis, cure, mitigation, treatment, or prevention of disease in man or other animals (C) Articles (other than food) intended to affect the structure or any function of the body of man or other animals (D) Articles intended for use as a component of any articles specified in Clause (A), (B), or (C)
Drug Administration Law (DAL)	(China, http://english.nmpa.gov.cn/2019-11/29/c_456284.htm) The DAL passed in China on 26 August 2019 and came into effect on 1 December 2019. It includes regulations, documents, and technical guidelines for drug Marketing Authorization Holders, drug clinical trials oversight, requirements for Drug GMP and GSP administration, review and approval of APIs, and investigation and prosecution of drug-related illegal activities.

Term	Definition
Drug-Device Combinations (DDCs)	(EU, EMA May 2019 Draft Guideline – https://www.medical-device-regulation.eu/wp-content/uploads/2019/06/draft-guideline-quality-requirements-drug-device-combinations_en-1.pdf) A medicinal product(s) with integral and/or non-integral medical device/device component(s) necessary for administration, correct dosing, or use of the medicinal product.
Drug Identification Number (DIN)	(Canada, https://www.canada.ca/en/health-canada/services/drugs-health-products/drug-products/fact-sheets/drug-identification-number.html) A computer-generated eight-digit number assigned by Health Canada to a drug product prior to being marketed in Canada. It uniquely identifies all drug products sold in a dosage form in Canada and is located on the label of prescription and over-the-counter drug products that have been evaluated and authorized for sale in Canada.
E	
eCTD	Electronic format for Common Technical Document (International Conference on Harmonization, https://www.ich.org/page/ctd) Under ICH, the CTD is a common format for assembling quality, safety, and efficacy information. The CTD is organized into five modules. Module 1 is region specific and Modules 2, 3, 4, and 5 are intended to be common for all regions. In July 2003, the CTD became the mandatory format for new drug applications in the EU and Japan, and the strongly recommended format of choice for NDAs submitted to United States FDA.
Essential Performance Requirements (EPRs)	Critical requirements, achieved at an acceptable level of risk, that are essential to the safe and effective operation/functioning and use of a medical product, i.e., those performance characteristics needed for safe and effective dose delivery (based on IEC 60601-1-11:2015); Critical requirements essential to the safe and effective operation/ functioning and use of the device constituent part and the combination product as a whole.
European Databank on Medical Devices (EUDAMED)	(EU, https://ec.europa.eu/health/md_eudamed/overview_en) The web-based portal developed by the European Commission to implement Regulation (EU) 2017/745 on medical devices and Regulation (EU) 2017/746 on in vitro diagnosis medical devices. It acts as a central hive for the exchange of information between national competent authorities and the European Commission.
European Medicines Agency (EMA)	(EU, https://www.ema.europa.eu/en) A decentralized agency of the European Union (EU) responsible for the scientific evaluation, supervision, and safety monitoring of medicines in the EU.
European Union (EU)	(EU, https://europa.eu/european-union/about-eu/countries/member-countries_en) An economic and political union between 27 EU countries that together cover much of the European continent. The predecessor of the EU was the European Economic Community (EEC), created in 1958 in the aftermath of WWII. What began as a purely economic union has evolved into an organization spanning policy areas, from climate, environment, and health to external relations and security, justice, and migration. European Economic Community (EEC) changed its name to the European Union (EU) in 1993 reflecting this.
Expectedness	For drugs, reportability takes into consideration the concept of "expectedness" based on the product label, accounting for the event in question. If the label specifies that such a serious event could be anticipated, then it is deemed "expected." This includes events that may be symptomatically and pathophysiologically related to an event listed in the labeling but, that differ from the event because of greater severity or specificity (see 21 CFR 314.80(a) and 600.80(a)). This contrasts with the interpretation of reportability for device malfunctions, where, for standalone devices, malfunctions are deemed reportable regardless of labeling expectedness.
Extractables	(USP <1663> Assessment of Extractables Associated with Pharmaceutical Packaging/Delivery Systems) Organic or inorganic chemical entities that are released from a pharmaceutical delivery system and into an extraction solvent under laboratory conditions.

Term	Definition
F	
Failure Mode	The specific manner or way by which a failure occurs, e.g., for a raw material, component, function, equipment, process, subsystem, sub-assembly, system, device, or combination product. If we say that there are several ways something can go wrong, we say it has multiple failure modes.
Failure Modes & Effects Analysis (FMEA)	A model/ approach used to identify and prioritize potential defects based on their severity, expected frequency, and likelihood of detection. A Failure Mode And Effects Analysis can be performed on a design (dFMEA), an intended design, an individual process (pFMEA) or a complete production process, including equipment. User FMEA or Application FMEA (uFMEA or aFMEA) can be conducted to identify use task failures that lead to unintended performance consequences. Critical Task Analysis uses Use-Related Risk Analysis, (URRA) a concept similar to uFMEA. Detectability is not a consideration in URRA.
Federal Drug and Cosmetic Act (FD&C Act or FDCA)	(US, https://www.fda.gov/regulatory-information/laws-enforced-fda/federal-food-drug-and -cosmetic-act-fdc-act) The FDCA and subsequent amending statutes are codified into Title 21 Chapter 9 of the United States Code. – The laws of the United States are organized by subject into the United States Code. The United States Code contains only the currently enacted statutory language. The official United States Code is maintained by the Office of the Law Revision Counsel in the United States House of Representatives. The Office of the Law Revision Counsel reviews enacted laws and determines where the statutory language should be codified related to its topic.
Field Alert Report (FAR) (or 3-Day Report)	(21 CFR §314.81 or) A report that is submitted for NDA/ANDA products, or those containing a drug constituent part, including • "[A]ny incident that causes the [product] or its labeling to be mistaken for, or applied to, another article," or "concerning any bacteriological contamination, or any significant chemical, physical, or other change or deterioration in the distributed [product]" or • "[A]ny failure of one or more distributed batches of the [product] to meet the specification established for it in the application." Importantly, per the PMSR Final Guidance, a FAR is required for "any of the issues described above that could have resulted from the manufacturing process for any of the constituent parts of the combination product or for the combination product (see 21 CFR 4.102(a) and 314.(81)." A FAR report is required even if the issue is the result of a material sourced from a third party.
Field Safety Corrective Action (FSCA)	(EU MDR 2017/745, Article 89) Action taken by a manufacturer to reduce the chance of death or serious deterioration in health with the use of a medical device on the market, e.g., recalls
Fifteen-Day Report	(21 CFR §314.80 or §600.80) A report that is submitted for NDA/ANDA-led or BLA-led products, or those containing a drug- or biologic-constituent part, for adverse experiences that are both serious and unexpected.
Fifteen-Day "Malfunction-only" Reporting	A traditional drug- or biologic- 15-day report would require the following four elements to be valid: 1) An identifiable patient 2) An identifiable reporter 3) A suspect product 4) An adverse experience A Malfunction report may also be required when a 15-day report is not. Fifteen-Day "Malfunction-only" reporting applies specifically to drug- or biologic-led PMSR types. A Fifteen-Day "Malfunction-only" is considered to be valid even in the absence of the above four elements. As a best practice, in the authors' experience, organizations will need to establish default values that can be used for the four elements.
Final Finished Combination Product	The combination product intended for market and submitted in the marketing application. This term applies to the combined final device, drug and/or biological product configuration, including all product user interfaces, such as proposed packaging, labels, and labeling, including training programs.

Term	Definition
Five-Day Report	(21 CFR §803.3, 803.53, 803.56) For products containing a device constituent part, these reports include: • A reportable event that "necessitates remedial action to prevent an unreasonable risk of substantial harm to the public health," or • FDA has "made a written request for the submission of a [Five-day] report" (21 CFR 803.53). • Remedial action includes "any action other than routine maintenance or servicing …where such action is necessary to prevent recurrence of a reportable event" (21 CFR 803.3(v)).
Fixed-dose Combination	Also referred to as combination therapy; fixed-dose combinations are treatments in which a patient is given two or more drugs, or two or more active pharmaceutical ingredients, or two or more biologics, for a single disease.
Follow-up Report	Under US Combination Products PMSR remit, there is flexibility to submit alternate constituent part-related report types as they become known. This can be done using a "Follow-up Report," one that captures any additional information gathered relative to an event for a given report type, after the initial report was submitted.
Food and Drug Administration (FDA)	(US, https://www.fda.gov/about-fda/transparency/fda-basics) A federal agency of the United States Department of Health and Human Services, one of the US federal executive departments. The FDA is responsible for protecting and promoting public health through the control and supervision of food safety, tobacco products, dietary supplements, prescription and over-the-counter pharmaceutical drugs (medications), vaccines, biopharmaceuti cals, blood transfusions, medical devices, electromagnetic radiation emitting devices (ERED), cosmetics, animal foods and feed, and veterinary products.
Formative Evaluation/Formative Study	A Human Factors study that is conducted on a combination products prototype user interface at one or more stages during the iterative product development process. The focus of a formative study is to assess user interaction with the product and to identify potential use errors. HF Formative studies are typically iterative, and inform the content of the HF Validation study, also known as Summative Study. (*see also Human Factors Study and Summative Study*)
G	
Global Harmonization Task Force (GHTF)	(International, http://www.imdrf.org/ghtf/ghtf-archives.asp) GHTF was conceived in 1992 in an effort to achieve greater uniformity between national medical device regulatory systems. This was done with two aims in mind: enhancing patient safety and increasing access to safe, effective, and clinically beneficial medical technologies around the world. Since 2011, The International Medical Device Regulators Forum (IMDRF) is continuing the work of the Global Harmonization Task Force (GHTF).
Good Clinical Practice (GCP)	(https://www.fda.gov/about-fda/center-drug-evaluation-and-research-cder/good-clinical-practice; *see also ISO 14155:2020, "Clinical investigation of medical devices for human subjects – Good Clinical Practice"*) Health authorities regulate scientific studies designed to develop "evidence to support the safety and effectiveness of investigational drugs (human and animal), biological products, and medical devices. Physicians and other qualified experts ('clinical investigators') who conduct these studies are required to comply with applicable statutes and regulations … intended to ensure the integrity of clinical data on which product approvals are based and to help protect the rights, safety, and welfare of human subjects."
Good Distribution Practice (GDP)	(EU, https://www.ema.europa.eu/en/glossary/good-distribution-practice) A code of standards ensuring that the quality of a medicine is maintained throughout the distribution network, so that authorized medicines are distributed to retail pharmacists and others selling medicines to the general public without any alteration of their properties.
Good Manufacturing Practice (GMP)	(International, https://www.who.int/medicines/areas/quality_safety/quality_assurance/gmp/en/) Regulations that require that manufacturers, processors, and packagers of drugs, medical devices, some food, and blood take proactive steps to ensure that their products are safe, pure, and effective.

Term	Definition
Good Pharmacovigilance Practices (GVP)	(EU, https://www.ema.europa.eu/en/human-regulatory/post-authorisation/pharmacovigilance/good-pharmacovigilance-practices) A set of measures drawn up to facilitate the performance of pharmacovigilance in the European Union (EU). GVP applies to marketing authorization holders, the European Medicines Agency (EMA), and medicines regulatory authorities in EU Member States. They cover medicines authorized centrally via the Agency as well as medicines authorized at national level.
General Safety and Performance Requirements (GSPRs)	(EU, https://www.medical-device-regulation.eu/category/chapter-1-general-requirements-2/) Minimum essential performance characteristics required of medical devices under EU MDR Annex 1. Fulfilling GSPRs is fundamental preconditions to placing any medical device on the European market. The "requirements" are a set of product characteristics, considered by the European authorities as being essential to ensuring that any new device will be safe and perform as intended throughout its life. The GSPRs are subdivided into general requirements, design and manufacturing requirements, and information that is supplied with the device.
H	
Harm	Injury or damage to the health of people, or damage to property or environment; injury includes physical or mental injury, including the damage that can occur from loss of product quality or availability (ISO/IEC 63:2019)
Hazard	Potential source of harm (ISO/IEC Guide 51)
Hazard Identification	The systematic use of information to identify potential sources of harm (hazards)
Hazardous Situation	Circumstances in which people, property, or the environment are exposed to one (or more) hazard(s) (ISO/IEC Guide 63:2019)
Health Canada (HC)	(Canada, https://www.canada.ca/en/health-canada.html) The Canadian health authority responsible for helping Canadians maintain and improve their health. It ensures that high-quality health services are accessible, and works to reduce health risks.
Health Products and Food Branch (HPFB)	(Canada, https://www.canada.ca/en/health-canada/corporate/about-health-canada/branches-agencies/health-products-food-branch.html) The branch of Health Canada that manages the health-related risks and benefits of health products and food by minimizing risk factors while maximizing the safety provided by the regulatory system.
Health Technology Assessment (HTA)	(International, https://www.who.int/medical_devices/assessment/en/) The systematic evaluation of properties, effects, and/or impacts of health technology. It is a multidisciplinary process to evaluate the social, economic, organizational, and ethical issues of a health intervention or health technology. The main purpose of conducting an assessment is to inform a policy decision-making.
Human Cell Tissue Therapy Product (HCT/P)	(fda.gov/vaccines-blood-biologics/tissue-tissue-products) "Human cells or tissue intended for implantation, transplantation, infusion, or transfer into a human recipient is regulated as a human cell, tissue, and cellular and tissue-based product or HCT/P." In the United States, these are regulated by the Center for Biologics Evaluation and Research (CBER) under 21 CFR Parts 1270 and 1271. Examples of such tissues include "bone, skin, corneas, ligaments, tendons, dura mater, heart valves, hematopoietic stem/progenitor cells derived from peripheral and cord blood, oocytes and semen. CBER does not regulate the transplantation of vascularized human organ transplants such as kidney, liver, heart, lung or pancreas. The Health Resources Services Administration (HRSA) oversees the transplantation of vascularized human organs."
Human Factors	"A body of knowledge about human abilities, human limitations, and other human characteristics that are relevant to design. Human factors engineering is the application of human factors information to the design of tools, machines, systems, tasks, jobs, and environments for safe, comfortable, and effective human use." (hfes.org)

Term	Definition
Human Factors Study (HF or UE)	Also called Usability Engineering Study; HF evaluates the ability of the user to perform critical tasks and their ability to understand the information in the packaging and labeling (e.g., labels or instructions for use) that inform the user's actions, and that are critical to the safe and effective use of the combination product. An HF Study is conducted with representative users to evaluate the adequacy of the combination product user interface design, to eliminate or mitigate potential use-related hazards. Typically HF studies are an iterative part of the design and development process, driven by the complexity of the combination product and the nature of the safety considerations. Example HF critical tasks might include product preparation/reconstitution steps, administration, maintenance and disposal, or the actions an individual takes in the event of an adverse reaction. *See also Formative Study and Summative Study.*
I	
Immunological	An action initiated by a substance or its metabolites on the human body and mediated or exerted (e.g. stimulation, modulation, blocking, replacement) by cells or molecules involved in the functioning of the immune system (e.g. lymphocytes, toll-like receptors, complement factors, cytokines, antibodies). Examples: • Presentation of an antigen to generate antibody response. • Modulation of an immune response (e.g. suppressing, blocking, activating/ enhancing), • Replacement, reconstitution or introduction of natural or modified immune cells or molecules, • Triggering an immune response against the targeted tissues, cells, or antigens by immune-specific recognition. • Targeting action of other linked or coupled substances (see Note 2) Examples of substances/products acting via immunological means: vaccine, tetanus anti-serum, monoclonal antibodies, CAR-T cells, anti-venom, C1 esterase-inhibitor.
In vitro Diagnostic (IVD)	(International, https://www.who.int/in-vitro-diagnostic/en/) IVDs are a subset of medical devices which are used for in vitro examination of specimens derived from the human body to provide information for screening, diagnosis, or treatment monitoring purposes. An IVD test may include reagents provided either in kit format or separately, as well as calibrators, and controls. In vitro testing may be performed on a variety of instruments ranging from small, hand-held tests to complex laboratory instruments.
Incident	(Article 87, EU MDR 2017/745) Any malfunction or deterioration in the characteristics or performance of a device, including use error
Individual Case Safety Reports (ICSRs)	FDA, EU, and World Health Organization (WHO) commonly refer to Individual Case Safety Reports (ICSRs) as the mechanism for capturing information needed to support the reporting of adverse events, product problems, and consumer complaints with respect to any product for the PMSR report types (e.g., 15-Day Reports, 5-Day Reports, Malfunction Reports, Follow-Up Reports). (Non-ICSR: FAR, BPDR, Correction & Removal Reports)
Information Sharing	Under 21 CFR §4 Subpart B and associated Final Guidance for Postmarketing Safety Reporting for Combination Products, information sharing between constituent part applicants of a combination product must occur: • No later than 5 days from receipt; • With the other Constituent Part Applicant(s) for the same combination product regarding an event associated with the combination product that involves: • A death or serious injury as described in 21 CFR §803.3, or • An adverse experience as described in 21 CFR §314.80(a) and §600.80(a). According to 21 C.F.R §4.103, this information must be shared regardless of whether the event is expected or unexpected and regardless of whether the Constituent Part Applicant believes it involves only its constituent part. If both constituent parts are registered to the same Market Authorization Holder (MAH), while the constituent parts may comprise a cross-label combination product, information-sharing requirements do not apply.
Installation	The process for making hardware and/or software ready for use.

Term	Definition
Installation Qualification (IQ)	Establishes objective evidence that all key aspects of the process equipment and ancillary system installation adhere to the manufacturer's approved specification as well as recommendations of the supplier of the equipment are appropriately considered. Software documentation is needed if the equipment or tool includes software.
Intellectual Property (IP)	(International, https://www.wipo.int/about-ip/en/) Creations of the mind, such as inventions; literary and artistic works; designs; and symbols, names, and images used in commerce.
Intended Use	Use for which a product, process or service is intended according to its specifications, instructions, and information provided by the manufacturer (ISO/IEC Guide 63:2019). Included in intended use includes intended users, target patient population, intended use environment, and operating principles of a given product.
International Council for Harmonization (ICH)	(International, https://www.ich.org/page/mission) The International Council for Harmonization of Technical Requirements for Pharmaceuticals for Human Use (ICH), established in 1990, brings together the regulatory authorities and pharmaceutical industry to discuss scientific and technical aspects of pharmaceuticals and develop ICH guidelines. ICH's mission is to achieve greater harmonization worldwide to ensure that safe, effective, and high-quality medicines are developed and registered in the most resource-efficient manner. Harmonization is achieved through the development of ICH Guidelines via a process of scientific consensus with regulatory and industry experts working side-by-side.
International Medical Device Regulators Forum (IMDRF)	(International, http://www.imdrf.org/about/about.asp) Established in October 2011, with representatives from global medical device authorities (Australia, Brazil, Canada, China, European Union, Japan, Russia, Singapore, South Korea, and the United States) as well as World Health Organization (WHO), this forum provides guidance on strategies, policies, directions for medical devices. IMDRF is a voluntary group of medical device regulators from around the world who have come together to build on the strong foundational work of the Global Harmonization Task Force on Medical Devices (GHTF) and aims to accelerate international medical device regulatory harmonization and convergence.
International Standards Organization (ISO)	(International, https://www.iso.org/about-us.html) An independent, non-governmental international organization with a membership of 165 national standards bodies. Through its members, it brings together experts to share knowledge and develop voluntary, consensus-based, market-relevant International Standards that support innovation and provide solutions to global challenges.
Issue-Related Summary Reports (IRSR)	(Canada, https://www.canada.ca/en/health-canada/services/drugs-health-products/reports-publications/medeffect-canada/preparing-submitting-summary-reports-marketed-drugs-natural-health-products-guidance-industry.html) A concise, critical analysis of the adverse drug reactions and serious adverse drug reactions to a drug that are known to the manufacturer with respect to a specific issue that Health Canada's Minister directs the manufacturer to analyze.

J

K

L

Term	Definition
Leachables	(USP <1664> Assessment of Drug Product Leachables Associated with Pharmaceutical Packaging/Delivery Systems) Organic or inorganic chemical entities that are present in a packaged drug product because they have migrated into the packaged drug product from a delivery system, related component, or material of construction under normal conditions of storage and use, or during drug product stability studies.
Leak Performance Testing	Test designed to observe leakage of drug-filled containers that have several interfaces, e.g., syringe barrel to Luer needle, syringe barrel to plunger, and cartridge to seal.

Term	Definition
Legal Manufacturer	The natural or legal person who is responsible for the design, manufacture, packaging, and labeling of a device before it is placed on the market under his own name, regardless of whether these operations are carried out by that person himself or on his behalf by a Third Party.
Letter of Designation	(21 CFR 3.2(i)) FDA's formal response to an RFD and is a binding determination with respect to classification and/or center assignment that may be changed under conditions specified in Section 563 of the FD&C Act and 21 CFR 3.9 in the regulations.
Life-threatening Adverse drug experience	[21 CFR §314.80(a)] Any adverse drug experience that places the patient, in the view of the initial reporter, at immediate risk of death from the adverse drug experience as it occurred, i.e., it does not include an adverse drug experience that, had it occurred in a more severe form, might have caused death. Examples: Anaphylactic shock; Miscarriage
M	
Major Clinical Study (or Major Clinical Trial)	A larger-scale clinical study that occurs during a later phase of combination product development. Major clinical studies provide the primary support for the safety and effectiveness of a product for a proposed indication. In addition to adequate and well-controlled studies, other types of later-phase, larger-scale clinical studies may also be considered major clinical studies, e.g., a long-term extension study. FDA Draft Guidance "Human Factors Studies and Related Clinical Study Considerations in Combination Product Design and Development" (Feb. 2016) https://www.fda.gov/media/96018/download
Malfunction	The failure of a device to meet its performance specifications or otherwise perform as intended. [21 CFR §803.3(k)] Performance specifications include all claims made in the labeling for the device. The intended performance of a device refers to the intended use for which the device is labeled or marketed (21 CFR §801.4). Examples: Bent needle, Plunger broken, Label damage, Seal integrity compromised
Malfunction Report (also called "30-Day Report")	(21CFR §803.50 and 803.56) A report filed under PMSR for products containing a device constituent part, for when information reasonably suggests: • The product has malfunctioned; and • The product, or a similar product marketed by the applicant, "would be ***likely*** to ***cause or contribute to*** a death or serious injury if the malfunction were to recur" [21 CFR § 803.3(o)(2)(ii) and 803.50].
Management with Executive Responsibility	(also Top Management) a senior employee and has the authority to establish and change the quality policy and quality system. FDA's definition or description of management with executive responsibility is consistent with the definition in the International Organization for Standardization document, ISO 9001/ 21 CFR 820.20
Mandatory Problem Reporting (MPR)	(Canada, https://www.canada.ca/en/health-canada/services/drugs-health-products/reports-publications/medeffect-canada/guidance-document-mandatory-problem-reporting-medical-devices-health-canada-2011.html#a22) An MPR is required under the Regulations in Canada for any incident involving a medical device that is sold in Canada when the incident: • occurs either within or outside Canada; • relates to a failure of the device or a deterioration in its effectiveness, or any inadequacy in its labeling or in its directions for use (section 59(1)(a)); and • has led to the death or a serious deterioration in the state of health of a patient, user, or other person, or could do so if it were to recur (section 59(1)(b)).
Manufacturer	**(Device)(21 CFR§820.3(o))** Any person who designs, manufactures, fabricates, assembles, or processes a finished device. Manufacturer includes but is not limited to those who perform the functions of contract sterilization, installation, relabeling, remanufacturing, repacking, or specification development, and initial distributors of foreign entities performing these functions.

Term	Definition
	(Drug) (21 CFR §210.3(12)) Manufacture, processing, packing, or holding of a drug product includes packaging and labeling operations, testing, and quality control of drug products. **(Biologic) (21 CFR §600.3(t))** means any legal person or entity engaged in the manufacture of a product subject to license under the act; "Manufacturer" also includes any legal person or entity who is an applicant for a license where the applicant assumes responsibility for compliance with the applicable product and establishment standards. **Combination Product Manufacturer* (US FDA Combination Products Compliance Program, June 2020):** An entity (facility) engaged in activities for a combination product that are considered within the scope of manufacturing for drugs, devices, biological products, and HCT/Ps. Such manufacturing **activities include, but are not limited to, designing, fabricating, assembling, filling, processing, sterilizing, testing, labeling, packaging, repackaging, holding, and storage, including a contract manufacturing facility** (see 21 CFR §4.2) **Note October 2019 guidance Identification of Manufacturing Establishments in Applications to CBER and CDER Q&A*
Marketing Authorization (MA)	(EU, https://www.ema.europa.eu/en/glossary/marketing-authorisation) The approval to market a medicine in one, several or all European Union Member States.
Marketing Authorization Application (MAA)	(EU, https://www.ema.europa.eu/en/glossary/marketing-authorisation-application) An application submitted by a drug manufacturer to the regulatory authority for approval to market a medicine within the EU. MAA is part of the official procedure before the Medicines and Healthcare Products Regulatory Agency in the UK and the Committee for Medicinal Products for Human Use of the EMA, a specialized agency of the EC. In the US, the equivalent process is the NDA.
Marketing Authorization Holder (MAH)	(EU, https://www.ema.europa.eu/en/glossary/marketing-authorisation-holder) MAH is the company or other legal entity that has the authorization to market a medicine in one, several, or all European Union Member States.
Medical Device	§201(h) of the FD&C Act (21 U.S.C. 321(h)) provides that the term "device" means: An instrument, apparatus, implement, machine, contrivance, implant, in vitro reagent, or other similar or related article, including any component, part, or accessory, which is 1. Recognized in the official National Formulary, or the US Pharmacopeia, or any supplement to them, 2. Intended for use in the diagnosis of disease or other conditions, or in the cure, mitigation, treatment, or prevention of disease, in man or animals, or Intended to affect the structure or any function of the body of man or other animals, and which does not achieve its primary intended purposes through chemical action within or on the body of man or other animals and *which is not dependent upon being metabolized for the achievement of its primary intended purposes*. Range from bedpans and tongue depressors to implantable cardiac defibrillators and surgical robots.
Medical Device Directives (MDD)	(EU, https://eur-lex.europa.eu/eli/reg/2017/745/oj) MDD or 93/42/EEC is a directive under the European Council intended to harmonize the laws relating to medical devices within the European Union. MDR (EU) 2017/745 supplants MDD in May 2021.
Medical Device Establishment License (MDEL)	(Canada, https://www.canada.ca/en/health-canada/services/drugs-health-products/compliance-enforcement/establishment-licences/forms/medical-device-establishment-licence-application-form-instructions-0292.html) If a company sells or imports any class of medical devices in Canada, the company must apply for and maintain a MDEL, unless exemption(s) are met in section 44 of the Medical Devices Regulations.

Term	Definition
Medical Device File	ISO 13485:2016, Clause 4.2.3; similar to Device Master Record (21 CFR 820.181) For each medical device type or medical device family, the organization shall establish and maintain one or more files either containing or referencing documents generated to demonstrate conformity to the requirement of this International Standard and compliance with applicable regulatory requirements.
Medical Device License (MDL)	(Canada, https://www.canada.ca/en/health-canada/services/drugs-health-products/compliance-enforcement/establishment-licences/annual-review-documents/frequently-asked-questions-medical-device-establishment-licensing-fees.html) A license issued to manufacturers authorizing them to import or sell their Class II, III, or IV medical devices in Canada.
MDR (Medical Device Report) Reportable Event	[US, 21 CFR §803.3(o)] (1) An event that user facilities become aware of that reasonably suggests that a device has or may have caused or contributed to a death or serious injury; or (2) An event that manufacturers or importers become aware of that reasonably suggests that one of their marketed devices: (i) May have caused or contributed to a death or serious injury, or (ii) Has malfunctioned and that the device or a similar device marketed by the manufacturer or importer would be **likely to cause or contribute** to a death or serious injury if the malfunction were to recur. Examples: Fractured implanted stent; Broken, embedded needle; Out-of-box pre-mature actuation of autoinjector; Out-of-box pre-mature deployed stent; Infected Insertion site for a drug delivery system; Oral dosing pipette with unexpected traces of latex in bulb, and same dosing pipette included in a different drug product.
Medical Device Reporting (MDR)	(US, https://www.fda.gov/medical-devices/medical-device-safety/medical-device-reporting-mdr-how-report-medical-device-problems) One of the post-market surveillance tools the FDA uses to monitor device performance, detect potential device-related safety issues, and contribute to benefit-risk assessments of these products.
Medical Device Single Audit Program (MDSAP)	(International, https://www.fda.gov/medical-devices/cdrh-international-programs/medical-device-single-audit-program-mdsap) MDSAP was developed by the IMDRF to allow an MDSAP-recognized Auditing Organization to conduct a single regulatory audit of a medical device manufacturer that satisfies the relevant requirements of the regulatory authorities participating in the program.
Medical Device User Fee and Modernization Act (MDUFMA)	(US, https://www.fda.gov/industry/fda-user-fee-programs/medical-device-user-fee-amendments-mdufa) Device user fees were first established in 2002 by the MDUFMA. Under the user fee system, medical device companies pay fees to the FDA when they register their establishments and list their devices with the agency, whenever they submit an application or a notification to market a new medical device in the US and for certain other types of submissions. These fees help the FDA increase the efficiency of regulatory processes with a goal of reducing the time it takes to bring safe and effective medical devices to the US market.
Medical Product	A generic term for any product used to diagnose, cure, mitigate, treat, prevent or manage a disease, including any medical device, drug, or biological product (or medicinal product).
Medication Error (also see "use error")	Any preventable event that may cause or lead to inappropriate medication use (use error) or patient harm while the medication is in the control of the healthcare professional, patient, or consumer. An unintended failure in the medication treatment process, that leads to or has the potential to lead to, harm to the patient, and is thereby preventable. A failure in the treatment process may not refer to lack of efficacy of the drug, rather to human or process-mediated failures. Different failures can happen in the process, e.g., abuse, misuse, medication error, and off-label use.

Term	Definition
Medicinal Product (MP)	(EU, https://www.ema.europa.eu/en/glossary/medicinal-product) A substance, or combination of substances, that is intended to treat, prevent or diagnose a disease, or to restore, correct, or modify physiological functions by exerting a pharmacological, immunological, or metabolic action.
MedRA coding	Drugs and biologics use MedRA coding for registration, documentation, and safety monitoring of medical products, both before and after the product has been authorized for sale. In the late 1990s, the International Council for Harmonization (ICH) of Technical Requirements for Pharmaceuticals for Human Use established a rich, highly specific standardized medical dictionary for regulatory activities, MedRA. MedRA was created to facilitate sharing of regulatory information across jurisdictions globally for medical products used by humans.
Metabolic	An action of a substance or its metabolites that involves an alteration, including stopping, starting or changing the rate, extent or nature of a physiological or pathological biochemical process participating in, and available for, function of the human body. "Biochemical processes" include anabolic and catabolic reactions and transport of substances between compartments. An interaction with a known receptor is not a prerequisite for the presence of metabolic means of action.
Ministry of Health, Labor and Welfare (MHLW)	(Japan, https://www.pmda.go.jp/files/000164006.pdf) A cabinet-level ministry of the Japanese government. The ministry provides services on health, labor, and welfare. MHLW has the authority to give marketing approval, to issue a license for a marketing authorization holder, and to issue a manufacturer license for pharmaceuticals, medical devices, and cosmetics. (see also PMDA).
Misuse (of a medicinal product)	Situations where the medicinal product is intentionally and inappropriately used not in accordance with the authorized product information.
Mode of Action (MOA)	(International; see 21 CFR §3.2(k)) A mode of action (MOA) is the means by which a product achieves its intended effect(s) or action(s). Each MOA of a combination product is based on the type of constituent part doing the action, whatever that constituent part may be doing. More specifically, under US FDA, MOA is defined as the means by which a product achieves an intended therapeutic effect or action. For purposes of this definition, "therapeutic" action or effect includes any effect or action of the combination product intended to diagnose, cure, mitigate, treat, or prevent disease, or affect the structure or any function of the body. When making assignments of combination products under this part, the US FDA considers three types of mode of action: The actions provided by a biological product, a device, and a drug. Because combination products are comprised of more than one type of regulated article (biological product, device, or drug), and each constituent part contributes a biological product, device, or drug mode of action, combination products will typically have more than one identifiable mode of action.
N	
National Medical Products Administration (NMPA)	(China, http://subsites.chinadaily.com.cn/nmpa/aboutNMPA.html) The NMPA of China has the responsibility to: • Supervise the safety of drugs (including traditional Chinese medicines (TCMs) and ethno-medicines, medical devices, and cosmetics; • Regulate the registration of drugs, medical devices, and cosmetics; and • Undertake standards management for drugs, medical devices, and cosmetics.
New Drug Submission (NDS)	(Canada, https://www.canada.ca/en/health-canada/services/drugs-health-products/drug-products/applications-submissions/guidance-documents/chemical-entity-products-quality/guidance-document-quality-chemistry-manufacturing-guidance-new-drug-submissions-ndss-abbreviated-new-drug-submissions.html) A written request to Health Canada to obtain marketing approval for a new drug. The NDS must contain sufficient information and material to allow an assessment of the safety and effectiveness of the new drug.

Term	Definition
Non-combination product	(www.fda.gov) A non-combination product is a product that is only either a drug, a device, or a biological product as each is defined in the FD&C Act. The term does not include combination products as defined in 21 CFR 3.2(e).
Nonconformity (or Nonconformance)	A non-fulfillment of a requirement. It is important to understand that requirements may relate to product, process or the Quality Management System (GHTF/SG3/N18:2010; 21 CFR 820.90; ISO 13485:2016 Clause 8.3)
Notice of Compliance (NOC)	(Canada, https://www.canada.ca/en/health-canada/services/drugs-health-products/drug-products/notice-compliance/database.html) A document issued to a manufacturer following the satisfactory review of a submission for a new drug, and signifies compliance with the Canadian Food and Drug Regulations.
Notified Body (NB)	In the European Union, an organization that has been designated by a member state to assess the conformity of certain products, e.g., medical devices, to applicable standards and technical requirements, before their placement on the EU market.
O	
Off-label Use	Situations where a medicinal product is intentionally used for a medical purpose not in accordance with the authorized product information. *Off-label use includes use in non-authorized pediatric age categories.
Office of Clinical Policy & Programs (OCPP)	(US, https://www.fda.gov/about-fda/office-commissioner/office-clinical-policy-and-programs) The OCPP in the Office of the Commissioner supports the development of clinical programs that address key policy issues across the FDA's medical product centers. OCPP is responsible for developing, fostering, and coordinating cross-center initiatives involving the design and implementation of FDA policy related to medical product development and evaluation. The Office collectively provides leadership and coordination on complex medical product policy issues across FDA, the Department, other agencies and external stakeholders, and international bodies.
Office of Combination Products (OCP)	(US, https://www.fda.gov/about-fda/office-clinical-policy-and-programs/office-combination-products) The US FDA OCP was established on Dec. 24, 2002, aligned to §204 of the Medical Device User Fee and Modernization Act of 2002 (MDUFMA). The law gives the OCP broad responsibilities covering the regulatory life cycle of combination products. OCP primarily: • Serves as a focal point for combination product issues for FDA staff and industry. • Develops guidance, regulations, and standard operating procedures to clarify the regulation of combination products. • Classifies products as drugs, devices, biological products, or combination products and assign an FDA center to have primary jurisdiction for pre-market review and post-market regulation where the jurisdiction is unclear or in dispute. • Ensures timely and effective pre-market review of combination products by overseeing the timeliness of and coordinating reviews involving more than one agency center. • Ensures consistency and appropriateness of post-market regulation of combination products. • Facilitates resolution of disputes regarding the timeliness of pre-market review of combination products. • Updates agreements, guidance documents, or practices specific to the assignment of combination products. • Develops annual reports to Congress on the Office's activities and impacts. • Provides training to FDA staff and regulated industry on combination product regulation.
Office of Submissions and Intellectual Property (OSIP)	(Canada, https://www.canada.ca/en/health-canada/services/drugs-health-products/drug-products/applications-submissions/guidance-documents/management-drug-submissions/management-drug-applications-2019/document.html#a24) The organization under Health Canada that is responsible for receiving, processing and forwarding submissions to the appropriate Centre/Bureau/Office for review.

Term	Definition
Operational Qualification	Operation is challenged at worst-case process settings (limits of acceptability criteria) for specific parameters, demonstrating that the output of the process is still within the defined specifications, even at the process limits. OQ therefore establishes objective evidence that the process control limits and action levels consistently result in product that meets all predetermined requirements/specifications.
P	
Packaging	Through lens of device manufacturer: The configuration of materials designed to: • Avoid physical damage, biological contamination, and any other external disturbance to the medical device or assembly components, from time of assembly to point of use, and • Support the proper identification and use of the device by means of labeling such that at any time point before the end of shelf life, the device can be used safely and efficaciously. Through the lens of drug manufacturer: The combination of components necessary to contain, preserve, protect, and deliver a safe, efficacious drug product, such that at any time point before the expiration date of the drug product, a safe and efficacious dosage form is available.
Periodic Safety Reports	In addition to the ICSR and Non-ICSR reporting, there is an expectation that post-launch reports be provided on a periodic basis to summarize, among other things, analysis of adverse events for drug and biologics. The frequency of these reports is variable. These periodic reports include: Periodic Adverse Drug Experience Reports(PADERs), Periodic Safety Update Reports (PSURs), and Periodic Benefit Risk Evaluation Reports (PBERs). These report types are recognized globally under ICH E2C(R2).
Pharmaceutical Affairs Law (PAL)	(Japan, https://www8.cao.go.jp/kiseikaikaku/oto/otodb/english/houseido/hou/lh_02070.html) This law regulates matters necessary for securing the quality, efficacy, and safety of pharmaceuticals, quasi-drugs, cosmetics, and medical devices, while taking necessary steps to promote research and development of pharmaceuticals and medical devices in high necessity, and thereby improve public better health and hygiene.
Pharmaceuticals and Medical Devices Agency (PMDA)	(Japan, https://www.emergobyul.com/resources/japan/pharmaceuticals-medical-devices-agency) The Pharmaceuticals and Medical Devices Agency is an Independent Administrative Institution responsible for ensuring the safety, efficacy, and quality of pharmaceuticals and medical devices in **Japan.** The PMDA is the government organization in Japan in charge of reviewing drugs and medical devices, overseeing post-market safety, and providing relief for adverse health effects. The current PMDA, established in 2004, incorporates the Pharmaceuticals and Medical Devices Evaluation Center of the National Institute of Health Sciences (PMDEC), the Organization for Pharmaceutical Safety and Research (OPSR/KIKO), and part of the Japan Association for the Advancement of Medical Equipment (JAAME). The PMDA is part of the Ministry of Health, Labor, and Welfare (MHLW). They both handle a gamut of activities, ranging from approval reviews to post-market surveillance. Within the PMDA, the Office of Medical Devices Evaluation provides oversight to the manufacturing of medical devices, enforces standards, and grants approval to manufacture and market devices.
Pharmacological	An interaction between a substance or its metabolites and a constituent of the human body (or a pathogenic agent of the human body) that results in a change in physiological or pathological characteristics, e.g. initiation, enhancement, reduction, or blockade. Examples of constituents of the human body may include, among others: cells and their constituents (cell membranes, intracellular structures, RNA, DNA, proteins, e.g. membrane proteins, enzymes), extracellular matrix, blood and its components, body fluids, and their components. Examples of pharmacological means: interaction between a ligand (e.g. agonist, antagonist) and a receptor; interaction between a substance and membrane lipids; interaction between a substance and components of the cytoskeleton.

Term	Definition
Pharmacovigilance	(International; https://www.ema.europa.eu/en/glossary/pharmacovigilance) Science and activities relating to the detection, assessment, understanding, and prevention of adverse effects or any other medicine-related problem.
Platform Approach	When a technology (drug constituent, biologic constituent or device technology or software) is used as a base upon which other applications, processes, or products are supported or developed. Example: A finished medical device, or device component (product/ device)/ system/ technology) could be developed as a platform for use in a combination product portfolio (including a Drug Delivery system), where the information generated to justify, verify or validate its design and development can be used repeatedly as a basis for the safety and/or effectiveness of other products (i.e. to deliver other drugs).
Platform Technology	(1) medical device or medical device system (e.g., Common materials, processes, components, or delivery devices may be considered platforms) that may be suitable for use with various medicinal products and whose pre-existing device data or information or both can be leveraged in whole or in part to support such use or (2) medicinal product formulation (e.g., steroid coating) that may be suitable as a platform for use with various medical devices.
Post-authorization Efficacy Studies (PAES)	(EU, EMA/PDCO/CAT/CMDh/PRAC/CHMP/261500/2015) Studies subsequently conducted within the authorized therapeutic indication to address well-reasoned scientific uncertainties identified by EU regulators on aspects of the evidence of benefits that should be, or can only be, addressed post-authorization.
Post Marketing Safety Reporting (PMSR)	A mechanism to report significant safety and efficacy issues to health authorities; PMSR includes collection, analysis, and reporting of relevant drug adverse experience data and device malfunctions and serious injuries. It is considered a reflection of product safety and performance.
Post-market Surveillance (PMS)	Monitoring safety and effectiveness of medicinal product, medical device, and/or combination product after it has been released on the market
Predictable Misuse	Use of a product or system in a way not intended by the manufacturer, but which can result from readily predicted human behavior. (ISO/IEC Guide 63:2019)
Preventive Action	Action to eliminate the cause of a potential nonconformity or other undesirable situation. There can be more than one cause for nonconformity. Preventive action is taken to prevent occurrence. (GHTF/SG3/N18:2010)
Primary Mode of Action (PMOA)	The single mode of action of a combination product that makes the greatest contribution to the combination product's overall intended use(s).
Process Validation	Process Validation answers the question: "Will the constituent parts and combination product production process (processes) *reliably and repeatedly produce conforming product*?" (21 CFR 820.75/ISO 13485:2016 Clauses 7.5.5, 7.5.6, 7.5.7, 8.2.5; 21 CFR 211.100(a), 211.110(a))
Product	Through the lens of device manufacturer: **Finished device** means any device or accessory to any device that is suitable for use or capable of functioning, whether or not it is packaged, labeled, or sterilized. Through the lens of drug manufacturer: A finished dosage form, generally, but not necessarily, in association with inactive ingredients.
Product Lifecycle	All phases in the life of the product from the initial development through marketing until product discontinuation and disposal. (ISO/IEC Guide 63:2019)
Purchasing Controls	Procedures that ensure you are only purchasing from suppliers who can meet your specifications and requirements. An Approved Supplier List is generally used to keep track of your qualified suppliers.(21 CFR 820.50/ ISO 13485:2016, Clause 7.4.1)
Q	
Quality	The degree to which a set of inherent properties of a product, system, or process fulfills requirements.

Term	Definition
Quality System	The organizational structure, responsibilities, procedures, processes, and resources for implementing the management of quality in an organization
Quasi-Drug	Japan: Products that fall in between cosmetics and pharmaceuticals.. Quasi-drugs are products with formulations that are recognized and licensed by the Ministry of Health, Labour and Welfare in Japan but are not as "strong" as drugs (pharmaceuticals) but their efficacy is recognized by authorizing constitutions. https://oem-cosmetic.com/en/blog/quasi-drugs-oem-japan South Korean Ministry of Food and Drug Safety (MFDS) considers quasi-drugs as one of two primary categories of beauty products, where their benefits are not as suitable as drugs. Examples include ointments and anti-inflammatory products for external use, contact lenses, and items used for sanitary purposes like bandages. https://www.freyrsolutions.com/what-are-quasi-drugs-in-south-korea
R	Medical devices that are obtained separately by the user for use with medicinal products.
Referenced Product	Medical devices that are obtained separately by the user for use with medicinal products
Reliability	"The probability that a product, system, or service will perform its intended function adequately for a specified period of time, or will operate in a defined environment without failure." (asq.org)
Request for Designation (RFD) (also called "Applicant's Letter of Request")	21 CFR 3.2(j) It is a written submission to US FDA Office of Combination Products to request a determination of (1) the regulatory identity or classification of a product as a drug, device, biological product, or combination product, and/or (2) either the component of FDA that will regulate the product if it is a non-combination product, or which Agency Center will have primary jurisdiction for pre-market review and regulation if it is a combination product.
Reserve Samples	21 CFR 211.170 requires retain sample to support any potential post-distribution drug investigations. Single-entity and co-packaged combination product manufacturers are expected to keep reserve samples of each lot of the active ingredient, if any, that they receive, in whatever form it arrives at their facility.
Residual Risk	Risk remaining after risk control measures have been implemented. (ISO/IEC Guide 63:2019)
Risk	Probability of the occurrence of harm and the severity of that harm
Risk Acceptance	The decision to accept risk (ISO Guide 73)
Risk Analysis	Systematic use of available information to identify hazards and to estimate the risk (ISO/IEC Guide 63:2019)
Risk Assessment	Overall process including risk analysis and risk evaluation. (ISO/IEC Guide 51:2014)
Risk-Benefit Analysis	Analysis that seeks to quantify the risk and benefits and hence their ratio. Analyzing a risk can be heavily dependent on the human factor. A certain level of risk in our lives is accepted as necessary to achieve certain benefits.
Risk Communication	Sharing information about risk and risk management between stakeholders and decision maker.
Risk Control	Process in which decisions are made and measures implemented to reduce or maintain risks within specified levels; actions implementing risk management decisions (ISO/IEC Guide 63:2019 and ISO Guide 73)
Risk Estimation	Process used to assign values to the probability of occurrence of harm and the severity of that harm (ISO/IEC Guide 63:2019)
Risk Evaluation	Comparison of estimated risk to given risk criteria to determine the acceptability of the risk (ISO/IEC Guide 63:2019); process of comparing the estimated risk against a given risk criteria to determine the significance of a risk, using a qualitative or quantitative scale (ICH Q9).
Risk Management	Systematic application of management policies, procedures, and practices to the tasks of analyzing, evaluating, controlling, and monitoring risk across the product lifecycle. (ISO/IEC Guide 63:2019). It is a framework that supports the identification and control of critical requirements essential to the safe and effective functioning and use of medical products.

Term	Definition
Risk Management File (RMF)	Set of records and other documents that are produced via risk management.
Risk Management Plan (RMP)	The iterative guide to risk management activities throughout the entire product lifecycle of a given product.
S	
Safety	Freedom from unacceptable risk (ISO/IEC Guide 63:2019)
Same-Similar	The concept of "same-similar" is defined under 21 CFR §803.50 by US FDA as a device with a) the same general purpose and function; and b) the same type (e.g., would have the same device product code); and c) the same basic design and performance characteristics, including slight modifications, as it relates to safety and effectiveness as the applicant's US marketed medical device. In general, an organization should determine, based on the three criteria above, if they would deem two products as same-similar. For combination products, the same-similar concept is applied to the device constituent part and to the active moiety of a combination product.
Serious Adverse Experience (SAE) (biologic)	[21 CFR §600.80 (a)] Any adverse experience occurring at any dose that results in any of the following outcomes: Death, a life-threatening adverse experience, inpatient hospitalization or prolongation of existing hospitalization, a persistent or significant disability/incapacity, or a congenital anomaly/birth defect. Important medical events that may not result in death, be life-threatening, or require hospitalization may be considered a serious adverse experience when, based upon appropriate medical judgment, they may jeopardize the patient or subject and may require medical or surgical intervention to prevent one of the outcomes listed in this definition. Examples: Hypoglycemic shock due to overdose of insulin; Allergic bronchospasm requiring intensive treatment in an emergency room or at home; Blood dyscrasias or convulsions that do not result in inpatient hospitalization; Development of drug dependency or drug abuse.
Serious Adverse Experience (SAE) (drug)	[21 CFR § 314.80(a)] Any adverse drug experience occurring at any dose that results in any of the following outcomes: Death, a life-threatening adverse drug experience, inpatient hospitalization or prolongation of existing hospitalization, a persistent or significant disability/incapacity, or a congenital anomaly/birth defect. Important medical events that may not result in death, be life-threatening, or require hospitalization may be considered a serious adverse drug experience when, based upon appropriate medical judgment, they may jeopardize the patient or subject and may require medical or surgical intervention to prevent one of the outcomes listed in this definition. Examples: Cancer; Congenital defects; Allergic bronchospasm requiring intensive; treatment in an emergency room or at home; Stroke; Blood dyscrasias or convulsions that do not result in inpatient hospitalization; Development of drug dependency or drug abuse
Serious Incident	(EU MDR 2017/745) Any incident that directly or indirectly led, or might have led, to any of the following: • Death of a patient, user, or other person • Temporary or permanent serious deterioration of a patient's, users, or other person's state of health • A serious public health threat

Term	Definition
Serious Injury (SI) (device)	[21 CFR §803.3(w)] An injury or illness that: (1) Is life-threatening; (2) Results in permanent impairment of a body function or permanent damage to a body structure; or (3) Necessitates medical or surgical intervention to preclude permanent impairment of a body function or permanent damage to a body structure. Permanent means irreversible impairment or damage to a body structure or function, excluding trivial impairment or damage. Examples: Patient suffers a stroke and paralysis after implantation of stent; Patient receives stitches due to explant of broken needle; Someone chokes on parts of product; Label damage on a laser, leading to improper use of device, leading to blindness
Seriousness	Patient/event outcome or action criteria usually associated with events that pose a threat to a patient's life or functioning (ICH-E2A)
Servicing	Where servicing is a specified requirement, each manufacturer shall establish and maintain instructions and procedures for performing and verifying that the servicing meets the specified requirements. (21 CFR 820.200/ ISO 13485:2016, Clause 7.5.4)
Severity	Measure of the possible consequences of a hazard (ISO/IEC Guide 63:2019); the intensity of a specific event (e.g., mild, moderate, or severe)
Shelf Life	The extent to which a product retains, within specified limits, and throughout its period of storage and use, the same properties and characteristics that it possessed at the time of manufacture
Single-Entity Combination Product	A product comprised of two or more regulated components that are physically, chemically, or otherwise combined or mixed and produced as a single entity. Sometimes referred to as "single integral."
Single Integral Combination Product	A product comprised of two or more regulated components that are physically, chemically, or otherwise combined or mixed and produced as a single entity. Sometimes referred to as "single entity."
Stability	The extent to which a product retains, within specified limits, and throughout its period of storage and use, the same properties and characteristics that it possessed at the time of manufacture. For combination products, one needs to consider the stability of the drug constituent part, the stability (typically referred to as "shelf life") of the device constituent part, and the stability of the combination product (e.g., interactions between the constituent parts and/ or components for the combined use configuration).
Standard Operating Procedure (SOP)	Step-by-step instructions used to help workers carry out specific procedures or routine operations.
Summative Study (Summative Validation	A human factors study conducted to demonstrate that the final finished combination product user interface can be used by intended users without serious use errors or problems, for the product's intended uses and under the expected use conditions. The study should demonstrate that use-related hazards for the final finished combination product have been eliminated or that the mitigation for residual risks is acceptable, i.e., the benefits of product use outweigh the residual risk of the product. The study participants are representative of the intended users and the study conditions are representative of expected use conditions. (US FDA Draft Guidance "Human Factors Studies and Related Clinical Study Considerations in Combination Product Design & Development" (Feb. 2016) https://www.fda.gov/media/96018/download.

Term	Definition
T	
Therapeutic Goods	Therapeutic Goods Administration (Australia) defines this term: "Therapeutic goods can comprise a broad range of things, such as bandages, pregnancy testing kits, herbal remedies, tissue grafts, and paracetamol. They generally fall under three main categories: • Medicines – including prescription, over-the-counter and complementary medicines, such as paracetamol and echinacea • Biologicals – something made from or containing human cells or tissues, such as human stem cells or skin • Medical devices – including instruments, implants and appliances, such as pacemakers and sterile bandages The TGA also regulates what are known as other therapeutic goods (OTGs), which include items such as tampons and disinfectants." https://www.tga.gov.au/what-are-therapeutic-goods
U	
Unexpected (adverse drug experience)	[21CFR §314.80(a)] Any adverse drug experience that is not listed in the current labeling for the drug product. This includes events that may be symptomatically and pathophysiologically related to an event listed in the labeling, but differ from the event because of greater severity or specificity. "Unexpected," as used in this definition, refers to an adverse drug experience that has not been previously observed (i.e., included in the labeling) rather than from the perspective of such experience not being anticipated from the pharmacological properties of the pharmaceutical product. **Examples:** Hepatic necrosis would be unexpected (by virtue of greater severity) if the labeling only referred to elevated hepatic enzymes or hepatitis; Cerebral thromboembolism and cerebral vasculitis would be unexpected (by virtue of greater specificity) if the labeling only listed cerebral vascular accidents.
Use error	A situation whereby the outcome of product use differs from that which was intended, but is not due to device malfunction. The error often highlights a problem with device labeling, user interface, or other aspects of device design; User action or lack of user action while using the medical product that leads to a different result than that intended by the manufacturer or that expected by the user. (ISO/IEC 62366-1:2015)
Use-Related Risk Assessment (URRA)	Use-related risk assessment of a combination product is the foundation for Human Factors study designs. It drives a risk-based approach to testing and evaluation. A URRA is a crucial step to help identify use-related hazards associated with the combination product, as well as to characterize high-risk hazards so they can be mitigated or eliminated through improved product interface design. The use-related risk assessment helps identify critical tasks that should be evaluated in an HF study, informs the priority of testing the tasks in an HF study, and determines if there are specific use scenarios to include in testing.
User interface	All components of a product with which a user interacts (e.g., sees and touches), such as labels and packaging, and the container/closure system interacts, such as controls and displays.
V	
Validation	Confirmation through provision of objective evidence that the requirements for a specific intended use or application have been fulfilled. The use conditions for validation can be real or simulated. (GHTF/SG3/N18:2010)
Validation Master Plan (VMP)	A summary document for a particular project or process undergoing validation. The VMP defines the process, strategy, key steps, and the roles and responsibilities of those engaged in the process validation activities. It generally follows a stage approach.

Term	Definition
Verification	Sometimes also referred to as "qualification"; Confirmation, with objective evidence, that specified requirements have been fulfilled. Confirmation can include activities such as performing alternative calculations; comparing a new design specification with a similar proven design specification; doing testing; performing demonstrations; and reviewing and approving documents prior to use. (GHTF/SG3/N18:2010)

NOTES

1. The definitions and applicable regulatory requirements differ from country to country and region to region. The organization needs to understand how the definitions in this Glossary will be interpreted in light of regulatory definitions in the jurisdictions in which their medical products are to be made available.
2. In some jurisdictions (for example, EMA, MHRA), the term "medicinal product" generally refers to any drug or biologic product or both. Generally speaking, medicinal products achieve their primary intended purpose through chemical, pharmacological, immunological, or metabolic action, whereas medical devices do not (https://www.ema.europa.eu/en/glossary/medicinal-product; EU MDR (2017/745) Article 2)). Other jurisdictions (for example, US FDA) treat drugs and biological products as separately regulated categories. Therefore, additional requirements may apply for drug/biological product combinations depending on the regional/country regulatory framework.
3. US FDA CDRH Draft Human Factors Engineering Guidance, "Applying Human Factors and Usability Engineering to Medical Devices" (February 2016), accessed July 31, 2022 at https://www.fda.gov/media/80481/download.
4. Note that CDER does not include the word "serious" in their definition of critical use task, so the scope of tasks that are considered critical is broader than in the definition from CDRH. US FDA CDER Draft Human Factors Engineering Guidance, "Human Factors Studies and Related Clinical Study Considerations in Combination Product Design and Development" (February 2016), accessed July 31, 2022 at https://www.fda.gov/regulatory-information/search-fda-guidance-documents/human-factors-studies-and-related-clinical-study-considerations-combination-product-design-and.
5. 21 CFR 820.30(h) or ISO 13485:2016 Clause 7.3.8 (results and conclusions are documented per Clause 4.2.5).

Index

Note: Locators in *italics* represent figures and **bold** indicate tables in the text.

Printed in the United States
by Baker & Taylor Publisher Services